The UK Pesticide Guide 2020

Editor: M.A. Lainsbury, BSc(Hons)

BCPC

BCPC (British Crop Production Council) promotes the use of good science and technology in understanding and applying effective and sustainable crop production. Its key objectives are to identify developing issues in the science and practice of crop protection and production and provide informed, independent analysis; to publish definitive information to growers, advisors and other stakeholders; to organise conferences and symposia, and to stimulate interest and learning. BCPC is a registered Charity and a Company limited by Guarantee. Further details available from BCPC, Garden Studio, 4 Hillside, Aldershot, Surrey, UK. GU11 3NB.

Telephone: (01252) 285223
e-mail: md@bcpc.org
Web: www.bcpc.org

ISBN: 978-1-9998966-3-8

Typeset and Printed in the United Kingdom by
Hobbs the Printers Ltd, Totton, Hampshire, SO40 3WX
Website: www.hobbs.uk.com

Contents

Editor's Note

This is the 33rd edition of *The UK Pesticide Guide* and 2019 has been a year which has seen the impending loss of a number of key active ingredients. The more significant losses are listed below but the loss that worries me the most if the loss of chlorothalonil. This multi-site fungicide is not only used alone but also in mixture with a large number of other fungicide actives. Because of the multi-site nature of its activity it means that it is virtually impossible for fungi to develop resistance to it. This is why it is used in so many co-formulations with other actives – it protects their mode of action from resistance. There are two other multi-site fungicides that I know of – mancozeb and folpet. I foresee a greater use of these actives as companies and growers alike seek to protect the fungicide armoury that they currently have.

Just like last year, as this book goes for preparation of the first proof news bulletins are filled with questions about whether we will get a deal from Europe over Brexit. It will not affect generic products but what impact will the final situation have on parallel approved products? We will still have to wait and see. On the more positive front, the 2020 edition has a number of new active ingredients, formulations and new products but we have also lost a lot of actives and I summarise the changes here with new actives shown in **bold**. BASF has launched its new powerful triazole mefentrifluconazole but it will need protecting against the development of fungal resistance if we are to keep that activity. Clearly this year there are many more losses than gains:

Herbicide gains are:
- **aclonifen straight**
- diflufenican + flufenacet + metribuzin
- iodosulfuron-methyl-sodium straight
- clopyralid + halauxifen-methyl
- dicamba + nicosulfuron
- diflufenican + iodosulfuron-methyl
- florasulam + halauxifen-methyl

Fungicide gains are:
- fluoxastrobin + tebuconazole
- laminarin
- **mefentrifluconazole**
- prochloraz
- azoxystrobin + difenoconazole
- **benzovindiflupyr**
- dimethomorph + propamocarb
- epoxiconazole + prochloraz
- isopyrazam + prothioconazole

Insecticide & PGR full entry gains are:
- **sulfoxaflor**
- Metarhizium anisopliae

Herbicides demoted to PAR or lost are:
- chloridazon
- chloridazon + metamitron
- chloridazon + quinmerac
- clodinafop-methyl + pinoxaden
- diflufenican + flufenacet + flurtamone
- diquat
- ethofumesate + metamitron + phenmedipham
- glufosinate ammonium
- mesotrione + nicosulfuron
- mesotrione + terbuthylazine
- quinoclamine

Insecticides & PGRs demoted to PAR or lost are:
- Bacillus thurigiensis israelensis AM65-52
- chlormequat + imazaquin
- chlopyrifos-methyl
- dazomet
- pymetrozine

Fungicides demoted to PAR or lost are:
- azoxystrobin + chlorothalonil
- azoxystrobin + propiconazole
- carboxin + thiram
- chlorothalonil
- chlorothalonil + cymoxanil
- chlorothalonil + cyproconazole
- cholorthalonil + cyproconazole + propiconazole
- chlorothalonil + fluxapyroxad
- chlorothalonil metalaxyl-M
- chlorothalonil + propiconazole
- chlorothalonil + proquinazid
- chlorothalonil + tebuconazole
- difenoconazole + propiconazole
- dithianon + pyraclostrobin
- epoxiconazole + fenpropimorph
- epoxiconazole + fenpropimorph + kresoxim-methyl
- epoxiconazole + fenpropimorph + metrafenone
- epoxiconazole + metrafenone
- fenamidone + fosetyl-aluminium
- fenamidone + propamocarb
- fenpropidin + prochloraz + tebuconazole
- fenpropimorph
- fenpropimorph + pyraclostrobin
- meptyldinocap

- prochloraz + propiconazole
- propiconazole
- quinoxyfen
- thiram

Insecticides & PGRs demoted to PAR or lost are:
- beta-cyfluthrin + clothianadin

- chlopyrifos
- clothianidin
- clothianidin + prothioconazole
- etofenprox
- fenoxycarb
- imidacloprid
- metam sodium
- methiocarb

One new entry, foramsulfuron + thiencarbazone-methyl is specifically for ALS-resistant sugarbeet. ALS-resistant varieties have been developed from naturally-occurring resistant examples of sugarbeet much like the ALS-resistant blackgrass that now dominates our cereal fields and these extremely rare alleles have been carefully bred into elite varieties that come through the variety selection trials. Besides excellent broad-leaved and grass weed control they offer an opportunity to deal with weed beet and may also help to compensate for the loss of desmedipham and chloridazon from the sugarbeet herbicide choice.

Many products have been demoted from Section 2 to Section 3 Products also Registered (PAR) as their approvals expire. As always, this book is a snapshot of the products with a valid approval on 16 October 2019 when the text went for initial typesetting. If you cannot see a product in the main section of the book, please check the index because it may be listed in Products also Registered. For those who must have the most up-to-date information I can recommend the on-line subscription version of the UKPG – ukpesticideguide.co.uk. This is a searchable database with much more information than we can squeeze into the book.

We express our thanks to all those who have provided the information for the compilation of this guide. Criticisms and suggestions for improvement are always welcome and if you notice any errors or omissions, please let me know via the email address below.

Martin Lainsbury, Editor
email: ukpg@bcpc.org

Disclaimer

Every effort has been made to ensure that the information in this book is correct at the time of going to press, but the Editor and the publishers do not accept liability for any error or omission in the content, or for any loss, damage or other accident arising from the use of the products listed herein. Omission of a product does not necessarily mean that it is not approved or that it is unavailable.

It is essential to follow the instructions on the approved label when handling, storing or using any crop protection product. Approved 'off-label' extensions of use now termed Extension of Authorisation for Minor Use (EAMUs) are undertaken entirely at the risk of the user.

The information in this publication has been selected from official sources, from the suppliers product labels and from the product manuals of the pesticides approved for use in the UK under EU Regulation 1107/2009.

Changes Since 2019 Edition

Pesticides and adjuvants added or deleted since the 2019 edition are listed below. Every effort has been made to ensure the accuracy of this list, but late notifications to the Editor are not included.

Products Added

Products new to this edition are shown here. In addition, products that were listed in the previous edition whose CRD/HSE registration number and/or supplier have changed are included.

Product	Reg. No.	Supplier	Product	Reg. No.	Supplier
Agate EW	19000	Nufarm UK	Dakota	19179	Dow (Corteva)
Agree 50 WG	18965	Certis	Deacon	19003	Nufarm UK
Alternator Met	18267	Bayer CropScience	Diamond	18832	Pan Amenity
			Difenostar	19118	Life Scientific
Angle	19119	Syngenta	Dime	18661	Pan Agriculture
Arixa	19034	Nurture			
Asbo	19186	Sphere	Dime	18825	Pan Agriculture
Asteroid	18958	FMC Agro			
Asteroid Pro	18754	FMC Agro	Dymid	19187	Dow (Corteva)
Asteroid Pro 450	18788	FMC Agro	Emerger	19056	Bayer CropScience
Axial Pro	19010	Syngenta			
Bactimos SC	UK15-0947	Resource Chemicals	Enervin SC	19095	BASF
			Enstar	18911	Headland Amenity
Bandit	18577	Harvest			
Betanal Tandem	19257	Bayer CropScience	Envision	18770	FMC Agro
			Ephon Top	18872	Nufarm UK
Bow	19054	Nufarm UK	Evito T	18671	Arysta
CCC 720	19008	Nurture	Fazor	19074	Arysta
Chime	18557	Harvest	Ferrex	19138	Cropco Ltd
Clayton Cob	18971	Clayton	Flagstaff	18586	De Sangosse
Clayton Occupy	18924	Clayton	Force ST	19042	Syngenta
Clayton Paragon	19051	Clayton	Frutogard	19105	Certis
Clayton Prius	18946	Clayton	Fusilade Forte	19019	Nufarm UK
Clayton Rift	18945	Clayton	Fusilade Max	19013	Nufarm UK
Clayton Tacit	19073	Clayton	Galaxy	18952	Dow (Corteva)
Cleancrop Peregrine	18999	Clayton	Giddo	19128	Bayer CropScience
Cleancrop Signifier	18998	Clayton			
Clozone	18979	UPL Europe	Grainstore 25EC	19218	Pan Agriculture
Concert SX	18806	FMC Agro			
Consul	18805	FMC Agro	Grainstore ULV	19195	Pan Agriculture
Conviso One	19036	Bayer CropScience			
			Headland Spruce	18259	FMC Agro
Croupier OD	18457	Certis	Icade	19069	Dow (Corteva)
Curb Liquid Crop Spray	18643	Sphere	Iconic	19094	Adama
			Integral Pro	18510	BASF
Cymax	19045	Belchim	Iodus	19163	Arysta
Cyper 500	18670	Nurture	Ironclad	18970	Gemini
			Kaltor	18749	Rotam

Product	Reg. No.	Supplier	Product	Reg. No.	Supplier
Laminator Flo	19104	Sumitomo	Robin	19209	Pan Agriculture
Lantern	19085	BASF	Romeo	19170	Fargro
Laquna	19232	Pan Agriculture	Rosan	18778	FMC Agro
Laya	19016	Life Scientific	Roundup Ultimate	12774	Monsanto
Lector Delta	19055	Nufarm UK	Shylock	18928	Nurture
Master Gly	19202	Ventura	Sigon	19014	De Sangosse
Metfin 90	18910	Clayton	Snapper	19037	Nufarm UK
Mitre	19002	Nufarm UK	Soriale	19166	BASF
Monkey	18997	Nufarm UK	Spindle	18949	Pan Agriculture
Myresa	19153	BASF	Steward	18792	FMC Agro
Nucleus	19026	FMC Agro	Storm Ultra	UK18-1164	BASF
Octavian Met	18266	Bayer CropScience	Storm Ultra Bait Blocks	UK18-1164	BASF
Omasa	19043	Nurture	Storm Ultra Secure	UK18-1164	BASF
Orius	18992	Nufarm UK			
Orius P	18994	Nufarm UK	Surrender T/F	19203	Agrovista
Pennant	18779	FMC Agro	Target Flo	19159	UPL Europe
Pest Expert Formula B+	UK17-1051	Killgerm	Tempest	A0887	Intracrop
			Terrane	18738	FMC Agro
Praxys	18953	Dow (Corteva)	Thiopron	19147	UPL Europe
Presite SX	18776	FMC Agro	Topsail	19024	Adama
Prizm	18940	Adama	Torres	19017	BASF
Probe	A0874	Life Scientific	Traciafin Plus	19212	Clayton
Proradix	18115	Fargro	V10	18969	Valto
Protefin	19126	Clayton	Vesuvius Green	19201	Ventura
Provalia LQM	18798	FMC Agro	Volcano	18926	UPL Europe
Pure Ace	19053	Pure Amenity			
Refine Max SX	18785	FMC Agro			
Robin	18915	Pan Agriculture			

Products Deleted

The appearance of a product name in the following list may not necessarily mean that it is no longer available. In cases of doubt, refer to Section 3 or the supplier.

Product	Reg. No.	Supplier	Product	Reg. No.	Supplier
Apache	14255	Syngenta	Fenomenal	15494	Bayer CropScience
Aubrac Xtra	18083	Harvest	Frozen	17691	Harvest
Cherokee	13251	Syngenta	Garrison	18174	Pan Agriculture
Chimpanzee	16043	Pan Agriculture	Mesostar	17410	Life Scientific
Consento	15150	Bayer CropScience	Mundua	16103	Agroquimicos
Difenag Fresh Bait	UK12-0436	Killgerm	Newman's T-80	A0192	De Sangosse
			potassium bicarbonate	00000	various
Difenag Wax Blocks	UK12-0435	Killgerm	Prima	A0864	De Sangosse
			Priori Xtra	11518	Syngenta

Product	Reg. No.	Supplier	Product	Reg. No.	Supplier
Prompto	15156	Bayer CropScience	Sumir	17974	Life Scientific
Pure Ace	18364	Pure Amenity	Tezate 220 SL	16979	UPL Europe
Pure Flazasul	15135	Pure Amenity	Torpedo-II	A0541	De Sangosse
Sceptre	18868	Syngenta	Vazor Cypermethrin 10	H10085	Killgerm
sodium hypochlorite	-	various	X-Wet	A0548	De Sangosse
Sorexa Gel	UK12-0364	BASF			

Introduction

Purpose

The primary aim of this book is to provide a practical handbook of pesticides, plant growth regulators and adjuvants that the farmer or grower can realistically and legally obtain in the UK and to identify the purposes for which they can be used. It is designed to help in the identification of products appropriate to a particular problem. In addition to uses recommended on product labels, details are provided of those uses which do not appear on product labels but which have been granted Extension of Authorisation for Minor Use (EAMUs). As well as identifying the products available, the book provides guidance on how to use them safely and effectively but without giving details of doses, volumes, spray schedules or approved tank mixtures. Sections 5 and 6 provide essential background information on a wide range of pesticide-related issues including: legislation, biodiversity, codes of practice, poisons and treatment of poisoning, products for use in special situations and weed and crop growth stage keys.

While we have tried to cover all other important factors, this book does not provide a full statement of product recommendations. Before using any pesticide product it is essential that the user should read the label carefully and comply strictly with the instructions it contains. **This Guide does not substitute for the product label.**

Scope

The *Guide* is confined to pesticides registered in the UK for professional use in arable agriculture, horticulture (including amenity use), forestry and areas in or near water. Within these fields of use there are about 3017 products in the UK with current approval. The *Guide* is a reference for virtually all of them given in two sections. Section 2 gives full details of the products notified to the editor as available on the market. Products are included in this section only if requested by the supplier and supported by evidence of approval. Section 3 gives brief details of all other registered products with extant approval.

Section 2 lists some 388 individual active ingredients and mixtures. For each entry a list is shown of the available approved products together with the name of the supplier from whom each may be obtained. All types of pesticide covered by Control of Pesticides Regulations or Plant Protection Products Regulations are included. This embraces acaricides, algicides, herbicides, fungicides, insecticides, lumbricides, molluscicides, nematicides and rodenticides together with plant growth regulators used as straw shorteners, sprout inhibitors and for various horticultural purposes. The total number of products included in section 2 (i.e. available in the market in 2020) is about 1193.

The *Guide* also gives information (in section 4) on around 149 authorised adjuvants which, although not active pesticides themselves, may be added to pesticide products to improve their effectiveness. The Guide does not include products solely approved for amateur, home and garden, domestic, food storage, public or animal health uses.

Sources of Information

The information in this edition has been drawn from these authoritative sources:

- approved labels and product manuals received from suppliers of pesticides up to October 2019.
- websites of the Chemicals Regulation Division (CRD) *www.hse.gov.uk/crd/index.htm* and the Health and Safety Executive (HSE), *www.hse.gov.uk*.

Criteria for Inclusion

To be included, a pesticide must meet the following conditions:

- it must have extant approval under UK pesticides legislation
- information on the approved uses must have been provided by the supplier
- it must be expected to be on the UK market during the currency of this edition.

When a company changes its name, whether by merger, takeover or joint venture, it is obliged to re-register its products in the name of the new company, and new MAPP numbers are normally assigned. Where stocks of the previously registered product remain legally available, both old and new identities are included in the *Guide and remain until approval of the former lapses or stocks are notified as exhausted.*

Products that have been withdrawn from the market and whose approval will finally lapse during 2020 are identified in each profile. After the indicated date, sale, storage or use of the product bearing that approval number becomes illegal. Where there is a direct replacement product, this is indicated.

The Voluntary Initiative

The Voluntary Initiative (VI) is a programme of measures, agreed by the crop protection industry and related organisations with Government, to minimise the environmental impacts of crop protection products. The programme has provided a framework of practices and principles to help Government achieve its objective of protection of water quality and enhancement of farmland biodiversity. Many of the environmental protection schemes launched under the VI represent current best practice.

The first five-year phase of the programme formally concluded in March 2006, but it is continuing with a rolling two-year review of proposals and targets; current proposals include an online crop protection management plan and greater involvement in catchment-sensitive farming (see below).

A key element of the VI has been the provision of environmental information on crop protection products. Members of the Crop Protection Association (CPA) have committed to do this by producing Environmental Information Sheets (EISs) for all their marketed professional products.

EISs reinforce and supplement information on a product label by giving specific environmental impact information in a standardised format. They highlight any situations where risk management is essential to ensure environmental protection. Their purpose is to provide user-friendly information to advisers (including those in the amenity sector), farmers and growers on the environmental impact of crop protection products, to allow planning with a better understanding of the practical implications.

Links are provided to give users rapid access to EISs on the VI website:
www.voluntaryinitiative.org.uk.

Catchment-sensitive Farming

Catchment-sensitive Farming (CSF) is the government's response to climate change and the European Water Directive. Rainfall events are likely to become heavier, with a greater risk of soil, nutrients and pesticides ending up in the waterways. The CSF programme is investigating the effects of specific targeted advice on 20 catchment areas perceived as being at risk, and new entries in this *Guide* can help identify the pesticides that pose the greatest risk of water contamination (see Environmental Safety).

Notes on Contents and Layout

The book consists of six main sections:

1 Crop/Pest Guide
2 Pesticide Profiles
3 Products Also Registered
4 Adjuvants
5 Useful Information
6 Appendices

1 Crop/Pest Guide

This section enables the user to identify which active ingredients are approved for a particular crop/pest combination. The crops are grouped as shown in the Crop Index. For convenience, some crops and pests have been grouped into generic units (for example, 'cereals' are covered as a group, not individually). Indications from the Crop/Pest Guide must always be checked against the specific entry in the pesticide profiles section because a product may not be approved on all crops in the group. Chemicals indicated as having uses in cereals, for example, may be approved for use only on winter wheat and winter barley, not for other cereals. Because of the difference in wording of product labels, it may sometimes be necessary to refer to broader categories of organism in addition to specific organisms (e.g. annual grasses and annual weeds in addition to annual meadow-grass).

2 Pesticide Profiles

Each active ingredient and mixture of ingredients has a separate numbered profile entry. The entries are arranged in alphabetical order using the common names approved by the British Standards Institution and used in *The Pesticide Manual* (18th Edition; BCPC, 2018) now available as an online subscription resource visit: *www.bcpc.org*. Where an active ingredient is available only in mixtures, this is stated. The ingredients of the mixtures are themselves ordered alphabetically, and the entries appear at the correct point for the first named active ingredient.

Below the profile title, a brief phrase describes the main use of the active ingredient(s). This is followed where appropriate by the mode of action code(s) as published by the Fungicide, Herbicide and Insecticide Resistance Action Committees.

Within each profile entry, a table lists the products approved and available on the market in the following style:

Product name	Main supplier	Active ingredient contents	Formulation type	Registration number
1 Liberator	Bayer CropScience	100:400g/l	SC	15206
2 Naceto	Belchim	200:400g/l	SC	18063

Many of the **product names** are registered trademarks, but no special indication of this is given. Individual products are indexed under their entry number in the Index of Proprietary Names at the back of the book. The main supplier indicates the marketing outlet through which the product may be purchased. Full addresses and contact details of suppliers are listed in Appendix 1. For mixtures, the **active ingredient contents** are given in the same order as they appear in the profile heading. The **formulation types** are listed in full in the Key to Abbreviations and Acronyms (Appendix 5). The **registration number** normally refers to the registration with the Chemicals Regulation Division (CRD). In cases where the products registered with the Health and Safety Executive are included, HSE numbers are quoted, e.g. H0462 and approved biocides are numbered UK12-XXXX.

Below the product table, a **Uses** section lists all approved uses notified by the suppliers to the editor by October 2019, giving both the principal target organisms and the recommended crops or situations (identified in ***bold italics***). Where there is an important condition of use, or where the approval is off-label, this is shown in parentheses as (*off-label*). Numbers in square brackets refer to the numbered products in the table above. Thus, in the example shown above, a use approved for Liberator (product 1) but not for Naceto (product 2) appears as:

- Annual grasses in ***winter oats*** (off-label) [1]

Any Extensions of Authorisation for Minor Uses (EAMUs), for products in the profile are detailed below the list of approved uses. Each EAMU has a separate entry and shows the crops to which it applies, the Notice of Approval number, the expiry date (if due to fall during the current edition), and the product reference number in square brackets. EAMUs may now appear on the product label but are undertaken entirely at the risk of the user.

Below the EAMU paragraph, **Notes** are listed under the following headings. Unless otherwise stated, any reference to dose is made in terms of product rather than active ingredient. Where a note refers to a particular product rather than to the entry generally, this is indicated by numbers in square brackets as described above.

Approval information	Information of a general nature about the approval status of the active ingredient or products in the profile is given here. Notes on approval for aerial or ULV application are given.
	Inclusion of the active ingredient in Annex I under EC Directive 1107/2009 is shown, as well as any acceptance from the British Beer and Pub Association (BBPA) for its use on barley or hops for brewing or malting.
	Where a product approval will finally expire in 2020, the expiry date is shown here.
Efficacy	This entry identifies factors important in making the most effective use of the treatment. Certain factors, such as the need to apply chemicals uniformly, the need to use appropriate volume rates, and the need to prevent settling out of the active ingredient in the spray tank, are assumed to be important for all treatments and are not emphasised in individual profiles.
Restrictions	Notes are included in this section where products are subject to the Poisons Law, and where the label warns about organophosphorus and/or anticholinesterase compounds. Factors important in optimising performance and minimising the risk of crop damage, including statutory conditions relating to the maximum permitted number of applications, are listed. Any restrictions on crop varieties that may not be sprayed are mentioned here.
Environmental safety	Where the label specifies an environmental hazard classification, it is noted together with the associated risk phrases. Any other special operator precautions and any conditions concerning withholding livestock from treated areas are specified. Where any of the products in the profile are subject to Category A, Category B or broadcast air-assisted buffer zone restrictions under the LERAP scheme, the relevant classification is shown. Where a LERAP requires a buffer zone greater than 5m, detail of the larger buffer zone is given here. Arthropod buffer zones and statutory drift reduction limitations are also listed here.
	Other environmental hazards are also noted here, including potential dangers to livestock, game, wildlife, bees and fish. The need to avoid drift onto neighbouring crops and the need to wash out equipment thoroughly after use are important with all pesticide treatments but may receive special mention here if of particular significance.
Crop-specific information	Instructions about the timing of application or cultivations that are specific to a particular crop, rather than generally applicable, are mentioned here. The latest permitted use and harvest intervals (if any) are shown for individual crops.

Following crops guidance	Any specific label instructions about what may be grown after a treated crop, whether harvested normally or after failure, are shown here.
Hazard classification and safety precautions	The label hazard classifications and precautions are shown using a series of letter and number codes which are explained in Appendix 4. Hazard warnings are now given in full, with the other precautions listed under the following subheadings:

- Risk phrases
- Operator protection
- Environmental protection
- Consumer protection
- Storage and disposal
- Treated seed
- Vertebrate/rodent control products
- Medical advice

This section is given for information only and should not be used for the purpose of making a COSHH assessment without reference to the actual label of the product to be used.

3 Products also Registered

Products with an extant approval for all or part of the year of the edition in which they appear are listed in this section if they have not been requested by their supplier or manufacturer for inclusion in Section 2. Details shown are the same (apart from formulation) as those in the product tables of Section 2 and the approval expiry date is shown. Not all products listed here will be available in the market, but this list, together with the products in Section 2, comprises a comprehensive listing of all approved products in the UK for uses within the scope of the *Guide at the time of going to print.*

4 Adjuvants

Adjuvants are listed in a table ordered alphabetically by product name. For each product, details are shown of the main supplier, the authorisation number and the mode of action (e.g. wetter, sticker, etc.) as shown on the label. A brief statement of the uses of the adjuvant is given. Protective clothing requirements and label precautions are listed using the codes from Appendix 4.

5 Useful Information

This section summarises legislation covering approval, storage, sale and use of pesticides in the UK. There are brief articles on the broader issues concerning the use of crop protection chemicals, including resistance management. Lists are provided of products approved for use in or near water,in forestry, as seed treatments and for aerial application. Chemicals subject to the Poisons Laws are listed, and there is a summary of first aid measures if pesticide poisoning should be suspected. Finally, this section provides guidance on environmental protection issues and covers the protection of bees and the use of pesticides in or near water.

6 Appendices

Appendix 1	Gives contact details of all companies listed as main suppliers in the pesticide profiles.
	Where a supplier is no longer trading under that name (usually following a merger or takeover), but products under that name are still listed in the Guide because they are still available in the supply chain, a cross-reference indicates the new 'parent' company from which technical or commercial information can be obtained.

Appendix 2   Gives contact details of useful contacts, including the National Poisons
                    Information Service.

Appendix 3   Gives details of the keys to crop and weed growth stages, including the
                    publication reference for each. The numeric codes are used in the descriptive
                    sections of the pesticide profiles (Section 2).

Appendix 4   Shows the full text for code letters and numbers used in the pesticide
                    profiles (Section 2) to indicate personal protective equipment requirements
                    and label precautions.

Appendix 5   Shows the full text for the formulation abbreviations used in the pesticide
                    profiles (Section 2). Other abbreviations and acronyms used in the Guide are
                    also explained here.

Appendix 6   Provides full definitions of officially agreed descriptive phrases for crops or
                    situations used in the pesticide profiles (Section 2) where misunderstandings
                    can occur.

Appendix 7   Shows a list of useful reference publications, which amplify the summarised
                    information provided in this section.

SECTION 1
CROP/PEST GUIDE

Crop/Pest Guide Index

Important note: The Crop/Pest Guide Index refers to pages on which the subject can be located.

Products that are not classed as pesticides (and therefore are not evaluated by Chemicals Regulation Division (CRD) do not have a MAPP No. They work by physical means and can be effective but are not governed by pesticide legislation and can be used in all appropriate crop/situations. These products are variously effective against aphids, mealy bugs, mites, mosquitoes, spider mites, suckers, whitefly and powdery mildew. For details see section 321 (Physical Pest Control) on page 477.

Crop/Pest Guide

Important note: For convenience, some crops and pests or targets have been brought together into genetic groups in this guide, e.g. 'cereals', 'annual grasses'. It is essential to check the profile entry in Section 2 *and* the label to ensure that a product is approved for a specific crop/pest combination, e.g. winter wheat/blackgrass.

Arable and vegetable crops

Agricultural herbage - Grass

Crop control	Desiccation	glyphosate
Pests	Aphids	esfenvalerate
	Flies	esfenvalerate
	Slugs/snails	ferric phosphate
Plant growth regulation	Growth control	gibberellins
Weeds	Broad-leaved weeds	2,4-D, 2,4-D + dicamba, 2,4-D + MCPA, 2,4-DB, 2,4-DB + MCPA, amidosulfuron, aminopyralid, aminopyralid + triclopyr, clopyralid, clopyralid + florasulam + fluroxypyr, clopyralid + triclopyr, florasulam + fluroxypyr, fluroxypyr, fluroxypyr (*off-label*), fluroxypyr + triclopyr, glyphosate, glyphosate (*wiper application*), MCPA, thifensulfuron-methyl, tribenuron-methyl
	Crops as weeds	florasulam + fluroxypyr, fluroxypyr (*off-label*)
	Grass weeds	glyphosate
	Weeds, miscellaneous	aminopyralid + triclopyr, clopyralid + florasulam + fluroxypyr, clopyralid + triclopyr, florasulam + fluroxypyr, fluroxypyr (*off-label*), glyphosate, MCPA
	Woody weeds/scrub	clopyralid + triclopyr

Agricultural herbage - Herbage legumes

Diseases	Powdery mildew	physical pest control
Pests	Aphids	physical pest control
	Mealybugs	physical pest control
	Scale insects	physical pest control
	Spider mites	physical pest control
	Suckers	physical pest control
	Whiteflies	physical pest control
Plant growth regulation	Growth control	trinexapac-ethyl (*off-label*)
Weeds	Broad-leaved weeds	2,4-DB, propyzamide
	Grass weeds	clethodim (*off-label*), propyzamide

Agricultural herbage - Herbage seed

Diseases	Crown rust	bixafen + prothioconazole (*off-label*), boscalid + epoxiconazole (*off-label*)
	Disease control	epoxiconazole (*off-label*), epoxiconazole + pyraclostrobin (*off-label*), tebuconazole (*off-label*)
	Drechslera leaf spot	bixafen + prothioconazole (*off-label*)

	Leaf spot	prothioconazole + tebuconazole (*off-label*)
	Net blotch	boscalid + epoxiconazole (*off-label*)
	Powdery mildew	bixafen + prothioconazole (*off-label*), cyflufenamid (*off-label*), epoxiconazole (*off-label*), prothioconazole + tebuconazole (*off-label*), sulphur (*off-label*)
	Rhynchosporium	bixafen + prothioconazole (*off-label*), boscalid + epoxiconazole (*off-label*), prothioconazole + tebuconazole (*off-label*)
	Rust	bixafen + prothioconazole (*off-label*), boscalid + epoxiconazole (*off-label*), epoxiconazole (*off-label*), prothioconazole + tebuconazole (*off-label*), pyraclostrobin (*off-label*), tebuconazole (*off-label*), trifloxystrobin (*off-label*)
Pests	Aphids	deltamethrin (*off-label*), esfenvalerate (*off-label*), tau-fluvalinate (*off-label*), zeta-cypermethrin (*off-label*)
	Caterpillars	lambda-cyhalothrin (*off-label*)
	Flies	esfenvalerate
	Pests, miscellaneous	deltamethrin (*off-label*), lambda-cyhalothrin (*off-label*), zeta-cypermethrin (*off-label*)
Plant growth regulation	Growth control	trinexapac-ethyl, trinexapac-ethyl (*off-label*)
Weeds	Broad-leaved weeds	bifenox (*off-label*), bromoxynil + diflufenican (*off-label*), clopyralid (*off-label*), clopyralid + florasulam + fluroxypyr, dicamba + MCPA + mecoprop-P, diflufenican (*off-label*), diflufenican + flufenacet (*off-label*), ethofumesate, florasulam (*off-label*), florasulam + fluroxypyr, florasulam + fluroxypyr (*off-label*), florasulam + halauxifen-methyl (*off-label*), fluroxypyr, MCPA, mecoprop-P, pendimethalin (*off-label*), propyzamide
	Crops as weeds	bromoxynil + diflufenican (*off-label*), clopyralid (*off-label*), diflufenican (*off-label*), florasulam + fluroxypyr
	Grass weeds	clodinafop-propargyl (*off-label*), diflufenican (*off-label*), diflufenican + flufenacet (*off-label*), ethofumesate, fenoxaprop-P-ethyl (*off-label*), pendimethalin (*off-label*), propyzamide
	Weeds, miscellaneous	clopyralid + florasulam + fluroxypyr, diflufenican (*off-label*), florasulam + fluroxypyr, MCPA

Brassicas - Brassica seed crops

Weeds	Broad-leaved weeds	propyzamide
	Grass weeds	propyzamide

Brassicas - Brassicas, general

Diseases	Alternaria	azoxystrobin + difenoconazole (*off-label*), cyprodinil + fludioxonil (*off-label*)
	Botrytis	boscalid + pyraclostrobin (*off-label*), cyprodinil + fludioxonil (*off-label*)
	Downy mildew	dimethomorph (*off-label*)
	Pythium	metalaxyl-M (*off-label*)
	Rhizoctonia	boscalid + pyraclostrobin (*off-label*), cyprodinil + fludioxonil (*off-label*)
	Ring spot	azoxystrobin + difenoconazole (*off-label*)

	Sclerotinia	boscalid + pyraclostrobin (*off-label*)
	Stem canker	cyprodinil + fludioxonil (*off-label*)
Pests	Aphids	thiacloprid (*off-label*)
	Caterpillars	indoxacarb (*off-label*)
	Flies	Metarhizium anisopliae (*off-label*)
Weeds	Broad-leaved weeds	clopyralid (*off-label*), pyridate (*off-label*)
	Crops as weeds	clopyralid (*off-label*)

Brassicas - Fodder brassicas

Diseases	Alternaria	azoxystrobin, azoxystrobin (*moderate control*), cyprodinil + fludioxonil (*off-label*), difenoconazole
	Anthracnose	Bacillus amyloliquefaciens D747 (*off-label*)
	Black rot	Bacillus amyloliquefaciens D747 (*off-label*)
	Botrytis	Bacillus amyloliquefaciens D747 (*off-label*), Bacillus subtilis (*off-label*), cyprodinil + fludioxonil (*off-label*), Gliocladium catenulatum (*off-label*)
	Damping off	Bacillus amyloliquefaciens D747 (*off-label*), Bacillus subtilis (*off-label*)
	Downy mildew	Bacillus amyloliquefaciens D747 (*off-label*), fluopicolide + propamocarb hydrochloride (*off-label*), metalaxyl-M (*off-label*)
	Fusarium	Bacillus amyloliquefaciens D747 (*off-label*), Gliocladium catenulatum (*off-label*)
	Phytophthora	Gliocladium catenulatum (*off-label*)
	Powdery mildew	azoxystrobin + difenoconazole, Bacillus amyloliquefaciens D747 (*off-label*), potassium bicarbonate (commodity substance) (*off-label*)
	Pythium	Bacillus amyloliquefaciens D747 (*off-label*), Bacillus subtilis (*off-label*), metalaxyl-M (*off-label*)
	Rhizoctonia	Bacillus amyloliquefaciens D747 (*off-label*), cyprodinil + fludioxonil (*off-label*), Gliocladium catenulatum (*off-label*)
	Ring spot	azoxystrobin, azoxystrobin (*moderate control*), difenoconazole
	Sclerotinia	Bacillus amyloliquefaciens D747 (*off-label*)
	Spear rot	Bacillus amyloliquefaciens D747 (*off-label*)
	Stem canker	cyprodinil + fludioxonil (*off-label*)
	White blister	azoxystrobin, azoxystrobin (*moderate control*), azoxystrobin + difenoconazole, boscalid + pyraclostrobin (*off-label*)
Pests	Aphids	esfenvalerate, spirotetramat, thiacloprid (*off-label*)
	Beetles	alpha-cypermethrin
	Caterpillars	alpha-cypermethrin, Bacillus thuringiensis, Bacillus thuringiensis (*off-label*), esfenvalerate, indoxacarb (*off-label*), spinosad (*off-label*)
	Flies	spinosad (*off-label*)
	Slugs/snails	ferric phosphate
	Whiteflies	spirotetramat
Weeds	Broad-leaved weeds	clomazone (*off-label*), clopyralid (*off-label*), dimethenamid-p + metazachlor (*off-label*), metazachlor (*off-label*), napropamide, pendimethalin (*off-label*), pyridate (*off-label*)

	Crops as weeds	clopyralid (*off-label*), fluazifop-P-butyl, fluazifop-P-butyl (*stockfeed only*)
	Grass weeds	dimethenamid-p + metazachlor (*off-label*), fluazifop-P-butyl, fluazifop-P-butyl (*stockfeed only*), metazachlor (*off-label*), napropamide, pendimethalin (*off-label*), S-metolachlor (*off-label*)

Brassicas - Leaf and flowerhead brassicas

Diseases	Alternaria	azoxystrobin, azoxystrobin (*moderate control*), boscalid + pyraclostrobin, cyprodinil + fludioxonil (*off-label*), difenoconazole, difenoconazole + fluxapyroxad (*reduction*), prothioconazole, tebuconazole, tebuconazole + trifloxystrobin
	Anthracnose	Bacillus amyloliquefaciens D747 (*off-label*)
	Black rot	azoxystrobin (*off-label*), Bacillus amyloliquefaciens D747 (*off-label*)
	Botrytis	Bacillus amyloliquefaciens D747 (*off-label*), Bacillus subtilis (*off-label*), cyprodinil + fludioxonil (*off-label*), Gliocladium catenulatum (*off-label*)
	Damping off	Bacillus amyloliquefaciens D747 (*off-label*), Bacillus subtilis (*off-label*)
	Downy mildew	Bacillus amyloliquefaciens D747 (*off-label*), fluopicolide + propamocarb hydrochloride (*off-label*), mandipropamid, metalaxyl-M (*off-label*)
	Fusarium	Bacillus amyloliquefaciens D747 (*off-label*), Gliocladium catenulatum (*off-label*)
	Light leaf spot	prothioconazole, tebuconazole, tebuconazole + trifloxystrobin
	Phoma leaf spot	tebuconazole + trifloxystrobin
	Phytophthora	Gliocladium catenulatum (*off-label*)
	Powdery mildew	azoxystrobin + difenoconazole, Bacillus amyloliquefaciens D747 (*off-label*), physical pest control, potassium bicarbonate (commodity substance) (*off-label*), prothioconazole, tebuconazole, tebuconazole + trifloxystrobin
	Pythium	Bacillus amyloliquefaciens D747 (*off-label*), Bacillus subtilis (*off-label*), metalaxyl-M, metalaxyl-M (*off-label*)
	Rhizoctonia	Bacillus amyloliquefaciens D747 (*off-label*), cyprodinil + fludioxonil (*off-label*), Gliocladium catenulatum (*off-label*)
	Ring spot	azoxystrobin, azoxystrobin (*moderate control*), boscalid + pyraclostrobin, difenoconazole, difenoconazole + fluxapyroxad (*reduction*), prothioconazole, tebuconazole, tebuconazole + trifloxystrobin
	Sclerotinia	Bacillus amyloliquefaciens D747 (*off-label*)
	Spear rot	Bacillus amyloliquefaciens D747 (*off-label*)
	Stem canker	cyprodinil + fludioxonil (*off-label*), prothioconazole
	Storage rots	1-methylcyclopropene (*off-label*)
	White blister	azoxystrobin, azoxystrobin (*moderate control*), azoxystrobin (*off-label*), azoxystrobin + difenoconazole, boscalid + pyraclostrobin, boscalid + pyraclostrobin (*off-label*), boscalid + pyraclostrobin

		(*qualified minor use*), mancozeb + metalaxyl-M (*off-label*), tebuconazole + trifloxystrobin
Pests	Aphids	acetamiprid (*off-label*), beta-cyfluthrin, cypermethrin, deltamethrin, esfenvalerate, flonicamid (*off-label*), physical pest control, pyrethrins, spirotetramat, spirotetramat (*off-label*), thiacloprid, thiacloprid (*off-label*)
	Beetles	alpha-cypermethrin, deltamethrin
	Birds/mammals	aluminium ammonium sulphate
	Caterpillars	alpha-cypermethrin, Bacillus thuringiensis, Bacillus thuringiensis (*off-label*), beta-cyfluthrin, cypermethrin, deltamethrin, esfenvalerate, indoxacarb, indoxacarb (*off-label*), lambda-cyhalothrin, pyrethrins, spinosad, spinosad (*off-label*)
	Flies	cyantraniliprole, spinosad (*off-label*)
	Mealybugs	physical pest control
	Scale insects	physical pest control
	Slugs/snails	ferric phosphate, metaldehyde
	Spider mites	physical pest control, pyrethrins
	Suckers	physical pest control
	Thrips	pyrethrins
	Whiteflies	acetamiprid (*off-label*), cypermethrin, deltamethrin, lambda-cyhalothrin, physical pest control, spirotetramat, spirotetramat (*off-label*)
Weeds	Broad-leaved weeds	clomazone (*off-label*), clopyralid, clopyralid (*off-label*), dimethenamid-p + metazachlor (*off-label*), dimethenamid-p + pendimethalin (*off-label*), metazachlor, metazachlor (*off-label*), napropamide, napropamide (*off-label*), pendimethalin, pendimethalin (*off-label*), propyzamide (*off-label*), pyridate, pyridate (*off-label*)
	Crops as weeds	clethodim (*off-label*), clopyralid (*off-label*), cycloxydim, cycloxydim (*off-label*), pendimethalin
	Grass weeds	clethodim (*off-label*), cycloxydim, cycloxydim (*off-label*), dimethenamid-p + metazachlor (*off-label*), dimethenamid-p + pendimethalin (*off-label*), metazachlor, metazachlor (*off-label*), napropamide, napropamide (*off-label*), pendimethalin, pendimethalin (*off-label*), propyzamide (*off-label*), S-metolachlor (*off-label*)

Brassicas - Mustard

Crop control	Desiccation	glyphosate
Diseases	Alternaria	Bacillus amyloliquefaciens D747, cyprodinil + fludioxonil (*off-label*), difenoconazole (*off-label*)
	Anthracnose	Bacillus amyloliquefaciens D747 (*off-label*)
	Bacterial blight	Bacillus amyloliquefaciens D747 (*off-label*)
	Black rot	azoxystrobin (*off-label*)
	Botrytis	Bacillus amyloliquefaciens D747 (*off-label*), Bacillus subtilis (*off-label*), cyprodinil + fludioxonil (*off-label*)
	Cercospora leaf spot	Bacillus amyloliquefaciens D747 (*off-label*)
	Damping off	Bacillus amyloliquefaciens D747 (*off-label*)
	Disease control	boscalid (*off-label*), mandipropamid, prothioconazole (*off-label*), tebuconazole (*off-label*)

	Downy mildew	Bacillus amyloliquefaciens D747 (*off-label*), fluopicolide + propamocarb hydrochloride (*off-label*), mandipropamid
	Fusarium	Bacillus amyloliquefaciens D747, Bacillus amyloliquefaciens D747 (*off-label*)
	Powdery mildew	Bacillus amyloliquefaciens D747, Bacillus amyloliquefaciens D747 (*off-label*), difenoconazole (*off-label*)
	Pythium	Bacillus amyloliquefaciens D747, Bacillus amyloliquefaciens D747 (*off-label*)
	Rhizoctonia	Bacillus amyloliquefaciens D747, Bacillus amyloliquefaciens D747 (*off-label*), cyprodinil + fludioxonil (*off-label*)
	Rust	tebuconazole (*off-label*)
	Sclerotinia	Bacillus amyloliquefaciens D747 (*off-label*), difenoconazole (*off-label*)
	Stem canker	cyprodinil + fludioxonil (*off-label*)
	Stemphylium	Bacillus amyloliquefaciens D747 (*off-label*)
	White blister	azoxystrobin (*off-label*)
Pests	Aphids	Beauveria bassiana (*off-label*), cypermethrin, tau-fluvalinate (*off-label*)
	Beetles	deltamethrin, indoxacarb (*off-label*), thiacloprid
	Caterpillars	deltamethrin, lambda-cyhalothrin (*off-label*)
	Pests, miscellaneous	cypermethrin, lambda-cyhalothrin (*off-label*)
	Slugs/snails	ferric phosphate
	Spider mites	Beauveria bassiana (*off-label*)
	Thrips	Beauveria bassiana (*off-label*)
	Weevils	deltamethrin
	Whiteflies	Beauveria bassiana (*off-label*)
Weeds	Broad-leaved weeds	bifenox (*off-label*), clopyralid (*off-label*), clopyralid + picloram (*off-label*), glyphosate, lenacil (*off-label*), metazachlor (*off-label*), napropamide (*off-label*)
	Crops as weeds	fluazifop-P-butyl, glyphosate, propaquizafop (*off-label*)
	Grass weeds	clethodim (*off-label*), fluazifop-P-butyl, glyphosate, metazachlor (*off-label*), napropamide (*off-label*), propaquizafop (*off-label*)
	Weeds, miscellaneous	glyphosate, lenacil (*off-label*), napropamide (*off-label*)

Brassicas - Root brassicas

Diseases	Alternaria	azoxystrobin (*off-label*), difenoconazole + fluxapyroxad (*off-label*), isopyrazam (*off-label*), prothioconazole, tebuconazole, tebuconazole + trifloxystrobin (*off-label*)
	Anthracnose	Bacillus amyloliquefaciens D747 (*off-label*)
	Black rot	Bacillus amyloliquefaciens D747 (*off-label*)
	Botrytis	Bacillus amyloliquefaciens D747 (*off-label*), Bacillus subtilis (*off-label*), boscalid + pyraclostrobin (*off-label*), Gliocladium catenulatum (*off-label*)
	Cercospora leaf spot	isopyrazam (*off-label*)
	Damping off	Bacillus amyloliquefaciens D747 (*off-label*)
	Disease control	boscalid + pyraclostrobin (*off-label*)

	Downy mildew	azoxystrobin (*off-label*), Bacillus amyloliquefaciens D747 (*off-label*), dimethomorph (*off-label*), fluopicolide + propamocarb hydrochloride (*off-label*), metalaxyl-M (*off-label*), propamocarb hydrochloride (*off-label*)
	Fungus diseases	azoxystrobin + difenoconazole (*off-label*)
	Fusarium	Bacillus amyloliquefaciens D747 (*off-label*), Gliocladium catenulatum (*off-label*)
	Leaf spot	tebuconazole + trifloxystrobin (*off-label*)
	Mycosphaerella	tebuconazole + trifloxystrobin (*off-label*)
	Phytophthora	Gliocladium catenulatum (*off-label*)
	Powdery mildew	Bacillus amyloliquefaciens D747 (*off-label*), difenoconazole + fluxapyroxad (*off-label*), isopyrazam (*off-label*), physical pest control, potassium bicarbonate (commodity substance) (*off-label*), prothioconazole, sulphur, tebuconazole, tebuconazole + trifloxystrobin (*off-label*)
	Pythium	Bacillus amyloliquefaciens D747 (*off-label*), metalaxyl-M, metalaxyl-M (*off-label*)
	Ramularia leaf spots	isopyrazam (*off-label*)
	Rhizoctonia	azoxystrobin (*off-label*), Bacillus amyloliquefaciens D747 (*off-label*), difenoconazole + fluxapyroxad (*off-label*), Gliocladium catenulatum (*off-label*)
	Rust	boscalid + pyraclostrobin (*off-label*)
	Sclerotinia	Bacillus amyloliquefaciens D747 (*off-label*), boscalid + pyraclostrobin (*off-label*), cyprodinil + fludioxonil (*off-label*), difenoconazole + fluxapyroxad (*off-label*), isopyrazam (*off-label*), tebuconazole + trifloxystrobin (*off-label*)
	Septoria	tebuconazole + trifloxystrobin (*off-label*)
	Spear rot	Bacillus amyloliquefaciens D747 (*off-label*)
	Stem canker	azoxystrobin + difenoconazole (*off-label*), isopyrazam (*off-label*)
	White blister	azoxystrobin (*off-label*), metalaxyl-M (*off-label*)
Pests	Aphids	cypermethrin, deltamethrin (*off-label*), esfenvalerate, physical pest control, pirimicarb (*off-label*), pyrethrins, spirotetramat, thiacloprid (*off-label*)
	Beetles	deltamethrin, deltamethrin (*off-label*), lambda-cyhalothrin (*off-label*)
	Caterpillars	Bacillus thuringiensis, Bacillus thuringiensis (*off-label*), Bacillus thuringiensis aizawai GC-91, chlorantraniliprole (*off-label*), deltamethrin, deltamethrin (*off-label*), esfenvalerate, indoxacarb (*off-label*), lambda-cyhalothrin (*off-label*), pyrethrins, spinosad (*off-label*)
	Cutworms	Bacillus thuringiensis (*off-label*), cypermethrin, deltamethrin (*off-label*), lambda-cyhalothrin (*off-label*)
	Flies	chlorantraniliprole (*off-label*), cypermethrin, deltamethrin (*off-label*), lambda-cyhalothrin (*off-label*), spinosad (*off-label*)
	Leaf miners	spinosad (*off-label*)
	Mealybugs	physical pest control
	Pests, miscellaneous	cypermethrin, deltamethrin (*off-label*), lambda-cyhalothrin (*off-label*)

	Scale insects	physical pest control
	Slugs/snails	ferric phosphate
	Spider mites	physical pest control, pyrethrins
	Suckers	physical pest control
	Thrips	deltamethrin (*off-label*), pyrethrins, spinosad (*off-label*)
	Weevils	deltamethrin (*off-label*), lambda-cyhalothrin (*off-label*)
	Whiteflies	physical pest control, spirotetramat
Plant growth regulation	Quality/yield control	sulphur
Weeds	Broad-leaved weeds	clomazone (*off-label*), clopyralid, dimethenamid-p + metazachlor (*off-label*), glyphosate, isoxaben (*off-label*), metamitron (*off-label*), metazachlor (*off-label*), napropamide (*off-label*), pelargonic acid (*off-label*), pendimethalin (*off-label*), prosulfocarb (*off-label*), S-metolachlor (*off-label*)
	Crops as weeds	cycloxydim, fluazifop-P-butyl, fluazifop-P-butyl (*stockfeed only*), glyphosate
	Grass weeds	clethodim (*off-label*), cycloxydim, cycloxydim (*off-label*), dimethenamid-p + metazachlor (*off-label*), fluazifop-P-butyl, fluazifop-P-butyl (*stockfeed only*), glyphosate, metamitron (*off-label*), metazachlor (*off-label*), napropamide (*off-label*), pelargonic acid (*off-label*), propaquizafop, prosulfocarb (*off-label*), S-metolachlor (*off-label*)
	Mosses	pelargonic acid (*off-label*)
	Weeds, miscellaneous	glyphosate, glyphosate (*off-label*), pelargonic acid (*off-label*)

Brassicas - Salad greens

Diseases	Alternaria	azoxystrobin + difenoconazole (*off-label*), cyprodinil + fludioxonil (*off-label*), difenoconazole (*off-label*)
	Anthracnose	Bacillus amyloliquefaciens D747 (*off-label*)
	Bacterial blight	Bacillus amyloliquefaciens D747 (*off-label*)
	Black rot	azoxystrobin (*off-label*), Bacillus amyloliquefaciens D747 (*off-label*)
	Botrytis	Bacillus amyloliquefaciens D747 (*off-label*), Bacillus subtilis (*off-label*), cyprodinil + fludioxonil (*off-label*), fenhexamid (*off-label - baby leaf production*), fluopyram + trifloxystrobin (*off-label*), Gliocladium catenulatum (*off-label*), pyrimethanil (*off-label*)
	Cercospora leaf spot	Bacillus amyloliquefaciens D747 (*off-label*)
	Damping off	Bacillus amyloliquefaciens D747 (*off-label*), Bacillus subtilis (*off-label*)
	Disease control	dimethomorph + mancozeb (*off-label*), mandipropamid
	Downy mildew	Bacillus amyloliquefaciens D747 (*off-label*), cos-oga (*off-label*), dimethomorph (*off-label*), dimethomorph + mancozeb (*off-label*), fluopicolide + propamocarb hydrochloride (*off-label*), mancozeb + metalaxyl-M (*off-label*), mandipropamid, metalaxyl-M (*off-label*), potassium bicarbonate (commodity substance) (*off-label*)

	Fusarium	Bacillus amyloliquefaciens D747 (*off-label*), Gliocladium catenulatum (*off-label*), Trichoderma asperellum (Strain T34) (*off-label*)
	Leaf spot	difenoconazole (*off-label*)
	Phytophthora	Gliocladium catenulatum (*off-label*)
	Powdery mildew	Ampelomyces quisqualis (Strain AQ10) (*off-label*), Bacillus amyloliquefaciens D747 (*off-label*), cos-oga (*off-label*), fluopyram + trifloxystrobin (*off-label*), potassium bicarbonate (commodity substance) (*off-label*)
	Pythium	Bacillus amyloliquefaciens D747 (*off-label*), metalaxyl-M, metalaxyl-M (*off-label*), Trichoderma asperellum (Strain T34) (*off-label*)
	Rhizoctonia	Bacillus amyloliquefaciens D747 (*off-label*), cyprodinil + fludioxonil (*off-label*), Gliocladium catenulatum (*off-label*)
	Ring spot	azoxystrobin + difenoconazole (*off-label*), difenoconazole (*off-label*)
	Sclerotinia	Bacillus amyloliquefaciens D747 (*off-label*), fluopyram + trifloxystrobin (*off-label*)
	Spear rot	Bacillus amyloliquefaciens D747 (*off-label*)
	Stem canker	cyprodinil + fludioxonil (*off-label*)
	Stemphylium	Bacillus amyloliquefaciens D747 (*off-label*)
	White blister	azoxystrobin (*off-label*), boscalid + pyraclostrobin (*off-label*)
Pests	Aphids	acetamiprid (*off-label*), Beauveria bassiana (*off-label*), esfenvalerate, pirimicarb (*off-label*), spirotetramat (*off-label*), thiacloprid (*off-label*), thiacloprid (*off-label - baby leaf production*)
	Caterpillars	Bacillus thuringiensis, Bacillus thuringiensis (*off-label*), chlorantraniliprole (*off-label*), esfenvalerate, indoxacarb (*off-label*), lambda-cyhalothrin (*off-label*), spinosad
	Cutworms	Bacillus thuringiensis (*off-label*)
	Pests, miscellaneous	lambda-cyhalothrin (*off-label*)
	Slugs/snails	ferric phosphate
	Spider mites	abamectin (*off-label*), Beauveria bassiana (*off-label*), Lecanicillium muscarium (*off-label*)
	Thrips	Beauveria bassiana (*off-label*), Lecanicillium muscarium (*off-label*)
	Whiteflies	Beauveria bassiana (*off-label*), Lecanicillium muscarium (*off-label*), spirotetramat (*off-label*)
Weeds	Broad-leaved weeds	chlorpropham, chlorpropham (*off-label*), clomazone (*off-label*), clopyralid (*off-label*), dimethenamid-p + metazachlor (*off-label*), dimethenamid-p + pendimethalin (*off-label*), lenacil (*off-label*), metazachlor (*off-label*), napropamide (*off-label*), pendimethalin (*off-label*), pendimethalin (*off-label - for baby leaf production*), propyzamide (*off-label*), S-metolachlor (*off-label*)
	Crops as weeds	clopyralid (*off-label*), cycloxydim (*off-label*), fluazifop-P-butyl
	Grass weeds	chlorpropham, chlorpropham (*off-label*), cycloxydim (*off-label*), dimethenamid-p + metazachlor (*off-label*), dimethenamid-p + pendimethalin (*off-label*), fluazifop-

P-butyl, metazachlor (*off-label*), napropamide (*off-label*), propyzamide (*off-label*), S-metolachlor (*off-label*)

| Weeds, miscellaneous | lenacil (*off-label*) |

Cereals - Barley

| Crop control | Desiccation | glyphosate |

| Diseases | Covered smut | fludioxonil (*seed treatment*), fludioxonil + tefluthrin, fluopyram + prothioconazole + tebuconazole, prothioconazole + tebuconazole |

| | Disease control | epoxiconazole + folpet, fluxapyroxad + metconazole, isopyrazam, isopyrazam + prothioconazole |

| | Ear blight | prothioconazole + tebuconazole |

| | Eyespot | azoxystrobin + cyproconazole (*reduction*), bixafen + prothioconazole (*reduction of incidence and severity*), boscalid + epoxiconazole (*moderate control*), boscalid + epoxiconazole + pyraclostrobin (*moderate control*), cyprodinil, epoxiconazole (*reduction*), epoxiconazole + metconazole (*reduction*), fluoxastrobin + prothioconazole (*reduction*), fluoxastrobin + prothioconazole + trifloxystrobin (*reduction*), prochloraz + proquinazid + tebuconazole, prothioconazole, prothioconazole + spiroxamine, prothioconazole + spiroxamine + tebuconazole (*reduction*), prothioconazole + tebuconazole, prothioconazole + tebuconazole (*reduction*), prothioconazole + trifloxystrobin (*reduction*), prothioconazole + trifloxystrobin (*reduction in severity*) |

| | Foot rot | fludioxonil (*seed treatment*), fludioxonil + tefluthrin, imazalil + ipconazole (*useful protection*) |

| | Fusarium | fluopyram + prothioconazole + tebuconazole, prothioconazole + tebuconazole |

| | Late ear diseases | bixafen + prothioconazole, fluoxastrobin + prothioconazole, fluoxastrobin + prothioconazole + trifloxystrobin, prothioconazole, prothioconazole + spiroxamine + tebuconazole, prothioconazole + tebuconazole |

| | Leaf stripe | fludioxonil (*seed treatment - reduction*), fludioxonil + tefluthrin (*partial control*), fluopyram + prothioconazole + tebuconazole, imazalil + ipconazole, prothioconazole + tebuconazole |

| | Loose smut | fluopyram + prothioconazole + tebuconazole, imazalil + ipconazole, prothioconazole + tebuconazole |

| | Net blotch | azoxystrobin, azoxystrobin + cyproconazole, benzovindiflupyr, benzovindiflupyr + prothioconazole, bixafen + prothioconazole, boscalid + epoxiconazole, boscalid + epoxiconazole + pyraclostrobin, cyprodinil, cyprodinil + isopyrazam, epoxiconazole, epoxiconazole (*moderate control*), epoxiconazole + fluxapyroxad, epoxiconazole + fluxapyroxad + pyraclostrobin, epoxiconazole + isopyrazam, epoxiconazole + metconazole, epoxiconazole + pyraclostrobin, fluopyram + prothioconazole + tebuconazole (*seed borne*), fluoxastrobin + prothioconazole, fluoxastrobin + prothioconazole + |

trifloxystrobin, fluxapyroxad, fluxapyroxad + metconazole, fluxapyroxad + pyraclostrobin, mefentrifluconazole (*reduction*), metconazole (*reduction*), prochloraz + proquinazid + tebuconazole (*moderate control*), prothioconazole, prothioconazole + spiroxamine, prothioconazole + spiroxamine + tebuconazole, prothioconazole + tebuconazole, prothioconazole + trifloxystrobin, pyraclostrobin, tebuconazole, trifloxystrobin

Powdery mildew   azoxystrobin, azoxystrobin (*moderate control*), azoxystrobin + cyproconazole, bixafen + prothioconazole, boscalid + epoxiconazole (*moderate control*), boscalid + epoxiconazole + pyraclostrobin (*moderate control*), cyflufenamid, cyprodinil, cyprodinil + isopyrazam, epoxiconazole, epoxiconazole + fluxapyroxad (*moderate control*), epoxiconazole + fluxapyroxad + pyraclostrobin (*moderate control*), epoxiconazole + metconazole (*moderate control*), epoxiconazole + pyraclostrobin, fluoxastrobin + prothioconazole, fluoxastrobin + prothioconazole + trifloxystrobin, flutriafol, fluxapyroxad (*reduction*), fluxapyroxad + metconazole (*moderate control*), fluxapyroxad + pyraclostrobin (*moderate control*), metconazole, metconazole (*moderate control*), metrafenone, prochloraz + proquinazid + tebuconazole, proquinazid, prothioconazole, prothioconazole + spiroxamine, prothioconazole + spiroxamine + tebuconazole, prothioconazole + tebuconazole, prothioconazole + trifloxystrobin, sulphur, tebuconazole, tebuconazole (*moderate control*)

Ramularia leaf spots   benzovindiflupyr , benzovindiflupyr + prothioconazole, bixafen + prothioconazole, boscalid + epoxiconazole (*moderate control*), boscalid + epoxiconazole + pyraclostrobin (*moderate control*), cyprodinil + isopyrazam, epoxiconazole (*reduction*), epoxiconazole + fluxapyroxad, epoxiconazole + fluxapyroxad + pyraclostrobin, epoxiconazole + isopyrazam, epoxiconazole + metconazole (*moderate control*), epoxiconazole + pyraclostrobin (*moderate control*), epoxiconazole + pyraclostrobin (*reduction*), fluxapyroxad + metconazole, fluxapyroxad + pyraclostrobin, mefentrifluconazole

Rhynchosporium   azoxystrobin, azoxystrobin (*reduction*), azoxystrobin + cyproconazole (*moderate control*), benzovindiflupyr (*moderate control*), benzovindiflupyr + prothioconazole (*moderate control*), bixafen + prothioconazole, boscalid + epoxiconazole (*moderate control*), boscalid + epoxiconazole + pyraclostrobin, cyprodinil, cyprodinil + isopyrazam, epoxiconazole, epoxiconazole + fluxapyroxad, epoxiconazole + fluxapyroxad + pyraclostrobin, epoxiconazole + folpet, epoxiconazole + isopyrazam, epoxiconazole + metconazole, epoxiconazole + pyraclostrobin, fluoxastrobin + prothioconazole, fluoxastrobin + prothioconazole + trifloxystrobin, flutriafol, fluxapyroxad, fluxapyroxad + metconazole, fluxapyroxad + pyraclostrobin, folpet (*reduction*), mefentrifluconazole, metconazole, metconazole (*reduction*), prochloraz + proquinazid + tebuconazole

		(*moderate control*), prothioconazole, prothioconazole + spiroxamine, prothioconazole + spiroxamine + tebuconazole, prothioconazole + tebuconazole, prothioconazole + trifloxystrobin, pyraclostrobin (*moderate control*), tebuconazole, tebuconazole (*moderate control*), trifloxystrobin
	Rust	azoxystrobin, azoxystrobin + cyproconazole, benzovindiflupyr, benzovindiflupyr + prothioconazole, bixafen + prothioconazole, boscalid + epoxiconazole, boscalid + epoxiconazole + pyraclostrobin, cyprodinil + isopyrazam, epoxiconazole, epoxiconazole + fluxapyroxad, epoxiconazole + fluxapyroxad + pyraclostrobin, epoxiconazole + isopyrazam, epoxiconazole + metconazole, epoxiconazole + pyraclostrobin, fluoxastrobin + prothioconazole, fluoxastrobin + prothioconazole + trifloxystrobin, flutriafol, fluxapyroxad, fluxapyroxad (*moderate control*), fluxapyroxad + metconazole, fluxapyroxad + pyraclostrobin, mefentrifluconazole, metconazole, prochloraz + proquinazid + tebuconazole, prothioconazole, prothioconazole + spiroxamine, prothioconazole + spiroxamine + tebuconazole, prothioconazole + tebuconazole, prothioconazole + trifloxystrobin, pyraclostrobin, tebuconazole, trifloxystrobin
	Seed-borne diseases	fludioxonil + tebuconazole, fluopyram + prothioconazole + tebuconazole, imazalil + ipconazole (*useful protection*), ipconazole
	Snow mould	fludioxonil (*seed treatment*), fludioxonil + tefluthrin
	Soil-borne diseases	ipconazole
	Sooty moulds	prothioconazole + tebuconazole
	Take-all	azoxystrobin (*reduction*), azoxystrobin + cyproconazole (*reduction*), fluoxastrobin + prothioconazole (*reduction*), silthiofam (*seed treatment*)
Pests	Aphids	alpha-cypermethrin, beta-cyfluthrin, cypermethrin, cypermethrin (*autumn sown*), deltamethrin, esfenvalerate, lambda-cyhalothrin, tau-fluvalinate, zeta-cypermethrin
	Beetles	lambda-cyhalothrin
	Birds/mammals	aluminium ammonium sulphate
	Caterpillars	lambda-cyhalothrin
	Flies	cypermethrin, cypermethrin (*autumn sown*), cypermethrin (*qualified recommendation*), cypermethrin (*reduction*), fludioxonil + tefluthrin
	Pests, miscellaneous	cypermethrin
	Slugs/snails	ferric phosphate
	Weevils	lambda-cyhalothrin
	Wireworms	fludioxonil + tefluthrin
Plant growth regulation	Growth control	chlormequat, chlormequat + 2-chloroethylphosphonic acid, ethephon, ethephon + mepiquat chloride, mepiquat chloride + prohexadione-calcium, prohexadione-calcium, prohexadione-calcium + trinexapac-ethyl, trinexapac-ethyl

	Quality/yield control	ethephon + mepiquat chloride (*low lodging situations*), mepiquat chloride + prohexadione-calcium, sulphur
Weeds	Broad-leaved weeds	2,4-D, 2,4-D + MCPA, 2,4-DB, 2,4-DB + MCPA, amidosulfuron, amidosulfuron + iodosulfuron-methyl-sodium, aminopyralid + halauxifen-methyl, bifenox, bromoxynil, bromoxynil + diflufenican, bromoxynil + diflufenican (*at 1.9 l/ha only*), carfentrazone-ethyl, carfentrazone-ethyl + mecoprop-P, chlorotoluron + diflufenican + pendimethalin, clopyralid, clopyralid + florasulam, clopyralid + florasulam + fluroxypyr, clopyralid + fluroxypyr + MCPA, dicamba + MCPA + mecoprop-P, dicamba + mecoprop-P, dichlorprop-P + MCPA + mecoprop-P, diflufenican, diflufenican + florasulam, diflufenican + flufenacet, diflufenican + flufenacet (*off-label*), diflufenican + flufenacet + metribuzin, diflufenican + metsulfuron-methyl, diflufenican + pendimethalin, florasulam, florasulam + fluroxypyr, florasulam + halauxifen-methyl, florasulam + tribenuron-methyl, flufenacet, flufenacet + pendimethalin, flufenacet + pendimethalin (*off-label*), flufenacet + picolinafen, fluroxypyr, fluroxypyr + halauxifen-methyl, fluroxypyr + metsulfuron-methyl, fluroxypyr + metsulfuron-methyl + thifensulfuron-methyl, glyphosate, isoxaben, MCPA, mecoprop-P, metsulfuron-methyl, metsulfuron-methyl + thifensulfuron-methyl, metsulfuron-methyl + tribenuron-methyl, pendimethalin, pendimethalin + picolinafen, pendimethalin + picolinafen (*off-label*), prosulfocarb, prosulfocarb (*off-label*), thifensulfuron-methyl, thifensulfuron-methyl + tribenuron-methyl, tribenuron-methyl
	Crops as weeds	amidosulfuron + iodosulfuron-methyl-sodium, bromoxynil, bromoxynil + diflufenican, chlorotoluron + diflufenican + pendimethalin, clopyralid + florasulam, diflufenican, diflufenican + florasulam, diflufenican + flufenacet, florasulam, florasulam + fluroxypyr, florasulam + halauxifen-methyl, flufenacet + picolinafen, fluroxypyr, glyphosate, metsulfuron-methyl, metsulfuron-methyl + tribenuron-methyl, pendimethalin
	Grass weeds	chlorotoluron + diflufenican + pendimethalin, diflufenican + flufenacet, diflufenican + flufenacet (*off-label*), diflufenican + flufenacet + metribuzin, diflufenican + pendimethalin, fenoxaprop-P-ethyl, flufenacet, flufenacet + pendimethalin, flufenacet + pendimethalin (*off-label*), flufenacet + picolinafen, glyphosate, pendimethalin, pendimethalin + picolinafen, pendimethalin + picolinafen (*off-label*), pinoxaden, prosulfocarb, prosulfocarb (*off-label*), tri-allate
	Weeds, miscellaneous	chlorotoluron + diflufenican + pendimethalin, clopyralid + florasulam, diflufenican, diflufenican + flufenacet + metribuzin, diflufenican + metsulfuron-methyl, florasulam, glyphosate, metsulfuron-methyl, metsulfuron-methyl + thifensulfuron-methyl

Cereals - Cereals, general

Crop control	Desiccation	glyphosate, glyphosate (*off-label*)

Diseases	Alternaria	bromuconazole + tebuconazole, difenoconazole (*off-label*)
	Blue mould	prothioconazole + tebuconazole
	Bunt	prothioconazole + tebuconazole
	Cladosporium	bromuconazole + tebuconazole, difenoconazole (*off-label*)
	Crown rust	boscalid + epoxiconazole, boscalid + epoxiconazole + pyraclostrobin, epoxiconazole, epoxiconazole + fluxapyroxad, epoxiconazole + fluxapyroxad + pyraclostrobin
	Disease control	benzovindiflupyr + prothioconazole, fluxapyroxad + metconazole, isopyrazam + prothioconazole
	Ear blight	metconazole (*reduction*)
	Eyespot	boscalid + epoxiconazole (*moderate control*), boscalid + epoxiconazole + pyraclostrobin (*moderate control*), epoxiconazole + fluxapyroxad (*moderate control*), epoxiconazole + fluxapyroxad + pyraclostrobin (*moderate control*), epoxiconazole + metconazole (*reduction*), fluoxastrobin + prothioconazole + trifloxystrobin (*reduction*), fluxapyroxad + metconazole (*good reduction*), prochloraz (*moderate control*), prothioconazole + trifloxystrobin
	Foot rot	fludioxonil (*seed treatment*)
	Fungus diseases	prothioconazole + tebuconazole (*qualified minor use*)
	Fusarium	bromuconazole + tebuconazole, prothioconazole + tebuconazole
	Late ear diseases	fluoxastrobin + prothioconazole + trifloxystrobin, prothioconazole + trifloxystrobin
	Loose smut	prothioconazole + tebuconazole
	Powdery mildew	azoxystrobin, azoxystrobin (*moderate control*), boscalid + epoxiconazole (*moderate control*), boscalid + epoxiconazole + pyraclostrobin (*moderate control*), bromuconazole + tebuconazole (*moderate control*), cyflufenamid (*off-label*), epoxiconazole, epoxiconazole + fluxapyroxad (*moderate control*), epoxiconazole + fluxapyroxad + pyraclostrobin (*moderate control*), epoxiconazole + metconazole (*moderate control*), epoxiconazole + pyraclostrobin, fluoxastrobin + prothioconazole + trifloxystrobin, fluxapyroxad (*reduction*), fluxapyroxad + metconazole (*moderate control*), fluxapyroxad + pyraclostrobin (*moderate control*), mefentrifluconazole (*moderate control*), metconazole (*moderate control*), physical pest control, prochloraz, prothioconazole + trifloxystrobin, sulphur, tebuconazole
	Ramularia leaf spots	epoxiconazole + pyraclostrobin (*moderate control*)
	Rhynchosporium	azoxystrobin, azoxystrobin (*reduction*), benzovindiflupyr (*moderate control*), benzovindiflupyr + prothioconazole (*moderate control*), boscalid + epoxiconazole (*moderate control*), boscalid + epoxiconazole + pyraclostrobin, epoxiconazole, epoxiconazole + fluxapyroxad + pyraclostrobin, epoxiconazole + metconazole (*moderate control*), epoxiconazole + pyraclostrobin, fluxapyroxad, fluxapyroxad + metconazole, fluxapyroxad + pyraclostrobin, mefentrifluconazole (*moderate*

		control), metconazole (*moderate control*), prochloraz (*moderate control*), tebuconazole
	Rust	azoxystrobin, benzovindiflupyr , benzovindiflupyr + prothioconazole, boscalid + epoxiconazole, boscalid + epoxiconazole + pyraclostrobin, bromuconazole + tebuconazole, difenoconazole (*off-label*), epoxiconazole, epoxiconazole + fluxapyroxad, epoxiconazole + fluxapyroxad + pyraclostrobin, epoxiconazole + metconazole, epoxiconazole + pyraclostrobin, fluoxastrobin + prothioconazole + trifloxystrobin, fluxapyroxad, fluxapyroxad (*moderate control*), fluxapyroxad + metconazole, fluxapyroxad + pyraclostrobin, mancozeb, mefentrifluconazole, mefentrifluconazole (*moderate control*), metconazole, prothioconazole + trifloxystrobin, pyraclostrobin (*off-label*), tebuconazole, trifloxystrobin (*off-label*)
	Seed-borne diseases	fludioxonil (*seed treatment*), silthiofam (*off-label*)
	Septoria	bromuconazole + tebuconazole (*moderate control*), difenoconazole (*off-label*), fluoxastrobin + prothioconazole + trifloxystrobin, mancozeb, mefentrifluconazole, mefentrifluconazole (*moderate control*), metconazole, prothioconazole + trifloxystrobin
	Snow mould	fludioxonil + sedaxane
	Sooty moulds	mancozeb
	Stripe smut	fludioxonil, fludioxonil + sedaxane
	Take-all	azoxystrobin (*reduction*), silthiofam (*off-label*)
	Tan spot	fluoxastrobin + prothioconazole + trifloxystrobin
Pests	Aphids	beta-cyfluthrin, cypermethrin, deltamethrin (*off-label*), dimethoate, lambda-cyhalothrin, physical pest control, tau-fluvalinate (*off-label*), zeta-cypermethrin (*off-label*)
	Beetles	deltamethrin (*off-label*), lambda-cyhalothrin
	Birds/mammals	aluminium ammonium sulphate
	Caterpillars	lambda-cyhalothrin, lambda-cyhalothrin (*off-label*)
	Flies	deltamethrin (*off-label*)
	Mealybugs	physical pest control
	Midges	beta-cyfluthrin, deltamethrin (*off-label*)
	Pests, miscellaneous	cypermethrin, deltamethrin (*off-label*), lambda-cyhalothrin (*off-label*), zeta-cypermethrin (*off-label*)
	Scale insects	physical pest control
	Slugs/snails	ferric phosphate
	Spider mites	physical pest control
	Suckers	physical pest control
	Weevils	lambda-cyhalothrin
	Whiteflies	physical pest control
Plant growth regulation	Growth control	chlormequat, ethephon, mepiquat chloride + prohexadione-calcium, prohexadione-calcium, trinexapac-ethyl, trinexapac-ethyl (*off-label*)
	Quality/yield control	mepiquat chloride + prohexadione-calcium, sulphur
Weeds	Broad-leaved weeds	2,4-DB + MCPA, aminopyralid + halauxifen-methyl, bromoxynil (*off-label*), carfentrazone-ethyl (*off-label*), clopyralid (*off-label*), clopyralid + florasulam, clopyralid + florasulam + fluroxypyr, clopyralid +

fluroxypyr + MCPA, dicamba, dicamba + MCPA + mecoprop-P (*off-label*), dicamba + mecoprop-P (*off-label*), dichlorprop-P + MCPA + mecoprop-P, diflufenican, diflufenican + florasulam, diflufenican + flufenacet (*off-label*), florasulam + fluroxypyr, florasulam + fluroxypyr (*off-label*), florasulam + halauxifen-methyl, flufenacet + picolinafen, fluroxypyr, fluroxypyr + halauxifen-methyl, mesosulfuron-methyl + propoxycarbazone-sodium, metsulfuron-methyl + tribenuron-methyl (*off-label*), pendimethalin + picolinafen, prosulfocarb (*off-label*), prosulfuron (*off-label*)

	Crops as weeds	clopyralid + florasulam, diflufenican + florasulam, florasulam + fluroxypyr, florasulam + halauxifen-methyl, flufenacet + picolinafen, mesosulfuron-methyl + propoxycarbazone-sodium, metsulfuron-methyl + tribenuron-methyl (*off-label*), prosulfocarb (*off-label*)
	Grass weeds	diflufenican + flufenacet (*off-label*), flufenacet (*off-label*), flufenacet + picolinafen, glyphosate, mesosulfuron-methyl + propoxycarbazone-sodium, pendimethalin + picolinafen, prosulfocarb (*off-label*)
	Weeds, miscellaneous	bromoxynil (*off-label*), clopyralid + florasulam, dicamba, dicamba + MCPA + mecoprop-P (*off-label*), glyphosate (*off-label*)

Cereals - Maize/sweetcorn

Diseases	Anthracnose	Bacillus amyloliquefaciens D747 (*off-label*)
	Bacterial blight	Bacillus amyloliquefaciens D747 (*off-label*)
	Botrytis	Bacillus amyloliquefaciens D747 (*off-label*), Bacillus subtilis (*off-label*), potassium bicarbonate (commodity substance) (*off-label*)
	Cercospora leaf spot	Bacillus amyloliquefaciens D747 (*off-label*)
	Damping off	fludioxonil + metalaxyl-M
	Disease control	epoxiconazole + pyraclostrobin (*off-label*), pyraclostrobin
	Downy mildew	Bacillus amyloliquefaciens D747 (*off-label*)
	Eyespot	epoxiconazole + pyraclostrobin, epoxiconazole + pyraclostrobin (*off-label*), pyraclostrobin, pyraclostrobin (*off-label*)
	Foliar diseases	epoxiconazole + pyraclostrobin, pyraclostrobin, pyraclostrobin (*moderate control*)
	Fusarium	Bacillus amyloliquefaciens D747 (*off-label*), fludioxonil + metalaxyl-M
	Powdery mildew	Bacillus amyloliquefaciens D747 (*off-label*), potassium bicarbonate (commodity substance) (*off-label*)
	Pythium	Bacillus amyloliquefaciens D747 (*off-label*), fludioxonil + metalaxyl-M
	Rhizoctonia	Bacillus amyloliquefaciens D747 (*off-label*)
	Rust	epoxiconazole + pyraclostrobin (*off-label*)
	Sclerotinia	Bacillus amyloliquefaciens D747 (*off-label*)
	Stemphylium	Bacillus amyloliquefaciens D747 (*off-label*)
Pests	Aphids	cypermethrin, spirotetramat (*off-label*)
	Caterpillars	Bacillus thuringiensis (*off-label*), indoxacarb (*off-label*)

	Cutworms	Bacillus thuringiensis (*off-label*)
	Flies	lambda-cyhalothrin (*off-label*)
	Pests, miscellaneous	cypermethrin
	Slugs/snails	ferric phosphate
Weeds	Bindweeds	dicamba + prosulfuron
	Broad-leaved weeds	2,4-D, bromoxynil, clopyralid, clopyralid (*off-label*), clopyralid + florasulam + fluroxypyr, dicamba, dicamba + nicosulfuron, dicamba + prosulfuron, dicamba + prosulfuron (*seedlings only*), dimethenamid-p + pendimethalin, dimethenamid-p + pendimethalin (*off-label*), flufenacet + isoxaflutole, fluroxypyr, fluroxypyr (*off-label*), foramsulfuron + iodosulfuron-methyl-sodium, mesotrione, mesotrione (*off-label*), nicosulfuron, nicosulfuron (*off-label*), pendimethalin, pendimethalin (*off-label*), pendimethalin (*off-label - under covers*), prosulfuron, pyridate, rimsulfuron, S-metolachlor, S-metolachlor (*moderately susceptible*), S-metolachlor (*off-label*)
	Crops as weeds	clopyralid (*off-label*), fluroxypyr (*off-label*), mesotrione, nicosulfuron, nicosulfuron (*off-label*), pendimethalin, rimsulfuron
	Grass weeds	dicamba + nicosulfuron, dimethenamid-p + pendimethalin, dimethenamid-p + pendimethalin (*off-label*), flufenacet + isoxaflutole, foramsulfuron + iodosulfuron-methyl-sodium, mesotrione, mesotrione (*off-label*), nicosulfuron, nicosulfuron (*off-label*), pendimethalin, S-metolachlor, S-metolachlor (*off-label*)
	Weed grasses	dicamba + nicosulfuron, foramsulfuron + iodosulfuron-methyl-sodium, nicosulfuron
	Weeds, miscellaneous	bromoxynil, dicamba, nicosulfuron (*off-label*)

Cereals - Oats

Crop control	Desiccation	glyphosate
Diseases	Blue mould	prothioconazole + tebuconazole
	Bunt	prothioconazole + tebuconazole
	Covered smut	prothioconazole + tebuconazole
	Crown rust	azoxystrobin, azoxystrobin + cyproconazole, bixafen + prothioconazole, boscalid + epoxiconazole, epoxiconazole + pyraclostrobin, fluoxastrobin + prothioconazole, fluxapyroxad, fluxapyroxad + pyraclostrobin, mefentrifluconazole, prochloraz + proquinazid + tebuconazole (*qualified minor use recommendation*), prothioconazole, prothioconazole + spiroxamine, prothioconazole + spiroxamine + tebuconazole, prothioconazole + tebuconazole, pyraclostrobin, tebuconazole, tebuconazole (*reduction*)
	Eyespot	bixafen + prothioconazole (*reduction of incidence and severity*), fluoxastrobin + prothioconazole, fluoxastrobin + prothioconazole (*reduction of incidence and severity*), prothioconazole, prothioconazole + spiroxamine, prothioconazole + spiroxamine + tebuconazole, prothioconazole + tebuconazole

	Foot rot	fludioxonil (*seed treatment*), fludioxonil + tefluthrin
	Fusarium	prothioconazole + tebuconazole
	Leaf spot	fludioxonil (*seed treatment*), fludioxonil + tefluthrin
	Loose smut	fludioxonil + sedaxane, prothioconazole + tebuconazole
	Net blotch	tebuconazole
	Powdery mildew	azoxystrobin, azoxystrobin (*moderate control*), azoxystrobin + cyproconazole, bixafen + prothioconazole, cyflufenamid, epoxiconazole, epoxiconazole + pyraclostrobin, fluoxastrobin + prothioconazole, fluxapyroxad (*reduction*), fluxapyroxad + pyraclostrobin (*moderate control*), mefentrifluconazole, metrafenone (*evidence of mildew control on oats is limited*), prochloraz + proquinazid + tebuconazole, proquinazid, prothioconazole, prothioconazole + spiroxamine, prothioconazole + spiroxamine + tebuconazole, prothioconazole + tebuconazole, sulphur, tebuconazole
	Rhynchosporium	tebuconazole
	Rust	tebuconazole
	Seed-borne diseases	fludioxonil + tebuconazole
	Snow mould	fludioxonil (*seed treatment*)
	Take-all	azoxystrobin (*reduction*)
Pests	Aphids	cypermethrin, deltamethrin, lambda-cyhalothrin, zeta-cypermethrin
	Beetles	lambda-cyhalothrin
	Birds/mammals	aluminium ammonium sulphate
	Caterpillars	lambda-cyhalothrin
	Flies	cypermethrin
	Pests, miscellaneous	cypermethrin
	Slugs/snails	ferric phosphate
	Weevils	lambda-cyhalothrin
	Wireworms	fludioxonil + tefluthrin
Plant growth regulation	Growth control	chlormequat, prohexadione-calcium + trinexapac-ethyl, trinexapac-ethyl
	Quality/yield control	sulphur
Weeds	Broad-leaved weeds	2,4-D, 2,4-D + MCPA, 2,4-DB, amidosulfuron, bromoxynil, carfentrazone-ethyl, carfentrazone-ethyl + mecoprop-P, clopyralid, clopyralid + florasulam, clopyralid + florasulam + fluroxypyr, clopyralid + fluroxypyr + MCPA, dicamba + MCPA + mecoprop-P, dicamba + mecoprop-P, dichlorprop-P + MCPA + mecoprop-P, diflufenican (*off-label*), diflufenican + flufenacet (*off-label*), florasulam, florasulam + fluroxypyr, florasulam + tribenuron-methyl, flumioxazin (*off-label*), fluroxypyr, glyphosate, isoxaben, MCPA, mecoprop-P, metsulfuron-methyl, metsulfuron-methyl + thifensulfuron-methyl, metsulfuron-methyl + tribenuron-methyl, thifensulfuron-methyl + tribenuron-methyl, tribenuron-methyl
	Crops as weeds	bromoxynil, clopyralid + florasulam, diflufenican (*off-label*), florasulam, flumioxazin (*off-label*), fluroxypyr,

		glyphosate, metsulfuron-methyl, metsulfuron-methyl + tribenuron-methyl
	Grass weeds	diflufenican (*off-label*), diflufenican + flufenacet (*off-label*), flumioxazin (*off-label*), glyphosate
	Weeds, miscellaneous	clopyralid + florasulam, diflufenican (*off-label*), florasulam, glyphosate, metsulfuron-methyl, metsulfuron-methyl + thifensulfuron-methyl

Cereals - Rye/triticale

Diseases	Alternaria	bromuconazole + tebuconazole, difenoconazole (*off-label*)
	Blue mould	prothioconazole + tebuconazole
	Bunt	prothioconazole + tebuconazole
	Cladosporium	bromuconazole + tebuconazole, difenoconazole (*off-label*)
	Disease control	epoxiconazole + folpet, fluxapyroxad + metconazole, isopyrazam + prothioconazole, tebuconazole (*off-label*)
	Ear blight	epoxiconazole + metconazole (*good reduction*), epoxiconazole + metconazole (*moderate control*), epoxiconazole + pyraclostrobin (*good reduction*), metconazole (*reduction*), tebuconazole, thiophanate-methyl (*reduction*)
	Eyespot	bixafen + prothioconazole (*reduction of incidence and severity*), bixafen + prothioconazole + spiroxamine, bixafen + prothioconazole + tebuconazole (*reduction of incidence and severity*), boscalid + epoxiconazole (*moderate control*), boscalid + epoxiconazole + pyraclostrobin (*moderate control*), epoxiconazole + fluxapyroxad (*moderate control*), epoxiconazole + fluxapyroxad + pyraclostrobin (*moderate control*), epoxiconazole + metconazole (*reduction*), fluoxastrobin + prothioconazole (*reduction*), fluoxastrobin + prothioconazole + trifloxystrobin (*reduction*), fluxapyroxad + metconazole (*good reduction*), prochloraz (*moderate control*), prochloraz + tebuconazole, prothioconazole, prothioconazole + spiroxamine, prothioconazole + spiroxamine + tebuconazole (*reduction*), prothioconazole + tebuconazole, prothioconazole + tebuconazole (*reduction*), prothioconazole + trifloxystrobin
	Foot rot	fludioxonil (*seed treatment*)
	Fusarium	bromuconazole + tebuconazole, prothioconazole + tebuconazole, thiophanate-methyl, thiophanate-methyl (*reduction*)
	Late ear diseases	bixafen + fluoxastrobin + prothioconazole, bixafen + prothioconazole, bixafen + prothioconazole + spiroxamine, bixafen + prothioconazole + tebuconazole, fluoxastrobin + prothioconazole + trifloxystrobin, prochloraz + tebuconazole, prothioconazole, prothioconazole + spiroxamine, prothioconazole + spiroxamine + tebuconazole, prothioconazole + trifloxystrobin
	Loose smut	prothioconazole + tebuconazole
	Mycosphaerella	benzovindiflupyr
	Net blotch	tebuconazole

Powdery mildew	azoxystrobin, azoxystrobin (*moderate control*), azoxystrobin + cyproconazole, bixafen + fluoxastrobin + prothioconazole, bixafen + prothioconazole, bixafen + prothioconazole + spiroxamine, bixafen + prothioconazole + tebuconazole, boscalid + epoxiconazole + pyraclostrobin (*moderate control*), bromuconazole + tebuconazole (*moderate control*), cyflufenamid, epoxiconazole, epoxiconazole + fluxapyroxad (*moderate control*), epoxiconazole + fluxapyroxad + pyraclostrobin (*moderate control*), epoxiconazole + metconazole (*moderate control*), fluoxastrobin + prothioconazole, fluoxastrobin + prothioconazole + trifloxystrobin, fluxapyroxad (*reduction*), fluxapyroxad + metconazole (*moderate control*), fluxapyroxad + pyraclostrobin (*moderate control*), mefentrifluconazole (*moderate control*), metconazole (*moderate control*), prochloraz + tebuconazole, proquinazid, prothioconazole, prothioconazole + spiroxamine, prothioconazole + spiroxamine + tebuconazole, prothioconazole + tebuconazole, prothioconazole + trifloxystrobin, sulphur, tebuconazole
Ramularia leaf spots	epoxiconazole + pyraclostrobin (*moderate control*)
Rhynchosporium	azoxystrobin, azoxystrobin (*reduction*), azoxystrobin + cyproconazole (*moderate control*), benzovindiflupyr + prothioconazole (*moderate control*), bixafen + fluoxastrobin + prothioconazole, bixafen + prothioconazole, bixafen + prothioconazole + spiroxamine, bixafen + prothioconazole + tebuconazole, epoxiconazole, epoxiconazole + folpet, epoxiconazole + isopyrazam, fluoxastrobin + prothioconazole, prochloraz (*moderate control*), prochloraz + tebuconazole, prothioconazole, prothioconazole + spiroxamine, prothioconazole + spiroxamine + tebuconazole, prothioconazole + tebuconazole, tebuconazole, tebuconazole (*moderate control*)
Rust	azoxystrobin, azoxystrobin + cyproconazole, benzovindiflupyr, benzovindiflupyr + prothioconazole, bixafen + fluoxastrobin + prothioconazole, bixafen + prothioconazole, bixafen + prothioconazole + spiroxamine, bixafen + prothioconazole + tebuconazole, boscalid + epoxiconazole, boscalid + epoxiconazole + pyraclostrobin, bromuconazole + tebuconazole, difenoconazole (*off-label*), epoxiconazole, epoxiconazole + fluxapyroxad, epoxiconazole + fluxapyroxad + pyraclostrobin, epoxiconazole + folpet, epoxiconazole + isopyrazam, epoxiconazole + metconazole, epoxiconazole + pyraclostrobin, fluoxastrobin + prothioconazole, fluoxastrobin + prothioconazole + trifloxystrobin, fluxapyroxad, fluxapyroxad (*moderate control*), fluxapyroxad + metconazole, fluxapyroxad + pyraclostrobin, mancozeb, mefentrifluconazole, metconazole, prochloraz + tebuconazole, prothioconazole, prothioconazole + spiroxamine, prothioconazole + spiroxamine + tebuconazole, prothioconazole + tebuconazole, prothioconazole + trifloxystrobin, pyraclostrobin (*off-label*),

		tebuconazole, tebuconazole (*off-label*), trifloxystrobin (*off-label*)
	Seed-borne diseases	fludioxonil, fludioxonil + tebuconazole, fludioxonil + tefluthrin (*off-label*), silthiofam (*off-label*)
	Septoria	benzovindiflupyr, benzovindiflupyr + prothioconazole, bixafen + fluoxastrobin + prothioconazole, bixafen + prothioconazole, bixafen + prothioconazole + spiroxamine, bixafen + prothioconazole + tebuconazole, boscalid + epoxiconazole, boscalid + epoxiconazole + pyraclostrobin, bromuconazole + tebuconazole (*moderate control*), difenoconazole (*off-label*), epoxiconazole, epoxiconazole + fluxapyroxad, epoxiconazole + fluxapyroxad + pyraclostrobin, epoxiconazole + folpet, epoxiconazole + isopyrazam, epoxiconazole + metconazole, epoxiconazole + pyraclostrobin, epoxiconazole + pyraclostrobin (*moderate control*), fluoxastrobin + prothioconazole + trifloxystrobin, fluxapyroxad, fluxapyroxad + metconazole, fluxapyroxad + pyraclostrobin, mancozeb, mefentrifluconazole, metconazole, prochloraz + tebuconazole, prothioconazole, prothioconazole + spiroxamine, prothioconazole + spiroxamine + tebuconazole, prothioconazole + trifloxystrobin, tebuconazole
	Snow mould	fludioxonil + sedaxane
	Sooty moulds	mancozeb, tebuconazole
	Take-all	azoxystrobin (*reduction*), azoxystrobin (*reduction in severity*), silthiofam (*off-label*)
	Tan spot	bixafen + fluoxastrobin + prothioconazole, bixafen + prothioconazole, bixafen + prothioconazole + spiroxamine, bixafen + prothioconazole + tebuconazole, epoxiconazole + pyraclostrobin (*moderate control*), fluoxastrobin + prothioconazole + trifloxystrobin, prothioconazole + spiroxamine
Pests	Aphids	beta-cyfluthrin, cypermethrin, deltamethrin (*off-label*), dimethoate, lambda-cyhalothrin, tau-fluvalinate (*off-label*), zeta-cypermethrin (*off-label*)
	Beetles	deltamethrin (*off-label*), lambda-cyhalothrin
	Birds/mammals	aluminium ammonium sulphate
	Caterpillars	lambda-cyhalothrin, lambda-cyhalothrin (*off-label*)
	Flies	cypermethrin, deltamethrin (*off-label*)
	Midges	beta-cyfluthrin, deltamethrin (*off-label*)
	Pests, miscellaneous	cypermethrin, deltamethrin (*off-label*), lambda-cyhalothrin (*off-label*), zeta-cypermethrin (*off-label*)
	Slugs/snails	ferric phosphate
	Weevils	lambda-cyhalothrin
Plant growth regulation	Growth control	chlormequat, ethephon, ethephon + mepiquat chloride, mepiquat chloride + prohexadione-calcium, prohexadione-calcium, prohexadione-calcium + trinexapac-ethyl, trinexapac-ethyl
	Quality/yield control	mepiquat chloride + prohexadione-calcium, sulphur
Weeds	Broad-leaved weeds	2,4-D, amidosulfuron, amidosulfuron + iodosulfuron-methyl-sodium, aminopyralid + halauxifen-methyl, bifenox, carfentrazone-ethyl, chlorotoluron + diflufenican + pendimethalin, clopyralid (*off-label*), clopyralid + florasulam, clopyralid + florasulam +

fluroxypyr, clopyralid + fluroxypyr + MCPA, dicamba + MCPA + mecoprop-P (*off-label*), dicamba + mecoprop-P (*off-label*), dichlorprop-P + MCPA + mecoprop-P, diflufenican, diflufenican + florasulam, diflufenican + flufenacet, diflufenican + flufenacet (*off-label*), diflufenican + pendimethalin, florasulam, florasulam + fluroxypyr, florasulam + fluroxypyr (*off-label*), florasulam + halauxifen-methyl, florasulam + pyroxsulam, florasulam + tribenuron-methyl, flufenacet + picolinafen, fluroxypyr, fluroxypyr + halauxifen-methyl, isoxaben, MCPA, mesosulfuron-methyl + propoxycarbazone-sodium, metsulfuron-methyl, metsulfuron-methyl + thifensulfuron-methyl, metsulfuron-methyl + tribenuron-methyl, pendimethalin, pendimethalin + picolinafen, pendimethalin + pyroxsulam, prosulfocarb (*off-label*), pyroxsulam, thifensulfuron-methyl + tribenuron-methyl, tribenuron-methyl

Crops as weeds · amidosulfuron + iodosulfuron-methyl-sodium, chlorotoluron + diflufenican + pendimethalin, clopyralid + florasulam, diflufenican, diflufenican + florasulam, diflufenican + flufenacet, florasulam, florasulam + fluroxypyr, florasulam + halauxifen-methyl, florasulam + pyroxsulam, flufenacet + picolinafen, mesosulfuron-methyl + propoxycarbazone-sodium, metsulfuron-methyl, metsulfuron-methyl + tribenuron-methyl, pendimethalin, pendimethalin + pyroxsulam (*from seed*), pyroxsulam

Grass weeds · chlorotoluron + diflufenican + pendimethalin, clodinafop-propargyl, diflufenican + flufenacet, diflufenican + flufenacet (*off-label*), diflufenican + pendimethalin, florasulam + pyroxsulam, flufenacet (*off-label*), flufenacet + picolinafen, mesosulfuron-methyl + propoxycarbazone-sodium, pendimethalin, pendimethalin + picolinafen, pendimethalin + pyroxsulam, pendimethalin + pyroxsulam (*MS only*), prosulfocarb (*off-label*), pyroxsulam, tri-allate (*off-label*)

Weeds, miscellaneous · chlorotoluron + diflufenican + pendimethalin, clopyralid + florasulam, dicamba + MCPA + mecoprop-P (*off-label*), diflufenican, florasulam, metsulfuron-methyl, metsulfuron-methyl + thifensulfuron-methyl

Cereals - Undersown cereals

Weeds · Broad-leaved weeds · 2,4-D, 2,4-DB, 2,4-DB + MCPA, clopyralid + florasulam + fluroxypyr, florasulam + fluroxypyr, MCPA, MCPA (*red clover or grass*), tribenuron-methyl

Crops as weeds · florasulam + fluroxypyr

Cereals - Wheat

Crop control · Desiccation · glyphosate

Diseases · Alternaria · bromuconazole + tebuconazole, tebuconazole

Blue mould · prothioconazole + tebuconazole

Bunt · fludioxonil (*seed treatment*), fludioxonil + sedaxane, fludioxonil + tefluthrin, imazalil + ipconazole, prothioconazole + tebuconazole

Cladosporium	bromuconazole + tebuconazole, fluxapyroxad + metconazole (*moderate control*), tebuconazole
Disease control	fluxapyroxad + metconazole, isopyrazam + prothioconazole
Ear blight	boscalid + epoxiconazole (*good reduction*), boscalid + epoxiconazole + pyraclostrobin (*good reduction*), dimoxystrobin + epoxiconazole, epoxiconazole, epoxiconazole (*good reduction*), epoxiconazole + fluxapyroxad (*good reduction*), epoxiconazole + fluxapyroxad + pyraclostrobin (*good reduction*), epoxiconazole + metconazole (*good reduction*), epoxiconazole + metconazole (*moderate control*), epoxiconazole + pyraclostrobin (*good reduction*), fludioxonil + sedaxane (*moderate control*), fluxapyroxad + metconazole (*good reduction*), metconazole (*reduction*), prothioconazole + tebuconazole, tebuconazole, thiophanate-methyl (*reduction*)
Eyespot	azoxystrobin + cyproconazole (*reduction*), bixafen + fluopyram + prothioconazole (*reduction of incidence and severity*), bixafen + prothioconazole (*reduction of incidence and severity*), bixafen + prothioconazole + spiroxamine, bixafen + prothioconazole + tebuconazole (*reduction of incidence and severity*), boscalid + epoxiconazole (*moderate control*), boscalid + epoxiconazole + pyraclostrobin (*moderate control*), epoxiconazole (*reduction*), epoxiconazole + fluxapyroxad (*moderate control*), epoxiconazole + fluxapyroxad + pyraclostrobin (*moderate control*), epoxiconazole + metconazole (*reduction*), fluoxastrobin + prothioconazole (*reduction*), fluoxastrobin + prothioconazole + trifloxystrobin (*reduction*), fluxapyroxad + metconazole (*good reduction*), metrafenone (*reduction*), prochloraz (*moderate control*), prochloraz + proquinazid + tebuconazole, prochloraz + tebuconazole, prothioconazole, prothioconazole + spiroxamine, prothioconazole + spiroxamine + tebuconazole (*reduction*), prothioconazole + tebuconazole, prothioconazole + tebuconazole (*reduction*), prothioconazole + trifloxystrobin
Foot rot	fludioxonil (*seed treatment*), fludioxonil + tefluthrin, fluoxastrobin + prothioconazole (*reduction*), imazalil + ipconazole (*useful protection*)
Fusarium	benzovindiflupyr + prothioconazole (*moderate control*), bromuconazole + tebuconazole, fludioxonil + sedaxane, prothioconazole + tebuconazole, tebuconazole, thiophanate-methyl, thiophanate-methyl (*reduction*)
Late ear diseases	azoxystrobin, bixafen + fluopyram + prothioconazole (*reduction of incidence and severity*), bixafen + fluoxastrobin + prothioconazole, bixafen + prothioconazole, bixafen + prothioconazole + spiroxamine, bixafen + prothioconazole + tebuconazole, fluoxastrobin + prothioconazole, fluoxastrobin + prothioconazole + trifloxystrobin, prochloraz + proquinazid + tebuconazole (*reduction only*), prochloraz + tebuconazole, prothioconazole, prothioconazole + spiroxamine, prothioconazole +

	spiroxamine + tebuconazole, prothioconazole + tebuconazole, prothioconazole + trifloxystrobin, tebuconazole
Leaf spot	prochloraz
Loose smut	fludioxonil + sedaxane, prothioconazole + tebuconazole
Mycosphaerella	benzovindiflupyr , benzovindiflupyr + prothioconazole
Net blotch	metconazole (*reduction*), tebuconazole
Powdery mildew	azoxystrobin + cyproconazole, bixafen + fluopyram + prothioconazole, bixafen + fluoxastrobin + prothioconazole, bixafen + prothioconazole, bixafen + prothioconazole + spiroxamine, bixafen + prothioconazole + tebuconazole, bromuconazole + tebuconazole (*moderate control*), cyflufenamid, epoxiconazole, epoxiconazole + fluxapyroxad (*moderate control*), epoxiconazole + fluxapyroxad + pyraclostrobin (*moderate control*), epoxiconazole + metconazole (*moderate control*), fluoxastrobin + prothioconazole, fluoxastrobin + prothioconazole + trifloxystrobin, flutriafol, fluxapyroxad (*reduction*), fluxapyroxad + metconazole (*moderate control*), fluxapyroxad + pyraclostrobin (*moderate control*), laminarin, mefentrifluconazole (*moderate control*), metconazole (*moderate control*), metrafenone, prochloraz + proquinazid + tebuconazole, prochloraz + tebuconazole, proquinazid, prothioconazole, prothioconazole + spiroxamine, prothioconazole + spiroxamine + tebuconazole, prothioconazole + tebuconazole, prothioconazole + trifloxystrobin, pyriofenone, sulphur, tebuconazole, tebuconazole (*moderate control*)
Rhynchosporium	tebuconazole
Rust	azoxystrobin, azoxystrobin + cyproconazole, benzovindiflupyr , benzovindiflupyr + prothioconazole, bixafen + fluopyram + prothioconazole, bixafen + fluoxastrobin + prothioconazole, bixafen + prothioconazole, bixafen + prothioconazole + spiroxamine, bixafen + prothioconazole + tebuconazole, boscalid + epoxiconazole, boscalid + epoxiconazole + pyraclostrobin, bromuconazole + tebuconazole, difenoconazole, dimoxystrobin + epoxiconazole, epoxiconazole, epoxiconazole + fluxapyroxad, epoxiconazole + fluxapyroxad + pyraclostrobin, epoxiconazole + folpet, epoxiconazole + isopyrazam, epoxiconazole + metconazole, epoxiconazole + pyraclostrobin, fluoxastrobin + prothioconazole, fluoxastrobin + prothioconazole + trifloxystrobin, flutriafol, fluxapyroxad, fluxapyroxad (*moderate control*), fluxapyroxad + metconazole, fluxapyroxad + pyraclostrobin, mancozeb, mancozeb (*useful control*), mefentrifluconazole, metconazole, prochloraz + proquinazid + tebuconazole, prochloraz + tebuconazole, prothioconazole, prothioconazole + spiroxamine, prothioconazole + spiroxamine + tebuconazole, prothioconazole + tebuconazole, prothioconazole + trifloxystrobin, pyraclostrobin, tebuconazole, trifloxystrobin

Seed-borne diseases	difenoconazole + fludioxonil, fludioxonil + tebuconazole, imazalil + ipconazole (*useful protection*), ipconazole
Septoria	azoxystrobin, azoxystrobin + cyproconazole, benzovindiflupyr , benzovindiflupyr (*moderate control*), benzovindiflupyr + prothioconazole, benzovindiflupyr + prothioconazole (*moderate control*), bixafen + fluopyram + prothioconazole, bixafen + fluoxastrobin + prothioconazole, bixafen + prothioconazole, bixafen + prothioconazole + spiroxamine, bixafen + prothioconazole + tebuconazole, boscalid + epoxiconazole, boscalid + epoxiconazole + pyraclostrobin, bromuconazole + tebuconazole (*moderate control*), difenoconazole, dimoxystrobin + epoxiconazole, epoxiconazole, epoxiconazole + fluxapyroxad, epoxiconazole + fluxapyroxad + pyraclostrobin, epoxiconazole + folpet, epoxiconazole + isopyrazam, epoxiconazole + metconazole, epoxiconazole + pyraclostrobin, epoxiconazole + pyraclostrobin (*moderate control*), fludioxonil (*seed treatment*), fludioxonil + sedaxane, fludioxonil + tefluthrin, fluoxastrobin + prothioconazole, fluoxastrobin + prothioconazole + trifloxystrobin, flutriafol, fluxapyroxad, fluxapyroxad + metconazole, fluxapyroxad + pyraclostrobin, fluxapyroxad + pyraclostrobin (*moderate control*), folpet (*reduction*), laminarin, mancozeb, mancozeb (*reduction*), mefentrifluconazole, metconazole, prochloraz, prochloraz + proquinazid + tebuconazole, prochloraz + proquinazid + tebuconazole (*moderate control*), prochloraz + tebuconazole, prothioconazole, prothioconazole + spiroxamine, prothioconazole + spiroxamine + tebuconazole, prothioconazole + tebuconazole, prothioconazole + trifloxystrobin, pyraclostrobin, tebuconazole, trifloxystrobin
Sharp eyespot	fluoxastrobin + prothioconazole, fluoxastrobin + prothioconazole (*reduction*)
Snow mould	fludioxonil (*seed treatment*), fludioxonil + sedaxane, fludioxonil + tefluthrin
Soil-borne diseases	ipconazole
Sooty moulds	boscalid + epoxiconazole, boscalid + epoxiconazole + pyraclostrobin (*good reduction*), epoxiconazole (*reduction*), epoxiconazole + fluxapyroxad (*reduction*), epoxiconazole + fluxapyroxad + pyraclostrobin (*reduction*), epoxiconazole + metconazole (*good reduction*), epoxiconazole + metconazole (*moderate control*), epoxiconazole + pyraclostrobin (*reduction*), fluoxastrobin + prothioconazole (*reduction*), mancozeb, prothioconazole + tebuconazole, tebuconazole
Take-all	azoxystrobin (*reduction*), azoxystrobin + cyproconazole (*reduction*), fluoxastrobin + prothioconazole (*reduction*), silthiofam (*seed treatment*)
Tan spot	bixafen + fluopyram + prothioconazole, bixafen + fluoxastrobin + prothioconazole, bixafen + prothioconazole, bixafen + prothioconazole + spiroxamine, bixafen + prothioconazole + tebuconazole, boscalid + epoxiconazole (*reduction*),

		boscalid + epoxiconazole + pyraclostrobin (*moderate control*), dimoxystrobin + epoxiconazole, epoxiconazole + fluxapyroxad (*moderate control*), epoxiconazole + fluxapyroxad + pyraclostrobin (*moderate control*), epoxiconazole + metconazole (*moderate control*), epoxiconazole + pyraclostrobin, epoxiconazole + pyraclostrobin (*moderate control*), fluoxastrobin + prothioconazole, fluoxastrobin + prothioconazole + trifloxystrobin, fluxapyroxad (*reduction*), fluxapyroxad + metconazole (*good reduction*), fluxapyroxad + pyraclostrobin, prothioconazole, prothioconazole + spiroxamine, prothioconazole + tebuconazole
Pests	Aphids	alpha-cypermethrin, beta-cyfluthrin, cypermethrin, cypermethrin (*autumn sown*), deltamethrin, dimethoate, esfenvalerate, flonicamid, lambda-cyhalothrin, tau-fluvalinate, thiacloprid, zeta-cypermethrin
	Beetles	lambda-cyhalothrin
	Birds/mammals	aluminium ammonium sulphate
	Caterpillars	lambda-cyhalothrin
	Flies	cypermethrin, cypermethrin (*autumn sown*), cypermethrin (*qualified recommendation*), cypermethrin (*reduction*), fludioxonil + tefluthrin, lambda-cyhalothrin
	Midges	beta-cyfluthrin, lambda-cyhalothrin, thiacloprid
	Pests, miscellaneous	cypermethrin
	Slugs/snails	ferric phosphate
	Suckers	lambda-cyhalothrin
	Weevils	lambda-cyhalothrin
	Wireworms	fludioxonil + tefluthrin
Plant growth regulation	Growth control	chlormequat, chlormequat + 2-chloroethylphosphonic acid, ethephon, ethephon + mepiquat chloride, mepiquat chloride + prohexadione-calcium, prohexadione-calcium, prohexadione-calcium + trinexapac-ethyl, trinexapac-ethyl
	Quality/yield control	ethephon + mepiquat chloride (*low lodging situations*), mepiquat chloride + prohexadione-calcium, sulphur
Weeds	Broad-leaved weeds	2,4-D, 2,4-D + MCPA, 2,4-DB, 2,4-DB + MCPA, amidosulfuron, amidosulfuron + iodosulfuron-methyl-sodium, amidosulfuron + iodosulfuron-methyl-sodium + mesosulfuron-methyl, aminopyralid + halauxifen-methyl, bifenox, bromoxynil, bromoxynil + diflufenican, carfentrazone-ethyl, carfentrazone-ethyl + mecoprop-P, chlorotoluron + diflufenican + pendimethalin, clopyralid, clopyralid + florasulam, clopyralid + florasulam + fluroxypyr, clopyralid + fluroxypyr + MCPA, dicamba + MCPA + mecoprop-P, dicamba + mecoprop-P, dichlorprop-P + MCPA + mecoprop-P, diflufenican, diflufenican + florasulam, diflufenican + flufenacet, diflufenican + flufenacet (*off-label*), diflufenican + flufenacet + metribuzin, diflufenican + iodosulfuron-methyl-sodium + mesosulfuron-methyl, diflufenican + metsulfuron-methyl, diflufenican + pendimethalin, ethofumesate, florasulam, florasulam + fluroxypyr, florasulam +

halauxifen-methyl, florasulam + pyroxsulam, florasulam + tribenuron-methyl, flufenacet, flufenacet + pendimethalin, flufenacet + picolinafen, flumioxazin, fluroxypyr, fluroxypyr + halauxifen-methyl, fluroxypyr + metsulfuron-methyl, fluroxypyr + metsulfuron-methyl + thifensulfuron-methyl, glyphosate, iodosulfuron-methyl-sodium + mesosulfuron-methyl, isoxaben, MCPA, mecoprop-P, mesosulfuron-methyl + propoxycarbazone-sodium, metsulfuron-methyl, metsulfuron-methyl + thifensulfuron-methyl, metsulfuron-methyl + tribenuron-methyl, pendimethalin, pendimethalin (*off-label*), pendimethalin + picolinafen, pendimethalin + picolinafen (*off-label*), pendimethalin + pyroxsulam, prosulfocarb, pyroxsulam, sulfosulfuron, thifensulfuron-methyl, thifensulfuron-methyl + tribenuron-methyl, tribenuron-methyl

Crops as weeds amidosulfuron + iodosulfuron-methyl-sodium, bromoxynil, bromoxynil + diflufenican, chlorotoluron + diflufenican + pendimethalin, clopyralid + florasulam, diflufenican, diflufenican + florasulam, diflufenican + flufenacet, diflufenican + iodosulfuron-methyl-sodium + mesosulfuron-methyl, florasulam, florasulam + fluroxypyr, florasulam + halauxifen-methyl, florasulam + pyroxsulam, flufenacet + picolinafen, flumioxazin, fluroxypyr, glyphosate, mesosulfuron-methyl + propoxycarbazone-sodium, metsulfuron-methyl, metsulfuron-methyl + tribenuron-methyl, pendimethalin, pendimethalin + pyroxsulam (*from seed*), pyroxsulam

Grass weeds amidosulfuron + iodosulfuron-methyl-sodium + mesosulfuron-methyl, chlorotoluron + diflufenican + pendimethalin, clodinafop-propargyl, clodinafop-propargyl + cloquintocet-mexyl, diflufenican + flufenacet, diflufenican + flufenacet (*off-label*), diflufenican + flufenacet + metribuzin, diflufenican + iodosulfuron-methyl-sodium + mesosulfuron-methyl, diflufenican + pendimethalin, ethofumesate, fenoxaprop-P-ethyl, florasulam + pyroxsulam, flufenacet, flufenacet + pendimethalin, flufenacet + picolinafen, flumioxazin, glyphosate, iodosulfuron-methyl-sodium + mesosulfuron-methyl, mesosulfuron-methyl + propoxycarbazone-sodium, pendimethalin, pendimethalin (*off-label*), pendimethalin + picolinafen, pendimethalin + picolinafen (*off-label*), pendimethalin + pyroxsulam, pendimethalin + pyroxsulam (*MS only*), pinoxaden, propoxycarbazone-sodium, prosulfocarb, pyroxsulam, sulfosulfuron, sulfosulfuron (*moderate control of barren brome*), sulfosulfuron (*moderate control only*), tri-allate

Weeds, miscellaneous chlorotoluron + diflufenican + pendimethalin, clopyralid + florasulam, diflufenican, diflufenican + flufenacet + metribuzin, diflufenican + metsulfuron-methyl, florasulam, glyphosate, metsulfuron-methyl, metsulfuron-methyl + thifensulfuron-methyl

Edible fungi - Mushrooms

Diseases Alternaria Bacillus amyloliquefaciens D747

	Cobweb	metrafenone, prochloraz
	Dry bubble	prochloraz
	Fusarium	Bacillus amyloliquefaciens D747
	Pythium	Bacillus amyloliquefaciens D747
	Rhizoctonia	Bacillus amyloliquefaciens D747
	Wet bubble	prochloraz
Pests	Flies	Metarhizium anisopliae (off-label)

Fruiting vegetables - Aubergines

Diseases	Alternaria	Bacillus amyloliquefaciens D747
	Botrytis	azoxystrobin (off-label), Bacillus subtilis (off-label)
	Downy mildew	cyazofamid (off-label)
	Foot rot	Trichoderma asperellum (Strain T34)
	Fusarium	Bacillus amyloliquefaciens D747
	Phytophthora	azoxystrobin (off-label), cyazofamid (off-label)
	Powdery mildew	azoxystrobin (off-label), Bacillus amyloliquefaciens D747, physical pest control
	Pythium	Bacillus amyloliquefaciens D747, Bacillus subtilis (off-label)
	Rhizoctonia	Bacillus amyloliquefaciens D747
Pests	Aphids	physical pest control
	Caterpillars	methoxyfenozide (off-label)
	Mealybugs	physical pest control
	Pests, miscellaneous	methoxyfenozide (off-label)
	Scale insects	physical pest control
	Slugs/snails	ferric phosphate
	Spider mites	physical pest control
	Suckers	physical pest control
	Whiteflies	physical pest control

Fruiting vegetables - Cucurbits

Diseases	Alternaria	Bacillus amyloliquefaciens D747
	Anthracnose	Bacillus amyloliquefaciens D747 (off-label)
	Bacterial blight	Bacillus amyloliquefaciens D747 (off-label)
	Botrytis	azoxystrobin (off-label), Bacillus amyloliquefaciens D747 (off-label), Bacillus subtilis (off-label), cyprodinil + fludioxonil (off-label), mepanipyrim (off-label), potassium bicarbonate (commodity substance) (off-label)
	Cercospora leaf spot	Bacillus amyloliquefaciens D747 (off-label)
	Cladosporium	boscalid + pyraclostrobin (off-label)
	Disease control	cyflufenamid
	Downy mildew	azoxystrobin (off-label), Bacillus amyloliquefaciens D747 (off-label), cyazofamid (off-label)
	Fusarium	Bacillus amyloliquefaciens D747, Bacillus amyloliquefaciens D747 (off-label)
	Leaf spot	boscalid + pyraclostrobin (off-label)
	Phytophthora	Bacillus subtilis (off-label), cyazofamid (off-label)
	Powdery mildew	azoxystrobin (off-label), Bacillus amyloliquefaciens D747, Bacillus amyloliquefaciens D747 (off-label),

		boscalid + pyraclostrobin (*off-label*), cyflufenamid, cyflufenamid (*off-label*), penconazole (*off-label*), physical pest control, potassium bicarbonate (commodity substance) (*off-label*), proquinazid (*off-label*)
	Pythium	Bacillus amyloliquefaciens D747, Bacillus amyloliquefaciens D747 (*off-label*), Bacillus subtilis (*off-label*)
	Rhizoctonia	Bacillus amyloliquefaciens D747, Bacillus amyloliquefaciens D747 (*off-label*)
	Sclerotinia	Bacillus amyloliquefaciens D747 (*off-label*)
	Stemphylium	Bacillus amyloliquefaciens D747 (*off-label*)
Pests	Aphids	flonicamid (*off-label*), physical pest control, pyrethrins, thiacloprid (*off-label*)
	Caterpillars	Bacillus thuringiensis, pyrethrins, spinosad (*off-label*)
	Leaf miners	spinosad (*off-label*)
	Mealybugs	physical pest control
	Pests, miscellaneous	thiacloprid (*off-label*)
	Scale insects	physical pest control
	Slugs/snails	ferric phosphate
	Spider mites	physical pest control, pyrethrins
	Suckers	physical pest control
	Thrips	pyrethrins, spinosad (*off-label*)
	Whiteflies	flonicamid (*off-label*), physical pest control
Weeds	Broad-leaved weeds	clomazone (*off-label*), dimethenamid-p + pendimethalin (*off-label*), isoxaben (*off-label*), propyzamide (*off-label*)
	Grass weeds	dimethenamid-p + pendimethalin (*off-label*), propyzamide (*off-label*)

Fruiting vegetables - Peppers

Diseases	Alternaria	Bacillus amyloliquefaciens D747
	Botrytis	Bacillus subtilis (*off-label*)
	Foot rot	Trichoderma asperellum (Strain T34)
	Fusarium	Bacillus amyloliquefaciens D747
	Powdery mildew	azoxystrobin (*off-label*), Bacillus amyloliquefaciens D747, physical pest control
	Pythium	Bacillus amyloliquefaciens D747, Bacillus subtilis (*off-label*)
	Rhizoctonia	Bacillus amyloliquefaciens D747
Pests	Aphids	physical pest control, pirimicarb (*off-label*)
	Caterpillars	methoxyfenozide (*off-label*)
	Mealybugs	physical pest control
	Pests, miscellaneous	methoxyfenozide (*off-label*)
	Scale insects	physical pest control
	Spider mites	physical pest control
	Suckers	physical pest control
	Whiteflies	physical pest control

Fruiting vegetables - Tomatoes

Diseases	Alternaria	Bacillus amyloliquefaciens D747

	Botrytis	azoxystrobin (*off-label*), Bacillus subtilis (*off-label*), fenhexamid (*off-label*)
	Downy mildew	cyazofamid (*off-label*)
	Foot rot	Trichoderma asperellum (Strain T34)
	Fungus diseases	1-methylcyclopropene
	Fusarium	Bacillus amyloliquefaciens D747
	Phytophthora	azoxystrobin (*off-label*), cyazofamid (*off-label*), mancozeb
	Powdery mildew	azoxystrobin (*off-label*), Bacillus amyloliquefaciens D747, physical pest control
	Pythium	Bacillus amyloliquefaciens D747, Bacillus subtilis (*off-label*)
	Rhizoctonia	Bacillus amyloliquefaciens D747
Pests	Aphids	physical pest control
	Caterpillars	methoxyfenozide (*off-label*)
	Mealybugs	physical pest control
	Pests, miscellaneous	methoxyfenozide (*off-label*)
	Scale insects	physical pest control
	Spider mites	physical pest control
	Suckers	physical pest control
	Whiteflies	physical pest control

Herb crops - Herbs

Diseases	Alternaria	azoxystrobin (*off-label*), Bacillus amyloliquefaciens D747, boscalid (*off-label*), boscalid + dimoxystrobin (*off-label*), boscalid + pyraclostrobin (*off-label*), cyprodinil + fludioxonil (*off-label*)
	Anthracnose	Bacillus amyloliquefaciens D747 (*off-label*)
	Bacterial blight	Bacillus amyloliquefaciens D747 (*off-label*)
	Bacterial canker	Bacillus amyloliquefaciens D747 (*off-label*)
	Black rot	azoxystrobin (*off-label*)
	Botrytis	Bacillus amyloliquefaciens D747 (*off-label*), Bacillus subtilis (*off-label*), boscalid (*off-label*), boscalid + pyraclostrobin (*off-label*), cyprodinil + fludioxonil (*off-label*), fenhexamid (*off-label*), Gliocladium catenulatum (*off-label*)
	Cercospora leaf spot	Bacillus amyloliquefaciens D747 (*off-label*)
	Damping off	Bacillus amyloliquefaciens D747 (*off-label*), Bacillus subtilis (*off-label*), fosetyl-aluminium + propamocarb hydrochloride (*off-label*)
	Disease control	boscalid (*off-label*), difenoconazole (*off-label*), dimethomorph + mancozeb (*off-label*), mandipropamid
	Downy mildew	Bacillus amyloliquefaciens D747 (*off-label*), cos-oga (*off-label*), dimethomorph (*off-label*), dimethomorph + mancozeb (*off-label*), fluopicolide + propamocarb hydrochloride (*off-label*), fosetyl-aluminium + propamocarb hydrochloride (*off-label*), mancozeb + metalaxyl-M (*off-label*), mandipropamid, metalaxyl-M (*off-label*), potassium bicarbonate (commodity substance) (*off-label*)
	Fire	boscalid (*off-label*), boscalid + pyraclostrobin (*off-label*)

	Fireblight	Bacillus amyloliquefaciens D747 (*off-label*)
	Fungus diseases	Bacillus amyloliquefaciens D747 (*off-label*)
	Fusarium	Bacillus amyloliquefaciens D747, Bacillus amyloliquefaciens D747 (*off-label*), Gliocladium catenulatum (*off-label*)
	Leaf spot	difenoconazole (*off-label*)
	Phytophthora	difenoconazole (*off-label*), Gliocladium catenulatum (*off-label*)
	Powdery mildew	azoxystrobin (*off-label*), azoxystrobin + difenoconazole, Bacillus amyloliquefaciens D747, Bacillus amyloliquefaciens D747 (*off-label*), cos-oga (*off-label*), physical pest control, potassium bicarbonate (commodity substance) (*off-label*), prothioconazole (*off-label*)
	Pythium	Bacillus amyloliquefaciens D747, Bacillus amyloliquefaciens D747 (*off-label*), metalaxyl-M (*off-label*)
	Rhizoctonia	Bacillus amyloliquefaciens D747, Bacillus amyloliquefaciens D747 (*off-label*), cyprodinil + fludioxonil (*off-label*), Gliocladium catenulatum (*off-label*)
	Rust	azoxystrobin + difenoconazole (*off-label*)
	Sclerotinia	Bacillus amyloliquefaciens D747 (*off-label*), boscalid (*off-label*), boscalid + dimoxystrobin (*off-label*), boscalid + pyraclostrobin (*off-label*), cyprodinil + fludioxonil (*off-label*), prothioconazole (*off-label*)
	Septoria	azoxystrobin + difenoconazole (*off-label*)
	Stem canker	cyprodinil + fludioxonil (*off-label*), prothioconazole (*off-label*)
	Stemphylium	Bacillus amyloliquefaciens D747 (*off-label*)
	White blister	azoxystrobin (*off-label*), boscalid + pyraclostrobin (*off-label*)
Pests	Aphids	acetamiprid (*off-label*), Beauveria bassiana (*off-label*), deltamethrin (*off-label*), flonicamid (*off-label*), Metarhizium anisopliae (*off-label*), physical pest control, pirimicarb (*off-label*), pyrethrins, spirotetramat (*off-label*), thiacloprid (*off-label*)
	Beetles	deltamethrin (*off-label*), lambda-cyhalothrin (*off-label*)
	Caterpillars	Bacillus thuringiensis, Bacillus thuringiensis (*off-label*), chlorantraniliprole (*off-label*), deltamethrin (*off-label*), lambda-cyhalothrin (*off-label*), pyrethrins, spinosad (*off-label*)
	Cutworms	deltamethrin (*off-label*), lambda-cyhalothrin (*off-label*)
	Flies	lambda-cyhalothrin (*off-label*), Metarhizium anisopliae (*off-label*), spinosad (*off-label*)
	Leaf miners	deltamethrin (*off-label*), spinosad (*off-label*)
	Mealybugs	physical pest control
	Midges	lambda-cyhalothrin (*off-label*), Metarhizium anisopliae (*off-label*)
	Pests, miscellaneous	deltamethrin (*off-label*), lambda-cyhalothrin (*off-label*)
	Scale insects	physical pest control
	Slugs/snails	ferric phosphate
	Spider mites	Beauveria bassiana (*off-label*), physical pest control, pyrethrins

	Springtails	lambda-cyhalothrin (*off-label*)
	Suckers	physical pest control
	Thrips	Beauveria bassiana (*off-label*), lambda-cyhalothrin (*off-label*), Metarhizium anisopliae (*off-label*), pyrethrins, spinosad (*off-label*), thiacloprid (*off-label*)
	Weevils	lambda-cyhalothrin (*off-label*)
	Whiteflies	Beauveria bassiana (*off-label*), deltamethrin (*off-label*), physical pest control, spirotetramat (*off-label*)
	Wireworms	deltamethrin (*off-label*)
Weeds	Broad-leaved weeds	aclonifen (*off-label*), amidosulfuron (*off-label*), bentazone (*off-label*), carbetamide (*off-label*), chlorpropham (*off-label*), clomazone (*off-label*), clopyralid (*off-label*), clopyralid + picloram (*off-label*), dimethenamid-p + metazachlor (*off-label*), dimethenamid-p + metazachlor + quinmerac (*off-label*), dimethenamid-p + pendimethalin (*off-label*), fluroxypyr (*off-label*), lenacil (*off-label*), mesotrione (*off-label*), metazachlor + quinmerac (*off-label*), napropamide (*off-label*), pendimethalin (*off-label*), phenmedipham (*off-label*), propyzamide (*off-label*), prosulfocarb (*off-label*), pyridate (*off-label*), S-metolachlor (*off-label*)
	Crops as weeds	aclonifen (*off-label*), carbetamide (*off-label*), clethodim (*off-label*), cycloxydim (*off-label*), fluazifop-P-butyl, fluroxypyr (*off-label*), mesotrione (*off-label*), propaquizafop (*off-label*)
	Grass weeds	carbetamide (*off-label*), chlorpropham (*off-label*), clethodim (*off-label*), cycloxydim (*off-label*), dimethenamid-p + metazachlor (*off-label*), dimethenamid-p + metazachlor + quinmerac (*off-label*), dimethenamid-p + pendimethalin (*off-label*), fluazifop-P-butyl, mesotrione (*off-label*), metazachlor + quinmerac (*off-label*), napropamide (*off-label*), pendimethalin (*off-label*), propaquizafop (*off-label*), propyzamide (*off-label*), prosulfocarb (*off-label*), S-metolachlor (*off-label*)
	Weeds, miscellaneous	fluroxypyr (*off-label*), lenacil (*off-label*)

Leafy vegetables - Endives

Diseases	Alternaria	Bacillus amyloliquefaciens D747, cyprodinil + fludioxonil (*off-label*)
	Botrytis	Bacillus subtilis (*off-label*), cyprodinil + fludioxonil (*off-label*), fenhexamid (*off-label*), Gliocladium catenulatum (*off-label*)
	Disease control	dimethomorph + mancozeb (*off-label*), mandipropamid
	Downy mildew	azoxystrobin, boscalid + pyraclostrobin (*off-label*), dimethomorph + mancozeb (*off-label*), mandipropamid
	Fusarium	Bacillus amyloliquefaciens D747, Gliocladium catenulatum (*off-label*)
	Phytophthora	Gliocladium catenulatum (*off-label*)
	Powdery mildew	Bacillus amyloliquefaciens D747
	Pythium	Bacillus amyloliquefaciens D747

	Rhizoctonia	Bacillus amyloliquefaciens D747, cyprodinil + fludioxonil (*off-label*), Gliocladium catenulatum (*off-label*)
	Stem canker	cyprodinil + fludioxonil (*off-label*)
Pests	Aphids	deltamethrin (*off-label*), spirotetramat (*off-label*)
	Beetles	deltamethrin (*off-label*)
	Caterpillars	Bacillus thuringiensis, Bacillus thuringiensis (*off-label*), deltamethrin (*off-label*), indoxacarb (*off-label*), lambda-cyhalothrin (*off-label*), spinosad (*off-label*)
	Pests, miscellaneous	deltamethrin (*off-label*), lambda-cyhalothrin (*off-label*)
	Slugs/snails	ferric phosphate
	Thrips	spinosad (*off-label*)
	Whiteflies	spirotetramat (*off-label*)
Weeds	Broad-leaved weeds	chlorpropham, pendimethalin (*off-label*), propyzamide (*off-label*), S-metolachlor (*off-label*), triflusulfuron-methyl (*off-label*)
	Crops as weeds	cycloxydim (*off-label*), fluazifop-P-butyl
	Grass weeds	chlorpropham, cycloxydim (*off-label*), fluazifop-P-butyl, propyzamide (*off-label*), S-metolachlor (*off-label*)

Leafy vegetables - Lettuce

Diseases	Alternaria	Bacillus amyloliquefaciens D747, cyprodinil + fludioxonil (*off-label*)
	Bacterial canker	Bacillus amyloliquefaciens D747 (*off-label*)
	Black rot	azoxystrobin (*off-label*)
	Botrytis	Bacillus amyloliquefaciens D747 (*off-label*), Bacillus subtilis (*off-label*), boscalid + pyraclostrobin, boscalid + pyraclostrobin (*off-label*), cyprodinil + fludioxonil (*off-label*), fenhexamid (*off-label*), fluopyram + trifloxystrobin (*off-label*), Gliocladium catenulatum (*off-label*)
	Bottom rot	boscalid + pyraclostrobin
	Damping off	Bacillus subtilis (*off-label*)
	Disease control	boscalid + pyraclostrobin (*off-label*), dimethomorph + mancozeb (*off-label*), mandipropamid
	Downy mildew	azoxystrobin, cos-oga (*off-label*), dimethomorph (*off-label*), dimethomorph + mancozeb (*off-label*), fluopicolide + propamocarb hydrochloride (*off-label*), fosetyl-aluminium + propamocarb hydrochloride, mancozeb + metalaxyl-M (*off-label*), mandipropamid, potassium bicarbonate (commodity substance) (*off-label*)
	Fireblight	Bacillus amyloliquefaciens D747 (*off-label*)
	Fungus diseases	Bacillus amyloliquefaciens D747 (*off-label*), Bacillus subtilis (*off-label*)
	Fusarium	Bacillus amyloliquefaciens D747, Gliocladium catenulatum (*off-label*)
	Phytophthora	Gliocladium catenulatum (*off-label*)
	Powdery mildew	Bacillus amyloliquefaciens D747, Bacillus amyloliquefaciens D747 (*off-label*), cos-oga (*off-label*), fluopyram + trifloxystrobin (*off-label*), physical pest control

	Pythium	Bacillus amyloliquefaciens D747, fosetyl-aluminium + propamocarb hydrochloride
	Rhizoctonia	Bacillus amyloliquefaciens D747, Bacillus subtilis (*off-label*), cyprodinil + fludioxonil (*off-label*), Gliocladium catenulatum (*off-label*)
	Sclerotinia	Bacillus amyloliquefaciens D747 (*off-label*), difenoconazole + fluxapyroxad, fluopyram + trifloxystrobin (*off-label*)
	Soft rot	boscalid + pyraclostrobin
	Stem canker	cyprodinil + fludioxonil (*off-label*)
	White blister	azoxystrobin (*off-label*)
Pests	Aphids	acetamiprid (*off-label*), Beauveria bassiana (*off-label*), deltamethrin, deltamethrin (*off-label*), lambda-cyhalothrin, physical pest control, pyrethrins, spirotetramat, spirotetramat (*off-label*)
	Beetles	deltamethrin, deltamethrin (*off-label*)
	Caterpillars	Bacillus thuringiensis, Bacillus thuringiensis (*off-label*), chlorantraniliprole (*off-label*), deltamethrin, deltamethrin (*off-label*), indoxacarb (*off-label*), lambda-cyhalothrin (*off-label*), pyrethrins, spinosad (*off-label*)
	Cutworms	deltamethrin, lambda-cyhalothrin
	Leafhoppers	deltamethrin
	Mealybugs	physical pest control
	Pests, miscellaneous	deltamethrin, deltamethrin (*off-label*), lambda-cyhalothrin (*off-label*)
	Scale insects	physical pest control
	Slugs/snails	ferric phosphate
	Spider mites	Beauveria bassiana (*off-label*), physical pest control, pyrethrins
	Suckers	physical pest control
	Thrips	Beauveria bassiana (*off-label*), deltamethrin, pyrethrins, spinosad (*off-label*)
	Whiteflies	Beauveria bassiana (*off-label*), physical pest control, spirotetramat (*off-label*)
Plant growth regulation	Growth control	trinexapac-ethyl (*off-label*)
Weeds	Broad-leaved weeds	chlorpropham, dimethenamid-p + pendimethalin (*off-label*), napropamide (*off-label*), pendimethalin (*off-label*), propyzamide, propyzamide (*off-label*), propyzamide (*outdoor crops*), S-metolachlor (*off-label*)
	Crops as weeds	cycloxydim (*off-label*), fluazifop-P-butyl
	Grass weeds	chlorpropham, cycloxydim (*off-label*), dimethenamid-p + pendimethalin (*off-label*), fluazifop-P-butyl, napropamide (*off-label*), pendimethalin (*off-label*), propyzamide, propyzamide (*off-label*), propyzamide (*outdoor crops*), S-metolachlor (*off-label*)

Leafy vegetables - Spinach

Diseases	Alternaria	Bacillus amyloliquefaciens D747, cyprodinil + fludioxonil (*off-label*)
	Anthracnose	Bacillus amyloliquefaciens D747 (*off-label*)
	Bacterial blight	Bacillus amyloliquefaciens D747 (*off-label*)

	Black rot	azoxystrobin (*off-label*)
	Botrytis	Bacillus amyloliquefaciens D747 (*off-label*), Bacillus subtilis (*off-label*), cyprodinil + fludioxonil (*off-label*), Gliocladium catenulatum (*off-label*)
	Cercospora leaf spot	Bacillus amyloliquefaciens D747 (*off-label*)
	Damping off	Bacillus amyloliquefaciens D747 (*off-label*), Bacillus subtilis (*off-label*)
	Disease control	mandipropamid
	Downy mildew	Bacillus amyloliquefaciens D747 (*off-label*), boscalid + pyraclostrobin (*off-label*), dimethomorph (*off-label*), mandipropamid, metalaxyl-M (*off-label*)
	Fusarium	Bacillus amyloliquefaciens D747, Bacillus amyloliquefaciens D747 (*off-label*), Gliocladium catenulatum (*off-label*)
	Phytophthora	Gliocladium catenulatum (*off-label*)
	Powdery mildew	Bacillus amyloliquefaciens D747, Bacillus amyloliquefaciens D747 (*off-label*)
	Pythium	Bacillus amyloliquefaciens D747, Bacillus amyloliquefaciens D747 (*off-label*), metalaxyl-M (*off-label*)
	Rhizoctonia	Bacillus amyloliquefaciens D747, Bacillus amyloliquefaciens D747 (*off-label*), cyprodinil + fludioxonil (*off-label*), Gliocladium catenulatum (*off-label*)
	Sclerotinia	Bacillus amyloliquefaciens D747 (*off-label*)
	Stem canker	cyprodinil + fludioxonil (*off-label*)
	Stemphylium	Bacillus amyloliquefaciens D747 (*off-label*)
	White blister	azoxystrobin (*off-label*)
Pests	Aphids	acetamiprid (*off-label*), Beauveria bassiana (*off-label*)
	Caterpillars	Bacillus thuringiensis, Bacillus thuringiensis (*off-label*)
	Slugs/snails	ferric phosphate
	Spider mites	Beauveria bassiana (*off-label*)
	Thrips	Beauveria bassiana (*off-label*)
	Whiteflies	Beauveria bassiana (*off-label*)
Weeds	Broad-leaved weeds	chlorpropham (*off-label*), clopyralid (*off-label*), phenmedipham (*off-label*)
	Crops as weeds	cycloxydim (*off-label*), fluazifop-P-butyl
	Grass weeds	chlorpropham (*off-label*), cycloxydim (*off-label*), fluazifop-P-butyl

Leafy vegetables - Watercress

Diseases	Alternaria	Bacillus amyloliquefaciens D747
	Anthracnose	Bacillus amyloliquefaciens D747 (*off-label*)
	Bacterial blight	Bacillus amyloliquefaciens D747 (*off-label*)
	Botrytis	Bacillus amyloliquefaciens D747 (*off-label*), Bacillus subtilis (*off-label*)
	Cercospora leaf spot	Bacillus amyloliquefaciens D747 (*off-label*)
	Damping off	Bacillus amyloliquefaciens D747 (*off-label*), metalaxyl-M (*off-label*)
	Downy mildew	Bacillus amyloliquefaciens D747 (*off-label*), metalaxyl-M (*off-label*), propamocarb hydrochloride (*off-label*)

	Fusarium	Bacillus amyloliquefaciens D747, Bacillus amyloliquefaciens D747 (*off-label*), Trichoderma asperellum (Strain T34) (*off-label*)
	Phytophthora	propamocarb hydrochloride (*off-label*)
	Powdery mildew	Ampelomyces quisqualis (Strain AQ10) (*off-label*), Bacillus amyloliquefaciens D747, Bacillus amyloliquefaciens D747 (*off-label*)
	Pythium	Bacillus amyloliquefaciens D747, Bacillus amyloliquefaciens D747 (*off-label*), propamocarb hydrochloride (*off-label*), Trichoderma asperellum (Strain T34) (*off-label*)
	Rhizoctonia	Bacillus amyloliquefaciens D747, Bacillus amyloliquefaciens D747 (*off-label*)
	Root rot	propamocarb hydrochloride (*off-label*)
	Sclerotinia	Bacillus amyloliquefaciens D747 (*off-label*)
	Stemphylium	Bacillus amyloliquefaciens D747 (*off-label*)
Pests	Aphids	Beauveria bassiana (*off-label*)
	Caterpillars	Bacillus thuringiensis (*off-label*)
	Cutworms	Bacillus thuringiensis (*off-label*)
	Spider mites	Beauveria bassiana (*off-label*)
	Thrips	Beauveria bassiana (*off-label*)
	Whiteflies	Beauveria bassiana (*off-label*)
Weeds	Crops as weeds	cycloxydim (*off-label*)
	Grass weeds	cycloxydim (*off-label*)

Legumes - Beans (Phaseolus)

Diseases	Ascochyta	azoxystrobin (*useful control*)
	Botrytis	azoxystrobin (*some control*), Bacillus subtilis (*off-label*), boscalid + pyraclostrobin (*off-label*), cyprodinil + fludioxonil (*moderate control*), difenoconazole + fluxapyroxad (*off-label*), Gliocladium catenulatum (*off-label*)
	Downy mildew	azoxystrobin (*reduction*)
	Fusarium	Gliocladium catenulatum (*off-label*)
	Leaf spot	difenoconazole + fluxapyroxad (*off-label*)
	Mycosphaerella	azoxystrobin (*some control*)
	Phytophthora	difenoconazole + fluxapyroxad (*off-label*), Gliocladium catenulatum (*off-label*)
	Powdery mildew	physical pest control
	Rhizoctonia	Gliocladium catenulatum (*off-label*)
	Rust	tebuconazole (*off-label*)
	Sclerotinia	cyprodinil + fludioxonil
Pests	Aphids	cypermethrin, deltamethrin (*off-label*), flonicamid (*off-label*), physical pest control, pirimicarb (*off-label*), pyrethrins
	Birds/mammals	aluminium ammonium sulphate
	Caterpillars	Bacillus thuringiensis, Bacillus thuringiensis (*off-label*), deltamethrin (*off-label*), lambda-cyhalothrin (*off-label*), pyrethrins, spinosad (*off-label*)
	Cutworms	Bacillus thuringiensis (*off-label*)
	Mealybugs	physical pest control

	Midges	deltamethrin (*off-label*)
	Pests, miscellaneous	cypermethrin, lambda-cyhalothrin (*off-label*)
	Scale insects	physical pest control
	Slugs/snails	ferric phosphate
	Spider mites	physical pest control, pyrethrins
	Suckers	physical pest control
	Thrips	deltamethrin (*off-label*), pyrethrins
	Weevils	deltamethrin (*off-label*)
	Whiteflies	physical pest control
Weeds	Broad-leaved weeds	bentazone, clomazone (*off-label*), pendimethalin, pendimethalin (*off-label*)
	Crops as weeds	clethodim (*off-label*), cycloxydim, cycloxydim (*off-label*), fluazifop-P-butyl
	Grass weeds	clethodim (*off-label*), cycloxydim, cycloxydim (*off-label*), fluazifop-P-butyl, pendimethalin (*off-label*), propaquizafop, S-metolachlor (*off-label*)

Legumes - Beans (Vicia)

Crop control	Desiccation	glyphosate
Diseases	Botrytis	Bacillus subtilis (*off-label*), boscalid + pyraclostrobin (*off-label*), cyprodinil + fludioxonil (*moderate control*), Gliocladium catenulatum (*off-label*)
	Chocolate spot	azoxystrobin + tebuconazole, azoxystrobin + tebuconazole (*moderate control*), boscalid + pyraclostrobin (*moderate control*), tebuconazole, tebuconazole (*moderate control*)
	Disease control	azoxystrobin + tebuconazole, azoxystrobin + tebuconazole (*moderate control*), metconazole (*off-label*), tebuconazole
	Downy mildew	metalaxyl-M (*off-label*)
	Fusarium	Gliocladium catenulatum (*off-label*)
	Phytophthora	Gliocladium catenulatum (*off-label*)
	Rhizoctonia	Gliocladium catenulatum (*off-label*)
	Rust	azoxystrobin, boscalid + pyraclostrobin, metconazole, tebuconazole
	Sclerotinia	cyprodinil + fludioxonil
Pests	Aphids	esfenvalerate, lambda-cyhalothrin, pirimicarb, pyrethrins, thiacloprid, zeta-cypermethrin (*off-label*)
	Beetles	lambda-cyhalothrin, lambda-cyhalothrin (*off-label*), thiacloprid
	Birds/mammals	aluminium ammonium sulphate
	Caterpillars	Bacillus thuringiensis (*off-label*), lambda-cyhalothrin, pyrethrins
	Cutworms	Bacillus thuringiensis (*off-label*)
	Pests, miscellaneous	zeta-cypermethrin (*off-label*)
	Slugs/snails	ferric phosphate
	Spider mites	pyrethrins
	Thrips	lambda-cyhalothrin (*off-label*), pyrethrins
	Weevils	alpha-cypermethrin, cypermethrin, deltamethrin, esfenvalerate, lambda-cyhalothrin, zeta-cypermethrin

Weeds	Broad-leaved weeds	bentazone, carbetamide, clomazone, clomazone (*off-label*), clomazone + pendimethalin, glyphosate, imazamox + pendimethalin, imazamox + pendimethalin (*off-label*), isoxaben (*off-label*), pendimethalin (*off-label*), propyzamide, prosulfocarb (*off-label*), S-metolachlor (*off-label*)
	Crops as weeds	carbetamide, clethodim (*off-label*), cycloxydim, cycloxydim (*off-label*), fluazifop-P-butyl, glyphosate, pendimethalin (*off-label*), propyzamide, quizalofop-P-ethyl, quizalofop-P-tefuryl
	Grass weeds	carbetamide, clethodim (*off-label*), cycloxydim, cycloxydim (*off-label*), fluazifop-P-butyl, glyphosate, pendimethalin (*off-label*), propaquizafop, propyzamide, prosulfocarb (*off-label*), quizalofop-P-ethyl, quizalofop-P-tefuryl, S-metolachlor (*off-label*)
	Weeds, miscellaneous	glyphosate, pendimethalin (*off-label*)

Legumes - Forage legumes, general

Weeds	Broad-leaved weeds	pendimethalin (*off-label*)
	Grass weeds	pendimethalin (*off-label*)

Legumes - Lupins

Crop control	Desiccation	glyphosate (*off-label*)
Diseases	Ascochyta	metconazole (*qualified minor use*)
	Botrytis	Bacillus subtilis (*off-label*), cyprodinil + fludioxonil (*off-label*), metconazole (*qualified minor use*)
	Powdery mildew	physical pest control
	Rust	azoxystrobin, metconazole (*qualified minor use*)
Pests	Aphids	cypermethrin, physical pest control, zeta-cypermethrin (*off-label*)
	Caterpillars	lambda-cyhalothrin (*off-label*)
	Mealybugs	physical pest control
	Pests, miscellaneous	cypermethrin, lambda-cyhalothrin (*off-label*), zeta-cypermethrin (*off-label*)
	Scale insects	physical pest control
	Slugs/snails	ferric phosphate
	Spider mites	physical pest control
	Suckers	physical pest control
	Whiteflies	physical pest control
Weeds	Broad-leaved weeds	clomazone (*off-label*), pendimethalin (*off-label*), pyridate (*off-label*)
	Crops as weeds	fluazifop-P-butyl
	Grass weeds	fluazifop-P-butyl, pendimethalin (*off-label*), propaquizafop (*off-label*)
	Weeds, miscellaneous	glyphosate (*off-label*)

Legumes - Peas

Crop control	Desiccation	glyphosate
Diseases	Ascochyta	azoxystrobin, azoxystrobin (*useful control*), boscalid + pyraclostrobin (*moderate control only*), cyprodinil +

		fludioxonil, difenoconazole + fluxapyroxad (*moderate control*), metconazole (*reduction*)
	Botrytis	azoxystrobin (*some control*), Bacillus subtilis (*off-label*), cyprodinil + fludioxonil (*moderate control*), difenoconazole + fluxapyroxad (*off-label*), Gliocladium catenulatum (*off-label*), metconazole (*reduction*)
	Downy mildew	azoxystrobin (*reduction*), cymoxanil + fludioxonil + metalaxyl-M
	Fusarium	Gliocladium catenulatum (*off-label*)
	Leaf spot	difenoconazole + fluxapyroxad (*off-label*)
	Mycosphaerella	azoxystrobin (*some control*), cyprodinil + fludioxonil, metconazole (*reduction*)
	Phytophthora	difenoconazole + fluxapyroxad (*off-label*), Gliocladium catenulatum (*off-label*)
	Powdery mildew	physical pest control, sulphur (*off-label*)
	Rhizoctonia	Gliocladium catenulatum (*off-label*)
	Rust	metconazole
	Sclerotinia	cyprodinil + fludioxonil
Pests	Aphids	alpha-cypermethrin (*reduction*), cypermethrin, deltamethrin (*off-label*), esfenvalerate, flonicamid (*off-label*), lambda-cyhalothrin, physical pest control, pirimicarb, pirimicarb (*off-label*), pyrethrins, thiacloprid, zeta-cypermethrin
	Beetles	lambda-cyhalothrin
	Birds/mammals	aluminium ammonium sulphate
	Caterpillars	alpha-cypermethrin, Bacillus thuringiensis, cypermethrin, deltamethrin, deltamethrin (*off-label*), lambda-cyhalothrin, pyrethrins, zeta-cypermethrin
	Mealybugs	physical pest control
	Midges	deltamethrin, deltamethrin (*off-label*), lambda-cyhalothrin, thiacloprid
	Pests, miscellaneous	aluminium phosphide, cypermethrin
	Scale insects	physical pest control
	Slugs/snails	ferric phosphate
	Spider mites	physical pest control, pyrethrins
	Suckers	physical pest control
	Thrips	deltamethrin (*off-label*), esfenvalerate, pyrethrins
	Weevils	alpha-cypermethrin, cypermethrin, deltamethrin, deltamethrin (*off-label*), esfenvalerate, lambda-cyhalothrin, zeta-cypermethrin
	Whiteflies	physical pest control
Weeds	Broad-leaved weeds	bentazone, clomazone, clomazone (*off-label*), clomazone + pendimethalin, flumioxazin (*off-label*), glyphosate, imazamox + pendimethalin, MCPB, pendimethalin, pendimethalin (*off-label*), prosulfocarb (*off-label*)
	Crops as weeds	clethodim (*off-label*), cycloxydim, cycloxydim (*off-label*), fluazifop-P-butyl, flumioxazin (*off-label*), glyphosate, pendimethalin, quizalofop-P-ethyl, quizalofop-P-tefuryl
	Grass weeds	clethodim (*off-label*), cycloxydim, cycloxydim (*off-label*), fluazifop-P-butyl, flumioxazin (*off-label*), glyphosate, pendimethalin, pendimethalin (*off-label*),

propaquizafop, quizalofop-P-ethyl, quizalofop-P-tefuryl, S-metolachlor (*off-label*)

| Weeds, miscellaneous | glyphosate, glyphosate (*off-label*) |

Miscellaneous arable - Industrial crops

Crop control	Desiccation	glyphosate (*off-label*)
Diseases	Powdery mildew	physical pest control
Pests	Aphids	lambda-cyhalothrin (*off-label*), physical pest control
	Caterpillars	lambda-cyhalothrin (*off-label*)
	Leaf miners	thiacloprid (*off-label*)
	Mealybugs	physical pest control
	Pests, miscellaneous	lambda-cyhalothrin (*off-label*)
	Scale insects	physical pest control
	Spider mites	physical pest control
	Suckers	physical pest control
	Whiteflies	physical pest control
Weeds	Broad-leaved weeds	2,4-D (*off-label*), bromoxynil (*off-label*), clopyralid (*off-label*), fluroxypyr (*off-label*), MCPA (*off-label*), mecoprop-P (*off-label*), metsulfuron-methyl + thifensulfuron-methyl (*off-label*), pendimethalin (*off-label*), prosulfocarb (*off-label*)
	Crops as weeds	fluroxypyr (*off-label*)
	Grass weeds	pendimethalin (*off-label*), prosulfocarb (*off-label*), tri-allate (*off-label*)
	Weeds, miscellaneous	bromoxynil (*off-label*), glyphosate (*off-label*)

Miscellaneous arable - Miscellaneous arable crops

Crop control	Desiccation	glyphosate (*off-label*)
	Miscellaneous non-selective situations	benzoic acid
Diseases	Alternaria	Bacillus amyloliquefaciens D747
	Anthracnose	Bacillus amyloliquefaciens D747 (*off-label*)
	Bacterial blight	Bacillus amyloliquefaciens D747 (*off-label*)
	Botrytis	azoxystrobin (*off-label*), Bacillus amyloliquefaciens D747 (*off-label*), Bacillus subtilis (*off-label*), potassium bicarbonate (commodity substance) (*off-label*)
	Cercospora leaf spot	Bacillus amyloliquefaciens D747 (*off-label*)
	Didymella stem rot	Gliocladium catenulatum (*moderate control*)
	Disease control	boscalid (*off-label*), boscalid + pyraclostrobin (*off-label*), potassium bicarbonate (commodity substance)
	Downy mildew	azoxystrobin (*off-label*), Bacillus amyloliquefaciens D747 (*off-label*), boscalid + pyraclostrobin (*off-label*), cyazofamid (*off-label*)
	Fusarium	Bacillus amyloliquefaciens D747, Bacillus amyloliquefaciens D747 (*off-label*), Gliocladium catenulatum (*moderate control*)
	Phytophthora	cyazofamid (*off-label*), Gliocladium catenulatum (*moderate control*)
	Powdery mildew	azoxystrobin (*off-label*), Bacillus amyloliquefaciens D747, Bacillus amyloliquefaciens D747 (*off-label*),

		cyflufenamid (*off-label*), potassium bicarbonate (commodity substance) (*off-label*)
	Pythium	Bacillus amyloliquefaciens D747, Bacillus amyloliquefaciens D747 (*off-label*), Gliocladium catenulatum (*moderate control*)
	Rhizoctonia	Bacillus amyloliquefaciens D747, Bacillus amyloliquefaciens D747 (*off-label*), Gliocladium catenulatum (*moderate control*)
	Sclerotinia	Bacillus amyloliquefaciens D747 (*off-label*), Coniothyrium minitans
	Stemphylium	Bacillus amyloliquefaciens D747 (*off-label*)
Pests	Aphids	flonicamid (*off-label*), lambda-cyhalothrin (*off-label*), maltodextrin, thiacloprid (*off-label*)
	Caterpillars	Bacillus thuringiensis, Bacillus thuringiensis (*off-label*), lambda-cyhalothrin (*off-label*)
	Pests, miscellaneous	lambda-cyhalothrin (*off-label*)
	Slugs/snails	ferric phosphate, metaldehyde
	Spider mites	maltodextrin
	Whiteflies	maltodextrin
Plant growth regulation	Growth control	chlormequat (*off-label*), trinexapac-ethyl (*off-label*)
Weeds	Broad-leaved weeds	2,4-D (*off-label*), amidosulfuron (*off-label*), bentazone (*off-label*), bromoxynil (*off-label*), carfentrazone-ethyl (*before planting*), clomazone (*off-label*), clopyralid (*off-label*), clopyralid + picloram (*off-label*), clopyralid + triclopyr (*off-label*), dicamba + MCPA + mecoprop-P (*off-label*), diflufenican (*off-label*), dimethenamid-p + metazachlor (*off-label*), dimethenamid-p + pendimethalin (*off-label*), florasulam (*off-label*), florasulam + fluroxypyr (*off-label*), flufenacet + pendimethalin (*off-label*), fluroxypyr (*off-label*), MCPA (*off-label*), mecoprop-P (*off-label*), mesotrione (*off-label*), napropamide (*off-label*), nicosulfuron (*off-label*), pendimethalin (*off-label*), pendimethalin + picolinafen (*off-label*), propyzamide (*off-label*), prosulfocarb (*off-label*), prosulfuron (*off-label*), pyridate (*off-label*), thifensulfuron-methyl + tribenuron-methyl (*off-label*)
	Crops as weeds	bromoxynil (*off-label*), carfentrazone-ethyl (*before planting*), diflufenican (*off-label*), fluroxypyr (*off-label*), nicosulfuron (*off-label*), pendimethalin (*off-label*)
	Grass weeds	clethodim (*off-label*), diflufenican (*off-label*), dimethenamid-p + metazachlor (*off-label*), dimethenamid-p + pendimethalin (*off-label*), flufenacet + pendimethalin (*off-label*), mesotrione (*off-label*), napropamide (*off-label*), nicosulfuron (*off-label*), pendimethalin (*off-label*), pendimethalin + picolinafen (*off-label*), propaquizafop (*off-label*), propyzamide (*off-label*), prosulfocarb (*off-label*), tri-allate (*off-label*)
	Weeds, miscellaneous	2,4-D + glyphosate, bromoxynil (*off-label*), carfentrazone-ethyl, diflufenican (*off-label*), glyphosate, glyphosate (*before planting*), glyphosate (*off-label*), glyphosate (*pre-sowing/planting*), napropamide (*off-label*)

Miscellaneous arable - Miscellaneous arable situations

Weeds	Broad-leaved weeds	citronella oil
	Crops as weeds	glyphosate
	Grass weeds	glyphosate
	Weeds, miscellaneous	cycloxydim, fluazifop-P-butyl, glyphosate, metsulfuron-methyl, thifensulfuron-methyl

Miscellaneous field vegetables - All vegetables

Diseases	Anthracnose	Bacillus amyloliquefaciens D747 (*off-label*)
	Bacterial blight	Bacillus amyloliquefaciens D747 (*off-label*)
	Botrytis	Bacillus amyloliquefaciens D747 (*off-label*), Bacillus subtilis (*off-label*), potassium bicarbonate (commodity substance) (*off-label*)
	Cercospora leaf spot	Bacillus amyloliquefaciens D747 (*off-label*)
	Downy mildew	Bacillus amyloliquefaciens D747 (*off-label*)
	Fusarium	Bacillus amyloliquefaciens D747 (*off-label*)
	Powdery mildew	Bacillus amyloliquefaciens D747 (*off-label*), potassium bicarbonate (commodity substance) (*off-label*)
	Pythium	Bacillus amyloliquefaciens D747 (*off-label*)
	Rhizoctonia	Bacillus amyloliquefaciens D747 (*off-label*)
	Sclerotinia	Bacillus amyloliquefaciens D747 (*off-label*)
	Stemphylium	Bacillus amyloliquefaciens D747 (*off-label*)
Weeds	Broad-leaved weeds	clomazone (*off-label*)

Oilseed crops - Linseed/flax

Crop control	Desiccation	glyphosate
Diseases	Alternaria	difenoconazole (*off-label*)
	Botrytis	tebuconazole, tebuconazole (*reduction*)
	Disease control	boscalid (*off-label*), metconazole (*off-label*), prothioconazole (*off-label*), tebuconazole
	Foliar diseases	difenoconazole + paclobutrazol (*off-label*), prothioconazole (*off-label*)
	Powdery mildew	difenoconazole (*off-label*), difenoconazole + paclobutrazol (*off-label*), tebuconazole
	Rust	tebuconazole
	Sclerotinia	difenoconazole (*off-label*)
	Septoria	difenoconazole + paclobutrazol (*off-label*), prothioconazole (*off-label*)
Pests	Aphids	cypermethrin, tau-fluvalinate (*off-label*)
	Beetles	zeta-cypermethrin
	Caterpillars	lambda-cyhalothrin (*off-label*)
	Pests, miscellaneous	cypermethrin, lambda-cyhalothrin (*off-label*)
	Slugs/snails	ferric phosphate
Weeds	Broad-leaved weeds	amidosulfuron, bentazone, bifenox (*off-label*), bromoxynil, carbetamide (*off-label*), clopyralid, glyphosate, mesotrione (*off-label*), metazachlor (*off-label*), metsulfuron-methyl, napropamide (*off-label*), prosulfocarb (*off-label*)
	Crops as weeds	carbetamide (*off-label*), clethodim (*off-label*), cycloxydim, fluazifop-P-butyl, glyphosate, mesotrione

	(*off-label*), metsulfuron-methyl, quizalofop-P-ethyl, quizalofop-P-tefuryl
Grass weeds	carbetamide (*off-label*), clethodim (*off-label*), cycloxydim, fluazifop-P-butyl, glyphosate, mesotrione (*off-label*), metazachlor (*off-label*), napropamide (*off-label*), propaquizafop, prosulfocarb (*off-label*), quizalofop-P-ethyl, quizalofop-P-tefuryl, tri-allate (*off-label*)
Weeds, miscellaneous	bromoxynil, glyphosate, metsulfuron-methyl, napropamide (*off-label*)

Oilseed crops - Miscellaneous oilseeds

Crop control	Desiccation	glyphosate (*off-label*)
Diseases	Alternaria	boscalid + pyraclostrobin (*off-label*), difenoconazole (*off-label*)
	Botrytis	boscalid (*off-label*), boscalid + pyraclostrobin (*off-label*)
	Disease control	boscalid (*off-label*), dimethomorph + mancozeb (*off-label*), tebuconazole (*off-label*)
	Downy mildew	dimethomorph + mancozeb (*off-label*), mancozeb + metalaxyl-M (*off-label*)
	Fire	boscalid + pyraclostrobin (*off-label*)
	Powdery mildew	difenoconazole (*off-label*), sulphur (*off-label*)
	Rust	tebuconazole (*off-label*)
	Sclerotinia	boscalid (*off-label*), difenoconazole (*off-label*)
Pests	Aphids	deltamethrin (*off-label*), lambda-cyhalothrin (*off-label*), tau-fluvalinate (*off-label*)
	Beetles	lambda-cyhalothrin (*off-label*)
	Caterpillars	deltamethrin (*off-label*), lambda-cyhalothrin (*off-label*)
	Pests, miscellaneous	deltamethrin (*off-label*), lambda-cyhalothrin (*off-label*)
Weeds	Broad-leaved weeds	bifenox (*off-label*), clomazone (*off-label*), clopyralid (*off-label*), clopyralid + picloram (*off-label*), fluroxypyr (*off-label*), mesotrione (*off-label*), metazachlor (*off-label*), napropamide (*off-label*), pendimethalin (*off-label*), prosulfocarb (*off-label*)
	Crops as weeds	clethodim (*off-label*), fluazifop-P-butyl, fluroxypyr (*off-label*), propaquizafop (*off-label*)
	Grass weeds	clethodim (*off-label*), fluazifop-P-butyl, metazachlor (*off-label*), napropamide (*off-label*), propaquizafop (*off-label*), prosulfocarb (*off-label*)
	Weeds, miscellaneous	glyphosate (*off-label*), napropamide (*off-label*)

Oilseed crops - Oilseed rape

Crop control	Desiccation	glyphosate
Diseases	Alternaria	azoxystrobin, azoxystrobin + cyproconazole, boscalid, boscalid + dimoxystrobin (*moderate control*), difenoconazole, metconazole, tebuconazole
	Black scurf and stem canker	Bacillus amyloliquefaciens strain MBI600 (*useful reduction*), bixafen + prothioconazole + tebuconazole, boscalid + dimoxystrobin (*moderate control*), difenoconazole, fluopyram + prothioconazole, prothioconazole, prothioconazole + tebuconazole, tebuconazole

	Disease control	azoxystrobin + difenoconazole, boscalid + dimoxystrobin, difenoconazole + paclobutrazol, fluopyram + prothioconazole
	Ear blight	thiophanate-methyl (*reduction*)
	Fusarium	thiophanate-methyl
	Light leaf spot	bixafen + prothioconazole + tebuconazole, boscalid + dimoxystrobin (*reduction*), difenoconazole, difenoconazole + paclobutrazol, fluopyram + prothioconazole, metconazole, metconazole (*reduction*), prothioconazole, prothioconazole + tebuconazole, prothioconazole + tebuconazole (*moderate control only*), tebuconazole
	Phoma leaf spot	prothioconazole + tebuconazole, tebuconazole
	Powdery mildew	fluopyram + prothioconazole
	Ring spot	tebuconazole, tebuconazole (*reduction*)
	Sclerotinia	azoxystrobin, azoxystrobin + cyproconazole, azoxystrobin + isopyrazam, azoxystrobin + tebuconazole (*moderate control*), bixafen + prothioconazole + tebuconazole, boscalid, boscalid + dimoxystrobin, boscalid + metconazole, fluopyram + prothioconazole, fluoxastrobin + tebuconazole, prothioconazole, prothioconazole + tebuconazole, tebuconazole, thiophanate-methyl
	Stem canker	bixafen + prothioconazole + tebuconazole, boscalid + dimoxystrobin (*moderate control*), difenoconazole + paclobutrazol, fluopyram + prothioconazole, metconazole (*reduction*), prothioconazole, prothioconazole + tebuconazole, tebuconazole
Pests	Aphids	acetamiprid, deltamethrin, flonicamid, lambda-cyhalothrin, tau-fluvalinate
	Beetles	alpha-cypermethrin, Bacillus amyloliquefaciens strain MBI600 (*stimulation of plants own defenses*), beta-cyfluthrin, cypermethrin, deltamethrin, indoxacarb, lambda-cyhalothrin, tau-fluvalinate, thiacloprid, zeta-cypermethrin
	Birds/mammals	aluminium ammonium sulphate
	Caterpillars	lambda-cyhalothrin
	Midges	beta-cyfluthrin, cypermethrin, lambda-cyhalothrin, zeta-cypermethrin
	Slugs/snails	ferric phosphate
	Weevils	alpha-cypermethrin, beta-cyfluthrin, cypermethrin, deltamethrin, lambda-cyhalothrin, zeta-cypermethrin
Plant growth regulation	Growth control	mepiquat chloride + metconazole, metconazole, tebuconazole, trinexapac-ethyl
	Quality/yield control	sulphur
Weeds	Broad-leaved weeds	aminopyralid + metazachlor + picloram, aminopyralid + propyzamide, bifenox (*off-label*), carbetamide, clomazone, clomazone + napropamide, clopyralid, clopyralid + picloram, clopyralid + picloram (*off-label*), dimethachlor, dimethenamid-p + metazachlor, dimethenamid-p + metazachlor + quinmerac, dimethenamid-p + quinmerac, glyphosate, halauxifen-methyl + picloram, imazamox + metazachlor, imazamox + quinmerac, imazamox + quinmerac (*cut-leaved*), metazachlor, metazachlor +

	quinmerac, napropamide, propyzamide, pyridate (*off-label*)
Crops as weeds	carbetamide, clethodim, cycloxydim, fluazifop-P-butyl, glyphosate, imazamox + quinmerac, propyzamide, quizalofop-P-ethyl, quizalofop-P-tefuryl
Grass weeds	aminopyralid + propyzamide, carbetamide, clethodim, clomazone + napropamide, cycloxydim, dimethachlor, dimethenamid-p + metazachlor + quinmerac, fluazifop-P-butyl, glyphosate, imazamox + metazachlor, metazachlor, metazachlor + quinmerac, napropamide, propaquizafop, propyzamide, quizalofop-P-ethyl, quizalofop-P-tefuryl
Weeds, miscellaneous	glyphosate

Oilseed crops - Soya

Diseases	Botrytis	Gliocladium catenulatum (*off-label*)
	Fusarium	Gliocladium catenulatum (*off-label*)
	Phytophthora	Gliocladium catenulatum (*off-label*)
	Rhizoctonia	Gliocladium catenulatum (*off-label*)
	Sclerotinia	azoxystrobin (*off-label*)
Pests	Aphids	pyrethrins
	Birds/mammals	aluminium ammonium sulphate
	Caterpillars	pyrethrins
	Flies	lambda-cyhalothrin (*off-label*)
	Slugs/snails	ferric phosphate
	Spider mites	pyrethrins
	Thrips	pyrethrins
Weeds	Broad-leaved weeds	bentazone, bentazone (*off-label*), clomazone (*off-label*), flufenacet + metribuzin (*off-label*), imazamox + pendimethalin (*off-label*), pendimethalin (*off-label*), thifensulfuron-methyl (*off-label*)
	Crops as weeds	cycloxydim (*off-label*)
	Grass weeds	cycloxydim (*off-label*), flufenacet + metribuzin (*off-label*), pendimethalin (*off-label*)
	Weeds, miscellaneous	imazamox + pendimethalin (*off-label*)

Oilseed crops - Sunflowers

Diseases	Powdery mildew	physical pest control
Pests	Aphids	physical pest control
	Mealybugs	physical pest control
	Scale insects	physical pest control
	Slugs/snails	ferric phosphate
	Spider mites	physical pest control
	Suckers	physical pest control
	Whiteflies	physical pest control
Weeds	Broad-leaved weeds	aclonifen (*off-label*), pendimethalin
	Crops as weeds	aclonifen (*off-label*), fluazifop-P-butyl, pendimethalin
	Grass weeds	clethodim (*off-label*), fluazifop-P-butyl, pendimethalin

Root and tuber crops - Beet crops

Diseases	Alternaria	Bacillus amyloliquefaciens D747, cyprodinil + fludioxonil (*off-label*), difenoconazole + fluxapyroxad (*off-label*), isopyrazam (*off-label*), prothioconazole (*off-label*)
	Anthracnose	Bacillus amyloliquefaciens D747 (*off-label*)
	Bacterial blight	Bacillus amyloliquefaciens D747 (*off-label*)
	Black leg	hymexazol (*seed treatment*)
	Black rot	azoxystrobin (*off-label*)
	Botrytis	Bacillus amyloliquefaciens D747 (*off-label*), Bacillus subtilis (*off-label*), cyprodinil + fludioxonil (*off-label*), Gliocladium catenulatum (*off-label*)
	Cercospora leaf spot	azoxystrobin + cyproconazole, Bacillus amyloliquefaciens D747 (*off-label*), cyproconazole + trifloxystrobin, cyproconazole + trifloxystrobin (*off-label*), epoxiconazole (*off-label*), epoxiconazole + pyraclostrobin, isopyrazam (*off-label*), prothioconazole (*off-label*)
	Damping off	Bacillus amyloliquefaciens D747 (*off-label*), Bacillus subtilis (*off-label*)
	Disease control	azoxystrobin + difenoconazole, boscalid + pyraclostrobin (*off-label*), mandipropamid
	Downy mildew	Bacillus amyloliquefaciens D747 (*off-label*), boscalid + pyraclostrobin (*off-label*), dimethomorph (*off-label*), fluopicolide + propamocarb hydrochloride (*off-label*), fosetyl-aluminium + propamocarb hydrochloride (*off-label*), mandipropamid, metalaxyl-M (*off-label*)
	Fusarium	Bacillus amyloliquefaciens D747, Bacillus amyloliquefaciens D747 (*off-label*), Gliocladium catenulatum (*off-label*)
	Phytophthora	Gliocladium catenulatum (*off-label*)
	Powdery mildew	azoxystrobin + cyproconazole, Bacillus amyloliquefaciens D747, Bacillus amyloliquefaciens D747 (*off-label*), cyproconazole + trifloxystrobin, cyproconazole + trifloxystrobin (*off-label*), difenoconazole + fluxapyroxad (*off-label*), epoxiconazole, epoxiconazole + pyraclostrobin, epoxiconazole + pyraclostrobin (*off-label*), isopyrazam (*off-label*), physical pest control, prothioconazole (*off-label*), sulphur, sulphur (*off-label*)
	Pythium	Bacillus amyloliquefaciens D747, Bacillus amyloliquefaciens D747 (*off-label*), fludioxonil + metalaxyl-M + sedaxane, metalaxyl-M, metalaxyl-M (*off-label*)
	Ramularia leaf spots	azoxystrobin + cyproconazole, cyproconazole + trifloxystrobin, cyproconazole + trifloxystrobin (*off-label*), epoxiconazole + pyraclostrobin, isopyrazam (*off-label*)
	Rhizoctonia	Bacillus amyloliquefaciens D747, Bacillus amyloliquefaciens D747 (*off-label*), cyprodinil + fludioxonil (*off-label*), difenoconazole + fluxapyroxad (*off-label*), fludioxonil + metalaxyl-M + sedaxane, Gliocladium catenulatum (*off-label*)
	Root malformation disorder	azoxystrobin (*off-label*), metalaxyl-M (*off-label*)

	Rust	azoxystrobin + cyproconazole, cyproconazole + trifloxystrobin, cyproconazole + trifloxystrobin (*off-label*), epoxiconazole, epoxiconazole + pyraclostrobin, epoxiconazole + pyraclostrobin (*off-label*)
	Sclerotinia	Bacillus amyloliquefaciens D747 (*off-label*), cyprodinil + fludioxonil (*off-label*), difenoconazole + fluxapyroxad (*off-label*), isopyrazam (*off-label*)
	Stem canker	cyprodinil + fludioxonil (*off-label*), fludioxonil + metalaxyl-M + sedaxane, isopyrazam (*off-label*), prothioconazole (*off-label*)
	Stemphylium	Bacillus amyloliquefaciens D747 (*off-label*)
	White blister	azoxystrobin (*off-label*)
Pests	Aphids	acetamiprid (*off-label*), Beauveria bassiana (*off-label*), beta-cyfluthrin, deltamethrin (*off-label*), flonicamid, lambda-cyhalothrin, physical pest control, pirimicarb (*off-label*), pyrethrins, spirotetramat (*off-label*), thiacloprid (*off-label*), zeta-cypermethrin (*off-label*)
	Beetles	deltamethrin, deltamethrin (*off-label*), lambda-cyhalothrin, lambda-cyhalothrin (*off-label*), tefluthrin (*seed treatment*)
	Birds/mammals	aluminium ammonium sulphate
	Caterpillars	Bacillus thuringiensis, Bacillus thuringiensis (*off-label*), Bacillus thuringiensis aizawai GC-91, chlorantraniliprole (*off-label*), deltamethrin (*off-label*), lambda-cyhalothrin, lambda-cyhalothrin (*off-label*), pyrethrins
	Cutworms	Bacillus thuringiensis (*off-label*), cypermethrin, lambda-cyhalothrin, lambda-cyhalothrin (*off-label*), zeta-cypermethrin
	Flies	chlorantraniliprole (*off-label*), deltamethrin (*off-label*)
	Free-living nematodes	garlic extract (*off-label*), oxamyl
	Leaf miners	lambda-cyhalothrin, lambda-cyhalothrin (*off-label*)
	Mealybugs	physical pest control
	Millipedes	tefluthrin (*seed treatment*)
	Pests, miscellaneous	lambda-cyhalothrin (*off-label*), zeta-cypermethrin (*off-label*)
	Scale insects	physical pest control
	Slugs/snails	ferric phosphate
	Spider mites	Beauveria bassiana (*off-label*), physical pest control, pyrethrins
	Springtails	tefluthrin (*seed treatment*)
	Suckers	physical pest control
	Symphylids	tefluthrin (*seed treatment*)
	Thrips	Beauveria bassiana (*off-label*), deltamethrin (*off-label*), pyrethrins
	Weevils	lambda-cyhalothrin
	Whiteflies	Beauveria bassiana (*off-label*), physical pest control, spirotetramat (*off-label*)
Plant growth regulation	Quality/yield control	sulphur
Weeds	Broad-leaved weeds	chlorpropham (*off-label*), clomazone (*off-label*), clopyralid, clopyralid (*off-label*), desmedipham + ethofumesate + lenacil + phenmedipham,

desmedipham + ethofumesate + lenacil + phenmedipham (*off-label*), desmedipham + ethofumesate + phenmedipham, desmedipham + phenmedipham, dimethenamid-p + quinmerac, ethofumesate, ethofumesate + metamitron, ethofumesate + phenmedipham, foramsulfuron + thiencarbazone-methyl, glyphosate, lenacil, lenacil (*off-label*), lenacil + triflusulfuron-methyl, lenacil + triflusulfuron-methyl (*off-label*), metamitron, metamitron + quinmerac, pelargonic acid (*off-label*), phenmedipham, phenmedipham (*off-label*), propyzamide, S-metolachlor (*off-label*), triflusulfuron-methyl, triflusulfuron-methyl (*off-label*)

Crops as weeds

clethodim, cycloxydim, cycloxydim (*off-label*), fluazifop-P-butyl, foramsulfuron + thiencarbazone-methyl, glyphosate, glyphosate (*wiper application*), lenacil + triflusulfuron-methyl, lenacil + triflusulfuron-methyl (*off-label*), propaquizafop (*off-label*), propyzamide, quizalofop-P-ethyl, quizalofop-P-tefuryl, triflusulfuron-methyl (*off-label*)

Grass weeds

chlorpropham (*off-label*), clethodim, clethodim (*off-label*), cycloxydim, cycloxydim (*off-label*), desmedipham + ethofumesate + lenacil + phenmedipham, desmedipham + ethofumesate + lenacil + phenmedipham (*off-label*), desmedipham + ethofumesate + phenmedipham, desmedipham + ethofumesate + phenmedipham (*moderately susceptible*), ethofumesate, ethofumesate + metamitron, ethofumesate + phenmedipham, fluazifop-P-butyl, glyphosate, lenacil, metamitron, metamitron + quinmerac, pelargonic acid (*off-label*), propaquizafop, propaquizafop (*off-label*), propyzamide, quizalofop-P-ethyl, quizalofop-P-tefuryl, S-metolachlor (*off-label*)

Mosses

pelargonic acid (*off-label*)

Weeds, miscellaneous

desmedipham + ethofumesate + lenacil + phenmedipham (*off-label*), desmedipham + ethofumesate + phenmedipham, glyphosate, glyphosate (*off-label*), lenacil (*off-label*), pelargonic acid (*off-label*)

Root and tuber crops - Carrots/parsnips

Diseases | Alternaria

azoxystrobin, azoxystrobin (*off-label*), azoxystrobin + difenoconazole, boscalid + pyraclostrobin (*moderate control*), cyprodinil + fludioxonil (*moderate control*), difenoconazole + fluxapyroxad (*off-label*), isopyrazam, isopyrazam (*off-label*), mancozeb, prothioconazole, tebuconazole, tebuconazole + trifloxystrobin

Botrytis

Bacillus subtilis (*off-label*), cyprodinil + fludioxonil (*moderate control*), Gliocladium catenulatum (*off-label*)

Cavity spot

Bacillus subtilis (*off-label*), metalaxyl-M, metalaxyl-M (*off-label*), metalaxyl-M (*off-label - reduction*), metalaxyl-M (*reduction only*)

Cercospora leaf spot

isopyrazam (*off-label*)

Damping off

Bacillus subtilis (*off-label*), cymoxanil + fludioxonil + metalaxyl-M (*off-label*)

Fungus diseases

azoxystrobin + difenoconazole (*off-label*)

	Fusarium	Gliocladium catenulatum (*off-label*)
	Leaf spot	tebuconazole + trifloxystrobin (*off-label*)
	Phytophthora	Gliocladium catenulatum (*off-label*)
	Powdery mildew	azoxystrobin, azoxystrobin (*off-label*), azoxystrobin + difenoconazole, boscalid + pyraclostrobin, difenoconazole + fluxapyroxad (*off-label*), isopyrazam, isopyrazam (*off-label*), prothioconazole, sulphur (*off-label*), tebuconazole, tebuconazole + trifloxystrobin, tebuconazole + trifloxystrobin (*off-label*)
	Pythium	Bacillus subtilis (*off-label*), cymoxanil + fludioxonil + metalaxyl-M (*off-label*)
	Ramularia leaf spots	isopyrazam (*off-label*)
	Rhizoctonia	difenoconazole + fluxapyroxad (*off-label*), Gliocladium catenulatum (*off-label*)
	Sclerotinia	boscalid + pyraclostrobin (*moderate control*), cyprodinil + fludioxonil (*moderate control*), cyprodinil + fludioxonil (*off-label*), difenoconazole + fluxapyroxad (*off-label*), isopyrazam (*off-label*), prothioconazole, tebuconazole, tebuconazole + trifloxystrobin, tebuconazole + trifloxystrobin (*off-label*)
	Stem canker	azoxystrobin + difenoconazole (*off-label*), isopyrazam (*off-label*)
Pests	Aphids	cypermethrin, pyrethrins, spirotetramat, thiacloprid
	Birds/mammals	aluminium ammonium sulphate
	Caterpillars	pyrethrins
	Cutworms	Bacillus thuringiensis (*off-label*), deltamethrin (*off-label*), lambda-cyhalothrin
	Flies	chlorantraniliprole (*off-label*), cypermethrin, deltamethrin (*off-label*), lambda-cyhalothrin (*off-label*)
	Free-living nematodes	garlic extract
	Pests, miscellaneous	cypermethrin, deltamethrin (*off-label*)
	Slugs/snails	ferric phosphate
	Spider mites	pyrethrins
	Stem nematodes	oxamyl
	Thrips	pyrethrins
Weeds	Broad-leaved weeds	aclonifen (*off-label*), clomazone, clomazone + pendimethalin, diflufenican (*off-label*), flumioxazin (*off-label*), isoxaben (*off-label*), metamitron (*off-label*), metribuzin (*off-label*), pelargonic acid (*off-label*), pendimethalin, pendimethalin (*off-label*), prosulfocarb (*off-label*)
	Crops as weeds	aclonifen (*off-label*), clethodim (*off-label*), cycloxydim, diflufenican (*off-label*), fluazifop-P-butyl, flumioxazin (*off-label*), metribuzin (*off-label*), pendimethalin
	Grass weeds	clethodim (*off-label*), cycloxydim, fluazifop-P-butyl, flumioxazin (*off-label*), metamitron (*off-label*), metribuzin (*off-label*), pelargonic acid (*off-label*), pendimethalin, propaquizafop, prosulfocarb (*off-label*)
	Mosses	pelargonic acid (*off-label*)
	Weeds, miscellaneous	glyphosate (*off-label*), pelargonic acid (*off-label*)

Root and tuber crops - Miscellaneous root crops

Diseases	Alternaria	difenoconazole + fluxapyroxad (*off-label*), isopyrazam (*off-label*), tebuconazole + trifloxystrobin (*off-label*)
	Bacterial canker	Bacillus amyloliquefaciens D747 (*off-label*)
	Botrytis	azoxystrobin + difenoconazole (*off-label*), Bacillus amyloliquefaciens D747 (*off-label*), Bacillus subtilis (*off-label*), cyprodinil + fludioxonil (*moderate control*), Gliocladium catenulatum (*off-label*)
	Cercospora leaf spot	isopyrazam (*off-label*)
	Damping off	Bacillus subtilis (*off-label*)
	Disease control	tebuconazole (*off-label*)
	Fireblight	Bacillus amyloliquefaciens D747 (*off-label*)
	Fungus diseases	azoxystrobin + difenoconazole (*off-label*), Bacillus amyloliquefaciens D747 (*off-label*)
	Fusarium	Gliocladium catenulatum (*off-label*)
	Leaf spot	boscalid + pyraclostrobin (*off-label*), tebuconazole + trifloxystrobin (*off-label*)
	Mycosphaerella	tebuconazole + trifloxystrobin (*off-label*)
	Phytophthora	difenoconazole (*off-label*), Gliocladium catenulatum (*off-label*)
	Powdery mildew	Bacillus amyloliquefaciens D747 (*off-label*), boscalid + pyraclostrobin (*off-label*), difenoconazole + fluxapyroxad (*off-label*), isopyrazam (*off-label*), physical pest control, tebuconazole + trifloxystrobin (*off-label*)
	Ramularia leaf spots	isopyrazam (*off-label*)
	Rhizoctonia	difenoconazole + fluxapyroxad (*off-label*), Gliocladium catenulatum (*off-label*)
	Rust	boscalid + pyraclostrobin (*off-label*), tebuconazole (*off-label*)
	Sclerotinia	azoxystrobin (*off-label*), azoxystrobin + difenoconazole (*off-label*), Bacillus amyloliquefaciens D747 (*off-label*), boscalid + pyraclostrobin (*off-label*), cyprodinil + fludioxonil (*off-label*), difenoconazole + fluxapyroxad (*off-label*), isopyrazam (*off-label*), tebuconazole + trifloxystrobin (*off-label*)
	Septoria	tebuconazole + trifloxystrobin (*off-label*)
	Stem canker	azoxystrobin + difenoconazole (*off-label*), isopyrazam (*off-label*)
Pests	Aphids	cypermethrin, deltamethrin (*off-label*), physical pest control, pirimicarb (*off-label*), pyrethrins, thiacloprid (*off-label*)
	Beetles	deltamethrin (*off-label*)
	Caterpillars	deltamethrin (*off-label*), pyrethrins, spinosad (*off-label*)
	Cutworms	Bacillus thuringiensis (*off-label*), lambda-cyhalothrin (*off-label*)
	Flies	chlorantraniliprole (*off-label*), cypermethrin, lambda-cyhalothrin (*off-label*)
	Leaf miners	spinosad (*off-label*)
	Mealybugs	physical pest control
	Pests, miscellaneous	cypermethrin, lambda-cyhalothrin (*off-label*)
	Scale insects	physical pest control

	Slugs/snails	ferric phosphate
	Spider mites	physical pest control, pyrethrins
	Suckers	physical pest control
	Thrips	deltamethrin (*off-label*), pyrethrins, spinosad (*off-label*)
	Weevils	deltamethrin (*off-label*), lambda-cyhalothrin (*off-label*)
	Whiteflies	physical pest control
Weeds	Broad-leaved weeds	aclonifen (*off-label*), clomazone (*off-label*), metribuzin (*off-label*), napropamide (*off-label*), pendimethalin (*off-label*), propyzamide (*off-label*), prosulfocarb (*off-label*), triflusulfuron-methyl (*off-label*)
	Crops as weeds	aclonifen (*off-label*), fluazifop-P-butyl, metribuzin (*off-label*)
	Grass weeds	clethodim (*off-label*), cycloxydim (*off-label*), fluazifop-P-butyl, metribuzin (*off-label*), napropamide (*off-label*), propyzamide (*off-label*), prosulfocarb (*off-label*)
	Weeds, miscellaneous	glyphosate (*off-label*), metribuzin (*off-label*), prosulfocarb (*off-label*)

Root and tuber crops - Potatoes

Crop control	Desiccation	carfentrazone-ethyl, pyraflufen-ethyl
Diseases	Alternaria	difenoconazole + fluxapyroxad (*moderate control*), difenoconazole + mandipropamid, mancozeb
	Bacterial blight	Bacillus subtilis (*off-label*)
	Black dot	azoxystrobin, azoxystrobin (*reduction*), fludioxonil (*some reduction*)
	Black scurf and stem canker	azoxystrobin, azoxystrobin (*reduction*), fludioxonil, flutolanil (*tuber treatment*), pencycuron (*tuber treatment*), penflufen, penflufen (*reduction*)
	Blight	ametoctradin + dimethomorph (*reduction*), amisulbrom, mandipropamid (*protection*), oxathiapiprolin
	Disease control	azoxystrobin + fluazinam, difenoconazole, fluxapyroxad
	Dry rot	imazalil, thiabendazole (*tuber treatment - post-harvest*)
	Gangrene	imazalil, thiabendazole (*tuber treatment - post-harvest*)
	Helminthosporium seedling rot	Bacillus subtilis (*off-label*)
	Phytophthora	ametoctradin, ametoctradin + dimethomorph, amisulbrom, azoxystrobin + fluazinam, benthiavalicarb-isopropyl + mancozeb, boscalid + pyraclostrobin (*off-label*), cyazofamid, cymoxanil, cymoxanil + fluazinam, cymoxanil + mancozeb, cymoxanil + mandipropamid, cymoxanil + propamocarb, cymoxanil + zoxamide, difenoconazole + mandipropamid, dimethomorph, dimethomorph + fluazinam, dimethomorph + mancozeb, dimethomorph + zoxamide, fluazinam, fluopicolide + propamocarb hydrochloride, mancozeb, mancozeb + metalaxyl-M, mancozeb + metalaxyl-M (*reduction*), mancozeb + zoxamide, mandipropamid, oxathiapiprolin
	Powdery mildew	physical pest control

	Powdery scab	fluazinam (*off-label*)
	Rhizoctonia	azoxystrobin, Bacillus subtilis (*off-label*), Pseudomonas SP (DSMZ 13134) (*reduction*)
	Scab	fludioxonil (*off-label*)
	Silver scurf	fludioxonil (*reduction*), imazalil, thiabendazole (*tuber treatment - post-harvest*)
	Skin spot	imazalil, thiabendazole (*tuber treatment - post-harvest*)
Pests	Aphids	acetamiprid, esfenvalerate, flonicamid, lambda-cyhalothrin, physical pest control, spirotetramat, thiacloprid
	Beetles	lambda-cyhalothrin
	Caterpillars	lambda-cyhalothrin
	Cutworms	cypermethrin, zeta-cypermethrin
	Cyst nematodes	fluopyram, fosthiazate, oxamyl
	Free-living nematodes	fosthiazate (*reduction*), oxamyl
	Mealybugs	physical pest control
	Scale insects	physical pest control
	Slugs/snails	ferric phosphate, metaldehyde
	Spider mites	physical pest control
	Suckers	physical pest control
	Weevils	lambda-cyhalothrin
	Whiteflies	physical pest control
	Wireworms	fosthiazate (*reduction*)
Plant growth regulation	Growth control	chlorpropham, chlorpropham (*thermal fog*), maleic hydrazide, spearmint oil
Weeds	Broad-leaved weeds	aclonifen, bentazone, carfentrazone-ethyl, clomazone, clomazone + metribuzin, clomazone + pendimethalin, flufenacet + metribuzin, glyphosate, metobromuron, metribuzin, pendimethalin, prosulfocarb, pyraflufen-ethyl, rimsulfuron
	Crops as weeds	carfentrazone-ethyl, cycloxydim, fluazifop-P-butyl, glyphosate, metribuzin, pendimethalin, quizalofop-P-ethyl, quizalofop-P-tefuryl, rimsulfuron
	Grass weeds	clomazone + metribuzin, cycloxydim, fluazifop-P-butyl, flufenacet + metribuzin, glyphosate, metobromuron, metribuzin, pendimethalin, propaquizafop, prosulfocarb, quizalofop-P-ethyl, quizalofop-P-tefuryl
	Weeds, miscellaneous	carfentrazone-ethyl, glyphosate

Stem and bulb vegetables - Asparagus

Diseases	Ascochyta	azoxystrobin + difenoconazole (*off-label*)
	Botrytis	Bacillus subtilis (*off-label*), boscalid + pyraclostrobin (*off-label*), cyprodinil + fludioxonil (*off-label*), Gliocladium catenulatum (*off-label*)
	Downy mildew	metalaxyl-M (*off-label*)
	Fusarium	Gliocladium catenulatum (*off-label*)
	Phytophthora	Gliocladium catenulatum (*off-label*)
	Powdery mildew	physical pest control
	Rhizoctonia	Gliocladium catenulatum (*off-label*)

	Rust	azoxystrobin, azoxystrobin (*moderate control*), azoxystrobin + difenoconazole (*off-label*), boscalid + pyraclostrobin (*off-label*), difenoconazole (*off-label*)
	Stemphylium	azoxystrobin
Pests	Aphids	cypermethrin, lambda-cyhalothrin (*off-label*), physical pest control, pyrethrins
	Beetles	cypermethrin (*off-label*), spinosad (*off-label*), thiacloprid (*off-label*)
	Caterpillars	pyrethrins
	Mealybugs	physical pest control
	Pests, miscellaneous	cypermethrin
	Scale insects	physical pest control
	Slugs/snails	ferric phosphate
	Spider mites	physical pest control, pyrethrins
	Suckers	physical pest control
	Thrips	pyrethrins, spinosad (*off-label*)
	Whiteflies	physical pest control
Weeds	Broad-leaved weeds	carfentrazone-ethyl (*off-label*), clomazone (*off-label*), clopyralid (*off-label*), isoxaben, mesotrione (*off-label*), metribuzin (*off-label*), metribuzin (*off-label - from seed*), pelargonic acid (*off-label*), pendimethalin (*off-label*), pyridate (*off-label*)
	Crops as weeds	carfentrazone-ethyl (*off-label*), fluazifop-P-butyl, metribuzin (*off-label*)
	Grass weeds	fluazifop-P-butyl, metribuzin (*off-label*), pelargonic acid (*off-label*)
	Weeds, miscellaneous	carfentrazone-ethyl (*off-label*), clomazone (*off-label*), glyphosate, glyphosate (*off-label*), metribuzin (*off-label*), pelargonic acid (*off-label*)

Stem and bulb vegetables - Celery/chicory

Diseases	Alternaria	Bacillus amyloliquefaciens D747, cyprodinil + fludioxonil (*off-label*)
	Bacterial canker	Bacillus amyloliquefaciens D747 (*off-label*)
	Botrytis	azoxystrobin (*off-label*), azoxystrobin + difenoconazole (*off-label*), Bacillus amyloliquefaciens D747 (*off-label*), Bacillus subtilis (*off-label*), cyprodinil + fludioxonil (*off-label*), Gliocladium catenulatum (*off-label*)
	Damping off	Bacillus subtilis (*off-label*)
	Disease control	difenoconazole (*off-label*), dimethomorph + mancozeb (*off-label*), mandipropamid
	Downy mildew	dimethomorph + mancozeb (*off-label*), mandipropamid
	Fireblight	Bacillus amyloliquefaciens D747 (*off-label*)
	Fungus diseases	Bacillus amyloliquefaciens D747 (*off-label*)
	Fusarium	Bacillus amyloliquefaciens D747, Gliocladium catenulatum (*off-label*)
	Leaf spot	azoxystrobin (*off-label*), cyprodinil + fludioxonil (*off-label*), difenoconazole (*off-label*)
	Phytophthora	difenoconazole (*off-label*), Gliocladium catenulatum (*off-label*)

	Powdery mildew	Bacillus amyloliquefaciens D747, Bacillus amyloliquefaciens D747 (*off-label*)
	Pythium	Bacillus amyloliquefaciens D747
	Rhizoctonia	azoxystrobin (*off-label*), Bacillus amyloliquefaciens D747, Gliocladium catenulatum (*off-label*)
	Sclerotinia	azoxystrobin (*off-label*), azoxystrobin + difenoconazole (*off-label*), Bacillus amyloliquefaciens D747 (*off-label*), cyprodinil + fludioxonil (*off-label*)
	Septoria	tebuconazole + trifloxystrobin (*off-label*)
Pests	Aphids	Beauveria bassiana (*off-label*), pirimicarb (*off-label*)
	Caterpillars	Bacillus thuringiensis, Bacillus thuringiensis (*off-label*), chlorantraniliprole (*off-label*), lambda-cyhalothrin (*off-label*), spinosad (*off-label*)
	Cutworms	lambda-cyhalothrin (*off-label*)
	Flies	lambda-cyhalothrin (*off-label*), spinosad (*off-label*)
	Leaf miners	spinosad (*off-label*)
	Pests, miscellaneous	lambda-cyhalothrin (*off-label*)
	Slugs/snails	ferric phosphate
	Spider mites	Beauveria bassiana (*off-label*)
	Thrips	Beauveria bassiana (*off-label*), spinosad (*off-label*)
	Whiteflies	Beauveria bassiana (*off-label*)
Weeds	Broad-leaved weeds	clomazone (*off-label*), isoxaben (*off-label*), pelargonic acid (*off-label*), pendimethalin (*off-label*), propyzamide (*off-label*), prosulfocarb (*off-label*), S-metolachlor (*off-label*)
	Crops as weeds	cycloxydim (*off-label*), fluazifop-P-butyl
	Grass weeds	clethodim (*off-label*), cycloxydim (*off-label*), fluazifop-P-butyl, pelargonic acid (*off-label*), pendimethalin (*off-label*), propyzamide (*off-label*), prosulfocarb (*off-label*), S-metolachlor (*off-label*)
	Mosses	pelargonic acid (*off-label*)

Stem and bulb vegetables - Globe artichokes/cardoons

Diseases	Botrytis	Bacillus subtilis (*off-label*), Gliocladium catenulatum (*off-label*)
	Disease control	difenoconazole (*off-label*)
	Fusarium	Gliocladium catenulatum (*off-label*)
	Leaf spot	difenoconazole (*off-label*)
	Phytophthora	difenoconazole (*off-label*), Gliocladium catenulatum (*off-label*)
	Rhizoctonia	Gliocladium catenulatum (*off-label*)
Pests	Slugs/snails	ferric phosphate
Weeds	Crops as weeds	fluazifop-P-butyl
	Grass weeds	fluazifop-P-butyl

Stem and bulb vegetables - Onions/leeks/garlic

Diseases	Blight	fluoxastrobin + prothioconazole (*useful reduction*)
	Botrytis	Bacillus amyloliquefaciens D747 (*off-label*), Bacillus subtilis (*off-label*), cyprodinil + fludioxonil (*off-label*), fluoxastrobin + prothioconazole (*useful reduction*), Gliocladium catenulatum (*off-label*)
	Damping off	Bacillus subtilis (*off-label*)

	Disease control	dimethomorph + mancozeb, dimethomorph + mancozeb (*off-label*)
	Downy mildew	azoxystrobin, azoxystrobin (*moderate control*), azoxystrobin (*off-label*), azoxystrobin (*reduction*), benthiavalicarb-isopropyl + mancozeb (*off-label*), dimethomorph + mancozeb, dimethomorph + mancozeb (*off-label*), fluopicolide + propamocarb hydrochloride (*off-label*), fluoxastrobin + prothioconazole (*control*), mancozeb, mancozeb + metalaxyl-M (*off-label*), mancozeb + metalaxyl-M (*useful control*), metalaxyl-M (*off-label*)
	Fusarium	boscalid + pyraclostrobin (*off-label*), Gliocladium catenulatum (*off-label*)
	Phytophthora	Gliocladium catenulatum (*off-label*)
	Powdery mildew	physical pest control
	Purple blotch	azoxystrobin, azoxystrobin (*moderate control*), azoxystrobin + difenoconazole (*moderate control*), prothioconazole
	Pythium	metalaxyl-M, metalaxyl-M (*off-label*)
	Rhizoctonia	Gliocladium catenulatum (*off-label*)
	Rhynchosporium	prothioconazole (*useful reduction*)
	Rust	azoxystrobin, azoxystrobin + difenoconazole, mancozeb, prothioconazole, tebuconazole, tebuconazole + trifloxystrobin (*off-label*)
	Stemphylium	prothioconazole (*useful reduction*)
	White rot	Bacillus subtilis (*off-label*), boscalid + pyraclostrobin (*off-label*), tebuconazole (*off-label*)
	White tip	ametoctradin, azoxystrobin, azoxystrobin + difenoconazole (*qualified minor use*), boscalid + pyraclostrobin (*off-label*), dimethomorph + mancozeb (*reduction*)
Pests	Aphids	lambda-cyhalothrin (*off-label*), physical pest control, pyrethrins, spirotetramat
	Caterpillars	Bacillus thuringiensis, Bacillus thuringiensis (*off-label*), pyrethrins
	Cutworms	Bacillus thuringiensis (*off-label*), deltamethrin (*off-label*)
	Mealybugs	physical pest control
	Pests, miscellaneous	deltamethrin (*off-label*), lambda-cyhalothrin (*off-label*)
	Root-knot nematodes	garlic extract (*off-label*)
	Scale insects	physical pest control
	Slugs/snails	ferric phosphate
	Spider mites	physical pest control, pyrethrins
	Stem nematodes	garlic extract (*off-label*)
	Suckers	physical pest control
	Thrips	lambda-cyhalothrin (*off-label*), pyrethrins, spinosad, thiacloprid (*off-label*)
	Whiteflies	physical pest control, spirotetramat
Plant growth regulation	Growth control	maleic hydrazide
Weeds	Broad-leaved weeds	aclonifen (*off-label*), bentazone (*off-label*), bromoxynil, bromoxynil (*off-label*), chlorpropham, chlorpropham (*off-label*), clopyralid (*off-label*), dimethenamid-p +

metazachlor (*off-label*), dimethenamid-p + pendimethalin (*off-label*), flumioxazin (*off-label*), fluroxypyr (*off-label*), glyphosate, isoxaben (*off-label*), pelargonic acid (*off-label*), pendimethalin, pendimethalin (*off-label*), pendimethalin (*pre + post emergence treatment*), prosulfocarb (*off-label*), pyridate, pyridate (*off-label*), S-metolachlor (*off-label*)

Crops as weeds	aclonifen (*off-label*), bromoxynil, bromoxynil (*off-label*), clethodim (*off-label*), clopyralid (*off-label*), cycloxydim, fluazifop-P-butyl, flumioxazin (*off-label*), fluroxypyr (*off-label*), glyphosate, pendimethalin
Grass weeds	chlorpropham, clethodim (*off-label*), cycloxydim, dimethenamid-p + metazachlor (*off-label*), dimethenamid-p + pendimethalin (*off-label*), fluazifop-P-butyl, flumioxazin (*off-label*), glyphosate, pelargonic acid (*off-label*), pendimethalin, pendimethalin (*off-label*), propaquizafop, propaquizafop (*off-label*), prosulfocarb (*off-label*), S-metolachlor (*off-label*)
Mosses	pelargonic acid (*off-label*)
Weeds, miscellaneous	bromoxynil (*off-label*), fluroxypyr (*off-label*), glyphosate, glyphosate (*off-label*), pelargonic acid (*off-label*)

Stem and bulb vegetables - Vegetables

Diseases	Alternaria	difenoconazole + fluxapyroxad (*off-label*)
	Botrytis	Bacillus subtilis (*off-label*), Gliocladium catenulatum (*off-label*)
	Fusarium	Gliocladium catenulatum (*off-label*)
	Phytophthora	Gliocladium catenulatum (*off-label*)
	Powdery mildew	difenoconazole + fluxapyroxad (*off-label*), penconazole (*off-label*)
	Rhizoctonia	difenoconazole + fluxapyroxad (*off-label*), Gliocladium catenulatum (*off-label*)
	Sclerotinia	difenoconazole + fluxapyroxad (*off-label*)
Pests	Aphids	deltamethrin (*off-label*), Metarhizium anisopliae (*off-label*), pyrethrins
	Beetles	deltamethrin (*off-label*)
	Caterpillars	Bacillus thuringiensis, deltamethrin (*off-label*), pyrethrins
	Flies	Metarhizium anisopliae (*off-label*)
	Midges	Metarhizium anisopliae (*off-label*)
	Pests, miscellaneous	deltamethrin (*off-label*)
	Slugs/snails	ferric phosphate
	Spider mites	pyrethrins
	Thrips	deltamethrin (*off-label*), Metarhizium anisopliae (*off-label*), pyrethrins
	Weevils	deltamethrin (*off-label*)
Weeds	Crops as weeds	fluazifop-P-butyl
	Grass weeds	cycloxydim (*off-label*), fluazifop-P-butyl

Flowers and ornamentals

Flowers - Bedding plants, general

Diseases	Powdery mildew	bupirimate, physical pest control
Pests	Aphids	physical pest control, thiacloprid
	Beetles	thiacloprid
	Mealybugs	physical pest control
	Scale insects	physical pest control
	Spider mites	physical pest control
	Suckers	physical pest control
	Weevils	thiacloprid
	Whiteflies	physical pest control, thiacloprid
Plant growth regulation	Flowering control	paclobutrazol
	Growth control	paclobutrazol

Flowers - Bulbs/corms

Crop control	Desiccation	carfentrazone-ethyl (*off-label*)
Diseases	Basal stem rot	cyprodinil + fludioxonil (*off-label*)
	Botrytis	boscalid + epoxiconazole (*off-label*), Gliocladium catenulatum (*off-label*), mancozeb
	Crown rot	metalaxyl-M (*off-label*)
	Downy mildew	metalaxyl-M (*off-label*)
	Fusarium	cyprodinil + fludioxonil (*off-label*), Gliocladium catenulatum (*off-label*)
	Phytophthora	Gliocladium catenulatum (*off-label*)
	Powdery mildew	bupirimate, physical pest control
	Rhizoctonia	Gliocladium catenulatum (*off-label*)
	White mould	boscalid + epoxiconazole (*off-label*)
Pests	Aphids	physical pest control
	Mealybugs	physical pest control
	Scale insects	physical pest control
	Spider mites	physical pest control
	Suckers	physical pest control
	Whiteflies	physical pest control
Plant growth regulation	Growth control	ethephon
Weeds	Broad-leaved weeds	bentazone, bromoxynil (*off-label*), metobromuron (*off-label*)
	Grass weeds	metobromuron (*off-label*), S-metolachlor (*off-label*)

Flowers - Miscellaneous flowers

Diseases	Anthracnose	Bacillus amyloliquefaciens D747 (*off-label*)
	Bacterial blight	Bacillus amyloliquefaciens D747 (*off-label*)
	Botrytis	Bacillus amyloliquefaciens D747 (*off-label*), Gliocladium catenulatum (*off-label*)
	Cercospora leaf spot	Bacillus amyloliquefaciens D747 (*off-label*)
	Damping off	Bacillus amyloliquefaciens D747 (*off-label*)

	Disease control	tebuconazole (*off-label*)
	Downy mildew	Bacillus amyloliquefaciens D747 (*off-label*)
	Fusarium	Bacillus amyloliquefaciens D747 (*off-label*), Gliocladium catenulatum (*off-label*)
	Phytophthora	Gliocladium catenulatum (*off-label*)
	Powdery mildew	Bacillus amyloliquefaciens D747 (*off-label*), bupirimate, physical pest control
	Pythium	Bacillus amyloliquefaciens D747 (*off-label*)
	Rhizoctonia	Bacillus amyloliquefaciens D747 (*off-label*), Gliocladium catenulatum (*off-label*)
	Rust	tebuconazole (*off-label*)
	Sclerotinia	Bacillus amyloliquefaciens D747 (*off-label*)
	Stemphylium	Bacillus amyloliquefaciens D747 (*off-label*)
Pests	Aphids	Beauveria bassiana (*off-label*), physical pest control
	Cutworms	deltamethrin (*off-label*)
	Flies	deltamethrin (*off-label*), lambda-cyhalothrin (*off-label*)
	Mealybugs	physical pest control
	Pests, miscellaneous	deltamethrin (*off-label*), lambda-cyhalothrin (*off-label*)
	Scale insects	physical pest control
	Slugs/snails	ferric phosphate
	Spider mites	Beauveria bassiana (*off-label*), physical pest control
	Suckers	physical pest control
	Thrips	Beauveria bassiana (*off-label*)
	Whiteflies	Beauveria bassiana (*off-label*), physical pest control
Plant growth regulation	Flowering control	paclobutrazol
	Growth control	paclobutrazol
Weeds	Broad-leaved weeds	lenacil (*off-label*), metribuzin (*off-label*), pendimethalin (*off-label*), prosulfocarb (*off-label*)
	Crops as weeds	fluazifop-P-butyl, metribuzin (*off-label*)
	Grass weeds	fluazifop-P-butyl, metribuzin (*off-label*)
	Weeds, miscellaneous	lenacil (*off-label*)

Flowers - Pot plants

Diseases	Powdery mildew	physical pest control
Pests	Aphids	physical pest control, thiacloprid
	Beetles	thiacloprid
	Mealybugs	physical pest control
	Scale insects	physical pest control
	Spider mites	physical pest control
	Suckers	physical pest control
	Weevils	thiacloprid
	Whiteflies	physical pest control, thiacloprid
Plant growth regulation	Flowering control	paclobutrazol
	Growth control	paclobutrazol

Flowers - Protected bulbs/corms

Diseases	Downy mildew	propamocarb hydrochloride (*off-label*)
	Fungus diseases	propamocarb hydrochloride (*off-label*)
	Phytophthora	propamocarb hydrochloride (*off-label*)
	Pythium	propamocarb hydrochloride (*off-label*)

Flowers - Protected flowers

Diseases	Crown rot	metalaxyl-M (*off-label*)
	Downy mildew	metalaxyl-M (*off-label*)
	Powdery mildew	bupirimate
Pests	Aphids	pirimicarb (*off-label*)
	Spider mites	tebufenpyrad
Plant growth regulation	Flowering control	benzyladenine

Miscellaneous flowers and ornamentals - Miscellaneous uses

Diseases	Botrytis	boscalid + pyraclostrobin (*off-label*), isopyrazam (*off-label*)
	Disease control	boscalid + pyraclostrobin (*off-label*), potassium bicarbonate (commodity substance)
	Leaf spot	boscalid + pyraclostrobin (*off-label*)
	Powdery mildew	boscalid + pyraclostrobin (*off-label*), isopyrazam (*off-label*), physical pest control
	Rust	boscalid + pyraclostrobin (*off-label*)
	Sclerotinia	Coniothyrium minitans, isopyrazam (*off-label*)
Pests	Aphids	maltodextrin, physical pest control
	Caterpillars	Bacillus thuringiensis (*off-label*)
	Mealybugs	physical pest control
	Pests, miscellaneous	diflubenzuron
	Scale insects	physical pest control
	Slugs/snails	ferric phosphate, metaldehyde
	Spider mites	maltodextrin, physical pest control
	Suckers	physical pest control
	Whiteflies	maltodextrin, physical pest control
Weeds	Broad-leaved weeds	carfentrazone-ethyl (*before planting*), propyzamide
	Crops as weeds	carfentrazone-ethyl (*before planting*)
	Grass weeds	propyzamide
	Weeds, miscellaneous	carfentrazone-ethyl, glyphosate, glyphosate (*before planting*), glyphosate (*pre-sowing/planting*)

Ornamentals - Nursery stock

Diseases	Anthracnose	Bacillus amyloliquefaciens D747 (*off-label*)
	Bacterial canker	Bacillus amyloliquefaciens D747 (*off-label*)
	Black root rot	azoxystrobin (*off-label*), tebuconazole + trifloxystrobin (*off-label*)
	Botrytis	azoxystrobin (*off-label*), Bacillus amyloliquefaciens D747 (*off-label*), Bacillus subtilis (*off-label*), boscalid + pyraclostrobin (*off-label*), captan (*off-label*), cyprodinil + fludioxonil, Gliocladium catenulatum (*off-label*),

		mepanipyrim (*off-label*), prohexadione-calcium (*off-label*), pyrimethanil (*off-label*)
	Cercospora leaf spot	Bacillus amyloliquefaciens D747 (*off-label*)
	Damping off	Bacillus amyloliquefaciens D747 (*off-label*)
	Disease control	dimethomorph + mancozeb (*off-label*), mancozeb, mancozeb (*off-label*), mancozeb + metalaxyl-M (*off-label*), mancozeb + zoxamide (*off-label*), prohexadione-calcium (*off-label*)
	Downy mildew	ametoctradin + dimethomorph (*off-label*), azoxystrobin (*off-label*), Bacillus amyloliquefaciens D747 (*off-label*), benthiavalicarb-isopropyl + mancozeb (*off-label*), captan (*off-label*), cos-oga (*off-label*), dimethomorph (*off-label*), dimethomorph + mancozeb (*off-label*), fluopicolide + propamocarb hydrochloride (*off-label*), fosetyl-aluminium + propamocarb hydrochloride (*off-label*), mancozeb, mandipropamid (*off-label*), propamocarb hydrochloride (*off-label*)
	Fireblight	Bacillus amyloliquefaciens D747 (*off-label*)
	Foliar diseases	pyraclostrobin (*off-label*), trifloxystrobin (*off-label*)
	Foot rot	Trichoderma asperellum (Strain T34)
	Fungus diseases	Bacillus amyloliquefaciens D747 (*off-label*), captan (*off-label*), propamocarb hydrochloride (*off-label*)
	Fusarium	azoxystrobin (*off-label*), Bacillus amyloliquefaciens D747 (*off-label*), Gliocladium catenulatum (*off-label*), thiophanate-methyl (*off-label*), Trichoderma asperellum (Strain T34) (*off-label*)
	Leaf spot	azoxystrobin (*off-label*)
	Phytophthora	Gliocladium catenulatum (*off-label*), metalaxyl-M, propamocarb hydrochloride (*off-label*)
	Powdery mildew	Ampelomyces quisqualis (Strain AQ10) (*off-label*), azoxystrobin (*off-label*), Bacillus amyloliquefaciens D747 (*off-label*), bupirimate, cos-oga (*off-label*), cyflufenamid (*off-label*), isopyrazam (*off-label*), kresoxim-methyl, mepanipyrim (*off-label*), penconazole (*off-label*), physical pest control, potassium bicarbonate (commodity substance) (*off-label*)
	Pythium	Bacillus amyloliquefaciens D747 (*off-label*), fosetyl-aluminium + propamocarb hydrochloride (*off-label*), metalaxyl-M, metalaxyl-M (*off-label*), propamocarb hydrochloride (*off-label*), Trichoderma asperellum (Strain T34) (*off-label*)
	Rhizoctonia	Bacillus amyloliquefaciens D747 (*off-label*), Gliocladium catenulatum (*off-label*)
	Root diseases	thiophanate-methyl (*off-label*)
	Rust	azoxystrobin (*off-label*), tebuconazole + trifloxystrobin (*off-label*)
	Scab	azoxystrobin (*off-label*), captan (*off-label*)
	Sclerotinia	Bacillus amyloliquefaciens D747 (*off-label*)
	Stemphylium	Bacillus amyloliquefaciens D747 (*off-label*)
	White blister	azoxystrobin (*off-label*)
Pests	Aphids	acetamiprid, Beauveria bassiana, cypermethrin, deltamethrin, dimethoate, esfenvalerate, lambda-cyhalothrin (*off-label*), physical pest control,

		pyrethrins, spirotetramat (*off-label*), sulfoxaflor, thiacloprid, thiacloprid (*off-label*)
	Beetles	Beauveria bassiana, thiacloprid, thiacloprid (*off-label*)
	Capsid bugs	deltamethrin
	Caterpillars	Bacillus thuringiensis, Bacillus thuringiensis aizawai GC-91, deltamethrin, esfenvalerate, indoxacarb, indoxacarb (*off-label*), lambda-cyhalothrin (*off-label*), pyrethrins, spinosad (*off-label*)
	Flies	Metarhizium anisopliae (*off-label*), thiacloprid (*off-label*)
	Free-living nematodes	fosthiazate (*off-label*)
	Gall, rust and leaf & bud mites	clofentezine (*off-label*)
	Leaf miners	deltamethrin (*off-label*), dimethoate, esfenvalerate, thiacloprid (*off-label*)
	Mealybugs	deltamethrin, physical pest control
	Midges	Metarhizium anisopliae (*off-label*)
	Mites	clofentezine (*off-label*)
	Pests, miscellaneous	chlorantraniliprole (*off-label*), cypermethrin, deltamethrin (*off-label*), diflubenzuron, esfenvalerate, indoxacarb (*off-label*), lambda-cyhalothrin (*off-label*), methoxyfenozide (*off-label*), spinosad (*off-label*), thiacloprid (*off-label*)
	Scale insects	deltamethrin, physical pest control
	Slugs/snails	ferric phosphate
	Spider mites	abamectin, clofentezine (*off-label*), dimethoate, etoxazole (*off-label*), fenazaquin, physical pest control, pyrethrins, spirodiclofen (*off-label*)
	Suckers	physical pest control
	Thrips	abamectin, azadirachtin, deltamethrin, deltamethrin (*off-label*), esfenvalerate, Metarhizium anisopliae (*off-label*), pyrethrins, spinosad, spirotetramat (*off-label*), thiacloprid (*off-label*)
	Weevils	chlorantraniliprole (*off-label*), Metarhizium anisopliae, thiacloprid, thiacloprid (*off-label*)
	Whiteflies	acetamiprid, Beauveria bassiana, buprofezin, deltamethrin, esfenvalerate, flonicamid (*off-label*), Lecanicillium muscarium, physical pest control, spirotetramat (*off-label*), sulfoxaflor, thiacloprid, thiacloprid (*off-label*)
Plant growth regulation	Flowering control	paclobutrazol, paclobutrazol (*off-label*)
	Growth control	4-indol-3-ylbutyric acid, benzyladenine, benzyladenine + gibberellin, chlormequat (*off-label*), daminozide, ethephon, ethephon (*off-label*), ethephon + mepiquat chloride (*off-label*), gibberellins, paclobutrazol, paclobutrazol (*off-label*), prohexadione-calcium (*off-label*), trinexapac-ethyl (*off-label*)
Weeds	Broad-leaved weeds	2,4-D (*off-label*), amidosulfuron (*off-label*), bentazone, bentazone (*off-label*), bromoxynil (*off-label*), chlorpropham, clomazone (*off-label*), clopyralid, clopyralid + picloram (*off-label*), diflufenican (*off-label*), dimethenamid-p + metazachlor (*off-label*), dimethenamid-p + pendimethalin (*off-label*), florasulam (*off-label*), florasulam + fluroxypyr (*off-*

label), flufenacet (*off-label*), flumioxazin (*off-label*),
fluroxypyr (*off-label*), imazamox + pendimethalin (*off-label*), isoxaben, metamitron (*off-label*), metazachlor,
metribuzin (*off-label*), nicosulfuron (*off-label*),
pelargonic acid, pendimethalin (*off-label*),
phenmedipham (*off-label*), propyzamide,
propyzamide (*off-label*), prosulfocarb (*off-label*),
rimsulfuron (*off-label*), S-metolachlor (*off-label*)

Crops as weeds	diflufenican (*off-label*), flumioxazin (*off-label*), fluroxypyr (*off-label*), metribuzin (*off-label*), nicosulfuron (*off-label*)
Grass weeds	chlorpropham, cycloxydim, cycloxydim (*off-label*), dimethenamid-p + metazachlor (*off-label*), dimethenamid-p + pendimethalin (*off-label*), flufenacet (*off-label*), flumioxazin (*off-label*), metamitron (*off-label*), metazachlor, metribuzin (*off-label*), nicosulfuron (*off-label*), pendimethalin (*off-label*), propyzamide, propyzamide (*off-label*), prosulfocarb (*off-label*), S-metolachlor (*off-label*)
Mosses	pelargonic acid
Weeds, miscellaneous	bromoxynil (*off-label*), carfentrazone-ethyl (*off-label*), diflufenican (*off-label*), glyphosate, glyphosate (*off-label - as a directed spray*), glyphosate + pyraflufen-ethyl, metribuzin (*off-label*), pelargonic acid

Ornamentals - Trees and shrubs

Crop control	Desiccation	glyphosate
Diseases	Botrytis	boscalid + pyraclostrobin (*off-label*)
	Downy mildew	benthiavalicarb-isopropyl + mancozeb (*off-label*)
	Leaf spot	boscalid + pyraclostrobin (*off-label*)
	Phytophthora	Bacillus subtilis (*off-label*), metalaxyl-M (*off-label*)
	Powdery mildew	boscalid + pyraclostrobin (*off-label*)
	Rust	boscalid + pyraclostrobin (*off-label*)
Pests	Aphids	deltamethrin, lambda-cyhalothrin (*off-label*)
	Birds/mammals	warfarin (*nut crops*)
	Capsid bugs	deltamethrin
	Caterpillars	Bacillus thuringiensis, Bacillus thuringiensis aizawai GC-91, deltamethrin
	Flies	Metarhizium anisopliae (*off-label*)
	Mealybugs	deltamethrin
	Midges	Metarhizium anisopliae (*off-label*)
	Pests, miscellaneous	diflubenzuron, lambda-cyhalothrin (*off-label*)
	Scale insects	deltamethrin
	Slugs/snails	ferric phosphate
	Thrips	deltamethrin, Metarhizium anisopliae (*off-label*)
	Weevils	Metarhizium anisopliae (*off-label*)
	Whiteflies	deltamethrin
Plant growth regulation	Growth control	glyphosate
	Plant growth regulation, miscellaneous	glyphosate (*off-label*)

Weeds	Broad-leaved weeds	clopyralid (*off-label*), isoxaben, pelargonic acid, propyzamide
	Grass weeds	propyzamide
	Mosses	pelargonic acid
	Weeds, miscellaneous	2,4-D + glyphosate, flazasulfuron, glyphosate, glyphosate + pyraflufen-ethyl, glyphosate + sulfosulfuron, pelargonic acid, propyzamide

Ornamentals - Woody ornamentals

| Weeds | Broad-leaved weeds | propyzamide |
| | Grass weeds | propyzamide |

Forestry

Forest nurseries, general - Forest nurseries

Crop control	Chemical stripping/ thinning	glyphosate
Diseases	Blight	copper oxychloride (*off-label*)
	Botrytis	boscalid + pyraclostrobin (*off-label*), cyprodinil + fludioxonil, fenhexamid (*off-label*), mepanipyrim (*off-label*), pyrimethanil (*off-label*)
	Damping off	fosetyl-aluminium + propamocarb hydrochloride (*off-label*)
	Disease control	cyprodinil (*off-label*), dimethomorph + mancozeb (*off-label*), epoxiconazole (*off-label*), mancozeb (*off-label*), mancozeb + metalaxyl-M (*off-label*), pyrimethanil (*off-label*)
	Downy mildew	benthiavalicarb-isopropyl + mancozeb (*off-label*), dimethomorph + mancozeb (*off-label*), fosetyl-aluminium + propamocarb hydrochloride (*off-label*), mancozeb (*off-label*), propamocarb hydrochloride (*off-label*)
	Fungus diseases	propamocarb hydrochloride (*off-label*)
	Fusarium	Trichoderma asperellum (Strain T34) (*off-label*)
	Leaf spot	difenoconazole (*off-label*)
	Phytophthora	Bacillus subtilis (*off-label*), difenoconazole (*off-label*), metalaxyl-M (*off-label*), propamocarb hydrochloride (*off-label*)
	Powdery mildew	Ampelomyces quisqualis (Strain AQ10) (*off-label*), proquinazid (*off-label*)
	Pythium	propamocarb hydrochloride (*off-label*), Trichoderma asperellum (Strain T34) (*off-label*)
	Root rot	Phlebiopsis gigantea
	Rust	difenoconazole (*off-label*)
Pests	Aphids	acetamiprid (*off-label*), Beauveria bassiana (*off-label*), deltamethrin (*off-label*), lambda-cyhalothrin (*off-label*), thiacloprid (*off-label*)
	Beetles	acetamiprid (*off-label*)
	Caterpillars	Bacillus thuringiensis (*off-label*), lambda-cyhalothrin (*off-label*)
	Cutworms	Bacillus thuringiensis (*off-label*)

	Flies	acetamiprid (*off-label*)
	Gall, rust and leaf & bud mites	clofentezine (*off-label*)
	Mites	bifenazate (*off-label*), clofentezine (*off-label*)
	Pests, miscellaneous	deltamethrin (*off-label*), lambda-cyhalothrin (*off-label*), thiacloprid (*off-label*)
	Spider mites	clofentezine (*off-label*)
	Wasps	acetamiprid (*off-label*)
	Weevils	acetamiprid (*off-label*)
	Whiteflies	Beauveria bassiana (*off-label*)
Plant growth regulation	Growth control	ethephon + mepiquat chloride (*off-label*), trinexapac-ethyl (*off-label*)
Weeds	Broad-leaved weeds	2,4-D (*off-label*), amidosulfuron (*off-label*), amidosulfuron + iodosulfuron-methyl-sodium (*off-label*), chlorpropham (*off-label*), clomazone (*off-label*), clopyralid (*off-label*), clopyralid + picloram (*off-label*), desmedipham + ethofumesate + lenacil + phenmedipham (*off-label*), diflufenican (*off-label*), dimethenamid-p + metazachlor (*off-label*), florasulam (*off-label*), flufenacet (*off-label*), fluroxypyr (*off-label*), isoxaben, nicosulfuron (*off-label*), pelargonic acid, pendimethalin (*off-label*), pendimethalin + picolinafen (*off-label*), propyzamide, propyzamide (*off-label*), prosulfocarb (*off-label*), rimsulfuron (*off-label*), S-metolachlor (*off-label*)
	Crops as weeds	amidosulfuron + iodosulfuron-methyl-sodium (*off-label*), clopyralid (*off-label*), diflufenican (*off-label*), fluroxypyr (*off-label*), nicosulfuron (*off-label*)
	Grass weeds	chlorpropham (*off-label*), cycloxydim, desmedipham + ethofumesate + lenacil + phenmedipham (*off-label*), dimethenamid-p + metazachlor (*off-label*), flufenacet (*off-label*), nicosulfuron (*off-label*), pendimethalin (*off-label*), pendimethalin + picolinafen (*off-label*), propaquizafop (*off-label*), propyzamide, propyzamide (*off-label*), prosulfocarb (*off-label*), quizalofop-P-ethyl (*off-label*), S-metolachlor (*off-label*)
	Weeds, miscellaneous	diflufenican (*off-label*), glyphosate
	Woody weeds/scrub	glyphosate

Forestry plantations, general - Forestry plantations

Crop control	Chemical stripping/ thinning	glyphosate
Diseases	Root rot	Phlebiopsis gigantea
Pests	Birds/mammals	warfarin
	Caterpillars	Bacillus thuringiensis (*off-label*)
	Pests, miscellaneous	diflubenzuron
	Weevils	acetamiprid (*off-label*), cypermethrin
Plant growth regulation	Growth control	1-naphthylacetic acid (*off-label*), glyphosate
	Plant growth regulation, miscellaneous	glyphosate (*off-label*)

Weeds	Broad-leaved weeds	clopyralid (*off-label*), isoxaben, pendimethalin (*off-label*), propyzamide
	Grass weeds	cycloxydim, glyphosate, propaquizafop (*off-label*), propyzamide
	Weeds, miscellaneous	carfentrazone-ethyl (*off-label*), glyphosate, glyphosate (*off-label*), propyzamide
	Woody weeds/scrub	glyphosate

Miscellaneous forestry situations - Cut logs/timber

| Pests | Beetles | cypermethrin |

Woodland on farms - Woodland

Crop control	Chemical stripping/ thinning	glyphosate
Diseases	Root rot	Phlebiopsis gigantea
Pests	Aphids	lambda-cyhalothrin (*off-label*)
	Beetles	lambda-cyhalothrin (*off-label*)
	Caterpillars	lambda-cyhalothrin (*off-label*)
	Pests, miscellaneous	lambda-cyhalothrin (*off-label*)
	Sawflies	lambda-cyhalothrin (*off-label*)
Plant growth regulation	Growth control	gibberellins
Weeds	Broad-leaved weeds	2,4-D (*off-label*), amidosulfuron (*off-label*), clopyralid (*off-label*), florasulam (*off-label*), fluroxypyr (*off-label*), lenacil (*off-label*), MCPA (*off-label*), pendimethalin (*off-label*), propyzamide
	Crops as weeds	fluazifop-P-butyl, fluroxypyr (*off-label*)
	Grass weeds	fluazifop-P-butyl, propyzamide
	Weeds, miscellaneous	glyphosate, glyphosate (*off-label*), lenacil (*off-label*)

Fruit and hops

All bush fruit - All currants

Diseases	Alternaria	Bacillus amyloliquefaciens D747
	Botrytis	Bacillus amyloliquefaciens D747 (*off-label*), Bacillus subtilis (*off-label*), boscalid + pyraclostrobin, boscalid + pyraclostrobin (*off-label*), cyprodinil + fludioxonil (*qualified minor use recommendation*), fenhexamid, Gliocladium catenulatum (*off-label*), pyrimethanil (*off-label*)
	Cane blight	Bacillus amyloliquefaciens D747 (*off-label*)
	Cladosporium	Bacillus amyloliquefaciens D747 (*off-label*)
	Disease control	mancozeb + metalaxyl-M (*off-label*), sulphur (*off-label*)
	Fusarium	Bacillus amyloliquefaciens D747, Gliocladium catenulatum (*off-label*)
	Leaf spot	boscalid + pyraclostrobin (*moderate control only*), boscalid + pyraclostrobin (*off-label*), dodine
	Phytophthora	Bacillus subtilis (*off-label*), Gliocladium catenulatum (*off-label*)

	Powdery mildew	Bacillus amyloliquefaciens D747, Bacillus amyloliquefaciens D747 (*off-label*), boscalid + pyraclostrobin (*moderate control only*), boscalid + pyraclostrobin (*off-label*), bupirimate, kresoxim-methyl, kresoxim-methyl (*off-label*), penconazole, sulphur
	Pythium	Bacillus amyloliquefaciens D747
	Rhizoctonia	Bacillus amyloliquefaciens D747, Gliocladium catenulatum (*off-label*)
Pests	Aphids	spirotetramat (*off-label*)
	Birds/mammals	aluminium ammonium sulphate
	Capsid bugs	deltamethrin (*off-label*)
	Caterpillars	Bacillus thuringiensis (*off-label*), chlorantraniliprole (*off-label*), deltamethrin (*off-label*), indoxacarb (*off-label*), lambda-cyhalothrin (*off-label*), spinosad (*off-label*)
	Cutworms	Bacillus thuringiensis (*off-label*)
	Flies	lambda-cyhalothrin (*off-label*), Metarhizium anisopliae (*off-label*)
	Gall, rust and leaf & bud mites	spirotetramat (*off-label*), sulphur
	Leatherjackets	Metarhizium anisopliae (*off-label*)
	Midges	lambda-cyhalothrin (*off-label*), Metarhizium anisopliae (*off-label*), spirotetramat (*off-label*)
	Pests, miscellaneous	Bacillus thuringiensis (*off-label*), indoxacarb (*off-label*), lambda-cyhalothrin (*off-label*), spinosad (*off-label*), thiacloprid (*off-label*)
	Sawflies	deltamethrin (*off-label*), lambda-cyhalothrin (*off-label*), spinosad (*off-label*)
	Scale insects	thiacloprid (*off-label*)
	Spider mites	Beauveria bassiana (*off-label*), spirodiclofen (*off-label*)
	Thrips	deltamethrin (*off-label*), lambda-cyhalothrin (*off-label*), Metarhizium anisopliae (*off-label*)
	Weevils	Metarhizium anisopliae
	Whiteflies	Beauveria bassiana (*off-label*)
Weeds	Broad-leaved weeds	carfentrazone-ethyl (*off-label*), clopyralid (*off-label*), flufenacet + metribuzin (*off-label*), isoxaben, pendimethalin, pendimethalin (*off-label*), propyzamide
	Crops as weeds	fluazifop-P-butyl, pendimethalin
	Grass weeds	clethodim, clethodim (*off-label*), fluazifop-P-butyl, flufenacet + metribuzin (*off-label*), pendimethalin, propyzamide
	Weeds, miscellaneous	carfentrazone-ethyl (*off-label*), glyphosate (*off-label*)

All bush fruit - All protected bush fruit

Diseases	Anthracnose	boscalid + pyraclostrobin (*off-label*), cyprodinil + fludioxonil (*off-label*)
	Botrytis	Bacillus amyloliquefaciens D747 (*off-label*), boscalid + pyraclostrobin (*off-label*), cyprodinil + fludioxonil (*off-label*), fenhexamid (*off-label*)
	Cane blight	Bacillus amyloliquefaciens D747 (*off-label*)
	Cladosporium	Bacillus amyloliquefaciens D747 (*off-label*)

	Disease control	mancozeb + metalaxyl-M (*off-label*)
	Fusarium	Trichoderma asperellum (Strain T34) (*off-label*)
	Leaf spot	boscalid + pyraclostrobin (*off-label*)
	Phytophthora	fluazinam (*off-label*)
	Powdery mildew	Ampelomyces quisqualis (Strain AQ10) (*off-label*), Bacillus amyloliquefaciens D747 (*off-label*)
	Pythium	Trichoderma asperellum (Strain T34) (*off-label*)
Pests	Caterpillars	Bacillus thuringiensis (*off-label*), indoxacarb (*off-label*), spinosad (*off-label*), thiacloprid (*off-label*)
	Flies	spinosad (*off-label*)
	Midges	thiacloprid (*off-label*)
	Pests, miscellaneous	indoxacarb (*off-label*)
	Spider mites	Beauveria bassiana (*off-label*), Lecanicillium muscarium (*off-label*)
	Thrips	Lecanicillium muscarium (*off-label*)
	Whiteflies	Beauveria bassiana (*off-label*), Lecanicillium muscarium (*off-label*)
Weeds	Weeds, miscellaneous	carfentrazone-ethyl (*off-label*)

All bush fruit - All vines

Diseases	Bacterial canker	Bacillus amyloliquefaciens D747 (*off-label*)
	Black rot	kresoxim-methyl (*off-label*)
	Botrytis	Bacillus amyloliquefaciens D747 (*off-label*), Bacillus subtilis (*off-label*), fenhexamid, fenpyrazamine, Gliocladium catenulatum (*off-label*), potassium bicarbonate (commodity substance) (*off-label*), prohexadione-calcium (*off-label*)
	Disease control	mancozeb, mancozeb + zoxamide, potassium phosphonates
	Downy mildew	ametoctradin + dimethomorph (*off-label*), amisulbrom (*off-label*), benthiavalicarb-isopropyl + mancozeb (*off-label*), copper oxychloride, cos-oga (*off-label*), cymoxanil (*off-label*), potassium bicarbonate (commodity substance) (*off-label*)
	Fireblight	Bacillus amyloliquefaciens D747 (*off-label*)
	Fungus diseases	Bacillus amyloliquefaciens D747 (*off-label*)
	Fusarium	Gliocladium catenulatum (*off-label*)
	Phytophthora	Gliocladium catenulatum (*off-label*), potassium phosphonates
	Powdery mildew	Ampelomyces quisqualis (Strain AQ10) (*off-label*), Bacillus amyloliquefaciens D747 (*off-label*), cos-oga (*off-label*), cyflufenamid (*off-label*), fluxapyroxad (*off-label*), kresoxim-methyl (*off-label*), metrafenone (*off-label*), penconazole, potassium bicarbonate (commodity substance) (*off-label*), sulphur, tebuconazole + trifloxystrobin (*off-label*)
	Pythium	potassium phosphonates
	Rhizoctonia	Gliocladium catenulatum (*off-label*)
	Sclerotinia	Bacillus amyloliquefaciens D747 (*off-label*)
Pests	Aphids	spirotetramat (*off-label*)

	Caterpillars	Bacillus thuringiensis, Bacillus thuringiensis (*off-label*), indoxacarb (*off-label*), methoxyfenozide (*off-label*)
	Flies	Metarhizium anisopliae (*off-label*)
	Leatherjackets	Metarhizium anisopliae (*off-label*)
	Midges	Metarhizium anisopliae (*off-label*)
	Pests, miscellaneous	indoxacarb (*off-label*)
	Scale insects	spirotetramat (*off-label*)
	Spider mites	Lecanicillium muscarium (*off-label*)
	Thrips	Lecanicillium muscarium (*off-label*), Metarhizium anisopliae (*off-label*)
	Whiteflies	Lecanicillium muscarium (*off-label*)
Plant growth regulation	Growth control	gibberellins
Weeds	Broad-leaved weeds	propyzamide (*off-label*)
	Crops as weeds	fluazifop-P-butyl
	Grass weeds	fluazifop-P-butyl, propyzamide (*off-label*)
	Weeds, miscellaneous	carfentrazone-ethyl (*off-label*), glyphosate

All bush fruit - Bilberries/blueberries/cranberries

Diseases	Alternaria	Bacillus amyloliquefaciens D747
	Anthracnose	boscalid + pyraclostrobin (*off-label*)
	Botrytis	Bacillus amyloliquefaciens D747 (*off-label*), Bacillus subtilis (*off-label*), boscalid + pyraclostrobin (*off-label*), cyprodinil + fludioxonil (*qualified minor use recommendation*), fenhexamid (*off-label*), Gliocladium catenulatum (*off-label*), pyrimethanil (*off-label*)
	Cane blight	Bacillus amyloliquefaciens D747 (*off-label*)
	Cladosporium	Bacillus amyloliquefaciens D747 (*off-label*)
	Disease control	mancozeb + metalaxyl-M (*off-label*), sulphur (*off-label*)
	Fusarium	Bacillus amyloliquefaciens D747, Gliocladium catenulatum (*off-label*)
	Leaf spot	boscalid + pyraclostrobin (*off-label*)
	Phytophthora	Bacillus subtilis (*off-label*), Gliocladium catenulatum (*off-label*)
	Powdery mildew	Bacillus amyloliquefaciens D747, Bacillus amyloliquefaciens D747 (*off-label*), kresoxim-methyl (*off-label*)
	Pythium	Bacillus amyloliquefaciens D747
	Rhizoctonia	Bacillus amyloliquefaciens D747, Gliocladium catenulatum (*off-label*)
Pests	Aphids	spirotetramat (*off-label*)
	Birds/mammals	aluminium ammonium sulphate
	Caterpillars	Bacillus thuringiensis (*off-label*), chlorantraniliprole (*off-label*), indoxacarb (*off-label*), lambda-cyhalothrin (*off-label*), spinosad (*off-label*)
	Cutworms	Bacillus thuringiensis (*off-label*)
	Flies	lambda-cyhalothrin (*off-label*), Metarhizium anisopliae (*off-label*), spinosad (*off-label*)

	Gall, rust and leaf & bud mites	spirotetramat (*off-label*)
	Leatherjackets	Metarhizium anisopliae (*off-label*)
	Midges	Metarhizium anisopliae (*off-label*), spirotetramat (*off-label*)
	Pests, miscellaneous	Bacillus thuringiensis (*off-label*), indoxacarb (*off-label*), lambda-cyhalothrin (*off-label*), thiacloprid (*off-label*)
	Scale insects	thiacloprid (*off-label*)
	Spider mites	Beauveria bassiana (*off-label*)
	Thrips	lambda-cyhalothrin (*off-label*), Metarhizium anisopliae (*off-label*)
	Weevils	Metarhizium anisopliae
	Whiteflies	Beauveria bassiana (*off-label*)
Weeds	Broad-leaved weeds	carfentrazone-ethyl (*off-label*), clopyralid (*off-label*), flufenacet + metribuzin (*off-label*), pendimethalin (*off-label*), propyzamide (*off-label*)
	Crops as weeds	fluazifop-P-butyl
	Grass weeds	clethodim, clethodim (*off-label*), fluazifop-P-butyl, flufenacet + metribuzin (*off-label*), propyzamide (*off-label*)
	Weeds, miscellaneous	carfentrazone-ethyl (*off-label*), glyphosate (*off-label*)

All bush fruit - Bush fruit, general

Diseases	Botrytis	Bacillus amyloliquefaciens D747 (*off-label*), Bacillus subtilis (*off-label*)
	Cane blight	Bacillus amyloliquefaciens D747 (*off-label*)
	Cladosporium	Bacillus amyloliquefaciens D747 (*off-label*)
	Powdery mildew	Bacillus amyloliquefaciens D747 (*off-label*)
Pests	Caterpillars	indoxacarb (*off-label*), spinosad (*off-label*)
	Pests, miscellaneous	indoxacarb (*off-label*), spinosad (*off-label*)
	Spider mites	Beauveria bassiana (*off-label*)
	Whiteflies	Beauveria bassiana (*off-label*)
Weeds	Broad-leaved weeds	clopyralid (*off-label*)
	Crops as weeds	fluazifop-P-butyl
	Grass weeds	fluazifop-P-butyl

All bush fruit - Gooseberries

Diseases	Alternaria	Bacillus amyloliquefaciens D747
	Anthracnose	boscalid + pyraclostrobin (*off-label*)
	Botrytis	Bacillus amyloliquefaciens D747 (*off-label*), Bacillus subtilis (*off-label*), boscalid + pyraclostrobin (*off-label*), cyprodinil + fludioxonil (*qualified minor use recommendation*), fenhexamid, Gliocladium catenulatum (*off-label*), pyrimethanil (*off-label*)
	Cane blight	Bacillus amyloliquefaciens D747 (*off-label*)
	Cladosporium	Bacillus amyloliquefaciens D747 (*off-label*)
	Disease control	mancozeb + metalaxyl-M (*off-label*)
	Fusarium	Bacillus amyloliquefaciens D747, Gliocladium catenulatum (*off-label*)
	Leaf spot	boscalid + pyraclostrobin (*off-label*)

	Phytophthora	Bacillus subtilis (*off-label*), Gliocladium catenulatum (*off-label*)
	Powdery mildew	Bacillus amyloliquefaciens D747, Bacillus amyloliquefaciens D747 (*off-label*), bupirimate, kresoxim-methyl (*off-label*), sulphur
	Pythium	Bacillus amyloliquefaciens D747
	Rhizoctonia	Bacillus amyloliquefaciens D747, Gliocladium catenulatum (*off-label*)
Pests	Aphids	spirotetramat (*off-label*)
	Birds/mammals	aluminium ammonium sulphate
	Capsid bugs	deltamethrin (*off-label*)
	Caterpillars	Bacillus thuringiensis (*off-label*), chlorantraniliprole (*off-label*), deltamethrin (*off-label*), indoxacarb (*off-label*), lambda-cyhalothrin (*off-label*), spinosad (*off-label*)
	Cutworms	Bacillus thuringiensis (*off-label*)
	Flies	spinosad (*off-label*)
	Gall, rust and leaf & bud mites	spirotetramat (*off-label*)
	Midges	lambda-cyhalothrin (*off-label*), spirotetramat (*off-label*)
	Pests, miscellaneous	Bacillus thuringiensis (*off-label*), indoxacarb (*off-label*), lambda-cyhalothrin (*off-label*), thiacloprid (*off-label*)
	Sawflies	deltamethrin (*off-label*), lambda-cyhalothrin (*off-label*)
	Scale insects	thiacloprid (*off-label*)
	Spider mites	Beauveria bassiana (*off-label*), spirodiclofen (*off-label*)
	Thrips	deltamethrin (*off-label*)
	Weevils	Metarhizium anisopliae
	Whiteflies	Beauveria bassiana (*off-label*)
Weeds	Broad-leaved weeds	carfentrazone-ethyl (*off-label*), clopyralid (*off-label*), flufenacet + metribuzin (*off-label*), isoxaben, pendimethalin, propyzamide
	Crops as weeds	fluazifop-P-butyl, pendimethalin
	Grass weeds	clethodim, clethodim (*off-label*), fluazifop-P-butyl, flufenacet + metribuzin (*off-label*), pendimethalin, propyzamide
	Weeds, miscellaneous	carfentrazone-ethyl (*off-label*), glyphosate (*off-label*)

All bush fruit - Miscellaneous bush fruit

Diseases	Botrytis	cyprodinil + fludioxonil (*off-label*), pyrimethanil (*off-label*)
	Downy mildew	metalaxyl-M (*off-label*)
	Powdery mildew	boscalid (*off-label*), physical pest control, proquinazid (*off-label*), sulphur
Pests	Aphids	physical pest control
	Capsid bugs	lambda-cyhalothrin (*off-label*)
	Caterpillars	indoxacarb (*off-label*), lambda-cyhalothrin (*off-label*)
	Mealybugs	physical pest control
	Pests, miscellaneous	spinosad (*off-label*)
	Scale insects	physical pest control
	Spider mites	physical pest control

	Suckers	physical pest control
	Wasps	lambda-cyhalothrin (*off-label*)
	Whiteflies	physical pest control
Weeds	Weeds, miscellaneous	glyphosate (*off-label*)

Cane fruit - All outdoor cane fruit

Crop control	Desiccation	carfentrazone-ethyl (*off-label*)
Diseases	Alternaria	Bacillus amyloliquefaciens D747
	Botrytis	Bacillus amyloliquefaciens D747 (*off-label*), Bacillus subtilis (*off-label*), fenhexamid, Gliocladium catenulatum (*off-label*), pyrimethanil, pyrimethanil (*off-label*)
	Botrytis fruit rot	cyprodinil + fludioxonil
	Cane blight	Bacillus amyloliquefaciens D747 (*off-label*), boscalid + pyraclostrobin (*off-label*)
	Cladosporium	Bacillus amyloliquefaciens D747 (*off-label*)
	Disease control	mancozeb + metalaxyl-M (*off-label*)
	Downy mildew	metalaxyl-M (*off-label*)
	Fusarium	Bacillus amyloliquefaciens D747, Gliocladium catenulatum (*off-label*)
	Phytophthora	Bacillus subtilis (*off-label*), fluazinam (*off-label*), Gliocladium catenulatum (*off-label*)
	Powdery mildew	azoxystrobin (*off-label*), Bacillus amyloliquefaciens D747, Bacillus amyloliquefaciens D747 (*off-label*), bupirimate, cos-oga (*off-label*)
	Purple blotch	boscalid + pyraclostrobin (*off-label*)
	Pythium	Bacillus amyloliquefaciens D747
	Rhizoctonia	Bacillus amyloliquefaciens D747, Gliocladium catenulatum (*off-label*)
	Root rot	dimethomorph, dimethomorph (*moderate control*)
Pests	Beetles	deltamethrin, deltamethrin (*off-label*)
	Birds/mammals	aluminium ammonium sulphate
	Capsid bugs	lambda-cyhalothrin (*off-label*), thiacloprid (*off-label*)
	Caterpillars	Bacillus thuringiensis, Bacillus thuringiensis (*off-label*), deltamethrin (*off-label*), indoxacarb (*off-label*), lambda-cyhalothrin (*off-label*), spinosad (*off-label*)
	Cutworms	Bacillus thuringiensis (*off-label*)
	Flies	Metarhizium anisopliae (*off-label*), spinosad (*off-label*)
	Leatherjackets	Metarhizium anisopliae (*off-label*)
	Midges	Metarhizium anisopliae (*off-label*)
	Pests, miscellaneous	Bacillus thuringiensis (*off-label*), deltamethrin (*off-label*), indoxacarb (*off-label*), lambda-cyhalothrin (*off-label*), spinosad (*off-label*), thiacloprid (*off-label*)
	Slugs/snails	metaldehyde
	Spider mites	Beauveria bassiana (*off-label*), clofentezine (*off-label*)
	Thrips	Metarhizium anisopliae (*off-label*), spinosad (*off-label*)
	Weevils	lambda-cyhalothrin (*off-label*), Metarhizium anisopliae
	Whiteflies	Beauveria bassiana (*off-label*)
Weeds	Broad-leaved weeds	isoxaben, pendimethalin, propyzamide, propyzamide (*England only*)

	Crops as weeds	fluazifop-P-butyl, pendimethalin
	Grass weeds	clethodim, clethodim (*off-label*), fluazifop-P-butyl, pendimethalin, propyzamide, propyzamide (*England only*)
	Weeds, miscellaneous	carfentrazone-ethyl (*off-label*), glyphosate (*off-label*)

Cane fruit - All protected cane fruit

Crop control	Desiccation	carfentrazone-ethyl (*off-label*)
Diseases	Botrytis	Bacillus amyloliquefaciens D747 (*off-label*), pyrimethanil (*off-label*)
	Cane blight	Bacillus amyloliquefaciens D747 (*off-label*), boscalid + pyraclostrobin (*off-label*)
	Cladosporium	Bacillus amyloliquefaciens D747 (*off-label*)
	Disease control	mancozeb + metalaxyl-M (*off-label*)
	Fusarium	Trichoderma asperellum (Strain T34) (*off-label*)
	Phytophthora	fluazinam (*off-label*)
	Powdery mildew	azoxystrobin (*off-label*), Bacillus amyloliquefaciens D747 (*off-label*), bupirimate, cos-oga (*off-label*)
	Purple blotch	boscalid + pyraclostrobin (*off-label*)
	Pythium	Trichoderma asperellum (Strain T34) (*off-label*)
	Root rot	dimethomorph (*moderate control*)
	Rust	boscalid + pyraclostrobin (*off-label*)
Pests	Capsid bugs	thiacloprid (*off-label*)
	Caterpillars	indoxacarb (*off-label*)
	Pests, miscellaneous	indoxacarb (*off-label*), thiacloprid (*off-label*)
	Spider mites	abamectin (*off-label*), Beauveria bassiana (*off-label*), Lecanicillium muscarium (*off-label*)
	Thrips	Lecanicillium muscarium (*off-label*), spinosad (*off-label*)
	Whiteflies	Beauveria bassiana (*off-label*), Lecanicillium muscarium (*off-label*)
Weeds	Weeds, miscellaneous	carfentrazone-ethyl (*off-label*)

Hops, general - Hops

Crop control	Desiccation	pyraflufen-ethyl (*off-label*)
Diseases	Botrytis	Bacillus subtilis (*off-label*), fenhexamid (*off-label*), Gliocladium catenulatum (*off-label*)
	Disease control	boscalid + pyraclostrobin (*off-label*), captan (*off-label*), cymoxanil + mancozeb (*off-label*), mancozeb + metalaxyl-M (*off-label*), thiophanate-methyl (*off-label*)
	Downy mildew	ametoctradin + dimethomorph (*off-label*), benthiavalicarb-isopropyl + mancozeb (*off-label*), cymoxanil (*off-label*), mandipropamid (*off-label*), metalaxyl-M (*off-label*)
	Fusarium	Gliocladium catenulatum (*off-label*)
	Phytophthora	Gliocladium catenulatum (*off-label*)
	Powdery mildew	cos-oga (*off-label*), fluopyram (*off-label*), metrafenone (*off-label*), penconazole (*off-label*), sulphur
	Rhizoctonia	Gliocladium catenulatum (*off-label*)

Pests	Aphids	acetamiprid (*off-label*), flonicamid (*off-label*), lambda-cyhalothrin (*off-label*), spirotetramat (*off-label*), tebufenpyrad, thiacloprid (*off-label*)
	Beetles	lambda-cyhalothrin (*off-label*)
	Capsid bugs	lambda-cyhalothrin (*off-label*)
	Caterpillars	Bacillus thuringiensis, chlorantraniliprole (*off-label*), indoxacarb (*off-label*), lambda-cyhalothrin (*off-label*), spinosad (*off-label*)
	Free-living nematodes	fosthiazate (*off-label*)
	Pests, miscellaneous	indoxacarb (*off-label*), lambda-cyhalothrin (*off-label*), methoxyfenozide (*off-label*), spinosad (*off-label*), thiacloprid (*off-label*)
	Spider mites	spirodiclofen (*off-label*), tebufenpyrad
Plant growth regulation	Growth control	prohexadione-calcium (*off-label*)
Weeds	Broad-leaved weeds	bentazone (*off-label*), clomazone (*off-label*), clopyralid (*off-label*), flumioxazin (*off-label*), imazamox + pendimethalin (*off-label*), isoxaben, pendimethalin (*off-label*), propyzamide (*off-label*), pyraflufen-ethyl (*off-label*)
	Crops as weeds	fluazifop-P-butyl, flumioxazin (*off-label*)
	Grass weeds	fluazifop-P-butyl, flumioxazin (*off-label*), propyzamide (*off-label*)
	Weeds, miscellaneous	glyphosate (*off-label*)

Miscellaneous fruit situations - Fruit crops, general

Diseases	Bacterial canker	Bacillus amyloliquefaciens D747 (*off-label*)
	Botrytis	Bacillus amyloliquefaciens D747 (*off-label*), Bacillus subtilis (*off-label*), fenhexamid (*off-label*), pyrimethanil (*off-label*)
	Damping off	fosetyl-aluminium + propamocarb hydrochloride (*off-label*)
	Disease control	boscalid + pyraclostrobin (*off-label*), pyrimethanil (*off-label*), thiophanate-methyl (*off-label*)
	Downy mildew	benthiavalicarb-isopropyl + mancozeb (*off-label*), fosetyl-aluminium + propamocarb hydrochloride (*off-label*), metalaxyl-M (*off-label*)
	Fireblight	Bacillus amyloliquefaciens D747 (*off-label*)
	Fungus diseases	Bacillus amyloliquefaciens D747 (*off-label*)
	Powdery mildew	Bacillus amyloliquefaciens D747 (*off-label*), kresoxim-methyl (*reduction*), physical pest control, proquinazid (*off-label*)
	Scab	kresoxim-methyl
	Sclerotinia	Bacillus amyloliquefaciens D747 (*off-label*)
Pests	Aphids	acetamiprid (*off-label*), lambda-cyhalothrin (*off-label*), physical pest control, thiacloprid (*off-label*)
	Caterpillars	Bacillus thuringiensis (*off-label*), indoxacarb (*off-label*)
	Flies	Metarhizium anisopliae (*off-label*)
	Free-living nematodes	fosthiazate (*off-label*)
	Mealybugs	physical pest control
	Midges	Metarhizium anisopliae (*off-label*)

	Pests, miscellaneous	aluminium phosphide, indoxacarb (*off-label*), lambda-cyhalothrin (*off-label*), methoxyfenozide (*off-label*), thiacloprid (*off-label*)
	Scale insects	physical pest control
	Spider mites	physical pest control
	Suckers	physical pest control
	Thrips	Metarhizium anisopliae (*off-label*)
	Weevils	Metarhizium anisopliae (*off-label*)
	Whiteflies	physical pest control
Weeds	Broad-leaved weeds	flumioxazin (*off-label*), imazamox + pendimethalin (*off-label*), nicosulfuron (*off-label*), pendimethalin (*off-label*), prosulfocarb (*off-label*)
	Crops as weeds	flumioxazin (*off-label*), nicosulfuron (*off-label*)
	Grass weeds	flumioxazin (*off-label*), nicosulfuron (*off-label*), pendimethalin (*off-label*), prosulfocarb (*off-label*)
	Weeds, miscellaneous	glyphosate (*off-label*)

Miscellaneous fruit situations - Fruit nursery stock

Diseases	Disease control	captan (*off-label*)
	Fungus diseases	potassium bicarbonate (commodity substance)
	Powdery mildew	penconazole (*off-label*)
	Sooty moulds	potassium bicarbonate (commodity substance)
Pests	Caterpillars	spinosad (*off-label*)
	Pests, miscellaneous	spinosad (*off-label*)
	Spider mites	spirodiclofen (*off-label*)
Plant growth regulation	Growth control	prohexadione-calcium (*off-label*)
Weeds	Broad-leaved weeds	clopyralid (*off-label*), metazachlor
	Grass weeds	metazachlor

Other fruit - All outdoor strawberries

Diseases	Alternaria	Bacillus amyloliquefaciens D747
	Anthracnose	azoxystrobin
	Bacterial canker	Bacillus amyloliquefaciens D747 (*off-label*)
	Black spot	boscalid + pyraclostrobin, cyprodinil + fludioxonil (*qualified minor use*)
	Botrytis	Bacillus amyloliquefaciens D747 (*off-label*), Bacillus subtilis, Bacillus subtilis (*off-label*), boscalid + pyraclostrobin, cyprodinil + fludioxonil, fenhexamid, fenpyrazamine, mepanipyrim, pyrimethanil
	Botrytis fruit rot	Bacillus subtilis
	Crown rot	dimethomorph (*moderate control*)
	Didymella stem rot	Gliocladium catenulatum (*moderate control*)
	Disease control	azoxystrobin + difenoconazole, difenoconazole + fluxapyroxad, mancozeb + metalaxyl-M (*off-label*)
	Fireblight	Bacillus amyloliquefaciens D747 (*off-label*)
	Fungus diseases	Bacillus amyloliquefaciens D747 (*off-label*)
	Fusarium	Bacillus amyloliquefaciens D747, Gliocladium catenulatum (*moderate control*)
	Phytophthora	Gliocladium catenulatum (*moderate control*)

	Powdery mildew	azoxystrobin, Bacillus amyloliquefaciens D747, Bacillus amyloliquefaciens D747 (*off-label*), boscalid + pyraclostrobin, bupirimate, cos-oga (*off-label*), cyflufenamid (*off-label*), kresoxim-methyl, penconazole, physical pest control, potassium bicarbonate (commodity substance) (*off-label*), sulphur
	Pythium	Bacillus amyloliquefaciens D747, Gliocladium catenulatum (*moderate control*)
	Rhizoctonia	Bacillus amyloliquefaciens D747, Gliocladium catenulatum (*moderate control*)
	Sclerotinia	Bacillus amyloliquefaciens D747 (*off-label*)
Pests	Aphids	physical pest control, spirotetramat
	Birds/mammals	aluminium ammonium sulphate
	Capsid bugs	indoxacarb (*off-label*), thiacloprid (*off-label*)
	Caterpillars	Bacillus thuringiensis, deltamethrin (*off-label*), indoxacarb (*off-label*), lambda-cyhalothrin (*off-label*), spinosad (*off-label*)
	Flies	spinosad (*off-label*)
	Mealybugs	physical pest control
	Pests, miscellaneous	indoxacarb (*off-label*), lambda-cyhalothrin (*off-label*), spinosad (*off-label*)
	Scale insects	physical pest control
	Slugs/snails	ferric phosphate
	Spider mites	clofentezine (*off-label*), physical pest control, spirodiclofen (*off-label*), tebufenpyrad
	Suckers	physical pest control
	Tarsonemid mites	spirotetramat
	Thrips	Beauveria bassiana (*off-label*), deltamethrin (*off-label*)
	Weevils	cyantraniliprole (*off-label*), Metarhizium anisopliae
	Whiteflies	Beauveria bassiana (*off-label*), physical pest control
Weeds	Broad-leaved weeds	carfentrazone-ethyl (*off-label*), clopyralid (*off-label*), dimethenamid-p + pendimethalin (*off-label*), isoxaben, metamitron (*off-label*), pendimethalin, propyzamide
	Crops as weeds	cycloxydim, fluazifop-P-butyl, pendimethalin
	Grass weeds	clethodim, clethodim (*off-label*), cycloxydim, dimethenamid-p + pendimethalin (*off-label*), fluazifop-P-butyl, metamitron (*off-label*), pendimethalin, propyzamide, S-metolachlor (*off-label*)
	Weeds, miscellaneous	carfentrazone-ethyl (*off-label*), glyphosate (*off-label*)

Other fruit - Protected miscellaneous fruit

Diseases	Anthracnose	azoxystrobin
	Black spot	boscalid + pyraclostrobin, cyprodinil + fludioxonil (*qualified minor use*)
	Botrytis	Bacillus subtilis (*reduction of damage to fruit*), boscalid + pyraclostrobin, cyprodinil + fludioxonil, fenpyrazamine, fluopyram + trifloxystrobin (*moderate control*), pyrimethanil, pyrimethanil (*moderate control*)

	Botrytis fruit rot	Bacillus subtilis (*reduction of damage to fruit*)
	Crown rot	dimethomorph (*moderate control*)
	Damping off	fosetyl-aluminium + propamocarb hydrochloride (*off-label*)
	Disease control	azoxystrobin + difenoconazole, boscalid + pyraclostrobin (*off-label*), difenoconazole + fluxapyroxad, mancozeb + metalaxyl-M (*off-label*)
	Downy mildew	cerevisane, fosetyl-aluminium + propamocarb hydrochloride (*off-label*)
	Fusarium	Trichoderma asperellum (Strain T34) (*off-label*)
	Powdery mildew	Ampelomyces quisqualis (Strain AQ10), azoxystrobin, boscalid + pyraclostrobin, bupirimate, cerevisane, cos-oga (*off-label*), fluopyram + trifloxystrobin, kresoxim-methyl, penconazole, penconazole (*off-label*), proquinazid (*off-label*)
	Pythium	Trichoderma asperellum (Strain T34) (*off-label*)
Pests	Aphids	acetamiprid (*off-label*), Beauveria bassiana, spirotetramat
	Beetles	Beauveria bassiana
	Capsid bugs	thiacloprid (*off-label*)
	Caterpillars	Bacillus thuringiensis (*off-label*), deltamethrin (*off-label*), indoxacarb (*off-label*), spinosad (*off-label*)
	Mites	bifenazate (*off-label*)
	Pests, miscellaneous	indoxacarb (*off-label*), lambda-cyhalothrin (*off-label*), spinosad, thiacloprid (*off-label*)
	Spider mites	abamectin, bifenazate, clofentezine (*off-label*), etoxazole (*off-label*), Lecanicillium muscarium (*off-label*), spirodiclofen (*off-label*)
	Tarsonemid mites	spirotetramat
	Thrips	abamectin, Beauveria bassiana (*off-label*), Lecanicillium muscarium (*off-label*), spinosad
	Weevils	cyantraniliprole (*off-label*)
	Whiteflies	Beauveria bassiana, Beauveria bassiana (*off-label*), Lecanicillium muscarium, Lecanicillium muscarium (*off-label*)
Plant growth regulation	Growth control	gibberellins (*off-label*)
Weeds	Weeds, miscellaneous	carfentrazone-ethyl (*off-label*)

Other fruit - Rhubarb

Diseases	Botrytis	Bacillus subtilis (*off-label*), Gliocladium catenulatum (*off-label*)
	Disease control	difenoconazole (*off-label*)
	Downy mildew	mancozeb + metalaxyl-M (*off-label*)
	Fusarium	Gliocladium catenulatum (*off-label*)
	Leaf spot	difenoconazole (*off-label*)
	Phytophthora	difenoconazole (*off-label*), Gliocladium catenulatum (*off-label*)
	Rhizoctonia	Gliocladium catenulatum (*off-label*)
Pests	Aphids	pirimicarb (*off-label*)
	Caterpillars	Bacillus thuringiensis (*off-label*)

	Cutworms	Bacillus thuringiensis (*off-label*)
	Slugs/snails	ferric phosphate
Plant growth regulation	Growth control	gibberellins
Weeds	Broad-leaved weeds	clomazone (*off-label*), clopyralid (*off-label*), isoxaben (*off-label*), mesotrione (*off-label*), pelargonic acid (*off-label*), pendimethalin (*off-label*), propyzamide, propyzamide (*outdoor*)
	Crops as weeds	fluazifop-P-butyl
	Grass weeds	fluazifop-P-butyl, pelargonic acid (*off-label*), pendimethalin (*off-label*), propyzamide, propyzamide (*outdoor*)
	Weeds, miscellaneous	glyphosate (*off-label*), mesotrione (*off-label*), metribuzin (*off-label*), pelargonic acid (*off-label*)

Tree fruit - All nuts

Diseases	Botrytis	Bacillus subtilis (*off-label*), Gliocladium catenulatum (*off-label*)
	Fusarium	Gliocladium catenulatum (*off-label*)
	Phytophthora	Gliocladium catenulatum (*off-label*)
	Rhizoctonia	Gliocladium catenulatum (*off-label*)
Pests	Aphids	lambda-cyhalothrin (*off-label*)
	Caterpillars	Bacillus thuringiensis (*off-label*), indoxacarb (*off-label*), lambda-cyhalothrin (*off-label*)
	Mites	thiacloprid (*off-label*)
	Pests, miscellaneous	aluminium phosphide, indoxacarb (*off-label*), lambda-cyhalothrin (*off-label*)
Weeds	Broad-leaved weeds	fluroxypyr (*off-label*), isoxaben (*off-label*), pendimethalin (*off-label*), propyzamide (*off-label*), prosulfocarb (*off-label*)
	Crops as weeds	fluazifop-P-butyl
	Grass weeds	fluazifop-P-butyl, propyzamide (*off-label*), prosulfocarb (*off-label*)
	Weeds, miscellaneous	glyphosate (*off-label*)

Tree fruit - All pome fruit

Crop control	Sucker/shoot control	glyphosate
Diseases	Alternaria	cyprodinil + fludioxonil, fludioxonil (*reduction*)
	Bacterial canker	Bacillus amyloliquefaciens D747 (*off-label*)
	Botrytis	Bacillus amyloliquefaciens D747 (*off-label*), Bacillus subtilis (*off-label*), fludioxonil (*reduction*), Gliocladium catenulatum (*off-label*), pyrimethanil (*off-label*)
	Botrytis fruit rot	cyprodinil + fludioxonil
	Disease control	captan (*off-label*), fludioxonil, fludioxonil + pyrimethanil, fluopyram, mancozeb, trifloxystrobin
	Fireblight	Bacillus amyloliquefaciens D747 (*off-label*)
	Fungus diseases	1-methylcyclopropene, Bacillus amyloliquefaciens D747 (*off-label*), fludioxonil (*reduction*), mancozeb + metalaxyl-M (*off-label*), potassium bicarbonate (commodity substance)
	Fusarium	cyprodinil + fludioxonil, Gliocladium catenulatum (*off-label*)

81

Penicillium rot	cyprodinil + fludioxonil, fludioxonil (*reduction*)
Phytophthora	Gliocladium catenulatum (*off-label*)
Powdery mildew	Bacillus amyloliquefaciens D747 (*off-label*), boscalid + pyraclostrobin, bupirimate, cyflufenamid, fluxapyroxad, kresoxim-methyl (*reduction*), penconazole, penthiopyrad (*suppression*), physical pest control, proquinazid (*off-label*), sulphur, tebuconazole (*off-label*)
Rhizoctonia	Gliocladium catenulatum (*off-label*)
Scab	boscalid + pyraclostrobin, captan, difenoconazole, dithianon, dithianon + potassium phosphonates, dodine, fluxapyroxad, kresoxim-methyl, mancozeb, penthiopyrad, potassium phosphonates, pyrimethanil, pyrimethanil (*off-label*), sulphur
Sclerotinia	Bacillus amyloliquefaciens D747 (*off-label*)
Sooty moulds	potassium bicarbonate (commodity substance), potassium bicarbonate (commodity substance) (*reduction*)
Stem canker	tebuconazole (*off-label*)
Storage rots	captan, cyprodinil + fludioxonil

Pests	Aphids	acetamiprid, deltamethrin, flonicamid, lambda-cyhalothrin, physical pest control, spirotetramat, spirotetramat (*off-label*), thiacloprid
	Capsid bugs	deltamethrin
	Caterpillars	Bacillus thuringiensis, Bacillus thuringiensis (*off-label*), chlorantraniliprole, Cydia pomonella GV, deltamethrin, dodecadienol + tetradecenyl acetate + tetradecylacetate, indoxacarb, indoxacarb (*off-label*), lambda-cyhalothrin, lambda-cyhalothrin (*off-label*), methoxyfenozide, spinosad
	Gall, rust and leaf & bud mites	spirodiclofen
	Mealybugs	physical pest control
	Midges	thiacloprid (*off-label*)
	Pests, miscellaneous	indoxacarb (*off-label*), lambda-cyhalothrin (*off-label*), potassium bicarbonate (commodity substance), thiacloprid (*off-label*)
	Sawflies	deltamethrin
	Scale insects	physical pest control, spirodiclofen, spirotetramat
	Spider mites	clofentezine, physical pest control, spirodiclofen, tebufenpyrad
	Suckers	deltamethrin, lambda-cyhalothrin, physical pest control, potassium bicarbonate (commodity substance) (*reduction*), spirodiclofen
	Whiteflies	acetamiprid, physical pest control
Plant growth regulation	Fruiting control	1-methylcyclopropene (*off-label - post harvest only*), 1-methylcyclopropene (*post-harvest use*), 1-naphthylacetic acid, benzyladenine, gibberellins, metamitron
	Growth control	benzyladenine, ethephon (*off-label*), gibberellins, metamitron, prohexadione-calcium, prohexadione-calcium (*off-label*)
	Quality/yield control	gibberellins, gibberellins (*off-label*)

Weeds	Broad-leaved weeds	2,4-D, clopyralid (*off-label*), fluroxypyr (*off-label*), isoxaben, pendimethalin, pendimethalin (*off-label*), propyzamide, propyzamide (*off-label*), prosulfocarb (*off-label*)
	Crops as weeds	fluazifop-P-butyl, pendimethalin
	Grass weeds	fluazifop-P-butyl, pendimethalin, propyzamide, propyzamide (*off-label*), prosulfocarb (*off-label*)
	Weeds, miscellaneous	2,4-D + glyphosate, glyphosate, glyphosate (*off-label*)

Tree fruit - All stone fruit

Crop control	Sucker/shoot control	glyphosate
Diseases	Bacterial canker	Bacillus amyloliquefaciens D747 (*off-label*)
	Blossom wilt	boscalid + pyraclostrobin (*off-label*), cyprodinil + fludioxonil (*off-label*)
	Botrytis	1-methylcyclopropene (*off-label*), Bacillus amyloliquefaciens D747 (*off-label*), Bacillus subtilis (*off-label*), cyprodinil + fludioxonil (*off-label*), fenhexamid (*off-label*), Gliocladium catenulatum (*off-label*)
	Fireblight	Bacillus amyloliquefaciens D747 (*off-label*)
	Fungus diseases	1-methylcyclopropene, Bacillus amyloliquefaciens D747 (*off-label*)
	Fusarium	Gliocladium catenulatum (*off-label*)
	Phytophthora	Gliocladium catenulatum (*off-label*)
	Powdery mildew	Bacillus amyloliquefaciens D747 (*off-label*), fluxapyroxad (*off-label*)
	Rhizoctonia	Gliocladium catenulatum (*off-label*)
	Scab	fluxapyroxad (*off-label*)
	Sclerotinia	Bacillus amyloliquefaciens D747 (*off-label*), boscalid + pyraclostrobin (*off-label*)
	Storage rots	1-methylcyclopropene (*off-label*)
Pests	Aphids	acetamiprid, spirotetramat, spirotetramat (*off-label*), thiacloprid (*off-label*)
	Capsid bugs	indoxacarb (*off-label*)
	Caterpillars	Bacillus thuringiensis (*off-label*), dodecadienol + tetradecenyl acetate + tetradecylacetate, indoxacarb (*off-label*), lambda-cyhalothrin (*off-label*), methoxyfenozide (*off-label*)
	Gall, rust and leaf & bud mites	spirodiclofen (*off-label*)
	Pests, miscellaneous	indoxacarb (*off-label*), lambda-cyhalothrin (*off-label*), thiacloprid (*off-label* - under temporary protective rain covers*)
	Spider mites	spirodiclofen (*off-label*)
	Whiteflies	acetamiprid
Plant growth regulation	Fruiting control	gibberellins (*off-label*)
	Growth control	gibberellins (*off-label*), prohexadione-calcium (*off-label*)
Weeds	Broad-leaved weeds	isoxaben, pendimethalin, propyzamide, propyzamide (*off-label*), prosulfocarb (*off-label*)
	Crops as weeds	fluazifop-P-butyl, pendimethalin

	Grass weeds	fluazifop-P-butyl, pendimethalin, propyzamide, propyzamide (*off-label*), prosulfocarb (*off-label*)
	Weeds, miscellaneous	glyphosate, glyphosate (*off-label*)

Tree fruit - Miscellaneous top fruit

Diseases	Bacterial canker	Bacillus amyloliquefaciens D747 (*off-label*)
	Botrytis	Bacillus amyloliquefaciens D747 (*off-label*), Bacillus subtilis (*off-label*), Gliocladium catenulatum (*off-label*)
	Fireblight	Bacillus amyloliquefaciens D747 (*off-label*)
	Fungus diseases	Bacillus amyloliquefaciens D747 (*off-label*)
	Fusarium	Gliocladium catenulatum (*off-label*)
	Phytophthora	Gliocladium catenulatum (*off-label*)
	Powdery mildew	Bacillus amyloliquefaciens D747 (*off-label*)
	Rhizoctonia	Gliocladium catenulatum (*off-label*)
	Sclerotinia	Bacillus amyloliquefaciens D747 (*off-label*)
Pests	Caterpillars	indoxacarb (*off-label*), lambda-cyhalothrin (*off-label*)
	Pests, miscellaneous	indoxacarb (*off-label*), lambda-cyhalothrin (*off-label*)
Weeds	Broad-leaved weeds	prosulfocarb (*off-label*)
	Grass weeds	prosulfocarb (*off-label*)
	Weeds, miscellaneous	glyphosate (*off-label*)

Tree fruit - Protected fruit

Diseases	Powdery mildew	Ampelomyces quisqualis (Strain AQ10) (*off-label*), penconazole (*off-label*)
Pests	Aphids	Beauveria bassiana
	Beetles	Beauveria bassiana
	Caterpillars	Bacillus thuringiensis (*off-label*), indoxacarb (*off-label*)
	Pests, miscellaneous	indoxacarb (*off-label*)
	Spider mites	Lecanicillium muscarium (*off-label*)
	Thrips	Lecanicillium muscarium (*off-label*)
	Whiteflies	Beauveria bassiana, Lecanicillium muscarium (*off-label*)

Grain/crop store uses

Stored produce - Food/produce storage

Pests	Food/grain storage pests	pirimiphos-methyl
	Pests, miscellaneous	aluminium phosphide, cypermethrin, magnesium phosphide

Stored seed - Stored grain/rape/linseed

Pests	Food/grain storage pests	phenothrin + tetramethrin, pirimiphos-methyl
	Mites	physical pest control, pirimiphos-methyl
	Pests, miscellaneous	aluminium phosphide, deltamethrin, magnesium phosphide, pirimiphos-methyl

Grass

Turf/amenity grass - Amenity grassland

Diseases	Anthracnose	fludioxonil (*reduction*), tebuconazole + trifloxystrobin
	Dollar spot	tebuconazole + trifloxystrobin
	Drechslera leaf spot	fludioxonil (*useful levels of control*)
	Fusarium	fludioxonil, tebuconazole + trifloxystrobin, trifloxystrobin
	Melting out	tebuconazole + trifloxystrobin
	Red thread	tebuconazole + trifloxystrobin, trifloxystrobin
	Rust	tebuconazole + trifloxystrobin
	Seed-borne diseases	fludioxonil
Pests	Birds/mammals	aluminium ammonium sulphate
	Slugs/snails	metaldehyde
Plant growth regulation	Growth control	trinexapac-ethyl
Weeds	Broad-leaved weeds	2,4-D, 2,4-D + dicamba, 2,4-D + dicamba + fluroxypyr, 2,4-D + florasulam, aminopyralid + fluroxypyr, aminopyralid + triclopyr, citronella oil, clopyralid + florasulam + fluroxypyr, clopyralid + triclopyr, dicamba + MCPA + mecoprop-P, dicamba + MCPA + mecoprop-P (*moderately susceptible*), dicamba + mecoprop-P, florasulam + fluroxypyr, fluroxypyr + triclopyr, mecoprop-P
	Crops as weeds	2,4-D + dicamba + fluroxypyr, 2,4-D + florasulam, dicamba + MCPA + mecoprop-P (*moderately susceptible*), florasulam + fluroxypyr
	Mosses	carfentrazone-ethyl + mecoprop-P, ferrous sulphate, pelargonic acid
	Weeds, miscellaneous	2,4-D + glyphosate, aminopyralid + triclopyr, carfentrazone-ethyl + mecoprop-P, clopyralid + florasulam + fluroxypyr, dicamba + MCPA + mecoprop-P, florasulam + fluroxypyr, pelargonic acid
	Woody weeds/scrub	aminopyralid + triclopyr, clopyralid + triclopyr

Turf/amenity grass - Managed amenity turf

Diseases	Anthracnose	azoxystrobin, azoxystrobin (*moderate control*), fludioxonil (*reduction*), tebuconazole + trifloxystrobin
	Brown patch	azoxystrobin
	Crown rust	azoxystrobin
	Disease control	difenoconazole + fludioxonil
	Dollar spot	fluopyram + trifloxystrobin, pyraclostrobin (*useful reduction*), tebuconazole + trifloxystrobin
	Drechslera leaf spot	fludioxonil (*useful levels of control*)
	Fairy rings	azoxystrobin, azoxystrobin (*reduction*)
	Fusarium	azoxystrobin, fludioxonil, pyraclostrobin (*moderate control*), pyraclostrobin (*moderate control only*), tebuconazole + trifloxystrobin, trifloxystrobin
	Melting out	azoxystrobin, tebuconazole + trifloxystrobin
	Red thread	pyraclostrobin, tebuconazole + trifloxystrobin, trifloxystrobin

	Rust	azoxystrobin, tebuconazole + trifloxystrobin
	Seed-borne diseases	fludioxonil, fluopyram + trifloxystrobin
	Take-all patch	azoxystrobin
Pests	Birds/mammals	aluminium ammonium sulphate
	Flies	esfenvalerate
	Slugs/snails	ferric phosphate, metaldehyde
Plant growth regulation	Growth control	trinexapac-ethyl
Weeds	Broad-leaved weeds	2,4-D, 2,4-D + dicamba + fluroxypyr, 2,4-D + dicamba + MCPA + mecoprop-P, 2,4-D + florasulam, clopyralid + 2,4-D + MCPA, clopyralid + florasulam + fluroxypyr, clopyralid + fluroxypyr + MCPA, dicamba + MCPA + mecoprop-P, dicamba + MCPA + mecoprop-P (*moderately susceptible*), dicamba + mecoprop-P, ferrous sulphate + MCPA + mecoprop-P, florasulam + fluroxypyr, MCPA + mecoprop-P, mecoprop-P
	Crops as weeds	2,4-D + dicamba + fluroxypyr, 2,4-D + dicamba + MCPA + mecoprop-P, 2,4-D + florasulam, clopyralid + 2,4-D + MCPA, dicamba + MCPA + mecoprop-P (*moderately susceptible*), florasulam + fluroxypyr
	Grass weeds	cycloxydim (*off-label*)
	Mosses	carfentrazone-ethyl + mecoprop-P, ferrous sulphate, ferrous sulphate + MCPA + mecoprop-P, pelargonic acid
	Weeds, miscellaneous	carfentrazone-ethyl + mecoprop-P, dicamba + MCPA + mecoprop-P, florasulam + fluroxypyr, glyphosate, glyphosate (*pre-establishment only*), pelargonic acid

Non-crop pest control

Farm buildings/yards - Farm buildings

Pests	Ants	cypermethrin + tetramethrin
	Bedbugs	cypermethrin + tetramethrin
	Birds/mammals	brodifacoum, bromadiolone, difenacoum, flocoumafen, warfarin
	Caterpillars	cypermethrin + tetramethrin
	Cockroaches	cypermethrin + tetramethrin
	Flies	cypermethrin + tetramethrin, phenothrin + tetramethrin
	Pests, miscellaneous	physical pest control, pyrethrins
	Silverfish	cypermethrin + tetramethrin
	Wasps	cypermethrin + tetramethrin, phenothrin + tetramethrin

Farm buildings/yards - Farmyards
Pests	Birds/mammals	brodifacoum, bromadiolone, difenacoum

Farmland pest control - Farmland situations
Pests	Birds/mammals	aluminium phosphide

Miscellaneous non-crop pest control - Manure/rubbish

Pests	Ants	cypermethrin + tetramethrin
	Bedbugs	cypermethrin + tetramethrin
	Caterpillars	cypermethrin + tetramethrin
	Cockroaches	cypermethrin + tetramethrin
	Flies	cypermethrin + tetramethrin
	Silverfish	cypermethrin + tetramethrin
	Wasps	cypermethrin + tetramethrin

Miscellaneous non-crop pest control - Miscellaneous pest control situations

Pests	Birds/mammals	carbon dioxide (commodity substance)

Protected salad and vegetable crops

Protected brassicas - Protected brassica vegetables

Diseases	Alternaria	boscalid + pyraclostrobin (*off-label*)
	Damping off	fosetyl-aluminium + propamocarb hydrochloride, propamocarb hydrochloride
	Downy mildew	fluopicolide + propamocarb hydrochloride (*off-label*), fosetyl-aluminium + propamocarb hydrochloride, propamocarb hydrochloride
	Fungus diseases	propamocarb hydrochloride
	Fusarium	Trichoderma asperellum (Strain T34) (*off-label*)
	Mycosphaerella	boscalid + pyraclostrobin (*off-label*)
	Phytophthora	propamocarb hydrochloride
	Pythium	propamocarb hydrochloride, Trichoderma asperellum (Strain T34) (*off-label*)
	Root rot	propamocarb hydrochloride
	White blister	boscalid + pyraclostrobin (*off-label*)
Pests	Aphids	acetamiprid (*off-label*), pyrethrins, spirotetramat (*off-label*)
	Beetles	lambda-cyhalothrin (*off-label*)
	Caterpillars	Bacillus thuringiensis (*off-label*), indoxacarb (*off-label*), lambda-cyhalothrin (*off-label*), pyrethrins
	Spider mites	pyrethrins
	Thrips	pyrethrins
	Whiteflies	spirotetramat (*off-label*)

Protected brassicas - Protected salad brassicas

Diseases	Damping off	fosetyl-aluminium + propamocarb hydrochloride
	Downy mildew	fosetyl-aluminium + propamocarb hydrochloride
	Fusarium	Trichoderma asperellum (Strain T34) (*off-label*)
	Pythium	Trichoderma asperellum (Strain T34) (*off-label*)
Pests	Aphids	lambda-cyhalothrin (*off-label*)
	Caterpillars	Bacillus thuringiensis (*off-label*)
	Pests, miscellaneous	lambda-cyhalothrin (*off-label*)

Protected crops, general - All protected crops

Pests	Aphids	Beauveria bassiana
	Caterpillars	Bacillus thuringiensis (*off-label*), indoxacarb (*off-label*), lambda-cyhalothrin (*off-label*)
	Pests, miscellaneous	indoxacarb (*off-label*), lambda-cyhalothrin (*off-label*)
	Whiteflies	Beauveria bassiana

Protected crops, general - Protected nuts

Pests	Caterpillars	indoxacarb (*off-label*)
	Pests, miscellaneous	indoxacarb (*off-label*)

Protected crops, general - Protected vegetables, general

Diseases	Anthracnose	Bacillus amyloliquefaciens D747 (*off-label*)
	Bacterial blight	Bacillus amyloliquefaciens D747 (*off-label*)
	Botrytis	Bacillus amyloliquefaciens D747 (*off-label*)
	Cercospora leaf spot	Bacillus amyloliquefaciens D747 (*off-label*)
	Downy mildew	Bacillus amyloliquefaciens D747 (*off-label*)
	Fusarium	Bacillus amyloliquefaciens D747 (*off-label*)
	Powdery mildew	Ampelomyces quisqualis (Strain AQ10) (*off-label*), Bacillus amyloliquefaciens D747 (*off-label*)
	Pythium	Bacillus amyloliquefaciens D747 (*off-label*)
	Rhizoctonia	Bacillus amyloliquefaciens D747 (*off-label*)
	Sclerotinia	Bacillus amyloliquefaciens D747 (*off-label*)
	Stemphylium	Bacillus amyloliquefaciens D747 (*off-label*)
Pests	Caterpillars	indoxacarb (*off-label*)
	Pests, miscellaneous	indoxacarb (*off-label*)
	Spider mites	Lecanicillium muscarium (*off-label*)
	Thrips	Lecanicillium muscarium (*off-label*)
	Whiteflies	Lecanicillium muscarium (*off-label*)

Protected fruiting vegetables - Protected aubergines

Diseases	Alternaria	difenoconazole + mandipropamid (*off-label*)
	Botrytis	boscalid + pyraclostrobin (*off-label*), cyprodinil + fludioxonil (*off-label*), fenhexamid (*off-label*), fenpyrazamine, mepanipyrim (*off-label*), potassium bicarbonate (commodity substance) (*off-label*), pyrimethanil (*off-label*)
	Damping off	fosetyl-aluminium + propamocarb hydrochloride (*off-label*)
	Downy mildew	cerevisane
	Phytophthora	difenoconazole + mandipropamid (*off-label*)
	Powdery mildew	Ampelomyces quisqualis (Strain AQ10), cerevisane, cos-oga, isopyrazam, mepanipyrim (*off-label*), potassium bicarbonate (commodity substance) (*off-label*), proquinazid (*off-label*), sulphur (*off-label*)
Pests	Aphids	acetamiprid, Beauveria bassiana, lambda-cyhalothrin (*off-label*), pirimicarb (*off-label*), sulfoxaflor
	Beetles	Beauveria bassiana
	Caterpillars	Bacillus thuringiensis, indoxacarb
	Gall, rust and leaf & bud mites	sulphur (*off-label*)

Leaf miners	deltamethrin (*off-label*), tetradecadienyl + trienyl acetate, thiacloprid (*off-label*)
Mealybugs	flonicamid (*off-label*)
Mites	etoxazole
Pests, miscellaneous	lambda-cyhalothrin (*off-label*)
Spider mites	abamectin, bifenazate (*off-label*)
Thrips	abamectin, deltamethrin (*off-label*), spinosad, thiacloprid (*off-label*)
Whiteflies	acetamiprid, Beauveria bassiana, buprofezin, cyantraniliprole (*off-label*), flonicamid (*off-label*), sulfoxaflor, thiacloprid (*off-label*)

Protected fruiting vegetables - Protected cucurbits

Diseases	Anthracnose	Bacillus amyloliquefaciens D747 (*off-label*)
	Bacterial blight	Bacillus amyloliquefaciens D747 (*off-label*)
	Botrytis	Bacillus amyloliquefaciens D747 (*off-label*), boscalid + pyraclostrobin (*off-label*), cyprodinil + fludioxonil (*off-label*), fenhexamid (*off-label*), fenpyrazamine, fluopyram + trifloxystrobin, mepanipyrim (*off-label*), potassium bicarbonate (commodity substance) (*off-label*)
	Cercospora leaf spot	Bacillus amyloliquefaciens D747 (*off-label*)
	Cladosporium	boscalid + pyraclostrobin (*off-label*)
	Damping off	fosetyl-aluminium + propamocarb hydrochloride, fosetyl-aluminium + propamocarb hydrochloride (*off-label*)
	Disease control	difenoconazole + fluxapyroxad, isopyrazam
	Downy mildew	Bacillus amyloliquefaciens D747 (*off-label*), cerevisane, fosetyl-aluminium + propamocarb hydrochloride, metalaxyl-M (*off-label*)
	Fungus diseases	boscalid + pyraclostrobin (*off-label*)
	Fusarium	Bacillus amyloliquefaciens D747 (*off-label*), cyprodinil + fludioxonil (*off-label*), Trichoderma asperellum (Strain T34) (*off-label*)
	Powdery mildew	Ampelomyces quisqualis (Strain AQ10), Ampelomyces quisqualis (Strain AQ10) (*off-label*), Bacillus amyloliquefaciens D747 (*off-label*), boscalid + pyraclostrobin (*off-label*), bupirimate, cerevisane, cos-oga, fluopyram + trifloxystrobin, isopyrazam, mepanipyrim (*off-label*), penconazole (*off-label*), potassium bicarbonate (commodity substance) (*off-label*), proquinazid (*off-label*), sulphur (*off-label*)
	Pythium	Bacillus amyloliquefaciens D747 (*off-label*), fosetyl-aluminium + propamocarb hydrochloride, Trichoderma asperellum (Strain T34) (*off-label*)
	Rhizoctonia	Bacillus amyloliquefaciens D747 (*off-label*)
	Sclerotinia	Bacillus amyloliquefaciens D747 (*off-label*)
	Stemphylium	Bacillus amyloliquefaciens D747 (*off-label*)
Pests	Aphids	acetamiprid, Beauveria bassiana, deltamethrin, flonicamid (*off-label*), lambda-cyhalothrin (*off-label*), pirimicarb (*off-label*), sulfoxaflor
	Beetles	Beauveria bassiana
	Caterpillars	Bacillus thuringiensis, Bacillus thuringiensis (*off-label*), deltamethrin, indoxacarb, spinosad (*off-label*)

	Cutworms	Bacillus thuringiensis (*off-label*)
	Gall, rust and leaf & bud mites	sulphur (*off-label*)
	Leaf miners	deltamethrin (*off-label*), spinosad (*off-label*), tetradecadienyl + trienyl acetate, thiacloprid (*off-label*)
	Mealybugs	deltamethrin
	Pests, miscellaneous	lambda-cyhalothrin (*off-label*)
	Scale insects	deltamethrin
	Spider mites	abamectin, abamectin (*off-label*), bifenazate (*off-label*), Lecanicillium muscarium (*off-label*), spirodiclofen (*off-label*)
	Thrips	abamectin, deltamethrin, deltamethrin (*off-label*), Lecanicillium muscarium (*off-label*), spinosad, spinosad (*off-label*), thiacloprid (*off-label*)
	Whiteflies	acetamiprid, Beauveria bassiana, buprofezin, deltamethrin, flonicamid (*off-label*), Lecanicillium muscarium, Lecanicillium muscarium (*off-label*), sulfoxaflor, thiacloprid (*off-label*)
Weeds	Broad-leaved weeds	clomazone (*off-label*)

Protected fruiting vegetables - Protected tomatoes

Crop control	Miscellaneous non-selective situations	Mild Pepino Mosaic Virus isolate VC1 + VX1
Diseases	Alternaria	difenoconazole + mandipropamid (*off-label*)
	Botrytis	boscalid + pyraclostrobin (*off-label*), cyprodinil + fludioxonil (*off-label*), fenhexamid (*off-label*), fenpyrazamine, mepanipyrim (*off-label*), potassium bicarbonate (commodity substance) (*off-label*), pyrimethanil (*off-label*)
	Damping off	propamocarb hydrochloride (*off-label*)
	Disease control	difenoconazole + fluxapyroxad
	Downy mildew	cerevisane
	Phytophthora	cerevisane, difenoconazole + mandipropamid (*off-label*), propamocarb hydrochloride (*off-label*)
	Powdery mildew	Ampelomyces quisqualis (Strain AQ10), bupirimate (*off-label*), cos-oga, cyflufenamid (*off-label*), isopyrazam, mepanipyrim (*off-label*), penconazole (*off-label*), potassium bicarbonate (commodity substance) (*off-label*), proquinazid (*off-label*), sulphur (*off-label*)
	Pythium	fosetyl-aluminium + propamocarb hydrochloride
	Root rot	propamocarb hydrochloride (*off-label*)
	Verticillium wilt	thiophanate-methyl (*off-label*)
Pests	Aphids	acetamiprid, Beauveria bassiana, deltamethrin, lambda-cyhalothrin (*off-label*), pirimicarb (*off-label*), pyrethrins, sulfoxaflor
	Beetles	Beauveria bassiana
	Caterpillars	Bacillus thuringiensis, chlorantraniliprole (*off-label*), deltamethrin, indoxacarb, pyrethrins
	Gall, rust and leaf & bud mites	sulphur (*off-label*)
	Leaf miners	tetradecadienyl + trienyl acetate, thiacloprid (*off-label*)
	Mealybugs	deltamethrin, flonicamid (*off-label*)

	Mites	etoxazole
	Pests, miscellaneous	lambda-cyhalothrin (*off-label*)
	Scale insects	deltamethrin
	Spider mites	abamectin, abamectin (*off-label*), bifenazate (*off-label*), pyrethrins, spirodiclofen (*off-label*)
	Thrips	abamectin, deltamethrin, pyrethrins, spinosad, spinosad (*off-label*), thiacloprid (*off-label*)
	Whiteflies	acetamiprid, Beauveria bassiana, buprofezin, cyantraniliprole (*off-label*), deltamethrin, flonicamid (*off-label*), Lecanicillium muscarium, sulfoxaflor, thiacloprid (*off-label*)
Plant growth regulation	Growth control	ethephon (*off-label*)

Protected herb crops - Protected herbs

Diseases	Alternaria	cyprodinil + fludioxonil (*off-label*)
	Botrytis	cyprodinil + fludioxonil (*off-label*), fenhexamid (*off-label*), fluopyram + trifloxystrobin, fluopyram + trifloxystrobin (*off-label*), potassium bicarbonate (commodity substance) (*off-label*), pyrimethanil (*off-label*)
	Damping off	fosetyl-aluminium + propamocarb hydrochloride (*off-label*)
	Disease control	difenoconazole + fluxapyroxad, mandipropamid
	Downy mildew	cos-oga (*off-label*), dimethomorph (*off-label*), mancozeb + metalaxyl-M (*off-label*), mandipropamid, metalaxyl-M (*off-label*)
	Fusarium	Trichoderma asperellum (Strain T34) (*off-label*)
	Powdery mildew	Ampelomyces quisqualis (Strain AQ10), Ampelomyces quisqualis (Strain AQ10) (*off-label*), cos-oga, cos-oga (*off-label*), fluopyram + trifloxystrobin, fluopyram + trifloxystrobin (*off-label*), isopyrazam, penconazole (*off-label*), potassium bicarbonate (commodity substance) (*off-label*), sulphur (*off-label*)
	Pythium	Trichoderma asperellum (Strain T34) (*off-label*)
	Rhizoctonia	cyprodinil + fludioxonil (*off-label*)
	Sclerotinia	fluopyram + trifloxystrobin (*off-label*)
	Stem canker	cyprodinil + fludioxonil (*off-label*)
	White blister	boscalid + pyraclostrobin (*off-label*)
Pests	Aphids	acetamiprid (*off-label*), Beauveria bassiana, deltamethrin, flonicamid (*off-label*), pirimicarb (*off-label*), spirotetramat (*off-label*), sulfoxaflor, thiacloprid (*off-label*)
	Beetles	Beauveria bassiana
	Caterpillars	Bacillus thuringiensis, Bacillus thuringiensis (*off-label*), deltamethrin, indoxacarb, indoxacarb (*off-label*)
	Cutworms	Bacillus thuringiensis (*off-label*)
	Gall, rust and leaf & bud mites	sulphur (*off-label*)
	Leaf miners	tetradecadienyl + trienyl acetate
	Mealybugs	deltamethrin
	Scale insects	deltamethrin

	Spider mites	abamectin, abamectin (*off-label*), bifenazate (*off-label*), Lecanicillium muscarium (*off-label*), spirodiclofen (*off-label*)
	Thrips	abamectin, deltamethrin, Lecanicillium muscarium (*off-label*)
	Whiteflies	Beauveria bassiana, buprofezin, deltamethrin, Lecanicillium muscarium (*off-label*), spirotetramat (*off-label*), sulfoxaflor
Weeds	Broad-leaved weeds	napropamide (*off-label*), propyzamide (*off-label*)
	Grass weeds	napropamide (*off-label*), propyzamide (*off-label*)

Protected leafy vegetables - Mustard and cress

Diseases	Alternaria	Bacillus amyloliquefaciens D747, cyprodinil + fludioxonil (*off-label*)
	Black rot	azoxystrobin (*off-label*)
	Botrytis	Bacillus subtilis (*off-label*), cyprodinil + fludioxonil (*off-label*), Gliocladium catenulatum (*off-label*)
	Disease control	dimethomorph + mancozeb (*off-label*), mandipropamid
	Downy mildew	dimethomorph + mancozeb (*off-label*), fluopicolide + propamocarb hydrochloride (*off-label*), mandipropamid
	Fusarium	Bacillus amyloliquefaciens D747, Gliocladium catenulatum (*off-label*)
	Phytophthora	Gliocladium catenulatum (*off-label*)
	Powdery mildew	Bacillus amyloliquefaciens D747
	Pythium	Bacillus amyloliquefaciens D747
	Rhizoctonia	Bacillus amyloliquefaciens D747, cyprodinil + fludioxonil (*off-label*), Gliocladium catenulatum (*off-label*)
	Stem canker	cyprodinil + fludioxonil (*off-label*)
	White blister	azoxystrobin (*off-label*)
Pests	Aphids	acetamiprid (*off-label*), Beauveria bassiana (*off-label*), deltamethrin (*off-label*), spirotetramat (*off-label*)
	Beetles	deltamethrin (*off-label*)
	Caterpillars	deltamethrin (*off-label*), spinosad (*off-label*)
	Pests, miscellaneous	deltamethrin (*off-label*)
	Spider mites	Beauveria bassiana (*off-label*)
	Thrips	Beauveria bassiana (*off-label*), spinosad (*off-label*)
	Whiteflies	Beauveria bassiana (*off-label*), spirotetramat (*off-label*)
Weeds	Broad-leaved weeds	pendimethalin (*off-label*), propyzamide (*off-label*)
	Crops as weeds	cycloxydim (*off-label*)
	Grass weeds	cycloxydim (*off-label*), propyzamide (*off-label*)

Protected leafy vegetables - Protected leafy vegetables

Diseases	Alternaria	cyprodinil + fludioxonil (*off-label*)
	Botrytis	boscalid + pyraclostrobin, boscalid + pyraclostrobin (*off-label*), cyprodinil + fludioxonil (*off-label*), fluopyram + trifloxystrobin (*off-label*), pyrimethanil (*off-label*)
	Bottom rot	boscalid + pyraclostrobin

	Disease control	mandipropamid
	Downy mildew	azoxystrobin, boscalid + pyraclostrobin (*off-label*), cerevisane, cos-oga (*off-label*), dimethomorph (*off-label*), fosetyl-aluminium + propamocarb hydrochloride, mandipropamid, potassium bicarbonate (commodity substance) (*off-label*)
	Fusarium	Trichoderma asperellum (Strain T34) (*off-label*)
	Powdery mildew	Ampelomyces quisqualis (Strain AQ10) (*off-label*), cerevisane, cos-oga (*off-label*), fluopyram + trifloxystrobin (*off-label*)
	Pythium	fosetyl-aluminium + propamocarb hydrochloride, Trichoderma asperellum (Strain T34) (*off-label*)
	Rhizoctonia	boscalid + pyraclostrobin (*off-label*), cyprodinil + fludioxonil (*off-label*)
	Sclerotinia	boscalid + pyraclostrobin (*off-label*), difenoconazole + fluxapyroxad (*moderate control*), fluopyram + trifloxystrobin (*off-label*)
	Soft rot	boscalid + pyraclostrobin
	Stem canker	cyprodinil + fludioxonil (*off-label*)
Pests	Aphids	acetamiprid (*off-label*), lambda-cyhalothrin (*off-label*), pirimicarb (*off-label*), pyrethrins, spirotetramat (*off-label*), thiacloprid (*off-label*)
	Caterpillars	Bacillus thuringiensis (*off-label*), pyrethrins
	Cutworms	Bacillus thuringiensis (*off-label*)
	Pests, miscellaneous	lambda-cyhalothrin (*off-label*)
	Spider mites	abamectin (*off-label*), pyrethrins
	Thrips	pyrethrins
	Whiteflies	spirotetramat (*off-label*)
Weeds	Broad-leaved weeds	napropamide (*off-label*), propyzamide (*off-label*)
	Grass weeds	napropamide (*off-label*), propyzamide (*off-label*)

Protected leafy vegetables - Protected spinach

Diseases	Alternaria	cyprodinil + fludioxonil (*off-label*)
	Botrytis	boscalid + pyraclostrobin (*off-label*), cyprodinil + fludioxonil (*off-label*)
	Disease control	mandipropamid
	Downy mildew	mandipropamid, metalaxyl-M (*off-label*)
	Powdery mildew	Ampelomyces quisqualis (Strain AQ10) (*off-label*)
	Rhizoctonia	boscalid + pyraclostrobin (*off-label*), cyprodinil + fludioxonil (*off-label*)
	Sclerotinia	boscalid + pyraclostrobin (*off-label*)
	Stem canker	cyprodinil + fludioxonil (*off-label*)
Pests	Aphids	spirotetramat (*off-label*)
	Caterpillars	Bacillus thuringiensis (*off-label*)
	Whiteflies	spirotetramat (*off-label*)

Protected legumes - Protected peas and beans

Diseases	Damping off	fosetyl-aluminium + propamocarb hydrochloride (*off-label*)
Pests	Caterpillars	Bacillus thuringiensis, Bacillus thuringiensis (*off-label*)
	Cutworms	Bacillus thuringiensis (*off-label*)

Protected root and tuber vegetables - Protected root brassicas

Diseases	Alternaria	cyprodinil + fludioxonil (*off-label*)
	Damping off	fosetyl-aluminium + propamocarb hydrochloride
	Disease control	boscalid + pyraclostrobin (*off-label*)
	Downy mildew	fosetyl-aluminium + propamocarb hydrochloride
	Leaf spot	cyprodinil + fludioxonil (*off-label*)
	Sclerotinia	cyprodinil + fludioxonil (*off-label*)
Pests	Caterpillars	chlorantraniliprole (*off-label*)
	Flies	chlorantraniliprole (*off-label*)

Protected stem and bulb vegetables - Protected asparagus

Diseases	Botrytis	cyprodinil + fludioxonil (*off-label*)
Pests	Aphids	lambda-cyhalothrin (*off-label*)

Protected stem and bulb vegetables - Protected celery/chicory

Diseases	Botrytis	azoxystrobin (*off-label*)
	Leaf spot	azoxystrobin (*off-label*)
	Powdery mildew	Ampelomyces quisqualis (Strain AQ10) (*off-label*)
	Rhizoctonia	azoxystrobin (*off-label*)
	Sclerotinia	azoxystrobin (*off-label*)
Pests	Caterpillars	Bacillus thuringiensis (*off-label*), deltamethrin (*off-label*)
	Cutworms	Bacillus thuringiensis (*off-label*)

Protected stem and bulb vegetables - Protected onions/leeks/garlic

Pests	Leaf miners	deltamethrin (*off-label*)
	Pests, miscellaneous	deltamethrin (*off-label*)

Total vegetation control

Aquatic situations, general - Aquatic situations

Pests	Aphids	Bacillus thuringiensis israelensis
	Flies	Bacillus thuringiensis israelensis, physical pest control
Plant growth regulation	Growth control	glyphosate
Weeds	Aquatic weeds	glyphosate
	Grass weeds	glyphosate
	Weeds, miscellaneous	glyphosate

Non-crop areas, general - Miscellaneous non-crop situations

Diseases	Botrytis	Bacillus subtilis (*off-label*)
Pests	Birds/mammals	difenacoum, flocoumafen
Weeds	Weeds, miscellaneous	2,4-D + glyphosate, flazasulfuron

Non-crop areas, general - Non-crop farm areas

Weeds	Broad-leaved weeds	metsulfuron-methyl
	Crops as weeds	metsulfuron-methyl
	Weeds, miscellaneous	2,4-D + glyphosate, glyphosate, glyphosate + pyraflufen-ethyl, metsulfuron-methyl

Non-crop areas, general - Paths/roads etc

Pests	Slugs/snails	metaldehyde
Weeds	Broad-leaved weeds	pelargonic acid
	Grass weeds	glyphosate
	Mosses	pelargonic acid, starch, protein, oil and water
	Weeds, miscellaneous	acetic acid, diflufenican + glyphosate, flazasulfuron, glyphosate, glyphosate + pyraflufen-ethyl, glyphosate + sulfosulfuron, pelargonic acid, starch, protein, oil and water

SECTION 2
PESTICIDE PROFILES

1   abamectin

A selective acaricide and insecticide for use in ornamentals and other protected crops
IRAC mode of action code: 6

Products

1 Amec	Nurture	18 g/l	EC	18691
2 Clayton Abba	Clayton	18 g/l	EC	18559
3 Dynamec	Syngenta	18 g/l	EC	18316
4 Killermite	Pan Agriculture	18 g/l	EC	18478
5 Smitten	Pan Agriculture	18 g/l	EC	18477

Uses

- Red spider mites in *protected baby leaf crops* (off-label), *protected blackberries* (off-label), *protected cayenne peppers* (off-label), *protected cherry tomatoes* (off-label), *protected herbs (see appendix 6)* (off-label), *protected lamb's lettuce* (off-label), *protected loganberries* (off-label), *protected raspberries* (off-label), *protected rubus hybrids* (off-label) [3]
- Two-spotted spider mite in *protected aubergines, protected chilli peppers, protected ornamentals, protected peppers, protected strawberries, protected tomatoes* [1-5]
- Western flower thrips in *protected aubergines, protected chilli peppers, protected ornamentals, protected peppers, protected strawberries, protected tomatoes* [1-5]

Extension of Authorisation for Minor Use (EAMUs)

- *protected baby leaf crops 20172832* [3]
- *protected blackberries 20172833* [3]
- *protected cayenne peppers 20172831* [3]
- *protected cherry tomatoes 20172831* [3]
- *protected herbs (see appendix 6) 20172832* [3]
- *protected lamb's lettuce 20172832* [3]
- *protected loganberries 20172833* [3]
- *protected raspberries 20172833* [3]
- *protected rubus hybrids 20172833* [3]

Approval information

- Abamectin included in Annex I under EC Regulation 1107/2009

Efficacy guidance

- Abamectin controls adults and immature stages of the two-spotted spider mite, the larval stages of leaf miners and the nymphs of Western Flower Thrips, plus a useful reduction in adults
- Treat at first sign of infestation. Repeat sprays may be required
- For effective control total cover of all plant surfaces is essential, but avoid run-off
- Target pests quickly become immobilised but 3-5 d may be required for maximum mortality
- Indoor applications should be made through a hydraulic nozzle applicator or a knapsack applicator. Outdoors suitable high volume hydraulic nozzle applicators should be used
- Limited data shows abamectin only slightly harmful to Anthocorid bugs and so compatible with biological control systems in which Anthocorid bugs are important

Restrictions

- Number of treatments 6 on protected tomatoes and cucumbers (only 4 of which can be made when flowers or fruit present); not restricted on flowers but rotation with other products advised
- Maximum concentration must not exceed 50 ml per 100 l water
- Do not mix with wetters, stickers or other adjuvants
- Do not use on ferns (*Adiantum* spp) or Shasta daisies
- Do not treat protected crops which are in flower or have set fruit between 1 Nov and 28 Feb
- Do not treat cherry tomatoes (but protected cherry tomatoes may be treated off-label)
- Consult manufacturer for list of plant varieties tested for safety
- There is insufficient evidence to support product compatibility with integrated and biological pest control programmes
- Unprotected persons must be kept out of treated areas until the spray has dried

SEE SECTION 3 FOR PRODUCTS ALSO REGISTERED

SECTION 2

Crop-specific information

- HI 3 d for protected edible crops
- On tomato or cucumber crops that are in flower, or have started to set fruit, treat only between 1 Mar and 31 Oct. Seedling tomatoes that have not started to flower or set fruit may be treated at any time
- Some spotting or staining may occur on carnation, kalanchoe and begonia foliage

Environmental safety

- Dangerous for the environment
- Very toxic to aquatic organisms
- High risk to bees. Do not apply to crops in flower or to those in which bees are actively foraging. Do not apply when flowering weeds are present
- Keep in original container, tightly closed, in a safe place, under lock and key
- Where bumble bees are used in tomatoes as pollinators, keep them out for 24 h after treatment

Hazard classification and safety precautions

Hazard Harmful, Dangerous for the environment, Harmful if swallowed, Very toxic to aquatic organisms
Transport code 9
Packaging group III
UN Number 3082
Risk phrases H319, H373
Operator protection A, C, D, H, K, M; U02a, U05a, U20a
Environmental protection E02a (until spray has dried); E12a, E12e, E15b, E34, E38, H410
Storage and disposal D01, D02, D05, D09b, D10c, D12a
Medical advice M04a

2   acetamiprid

A neonicotinoid insecticide for use in top fruit and horticulture
IRAC mode of action code: 4A

Products

1	Clayton Vault	Clayton	20% w/w	SG	18181
2	Gazelle SG	Certis	20% w/w	SG	13725
3	Insyst	Certis	20% w/w	SP	13414
4	Pure Ace	Pure Amenity	20% w/w	SG	19053
5	Vulcan	Pan Agriculture	20% w/w	SG	16689
6	Vulcan SP	Pan Agriculture	20% w/w	SG	18664

Uses

- Aphids in **apples, cherries, ornamental plant production, pears, plums, protected aubergines, protected ornamentals, protected peppers, protected tomatoes** [1, 2, 4-6]; **brussels sprouts** (off-label), **seed potatoes, spring oilseed rape, ware potatoes, winter oilseed rape** [3]; **chard** (off-label), **cress** (off-label), **forest nurseries** (off-label), **herbs (see appendix 6)** (off-label), **hops** (off-label), **lamb's lettuce** (off-label), **leaf brassicas** (off-label), **lettuce** (off-label), **parsley** (off-label), **protected cress** (off-label), **protected forest nurseries** (off-label), **protected herbs (see appendix 6)** (off-label), **protected hops** (off-label), **protected lamb's lettuce** (off-label), **protected leaf brassicas** (off-label), **protected parsley** (off-label), **protected rocket** (off-label), **protected soft fruit** (off-label), **rocket** (off-label), **soft fruit** (off-label), **spinach** (off-label), **spinach beet** (off-label), **top fruit** (off-label) [2]
- Common oak thelaxid in **forest nurseries** (off-label), **protected forest nurseries** (off-label) [2]
- Gall wasp in **forest nurseries** (off-label), **protected forest nurseries** (off-label) [2]
- Great spruce bark beetle in **forest nurseries** (off-label), **protected forest nurseries** (off-label) [2]
- Green aphid in **forest nurseries** (off-label), **protected forest nurseries** (off-label) [2]
- Large pine weevil in **forest** (off-label) [2]
- Pine weevil in **forest nurseries** (off-label) [2]
- Whitefly in **apples, cherries, ornamental plant production, pears, plums, protected aubergines, protected ornamentals, protected peppers, protected tomatoes** [1, 2, 4-6]; **brussels sprouts** (off-label) [3]

- Woolly aphid in *forest nurseries* (off-label), *protected forest nurseries* (off-label) [2]

Extension of Authorisation for Minor Use (EAMUs)
- *brussels sprouts* 20072866 [3]
- *chard* 20113144 [2]
- *cress* 20192251 [2]
- *forest* 20121068 [2]
- *forest nurseries* 20111313 [2], 20160084 [2], 20162653 [2]
- *herbs (see appendix 6)* 20192251 [2]
- *hops* 20082857 [2]
- *lamb's lettuce* 20192250 [2]
- *leaf brassicas* 20192250 [2]
- *lettuce* 20192250 [2]
- *parsley* 20192250 [2]
- *protected cress* 20192251 [2]
- *protected forest nurseries* 20162653 [2]
- *protected herbs (see appendix 6)* 20192251 [2]
- *protected hops* 20082857 [2]
- *protected lamb's lettuce* 20192250 [2]
- *protected leaf brassicas* 20192250 [2]
- *protected parsley* 20192250 [2]
- *protected rocket* 20192250 [2]
- *protected soft fruit* 20082857 [2]
- *rocket* 20192250 [2]
- *soft fruit* 20082857 [2]
- *spinach* 20192251 [2]
- *spinach beet* 20113144 [2]
- *top fruit* 20082857 [2]

Approval information
- Acetamiprid included in Annex I under EC Regulation 1107/2009

Efficacy guidance
- Best results obtained from application at the first sign of pest attack or when appropriate thresholds are reached
- Thorough coverage of foliage is essential to ensure best control. Acetamiprid has contact, systemic and translaminar activity

Restrictions
- Maximum number of treatments 1 per yr for cherries, ware potatoes; 2 per yr or crop for apples, pears, plums, ornamental plant production, protected crops, seed potatoes
- Do not use more than two applications of any neonicotinoid insecticide (e.g. acetamiprid, clothianidin, imidacloprid, thiacloprid) on any crop. Previous soil or seed treatment with a neonicotinoid counts as one such treatment
- 2 treatments are permitted on seed potatoes but must not be used consecutively [3]

Crop-specific information
- HI 3 d for protected crops; 14 d for apples, cherries, pears, plums, potatoes

Environmental safety
- Harmful to aquatic organisms
- Acetamiprid is slightly toxic to predatory mites and generally slightly toxic to other beneficials
- Broadcast air-assisted LERAP [1, 2, 4-6] (18 m); LERAP Category B [1-6]

Hazard classification and safety precautions
    Hazard Dangerous for the environment [1-6]; Harmful if swallowed [1, 2, 4-6]
    UN Number N/C
    Operator protection A [1-6]; H [1, 2, 4-6]; U02a, U19a [1-6]; U20a [3]; U20b [1, 2, 4-6]
    Environmental protection E15b, E16a, E16b [1-6]; E17b [1, 2, 4-6] (18 m); E38 [3]; H410 [1, 2, 4, 6]; H412 [3, 5]
    Storage and disposal D05 [1, 2, 4-6]; D09a, D10b [1-6]; D12a [3]

3   acetic acid

A non-selective herbicide for non-crop situations

Products

New-Way Weed Spray	Headland Amenity	240 g/l	SL	15319

Uses

- General weed control in **hard surfaces**, **natural surfaces not intended to bear vegetation**, **permeable surfaces overlying soil**

Approval information

- Acetic acid is included in Annex 1 under EC Regulation 1107/2009

Efficacy guidance

- Best results obtained from treatment of young tender weeds less than 10 cm high
- Treat in spring and repeat as necessary throughout the growing season
- Treat survivors as soon as fresh growth is seen
- Ensure complete coverage of foliage to the point of run-off
- Rainfall after treatment may reduce efficacy

Restrictions

- No restriction on number of treatments

Following crops guidance

- There is no residual activity in the soil and sowing or planting may take place as soon as treated weeds have died

Environmental safety

- Harmful to aquatic organisms
- High risk to bees
- Do not apply to crops in flower or to those in which bees are actively foraging. Do not apply when flowering weeds are present
- Keep people and animals off treated dense weed patches until spray has dried. This is not necessary for areas with occasional low growing prostrate weeds such as on pathways

Hazard classification and safety precautions

  Hazard Irritant
  UN Number N/C
  Risk phrases R36, R37, R38, R52
  Operator protection A, C, D, H; U05a, U09a, U19a, U20b
  Environmental protection E12a, E12e, E15a
  Storage and disposal D01, D02, D09a, D12a

4   aclonifen

A pre-emergence herbicide for broad-leaf and grass weed control in potatoes
HRAC mode of action code: F3

Products

Emerger	Bayer CropScience	600 g/l	SC	19056

Uses

- Annual dicotyledons in **potatoes**
- Black bindweed in **caraway** *(off-label)*, **carrots** *(off-label)*, **dill** *(off-label)*, **garlic** *(off-label)*, **onions** *(off-label)*, **parsley** *(off-label)*, **parsley root** *(off-label)*, **parsnips** *(off-label)*, **shallots** *(off-label)*, **sunflowers** *(off-label)*
- Chickweed in **caraway** *(off-label)*, **carrots** *(off-label)*, **dill** *(off-label)*, **garlic** *(off-label)*, **onions** *(off-label)*, **parsley** *(off-label)*, **parsley root** *(off-label)*, **parsnips** *(off-label)*, **shallots** *(off-label)*, **sunflowers** *(off-label)*

- Cleavers in **caraway** *(off-label)*, **carrots** *(off-label)*, **dill** *(off-label)*, **garlic** *(off-label)*, **onions** *(off-label)*, **parsley** *(off-label)*, **parsley root** *(off-label)*, **parsnips** *(off-label)*, **shallots** *(off-label)*, **sunflowers** *(off-label)*
- Crane's-bill in **caraway** *(off-label)*, **carrots** *(off-label)*, **dill** *(off-label)*, **garlic** *(off-label)*, **onions** *(off-label)*, **parsley** *(off-label)*, **parsley root** *(off-label)*, **parsnips** *(off-label)*, **shallots** *(off-label)*, **sunflowers** *(off-label)*
- Fat hen in **caraway** *(off-label)*, **carrots** *(off-label)*, **dill** *(off-label)*, **garlic** *(off-label)*, **onions** *(off-label)*, **parsley** *(off-label)*, **parsley root** *(off-label)*, **parsnips** *(off-label)*, **potatoes**, **shallots** *(off-label)*, **sunflowers** *(off-label)*
- Knotgrass in **caraway** *(off-label)*, **carrots** *(off-label)*, **dill** *(off-label)*, **garlic** *(off-label)*, **onions** *(off-label)*, **parsley** *(off-label)*, **parsley root** *(off-label)*, **parsnips** *(off-label)*, **shallots** *(off-label)*, **sunflowers** *(off-label)*
- Mayweeds in **caraway** *(off-label)*, **carrots** *(off-label)*, **dill** *(off-label)*, **garlic** *(off-label)*, **onions** *(off-label)*, **parsley** *(off-label)*, **parsley root** *(off-label)*, **parsnips** *(off-label)*, **potatoes**, **shallots** *(off-label)*, **sunflowers** *(off-label)*
- Redshank in **caraway** *(off-label)*, **carrots** *(off-label)*, **dill** *(off-label)*, **garlic** *(off-label)*, **onions** *(off-label)*, **parsley** *(off-label)*, **parsley root** *(off-label)*, **parsnips** *(off-label)*, **shallots** *(off-label)*, **sunflowers** *(off-label)*
- Shepherd's purse in **caraway** *(off-label)*, **carrots** *(off-label)*, **dill** *(off-label)*, **garlic** *(off-label)*, **onions** *(off-label)*, **parsley** *(off-label)*, **parsley root** *(off-label)*, **parsnips** *(off-label)*, **shallots** *(off-label)*, **sunflowers** *(off-label)*
- Small nettle in **caraway** *(off-label)*, **carrots** *(off-label)*, **dill** *(off-label)*, **garlic** *(off-label)*, **onions** *(off-label)*, **parsley** *(off-label)*, **parsley root** *(off-label)*, **parsnips** *(off-label)*, **shallots** *(off-label)*, **sunflowers** *(off-label)*
- Sowthistle in **caraway** *(off-label)*, **carrots** *(off-label)*, **dill** *(off-label)*, **garlic** *(off-label)*, **onions** *(off-label)*, **parsley** *(off-label)*, **parsley root** *(off-label)*, **parsnips** *(off-label)*, **shallots** *(off-label)*, **sunflowers** *(off-label)*
- Speedwells in **caraway** *(off-label)*, **carrots** *(off-label)*, **dill** *(off-label)*, **garlic** *(off-label)*, **onions** *(off-label)*, **parsley** *(off-label)*, **parsley root** *(off-label)*, **parsnips** *(off-label)*, **shallots** *(off-label)*, **sunflowers** *(off-label)*
- Volunteer oilseed rape in **caraway** *(off-label)*, **carrots** *(off-label)*, **dill** *(off-label)*, **garlic** *(off-label)*, **onions** *(off-label)*, **parsley** *(off-label)*, **parsley root** *(off-label)*, **parsnips** *(off-label)*, **shallots** *(off-label)*, **sunflowers** *(off-label)*

Extension of Authorisation for Minor Use (EAMUs)
- **caraway** *20191617*
- **carrots** *20191601*
- **dill** *20191617*
- **garlic** *20191616*
- **onions** *20191616*
- **parsley** *20191617*
- **parsley root** *20191601*
- **parsnips** *20191601*
- **shallots** *20191616*
- **sunflowers** *20191615*

Approval information
- Aclonifen included in Annex I under EC Regulation 1107/2009

Efficacy guidance
- DO NOT disrupt the soil surface after application as this will reduce the level of weed control provided

Restrictions
- Horizontal boom sprayers must be fitted with three star drift reduction technology for all uses
- Low drift spraying equipment must be operated according to the specific conditions stated in the official three star rating for that equipment as published on HSE Chemicals Regulation Division's website. These operating conditions must be maintained until the operator is 30m from the top of the bank of any surface water bodies

SEE SECTION 3 FOR PRODUCTS ALSO REGISTERED

SECTION 2

Following crops guidance
- Oilseed rape, field pea, dwarf french bean, wheat, barley, oat, winter triticale, winter rye, sugar beet, maize, and sunflower can be established in the normal rotation or in the event of failure of the treated crop, provided that cultivation (using cultivator, disc harrow or similar) and thorough mixing of treated soil has been conducted to at least 10 cm depth.
- Radish and phacelia require a period of at least 120 days after application, or plough to at least 20 cm before drilling.

Environmental safety
- Aquatic buffer zone requirement 6 m
- LERAP Category B

Hazard classification and safety precautions
Hazard Very toxic to aquatic organisms
Transport code 9
Packaging group III
UN Number 3082
Risk phrases H351
Operator protection A; U05a, U20b
Environmental protection E16a, H410
Storage and disposal D01, D02, D05, D09a, D10b

5   alpha-cypermethrin

A contact and ingested pyrethroid insecticide for use in arable crops and agricultural buildings
IRAC mode of action code: 3

Products

Al-cyper Ec	Harvest	100 g/l	EC	17801

Uses
- Cabbage seed weevil in *spring oilseed rape, winter oilseed rape*
- Caterpillars in *broccoli, brussels sprouts, cabbages, calabrese, cauliflowers, kale*
- Cereal aphid in *spring barley, spring wheat, winter barley, winter wheat*
- Flea beetle in *broccoli, brussels sprouts, cabbages, calabrese, cauliflowers, kale*
- Pea and bean weevil in *broad beans, combining peas, spring field beans, vining peas, winter field beans*
- Pea aphid in *combining peas* (reduction), *vining peas* (reduction)
- Pea moth in *combining peas, vining peas*
- Pollen beetle in *spring oilseed rape, winter oilseed rape*

Approval information
- Alpha-cypermethrin included in Annex I under EC Regulation 1107/2009
- Accepted by BBPA for use on malting barley

Efficacy guidance
- For cabbage stem flea beetle control spray oilseed rape when adult or larval damage first seen and about 1 mth later
- For flowering pests on oilseed rape apply at any time during flowering, against pollen beetle best results achieved at green to yellow bud stage (GS 3.3-3.7), against seed weevil between 20 pods set and 80% petal fall (GS 4.7-5.8)
- Spray cereals in autumn for control of cereal aphids, in spring/summer for grain aphids. (See label for details)
- For flea beetle, caterpillar and cabbage aphid control on brassicas apply when the pest or damage first seen or as a preventive spray. Repeat if necessary
- For pea and bean weevil control in peas and beans apply when pest attack first seen and repeat as necessary

Restrictions
- Apply up to 2 sprays on cereals in autumn and spring, 1 in summer between 1 Apr and 31 Aug. See label for details of rates and maximum total dose

FOR FULL CONDITIONS OF USE ALWAYS READ THE PRODUCT LABEL

- Only 1 aphicide treatment may be applied in cereals between 1 Apr and 31 Aug and spray volume must not be reduced in this period
- Do not apply to a cereal crop if any product containing a pyrethroid or dimethoate has been applied after the start of ear emergence (GS 51)
- Use low drift spraying equipment up to 30m from the top of the bank of any surface water bodies [1]
- Horizontal boom sprayers must be fitted with three star drift reduction technology for all uses

Crop-specific information
- Latest use: before the end of flowering for oilseed rape; before 31 Aug for cereals
- HI vining peas 1 d; brassicas, combining peas, broad beans, field beans 7 d
- For summer cereal application do not spray within 6 m from edge of crop and do not reduce volume when used after 31 Mar

Environmental safety
- Very toxic to aquatic organisms
- Extremely dangerous to fish or other aquatic life. Do not contaminate surface waters or ditches with chemical or used container
- Dangerous to bees. Where possible spray oilseed rape crops in the late evening or early morning or in dull weather. Give local beekeepers warning when using in flowering crops
- Do not spray within 6 m of the edge of a cereal crop after 31 Mar in yr of harvest
- Buffer zone requirement 12m [1]
- To protect non target insects/arthropods respect an unsprayed buffer zone of 5 m to non-crop land [1]
- Horizontal boom sprayers must be fitted with three star drift reduction technology for all uses
- LERAP Category A

Hazard classification and safety precautions
Hazard Harmful, Flammable, Dangerous for the environment, Flammable liquid and vapour, Toxic if swallowed, Harmful if inhaled, Very toxic to aquatic organisms
Transport code 6.1
Packaging group III
UN Number 2903
Risk phrases H304, H315, H317, H318, H335, H336, H373
Operator protection A, H; U05a, U10, U19c, U20b
Environmental protection E12c, E15a, E16c, E16d, E38, H410
Storage and disposal D01, D02, D05, D09a, D10b, D12a
Medical advice M05a, M05b

6   aluminium ammonium sulphate

An inorganic bird and animal repellent

Products

1	Asbo	Sphere	83 g/l	SC	19186
2	Curb Liquid Crop Spray	Sphere	83.3 g/l	LI	18643

Uses
- Animal repellent in *amenity grassland, beans without pods (dry), beans without pods (fresh), bilberries, blackberries, blackcurrants, blueberries, broad beans, broccoli, brussels sprouts, cabbages, calabrese, carrots, cauliflowers, combining peas, cranberries, durum wheat, dwarf beans, french beans, gooseberries, loganberries, managed amenity turf, raspberries, redcurrants, rubus hybrids, runner beans, rye, soya beans, spring barley, spring field beans, spring oats, spring oilseed rape, spring wheat, strawberries, sugar beet, triticale, vining peas, winter barley, winter field beans, winter oats, winter oilseed rape, winter wheat*
- Bird repellent in *amenity grassland, beans without pods (dry), beans without pods (fresh), bilberries, blackberries, blackcurrants, blueberries, broad beans, broccoli, brussels sprouts, cabbages, calabrese, carrots, cauliflowers, combining peas, cranberries, durum wheat, dwarf beans, french beans, gooseberries, loganberries, managed amenity turf,*

raspberries, redcurrants, rubus hybrids, runner beans, rye, soya beans, spring barley, spring field beans, spring oats, spring oilseed rape, spring wheat, strawberries, sugar beet, triticale, vining peas, winter barley, winter field beans, winter oats, winter oilseed rape, winter wheat

Approval information
- aluminium ammonium sulphate is included in Annex 1 under EC Regulation 1107/2009

Efficacy guidance
- Apply as overall spray to growing crops before damage starts or mix powder with seed depending on type of protection required
- Spray deposit protects growth present at spraying but gives little protection to new growth
- Product must be sprayed onto dry foliage to be effective and must dry completely before dew or frost forms. In winter this may require some wind

Crop-specific information
- Latest use: no restriction

Hazard classification and safety precautions
UN Number N/C
Operator protection A; U05a, U20a
Environmental protection E15a, E19b
Storage and disposal D01, D02, D05, D11a, D12a
Medical advice M03

7   aluminium phosphide

A phosphine generating compound used against vertebrates and grain store pests. Under new regulations any person wishing to purchase or use Aluminium Phosphide must be suitably qualified and licensed by January 2015. A new licensing body has been set up.
IRAC mode of action code: 24A

Products

1	Degesch Fumigation Tablets	Rentokil	56% w/w	GE 17035
2	Detia Gas-Ex-B	Rentokil	57% w/w	GE 17097
3	Phostoxin	Rentokil	56% w/w	GE 17000
4	Talunex	Killgerm	56% w/w	GE 17001

Uses
- Insect pests in *carob, cocoa, coffee, herbal infusions, stored dried spices, stored grain, stored linseed, stored nuts, stored oilseed rape, stored pulses, tea, tobacco* [1]; *coconuts, dried fruit, grain stores, nuts, pulses* [2]; *crop handling & storage structures, processed consumable products* [1, 2]
- Moles in *all situations* [3, 4]; *farmland* [3]
- Rabbits in *all situations* [3, 4]; *farmland* [3]
- Rats in *all situations* [3, 4]; *farmland* [3]

Approval information
- Aluminium phosphide is included in Annex 1 under EC Regulation 1107/2009
- Accepted by BBPA for use in stores for malting barley

Efficacy guidance
- Product releases poisonous hydrogen phosphide gas in contact with moisture
- Place fumigation tablets in grain stores as directed [1, 2]
- Place pellets in burrows or runs and seal hole by heeling in or covering with turf. Do not cover pellets with soil. Inspect daily and treat any new or re-opened holes [3, 4]

Restrictions
- Aluminium phosphide is subject to the Poisons Rules 1982 and the Poisons Act 1972. See Section 5 for more information

FOR FULL CONDITIONS OF USE ALWAYS READ THE PRODUCT LABEL

- Only to be used by operators instructed or trained in the use of aluminium phosphide and familiar with the precautionary measures to be taken. See label and HSE Guidance Notes for full precautions
- Only open container outdoors [3, 4] and for immediate use. Keep away from liquid or water as this causes immediate release of gas. Do not use in wet weather
- Do not use within 3 m of human or animal habitation. Before application ensure that no humans or domestic animals are in adjacent buildings or structures. Allow a minimum airing-off period of 4 h before re-admission

Environmental safety
- Product liberates very toxic, highly flammable gas
- Dangerous to fish or other aquatic life. Do not contaminate surface waters or ditches with chemical or used container
- Prevent access by livestock, pets and other non-target mammals and birds to buildings under fumigation and ventilation
- Pellets must never be placed or allowed to remain on ground surface
- Do not use adjacent to watercourses
- Take particular care to avoid gassing non-target animals, especially those protected under the Wildlife and Countryside Act (e.g. badgers, polecat, reptiles, natterjack toads, most birds). Do not use in burrows where there is evidence of badger or fox activity, or when burrows might be occupied by birds
- Dust remaining after decomposition is harmless and of no environmental hazard
- Keep in original container, tightly closed, in a safe place, under lock and key
- Dispose of empty containers as directed on label

Hazard classification and safety precautions
  Hazard Very toxic, In contact with water releases flammable gases which may ignite spontaneously, Fatal if swallowed, Toxic in contact with skin, Fatal if inhaled, Very toxic to aquatic organisms [1-4]; Harmful [2-4]; Highly flammable [2, 3]; Dangerous for the environment [1-3]
  Transport code 4.1 6.1 [4]; 4.3 [1-3]
  Packaging group I
  UN Number 1397
  Risk phrases H315, H318 [1, 3, 4]
  Operator protection A, D, H; U01, U07, U13, U19a, U20a [1-4]; U05a [2-4]; U05b, U18 [1]
  Environmental protection E02a [1] (4 h min); E02a [2-4] (4 h); E02b [1]; E13b, E34 [1-4]
  Storage and disposal D01, D02, D07, D09b [1-4]; D11b [2-4]; D11c [1]
  Vertebrate/rodent control products V04a [2-4]
  Medical advice M04a

8   ametoctradin

A potato blight fungicide
FRAC mode of action code: 45

See also ametoctradin + dimethomorph
*   ametoctradin + mancozeb*

Products

Enervin SC	BASF	200 g/l	SC	19095

Uses
- Late blight in **potatoes**
- White tip in **leeks**

Approval information
- Ametoctradin is included in Annex 1 under EC Regulation 1107/2009

Efficacy guidance
- Most effective when applied as a protectant spray before the risk of late blight in potato or white tip in leek occurs.
- Can be used on all varieties of potato, including seed crops

- Will currently give effective protection against phenylamide resistant strains of late blight.

Restrictions
- In potato apply no more than 3 consecutive applications and no more than a total of 4 applications of Enervin SC per season should be applied.
- In leek no more than a total of 2 applications per season should be applied.
- No more than 6 applications of a QoSI fungicide, making up no more than half of the total number of intended late blight sprays per season, should be made.

Hazard classification and safety precautions
    Transport code 9
    Packaging group III
    UN Number 3082
    Operator protection U05a
    Environmental protection E15b, E34, H411
    Storage and disposal D01, D02, D09a, D10c, D12c
    Medical advice M03

9   ametoctradin + dimethomorph

A systemic and protectant fungicide for potato blight control
FRAC mode of action code: 40 + 45

Products

1 Percos	BASF	300:225 g/l	SC	15248
2 Zampro DM	BASF	300:225 g/l	SC	15013

Uses
- Blight in **potatoes** [1, 2]
- Downy mildew in **hops** *(off-label)*, **ornamental plant production** *(off-label)*, **protected ornamentals** *(off-label)*, **table grapes** *(off-label)*, **wine grapes** *(off-label)* [1]
- Tuber blight in **potatoes** *(reduction)* [1, 2]

Extension of Authorisation for Minor Use (EAMUs)
- **hops** 20170556 [1]
- **ornamental plant production** 20130819 [1]
- **protected ornamentals** 20130819 [1]
- **table grapes** 20150254 [1]
- **wine grapes** 20150254 [1]

Approval information
- Ametoctradin and dimethomorph are included in Annex 1 under EC Regulation 1107/2009.

Efficacy guidance
- Application to dry foliage is rainfast within 1 hour of drying on the leaf.
- Will give effective control of phenylamide-resistant strains of blight

Crop-specific information
- HI 7 days for potatoes

Hazard classification and safety precautions
    Hazard Harmful, Dangerous for the environment, Harmful if swallowed
    UN Number N/C
    Operator protection U05a
    Environmental protection E15b, E34, E38, H410
    Storage and disposal D01, D02, D09a, D10c
    Medical advice M03, M05a

10   amidosulfuron

A post-emergence sulfonylurea herbicide for cleavers and other broad-leaved weed control in cereals
HRAC mode of action code: B

Products

| 1 | Eagle | Sumitomo | 75% w/w | WG | 16490 |
| 2 | Squire Ultra | Sumitomo | 75% w/w | WG | 16491 |

Uses
- Annual dicotyledons in *durum wheat, farm forestry* (off-label), *forest nurseries* (off-label), *game cover* (off-label), *linseed, spring barley, spring oats, spring rye, spring wheat, triticale, winter barley, winter oats, winter rye, winter wheat* [1]; *grassland* [2]; *ornamental plant production* (off-label) [1, 2]
- Charlock in *corn gromwell* (off-label), *durum wheat, farm forestry* (off-label), *forest nurseries* (off-label), *game cover* (off-label), *linseed, spring barley, spring oats, spring rye, spring wheat, triticale, winter barley, winter oats, winter rye, winter wheat* [1]; *ornamental plant production* (off-label) [1, 2]
- Cleavers in *corn gromwell* (off-label), *durum wheat, farm forestry* (off-label), *forest nurseries* (off-label), *game cover* (off-label), *linseed, spring barley, spring oats, spring rye, spring wheat, triticale, winter barley, winter oats, winter rye, winter wheat* [1]; *ornamental plant production* (off-label) [1, 2]
- Docks in *grassland* [2]
- Forget-me-not in *corn gromwell* (off-label), *durum wheat, farm forestry* (off-label), *forest nurseries* (off-label), *game cover* (off-label), *linseed, spring barley, spring oats, spring rye, spring wheat, triticale, winter barley, winter oats, winter rye, winter wheat* [1]; *ornamental plant production* (off-label) [1, 2]
- Shepherd's purse in *corn gromwell* (off-label), *durum wheat, farm forestry* (off-label), *forest nurseries* (off-label), *game cover* (off-label), *linseed, spring barley, spring oats, spring rye, spring wheat, triticale, winter barley, winter oats, winter rye, winter wheat* [1]; *ornamental plant production* (off-label) [1, 2]

Extension of Authorisation for Minor Use (EAMUs)
- *corn gromwell* 20150903 [1]
- *farm forestry* 20142508 [1]
- *forest nurseries* 20142508 [1]
- *game cover* 20142508 [1]
- *ornamental plant production* 20142508 [1], 20142511 [2]

Approval informationAccepted by BBPA for use on malting barley
- Amidosulfuron included in Annex I under EC Regulation 1107/2009

Efficacy guidance
- For best results apply in spring (from 1 Feb) in warm weather when soil moist and weeds growing actively. When used in grassland following cutting or grazing, docks should be allowed to regrow before treatment
- Weed kill is slow, especially under cool, dry conditions. Weeds may sometimes only be stunted but will have little or no competitive effect on crop
- May be used on all soil types unless certain sequences are used on linseed. See label
- Spray is rainfast after 1 h
- Cleavers controlled from emergence to flower bud stage. If present at application charlock (up to flower bud), shepherds purse (up to flower bud) and field forget-me-not (up to 6 leaves) will also be controlled
- Amidosulfuron is a member of the ALS-inhibitor group of herbicides and products should be used in a planned Resistance Management strategy. See Section 5 for more information

Restrictions
- Maximum number of treatments 1 per crop [1]
- Use after 1 Feb and do not apply to rotational grass after 30 Jun, or to permanent grassland after 15 Oct [2]

SEE SECTION 3 FOR PRODUCTS ALSO REGISTERED

- Do not apply to crops undersown or due to be undersown with clover or alfalfa [1]
- Do not spray crops under stress, suffering drought, waterlogged, grazed, lacking nutrients or if soil compacted
- Do not spray if frost expected
- Do not roll or harrow within 1 wk of spraying
- Specific restrictions apply to use in sequence or tank mixture with other sulfonylurea or ALS-inhibiting herbicides. See label for details. There are no recommendations for mixtures with metsulfuron-methyl products on linseed
- Certain mixtures with fungicides are expressly forbidden. See label for details
- Livestock must be kept out of treated areas for at least 7 days weeks following treatment and until poisonous weeds such as ragwort have died and become unpalatable

Crop-specific information
- Latest use: before first spikelets just visible (GS 51) for cereals; before flower buds visible for linseed; 15 Oct for grassland
- Broadcast cereal crops should be sprayed post-emergence after plants have a well established root system

Following crops guidance
- If a treated crop fails cereals may be sown after 15 d and thorough cultivation
- After normal harvest of a treated crop only cereals, winter oilseed rape, mustard, turnips, winter field beans or vetches may be sown in the same year as treatment and these must be preceded by ploughing or thorough cultivation
- Only cereals may be sown within 12 mth of application to grassland [2]
- Cereals or potatoes must be sown as the following crop after use of permitted mixtures or sequences with other sulfonylurea herbicides in cereals. Only cereals may be sown after the use of such sequences in linseed

Environmental safety
- Take care to wash out sprayers thoroughly. See label for details
- Avoid drift onto neighbouring broad-leaved plants or onto surface waters or ditches

Hazard classification and safety precautions
Hazard Dangerous for the environment
Transport code 9
Packaging group III
UN Number 3077
Operator protection U20a [1]; U20b [2]
Environmental protection E07b [2] (1 week); E15a, E38, H410 [1, 2]; E41 [2]
Storage and disposal D10a, D12a

11   amidosulfuron + iodosulfuron-methyl-sodium

A post-emergence sulfonylurea herbicide mixture for cereals
HRAC mode of action code: B + B

See also iodosulfuron-methyl-sodium

Products
1 Chekker	Bayer CropScience	12.5:1.25% w/w	WG	16495
2 Sekator OD	Sumitomo	100:25 g/l	OD	16494

Uses
- Annual dicotyledons in *forest nurseries* (off-label) [1]; **spring barley, spring rye, spring wheat, triticale, winter barley, winter rye, winter wheat** [1, 2]
- Chickweed in *forest nurseries* (off-label) [1]; **spring barley, spring rye, spring wheat, triticale, winter barley, winter rye, winter wheat** [1, 2]
- Cleavers in *forest nurseries* (off-label) [1]; **spring barley, spring rye, spring wheat, triticale, winter barley, winter rye, winter wheat** [1, 2]
- Mayweeds in *forest nurseries* (off-label) [1]; **spring barley, spring rye, spring wheat, triticale, winter barley, winter rye, winter wheat** [1, 2]

FOR FULL CONDITIONS OF USE ALWAYS READ THE PRODUCT LABEL

- Volunteer oilseed rape in **forest nurseries** *(off-label)* [1]; **spring barley**, **spring rye**, **spring wheat**, **triticale**, **winter barley**, **winter rye**, **winter wheat** [1, 2]

Extension of Authorisation for Minor Use (EAMUs)
- **forest nurseries** *20142722* [1]

Approval information
- Amidosulfuron and iodosulfuron-methyl-sodium included in Annex I under EC Regulation 1107/2009
- Accepted by BBPA for use on malting barley

Efficacy guidance
- Best results obtained from treatment in warm weather when soil is moist and the weeds are growing actively
- Weeds must be present at application to be controlled
- Dry conditions resulting in moisture stress may reduce effectiveness
- Weed control is slow especially under cool dry conditions
- Occasionally weeds may only be stunted but they will normally have little or no competitive effect on the crop
- Amidosulfuron and iodosulfuron are members of the ALS-inhibitor group of herbicides and products should be used in a planned Resistance Management strategy. See Section 5 for more information

Restrictions
- Maximum number of treatments 1 per crop
- Must only be applied between 1 Feb in yr of harvest and specified latest time of application
- Do not apply to crops undersown or to be undersown with grass, clover or alfalfa
- Do not roll or harrow within 1 wk of spraying
- Do not spray crops under stress from any cause or if the soil is compacted
- Do not spray if rain or frost expected
- Do not apply in mixture or in sequence with any other ALS inhibitor

Crop-specific information
- Latest use: before first spikelet of inflorescence just visible (GS 51)
- Treat drilled crops after the 2-leaf stage; treat broadcast crops after the plants have a well-established root system
- Applications to spring barley may cause transient crop yellowing

Following crops guidance
- Cereals, winter oilseed rape and winter field beans may be sown in the same yr as treatment provided they are preceded by ploughing or thorough cultivation. Any crop may be sown in the spring of the yr following treatment
- A minimum of 3 mth must elapse between treatment and sowing winter oilseed rape

Environmental safety
- Dangerous for the environment
- Toxic to aquatic organisms
- Take extreme care to avoid damage by drift onto broad-leaved plants outside the target area or onto ponds, waterways and ditches
- Observe carefully label instructions for sprayer cleaning
- LERAP Category B

Hazard classification and safety precautions
Hazard Irritant, Dangerous for the environment [1, 2]; Very toxic to aquatic organisms [1]
Transport code 9
Packaging group III
UN Number 3077 [1]; 3082 [2]
Risk phrases H317 [2]; H319 [1, 2]
Operator protection A, C, H; U05a, U08, U11, U14, U15, U20b
Environmental protection E15a, E16a, E16b, E38, H410
Storage and disposal D01, D02, D10a, D12a

SEE SECTION 3 FOR PRODUCTS ALSO REGISTERED

12 amidosulfuron + iodosulfuron-methyl-sodium + mesosulfuron-methyl

A herbicide mixture for weed control in winter wheat
HRAC mode of action code: B + B + B

See also amidosulfuron
amidosulfuron + iodosulfuron-methyl-sodium
iodosulfuron-methyl-sodium
mesosulfuron-methyl

Products

Pacifica Plus	Bayer CropScience	0.5:0.1:0.3% w/w	WG	17272

Uses
- Annual dicotyledons in **winter wheat**
- Annual grasses in **winter wheat**
- Blackgrass in **winter wheat**

Approval information
- Amidosulfuron, iodosulfuron-methyl-sodium and mesosulfuron-methyl included in Annex I under EC Regulation 1107/2009
- Accepted by BBPA for use on malting barley

Efficacy guidance
- Apply as early as possible in spring and before stem extension stage of any grass weeds
- Monitor efficacy and investigate areas of poor control

Restrictions
- Do not apply in tank-mixture or sequence with any product containing any other sulfonylurea or ALS inhibiting herbicide
- Do not use on crops undersown with grasses, clover, legumes or any other broad-leaved crop
- Take care to avoid drift on to non-target plants

Following crops guidance
- Winter wheat and winter barley may be sown in the same year as harvest of a crop treated with 0.5 kg/ha. Winter oilseed rape may be sown after ploughing following a crop treated with 0.4 kg/ha. In the event of crop failure, sow only winter or spring wheat.

Environmental safety
- LERAP Category B

Hazard classification and safety precautions
Hazard Very toxic to aquatic organisms
Transport code 9
Packaging group III
UN Number 3077
Risk phrases H317, H318
Operator protection A, C, H; U05a, U11, U20a
Environmental protection E15a, E16a, H410
Storage and disposal D01, D02, D05, D09a, D10b

13 aminopyralid

A pyridine carboxylic acid herbicide for weed control in grassland
HRAC mode of action code: O

Products

Pro-Banish	Dow (Corteva)	30 g/l	SL	18339

Uses
- Annual dicotyledons in **grassland**

FOR FULL CONDITIONS OF USE ALWAYS READ THE PRODUCT LABEL

Approval information
- Approvals of aminopyralid products have now been restored with extra advice stating that manure from animals fed on pasture treated with aminopyralid should not leave the farm.
- Aminopyralid is included in Annex 1 under 1107/2009.

Restrictions
- Livestock must be kept out of treated areas for at least 1 week following treatment and until poisonous weeds such as ragwort have died and become unpalatable.
- To protect groundwater do not apply to leys less than 1 year old.
- Users must have received adequate instruction, training and guidance in the safe and efficient use of the product and must take all reasonable precautions to protect the health of human beings, creatures and plants and safeguard the environment.
- This product must not be used on grassland grazed by animals other than cattle and sheep.
- The product must not be used on land where vegetation will be cut for animal feed, fodder or bedding nor for composting or mulching within one calendar year of treatment.

Following crops guidance
- In the event of failure of newly seeded treated grassland, grass may be re-seeded immediately or wheat may be sown provided 4 mth have elapsed since application
- Residues in plant tissues, including manure, may affect succeeding susceptible crops of peas, beans, other legumes, carrots, other Umbelliferae, potatoes, tomatoes, lettuce and other Compositae. These crops should not be sown within 3 mth of ploughing up treated grassland

Hazard classification and safety precautions
Hazard Dangerous for the environment
UN Number N/C
Operator protection A, H; U05a, U08, U11, U14, U23a
Environmental protection E07a, E15b, E38
Consumer protection C01
Storage and disposal D01, D02, D05, D09a, D10b, D12a

14   aminopyralid + fluroxypyr

A foliar acting herbicide mixture for use in grassland
HRAC mode of action code: O + O

See also fluroxypyr

Products
Synero	Dow (Corteva)	30:100 g/l	EW	18578

Uses
- Buttercups in *amenity grassland*
- Chickweed in *amenity grassland*
- Dandelions in *amenity grassland*
- Docks in *amenity grassland*
- Stinging nettle in *amenity grassland*
- Thistles in *amenity grassland*

Approval information
- Aminopyralid and fluroxypyr are included in Annex I under EC Regulation 1107/2009.

Efficacy guidance
- For best results and to avoid crop check, grass and weeds must be growing actively
- Allow 2-3 wk after cutting for hay or silage for sufficient regrowth to occur before spraying and leave 7 d afterwards to allow maximum translocation
- Where there is a high reservoir of weed seed or a historically high weed population a programmed approach may be needed involving a second treatment in the following yr
- Control may be reduced if rain falls within 1 h of spraying

Restrictions
- Maximum number of treatments 1 per yr.

SEE SECTION 3 FOR PRODUCTS ALSO REGISTERED

SECTION 2

- Do not apply to leys less than 1 year old.
- Do not apply by hand-held equipment.
- Do not use on grassland that will be used for animal feed, bedding, composting or mulching within 1 calender year of application.
- Do not use on grassland that will be grazed by animals other than cattle or sheep.
- Use of an antifoam is compulsory
- Do not use any treated plant material, or manure from animals fed on treated crops, for composting or mulching
- Do not use on crops grown for seed
- Manure from animals fed on pasture or silage treated with this herbicide should not leave the farm
- Do not use between 31st July and 1st March [1]
- For applications made between 1st March and 31st May, 1 application may be made per calendar year [1]
- For applications made between 1st June and 31st July, only 1 application may be made in a 2 year period [1]
- Do not apply spot applications to more than 20% of the area: if infestation levels exceed this figure then a broadcast application should be made [1]

Crop-specific information
- Treatment will kill clover
- Treatment may occasionally cause transient yellowing of the sward which is quickly outgrown
- Late treatments may lead to a slight transient leaning of grass that does not affect yield

Following crops guidance
- Do not drill clover or other legumes within 4 mth of treatment, or potatoes in the spring following treatment in the previous autumn
- In the event of failure of newly seeded treated grassland, grass may be re-seeded immediately, or wheat may be sown provided 4 mth have elapsed since application
- Residues in plant tissues, including manure, may affect succeeding susceptible crops of peas, beans, other legumes, carrots, other Umbelliferae, potatoes, tomatoes, lettuce and other Compositae. These crops should not be sown within 3 mth of ploughing up treated grassland

Environmental safety
- Dangerous for the environment
- Toxic to aquatic organisms
- Keep livestock out of treated areas for at least 7 d following treatment and until poisonous weeds, such as ragwort, have died down and become unpalatable
- Avoid damage by drift onto susceptible crops, non-target plants or waterways
- LERAP Category B

Hazard classification and safety precautions
Hazard Irritant, Dangerous for the environment, Very toxic to aquatic organisms
Transport code 9
Packaging group III
UN Number 3082
Risk phrases H304, H315, H318, H336
Operator protection A, C, H; U05a, U08, U11, U14, U23a
Environmental protection E07a, E15b, E16a, E38, H410
Consumer protection C01
Storage and disposal D01, D02, D05, D09a, D10b, D12a

15   aminopyralid + halauxifen-methyl

A herbicide mixture for use in cereals
HRAC mode of action code: O + O

See also aminopyralid
halauxifen-methyl

Products

Trezac	Dow (Corteva)	25:31.3 g/l	EC	18253

Uses
- Annual dicotyledons in **durum wheat**, **spelt**, **spring barley**, **spring wheat**, **triticale**, **winter barley**, **winter rye**, **winter wheat**

Approval information
- Aminopyralid and halauxifen-methyl included in Annex 1 under EC Regulation 1107/2009

Restrictions
- Applications to winter cereals must only be carried out after 1st March annual to prevent contamination of groundwater.
- Application must be made between February 1st and 31st May in the year of harvest for spring cereals
- This product must not be used on cereal crops where the straw is intended for animal feed, fodder or bedding, nor for composting or mulching.
- This product must only be used on cereal crops where the straw is to be fully incorporated back into the soil and will not leave the farm

Environmental safety
- LERAP Category B

Hazard classification and safety precautions
Hazard Harmful if swallowed, Very toxic to aquatic organisms
Transport code 9
Packaging group III
UN Number 3082
Risk phrases H315, H317, H318, H335
Operator protection A, H
Environmental protection E16a, H410

16   aminopyralid + metazachlor + picloram

A herbicide mixture for weed control in oilseed rape.
HRAC mode of action code: K3 + O

See also metazachlor

Products

Ralos	Dow (Corteva)	5.3:500:13.3 g/l	SC	16737

Uses
- Chickweed in **winter oilseed rape**
- Dead nettle in **winter oilseed rape**
- Fat hen in **winter oilseed rape**
- Penny cress in **winter oilseed rape**
- Poppies in **winter oilseed rape**
- Scentless mayweed in **winter oilseed rape**

Approval information
- Aminopyralid, metazachlor and picloram are included in Annex 1 under 1107/2009.

Restrictions
- Applications shall be limited to a total dose of not more than 1.0 kg metazachlor/ha in a three year period on the same field.
- Do not use between 1st October and 31st July
- Livestock must be kept out of treated areas for at least 1 week following treatment and until poisonous weeds such as ragwort have died and become unpalatable
- Do not use on sands and very light soils or on soils with more than 10% organic matter.
- Straw from treated crops must not be baled and must stay on the field.
- Do not use treated plant material for composting or mulching.
- Do not use digestate from anaerobic digesters on crops such as peas, beans and other legumes, carrots and other Umbelliferae, potatoes, lettuce or other Compositae, glasshouse or protected crops.

Following crops guidance
- Only wheat, barley, oats, maize or oilseed rape should be planted within 4 months (120 days) of application. All other crops should wait until 12 months after application.
- In the event of crop failure, only spring wheat, spring barley, spring oats, maize or ryegrass should be planted after ploughing or thorough cultivation of the soil.

Environmental safety
- Metazachlor stewardship guidelines advise a maximum dose of 750 g.a.i/ha/annum. Applications to drained fields should be complete by 15th Oct but, if drains are flowing, complete applications by 1st Oct.
- LERAP Category B

Hazard classification and safety precautions

Hazard Dangerous for the environment
Transport code 9
Packaging group III
UN Number 3082
Risk phrases H351
Operator protection A; U05a, U20a
Environmental protection E06c (1 week); E15b, E16a, E34, H410
Storage and disposal D01, D02, D09a, D10c
Medical advice M03

17   aminopyralid + propyzamide

A herbicide mixture for weed control in winter oilseed rape
HRAC mode of action code: K1 + O

Products

1	AstroKerb	Dow (Corteva)	6.3:500 g/l	SC	16184
2	Dymid	Dow (Corteva)	5.3:500 g/l	SC	19187
3	Pamino	Harvest	5.3:500 g/l	SC	18069

Uses
- Annual dicotyledons in *winter oilseed rape*
- Annual grasses in *winter oilseed rape*
- Mayweeds in *winter oilseed rape*
- Poppies in *winter oilseed rape*

Approval information
- Aminopyralid and propyzamide are included in Annex 1 under EC Regulation 1107/2009.

Efficacy guidance
- Allow soil temperatures to drop before application to winter oilseed rape to optimise activity of propyzamide component

Restrictions
- Users must have received adequate instruction, training and guidance on the safe and efficient use of the product.
- Must not be used on land where vegetation will be cut for animal feed, fodder, bedding nor for composting or mulching within one calendar year of treatment.
- Do not remove oilseed rape straw from the field unless it is for burning for heat or electricity production. Do not use the oilseed rape straw for animal feed, animal bedding, composting or mulching.
- Livestock must be kept out of treated areas for at least 1 week following treatment and until poisonous weeds such as ragwort have died and become unpalatable

Crop-specific information
- Only cereals may follow application of aminopyralid + propyzamide and these should not be planted within 30 weeks of treatment.

FOR FULL CONDITIONS OF USE ALWAYS READ THE PRODUCT LABEL

- Aminopyralid residues in plant tissue which have not completely decayed may affect susceptible crops such as legumes (e.g. peas and beans), sugar and fodder beet, carrots and other Umbelliferae, potatoes, tomatoes, lettuce and other Compositae.

Following crops guidance
- May only be followed by spring or winter cereals. Treated land must be mouldboard ploughed to a depth of at least 15 cm before a following cereal crop is planted.
- Following crops (winter and spring wheat) should not be planted within 30 weeks of application [2]

Hazard classification and safety precautions
    Hazard Harmful, Dangerous for the environment
    Transport code 9
    Packaging group III
    UN Number 3082
    Risk phrases H351
    Operator protection A, H; U20c
    Environmental protection E07b [1-3] (1); E15b, E38, E39 [1-3]; H410 [2]; H411 [1, 3]
    Storage and disposal D09a, D11a, D12a

18   aminopyralid + triclopyr

A foliar acting herbicide mixture for broad-leaved weed control in grassland
HRAC mode of action code: O + O

See also triclopyr

Products

1	Forefront T	Dow (Corteva)	30:240g/l	EW	15568
2	Garlon Ultra	Nomix Enviro	12:120 g/l	SL	16211
3	Icade	Dow (Corteva)	12:120 g/l	SL	16182
4	Icade	Dow (Corteva)	12:120 g/l	SL	19069

Uses
- Annual dicotyledons in *amenity grassland* [2-4]
- Brambles in *amenity grassland* [3, 4]
- Broom in *amenity grassland* [3, 4]
- Buddleia in *amenity grassland* [3, 4]
- Buttercups in *grassland* [1]
- Common mugwort in *amenity grassland* [3, 4]
- Common nettle in *amenity grassland* [3, 4]; *grassland* [1]
- Creeping thistle in *amenity grassland* [3, 4]
- Dandelions in *grassland* [1]
- Docks in *grassland* [1]
- Gorse in *amenity grassland* [3, 4]
- Hogweed in *amenity grassland* [3, 4]
- Japanese knotweed in *amenity grassland* [3, 4]
- Perennial dicotyledons in *amenity grassland* [3, 4]
- Rosebay willowherb in *amenity grassland* [3, 4]
- Thistles in *grassland* [1]

Approval information
- Aminopyralid and triclopyr are included in Annex I under EC Regulation 1107/2009.
- Approval expiry 31 Dec 2020 [1, 2]

Efficacy guidance
- For best results and to avoid crop check, grass and weeds must be growing actively
- Use adeqate water volume to ensure good weed coverage. Increase water volume if necessary where the weed population is high and where the grass is dense
- Allow 2-3 wk after cutting for hay or silage for sufficient regrowth to occur before spraying and leave 7 d afterwards to allow maximum translocation

SEE SECTION 3 FOR PRODUCTS ALSO REGISTERED

- Where there is a high reservoir of weed seed or a historically high weed population a programmed approach may be needed involving a second treatment in the following yr
- Control may be reduced if rain falls within 1 h of spraying

Restrictions
- Maximum number of treatments 1 per yr.
- Do not apply to leys less than 1 year old.
- Do not apply by hand-held equipment [1]
- Do not use on grassland that will be used for animal feed, bedding, composting or mulching within 1 calender year of application.
- Do not use on grassland that will be grazed by animals other than cattle or sheep.
- Do not use any treated plant material, or manure from animals fed on treated crops, for composting or mulching
- Do not use on crops grown for seed
- Manure from animals fed on pasture or silage from treated crops should not leave the farm
- Do not apply spot applications to more than 20% of the area

Crop-specific information
- Latest use: 7 d before grazing or harvest for grassland
- Late applications may lead to transient leaning of grass which does not affect final yield

Following crops guidance
- Ensure that all plant remains of a treated crop have completely decayed before planting susceptible crops such as peas, beans and other legumes, sugar beet, carrots and other Umbelliferae, potatoes and tomatoes, lettuce and other Compositae
- Do not plant potatoes, sugar beet, vegetables, beans or other leguminous crops in the calendar yr following application

Environmental safety
- Dangerous for the environment
- Toxic to aquatic organisms
- Keep livestock out of treated areas for at least 7 d after treatment or until foliage of any poisonous weeds such as ragwort has died and become unpalatable
- To protect groundwater do not apply to grass leys less than 1 yr old
- Take extreme care to avoid drift onto susceptible crops, non-target plants or waterways. All conifers, especially pine and larch, are very sensitive and may be damaged by vapour drift in hot conditions
- LERAP Category B [1]

Hazard classification and safety precautions
Hazard Harmful [2-4]; Irritant [1]; Dangerous for the environment [1-4]
Transport code 9 [1]
Packaging group III [1]
UN Number 3082 [1]; N/C [2-4]
Risk phrases H317 [1]; H319, H335 [2-4]
Operator protection A, H [1-4]; M [2-4]; U05a, U08 [1-4]; U14, U23a [1]
Environmental protection E06a [1] (7 days); E07c [2-4] (7 days); E15b, E38 [1-4]; E15c, E16a, H411 [1]; E39 [2-4]
Consumer protection C01
Storage and disposal D01, D02, D09a [1-4]; D05, D10c [2-4]; D10b, D12a [1]
Medical advice M05a [2-4]

19   amisulbrom

A fungicide for use in potatoes
FRAC mode of action code: 21

Products

1 Gachinko	DuPont (Corteva)	200 g/l	SC	18219
2 Shinkon	Gowan	200 g/l	SC	17498

SECTION 2

Uses
- Downy mildew in **wine grapes** *(off-label)* [2]
- Late blight in **potatoes** [1, 2]
- Tuber blight in **potatoes** [1, 2]

Extension of Authorisation for Minor Use (EAMUs)
- **wine grapes** *20193119* [2]

Approval information
- Amisulbrom is included in Annex 1 under EC Regulation 1107/2009

Efficacy guidance
- Make no more than 3 consecutive applications before switching to a blight fungicide with a different mode of action to protect against the risk of resistance.
- Do not use if blight is already visible on 1% of leaves.

Crop-specific information
- Rainfast within 3 hours of application but apply to dry foliage.

Following crops guidance
- Plough before planting new crops in the same soil.

Environmental safety
- LERAP Category B

Hazard classification and safety precautions
Hazard Dangerous for the environment, Very toxic to aquatic organisms
Transport code 9
Packaging group III
UN Number 3082
Risk phrases H351
Operator protection A, C; U05a, U15, U20b
Environmental protection E13b, E15a, E16a, E34, E38, H410
Storage and disposal D01, D02, D05, D09a, D10b, D12a

20   Ampelomyces quisqualis (Strain AQ10)

A biological fungicide that combats powdery mildew in fruit and ornamentals
FRAC mode of action code: BM 02

Products

AQ 10	Fargro	58% w/w	WG	17102

Uses
- Powdery mildew in **protected apple** *(off-label)*, **protected aubergines**, **protected baby leaf crops** *(off-label)*, **protected blackcurrants** *(off-label)*, **protected blueberry** *(off-label)*, **protected chicory** *(off-label)*, **protected chilli peppers**, **protected chilli peppers** *(off-label)*, **protected courgettes**, **protected courgettes** *(off-label)*, **protected crabapple** *(off-label)*, **protected cress** *(off-label)*, **protected cucumbers**, **protected endives** *(off-label)*, **protected forest nurseries** *(off-label)*, **protected gooseberries** *(off-label)*, **protected herbs (see appendix 6)** *(off-label)*, **protected lamb's lettuce** *(off-label)*, **protected land cress** *(off-label)*, **protected lettuce** *(off-label)*, **protected melons**, **protected ornamentals** *(off-label)*, **protected other small fruit and berries** *(off-label)*, **protected pear** *(off-label)*, **protected peppers**, **protected peppers** *(off-label)*, **protected pumpkins**, **protected purslane** *(off-label)*, **protected quince** *(off-label)*, **protected red mustard** *(off-label)*, **protected redcurrants** *(off-label)*, **protected rocket** *(off-label)*, **protected spinach** *(off-label)*, **protected spinach beet** *(off-label)*, **protected squashes**, **protected strawberries**, **protected summer squash** *(off-label)*, **protected table grapes** *(off-label)*, **protected tomatoes**, **protected watercress** *(off-label)*, **protected watermelon** *(off-label)*, **protected wine grapes** *(off-label)*, **protected winter squash**

Extension of Authorisation for Minor Use (EAMUs)
- **protected apple** *20152646*

SEE SECTION 3 FOR PRODUCTS ALSO REGISTERED

- *protected baby leaf crops* 20180375
- *protected blackcurrants* 20152646
- *protected blueberry* 20152646
- *protected chicory* 20180375
- *protected chilli peppers* 20152646
- *protected courgettes* 20152646
- *protected crabapple* 20152646
- *protected cress* 20180375
- *protected endives* 20180375
- *protected forest nurseries* 20152646
- *protected gooseberries* 20152646
- *protected herbs (see appendix 6)* 20152646
- *protected lamb's lettuce* 20180375
- *protected land cress* 20180375
- *protected lettuce* 20180375
- *protected ornamentals* 20152646
- *protected other small fruit and berries* 20152646
- *protected pear* 20152646
- *protected peppers* 20152646
- *protected purslane* 20180375
- *protected quince* 20152646
- *protected red mustard* 20180375
- *protected redcurrants* 20152646
- *protected rocket* 20180375
- *protected spinach* 20180375
- *protected spinach beet* 20180375
- *protected summer squash* 20152646
- *protected table grapes* 20152646
- *protected watercress* 20180375
- *protected watermelon* 20152646
- *protected wine grapes* 20152646

Approval information
- Ampelomyces quisqualis is included in Annex 1 under EC Regulation 1107/2009

Hazard classification and safety precautions
  UN Number N/C
  Operator protection A, D, H; U14, U20b
  Environmental protection E15b, E34
  Storage and disposal D01, D02, D05, D09a, D10a, D16, D20
  Medical advice M03

21   azadirachtin

An insecticide for use on ornamentals under full enclosure acting as a feeding inhibitor and a growth disruptor
IRAC mode of action code: Unknown

Products

Azatin	Certis	217 g/l	EC	18301

Uses
- Onion thrips in *ornamental plant production*
- Western flower thrips in *ornamental plant production*

Approval information
- Azadirachtin included in Annex I under EC Regulation 1107/2009

Efficacy guidance

- A minimum interval of 7 days must be observed between applications within a block (i.e. 4 consecutive sprays represents a block) and 42 days between blocks
- Soil bound systems may not be treated until BBCH 40 or later
- Applications are carried out at a dose rate of 1.4 l/ha for ornamental plants grown under cover and 1.68 l/ha for roses grown on substrate under cover in 0.14% concentration and a minimum interval of 7 days between applications.
- The total number of applications per year is a maximum of 20 applications, divided over 5 blocks. 42 days between each block of 4 applications must be observed.

Restrictions

- Reasonable precautions must be taken to prevent access of birds, wild mammals and bees to treated crops
- To minimise airborne environmental exposure, vents, doors and other openings must be closed during and after application until the applied product has fully settled
- To protect ground water, a minimum interval of 39 days after final treatment must be observed before planting out of containerised forest nursery plants
- To protect the environment, plants grown in synthetic media treated with azadirachtin must not be planted out into soil.

Following crops guidance

- Because of the diversity of crops that may be treated, users are advised to carry out a small scale trial application to crops to ascertain safety to the variety being treated before large scale applications

Hazard classification and safety precautions

Transport code 9
Packaging group III
UN Number 3082
Risk phrases H319
Operator protection A, C; U11
Environmental protection E15b, E34, H410
Storage and disposal D01, D06e, D07, D09a

22   azoxystrobin

A systemic translaminar and protectant strobilurin fungicide for a wide range of crops
FRAC mode of action code: 11

See also azoxystrobin + cyproconazole
azoxystrobin + difenoconazole
azoxystrobin + fluazinam
azoxystrobin + isopyrazam
azoxystrobin + tebuconazole

Products

1	Affix	UPL Europe	250 g/l	SC	18324
2	Amistar	Syngenta	250 g/l	SC	18039
3	Arixa	Nurture	250 g/l	SC	19034
4	Azaka	FMC Agro	250 g/l	SC	18731
5	Azofin Plus	Clayton	250 g/l	SC	18552
6	Azoxystar	Life Scientific	250 g/l	SC	17407
7	Chamane	UPL Europe	250 g/l	SC	15922
8	Clayton Belfry	Clayton	250 g/l	SC	18154
9	Conclude AZT 250SC	Belchim	250 g/l	SC	18440
10	Heritage	Syngenta	50% w/w	WG	13536
11	Heritage Maxx	Syngenta	95 g/l	DC	18246
12	Hill-Star	Stefes	250 g/l	SC	18150
13	Pure Azoxy	Pure Amenity	50% w/w	WG	18065
14	Sinstar	Agrii	250 g/l	SC	16852

SECTION 2

SEE SECTION 3 FOR PRODUCTS ALSO REGISTERED

Products – continued

15 Stunner	Harvest	250 g/l	SC	18479
16 Tazer	Nufarm UK	250 g/l	SC	15495
17 Toran	Becesane	250 g/l	SC	18239
18 Valiant	FMC Agro	250 g/l	SC	18742
19 Zoxis	Arysta	250 g/l	SC	18438

Uses

- Alternaria in *broccoli, brussels sprouts, cabbages, calabrese, carrots, cauliflowers, collards, kale* [1-3, 5, 8, 9, 12, 17, 19]; *broccoli (moderate control), brussels sprouts (moderate control), cabbages (moderate control), calabrese (moderate control), cauliflowers (moderate control), collards (moderate control), kale (moderate control)* [7]; *horseradish (off-label), parsnips (off-label), poppies (off-label), poppies grown for seed production (off-label)* [2]; *spring oilseed rape, winter oilseed rape* [1-3, 5-9, 12, 14-17, 19]
- Alternaria blight in *carrots* [7, 15, 16]
- Anthracnose in *managed amenity turf* [10, 13]; *managed amenity turf (moderate control)* [11]; *protected strawberries, strawberries* [2, 3, 5, 8, 9, 12, 17]
- Ascochyta in *combining peas, vining peas* [6, 7]; *combining peas (useful control), vining peas (useful control)* [1-3, 5, 8, 9, 12, 17, 19]; *dwarf beans (useful control), edible podded peas (useful control), french beans (useful control), mange-tout peas (useful control), sugar snap peas (useful control)* [2, 3, 5, 8, 9, 12, 17]
- Black dot in *potatoes* [1-3, 5, 7-9, 12, 17, 19]; *potatoes (reduction)* [6]
- Black root rot in *container-grown ornamentals (off-label), ornamental plant production (off-label), protected ornamentals (off-label)* [2]
- Black rot in *baby leaf crops (off-label), cress (off-label), herbs (see appendix 6) (off-label), lamb's lettuce (off-label), land cress (off-label), oriental cabbage (off-label), purslane (off-label), red mustard (off-label), rocket (off-label), spinach (off-label), spinach beet (off-label)* [2]
- Black scurf in *potatoes* [7]; *potatoes (reduction)* [6]
- Black scurf and stem canker in *potatoes* [1-3, 5, 8, 9, 12, 17, 19]
- Botrytis in *celery (outdoor) (off-label), container-grown ornamentals (off-label), ornamental plant production (off-label), protected celery (off-label), protected ornamentals (off-label)* [2]
- Brown patch in *managed amenity turf* [10, 11, 13]
- Brown rust in *rye* [4, 15, 16, 18]; *spring barley, spring wheat, winter barley, winter wheat* [1-9, 12, 14-19]; *spring rye, winter rye* [1-3, 5-9, 12, 17, 19]; *triticale* [1-9, 12, 15-19]
- Crown rust in *managed amenity turf* [10, 13]; *spring oats, winter oats* [1-6, 8, 9, 12, 15-19]
- Dark leaf and pod spot in *spring oilseed rape, winter oilseed rape* [7]
- Dark leaf spot in *spring oilseed rape, winter oilseed rape* [4, 15, 16, 18]
- Downy mildew in *bulb onions* [7]; *bulb onions (moderate control)* [15, 16]; *bulb onions (reduction)* [1-3, 5, 8, 9, 12, 17, 19]; *combining peas (reduction), vining peas (reduction)* [1-3, 5, 8, 9, 12, 15-17, 19]; *container-grown ornamentals (off-label), courgettes (off-label), cucumbers (off-label), gherkins (off-label), melons (off-label), ornamental plant production (off-label), protected ornamentals (off-label), pumpkins (off-label), radishes (off-label), salad onions (off-label), summer squash (off-label), watermelons (off-label), winter squash (off-label)* [2]; *dwarf beans (reduction), edible podded peas (reduction), endives, french beans (reduction), garlic (reduction), lettuce, mange-tout peas (reduction), protected endives, protected lettuce, shallots (reduction), sugar snap peas (reduction)* [2, 3, 5, 8, 9, 12, 17]
- Ear diseases in *spring wheat, winter wheat* [4, 15, 16, 18]
- Fairy rings in *managed amenity turf* [10, 13]; *managed amenity turf (reduction)* [11]
- Fusarium in *container-grown ornamentals (off-label), ornamental plant production (off-label), protected ornamentals (off-label)* [2]
- Fusarium patch in *managed amenity turf* [10, 11, 13]
- Glume blotch in *spring wheat, winter wheat* [1-9, 12, 14-19]
- Grey mould in *aubergines (off-label), courgettes (off-label), melons (off-label), pumpkins (off-label), summer squash (off-label), tomatoes (off-label), watermelons (off-label), winter squash (off-label)* [2]; *combining peas (some control), vining peas (some control)* [1-3, 5, 8, 9, 12, 15-17, 19]; *dwarf beans (some control), edible podded peas (some control), french beans (some control), mange-tout peas (some control), sugar snap peas (some control)* [2, 3, 5, 8, 9, 12, 17]

- Late blight in **aubergines** *(off-label)*, **tomatoes** *(off-label)* [2]
- Late ear diseases in **spring wheat**, **winter wheat** [1-3, 5-9, 12, 14, 17, 19]
- Leaf and pod spot in **combining peas** *(useful control)*, **vining peas** *(useful control)* [1-3, 5, 8, 9, 12, 15-17, 19]; **dwarf beans** *(useful control)*, **edible podded peas** *(useful control)*, **french beans** *(useful control)*, **mange-tout peas** *(useful control)*, **sugar snap peas** *(useful control)* [2, 3, 5, 8, 9, 12, 17]
- Leaf spot in **celery (outdoor)** *(off-label)*, **container-grown ornamentals** *(off-label)*, **ornamental plant production** *(off-label)*, **protected celery** *(off-label)*, **protected ornamentals** *(off-label)* [2]
- Melting out in **managed amenity turf** [10, 11, 13]
- Mycosphaerella in **combining peas** *(some control)*, **vining peas** *(some control)* [1-3, 5, 8, 9, 12, 15-17, 19]; **dwarf beans** *(some control)*, **edible podded peas** *(some control)*, **french beans** *(some control)*, **mange-tout peas** *(some control)*, **sugar snap peas** *(some control)* [2, 3, 5, 8, 9, 12, 17]
- Needle casts in **container-grown ornamentals** *(off-label)*, **ornamental plant production** *(off-label)*, **protected ornamentals** *(off-label)* [2]
- Net blotch in **spring barley**, **winter barley** [1-9, 12, 14-19]
- Powdery mildew in **aubergines** *(off-label)*, **blackberries** *(off-label)*, **chillies** *(off-label)*, **container-grown ornamentals** *(off-label)*, **courgettes** *(off-label)*, **cucumbers** *(off-label)*, **gherkins** *(off-label)*, **loganberries** *(off-label)*, **melons** *(off-label)*, **ornamental plant production** *(off-label)*, **parsnips** *(off-label)*, **peppers** *(off-label)*, **poppies** *(off-label)*, **poppies grown for seed production** *(off-label)*, **protected blackberries** *(off-label)*, **protected loganberries** *(off-label)*, **protected ornamentals** *(off-label)*, **protected raspberries** *(off-label)*, **protected rubus hybrids** *(off-label)*, **pumpkins** *(off-label)*, **raspberries** *(off-label)*, **rubus hybrids** *(off-label)*, **summer squash** *(off-label)*, **tomatoes** *(off-label)*, **watermelons** *(off-label)*, **winter squash** *(off-label)* [2]; **carrots** [1-3, 5, 7-9, 12, 15-17, 19]; **protected strawberries**, **strawberries** [2, 3, 5, 8, 9, 12, 17]; **rye** [4]; **rye** *(moderate control)*, **triticale** *(moderate control)* [15, 16, 18]; **spring barley**, **winter barley** [6, 7, 14]; **spring barley** *(moderate control)*, **winter barley** *(moderate control)* [4, 15, 16, 18]; **spring oats**, **winter oats** [6]; **spring oats** *(moderate control)*, **winter oats** *(moderate control)* [4, 18]; **triticale** [4, 6, 7]; **winter rye** [1-3, 5-9, 12, 17, 19]
- Purple blotch in **leeks** [1-3, 5, 8, 9, 12, 15-17, 19]; **leeks** *(moderate control)* [7]
- Rhizoctonia in **celery (outdoor)** *(off-label)*, **protected celery** *(off-label)*, **radishes** *(off-label)*, **swedes** *(off-label)*, **turnips** *(off-label)* [2]; **potatoes** [7]
- Rhynchosporium in **rye** [4]; **rye** *(reduction)*, **triticale** *(reduction)* [15, 16, 18]; **spring barley**, **winter barley** [1-3, 5-9, 12, 14, 17, 19]; **spring barley** *(reduction)*, **winter barley** *(reduction)* [4, 15, 16, 18]; **spring rye**, **winter rye** [1-3, 5-9, 12, 17, 19]; **triticale** [1-9, 12, 17, 19]
- Ring spot in **broccoli**, **brussels sprouts**, **cabbages**, **calabrese**, **cauliflowers**, **collards**, **kale** [1-3, 5, 8, 9, 12, 17, 19]; **broccoli** *(moderate control)*, **brussels sprouts** *(moderate control)*, **cabbages** *(moderate control)*, **calabrese** *(moderate control)*, **cauliflowers** *(moderate control)*, **collards** *(moderate control)*, **kale** *(moderate control)* [7]
- Root malformation disorder in **red beet** *(off-label)* [2]
- Rust in **asparagus** [1-3, 5, 8, 9, 12, 15-17, 19]; **asparagus** *(moderate control)* [7]; **broad beans**, **lupins** [2, 3, 5, 8, 9, 12, 17]; **container-grown ornamentals** *(off-label)*, **ornamental plant production** *(off-label)*, **protected ornamentals** *(off-label)* [2]; **leeks**, **spring field beans**, **winter field beans** [1-3, 5, 7-9, 12, 15-17, 19]; **managed amenity turf** [11]
- Scab in **container-grown ornamentals** *(off-label)*, **ornamental plant production** *(off-label)*, **protected ornamentals** *(off-label)* [2]
- Sclerotinia in **celeriac** *(off-label)*, **celery (outdoor)** *(off-label)*, **protected celery** *(off-label)* [2]; **soya beans** *(off-label)* [6]; **spring oilseed rape**, **winter oilseed rape** [7, 15, 16]
- Sclerotinia stem rot in **spring oilseed rape**, **winter oilseed rape** [1-6, 8, 9, 12, 14, 17-19]
- Septoria leaf blotch in **spring wheat** [4, 6, 7, 14-16, 18]; **winter wheat** [1-9, 12, 14-19]
- Stem canker in **potatoes** [7]; **potatoes** *(reduction)* [6]
- Stemphylium in **asparagus** [1-3, 5, 7-9, 12, 15-17, 19]
- Take-all in **rye** *(reduction)* [4, 15, 16, 18]; **spring barley** *(reduction)*, **spring wheat** *(reduction)*, **winter barley** *(reduction)*, **winter wheat** *(reduction)* [1-9, 12, 14-19]; **spring oats** *(reduction)*, **winter oats** *(reduction)* [4]; **spring rye** *(reduction in severity)* [1-3, 5, 8, 9, 12, 17, 19]; **triticale** *(reduction)* [1-5, 8, 9, 12, 15-19]
- Take-all patch in **managed amenity turf** [10, 11, 13]
- White blister in **baby leaf crops** *(off-label)*, **container-grown ornamentals** *(off-label)*, **cress** *(off-label)*, **herbs (see appendix 6)** *(off-label)*, **horseradish** *(off-label)*, **lamb's lettuce** *(off-label)*, **land**

SEE SECTION 3 FOR PRODUCTS ALSO REGISTERED

cress (off-label), **oriental cabbage** (off-label), **ornamental plant production** (off-label), **protected ornamentals** (off-label), **purslane** (off-label), **red mustard** (off-label), **rocket** (off-label), **spinach** (off-label), **spinach beet** (off-label) [2]; **broccoli, brussels sprouts, cabbages**, **calabrese, cauliflowers, collards, kale** [1-3, 5, 8, 9, 12, 17, 19]; **broccoli** (moderate control), **brussels sprouts** (moderate control), **cabbages** (moderate control), **calabrese** (moderate control), **cauliflowers** (moderate control), **collards** (moderate control), **kale** (moderate control) [7]

- White tip in **leeks** [1-3, 5, 8, 9, 12, 15-17, 19]
- Yellow rust in **spring wheat**, **winter wheat** [1-9, 12, 14-19]

Extension of Authorisation for Minor Use (EAMUs)

- **aubergines** *20170894* [2]
- **baby leaf crops** *20172069* [2]
- **blackberries** *20170895* [2]
- **celeriac** *20192198* [2]
- **celery (outdoor)** *20170891* [2]
- **chillies** *20170894* [2]
- **container-grown ornamentals** *20183388* [2]
- **courgettes** *20170893* [2], *20170894* [2]
- **cress** *20172069* [2]
- **cucumbers** *20170894* [2]
- **gherkins** *20170894* [2]
- **herbs (see appendix 6)** *20172069* [2]
- **horseradish** *20192198* [2]
- **lamb's lettuce** *20172069* [2]
- **land cress** *20172069* [2]
- **loganberries** *20170895* [2]
- **melons** *20170893* [2], *20170894* [2]
- **oriental cabbage** *20170889* [2]
- **ornamental plant production** *20183388* [2]
- **parsnips** *20192198* [2]
- **peppers** *20170894* [2]
- **poppies** *20180713* [2]
- **poppies grown for seed production** *20180713* [2]
- **protected blackberries** *20170895* [2]
- **protected celery** *20170891* [2]
- **protected loganberries** *20170895* [2]
- **protected ornamentals** *20183388* [2]
- **protected raspberries** *20170895* [2]
- **protected rubus hybrids** *20170895* [2]
- **pumpkins** *20170893* [2], *20170894* [2]
- **purslane** *20172069* [2]
- **radishes** *20192198* [2]
- **raspberries** *20170895* [2]
- **red beet** *20192198* [2]
- **red mustard** *20172069* [2]
- **rocket** *20172069* [2]
- **rubus hybrids** *20170895* [2]
- **salad onions** *20170890* [2]
- **soya beans** *20180786* [6]
- **spinach** *20172069* [2]
- **spinach beet** *20172069* [2]
- **summer squash** *20170893* [2], *20170894* [2]
- **swedes** *20192198* [2]
- **tomatoes** *20170894* [2]
- **turnips** *20192198* [2]
- **watermelons** *20170893* [2], *20170894* [2]
- **winter squash** *20170893* [2], *20170894* [2]

FOR FULL CONDITIONS OF USE ALWAYS READ THE PRODUCT LABEL

Approval information
- Azoxystrobin included in Annex I under EC Regulation 1107/2009
- Accepted by BBPA for use on malting barley

Efficacy guidance
- Best results obtained from use as a protectant or during early stages of disease establishment or when a predictive assessment indicates a risk of disease development
- Azoxystrobin inhibits fungal respiration and should always be used in mixture with fungicides with other modes of action
- Treatment under poor growing conditions may give less reliable results
- For good control of *Fusarium* patch in amenity turf and grass repeat treatment at minimum intervals of 2 wk
- Azoxystrobin is a member of the QoI cross resistance group. Product should be used preventatively and not relied on for its curative potential
- Use product in cereals as part of an Integrated Crop Management strategy incorporating other methods of control, including where appropriate other fungicides with a different mode of action. Do not apply more than two foliar applications of QoI containing products to any cereal crop
- There is a significant risk of widespread resistance occurring in *Septoria tritici* populations in UK. Failure to follow resistance management action may result in reduced levels of disease control
- On cereal crops product must always be used in mixture with another product, recommended for control of the same target disease, that contains a fungicide from a different cross resistance group and is applied at a dose that will give robust control
- Strains of barley powdery mildew resistant to QoIs are common in the UK

Restrictions
- Maximum number of treatments 1 per crop for potatoes; 2 per crop for brassicas, peas, cereals, oilseed rape; 4 per crop or yr for onions, carrots, leeks, amenity turf
- Maximum total dose ranges from 2-4 times the single full dose depending on crop and product. See labels for details
- On turf the maximum number of treatments is 4 per yr but they must not exceed one third of the total number of fungicide treatments applied
- Do not use where there is risk of spray drift onto neighbouring apple crops
- The same spray equipment should not be used to treat apples
- To reduce the risk of resistance developing on target diseases the total number of applications of products containing QoI fungicides made to any cereal crop must not exceed two
- To protect aquatic life, the maximum total dose applied to crops grown outdoors must not exceed 500 g a.i/ha per year [1, 2]
- To protect aquatic life, for uses on crops of broccoli, calabrese, Brussels sprouts, cabbage, cauliflower, collards, lettuce and kale, the maximum total dose applied must not exceed 500 g azoxystrobin per hectare per year [19]
- Do not apply to protected or outdoor ornamentals at temperatures above 30°C or below 10°C [2]

Crop-specific information
- Latest use: at planting for potatoes; grain watery ripe (GS 71) for cereals; before senescence for asparagus
- HI: 10 d for carrots;14 d for broccoli, Brussels sprouts, bulb onions, cabbages, calabrese, cauliflowers, collards, kale, vining peas; 21 d for leeks, spring oilseed rape, winter oilseed rape; 36 d for combining peas, 35 d for field beans
- In cereals control of established infections can be improved by appropriate tank mixtures or application as part of a programme. Always use in mixture with another product from a different cross-resistance group
- In turf use product at full dose rate in a disease control programme, alternating with fungicides of different modes of action
- In potatoes when used as incorporated treatment apply overall to the entire area to be planted, incorporate to 15 cm and plant on the same day. In-furrow spray should be directed at the furrow and not the seed tubers
- Applications to brassica crops must only be made to a developed leaf canopy and not before growth stages specified on the label
- Heavy disease pressure in brassicae and oilseed rape may require a second treatment
- All crops should be treated when not under stress. Check leaf wax on peas if necessary

SECTION 2

SEE SECTION 3 FOR PRODUCTS ALSO REGISTERED

- Consult processor before treating any crops for processing
- Treat asparagus after the harvest season. Where a new bed is established do not treat within 3 wk of transplanting out the crowns
- Do not apply to turf when ground is frozen or during drought

Environmental safety
- Dangerous for the environment
- Very toxic to aquatic organisms
- Avoid spray drift onto surrounding areas or crops, especially apples, plums or privet
- Buffer zone requirement 5 m in winter wheat, winter barley and winter oilseed rape [14]
- Buffer zone requirement 10 m in blackberry, loganberry, raspberry and rubus hybrid [2]
- Buffer zone requirement 6 m in bulb onion, carrots and leeks [16]
- LERAP Category B [1-17, 19]

Hazard classification and safety precautions
Hazard Harmful [7, 15, 16]; Dangerous for the environment [1-19]; Harmful if swallowed [11]; Harmful if inhaled [9, 12, 19]; Very toxic to aquatic organisms [6, 10, 13, 15, 16]
Transport code 9
Packaging group III
UN Number 3077 [10, 13]; 3082 [1-9, 11, 12, 14-19]
Risk phrases H317 [19]; H360 [11]; R50, R53a [18]
Operator protection A [1-9, 11, 12, 14-19]; H [4]; U05a, U20b [1-19]; U09a, U19a [1-9, 12, 14-19]
Environmental protection E15a [1-3, 5, 6, 8, 9, 12, 14, 17, 19]; E15b [4, 7, 10, 11, 13, 15, 16, 18]; E16a, H410 [1-17, 19]; E23 [7]; E38 [1-19]
Storage and disposal D01, D02, D09a, D12a [1-19]; D03, D10b [10, 11, 13]; D05 [1-3, 5-14, 17, 19]; D10c [1-9, 12, 14-19]
Medical advice M03 [7]; M05a [10, 11, 13]

23   azoxystrobin + cyproconazole

A contact and systemic broad spectrum fungicide mixture for cereals, beet crops and oilseed rape
FRAC mode of action code: 11 + 3

See also cyproconazole

Products

Mirador Xtra	Adama	200:80 g/l	SC	18542

Uses
- Alternaria in *spring oilseed rape, winter oilseed rape*
- Brown rust in *spring barley, spring rye, spring wheat, winter barley, winter rye, winter wheat*
- Cercospora leaf spot in *fodder beet, sugar beet*
- Crown rust in *spring oats, winter oats*
- Eyespot in *spring barley* (reduction), *spring wheat* (reduction), *winter barley* (reduction), *winter wheat* (reduction)
- Glume blotch in *spring wheat, winter wheat*
- Net blotch in *spring barley, winter barley*
- Powdery mildew in *fodder beet, spring barley, spring oats, spring rye, spring wheat, sugar beet, winter barley, winter oats, winter rye, winter wheat*
- Ramularia leaf spots in *fodder beet, sugar beet*
- Rhynchosporium in *spring barley* (moderate control), *spring rye* (moderate control), *winter barley* (moderate control), *winter rye* (moderate control)
- Rust in *fodder beet, sugar beet*
- Sclerotinia stem rot in *spring oilseed rape, winter oilseed rape*
- Septoria leaf blotch in *spring wheat, winter wheat*
- Take-all in *spring barley* (reduction), *spring wheat* (reduction), *winter barley* (reduction), *winter wheat* (reduction)
- Yellow rust in *spring wheat, winter wheat*

FOR FULL CONDITIONS OF USE ALWAYS READ THE PRODUCT LABEL

Approval information
- Azoxystrobin and cyproconazole included in Annex I under EC Regulation 1107/2009
- Accepted by BBPA for use on malting barley

Efficacy guidance
- Best results obtained from treatment during the early stages of disease development
- A second application may be needed if disease attack is prolonged
- Azoxystrobin is a member of the QoI cross resistance group. Product should be used preventatively and not relied on for its curative potential
- Use product as part of an Integrated Crop Management strategy incorporating other methods of control, including where appropriate other fungicides with a different mode of action. Do not apply more than two foliar applications of QoI containing products to any cereal crop
- There is a significant risk of widespread resistance occurring in *Septoria tritici* populations in UK. Failure to follow resistance management action may result in reduced levels of disease control
- Strains of wheat and barley powdery mildew resistant to QoIs are common in the UK. Control of wheat powdery mildew can only be relied upon from the triazole component
- Where specific control of wheat mildew is required this should be achieved through a programme of measures including products recommended for the control of mildew that contain a fungicide from a different cross-resistance group and applied at a dose that will give robust control
- Cyproconazole is a DMI fungicide. Resistance to some DMI fungicides has been identified in Septoria leaf blotch which may seriously affect performance of some products. For further advice contact a specialist advisor and visit the Fungicide Resistance Action Group (FRAG)-UK website

Restrictions
- Maximum total dose equivalent to two full dose treatments
- Do not use where there is risk of spray drift onto neighbouring apple crops
- The same spray equipment should not be used to treat apples

Crop-specific information
- Latest use: up to and including anthesis complete (GS 69) for rye and wheat; up to and including emergence of ear complete (GS 59) for barley and oats; BBCH79 (nearly all pods at final size) or 30 days before harvest, whichever is sooner for oilseed rape

Environmental safety
- Dangerous for the environment
- Very toxic to aquatic organisms

Hazard classification and safety precautions
Hazard Harmful, Dangerous for the environment, Harmful if swallowed, Harmful if inhaled, Very toxic to aquatic organisms
Transport code 9
Packaging group III
UN Number 3082
Risk phrases H361
Operator protection A; U05a, U09a, U19a, U20b
Environmental protection E15b, E34, E38, H410
Storage and disposal D01, D02, D05, D09a, D10c, D12a
Medical advice M03

24   azoxystrobin + difenoconazole

A broad spectrum fungicide mixture for field crops
FRAC mode of action code: 11 + 3

See also difenoconazole

Products
1	Amistar Top	Syngenta	200:125 g/l	SC	18050
2	Angle	Syngenta	125:125 g/l	SC	19119

SEE SECTION 3 FOR PRODUCTS ALSO REGISTERED

Uses

- Alternaria in **choi sum** *(off-label)*, **oriental brassicas** *(off-label)* [1]
- Alternaria blight in **carrots** [1]
- Black canker in **horseradish** *(off-label)*, **parsley root** *(off-label)*, **parsnips** *(off-label)*, **salsify** *(off-label)* [1]
- Botrytis in **chicory** *(off-label)*, **chicory grown outside for forcing** *(off-label)* [1]
- Disease control in **fodder beet, spring oilseed rape, sugar beet, winter oilseed rape** [2]; **protected strawberries, strawberries** [1]
- Phoma in **horseradish** *(off-label)*, **parsley root** *(off-label)*, **parsnips** *(off-label)*, **salsify** *(off-label)* [1]
- Powdery mildew in **broccoli, brussels sprouts, cabbages, calabrese, carrots, cauliflowers, collards, kale, rocket** [1]
- Purple blotch in **leeks** *(moderate control)* [1]
- Purple spot in **asparagus** *(off-label)* [1]
- Ring spot in **choi sum** *(off-label)*, **oriental brassicas** *(off-label)* [1]
- Rust in **asparagus** *(off-label)*, **herbs (see appendix 6)** *(off-label)*, **leeks** [1]
- Sclerotinia in **chicory** *(off-label)*, **chicory grown outside for forcing** *(off-label)* [1]
- Septoria seedling blight in **herbs (see appendix 6)** *(off-label)* [1]
- White blister in **broccoli, brussels sprouts, cabbages, calabrese, cauliflowers, collards, kale** [1]
- White tip in **leeks** *(qualified minor use)* [1]

Extension of Authorisation for Minor Use (EAMUs)

- **asparagus** *20171501* [1]
- **chicory** *20171503* [1]
- **chicory grown outside for forcing** *20171503* [1]
- **choi sum** *20171502* [1]
- **herbs (see appendix 6)** *20182671* [1]
- **horseradish** *20171340* [1]
- **oriental brassicas** *20171502* [1]
- **parsley root** *20171340* [1]
- **parsnips** *20171340* [1]
- **salsify** *20171340* [1]

Approval information

- Azoxystrobin and difenoconazole included in Annex I under EC Regulation 1107/2009

Efficacy guidance

- Best results obtained from applications made in the earliest stages of disease development or as a protectant treatment following a disease risk assessment
- Ensure the crop is free from any stress caused by environmental or agronomic effects
- Azoxystrobin is a member of the QoI cross resistance group. Product should be used preventatively and not relied on for its curative potential
- Use as part of an Integrated Crop Management strategy incorporating other methods of control, including where appropriate other fungicides with a different mode of action. Do not apply more than two foliar applications of QoI containing products

Restrictions

- Maximum number of treatments 2 per crop
- Do not apply where there is a risk of spray drift onto neighbouring apple crops
- Consult processors before treating a crop destined for processing

Crop-specific information

- HI 14 d for carrots; 21 d for brassicas, leeks
- Minimum spray interval of 14 d must be observed on brassicas

Environmental safety

- Dangerous for the environment
- Very toxic to aquatic organisms
- LERAP Category B

FOR FULL CONDITIONS OF USE ALWAYS READ THE PRODUCT LABEL

Hazard classification and safety precautions

Hazard Irritant, Dangerous for the environment, Harmful if swallowed, Harmful if inhaled
Transport code 9
Packaging group III
UN Number 3082
Risk phrases H317 [1]
Operator protection A, H; U05a, U09a, U19a, U20b
Environmental protection E15b, E16a, E38, H410
Storage and disposal D01, D02, D05, D09a, D10c, D12a

25   azoxystrobin + fluazinam

A blight fungicide for use in potatoes
FRAC mode of action code: 11 + 29

Products

Vendetta	FMC Agro	150:375 g/l	SC	18709

Uses
- Blight in **potatoes**
- Leaf blight in **potatoes**

Approval information
- Azoxystrobin and fluazinam included in Annex I under EC Regulation 1107/2009

Efficacy guidance
- Applications must begin prior to blight development. The first application should be applied at the first blight warning or when local weather conditions are favourable for disease development, whichever is the sooner. In the absence of weather conducive to disease development, the first application should be made just before the crop meets in the rows

Restrictions
- Do not use more than 3 consecutive QoI-containing sprays
- Certain apple varieties are highly sensitive. As a precaution , do not apply when there is a risk of spray drift onto neighbouring apple crops. Spray equipment used for application should not be used to treat apples

Environmental safety
- Buffer zone requirement 7m

Hazard classification and safety precautions

Hazard Very toxic to aquatic organisms
Transport code 9
Packaging group III
UN Number 3082
Risk phrases H317, H361
Operator protection A, H
Environmental protection H410
Storage and disposal D01, D02, D05

26   azoxystrobin + isopyrazam

A fungicide mixture for use in oilseed rape
FRAC mode of action code: 11 + 7

See also azoxystrobin

Products

1	Symetra	Syngenta	200:125 g/l	SC	16701
2	Symetra	Adama	200:125 g/l	SC	18556

SEE SECTION 3 FOR PRODUCTS ALSO REGISTERED

Uses
- Sclerotinia stem rot in *spring oilseed rape*, *winter oilseed rape*

Approval information
- Azoxystrobin and isopyrazam are included in Appendix 1 under EC Regulation 1107/2009

Efficacy guidance
- Rainfast within 1 hour of application.

Restrictions
- Do not use application equipment used to apply Symetra on apples or damage will occur.
- Contains a member of the QoI cross resistance group and a member of the SDHI cross resistance group. Should be used preventatively and should not be relied on for its curative potential.

Environmental safety
- LERAP Category B

Hazard classification and safety precautions
> **Hazard** Toxic, Dangerous for the environment, Harmful if swallowed, Toxic if inhaled, Very toxic to aquatic organisms
> **Transport code** 9
> **Packaging group** III
> **UN Number** 3082
> **Risk phrases** H361
> **Operator protection** A, H; U05a, U19a
> **Environmental protection** E16a, E38, H410
> **Storage and disposal** D01, D02, D12a

27   azoxystrobin + tebuconazole

A strobilurin/triazole fungicide mixture for disease control in oilseed rape
FRAC mode of action code: 11 + 3

Products

Custodia	Adama	120:200 g/l	SC	16393

Uses
- Chocolate spot in *spring field beans* (moderate control), *winter field beans*
- Disease control in *spring field beans*, *winter field beans* (moderate control)
- Sclerotinia in *spring oilseed rape* (moderate control), *winter oilseed rape* (moderate control)

Approval information
- Azoxystrobin and tebuconazole included in Annex I under EC Regulation 1107/2009

Efficacy guidance
- Azoxystrobin is a member of the QoI cross resistance group. Product should be used preventatively and not relied on for its curative potential
- Use product as part of an Integrated Crop Management strategy incorporating other methods of control, including where appropriate other fungicides with a different mode of action. Do not apply more than two foliar applications of QoI containing products to any crop

Restrictions
- Avoid drift on to neighbouring crops since damage may occur especially to broad-leaved plants
- Certain apple varieties are highly sensitive to azoxystrobin. As a precaution, azoxystrobin should not be applied when there is a risk of spray drift onto neighbouring apple crops. Spray equipment used to apply azoxystrobin to other crops should not be used to treat apples.
- Newer authorisations for tebuconazole products require application to cereals only after GS 30 and applications to oilseed rape and linseed after GS20 - check label

Environmental safety
- LERAP Category B

Hazard classification and safety precautions

Hazard Harmful, Dangerous for the environment, Harmful if swallowed, Very toxic to aquatic organisms

Transport code 9

Packaging group III

UN Number 3082

Risk phrases H361

Operator protection A, H; U02a, U04a, U05a, U20b

Environmental protection E15b, E16a, E34, E38, H410

Storage and disposal D01, D02, D09a, D10b, D12a

Medical advice M03, M05a

28   Bacillus amyloliquefaciens D747

A fungicide for use on a range of horticultural crops
FRAC mode of action code: 44

Products

Amylo X WG	Certis	25% w/w	PO	17978

Uses

*   Alternaria in *aubergines, blackberries, blackcurrants, blueberries, chicory, chillies, courgettes, cress, cucumbers, endives, gooseberries, lamb's lettuce, land cress, lettuce, loganberries, melons, mushrooms, peppers, pumpkins, raspberries, red mustard, redcurrants, rocket, rubus hybrids, spinach, spinach beet, strawberries, summer squash, tomatoes, watercress, watermelons, winter squash*
*   Anthracnose in *baby leaf crops* (off-label), *broccoli* (off-label), *brussels sprouts* (off-label), *cabbages* (off-label), *calabrese* (off-label), *cauliflowers* (off-label), *choi sum* (off-label), *collards* (off-label), *courgettes* (off-label), *edible flowers* (off-label), *gherkins* (off-label), *herbs (see appendix 6)* (off-label), *kale* (off-label), *kohlrabi* (off-label), *mustard* (off-label), *okra* (off-label), *oriental cabbage* (off-label), *ornamental plant production* (off-label), *parsley* (off-label), *protected gherkins* (off-label), *protected okra* (off-label), *pumpkins* (off-label), *spinach* (off-label), *spinach beet* (off-label), *summer squash* (off-label), *sweetcorn* (off-label), *watercress* (off-label), *winter squash* (off-label)
*   Bacterial canker in *apples* (off-label), *apricots* (off-label), *cherries* (off-label), *chicory* (off-label), *chicory root* (off-label), *chinese dates* (off-label), *crab apples* (off-label), *lamb's lettuce* (off-label), *lettuce* (off-label), *medlar* (off-label), *ornamental plant production* (off-label), *peaches* (off-label), *pears* (off-label), *plums* (off-label), *quinces* (off-label), *rocket* (off-label), *strawberries* (off-label), *wine grapes* (off-label), *witloof* (off-label)
*   Black rot in *broccoli* (off-label), *brussels sprouts* (off-label), *cabbages* (off-label), *calabrese* (off-label), *cauliflowers* (off-label), *choi sum* (off-label), *collards* (off-label), *kale* (off-label), *kohlrabi* (off-label), *oriental cabbage* (off-label)
*   Botrytis in *apples* (off-label), *apricots* (off-label), *baby leaf crops* (off-label), *bilberries* (off-label), *blackberries* (off-label), *blackcurrants* (off-label), *blueberries* (off-label), *broccoli* (off-label), *brussels sprouts* (off-label), *cabbages* (off-label), *calabrese* (off-label), *cauliflowers* (off-label), *cherries* (off-label), *chicory* (off-label), *chicory root* (off-label), *chinese dates* (off-label), *choi sum* (off-label), *collards* (off-label), *courgettes* (off-label), *crab apples* (off-label), *cranberries* (off-label), *edible flowers* (off-label), *gherkins* (off-label), *gooseberries* (off-label), *herbs (see appendix 6)* (off-label), *kale* (off-label), *kohlrabi* (off-label), *lamb's lettuce* (off-label), *lettuce* (off-label), *loganberries* (off-label), *medlar* (off-label), *mustard* (off-label), *okra* (off-label), *oriental cabbage* (off-label), *ornamental plant production* (off-label), *parsley* (off-label), *peaches* (off-label), *pears* (off-label), *plums* (off-label), *protected bilberries* (off-label), *protected blackberries* (off-label), *protected blackcurrants* (off-label), *protected blueberry* (off-label), *protected cranberries* (off-label), *protected gherkins* (off-label), *protected gooseberries* (off-label), *protected loganberries* (off-label), *protected okra* (off-label), *protected raspberries* (off-label), *protected redcurrants* (off-label), *protected rose hips* (off-label), *protected rubus hybrids* (off-label), *pumpkins* (off-label), *quinces* (off-label), *raspberries* (off-label), *redcurrants* (off-label), *rocket* (off-label), *rose hips* (off-label), *rubus hybrids* (off-label), *spinach* (off-label), *spinach beet* (off-label), *strawberries* (off-label),

summer squash *(off-label)*, **sweetcorn** *(off-label)*, **watercress** *(off-label)*, **wine grapes** *(off-label)*, **winter squash** *(off-label)*, **witloof** *(off-label)*

- Brown rot in **apples** *(off-label)*, **apricots** *(off-label)*, **cherries** *(off-label)*, **chicory** *(off-label)*, **chicory root** *(off-label)*, **chinese dates** *(off-label)*, **crab apples** *(off-label)*, **lamb's lettuce** *(off-label)*, **lettuce** *(off-label)*, **medlar** *(off-label)*, **peaches** *(off-label)*, **pears** *(off-label)*, **plums** *(off-label)*, **quinces** *(off-label)*, **rocket** *(off-label)*, **strawberries** *(off-label)*, **wine grapes** *(off-label)*, **witloof** *(off-label)*
- Cane blight in **bilberries** *(off-label)*, **blackberries** *(off-label)*, **blackcurrants** *(off-label)*, **blueberries** *(off-label)*, **cranberries** *(off-label)*, **gooseberries** *(off-label)*, **loganberries** *(off-label)*, **protected bilberries** *(off-label)*, **protected blackberries** *(off-label)*, **protected blackcurrants** *(off-label)*, **protected blueberry** *(off-label)*, **protected cranberries** *(off-label)*, **protected gooseberries** *(off-label)*, **protected loganberries** *(off-label)*, **protected raspberries** *(off-label)*, **protected redcurrants** *(off-label)*, **protected rose hips** *(off-label)*, **protected rubus hybrids** *(off-label)*, **raspberries** *(off-label)*, **redcurrants** *(off-label)*, **rose hips** *(off-label)*, **rubus hybrids** *(off-label)*
- Cercospora leaf spot in **baby leaf crops** *(off-label)*, **courgettes** *(off-label)*, **edible flowers** *(off-label)*, **gherkins** *(off-label)*, **herbs (see appendix 6)** *(off-label)*, **mustard** *(off-label)*, **okra** *(off-label)*, **ornamental plant production** *(off-label)*, **parsley** *(off-label)*, **protected gherkins** *(off-label)*, **protected okra** *(off-label)*, **pumpkins** *(off-label)*, **spinach** *(off-label)*, **spinach beet** *(off-label)*, **summer squash** *(off-label)*, **sweetcorn** *(off-label)*, **watercress** *(off-label)*, **winter squash** *(off-label)*
- Cladosporium in **bilberries** *(off-label)*, **blackberries** *(off-label)*, **blackcurrants** *(off-label)*, **blueberries** *(off-label)*, **cranberries** *(off-label)*, **gooseberries** *(off-label)*, **loganberries** *(off-label)*, **protected bilberries** *(off-label)*, **protected blackberries** *(off-label)*, **protected blackcurrants** *(off-label)*, **protected blueberry** *(off-label)*, **protected cranberries** *(off-label)*, **protected gooseberries** *(off-label)*, **protected loganberries** *(off-label)*, **protected raspberries** *(off-label)*, **protected redcurrants** *(off-label)*, **protected rose hips** *(off-label)*, **protected rubus hybrids** *(off-label)*, **raspberries** *(off-label)*, **redcurrants** *(off-label)*, **rose hips** *(off-label)*, **rubus hybrids** *(off-label)*
- Damping off in **baby leaf crops** *(off-label)*, **broccoli** *(off-label)*, **brussels sprouts** *(off-label)*, **cabbages** *(off-label)*, **calabrese** *(off-label)*, **cauliflowers** *(off-label)*, **choi sum** *(off-label)*, **collards** *(off-label)*, **edible flowers** *(off-label)*, **herbs (see appendix 6)** *(off-label)*, **kale** *(off-label)*, **kohlrabi** *(off-label)*, **mustard** *(off-label)*, **oriental cabbage** *(off-label)*, **ornamental plant production** *(off-label)*, **parsley** *(off-label)*, **spinach** *(off-label)*, **spinach beet** *(off-label)*, **watercress** *(off-label)*
- Downy mildew in **baby leaf crops** *(off-label)*, **broccoli** *(off-label)*, **brussels sprouts** *(off-label)*, **cabbages** *(off-label)*, **calabrese** *(off-label)*, **cauliflowers** *(off-label)*, **choi sum** *(off-label)*, **collards** *(off-label)*, **courgettes** *(off-label)*, **edible flowers** *(off-label)*, **gherkins** *(off-label)*, **herbs (see appendix 6)** *(off-label)*, **kale** *(off-label)*, **kohlrabi** *(off-label)*, **mustard** *(off-label)*, **okra** *(off-label)*, **oriental cabbage** *(off-label)*, **ornamental plant production** *(off-label)*, **parsley** *(off-label)*, **protected gherkins** *(off-label)*, **protected okra** *(off-label)*, **pumpkins** *(off-label)*, **spinach** *(off-label)*, **spinach beet** *(off-label)*, **summer squash** *(off-label)*, **sweetcorn** *(off-label)*, **watercress** *(off-label)*, **winter squash** *(off-label)*
- Fireblight in **apples** *(off-label)*, **apricots** *(off-label)*, **cherries** *(off-label)*, **chicory** *(off-label)*, **chicory root** *(off-label)*, **chinese dates** *(off-label)*, **crab apples** *(off-label)*, **lamb's lettuce** *(off-label)*, **lettuce** *(off-label)*, **medlar** *(off-label)*, **ornamental plant production** *(off-label)*, **peaches** *(off-label)*, **pears** *(off-label)*, **plums** *(off-label)*, **quinces** *(off-label)*, **rocket** *(off-label)*, **strawberries** *(off-label)*, **wine grapes** *(off-label)*, **witloof** *(off-label)*
- Fusarium in **aubergines**, **baby leaf crops** *(off-label)*, **blackberries**, **blackcurrants**, **blueberries**, **broccoli** *(off-label)*, **brussels sprouts** *(off-label)*, **cabbages** *(off-label)*, **calabrese** *(off-label)*, **cauliflowers** *(off-label)*, **chicory**, **chillies**, **choi sum** *(off-label)*, **collards** *(off-label)*, **courgettes**, **courgettes** *(off-label)*, **cress**, **cucumbers**, **edible flowers** *(off-label)*, **endives**, **gherkins** *(off-label)*, **gooseberries**, **herbs (see appendix 6)** *(off-label)*, **kale** *(off-label)*, **kohlrabi** *(off-label)*, **lamb's lettuce**, **land cress**, **lettuce**, **loganberries**, **melons**, **mushrooms**, **mustard** *(off-label)*, **okra** *(off-label)*, **oriental cabbage** *(off-label)*, **ornamental plant production** *(off-label)*, **parsley** *(off-label)*, **peppers**, **protected gherkins** *(off-label)*, **protected okra** *(off-label)*, **pumpkins**, **pumpkins** *(off-label)*, **raspberries**, **red mustard**, **redcurrants**, **rocket**, **rubus hybrids**, **spinach**, **spinach** *(off-label)*, **spinach beet**, **spinach beet** *(off-label)*, **strawberries**, **summer squash**, **summer squash** *(off-label)*, **sweetcorn** *(off-label)*, **tomatoes**, **watercress**, **watercress** *(off-label)*, **watermelons**, **winter squash**, **winter squash** *(off-label)*

- Monilinia spp. in **apples** *(off-label)*, **apricots** *(off-label)*, **cherries** *(off-label)*, **chicory** *(off-label)*, **chicory root** *(off-label)*, **chinese dates** *(off-label)*, **crab apples** *(off-label)*, **lamb's lettuce** *(off-label)*, **lettuce** *(off-label)*, **medlar** *(off-label)*, **ornamental plant production** *(off-label)*, **peaches** *(off-label)*, **pears** *(off-label)*, **plums** *(off-label)*, **quinces** *(off-label)*, **rocket** *(off-label)*, **strawberries** *(off-label)*, **wine grapes** *(off-label)*, **witloof** *(off-label)*
- Neck rot in **bulb onions** *(off-label)*, **shallots** *(off-label)*
- Powdery mildew in **apples** *(off-label)*, **apricots** *(off-label)*, **aubergines**, **baby leaf crops** *(off-label)*, **bilberries** *(off-label)*, **blackberries**, **blackberries** *(off-label)*, **blackcurrants**, **blackcurrants** *(off-label)*, **blueberries**, **blueberries** *(off-label)*, **broccoli** *(off-label)*, **brussels sprouts** *(off-label)*, **cabbages** *(off-label)*, **calabrese** *(off-label)*, **cauliflowers** *(off-label)*, **cherries** *(off-label)*, **chicory**, **chicory** *(off-label)*, **chicory root** *(off-label)*, **chillies**, **chinese dates** *(off-label)*, **choi sum** *(off-label)*, **collards** *(off-label)*, **courgettes**, **courgettes** *(off-label)*, **crab apples** *(off-label)*, **cranberries** *(off-label)*, **cress**, **cucumbers**, **edible flowers** *(off-label)*, **endives**, **gherkins** *(off-label)*, **gooseberries**, **gooseberries** *(off-label)*, **herbs (see appendix 6)** *(off-label)*, **kale** *(off-label)*, **kohlrabi** *(off-label)*, **lamb's lettuce**, **lamb's lettuce** *(off-label)*, **land cress**, **lettuce**, **lettuce** *(off-label)*, **loganberries**, **loganberries** *(off-label)*, **medlar** *(off-label)*, **melons**, **mustard** *(off-label)*, **okra** *(off-label)*, **oriental cabbage** *(off-label)*, **ornamental plant production** *(off-label)*, **parsley** *(off-label)*, **peaches** *(off-label)*, **pears** *(off-label)*, **peppers**, **plums** *(off-label)*, **protected bilberries** *(off-label)*, **protected blackberries** *(off-label)*, **protected blackcurrants** *(off-label)*, **protected blueberry** *(off-label)*, **protected cranberries** *(off-label)*, **protected gherkins** *(off-label)*, **protected gooseberries** *(off-label)*, **protected loganberries** *(off-label)*, **protected okra** *(off-label)*, **protected raspberries** *(off-label)*, **protected redcurrants** *(off-label)*, **protected rose hips** *(off-label)*, **protected rubus hybrids** *(off-label)*, **pumpkins**, **pumpkins** *(off-label)*, **quinces** *(off-label)*, **raspberries**, **raspberries** *(off-label)*, **red mustard**, **redcurrants**, **redcurrants** *(off-label)*, **rocket**, **rocket** *(off-label)*, **rose hips** *(off-label)*, **rubus hybrids**, **rubus hybrids** *(off-label)*, **spinach**, **spinach** *(off-label)*, **spinach beet**, **spinach beet** *(off-label)*, **strawberries**, **strawberries** *(off-label)*, **summer squash**, **summer squash** *(off-label)*, **sweetcorn** *(off-label)*, **tomatoes**, **watercress**, **watercress** *(off-label)*, **watermelons**, **wine grapes** *(off-label)*, **winter squash**, **winter squash** *(off-label)*, **witloof** *(off-label)*
- Pseudomonas in **baby leaf crops** *(off-label)*, **courgettes** *(off-label)*, **edible flowers** *(off-label)*, **gherkins** *(off-label)*, **herbs (see appendix 6)** *(off-label)*, **mustard** *(off-label)*, **okra** *(off-label)*, **parsley** *(off-label)*, **protected gherkins** *(off-label)*, **protected okra** *(off-label)*, **pumpkins** *(off-label)*, **spinach** *(off-label)*, **spinach beet** *(off-label)*, **summer squash** *(off-label)*, **sweetcorn** *(off-label)*, **watercress** *(off-label)*, **winter squash** *(off-label)*
- Pythium in **aubergines**, **baby leaf crops** *(off-label)*, **blackberries**, **blackcurrants**, **blueberries**, **broccoli** *(off-label)*, **brussels sprouts** *(off-label)*, **cabbages** *(off-label)*, **calabrese** *(off-label)*, **cauliflowers** *(off-label)*, **chicory**, **chillies**, **choi sum** *(off-label)*, **collards** *(off-label)*, **courgettes**, **courgettes** *(off-label)*, **cress**, **cucumbers**, **edible flowers** *(off-label)*, **endives**, **gherkins** *(off-label)*, **gooseberries**, **herbs (see appendix 6)** *(off-label)*, **kale** *(off-label)*, **kohlrabi** *(off-label)*, **lamb's lettuce**, **land cress**, **lettuce**, **loganberries**, **melons**, **mushrooms**, **mustard** *(off-label)*, **okra** *(off-label)*, **oriental cabbage** *(off-label)*, **ornamental plant production** *(off-label)*, **parsley** *(off-label)*, **peppers**, **protected gherkins** *(off-label)*, **protected okra** *(off-label)*, **pumpkins**, **pumpkins** *(off-label)*, **raspberries**, **red mustard**, **redcurrants**, **rocket**, **rubus hybrids**, **spinach**, **spinach** *(off-label)*, **spinach beet**, **spinach beet** *(off-label)*, **strawberries**, **summer squash**, **summer squash** *(off-label)*, **sweetcorn** *(off-label)*, **tomatoes**, **watercress**, **watercress** *(off-label)*, **watermelons**, **winter squash**, **winter squash** *(off-label)*
- Rhizoctonia in **aubergines**, **baby leaf crops** *(off-label)*, **blackberries**, **blackcurrants**, **blueberries**, **broccoli** *(off-label)*, **brussels sprouts** *(off-label)*, **cabbages** *(off-label)*, **calabrese** *(off-label)*, **cauliflowers** *(off-label)*, **chicory**, **chillies**, **choi sum** *(off-label)*, **collards** *(off-label)*, **courgettes**, **courgettes** *(off-label)*, **cress**, **cucumbers**, **edible flowers** *(off-label)*, **endives**, **gherkins** *(off-label)*, **gooseberries**, **herbs (see appendix 6)** *(off-label)*, **kale** *(off-label)*, **kohlrabi** *(off-label)*, **lamb's lettuce**, **land cress**, **lettuce**, **loganberries**, **melons**, **mushrooms**, **mustard** *(off-label)*, **okra** *(off-label)*, **oriental cabbage** *(off-label)*, **ornamental plant production** *(off-label)*, **parsley** *(off-label)*, **peppers**, **protected gherkins** *(off-label)*, **protected okra** *(off-label)*, **pumpkins**, **pumpkins** *(off-label)*, **raspberries**, **red mustard**, **redcurrants**, **rocket**, **rubus hybrids**, **spinach**, **spinach** *(off-label)*, **spinach beet**, **spinach beet** *(off-label)*, **strawberries**, **summer squash**, **summer squash** *(off-label)*, **sweetcorn** *(off-label)*, **tomatoes**, **watercress**, **watercress** *(off-label)*, **watermelons**, **winter squash**, **winter squash** *(off-label)*

SEE SECTION 3 FOR PRODUCTS ALSO REGISTERED

- Sclerotinia in **baby leaf crops** *(off-label)*, **broccoli** *(off-label)*, **brussels sprouts** *(off-label)*, **cabbages** *(off-label)*, **calabrese** *(off-label)*, **cauliflowers** *(off-label)*, **choi sum** *(off-label)*, **collards** *(off-label)*, **courgettes** *(off-label)*, **edible flowers** *(off-label)*, **gherkins** *(off-label)*, **herbs (see appendix 6)** *(off-label)*, **kale** *(off-label)*, **kohlrabi** *(off-label)*, **mustard** *(off-label)*, **okra** *(off-label)*, **oriental cabbage** *(off-label)*, **ornamental plant production** *(off-label)*, **parsley** *(off-label)*, **protected gherkins** *(off-label)*, **protected okra** *(off-label)*, **pumpkins** *(off-label)*, **spinach** *(off-label)*, **spinach beet** *(off-label)*, **summer squash** *(off-label)*, **sweetcorn** *(off-label)*, **watercress** *(off-label)*, **winter squash** *(off-label)*
- Sclerotinia rot in **apples** *(off-label)*, **apricots** *(off-label)*, **cherries** *(off-label)*, **chicory** *(off-label)*, **chicory root** *(off-label)*, **chinese dates** *(off-label)*, **crab apples** *(off-label)*, **lamb's lettuce** *(off-label)*, **lettuce** *(off-label)*, **medlar** *(off-label)*, **peaches** *(off-label)*, **pears** *(off-label)*, **plums** *(off-label)*, **quinces** *(off-label)*, **rocket** *(off-label)*, **strawberries** *(off-label)*, **wine grapes** *(off-label)*, **witloof** *(off-label)*
- Spear rot in **broccoli** *(off-label)*, **brussels sprouts** *(off-label)*, **cabbages** *(off-label)*, **calabrese** *(off-label)*, **cauliflowers** *(off-label)*, **choi sum** *(off-label)*, **collards** *(off-label)*, **kale** *(off-label)*, **kohlrabi** *(off-label)*, **oriental cabbage** *(off-label)*
- Stemphylium in **baby leaf crops** *(off-label)*, **courgettes** *(off-label)*, **edible flowers** *(off-label)*, **gherkins** *(off-label)*, **herbs (see appendix 6)** *(off-label)*, **mustard** *(off-label)*, **okra** *(off-label)*, **ornamental plant production** *(off-label)*, **parsley** *(off-label)*, **protected gherkins** *(off-label)*, **protected okra** *(off-label)*, **pumpkins** *(off-label)*, **spinach** *(off-label)*, **spinach beet** *(off-label)*, **summer squash** *(off-label)*, **sweetcorn** *(off-label)*, **watercress** *(off-label)*, **winter squash** *(off-label)*
- Xanthomonas in **baby leaf crops** *(off-label)*, **courgettes** *(off-label)*, **edible flowers** *(off-label)*, **gherkins** *(off-label)*, **herbs (see appendix 6)** *(off-label)*, **mustard** *(off-label)*, **okra** *(off-label)*, **parsley** *(off-label)*, **protected gherkins** *(off-label)*, **protected okra** *(off-label)*, **pumpkins** *(off-label)*, **spinach** *(off-label)*, **spinach beet** *(off-label)*, **summer squash** *(off-label)*, **sweetcorn** *(off-label)*, **watercress** *(off-label)*, **winter squash** *(off-label)*

Extension of Authorisation for Minor Use (EAMUs)
- **apples** *20180469*
- **apricots** *20180469*
- **baby leaf crops** *20183116*
- **bilberries** *20193147*
- **blackberries** *20193147*
- **blackcurrants** *20193147*
- **blueberries** *20193147*
- **broccoli** *20190427*
- **brussels sprouts** *20190427*
- **bulb onions** *20181580*
- **cabbages** *20190427*
- **calabrese** *20190427*
- **cauliflowers** *20190427*
- **cherries** *20180469*
- **chicory** *20180469*
- **chicory root** *20180469*
- **chinese dates** *20180469*
- **choi sum** *20190427*
- **collards** *20190427*
- **courgettes** *20192204*
- **crab apples** *20180469*
- **cranberries** *20193147*
- **edible flowers** *20183116*
- **gherkins** *20192204*
- **gooseberries** *20193147*
- **herbs (see appendix 6)** *20183116*
- **kale** *20190427*
- **kohlrabi** *20190427*
- **lamb's lettuce** *20180469*
- **lettuce** *20180469*
- **loganberries** *20193147*

- *medlar* 20180469
- *mustard* 20183116
- *okra* 20192204
- *oriental cabbage* 20190427
- *ornamental plant production* 20190428
- *parsley* 20183116
- *peaches* 20180469
- *pears* 20180469
- *plums* 20180469
- *protected bilberries* 20193147
- *protected blackberries* 20193147
- *protected blackcurrants* 20193147
- *protected blueberry* 20193147
- *protected cranberries* 20193147
- *protected gherkins* 20192204
- *protected gooseberries* 20193147
- *protected loganberries* 20193147
- *protected okra* 20192204
- *protected raspberries* 20193147
- *protected redcurrants* 20193147
- *protected rose hips* 20193147
- *protected rubus hybrids* 20193147
- *pumpkins* 20192204
- *quinces* 20180469
- *raspberries* 20193147
- *redcurrants* 20193147
- *rocket* 20180469
- *rose hips* 20193147
- *rubus hybrids* 20193147
- *shallots* 20181580
- *spinach* 20183116
- *spinach beet* 20183116
- *strawberries* 20180469
- *summer squash* 20192204
- *sweetcorn* 20192586
- *watercress* 20183116
- *wine grapes* 20180469
- *winter squash* 20192204
- *witloof* 20180469

Approval information
- Bacillus amyloliquefaciens included in Annex I under EC Regulation 1107/2009

Restrictions
- Do not apply before BBCH 14 on leafy vegetables and BBCH 10 on pome fruit, stone fruit, vine and strawberry

Hazard classification and safety precautions
 UN Number N/C
 Operator protection A, D, H; U05b
 Environmental protection E15b, E34
 Storage and disposal D01, D02, D10c

29   Bacillus amyloliquefaciens strain MBI600

A flowable seed treatment for use in winter oilseed rape
FRAC mode of action code: 44

Products

Integral Pro	BASF	6.85% w/v	FS	18510

SEE SECTION 3 FOR PRODUCTS ALSO REGISTERED

Uses
- Cabbage stem flea beetle in **winter oilseed rape** *(stimulation of plants own defenses)*
- Stem canker in **winter oilseed rape** *(useful reduction)*

Approval information
- Bacillus amyloliquefaciens MBI600 included in Annex I under EC Regulation 1107/2009

Efficacy guidance
- It is important to achieve good even coverage on the seed to obtain reliable disease control.
- The germination capacity should be checked prior to use and in the event of any minor effects a small change in sowing rate may be made
- Some slight delay in emergence may occur, but this will be outgrown with no lasting effects

Restrictions
- Do not treat grain with a moisture content exceeding 16% and do not allow the moisture content of treated seed to exceed 16%.
- Do not apply to cracked, split or sprouted seed.

Hazard classification and safety precautions
UN Number N/C
Operator protection A, D, H; U05a
Environmental protection E03, E15b, E34
Storage and disposal D01, D02, D09a, D11a, D21
Treated seed S01, S02, S04b, S05, S07

30 Bacillus subtilis

A bacterial fungicide for the control of Botrytis cinerea
FRAC mode of action code: 44

Products

Serenade ASO	Bayer CropScience	13.96 g/l	SC	16139

Uses
- Botrytis in **almonds** *(off-label)*, **apples** *(off-label)*, **apricots** *(off-label)*, **asparagus** *(off-label)*, **aubergines** *(off-label)*, **baby leaf crops** *(off-label)*, **beans without pods (dry)** *(off-label)*, **beans without pods (fresh)** *(off-label)*, **bilberries** *(off-label)*, **blackberries** *(off-label)*, **blackcurrants** *(off-label)*, **blueberries** *(off-label)*, **broad beans** *(off-label)*, **broccoli** *(off-label)*, **brussels sprouts** *(off-label)*, **bulb onions** *(off-label)*, **cabbages** *(off-label)*, **calabrese** *(off-label)*, **canary grass** *(off-label)*, **cardoons** *(off-label)*, **carrots** *(off-label)*, **cauliflowers** *(off-label)*, **celeriac** *(off-label)*, **celery (outdoor)** *(off-label)*, **cherries** *(off-label)*, **chestnuts** *(off-label)*, **chickpeas** *(off-label)*, **chicory** *(off-label)*, **chicory root** *(off-label)*, **chillies** *(off-label)*, **choi sum** *(off-label)*, **collards** *(off-label)*, **combining peas** *(off-label)*, **courgettes** *(off-label)*, **cranberries** *(off-label)*, **cress** *(off-label)*, **cucumbers** *(off-label)*, **dwarf beans** *(off-label)*, **edible podded peas** *(off-label)*, **elderberries** *(off-label)*, **endives** *(off-label)*, **figs** *(off-label)*, **florence fennel** *(off-label)*, **french beans** *(off-label)*, **garlic** *(off-label)*, **gherkins** *(off-label)*, **globe artichoke** *(off-label)*, **gooseberries** *(off-label)*, **hazel nuts** *(off-label)*, **herbs (see appendix 6)** *(off-label)*, **hops** *(off-label)*, **horseradish** *(off-label)*, **jerusalem artichokes** *(off-label)*, **kale** *(off-label)*, **kohlrabi** *(off-label)*, **lamb's lettuce** *(off-label)*, **land clearance** *(off-label)*, **leeks** *(off-label)*, **lentils** *(off-label)*, **lettuce** *(off-label)*, **loganberries** *(off-label)*, **lupins** *(off-label)*, **medlar** *(off-label)*, **melons** *(off-label)*, **mulberries** *(off-label)*, **nectarines** *(off-label)*, **okra** *(off-label)*, **oriental cabbage** *(off-label)*, **ornamental plant production** *(off-label)*, **parsley root** *(off-label)*, **parsnips** *(off-label)*, **peaches** *(off-label)*, **pears** *(off-label)*, **peppers** *(off-label)*, **plums** *(off-label)*, **pumpkins** *(off-label)*, **quinces** *(off-label)*, **radishes** *(off-label)*, **raspberries** *(off-label)*, **red beet** *(off-label)*, **red mustard** *(off-label)*, **redcurrants** *(off-label)*, **rhubarb** *(off-label)*, **rocket** *(off-label)*, **rose hips** *(off-label)*, **rubus hybrids** *(off-label)*, **runner beans** *(off-label)*, **salad onions** *(off-label)*, **salsify** *(off-label)*, **shallots** *(off-label)*, **spinach** *(off-label)*, **spinach beet** *(off-label)*, **strawberries** *(off-label)*, **summer squash** *(off-label)*, **swedes** *(off-label)*, **sweetcorn** *(off-label)*, **table grapes** *(off-label)*, **tomatoes (outdoor)** *(off-label)*, **turnips** *(off-label)*, **vining peas** *(off-label)*, **walnuts** *(off-label)*, **watercress** *(off-label)*, **watermelons** *(off-label)*, **wine grapes** *(off-label)*, **winter squash** *(off-label)*, **witloof** *(off-label)*

- Botrytis fruit rot in **protected strawberries** *(reduction of damage to fruit)*, **strawberries**
- Butt rot in **lettuce** *(off-label)*
- Cavity spot in **carrots** *(off-label)*, **parsnips** *(off-label)*
- Damping off in **baby leaf crops** *(off-label)*, **broccoli** *(off-label)*, **brussels sprouts** *(off-label)*, **bulb onions** *(off-label)*, **cabbages** *(off-label)*, **calabrese** *(off-label)*, **carrots** *(off-label)*, **cauliflowers** *(off-label)*, **celeriac** *(off-label)*, **celery (outdoor)** *(off-label)*, **chard** *(off-label)*, **collards** *(off-label)*, **garlic** *(off-label)*, **herbs (see appendix 6)** *(off-label)*, **kale** *(off-label)*, **leeks** *(off-label)*, **lettuce** *(off-label)*, **parsnips** *(off-label)*, **salad onions** *(off-label)*, **shallots** *(off-label)*, **spinach** *(off-label)*
- Grey mould in **protected strawberries** *(reduction of damage to fruit)*, **strawberries**
- Helminthosporium in **potatoes** *(off-label)*
- Phytophthora in **amenity vegetation** *(off-label)*, **blackberries** *(off-label)*, **blackcurrants** *(off-label)*, **blueberries** *(off-label)*, **courgettes** *(off-label)*, **cranberries** *(off-label)*, **cucumbers** *(off-label)*, **forest nurseries** *(off-label)*, **gooseberries** *(off-label)*, **loganberries** *(off-label)*, **marrows** *(off-label)*, **pumpkins** *(off-label)*, **raspberries** *(off-label)*, **redcurrants** *(off-label)*, **rubus hybrids** *(off-label)*, **squashes** *(off-label)*
- Pythium in **aubergines** *(off-label)*, **broccoli** *(off-label)*, **brussels sprouts** *(off-label)*, **cabbages** *(off-label)*, **calabrese** *(off-label)*, **carrots** *(off-label)*, **cauliflowers** *(off-label)*, **chillies** *(off-label)*, **collards** *(off-label)*, **courgettes** *(off-label)*, **cucumbers** *(off-label)*, **kale** *(off-label)*, **marrows** *(off-label)*, **parsnips** *(off-label)*, **peppers** *(off-label)*, **pumpkins** *(off-label)*, **squashes** *(off-label)*, **tomatoes (outdoor)** *(off-label)*
- Rhizoctonia in **lettuce** *(off-label)*, **potatoes** *(off-label)*
- Streptomyces in **potatoes** *(off-label)*
- White rot in **bulb onions** *(off-label)*, **garlic** *(off-label)*, **leeks** *(off-label)*, **salad onions** *(off-label)*, **shallots** *(off-label)*

Extension of Authorisation for Minor Use (EAMUs)
- **almonds** *20182358*
- **amenity vegetation** *20130704*
- **apples** *20182360*
- **apricots** *20182354*
- **asparagus** *20182353*
- **aubergines** *20150306, 20182343*
- **baby leaf crops** *20150306, 20182363*
- **beans without pods (dry)** *20182352*
- **beans without pods (fresh)** *20182365*
- **bilberries** *20182346*
- **blackberries** *20150306, 20182336*
- **blackcurrants** *20150306, 20182346*
- **blueberries** *20150306, 20182346*
- **broad beans** *20182365*
- **broccoli** *20150306, 20182357*
- **brussels sprouts** *20150306, 20182357*
- **bulb onions** *20150306, 20182351*
- **cabbages** *20150306, 20182357, 20182825*
- **calabrese** *20150306, 20182357*
- **canary grass** *20182344*
- **cardoons** *20182353*
- **carrots** *20150306, 20182359*
- **cauliflowers** *20150306, 20182357*
- **celeriac** *20150306, 20182359*
- **celery (outdoor)** *20150306, 20182353*
- **chard** *20150306*
- **cherries** *20182354*
- **chestnuts** *20182358*
- **chickpeas** *20182352*
- **chicory** *20182363*
- **chicory root** *20182359*
- **chillies** *20150306, 20182343*
- **choi sum** *20182357*

SECTION 2

- **collards** *20150306, 20182357*
- **combining peas** *20182352*
- **courgettes** *20150306, 20182343*
- **cranberries** *20150306, 20182346*
- **cress** *20182363*
- **cucumbers** *20150306, 20182343*
- **dwarf beans** *20182365*
- **edible podded peas** *20182365*
- **elderberries** *20182346*
- **endives** *20182363*
- **figs** *20182341*
- **florence fennel** *20182353*
- **forest nurseries** *20130704*
- **french beans** *20182365*
- **garlic** *20150306, 20182351*
- **gherkins** *20182343*
- **globe artichoke** *20182353*
- **gooseberries** *20150306, 20182346*
- **hazel nuts** *20182358*
- **herbs (see appendix 6)** *20150306, 20182345*
- **hops** *20182355*
- **horseradish** *20182359*
- **jerusalem artichokes** *20182359*
- **kale** *20150306, 20182357*
- **kohlrabi** *20182357*
- **lamb's lettuce** *20182363*
- **land clearance** *20182363*
- **leeks** *20150306, 20182353*
- **lentils** *20182352, 20182365*
- **lettuce** *20150306, 20182363*
- **loganberries** *20150306, 20182336*
- **lupins** *20182352*
- **marrows** *20150306*
- **medlar** *20182360*
- **melons** *20182343*
- **mulberries** *20182346*
- **nectarines** *20182354*
- **okra** *20182343*
- **oriental cabbage** *20182357*
- **ornamental plant production** *20182364*
- **parsley root** *20182359*
- **parsnips** *20150306, 20182359*
- **peaches** *20182354*
- **pears** *20182360*
- **peppers** *20150306, 20182343*
- **plums** *20182354*
- **potatoes** *20150306*
- **pumpkins** *20150306, 20182343*
- **quinces** *20182360*
- **radishes** *20182359*
- **raspberries** *20150306, 20182336*
- **red beet** *20182359*
- **red mustard** *20182363*
- **redcurrants** *20150306, 20182346*
- **rhubarb** *20182353*
- **rocket** *20182363*
- **rose hips** *20182346*
- **rubus hybrids** *20150306, 20182336*
- **runner beans** *20182365*

- **salad onions** *20150306, 20182351*
- **salsify** *20182359*
- **shallots** *20150306, 20182351*
- **spinach** *20150306, 20182363*
- **spinach beet** *20182363*
- **squashes** *20150306*
- **strawberries** *20182356*
- **summer squash** *20182343*
- **swedes** *20182359*
- **sweetcorn** *20182343*
- **table grapes** *20182342*
- **tomatoes (outdoor)** *20150306, 20182343*
- **turnips** *20182359*
- **vining peas** *20182365*
- **walnuts** *20182358*
- **watercress** *20182363*
- **watermelons** *20182343*
- **wine grapes** *20182342*
- **winter squash** *20182343*
- **witloof** *20182363*

Approval information
- Bacillus subtilis included in Annex I under EC Regulation 1107/2009

Efficacy guidance
- Alternating applications with fungicides using a different mode of action is recommended for resistance management.
- Do not apply using irrigation equipment.
- For maximum effectiveness, start applications before disease development.
- Apply in a minimum water volume of 400 l/ha.

Restrictions
- Consult processor before using on crops grown for processing

Hazard classification and safety precautions
Hazard Irritant
UN Number N/C
Operator protection A, D, H; U05a, U11, U14, U20b
Environmental protection E15b
Storage and disposal D01, D02, D05, D10a, D16

31   Bacillus thuringiensis

A bacterial insecticide for control of caterpillars
IRAC mode of action code: 11

Products

1	Dipel DF	Sumitomo	54% w/w	WG	17499
2	Lepinox Plus	Fargro	37.5% w/w	WP	16269

Uses
- Bramble shoot moth in **bilberries** *(off-label)*, **blackberries** *(off-label)*, **blackcurrants** *(off-label)*, **blueberries** *(off-label)*, **cranberries** *(off-label)*, **gooseberries** *(off-label)*, **loganberries** *(off-label)*, **redcurrants** *(off-label)*, **rubus hybrids** *(off-label)* [2]
- Cabbage moth in **baby leaf crops** *(off-label)*, **choi sum** *(off-label)*, **collards** *(off-label)*, **kohlrabi** *(off-label)*, **oriental cabbage** *(off-label)*, **protected chicory** *(off-label)*, **protected cress** *(off-label)*, **protected lamb's lettuce** *(off-label)*, **protected rocket** *(off-label)*, **protected watercress** *(off-label)* [2]
- Cabbage white butterfly in **baby leaf crops** *(off-label)*, **choi sum** *(off-label)*, **collards** *(off-label)*, **kohlrabi** *(off-label)*, **oriental cabbage** *(off-label)*, **protected chicory** *(off-label)*, **protected cress**

(off-label), **protected lamb's lettuce** *(off-label)*, **protected rocket** *(off-label)*, **protected watercress** *(off-label)* [2]

- Caterpillars in **all edible seed crops grown outdoors** *(off-label)*, **all non-edible crops (outdoor)** *(off-label)*, **all protected non-edible crops** *(off-label)*, **amenity vegetation**, **apples** *(off-label)*, **apricots** *(off-label)*, **baby leaf crops** *(off-label)*, **calabrese** *(off-label)*, **cauliflowers**, **celery (outdoor)** *(off-label)*, **cherries** *(off-label)*, **chestnuts** *(off-label)*, **choi sum** *(off-label)*, **cob nuts** *(off-label)*, **collards** *(off-label)*, **edible podded peas**, **endives** *(off-label)*, **fennel** *(off-label)*, **filberts** *(off-label)*, **forest** *(off-label)*, **hazel nuts** *(off-label)*, **herbs (see appendix 6)** *(off-label)*, **kale** *(off-label)*, **kiwi fruit** *(off-label)*, **kohlrabi** *(off-label)*, **leeks**, **lettuce** *(off-label)*, **nectarines** *(off-label)*, **olives** *(off-label)*, **oriental cabbage** *(off-label)*, **ornamental plant production**, **peaches** *(off-label)*, **pears** *(off-label)*, **plums** *(off-label)*, **protected apricots** *(off-label)*, **protected baby leaf crops** *(off-label)*, **protected broad beans**, **protected broccoli** *(off-label)*, **protected brussels sprouts** *(off-label)*, **protected cabbages** *(off-label)*, **protected calabrese** *(off-label)*, **protected cauliflowers** *(off-label)*, **protected celery** *(off-label)*, **protected chestnuts** *(off-label)*, **protected chilli peppers**, **protected choi sum** *(off-label)*, **protected collards** *(off-label)*, **protected cucumbers**, **protected dwarf french beans**, **protected edible crops** *(off-label)*, **protected endives** *(off-label)*, **protected hazelnuts** *(off-label)*, **protected herbs (see appendix 6)** *(off-label)*, **protected kale** *(off-label)*, **protected kiwi fruit** *(off-label)*, **protected lettuce** *(off-label)*, **protected nectarines** *(off-label)*, **protected olives** *(off-label)*, **protected oriental cabbage** *(off-label)*, **protected peaches** *(off-label)*, **protected plums** *(off-label)*, **protected quince** *(off-label)*, **protected rhubarb** *(off-label)*, **protected runner beans**, **protected spinach** *(off-label)*, **protected spinach beet** *(off-label)*, **protected tatsoi** *(off-label)*, **protected walnuts** *(off-label)*, **protected watercress** *(off-label)*, **protected wine grapes** *(off-label)*, **quinces** *(off-label)*, **raspberries**, **rhubarb** *(off-label)*, **spinach** *(off-label)*, **spinach beet** *(off-label)*, **spring cabbage** *(off-label)*, **tatsoi** *(off-label)*, **vining peas**, **walnuts** *(off-label)*, **watercress** *(off-label)*, **wine grapes** *(off-label)* [1]; **apples, celery (outdoor), chicory, chinese cabbage, courgettes, dwarf beans, endives, french beans, herbs (see appendix 6), hops, kale, lettuce, pears, protected melons, protected watermelon, pumpkins, radishes, spinach, spinach beet, turnips, wine grapes, winter squash** [2]; **broccoli, brussels sprouts, cabbages, calabrese, combining peas, globe artichoke, protected aubergines, protected peppers, protected tomatoes, strawberries** [1, 2]
- Cherry fruit moth in **apricots** *(off-label)*, **cherries** *(off-label)*, **nectarines** *(off-label)*, **peaches** *(off-label)*, **plums** *(off-label)* [2]
- Cutworms in **baby leaf crops** *(off-label)*, **bilberries** *(off-label)*, **blackberries** *(off-label)*, **blackcurrants** *(off-label)*, **blueberries** *(off-label)*, **broad beans** *(off-label)*, **cranberries** *(off-label)*, **forest nurseries** *(off-label)*, **gooseberries** *(off-label)*, **loganberries** *(off-label)*, **protected chicory** *(off-label)*, **protected chilli peppers** *(off-label)*, **protected courgettes** *(off-label)*, **protected cress** *(off-label)*, **protected dwarf french beans** *(off-label)*, **protected edible podded peas** *(off-label)*, **protected lamb's lettuce** *(off-label)*, **protected peppers** *(off-label)*, **protected rocket** *(off-label)*, **protected runner beans** *(off-label)*, **protected summer squash** *(off-label)*, **protected watercress** *(off-label)*, **redcurrants** *(off-label)*, **rhubarb** *(off-label)*, **rubus hybrids** *(off-label)*, **sweetcorn** *(off-label)* [2]; **bulb onions** *(off-label)*, **carrots** *(off-label)*, **celeriac** *(off-label)*, **garlic** *(off-label)*, **horseradish** *(off-label)*, **mooli** *(off-label)*, **parsley root** *(off-label)*, **parsnips** *(off-label)*, **radishes** *(off-label)*, **red beet** *(off-label)*, **salsify** *(off-label)*, **shallots** *(off-label)*, **swedes** *(off-label)*, **turnips** *(off-label)* [1]; **leeks** *(off-label)* [1, 2]
- Diamond-back moth in **baby leaf crops** *(off-label)*, **broccoli, brussels sprouts, cabbages, calabrese, chinese cabbage, choi sum** *(off-label)*, **collards** *(off-label)*, **kohlrabi** *(off-label)*, **oriental cabbage** *(off-label)*, **protected chicory** *(off-label)*, **protected cress** *(off-label)*, **protected lamb's lettuce** *(off-label)*, **protected rocket** *(off-label)*, **protected watercress** *(off-label)*, **turnips** [2]
- Flax tortrix moth in **broad beans** *(off-label)*, **protected dwarf french beans** *(off-label)*, **protected edible podded peas** *(off-label)*, **protected runner beans** *(off-label)* [2]
- Leek moth in **leeks** *(off-label)*, **rhubarb** *(off-label)* [2]
- Light brown apple moth in **apricots** *(off-label)*, **cherries** *(off-label)*, **nectarines** *(off-label)*, **peaches** *(off-label)*, **plums** *(off-label)* [2]
- Oak Processionary Moth in **forest** *(off-label)* [1]
- Plum fruit moth in **apricots** *(off-label)*, **cherries** *(off-label)*, **nectarines** *(off-label)*, **peaches** *(off-label)*, **plums** *(off-label)* [2]

- Plum tortrix moth in **apricots** *(off-label)*, **cherries** *(off-label)*, **nectarines** *(off-label)*, **peaches** *(off-label)*, **plums** *(off-label)* [2]
- Plutella xylostella in **broccoli**, **brussels sprouts**, **cabbages**, **calabrese**, **chinese cabbage**, **turnips** [2]
- Raspberry moth in **bilberries** *(off-label)*, **blackberries** *(off-label)*, **blackcurrants** *(off-label)*, **blueberries** *(off-label)*, **cranberries** *(off-label)*, **gooseberries** *(off-label)*, **loganberries** *(off-label)*, **redcurrants** *(off-label)*, **rubus hybrids** *(off-label)* [2]
- Silver Y moth in **baby leaf crops** *(off-label)*, **protected chicory** *(off-label)*, **protected chilli peppers** *(off-label)*, **protected courgettes** *(off-label)*, **protected cress** *(off-label)*, **protected dwarf french beans** *(off-label)*, **protected edible podded peas** *(off-label)*, **protected lamb's lettuce** *(off-label)*, **protected peppers** *(off-label)*, **protected rocket** *(off-label)*, **protected runner beans** *(off-label)*, **protected summer squash** *(off-label)*, **protected watercress** *(off-label)* [2]; **broad beans** *(off-label)*, **sweetcorn** *(off-label)* [1, 2]; **dwarf beans** *(off-label)*, **french beans** *(off-label)*, **runner beans** *(off-label)* [1]
- Strawberry tortrix in **bilberries** *(off-label)*, **blackberries** *(off-label)*, **blackcurrants** *(off-label)*, **blueberries** *(off-label)*, **cranberries** *(off-label)*, **gooseberries** *(off-label)*, **loganberries** *(off-label)*, **redcurrants** *(off-label)*, **rubus hybrids** *(off-label)* [2]
- Summer-fruit tortrix moth in **apricots** *(off-label)*, **cherries** *(off-label)*, **nectarines** *(off-label)*, **peaches** *(off-label)*, **plums** *(off-label)* [2]
- Tomato moth in **protected chilli peppers** *(off-label)*, **protected courgettes** *(off-label)*, **protected peppers** *(off-label)*, **protected summer squash** *(off-label)*, **sweetcorn** *(off-label)* [2]
- Tortrix moths in **apples**, **pears**, **strawberries** [2]; **raspberries** [1]
- Vapourer moth in **forest nurseries** *(off-label)* [2]
- White ermine moth in **forest nurseries** *(off-label)* [2]
- Winter moth in **apricots** *(off-label)*, **cherries** *(off-label)*, **nectarines** *(off-label)*, **peaches** *(off-label)*, **plums** *(off-label)* [2]; **bilberries** *(off-label)*, **blackcurrants** *(off-label)*, **blueberries** *(off-label)*, **cranberries** *(off-label)*, **protected bilberries** *(off-label)*, **protected blackcurrants** *(off-label)*, **protected blueberry** *(off-label)*, **protected cranberries** *(off-label)*, **protected gooseberries** *(off-label)*, **protected redcurrants** *(off-label)*, **protected vaccinium spp.** *(off-label)*, **redcurrants** *(off-label)*, **vaccinium spp.** *(off-label)* [1]

Extension of Authorisation for Minor Use (EAMUs)
- *all edible seed crops grown outdoors* *20162634* [1]
- *all non-edible crops (outdoor)* *20162634* [1]
- *all protected non-edible crops* *20162634* [1]
- *apples* *20162635* [1]
- *apricots* *20162632* [1], *20142700* [2]
- *baby leaf crops* *20162627* [1], *20142704* [2]
- *bilberries* *20162633* [1], *20142706* [2]
- *blackberries* *20142706* [2]
- *blackcurrants* *20162633* [1], *20142706* [2]
- *blueberries* *20162633* [1], *20142706* [2]
- *broad beans* *20162625* [1], *20142702* [2]
- *bulb onions* *20162623* [1]
- *calabrese* *20162626* [1]
- *carrots* *20162631* [1]
- *celeriac* *20162631* [1]
- *celery (outdoor)* *20162626* [1]
- *cherries* *20162635* [1], *20142700* [2]
- *chestnuts* *20162632* [1]
- *choi sum* *20162626* [1], *20142701* [2]
- *cob nuts* *20162632* [1]
- *collards* *20162626* [1], *20142701* [2]
- *cranberries* *20162633* [1], *20142706* [2]
- *dwarf beans* *20162625* [1]
- *endives* *20162627* [1]
- *fennel* *20162626* [1]
- *filberts* *20162632* [1]
- *forest* *20160931* [1]

SEE SECTION 3 FOR PRODUCTS ALSO REGISTERED

- **forest nurseries** *20142705* [2]
- **french beans** *20162625* [1]
- **garlic** *20162623* [1]
- **gooseberries** *20142706* [2]
- **hazel nuts** *20162632* [1]
- **herbs (see appendix 6)** *20162627* [1]
- **horseradish** *20162631* [1]
- **kale** *20162626* [1]
- **kiwi fruit** *20162632* [1]
- **kohlrabi** *20162626* [1], *20142701* [2]
- **leeks** *20162629* [1], *20142703* [2]
- **lettuce** *20162627* [1]
- **loganberries** *20142706* [2]
- **mooli** *20162631* [1]
- **nectarines** *20162632* [1], *20142700* [2]
- **olives** *20162632* [1]
- **oriental cabbage** *20162626* [1], *20142701* [2]
- **parsley root** *20162631* [1]
- **parsnips** *20162631* [1]
- **peaches** *20162632* [1], *20142700* [2]
- **pears** *20162635* [1]
- **plums** *20162632* [1], *20142700* [2]
- **protected apricots** *20162632* [1]
- **protected baby leaf crops** *20162627* [1]
- **protected bilberries** *20162633* [1]
- **protected blackcurrants** *20162633* [1]
- **protected blueberry** *20162633* [1]
- **protected broccoli** *20162626* [1]
- **protected brussels sprouts** *20162626* [1]
- **protected cabbages** *20162626* [1]
- **protected calabrese** *20162626* [1]
- **protected cauliflowers** *20162626* [1]
- **protected celery** *20162626* [1]
- **protected chestnuts** *20162632* [1]
- **protected chicory** *20142704* [2]
- **protected chilli peppers** *20142707* [2]
- **protected choi sum** *20162626* [1]
- **protected collards** *20162626* [1]
- **protected courgettes** *20142707* [2]
- **protected cranberries** *20162633* [1]
- **protected cress** *20142704* [2]
- **protected dwarf french beans** *20142702* [2]
- **protected edible crops** *20162634* [1]
- **protected edible podded peas** *20142702* [2]
- **protected endives** *20162627* [1]
- **protected gooseberries** *20162633* [1]
- **protected hazelnuts** *20162632* [1]
- **protected herbs (see appendix 6)** *20162627* [1]
- **protected kale** *20162626* [1]
- **protected kiwi fruit** *20162632* [1]
- **protected lamb's lettuce** *20142704* [2]
- **protected lettuce** *20162627* [1]
- **protected nectarines** *20162632* [1]
- **protected olives** *20162632* [1]
- **protected oriental cabbage** *20162626* [1]
- **protected peaches** *20162632* [1]
- **protected peppers** *20142707* [2]
- **protected plums** *20162632* [1]
- **protected quince** *20162632* [1]

FOR FULL CONDITIONS OF USE ALWAYS READ THE PRODUCT LABEL

- *protected redcurrants* 20162633 [1]
- *protected rhubarb* 20162626 [1]
- *protected rocket* 20142704 [2]
- *protected runner beans* 20142702 [2]
- *protected spinach* 20162627 [1]
- *protected spinach beet* 20162627 [1]
- *protected summer squash* 20142707 [2]
- *protected tatsoi* 20162626 [1]
- *protected vaccinium spp.* 20162633 [1]
- *protected walnuts* 20162632 [1]
- *protected watercress* 20162628 [1], 20142704 [2]
- *protected wine grapes* 20162632 [1]
- *quinces* 20162632 [1]
- *radishes* 20162631 [1]
- *red beet* 20162624 [1]
- *redcurrants* 20162633 [1], 20142706 [2]
- *rhubarb* 20162626 [1], 20142703 [2]
- *rubus hybrids* 20142706 [2]
- *runner beans* 20162625 [1]
- *salsify* 20162631 [1]
- *shallots* 20162623 [1]
- *spinach* 20162627 [1]
- *spinach beet* 20162627 [1]
- *spring cabbage* 20162626 [1]
- *swedes* 20162631 [1]
- *sweetcorn* 20162630 [1], 20142707 [2]
- *tatsoi* 20162626 [1]
- *turnips* 20162631 [1]
- *vaccinium spp.* 20162633 [1]
- *walnuts* 20162632 [1]
- *watercress* 20162628 [1]
- *wine grapes* 20162632 [1]

Approval information
- Bacillus thuringiensis included in Annex I under EC Regulation 1107/2009

Efficacy guidance
- Pest control achieved by ingestion by caterpillars of the treated plant vegetation. Caterpillars cease feeding and die in 1-3 d
- Apply as soon as larvae appear on crop and repeat every 7-10 d until the end of the hatching period
- Good coverage is essential, especially of undersides of leaves. Spray onto dry foliage and do not apply if rain expected within 6 h

Restrictions
- Apply spray mixture as soon as possible after preparation

Crop-specific information
- HI zero

Environmental safety
- Store out of direct sunlight

Hazard classification and safety precautions
  UN Number N/C
  Operator protection A, C, D, H; U05a, U15, U20c [1]; U11, U12, U14, U16b, U20b [2]; U19a [1, 2]
  Environmental protection E15a [1]; E15b, E34 [2]
  Storage and disposal D01, D02, D09a [1, 2]; D05, D11a [1]; D10b [2]
  Medical advice M03, M05a [2]

SEE SECTION 3 FOR PRODUCTS ALSO REGISTERED

32   Bacillus thuringiensis aizawai GC-91

An insecticide for use in amenity, ornamentals and some crops
IRAC mode of action code: 11

Products

Agree 50 WG	Certis	50% w/w	WG	18965

Uses

- Caterpillars in *amenity vegetation*, *ornamental plant production*, *radishes*, *red beet*, *swedes*

Approval information

- Bacillus thuringiensis aizawai GC-91 included in Annex I under EC Regulation 1107/2009

Hazard classification and safety precautions
     UN Number N/C
     Risk phrases H317
     Operator protection A, D, H; U05a, U08, U14, U19a, U20a
     Environmental protection E15b
     Storage and disposal D01, D02, D05, D09a, D10b

33   Bacillus thuringiensis israelensis

A bacterial insecticide for control of larvae of chironomid midges
IRAC mode of action code: 11

Products

1	Bactimos SC	Resource Chemicals	123 g/l	SC	UK15-0947
2	Vectobac AS	Resource Chemicals	123 g/l	AS	UK19-1198

Uses

- Blackfly in *open waters* [2]
- Filter fly in *open waters* [1]
- Mosquitoes in *open waters* [2]

Approval information

- Bacillus thuringiensis israelensis included in Annex I under EC Regulation 1107/2009

Hazard classification and safety precautions
     UN Number N/C
     Operator protection A, H

34   Beauveria bassiana

It is an entomopathogenic fungus causing white muscardine disease. It can be used as a
biological insecticide to control a number of pests such as termites, whitefly and some beetles
IRAC mode of action code: UNF

Products

1	Botanigard WP	Certis	22% w/w	WP	17054
2	Naturalis-L	Fargro	7.16% w/w	OD	17526

Uses

- Aphids in *angelica* (off-label), *baby leaf crops* (off-label), *balm* (off-label), *basil* (off-label), *bay* (off-label), *caraway* (off-label), *celery leaves* (off-label), *chervil* (off-label), *chicory* (off-label), *chives* (off-label), *coriander* (off-label), *cress* (off-label), *dill* (off-label), *edible flowers* (off-label), *fennel leaves* (off-label), *herbs (see appendix 6)* (off-label), *hyssop* (off-label), *lamb's lettuce* (off-label), *land cress* (off-label), *lettuce* (off-label), *lovage* (off-label), *marjoram* (off-label), *mint* (off-label), *oregano* (off-label), *protected aubergines*, *protected chilli peppers*, *protected courgettes*, *protected cucumbers*, *protected melons*, *protected nursery fruit trees*,

FOR FULL CONDITIONS OF USE ALWAYS READ THE PRODUCT LABEL

protected peppers, **protected strawberries**, **protected summer squash**, **protected tomatoes**, **purslane** *(off-label)*, **red mustard** *(off-label)*, **rocket** *(off-label)*, **rosemary** *(off-label)*, **sage** *(off-label)*, **salad burnet** *(off-label)*, **savory** *(off-label)*, **spinach** *(off-label)*, **spinach beet** *(off-label)*, **sweet ciceley** *(off-label)*, **tarragon** *(off-label)*, **thyme** *(off-label)*, **watercress** *(off-label)* [1]; **protected edible crops**, **protected forest nurseries** *(off-label)* [2]; **protected ornamentals** [1, 2]

- Beetles in **protected aubergines**, **protected chilli peppers**, **protected courgettes**, **protected cucumbers**, **protected melons**, **protected nursery fruit trees**, **protected ornamentals**, **protected peppers**, **protected strawberries**, **protected summer squash**, **protected tomatoes** [1]

- Thrips in **angelica** *(off-label)*, **baby leaf crops** *(off-label)*, **balm** *(off-label)*, **basil** *(off-label)*, **bay** *(off-label)*, **caraway** *(off-label)*, **celery leaves** *(off-label)*, **chervil** *(off-label)*, **chicory** *(off-label)*, **chives** *(off-label)*, **coriander** *(off-label)*, **cress** *(off-label)*, **dill** *(off-label)*, **edible flowers** *(off-label)*, **fennel leaves** *(off-label)*, **herbs (see appendix 6)** *(off-label)*, **hyssop** *(off-label)*, **lamb's lettuce** *(off-label)*, **land cress** *(off-label)*, **lettuce** *(off-label)*, **lovage** *(off-label)*, **marjoram** *(off-label)*, **mint** *(off-label)*, **oregano** *(off-label)*, **protected strawberries** *(off-label)*, **purslane** *(off-label)*, **red mustard** *(off-label)*, **rocket** *(off-label)*, **rosemary** *(off-label)*, **sage** *(off-label)*, **salad burnet** *(off-label)*, **savory** *(off-label)*, **spinach** *(off-label)*, **spinach beet** *(off-label)*, **strawberries** *(off-label)*, **sweet ciceley** *(off-label)*, **tarragon** *(off-label)*, **thyme** *(off-label)*, **watercress** *(off-label)* [1]

- Two-spotted spider mite in **angelica** *(off-label)*, **baby leaf crops** *(off-label)*, **balm** *(off-label)*, **basil** *(off-label)*, **bay** *(off-label)*, **bilberries** *(off-label)*, **blackberries** *(off-label)*, **blackcurrants** *(off-label)*, **blueberries** *(off-label)*, **caraway** *(off-label)*, **celery leaves** *(off-label)*, **chervil** *(off-label)*, **chicory** *(off-label)*, **chives** *(off-label)*, **coriander** *(off-label)*, **cranberries** *(off-label)*, **cress** *(off-label)*, **dill** *(off-label)*, **edible flowers** *(off-label)*, **elderberries** *(off-label)*, **fennel leaves** *(off-label)*, **gooseberries** *(off-label)*, **herbs (see appendix 6)** *(off-label)*, **hyssop** *(off-label)*, **lamb's lettuce** *(off-label)*, **land cress** *(off-label)*, **lettuce** *(off-label)*, **loganberries** *(off-label)*, **lovage** *(off-label)*, **marjoram** *(off-label)*, **mint** *(off-label)*, **mulberries** *(off-label)*, **oregano** *(off-label)*, **protected bilberries** *(off-label)*, **protected blackberries** *(off-label)*, **protected blackcurrants** *(off-label)*, **protected blueberry** *(off-label)*, **protected cranberries** *(off-label)*, **protected elderberries** *(off-label)*, **protected gooseberries** *(off-label)*, **protected loganberries** *(off-label)*, **protected mulberry** *(off-label)*, **protected raspberries** *(off-label)*, **protected redcurrants** *(off-label)*, **protected rose hips** *(off-label)*, **protected rubus hybrids** *(off-label)*, **purslane** *(off-label)*, **raspberries** *(off-label)*, **red mustard** *(off-label)*, **redcurrants** *(off-label)*, **rocket** *(off-label)*, **rose hips** *(off-label)*, **rosemary** *(off-label)*, **rubus hybrids** *(off-label)*, **sage** *(off-label)*, **salad burnet** *(off-label)*, **savory** *(off-label)*, **spinach** *(off-label)*, **spinach beet** *(off-label)*, **sweet ciceley** *(off-label)*, **tarragon** *(off-label)*, **thyme** *(off-label)*, **watercress** *(off-label)* [1]

- Whitefly in **angelica** *(off-label)*, **baby leaf crops** *(off-label)*, **balm** *(off-label)*, **basil** *(off-label)*, **bay** *(off-label)*, **bilberries** *(off-label)*, **blackberries** *(off-label)*, **blackcurrants** *(off-label)*, **blueberries** *(off-label)*, **caraway** *(off-label)*, **celery leaves** *(off-label)*, **chervil** *(off-label)*, **chicory** *(off-label)*, **chives** *(off-label)*, **coriander** *(off-label)*, **cranberries** *(off-label)*, **cress** *(off-label)*, **dill** *(off-label)*, **edible flowers** *(off-label)*, **elderberries** *(off-label)*, **fennel leaves** *(off-label)*, **gooseberries** *(off-label)*, **herbs (see appendix 6)** *(off-label)*, **hyssop** *(off-label)*, **lamb's lettuce** *(off-label)*, **land cress** *(off-label)*, **lettuce** *(off-label)*, **loganberries** *(off-label)*, **lovage** *(off-label)*, **marjoram** *(off-label)*, **mint** *(off-label)*, **mulberries** *(off-label)*, **oregano** *(off-label)*, **protected aubergines**, **protected bilberries** *(off-label)*, **protected blackberries** *(off-label)*, **protected blackcurrants** *(off-label)*, **protected blueberry** *(off-label)*, **protected chilli peppers**, **protected courgettes**, **protected cranberries** *(off-label)*, **protected cucumbers**, **protected elderberries** *(off-label)*, **protected gooseberries** *(off-label)*, **protected loganberries** *(off-label)*, **protected melons**, **protected mulberry** *(off-label)*, **protected nursery fruit trees**, **protected peppers**, **protected raspberries** *(off-label)*, **protected redcurrants** *(off-label)*, **protected rose hips** *(off-label)*, **protected rubus hybrids** *(off-label)*, **protected strawberries**, **protected strawberries** *(off-label)*, **protected summer squash**, **protected tomatoes**, **purslane** *(off-label)*, **raspberries** *(off-label)*, **red mustard** *(off-label)*, **redcurrants** *(off-label)*, **rocket** *(off-label)*, **rose hips** *(off-label)*, **rosemary** *(off-label)*, **rubus hybrids** *(off-label)*, **sage** *(off-label)*, **salad burnet** *(off-label)*, **savory** *(off-label)*, **spinach** *(off-label)*, **spinach beet** *(off-label)*, **strawberries** *(off-label)*, **sweet ciceley** *(off-label)*, **tarragon** *(off-label)*, **thyme** *(off-label)*, **watercress** *(off-label)* [1]; **protected edible crops**, **protected forest nurseries** *(off-label)* [2]; **protected ornamentals** [1, 2]

SEE SECTION 3 FOR PRODUCTS ALSO REGISTERED

Extension of Authorisation for Minor Use (EAMUs)

- *angelica* 20181792 [1]
- *baby leaf crops* 20181792 [1]
- *balm* 20181792 [1]
- *basil* 20181792 [1]
- *bay* 20181792 [1]
- *bilberries* 20181507 [1]
- *blackberries* 20181507 [1]
- *blackcurrants* 20181507 [1]
- *blueberries* 20181507 [1]
- *caraway* 20181792 [1]
- *celery leaves* 20181792 [1]
- *chervil* 20181792 [1]
- *chicory* 20181792 [1]
- *chives* 20181792 [1]
- *coriander* 20181792 [1]
- *cranberries* 20181507 [1]
- *cress* 20181792 [1]
- *dill* 20181792 [1]
- *edible flowers* 20181792 [1]
- *elderberries* 20181507 [1]
- *fennel leaves* 20181792 [1]
- *gooseberries* 20181507 [1]
- *herbs (see appendix 6)* 20181792 [1]
- *hyssop* 20181792 [1]
- *lamb's lettuce* 20181792 [1]
- *land cress* 20181792 [1]
- *lettuce* 20181792 [1]
- *loganberries* 20181507 [1]
- *lovage* 20181792 [1]
- *marjoram* 20181792 [1]
- *mint* 20181792 [1]
- *mulberries* 20181507 [1]
- *oregano* 20181792 [1]
- *protected bilberries* 20181507 [1]
- *protected blackberries* 20181507 [1]
- *protected blackcurrants* 20181507 [1]
- *protected blueberry* 20181507 [1]
- *protected cranberries* 20181507 [1]
- *protected elderberries* 20181507 [1]
- *protected forest nurseries* 20162195 [2]
- *protected gooseberries* 20181507 [1]
- *protected loganberries* 20181507 [1]
- *protected mulberry* 20181507 [1]
- *protected raspberries* 20181507 [1]
- *protected redcurrants* 20181507 [1]
- *protected rose hips* 20181507 [1]
- *protected rubus hybrids* 20181507 [1]
- *protected strawberries* 20160110 [1]
- *purslane* 20181792 [1]
- *raspberries* 20181507 [1]
- *red mustard* 20181792 [1]
- *redcurrants* 20181507 [1]
- *rocket* 20181792 [1]
- *rose hips* 20181507 [1]
- *rosemary* 20181792 [1]
- *rubus hybrids* 20181507 [1]
- *sage* 20181792 [1]
- *salad burnet* 20181792 [1]

- *savory* 20181792 [1]
- *spinach* 20181792 [1]
- *spinach beet* 20181792 [1]
- *strawberries* 20160110 [1]
- *sweet ciceley* 20181792 [1]
- *tarragon* 20181792 [1]
- *thyme* 20181792 [1]
- *watercress* 20181792 [1]

Approval information
- Beauveria bassiana included in Annex 1 under EC Regulation 1107/2009

Environmental safety
- Buffer zone requirement 10 m for bilberries, blackberry, blackcurrants, blueberry, cranberry, elderberry, gooseberry, loganberry, rubus hybrid, raspberry, redcurrant, mulberry and rosehips [1]

Hazard classification and safety precautions
    UN Number N/C
    Risk phrases H317, H334 [1]
    Operator protection A, D, H; U05a, U11, U14, U15, U16b, U19a, U20b
    Environmental protection E15b, E34
    Storage and disposal D01, D02, D05, D09a, D10a
    Medical advice M03, M04a

35   bentazone

A post-emergence contact benzothiadiazinone herbicide
HRAC mode of action code: C3

Products

1	Basagran SG	BASF	87% w/w	SG	08360
2	Benta 480 SL	Nufarm UK	480 g/l	SL	17355
3	Tanaru	Agroquimicos	480 g/l	SL	15726
4	Troy 480	UPL Europe	480 g/l	SL	16954

Uses
- Annual dicotyledons in *beans without pods (dry)*, *beans without pods (fresh)*, *bulb onions* (off-label), *chives* (off-label), *garlic* (off-label), *herbs (see appendix 6)* (off-label), *leeks* (off-label), *salad onions* (off-label), *shallots* (off-label), *soya beans*, *soya beans* (off-label) [1]; *broad beans, linseed, potatoes, runner beans, spring field beans, winter field beans* [1-4]; *combining peas, dwarf beans, vining peas* [1, 3, 4]; *french beans* [1, 2]; *narcissi* [2, 3]; *navy beans* [3]; *ornamental plant production* [1, 4]; *peas* [2]
- Chickweed in *game cover* (off-label), *hops* (off-label), *ornamental plant production* (off-label) [1]
- Cleavers in *game cover* (off-label), *hops* (off-label), *ornamental plant production* (off-label) [1]
- Common storksbill in *leeks* (off-label) [1]
- Fool's parsley in *leeks* (off-label) [1]
- Groundsel in *game cover* (off-label), *herbs (see appendix 6)* (off-label), *hops* (off-label), *ornamental plant production* (off-label) [1]
- Mayweeds in *bulb onions* (off-label), *game cover* (off-label), *garlic* (off-label), *herbs (see appendix 6)* (off-label), *hops* (off-label), *ornamental plant production* (off-label), *shallots* (off-label) [1]

Extension of Authorisation for Minor Use (EAMUs)
- *bulb onions* 20061631 [1]
- *chives* 20102994 [1]
- *game cover* 20082819 [1]
- *garlic* 20061631 [1]
- *herbs (see appendix 6)* 20120744 [1]
- *hops* 20082819 [1]

- **leeks** *20061630* [1]
- **ornamental plant production** *20082819* [1]
- **salad onions** *20102994* [1]
- **shallots** *20061631* [1]
- **soya beans** *20061629* [1]

Approval information
- Bentazone included in Annex I under EC Regulation 1107/2009
- Approval expiry 31 Dec 2020 [3, 4]

Efficacy guidance
- Most effective control obtained when weeds are growing actively and less than 5 cm high or across. Good spray cover is essential
- The addition of specified adjuvant oils is recommended for use on some crops to improve fat hen control. Do not use under hot or humid conditions. See label for details
- Split dose application may be made in all recommended crops except peas and generally gives better weed control. See label for details
- See label for guidance on water stewardship

Restrictions
- Maximum number of treatments normally 2 per crop but check label
- Crops must be treated at correct stage of growth to avoid danger of scorch. See label for details
- Not all varieties of recommended crops are fully tolerant. Use only on tolerant varieties named in label. Do not use on forage or mange-tout varieties of peas
- Do not use on crops which have been affected by drought, waterlogging, frost or other stress conditions
- Do not apply insecticides within 7 d of treatment
- Leave 14 d after using a post-emergence grass herbicide and carry out a leaf wax test where relevant or 7 d where treatment precedes the grass herbicide
- Do not spray at temperatures above 21°C. Delay spraying until evening if necessary
- Do not apply if rain or frost expected, if unseasonably cold, if foliage wet or in drought
- A minimum of 6 h (preferably 12 h) free from rain is required after application
- May be used on selected varieties of maincrop and second early potatoes (see label for details), not on seed crops or first earlies

Crop-specific information
- Latest use: before shoots exceed 15 cm high for potatoes and spring field beans (or 6-7 leaf pairs); 4 leaf pairs (6 pairs or 15 cm high with split dose) for broad beans; before flower buds visible for French, navy, runner and winter field beans, and linseed; before flower buds can be found enclosed in terminal shoot for peas
- Best results in narcissi obtained by using a suitable pre-emergence herbicide first
- Consult processor before using on crops for processing
- A satisfactory wax test must be carried out before use on peas
- Do not treat narcissi during flower bud formation

Environmental safety
- Dangerous for the environment
- Harmful to aquatic organisms
- Some pesticides pose a greater threat of contamination of water than others and bentazone is one of these pesticides. Take special care when applying bentazone near water and do not apply if heavy rain is forecast

Hazard classification and safety precautions
Hazard Harmful, Harmful if swallowed [1, 2]; Irritant [3, 4]
Packaging group III [1]
UN Number 2588 [1]; N/C [2-4]
Risk phrases H317 [1, 2, 4]; H318 [1]; H320 [4]; R36, R43 [3]
Operator protection A [1-4]; C [1, 3, 4]; H [3, 4]; U05a, U08, U19a [1-4]; U11, U20b [1, 3, 4]; U14 [1]; U20a [2]
Environmental protection E15a, H412 [2]; E15b, E34 [1, 3, 4]; E38 [4]
Storage and disposal D01, D02, D09a [1-4]; D10b [1]; D10c [2-4]
Medical advice M03 [1, 3, 4]; M05a [1]

FOR FULL CONDITIONS OF USE ALWAYS READ THE PRODUCT LABEL

36   benthiavalicarb-isopropyl

An amino acid amide carbamate fungicide available only in mixtures
FRAC mode of action code: 40

37   benthiavalicarb-isopropyl + mancozeb

A fungicide mixture for potatoes
FRAC mode of action code: 40 + M3

See also mancozeb

Products

Valbon	Certis	1.75:70% w/w	WG	14868

Uses
* Blight in **potatoes**
* Downy mildew in **amenity vegetation** *(off-label)*, **bulb onions** *(off-label)*, **forest nurseries** *(off-label)*, **garlic** *(off-label)*, **hops** *(off-label)*, **ornamental plant production** *(off-label)*, **shallots** *(off-label)*, **soft fruit** *(off-label)*, **table grapes** *(off-label)*, **wine grapes** *(off-label)*

Extension of Authorisation for Minor Use (EAMUs)
* **amenity vegetation** *20142325*
* **bulb onions** *20142326*
* **forest nurseries** *20142328*
* **garlic** *20142326*
* **hops** *20142328*
* **ornamental plant production** *20142325*
* **shallots** *20142326*
* **soft fruit** *20142328*
* **table grapes** *20142327*
* **wine grapes** *20142327*

Approval information
* Benthiavalicarb-isopropyl and mancozeb included in Annex I under EC Regulation 1107/2009

Efficacy guidance
* Apply as a protectant spray commencing before blight enters the crop irrespective of growth stage following a blight warning or where there is a local source of infection
* In the absence of a warning commence the spray programme before the foliage meets in the rows
* Repeat treatment every 7-10 d depending on disease pressure
* Increase water volume as necessary to ensure thorough coverage of the plant
* Use of air assisted sprayers or drop legs may help to improve coverage

Restrictions
* Maximum number of treatments 6 per crop
* Do not use more than three consecutive treatments. Include products with a different mode of action in the programme

Crop-specific information
* HI 7 d for potatoes

Environmental safety
* Very toxic to aquatic organisms
* Risk to non-target insects or other arthropods. Avoid spraying within 6 m of field boundary
* LERAP Category B

Hazard classification and safety precautions
   Hazard Irritant
   Transport code 9

Packaging group III
UN Number 3077
Risk phrases R40, R43, R50, R53a
Operator protection A, D, H; U14
Environmental protection E15a, E16a, E22c
Storage and disposal D01, D02, D05, D09a, D11a, D12b

38   benzoic acid

An organic horticultural disinfectant

Products

Menno Florades	Fargro	90 g/kg	SL	13985

Uses

* Bacteria in *all protected non-edible crops*
* Fungal spores in *all protected non-edible crops*
* Viroids in *all protected non-edible crops*
* Virus in *all protected non-edible crops*

Approval information

* Benzoic acid is included in Annex 1 under EC Regulation 1107/2009

Hazard classification and safety precautions

Hazard Corrosive
Transport code 3
Packaging group III
UN Number 1987
Risk phrases H318, H336
Operator protection A, C, H; U05a, U11, U13, U14, U15
Environmental protection E15a
Storage and disposal D01, D02, D09a, D10a
Medical advice M05a

39   benzovindiflupyr

An SDHI fungicide for disease control in cereals
FRAC mode of action code: 7

See also benzovindiflupyr + prothioconazole

Products

1	Ceravato Plus	Syngenta	100 g/l	EC	17865
2	Elatus Plus	Syngenta	100 g/l	EC	17841
3	Velogy Plus	Syngenta	100 g/l	EC	17866

Uses

* Brown rust in *rye, spring barley, spring wheat, triticale, winter barley, winter wheat*
* Glume blotch in *spring wheat* (moderate control), *winter wheat* (moderate control)
* Mycosphaerella in *spring wheat, triticale, winter wheat*
* Net blotch in *spring barley, winter barley*
* Ramularia leaf spots in *spring barley, winter barley*
* Rhynchosporium in *rye* (moderate control), *spring barley* (moderate control), *winter barley* (moderate control)
* Septoria leaf blotch in *spring wheat, triticale, winter wheat*
* Yellow rust in *spring wheat, triticale, winter wheat*

Approval information

* Benzovindiflupyr is included in Annex 1 under EC Regulation 1107/2009

FOR FULL CONDITIONS OF USE ALWAYS READ THE PRODUCT LABEL

Efficacy guidance
- Benzovindiflupyr works predominately via protective action

Restrictions
- Do not apply via hand-held equipment
- No more than 2 applications of SDHI fungicides should be applied to any cereal crop

Following crops guidance
- There are no restrictions on succeeding crops in a normal rotation

Environmental safety
- Buffer zone requirement 6 m [1-3]
- LERAP Category B

Hazard classification and safety precautions
    Hazard Harmful if swallowed, Harmful if inhaled, Very toxic to aquatic organisms
    Transport code 9
    Packaging group III
    UN Number 3082
    Risk phrases H317, H318
    Operator protection A, C, H; U09a, U10, U11, U19a, U20a
    Environmental protection E15b, E16a, H410
    Storage and disposal D01, D09a, D10c, D12a, D12b, D22

40   benzovindiflupyr + prothioconazole

An SDHI and triazole fungicide mixture for disease control in cereals
FRAC mode of action code: 3 + 7

See also benzovindiflupyr

Products

Elatus Era	Syngenta	75:150 g/l	EC	17889

Uses
- Brown rust in *rye*, *spring barley*, *spring wheat*, *triticale*, *winter barley*, *winter wheat*
- Disease control in *rye*
- Fusarium in *spring wheat* (moderate control), *winter wheat* (moderate control)
- Glume blotch in *spring wheat* (moderate control), *winter wheat* (moderate control)
- Mycosphaerella in *spring wheat*, *winter wheat*
- Net blotch in *spring barley*, *winter barley*
- Ramularia leaf spots in *spring barley*, *winter barley*
- Rhynchosporium in *rye* (moderate control), *spring barley* (moderate control), *triticale* (moderate control), *winter barley* (moderate control)
- Septoria leaf blotch in *spring wheat*, *triticale*, *winter wheat*
- Yellow rust in *spring wheat*, *winter wheat*

Approval information
- Benzovindiflupyr and prothioconazole are included in Annex 1 under EC Regulation 1107/2009
- Accepted by BBPA for use on malting barley
- Approval expiry 13 Dec 2020 [1]

Efficacy guidance
- Benzovindiflupyr is predominantly protectant while prothioconazole is a triazole with systemic and protectant activity. Use as a protectant treatment or in the earliest stages of disease development.

Restrictions
- The earliest time of application is GS31
- No more than 2 applications of SDHI fungicides should be applied to any cereal crop
- Do not apply via hand-held equipment

- Low drift spraying equipment must be operated according to the specific conditions stated in the official three star rating for that equipment as published on HSE Chemicals Regulation Division's website. These operating conditions must be maintained until the operator is 30m from the top of the bank of any surface water bodies.

Following crops guidance
- There are no restrictions on succeeding crops in a normal rotation

Environmental safety
- Buffer zone requirement 6m [1]
- LERAP Category B

Hazard classification and safety precautions
    Hazard Harmful if swallowed, Harmful if inhaled
    Transport code 9
    Packaging group III
    UN Number 3082
    Risk phrases H315, H317, H318
    Operator protection A, C, H; U09a, U10, U11, U19a, U20a
    Environmental protection E15b, E16a, H410
    Storage and disposal D01, D09a, D10c, D12a, D12b
    Medical advice M04a

41   benzyladenine

A cytokinin plant growth regulator for use in apples and pears

Products

1	Configure	Fargro	20 g/l	SC	17523
2	Exilis	Fine	20 g/l	SL	15706
3	Globaryll 100	Belchim	100 g/l	SL	16097
4	MaxCel	Sumitomo	20 g/l	SL	15708

Uses
- Fruit thinning in *apples* [2, 4]; *pears* [2]
- Growth regulation in *apples*, *pears* [3]; *ornamental plant production* [1]
- Increasing flowering in *phalaenopsis spp.*, *schlumbergera spp.*, *sempervivium spp.* [1]

Approval information
- Benzyladenine is included in Annex 1 under EC Directive 1107/2009

Efficacy guidance
- A post-bloom thinner for apples and pears with excessive blossom [2, 3]
- Apply when the temperature will exceed 15°C on day of application but note that temperatures above 28°C may result in excessive thinning [2-4]

Restrictions
- Use in pears is a Qualified Minor Use recommendation. Evidence of efficacy and crop safety in pears is limited to the pear variety 'Conference' [2]

Environmental safety
- Broadcast air-assisted LERAP [4] (5 m)

Hazard classification and safety precautions
    Hazard Harmful, Dangerous for the environment [3]
    UN Number N/C
    Risk phrases H317 [1]; H318, H361 [3]
    Operator protection A [1, 2, 4]; C [3]; H [1]; U02a, U09a, U19a, U20b [2, 4]; U05a [2-4]; U11, U14, U15, U20c [3]
    Environmental protection E15a [3]; E15b [1, 2, 4]; E17b [4] (5 m); E38, H412 [2-4]
    Storage and disposal D01 [1, 3]; D02, D09a [1-4]; D03, D05, D12a [2, 4]; D10a [1]; D10b [2-4]
    Medical advice M04a [3]; M05a [2, 4]

42   benzyladenine + gibberellin

A plant growth regulator mixture for use on ornamental plants

See also benzyladenine
*   gibberellins*

Products

Chrysal BVB	Chrysal	19:19 g/l	SL	17780

Uses
* Growth regulation in **ornamental plant production**

Approval information
* Benzyladenine and gibberellins are included in Annex 1 under EC Directive 1107/2009

Hazard classification and safety precautions
  UN Number N/C
  Operator protection A
  Environmental protection H412

43   beta-cyfluthrin

A non-systemic pyrethroid insecticide for insect control in cereals, oilseed rape, cabbages, cauliflowers and sugar beet
IRAC mode of action code: 3

Products

Gandalf	Nufarm UK	25 g/l	EC	18642

Uses
* Aphids in **oats**, **rye**, **spring barley**, **spring wheat**, **sugar beet**, **triticale**, **winter barley**, **winter wheat**
* Brassica pod midge in **oilseed rape**
* Cabbage moth in **cabbages**, **cauliflowers**
* Cabbage seed weevil in **oilseed rape**
* Cabbage stem flea beetle in **oilseed rape**
* Cabbage stem weevil in **oilseed rape**
* Peach-potato aphid in **cabbages**, **cauliflowers**
* Wheat-blossom midge in **rye**, **spring wheat**, **triticale**, **winter wheat**

Approval information
* Beta-cyfluthrin included in Annex I under EC Regulation 1107/2009

Environmental safety
* Dangerous to bees
* To protect non target insects/arthropods respect an unsprayed buffer zone of 5 m to non-crop land

Hazard classification and safety precautions
  Hazard Harmful, Dangerous for the environment, Flammable liquid and vapour, Harmful if swallowed, Harmful if inhaled, Very toxic to aquatic organisms
  Transport code 3
  Packaging group III
  UN Number 1993
  Risk phrases H304, H317, H319, H336
  Operator protection A, C, H; U04a, U05a, U10, U12, U14, U19a, U20a
  Environmental protection E12c, E12e, E16e, E34, E38, H410
  Storage and disposal D01, D02, D09a, D10a, D12a
  Medical advice M03, M05b

44   bifenazate

A bifenazate acaricide for use on protected strawberries
IRAC mode of action code: 25

Products

Floramite 240 SC	Arysta	240 g/l	SC	17958

Uses

- Mites in *protected forest nurseries* (off-label), *protected hops* (off-label)
- Two-spotted spider mite in *protected aubergines* (off-label), *protected chilli peppers* (off-label), *protected cucumbers* (off-label), *protected peppers* (off-label), *protected strawberries*, *protected tomatoes* (off-label)

Extension of Authorisation for Minor Use (EAMUs)

- *protected aubergines* 20182999
- *protected chilli peppers* 20182999
- *protected cucumbers* 20182999
- *protected forest nurseries* 20171472
- *protected hops* 20171472
- *protected peppers* 20182999
- *protected tomatoes* 20182999

Approval information

- Bifenazate included in Annex I under EC Regulation 1107/2009

Efficacy guidance

- Bifenazate acts by contact knockdown after approximately 4 d and a subsequent period of residual control
- All mobile stages of mites are controlled with occasionally some ovicidal activity
- Best results obtained from a programme of two treatments started as soon as first spider mites are seen
- To minimise possible development of resistance use in a planned Resistance Management strategy. Alternate at least two products with different modes of action between treatment programmes of bifenazate. See Section 5 for more information

Restrictions

- Maximum number of treatments 2 per yr for protected strawberries
- After use in protected environments avoid re-entry for at least 2 h while full ventilation is carried out allowing spray to dry on leaves
- Personnel working among treated crops should wear suitable long-sleeved garments and gloves for 14 d after treatment
- Do not apply as a low volume spray

Crop-specific information

- HI: 7 d for protected strawberries
- Carry out small scale tolerance test on the strawberry cultivar before large scale use

Environmental safety

- Dangerous for the environment
- Toxic to aquatic organisms
- Harmful to non-target predatory mites. Avoid spray drift onto field margins, hedges, ditches, surface water and neighbouring crops

Hazard classification and safety precautions

    Hazard Irritant, Dangerous for the environment
    Transport code 9
    Packaging group III
    UN Number 3082
    Risk phrases H317, H373
    Operator protection A, H; U02a, U04a, U05a, U10, U14, U15, U19a, U20b
    Environmental protection E15b, E38, H411

FOR FULL CONDITIONS OF USE ALWAYS READ THE PRODUCT LABEL

Storage and disposal D01, D02, D05, D09a, D10b, D12b
Medical advice M03

45   bifenox

A diphenyl ether herbicide for use in cereals and (off-label) in oilseed rape.
HRAC mode of action code: E

Products

1	Clayton Belstone	Clayton	480 g/l	SC	14033
2	Fox	Adama	480 g/l	SC	11981

Uses

- Annual dicotyledons in **canary flower (echium spp.)** *(off-label)*, **grass seed crops** *(off-label)* [1, 2]; **evening primrose** *(off-label)*, **honesty** *(off-label)*, **linseed** *(off-label)*, **mustard** *(off-label)*, **oilseed rape** *(off-label)* [2]; **triticale, winter barley, winter rye, winter wheat** [1]
- Charlock in **grass seed crops** *(off-label)* [2]
- Cleavers in **triticale, winter barley, winter rye, winter wheat** [2]
- Field pansy in **triticale, winter barley, winter rye, winter wheat** [2]
- Field speedwell in **triticale, winter barley, winter rye, winter wheat** [2]
- Forget-me-not in **triticale, winter barley, winter rye, winter wheat** [2]
- Ivy-leaved speedwell in **triticale, winter barley, winter rye, winter wheat** [2]
- Poppies in **triticale, winter barley, winter rye, winter wheat** [2]
- Red dead-nettle in **triticale, winter barley, winter rye, winter wheat** [2]

Extension of Authorisation for Minor Use (EAMUs)

- **canary flower (echium spp.)** *20121422* [1], *20142318* [2]
- **evening primrose** *20142318* [2]
- **grass seed crops** *20193238* [1], *20193257* [2]
- **honesty** *20142318* [2]
- **linseed** *20142318* [2]
- **mustard** *20142318* [2]
- **oilseed rape** *20142318* [2]

Approval information

- Accepted by BBPA for use on malting barley
- Bifenox included in Annex I under EC Regulation 1107/2009

Efficacy guidance

- Bifenox is absorbed by foliage and emerging roots of susceptible species
- Best results obtained when weeds are growing actively with adequate soil moisture

Restrictions

- Maximum number of treatments: one per crop
- Do not apply to crops suffering from stress from whatever cause
- Do not apply if the crop is wet or if rain or frost is expected
- Avoid drift onto broad-leaved plants outside the target area

Crop-specific information

- Latest use: before 2nd node detectable (GS 32) for all crops

Environmental safety

- Dangerous for the environment
- Very toxic to aquatic organisms
- Do not empty into drains

Hazard classification and safety precautions

Hazard Dangerous for the environment
Transport code 9
Packaging group III
UN Number 3082

SEE SECTION 3 FOR PRODUCTS ALSO REGISTERED

Operator protection A; U05a, U08, U20b
Environmental protection E15a, E19b, E34, E38, H410
Storage and disposal D01, D02, D09a, D12a
Medical advice M03

46 bixafen

A succinate dehydrogenase inhibitor (SDHI) fungicide available only in mixtures
FRAC mode of action code: 7

47 bixafen + fluopyram + prothioconazole

A fungicide mixture for disease control in wheat
FRAC mode of action code: 7 + 3

See also bixafen + prothioconazole

Products

Ascra Xpro	Bayer CropScience	65:65:130 g/l	EC	17623

Uses

- Brown rust in **spring wheat**, **winter wheat**
- Ear diseases in **spring wheat** *(reduction of incidence and severity)*, **winter wheat** *(reduction of incidence and severity)*
- Eyespot in **spring wheat** *(reduction of incidence and severity)*, **winter wheat** *(reduction of incidence and severity)*
- Glume blotch in **spring wheat**, **winter wheat**
- Powdery mildew in **spring wheat**, **winter wheat**
- Septoria leaf blotch in **spring wheat**, **winter wheat**
- Tan spot in **spring wheat**, **winter wheat**
- Yellow rust in **spring wheat**, **winter wheat**

Approval information

- Bixafen, fluopyram and prothioconazole are included in Annex 1 under EC Regulation 1107/2009.

Efficacy guidance

- Applications to upper leaves where *S. tritici* symptoms are present are likely to be less effective

Restrictions

- Bixafen is an SDH respiration inhibitor; Do not apply more than two foliar applications of products containing an SDH inhibitor to any cereal crop
- Resistance to some DMI fungicides has been identified in Septoria leaf blotch (*Mycosphaerella graminicola*) which may seriously affect the performance of some products. For further advice on resistance management in DMIs contact your agronomist or specialist advisor, and visit the FRAG-UK website.

Environmental safety

- LERAP Category B

Hazard classification and safety precautions
  Hazard Harmful if swallowed
  Transport code 9
  Packaging group III
  UN Number 3082
  Risk phrases H317, H318
  Operator protection A, C, H; U05a, U09b, U11, U20a
  Environmental protection E15b, E16a, E34, H410
  Storage and disposal D01, D02, D05, D09a, D10b, D12b

FOR FULL CONDITIONS OF USE ALWAYS READ THE PRODUCT LABEL

48   bixafen + fluoxastrobin + prothioconazole

An SDHI + strobilurin + triazole mixture for disease control in cereals
FRAC mode of action code: 7 + 11 + 3

Products

Variano Xpro	Bayer CropScience	40:50:100 g/l	EC	15218

Uses

- Brown rust in *spring wheat, triticale, winter rye, winter wheat*
- Ear diseases in *spring wheat, triticale, winter rye, winter wheat*
- Glume blotch in *spring wheat, triticale, winter rye, winter wheat*
- Leaf blotch in *triticale, winter rye*
- Powdery mildew in *spring wheat, triticale, winter rye, winter wheat*
- Rhynchosporium in *winter rye*
- Tan spot in *spring wheat, triticale, winter rye, winter wheat*
- Yellow rust in *spring wheat, triticale, winter rye, winter wheat*

Approval information

- Bixafen, fluroxystrobin and prothioconazole are included in Annex 1 under EC Regulation 1107/2009.

Restrictions

- Bixafen is an SDH respiration inhibitor; Do not apply more than two foliar applications of products containing an SDH inhibitor to any cereal crop

Environmental safety

- LERAP Category B

Hazard classification and safety precautions

Hazard Irritant, Dangerous for the environment
Transport code 9
Packaging group III
UN Number 3082
Risk phrases H317, H319
Operator protection A, H; U05b, U08, U14, U20b
Environmental protection E15a, E16a, E16b, E34, E38, H410
Storage and disposal D01, D02, D05, D09a, D10b, D12a
Medical advice M03

49   bixafen + prothioconazole

Succinate dehydrogenase inhibitor (SDHI) + triazole fungicide mixture for disease control in cereals
FRAC mode of action code: 3 + 7

Products

1	Aviator 235 Xpro	Bayer CropScience	75:160 g/l	EC	15026
2	Siltra Xpro	Bayer CropScience	60:200 g/l	EC	15082

Uses

- Brown rust in *grass seed crops* (off-label), *spring barley, winter barley* [2]; *spring wheat, triticale, winter rye, winter wheat* [1]
- Crown rust in *grass seed crops* (off-label), *spring oats, winter oats* [2]
- Drechslera leaf spot in *grass seed crops* (off-label) [2]
- Ear diseases in *spring barley, winter barley* [2]; *spring wheat, triticale, winter wheat* [1]
- Eyespot in *spring barley* (reduction of incidence and severity), *spring oats* (reduction of incidence and severity), *winter barley* (reduction of incidence and severity), *winter oats* (reduction of incidence and severity) [2]; *spring wheat* (reduction of incidence and severity), *triticale* (reduction of incidence and severity), *winter wheat* (reduction of incidence and severity) [1]

SEE SECTION 3 FOR PRODUCTS ALSO REGISTERED

- Glume blotch in *spring wheat, triticale, winter wheat* [1]
- Mildew in *grass seed crops (off-label)* [2]
- Net blotch in *spring barley, winter barley* [2]
- Powdery mildew in *spring barley, spring oats, winter barley, winter oats* [2]; *spring wheat, triticale, winter rye, winter wheat* [1]
- Ramularia leaf spots in *spring barley, winter barley* [2]
- Rhynchosporium in *grass seed crops (off-label), spring barley, winter barley* [2]; *winter rye* [1]
- Septoria leaf blotch in *spring wheat, triticale, winter wheat* [1]
- Tan spot in *spring wheat, triticale, winter wheat* [1]
- Yellow rust in *spring barley, winter barley* [2]; *spring wheat, triticale, winter wheat* [1]

Extension of Authorisation for Minor Use (EAMUs)
- *grass seed crops 20171034* [2]

Approval information
- Bixafen and prothioconazole are included in Annex I under EC Regulation 1107/2009.
- Accepted by BBPA for use on malting barley

Efficacy guidance
- Resistance to some DMI fungicides has been identified in Septoria leaf blotch (*Mycosphaerella graminicola*) which may seriously affect the performance of some products. For further advice on resistance management in DMIs contact your agronomist or specialist advisor, and visit the FRAG-UK website.

Restrictions
- Bixafen is an SDH respiration inhibitor; Do not apply more than two foliar applications of products containing an SDH inhibitor to any cereal crop

Environmental safety
- LERAP Category B

Hazard classification and safety precautions
    Hazard Harmful, Dangerous for the environment [1, 2]; Harmful if swallowed [2]
    Transport code 9
    Packaging group III
    UN Number 3082
    Risk phrases H317 [2]; H319 [1, 2]
    Operator protection U05a, U09a, U20b
    Environmental protection E15b, E16a, E16b, E34, E38, H410
    Storage and disposal D01, D02, D09a, D10b, D12a

50   bixafen + prothioconazole + spiroxamine

A broad-spectrum fungicide mixture for disease control in cereals
FRAC mode of action code: 3 + 5 + 7

Products

Boogie Xpro	Bayer CropScience	50:100:250 g/l	EC	15061

Uses
- Brown rust in *spring wheat, triticale, winter rye, winter wheat*
- Ear diseases in *spring wheat, triticale, winter rye, winter wheat*
- Eyespot in *spring wheat, triticale, winter rye, winter wheat*
- Glume blotch in *spring wheat, triticale, winter rye, winter wheat*
- Powdery mildew in *spring wheat, triticale, winter rye, winter wheat*
- Rhynchosporium in *triticale, winter rye*
- Septoria leaf blotch in *spring wheat, triticale, winter rye, winter wheat*
- Tan spot in *spring wheat, triticale, winter rye, winter wheat*
- Yellow rust in *spring wheat, triticale, winter rye, winter wheat*

FOR FULL CONDITIONS OF USE ALWAYS READ THE PRODUCT LABEL

Approval information
- Bixafen, prothioconazole and spiroxamine included in Annex I under EC Regulation 1107/2009.

Efficacy guidance
- Resistance to some DMI fungicides has been identified in Septoria leaf blotch (*Mycosphaerella graminicola*) which may seriously affect the performance of some products. For further advice on resistance management in DMIs contact your agronomist or specialist advisor, and visit the FRAG-UK website.

Restrictions
- No more than two applications of SDH inhibitors must be applied to the same cereal crop.
- Maintain an interval of at least 21 days between applications.
- Horizontal boom sprayers must be fitted with three star drift reduction technology for all uses.
- Low drift spraying equipment must be operated according to the specific conditions stated in the official three star rating for that equipment as published on HSE Chemicals Regulation Division's website. These operating conditions must be maintained until the operator is 30m from the top of the bank of any surface water bodies
- Only one application may be made before 30 April (BBCH 30-37) followed by a second application after 1 May (BBCH 37-69). Alternatively two applications can be made after 1 May (BBCH 37-69).

Environmental safety
- Maintain an aquatic buffer zone of 6 m
- LERAP Category B

Hazard classification and safety precautions
Hazard Harmful, Dangerous for the environment, Harmful if swallowed, Harmful if inhaled, Very toxic to aquatic organisms
Transport code 9
Packaging group III
UN Number 3082
Risk phrases H318, H335, H361, H373
Operator protection A, C, H; U05a, U09b, U11, U20a, U26
Environmental protection E15b, E16a, E16b, E34, E38, H410
Storage and disposal D01, D02, D05, D09a, D10b, D12a
Medical advice M03

51   bixafen + prothioconazole + tebuconazole

A succinate dehydrogenase inhibitor (SDHI) + triazole fungicide mixture for disease control in cereals
FRAC mode of action code: 3 + 7

Products

1	Skyway 285 Xpro	Bayer CropScience	75:110:100 g/l	EC	15028
2	Sparticus Xpro	Bayer CropScience	75:110:90 g/l	EC	15162

Uses
- Brown rust in *spring wheat, triticale, winter rye, winter wheat*
- Ear diseases in *spring wheat, triticale, winter wheat*
- Eyespot in *spring wheat* (reduction of incidence and severity), *triticale* (reduction of incidence and severity), *winter wheat* (reduction of incidence and severity)
- Glume blotch in *spring wheat, triticale, winter wheat*
- Light leaf spot in *winter oilseed rape*
- Phoma in *winter oilseed rape*
- Powdery mildew in *spring wheat, triticale, winter rye, winter wheat*
- Rhynchosporium in *winter rye*
- Sclerotinia in *winter oilseed rape*
- Septoria leaf blotch in *spring wheat, triticale, winter wheat*
- Stem canker in *winter oilseed rape*
- Tan spot in *spring wheat, triticale, winter wheat*
- Yellow rust in *spring wheat, triticale, winter wheat*

SEE SECTION 3 FOR PRODUCTS ALSO REGISTERED

Approval information
- Bixafen, prothioconazole and tebuconazole included in Annex I under EC Regulation 1107/2009.

Efficacy guidance
- Resistance to some DMI fungicides has been identified in Septoria leaf blotch (*Mycosphaerella graminicola*) which may seriously affect the performance of some products. For further advice on resistance management in DMIs contact your agronomist or specialist advisor, and visit the FRAG-UK website.

Restrictions
- Bixafen is an SDH respiration inhibitor; Do not apply more than two foliar applications of products containing an SDH inhibitor to any cereal crop
- Newer authorisations for tebuconazole products require application to cereals only after GS 30 and applications to oilseed rape and linseed after GS20 - check label

Environmental safety
- LERAP Category B

Hazard classification and safety precautions
    Hazard Harmful, Dangerous for the environment, Harmful if swallowed
    Transport code 9
    Packaging group III
    UN Number 3082
    Risk phrases H315, H317, H361
    Operator protection A, H; U05a, U09a, U20b
    Environmental protection E15b, E16a, E16b, E34, E38, H410
    Storage and disposal D01, D02, D09a, D10b, D12a

52   boscalid

A translocated and translaminar anilide fungicide
FRAC mode of action code: 7

Products

1	Filan	BASF	50% w/w	WG	11449
2	Fulmar	AgChem Access	50% w/w	WG	15767

Uses
- Alternaria in *spring oilseed rape*, *winter oilseed rape* [1, 2]
- Botrytis in *poppies* (off-label), *poppies for morphine production* (off-label), *poppies grown for seed production* (off-label) [1]
- Dark leaf and pod spot in *poppies* (off-label), *poppies grown for seed production* (off-label) [1]
- Disease control in *all edible seed crops grown outdoors* (off-label), *all non-edible seed crops grown outdoors* (off-label), *borage for oilseed production* (off-label), *canary flower (echium spp.)* (off-label), *corn gromwell* (off-label), *evening primrose* (off-label), *honesty* (off-label), *linseed* (off-label), *mustard* (off-label) [1]
- Poppy fire in *poppies* (off-label), *poppies grown for seed production* (off-label) [1]
- Powdery mildew in *grapevines* (off-label) [1]
- Sclerotinia in *poppies for morphine production* (off-label) [1]
- Sclerotinia stem rot in *poppies* (off-label), *poppies grown for seed production* (off-label) [1]; *spring oilseed rape*, *winter oilseed rape* [1, 2]

Extension of Authorisation for Minor Use (EAMUs)
- *all edible seed crops grown outdoors* 20093231 [1]
- *all non-edible seed crops grown outdoors* 20093231 [1]
- *borage for oilseed production* 20131293 [1]
- *canary flower (echium spp.)* 20131293 [1]
- *corn gromwell* 20131293 [1]
- *evening primrose* 20131293 [1]
- *grapevines* 20131947 [1]
- *honesty* 20131293 [1]

- **linseed** *20131293* [1]
- **mustard** *20131293* [1]
- **poppies** *20180175* [1]
- **poppies for morphine production** *20093222* [1]
- **poppies grown for seed production** *20180175* [1]

Approval information
- Boscalid included in Annex I under EC Regulation 1107/2009
- Accepted by BBPA on malting barley before ear emergence only and for use on hops

Efficacy guidance
- For best results apply as a protectant spray before symptoms are visible
- Applications against *Sclerotinia* should be made in high disease risk situations at early to full flower
- *Sclerotinia* control may be reduced when high risk conditions occur after flowering, leading to secondary disease spread
- Applications against *Alternaria* should be made at full flowering. Later applications may result in reduced levels of control
- Ensure adequate spray penetration and good coverage
- Use with an effective tank-mix partner to reduce the risk of resistance developing

Restrictions
- Maximum total dose equivalent to two full dose treatments
- Do no treat crops to be used for seed production later than full flowering stage
- Avoid spray drift onto neighbouring crops

Crop-specific information
- Latest use: up to and including when 50% of pods have reached final size (GS 75) for oilseed rape

Environmental safety
- Dangerous for the environment
- Toxic to aquatic organisms
- Product represents minimal hazard to bees when used as directed. However local bee-keepers should be notified if crops are to be sprayed when in flower

Hazard classification and safety precautions
Hazard Dangerous for the environment
Transport code 9
Packaging group III
UN Number 3077
Operator protection U05a, U20c
Environmental protection E15a, E38, H411
Storage and disposal D01, D02, D05, D09a, D10c, D12a

53   boscalid + dimoxystrobin

A fungicide mixture for disease control in oilseed rape
FRAC mode of action code: 7 + 11

See also boscalid

Products
Pictor	BASF	200:200 g/l	SC	16783

Uses
- Alternaria in **corn gromwell** *(off-label)*, **winter oilseed rape** *(moderate control)*
- Dark leaf and pod spot in **corn gromwell** *(off-label)*
- Dark leaf spot in **winter oilseed rape** *(moderate control)*
- Disease control in **winter oilseed rape**
- Light leaf spot in **winter oilseed rape** *(reduction)*

- Phoma in **winter oilseed rape** *(moderate control)*
- Sclerotinia in **corn gromwell** *(off-label)*, **winter oilseed rape**
- Stem canker in **winter oilseed rape** *(moderate control)*

Extension of Authorisation for Minor Use (EAMUs)
- **corn gromwell** *20151106*

Approval information
- Boscalid and dimoxystrobin included in Annex I under EC Regulation 1107/2009

Efficacy guidance
- Treat light leaf spot and phoma/stem canker at first sign of infection in the spring. Treat sclerotinia at early - full flower in high disease risk situations. Treat altenaria from full flower to when 50% of pods have reached full size.

Environmental safety
- LERAP Category B

Hazard classification and safety precautions
Hazard Harmful if swallowed, Harmful if inhaled, Very toxic to aquatic organisms
Transport code 9
Packaging group III
UN Number 3082
Risk phrases H317, H351, H361
Operator protection A, H; U05a
Environmental protection E15b, E16a, E34, H410
Storage and disposal D01, D02, D09a, D10c, D12b
Medical advice M05a

54   boscalid + epoxiconazole

A broad spectrum fungicide mixture for cereals
FRAC mode of action code: 7 + 3

See also epoxiconazole

Products

1	Chord	BASF	210:75 g/l	SC	16065
2	Enterprise	BASF	140:50 g/l	OD	15228
3	Kingdom	BASF	140:50 g/l	OD	15240
4	Tracker	BASF	233:67 g/l	SC	16048

Uses
- Botrytis in **narcissi** *(off-label)* [4]
- Brown rust in **durum wheat** [1, 4]; **grass seed crops** *(off-label)* [4]; **rye, spring barley, spring wheat, triticale, winter barley, winter wheat** [1-4]
- Crown rust in **grass seed crops** *(off-label)* [4]; **oats** [2, 3]; **spring oats, winter oats** [1, 4]
- Eyespot in **durum wheat** *(moderate control)* [1, 4]; **rye** *(moderate control)*, **spring barley** *(moderate control)*, **spring wheat** *(moderate control)*, **winter barley** *(moderate control)*, **winter wheat** *(moderate control)* [1-4]; **triticale** *(moderate control)* [1-3]
- Fusarium ear blight in **spring wheat** *(good reduction)*, **winter wheat** *(good reduction)* [2, 3]
- Glume blotch in **spring wheat, triticale, winter wheat** [2, 3]
- Net blotch in **grass seed crops** *(off-label)* [4]; **spring barley, winter barley** [1-4]
- Powdery mildew in **oats** *(moderate control)*, **rye** *(moderate control)*, **spring barley** *(moderate control)*, **winter barley** *(moderate control)* [2, 3]
- Ramularia leaf spots in **spring barley** *(moderate control)*, **winter barley** *(moderate control)* [1-4]
- Rhynchosporium in **grass seed crops** *(off-label)* [4]; **rye** *(moderate control)*, **spring barley** *(moderate control)*, **winter barley** *(moderate control)* [1-4]
- Septoria leaf blotch in **durum wheat** [1, 4]; **spring wheat, triticale, winter wheat** [1-4]
- Smoulder in **narcissi** *(off-label)* [4]
- Sooty moulds in **spring wheat, winter wheat** [2, 3]

- Tan spot in **durum wheat** *(reduction)* [1, 4]; **spring wheat** *(reduction)*, **winter wheat** *(reduction)* [1-4]
- White mould in **narcissi** *(off-label)* [4]
- Yellow rust in **rye**, **spring barley**, **spring wheat**, **triticale**, **winter barley**, **winter wheat** [2, 3]

Extension of Authorisation for Minor Use (EAMUs)
- **grass seed crops** *20150655* [4]
- **narcissi** *20132458* [4]

Approval information
- Boscalid and epoxiconazole included in Annex I under EC Regulation 1107/2009
- Accepted by BBPA for use on malting barley (before ear emergence only)

Efficacy guidance
- Apply at the start of foliar or stem based disease attack
- Optimum effect against eyespot achieved by spraying between leaf-sheath erect and second node detectable stages (GS 30-32)
- Epoxiconazole is a DMI fungicide. Resistance to some DMI fungicides has been identified in Septoria leaf blotch which may seriously affect performance of some products. For further advice contact a specialist advisor and visit the Fungicide Resistance Action Group (FRAG)-UK website

Crop-specific information
- Latest use: before cereal ear emergence

Environmental safety
- Dangerous for the environment
- Toxic to aquatic organisms
- Avoid drift onto neighbouring crops. May cause damage to broad-leaved plant species
- LERAP Category B

Hazard classification and safety precautions
Hazard Harmful, Dangerous for the environment [1-4]; Harmful if inhaled [2-4]
Transport code 9
Packaging group III
UN Number 3082
Risk phrases H318 [2-4]; H319 [1]; H351, H360 [1-4]
Operator protection A, C [1-4]; H [1, 4]; U05a [1-4]; U14, U20a [2, 3]; U20b [1, 4]
Environmental protection E15a [1, 4]; E15b [2, 3]; E16a, E38, H410 [1-4]
Storage and disposal D01, D02, D10c, D12a [1-4]; D05, D09a [2, 3]; D08 [1, 4]
Medical advice M03 [1-4]; M05a [2, 3]

55   boscalid + epoxiconazole + pyraclostrobin

A systemic and protectant fungicide mixture for disease control in cereals
FRAC mode of action code: 3 + 7 + 11

Products

Nebula XL	BASF	140:50:60 g/l	OD	15555

Uses
- Brown rust in **rye**, **spring wheat**, **triticale**, **winter barley**, **winter wheat**
- Crown rust in **oats**
- Eyespot in **rye** *(moderate control)*, **spring wheat** *(moderate control)*, **triticale** *(moderate control)*, **winter barley** *(moderate control)*, **winter wheat** *(moderate control)*
- Fusarium ear blight in **spring wheat** *(good reduction)*, **winter wheat** *(good reduction)*
- Glume blotch in **spring wheat**, **triticale**, **winter wheat**
- Net blotch in **winter barley**
- Powdery mildew in **oats** *(moderate control)*, **rye** *(moderate control)*, **triticale** *(moderate control)*, **winter barley** *(moderate control)*
- Ramularia leaf spots in **winter barley** *(moderate control)*

- Rhynchosporium in *rye*, *winter barley*
- Septoria leaf blotch in *spring wheat*, *triticale*, *winter wheat*
- Sooty moulds in *spring wheat* (good reduction), *winter wheat* (good reduction)
- Tan spot in *spring wheat* (moderate control), *winter wheat* (moderate control)
- Yellow rust in *rye*, *spring wheat*, *triticale*, *winter barley*, *winter wheat*

Approval information
- Boscalid, epoxiconazole and pyraclostrobin included in Annex I under EC Regulation 1107/2009

Following crops guidance
- Onions, oats, sugar beet, oilseed rape, cabbages, carrots, sunflowers, barley, lettuce, ryegrass, dwarf french beans, peas, potatoes, clover, wheat, field beans and maize may be sown as the following crop.

Environmental safety
- LERAP Category B

Hazard classification and safety precautions
Hazard Harmful, Dangerous for the environment, Harmful if inhaled, Very toxic to aquatic organisms
Transport code 9
Packaging group III
UN Number 3082
Risk phrases H317, H319, H351, H360, R53a
Operator protection A, H; U05a, U14, U20a
Environmental protection E15b, E16a, E34, E38, H410
Storage and disposal D01, D02, D05, D09a, D10c, D12a
Medical advice M03, M05a

56 boscalid + metconazole

An anilide and triazole fungicide mixture for disease control in oilseed rape
FRAC mode of action code: 3 + 7

Products

1	Highgate	BASF	133:60 g/l	SC	15251
2	Tectura	BASF	133:60 g/l	SC	15232

Uses
- Sclerotinia stem rot in *oilseed rape*

Approval information
- Boscalid and metconazole included in Annex I under EC Regulation 1107/2009

Following crops guidance
- Any crop can follow normally-harvested oilseed rape treated with [1] or [2]

Environmental safety
- LERAP Category B

Hazard classification and safety precautions
Hazard Harmful
Transport code 9
Packaging group III
UN Number 3082
Risk phrases H361
Operator protection A, H; U05a, U20a
Environmental protection E15b, E16a, E16b, E34, E38, H412
Storage and disposal D01, D02, D05, D09a, D10c, D12a
Medical advice M03, M05a

57   boscalid + pyraclostrobin

A protectant and systemic fungicide mixture
FRAC mode of action code: 7 + 11

See also pyraclostrobin

Products

1	Bellis	BASF	25.2:12.8% w/w	WG	12522
2	Signum	BASF	26.7:6.7% w/w	WG	11450

SECTION 2

Uses
- Alternaria in *cabbages*, *calabrese*, *carrots* (moderate control), *cauliflowers*, *poppies* (off-label), *poppies for morphine production* (off-label), *poppies grown for seed production* (off-label), *protected oriental cabbage* (off-label) [2]
- American gooseberry mildew in *blackcurrants* (moderate control only) [2]
- Anthracnose in *blueberries* (off-label), *gooseberries* (off-label), *protected blueberry* (off-label), *protected gooseberries* (off-label) [2]
- Black spot in *protected strawberries*, *strawberries* [2]
- Blight in *potatoes* (off-label) [2]
- Blossom wilt in *cherries* (off-label), *mirabelles* (off-label), *plums* (off-label), *protected cherries* (off-label), *protected mirabelles* (off-label), *protected plums* (off-label) [2]
- Botrytis in *amenity vegetation* (off-label), *asparagus* (off-label), *beans without pods (fresh)* (off-label), *blueberries* (off-label), *broad beans* (off-label), *forest nurseries* (off-label), *gooseberries* (off-label), *interior landscapes* (off-label), *lamb's lettuce* (off-label), *ornamental plant production* (off-label), *poppies* (off-label), *poppies for morphine production* (off-label), *poppies grown for seed production* (off-label), *protected aubergines* (off-label), *protected blueberry* (off-label), *protected chard* (off-label), *protected forest nurseries* (off-label), *protected gooseberries* (off-label), *protected lettuce* (off-label), *protected ornamentals* (off-label), *protected peppers* (off-label), *protected spinach beet* (off-label), *protected tomatoes* (off-label), *redcurrants* (off-label), *swedes* (off-label), *turnips* (off-label), *whitecurrants* (off-label) [2]
- Bottom rot in *lettuce*, *protected lettuce* [2]
- Brown rot in *apricots* (off-label), *nectarines* (off-label), *peaches* (off-label) [2]
- Cane blight in *blackberries* (off-label), *protected blackberries* (off-label), *raspberries* (off-label) [2]
- Chocolate spot in *spring field beans* (moderate control), *winter field beans* (moderate control) [2]
- Cladosporium in *protected cucumbers* (off-label) [2]
- Cladosporium leaf blotch in *courgettes* (off-label), *summer squash* (off-label) [2]
- Colletotrichum orbiculare in *courgettes* (off-label), *summer squash* (off-label) [2]
- Disease control in *all edible crops (outdoor and protected)* (off-label), *all non-edible crops (outdoor)* (off-label), *all protected non-edible crops* (off-label), *beetroot* (off-label), *kohlrabi* (off-label), *lamb's lettuce* (off-label), *protected hops* (off-label), *protected radishes* (off-label), *protected soft fruit* (off-label), *protected top fruit* (off-label), *radishes* (off-label), *red beet* (off-label), *soft fruit* (off-label), *top fruit* (off-label) [2]; *hops* (off-label) [1, 2]
- Downy mildew in *bulls blood* (off-label), *chard* (off-label), *endives* (off-label), *protected endives* (off-label), *spinach* (off-label), *spinach beet* (off-label) [2]
- Fusarium in *bulb onion sets* (off-label) [2]
- Grey mould in *blackcurrants*, *lettuce*, *protected lettuce*, *protected strawberries*, *strawberries* [2]
- Gummy stem blight in *protected cucumbers* (off-label) [2]
- Late blight in *potatoes* (off-label) [2]
- Leaf and pod spot in *combining peas* (moderate control only), *vining peas* (moderate control only) [2]
- Leaf spot in *amenity vegetation* (off-label), *black salsify* (off-label), *blackcurrants* (moderate control only), *blueberries* (off-label), *gooseberries* (off-label), *interior landscapes* (off-label), *parsley root* (off-label), *protected blueberry* (off-label), *protected gooseberries* (off-label), *redcurrants* (off-label), *salsify* (off-label), *whitecurrants* (off-label) [2]
- Mycosphaerella in *protected oriental cabbage* (off-label) [2]

- Poppy fire in **poppies** *(off-label)*, **poppies for morphine production** *(off-label)*, **poppies grown for seed production** *(off-label)* [2]
- Powdery mildew in **amenity vegetation** *(off-label)*, **black salsify** *(off-label)*, **carrots, interior landscapes** *(off-label)*, **parsley root** *(off-label)*, **protected cucumbers** *(off-label)*, **protected strawberries, pumpkins** *(off-label)*, **redcurrants** *(off-label)*, **salsify** *(off-label)*, **strawberries, whitecurrants** *(off-label)* [2]; **apples** [1]
- Purple blotch in **blackberries** *(off-label)*, **protected blackberries** *(off-label)*, **raspberries** *(off-label)* [2]
- Rhizoctonia in **protected chard** *(off-label)*, **protected lettuce** *(off-label)*, **protected spinach beet** *(off-label)* [2]
- Ring spot in **broccoli, brussels sprouts, cabbages, calabrese, cauliflowers** [2]
- Rust in **amenity vegetation** *(off-label)*, **asparagus** *(off-label)*, **black salsify** *(off-label)*, **interior landscapes** *(off-label)*, **kohlrabi** *(off-label)*, **parsley root** *(off-label)*, **protected blackberries** *(off-label)*, **salsify** *(off-label)*, **spring field beans, winter field beans** [2]
- Scab in **apples, pears** [1]
- Sclerotinia in **black salsify** *(off-label)*, **carrots** *(moderate control)*, **celeriac** *(off-label)*, **horseradish** *(off-label)*, **parsley root** *(off-label)*, **poppies** *(off-label)*, **poppies grown for seed production** *(off-label)*, **protected chard** *(off-label)*, **protected lettuce** *(off-label)*, **protected spinach beet** *(off-label)*, **salsify** *(off-label)* [2]
- Soft rot in **lettuce, protected lettuce** [2]
- White blister in **broccoli, brussels sprouts, cabbages** *(qualified minor use)*, **calabrese** *(qualified minor use)*, **chinese cabbage** *(off-label)*, **choi sum** *(off-label)*, **collards** *(off-label)*, **herbs (see appendix 6)** *(off-label)*, **kale** *(off-label)*, **leaf brassicas** *(off-label)*, **pak choi** *(off-label)*, **protected herbs (see appendix 6)** *(off-label)*, **protected leaf brassicas** *(off-label)*, **tatsoi** *(off-label)* [2]
- White rot in **bulb onions** *(off-label)*, **garlic** *(off-label)*, **salad onions** *(off-label)*, **shallots** *(off-label)* [2]
- White rust in **kohlrabi** *(off-label)* [2]
- White tip in **leeks** *(off-label)* [2]

Extension of Authorisation for Minor Use (EAMUs)
- **all edible crops (outdoor and protected)** *20102111* [2]
- **all non-edible crops (outdoor)** *20102111* [2]
- **all protected non-edible crops** *20102111* [2]
- **amenity vegetation** *20122317* [2]
- **apricots** *20121721* [2]
- **asparagus** *20102105* [2]
- **beans without pods (fresh)** *20121009* [2]
- **beetroot** *20121717* [2]
- **black salsify** *20121720* [2]
- **blackberries** *20102110* [2]
- **blueberries** *20121722* [2]
- **broad beans** *20121009* [2]
- **bulb onion sets** *20103122* [2]
- **bulb onions** *20102108* [2]
- **bulls blood** *20102136* [2]
- **celeriac** *20141059* [2]
- **chard** *20102136* [2]
- **cherries** *20102109* [2]
- **chinese cabbage** *20130285* [2]
- **choi sum** *20130285* [2]
- **collards** *20130285* [2]
- **courgettes** *20142855* [2]
- **endives** *20102136* [2]
- **forest nurseries** *20102119* [2]
- **garlic** *20102108* [2]
- **gooseberries** *20121722* [2]
- **herbs (see appendix 6)** *20102115* [2]
- **hops** *20112732* [1], *20102111* [2]
- **horseradish** *20093375* [2]

FOR FULL CONDITIONS OF USE ALWAYS READ THE PRODUCT LABEL

- *interior landscapes* 20122317 [2]
- *kale* 20130285 [2]
- *kohlrabi* 20121719 [2]
- *lamb's lettuce* 20121718 [2]
- *leaf brassicas* 20102115 [2]
- *leeks* 20102134 [2]
- *mirabelles* 20102109 [2]
- *nectarines* 20121721 [2]
- *ornamental plant production* 20122141 [2]
- *pak choi* 20130285 [2]
- *parsley root* 20121720 [2]
- *peaches* 20121721 [2]
- *plums* 20102109 [2]
- *poppies* 20180176 [2]
- *poppies for morphine production* 20101233 [2]
- *poppies grown for seed production* 20180176 [2]
- *potatoes* 20110394 [2]
- *protected aubergines* 20120427 [2]
- *protected blackberries* 20102102 [2]
- *protected blueberry* 20121722 [2]
- *protected chard* 20131807 [2], 20181744 [2]
- *protected cherries* 20102109 [2]
- *protected cucumbers* 20151178 [2]
- *protected endives* 20102136 [2]
- *protected forest nurseries* 20102119 [2]
- *protected gooseberries* 20121722 [2]
- *protected herbs (see appendix 6)* 20102115 [2]
- *protected hops* 20102111 [2]
- *protected leaf brassicas* 20102115 [2]
- *protected lettuce* 20131807 [2]
- *protected mirabelles* 20102109 [2]
- *protected oriental cabbage* 20183055 [2]
- *protected ornamentals* 20122141 [2]
- *protected peppers* 20120427 [2]
- *protected plums* 20102109 [2]
- *protected radishes* 20121717 [2]
- *protected soft fruit* 20102111 [2]
- *protected spinach beet* 20131807 [2]
- *protected tomatoes* 20120427 [2]
- *protected top fruit* 20102111 [2]
- *pumpkins* 20152651 [2]
- *radishes* 20121717 [2]
- *raspberries* 20102110 [2]
- *red beet* 20121717 [2]
- *redcurrants* 20102114 [2]
- *salad onions* 20102107 [2]
- *salsify* 20121720 [2]
- *shallots* 20102108 [2]
- *soft fruit* 20102111 [2]
- *spinach* 20102136 [2]
- *spinach beet* 20102136 [2]
- *summer squash* 20142855 [2]
- *swedes* 20151961 [2]
- *tatsoi* 20130285 [2]
- *top fruit* 20102111 [2]
- *turnips* 20151961 [2]
- *whitecurrants* 20102114 [2]

SEE SECTION 3 FOR PRODUCTS ALSO REGISTERED

Approval information
- Boscalid and pyraclostrobin included in Annex I under EC Regulation 1107/2009
- Accepted by BBPA on malting barley before ear emergence only and for use on hops

Efficacy guidance
- On brassicas apply as a protectant spray or at the first sign of disease and repeat at 3-4 wk intervals depending on disease pressure [2]
- Ensure adequate spray penetration and coverage by increasing water volume in dense crops [2]
- For best results on strawberries apply as a protectant spray at the white bud stage. Applications should be made in sequence with other products as part of a fungicide spray programme during flowering at 7-10 day intervals [2]
- On carrots and field beans apply as a protectant spray or at the first sign of disease with a repeat treatment if needed, as directed on the label [2]
- On lettuce apply as a protectant spray 1-2 wk after planting [2]
- Optimum results on apples and pears obtained from a protectant treatment from bud burst [1]
- Application as the final 2 sprays on apples and pears gives reduction in storage rots [1]
- Pyraclostrobin is a member of the QoI cross-resistance group of fungicides and should be used in programmes with fungicides with a different mode of action

Restrictions
- Maximum total dose equivalent to three full dose treatments on brassicas; two full dose treatments on all other field crops [2]
- Maximum number of treatments (including other QoI treatments) on apples and pears 4 per yr if total number of applications is 12 or more, or 3 per yr if the total number is fewer than 12 [1]
- Do not use more than 2 consecutive treatments on apples and pears, and these must be separated by a minimum of 2 applications of a fungicide with a different mode of action [1]
- Use a maximum of three applications per yr on brassicas and no more than two per yr on other field crops [2]
- Do not use consecutive treatments; apply in alternation with fungicides from a different cross resistance group and effective against the target diseases [2]
- Do not apply more than 6 kg/ha to the same area of land per yr [2]
- Consult processor before use on crops for processing
- Applications to lettuce and protected salad crops and to crops grown under temporary structures or polytunnels must only be made between 1 April and 31 October [2]

Crop-specific information
- HI 21 d for field beans; 14 d for brassica crops, carrots, lettuce; 7 d for apples, pears; 3 d for strawberries

Environmental safety
- Dangerous for the environment
- Very toxic to aquatic organisms
- Broadcast air-assisted LERAP [1] (40 m); LERAP Category B [2]

Hazard classification and safety precautions

Hazard Harmful, Dangerous for the environment, Harmful if swallowed, Very toxic to aquatic organisms

Transport code 9

Packaging group III

UN Number 3077

Operator protection A; U05a, U20c [1, 2]; U13 [1]

Environmental protection E15a, E38, H410 [1, 2]; E16a [2]; E17b [1] (40 m); E34 [1]

Storage and disposal D01, D02, D10c, D12a [1, 2]; D08 [1]; D09a [2]

Medical advice M03 [1]; M05a [1, 2]

58 brodifacoum

An anticoagulant coumarin rodenticide

Products

| 1 | Brodiag Blocks | Killgerm | 0.0029% w/w | RB | UK15-0858 |

FOR FULL CONDITIONS OF USE ALWAYS READ THE PRODUCT LABEL

Products – continued

2	Brodiag Fresh Bait	Killgerm	0.0029% w/w	RB	UK15-0857
3	Brodiag Whole Wheat	Killgerm	0.0029% w/w	RB	UK17-1051
4	Pest Expert Formula B+	Killgerm	0.0029% w/w	RB	UK17-1051
5	Sakarat Brodikill Whole Wheat	Killgerm	0.0029% w/w	RB	UK17-1051

Uses
* Mice in *farm buildings*, *farmyards* [1-3]
* Rats in *farm buildings*, *farmyards* [1-3]

Approval information
* Brodifacoum is not included in Annex 1 under EC Regulation 1107/2009

Efficacy guidance
* Product is a self-contained control device which requires no bait handling
* Best results obtained by placing tubes where mice are active and will readily enter the device
* To acquire a lethal dose mice must pass through the tube a number of times
* Where a continuous source of infestation is present, or where sustained protection is needed, tubes can be installed permanently and replaced at monthly intervals
* For complete eradication other control methods should also be used bearing in mind the resistance status of the target population

Restrictions
* For use only by professional pest contractors
* Do not use outdoors. Products must be used in situations where baits are placed within a building or other enclosed structure, and the target is living or feeding predominantly within that building or structure
* Do not attempt to open or recharge the tubes

Environmental safety
* Prevent access to baits by children, birds and other animals
* Do not place tubes where food, feed or water could become contaminated
* Remove all remains of tubes after use and burn or bury
* Search for and burn or bury all rodent bodies. Do not place in refuse bins or on open rubbish tips
* Keep in original container, tightly closed, in a safe place, under lock and key

Hazard classification and safety precautions
Operator protection U13, U20b [1-3]
Storage and disposal D05, D07, D09a, D11a [1-3]
Vertebrate/rodent control products V01b, V02, V03b, V04b [1-3]
Medical advice M03 [1-3]

59   bromadiolone

An anti-coagulant coumarin-derivative rodenticide

Products

1	Bromag Fresh Bait	Killgerm	0.005% w/w	RB	UK13-0754
2	Bromag Wax Blocks	Killgerm	0.005% w/w	RB	UK13-0755
3	Bromag Whole Wheat	Killgerm	0.005% w/w	RB	UK15-0932
4	Formula B Rat & Mouse Killer	Killgerm	0.005% w/w	RB	UK15-0934
5	Sakarat Bromabait	Killgerm	0.005% w/w	RB	UK15-0931
6	Sakarat Bromakill	Killgerm	0.005% w/w	RB	UK15-0935

Uses
* Mice in *farm buildings*, *farmyards* [2, 5, 6]; *farm buildings/yards* [1, 3, 4]
* Rats in *farm buildings*, *farmyards* [2, 5, 6]; *farm buildings/yards* [1, 3, 4]

SEE SECTION 3 FOR PRODUCTS ALSO REGISTERED

Approval information
- Bromadiolone included in Annex I under EC Directive 1107/2009
- Approval expiry 31 Aug 2020 [1, 3-6]

Efficacy guidance
- Ready-to-use baits are formulated on a mould-resistant, whole-wheat base
- Use in baiting programme. Place baits in protected situations, sufficient for continuous feeding between treatments
- Chemical is effective against warfarin- and coumatetralyl-resistant rats and mice and does not induce bait shyness
- Use bait bags where loose baiting inconvenient (eg behind ricks, silage clamps etc)
- The resistance status of the rodent population should be assessed when considering the choice of product to use. Resistance to bromadiolone in rats is now widespread in the UK.

Restrictions
- For use only by professional operators

Environmental safety
- Access to baits by children, birds and animals, particularly cats, dogs, pigs and poultry, must be prevented
- Baits must not be placed where food, feed or water could become contaminated
- Remains of bait and bait containers must be removed after treatment and burned or buried
- Rodent bodies must be searched for and burned or buried. They must not be placed in refuse bins or on rubbish tips
- Take extreme care to prevent domestic animals having access to the bait

Hazard classification and safety precautions
Operator protection A [1, 3]; U13, U20b
Storage and disposal D05, D07, D09a, D11a
Vertebrate/rodent control products V01b, V02, V03b, V04b
Medical advice M03

60   bromoxynil

A contact acting HBN herbicide
HRAC mode of action code: C3

Products

1 Butryflow	Nufarm UK	402 g/l	SC	14056
2 Maya	Nufarm UK	402 g/l	SC	16760

Uses
- Annual dicotyledons in **bulb onion sets** *(off-label)*, **leeks** *(off-label)*, **millet** *(off-label)*, **salad onions** *(off-label)*, **shallots** *(off-label)* [1]; **bulb onions, forage maize, grain maize, linseed, ornamental bulbs** *(off-label)*, **shallots, spring barley, spring oats, spring wheat, sweetcorn, winter barley, winter oats, winter wheat** [1, 2]
- Black bindweed in **bulb onion sets** *(off-label)*, **canary grass** *(off-label)*, **forage maize, grain maize, leeks** *(off-label)*, **linseed, salad onions** *(off-label)*, **shallots** *(off-label)*, **sweetcorn** [2]; **game cover** *(off-label)*, **millet** *(off-label)*, **miscanthus** *(off-label)*, **ornamental bulbs** *(off-label)*, **ornamental plant production** *(off-label)* [1, 2]
- Black nightshade in **bulb onion sets** *(off-label)*, **canary grass** *(off-label)*, **forage maize, grain maize, leeks** *(off-label)*, **linseed, salad onions** *(off-label)*, **shallots** *(off-label)*, **sweetcorn** [2]; **game cover** *(off-label)*, **millet** *(off-label)*, **miscanthus** *(off-label)*, **ornamental plant production** *(off-label)* [1, 2]
- Chickweed in **bulb onion sets** *(off-label)*, **canary grass** *(off-label)*, **leeks** *(off-label)*, **salad onions** *(off-label)*, **shallots** *(off-label)* [2]; **game cover** *(off-label)*, **millet** *(off-label)*, **miscanthus** *(off-label)*, **ornamental plant production** *(off-label)* [1, 2]
- Common orache in **bulb onion sets** *(off-label)*, **forage maize, grain maize, leeks** *(off-label)*, **linseed, millet** *(off-label)*, **salad onions** *(off-label)*, **shallots** *(off-label)*, **sweetcorn** [2]; **game cover** *(off-label)*, **miscanthus** *(off-label)*, **ornamental plant production** *(off-label)* [1, 2]

FOR FULL CONDITIONS OF USE ALWAYS READ THE PRODUCT LABEL

- Docks in **bulb onion sets** *(off-label)*, **leeks** *(off-label)*, **salad onions** *(off-label)*, **shallots** *(off-label)* [1, 2]; **millet** *(off-label)* [2]
- Fat hen in **bulb onion sets** *(off-label)*, **bulb onions**, **canary grass** *(off-label)*, **forage maize**, **grain maize**, **leeks** *(off-label)*, **linseed**, **salad onions** *(off-label)*, **shallots**, **shallots** *(off-label)*, **spring barley**, **spring oats**, **spring wheat**, **sweetcorn**, **winter barley**, **winter oats**, **winter wheat** [2]; **game cover** *(off-label)*, **millet** *(off-label)*, **miscanthus** *(off-label)*, **ornamental bulbs** *(off-label)*, **ornamental plant production** *(off-label)* [1, 2]
- Field bindweed in **bulb onion sets** *(off-label)*, **forage maize**, **grain maize**, **leeks** *(off-label)*, **linseed**, **millet** *(off-label)*, **salad onions** *(off-label)*, **shallots** *(off-label)*, **sweetcorn** [2]; **game cover** *(off-label)*, **miscanthus** *(off-label)*, **ornamental plant production** *(off-label)* [1, 2]
- Groundsel in **bulb onion sets** *(off-label)*, **bulb onions**, **canary grass** *(off-label)*, **leeks** *(off-label)*, **salad onions** *(off-label)*, **shallots**, **shallots** *(off-label)*, **spring barley**, **spring oats**, **spring wheat**, **winter barley**, **winter oats**, **winter wheat** [2]; **ornamental bulbs** *(off-label)* [1, 2]
- Knotgrass in **bulb onion sets** *(off-label)*, **forage maize**, **grain maize**, **leeks** *(off-label)*, **linseed**, **millet** *(off-label)*, **salad onions** *(off-label)*, **shallots** *(off-label)*, **sweetcorn** [2]; **game cover** *(off-label)*, **miscanthus** *(off-label)*, **ornamental plant production** *(off-label)* [1, 2]
- Redshank in **bulb onion sets** *(off-label)*, **leeks** *(off-label)*, **salad onions** *(off-label)*, **shallots** *(off-label)* [2]; **game cover** *(off-label)*, **millet** *(off-label)*, **miscanthus** *(off-label)*, **ornamental plant production** *(off-label)* [1, 2]
- Scentless mayweed in **bulb onion sets** *(off-label)*, **canary grass** *(off-label)*, **leeks** *(off-label)*, **salad onions** *(off-label)*, **shallots** *(off-label)* [2]
- Small nettle in **ornamental bulbs** *(off-label)* [2]
- Volunteer oilseed rape in **bulb onion sets** *(off-label)*, **bulb onions**, **canary grass** *(off-label)*, **leeks** *(off-label)*, **salad onions** *(off-label)*, **shallots**, **shallots** *(off-label)*, **spring barley**, **spring oats**, **spring wheat**, **winter barley**, **winter oats**, **winter wheat** [2]

Extension of Authorisation for Minor Use (EAMUs)
- **bulb onion sets** *20162683* [1], *20162882* [2]
- **canary grass** *20171860* [2]
- **game cover** *20131620* [1], *20171044* [2]
- **leeks** *20162683* [1], *20162882* [2]
- **millet** *20120593* [1], *20162881* [2]
- **miscanthus** *20131620* [1], *20171044* [2]
- **ornamental bulbs** *20130517* [1], *20171045* [2]
- **ornamental plant production** *20140561* [1], *20171046* [2]
- **salad onions** *20162683* [1], *20162882* [2]
- **shallots** *20162683* [1], *20162882* [2]

Approval information
- Bromoxynil included in Annex I under EC Regulation 1107/2009
- Accepted by BBPA for use on malting barley

Efficacy guidance
- Spray when main weed flush has germinated and the largest are at the 4 leaf stage
- Weed control can be enhanced by using a split treatment spraying each application when the weeds are seedling to 2 true leaves.

Restrictions
- Maximum number of treatments 1 per crop or yr or maximum total dose equivalent to one full dose treatment
- Do not apply with oils or other adjuvants
- Do not apply to bulb onion or shallot grown from sets [1]
- Do not apply using hand-held equipment or at concentrations higher than those recommended
- Do not apply during frosty weather, drought, when soil is waterlogged, when rain expected within 4 h or to crops under any stress
- Take particular care to avoid drift onto neighbouring susceptible crops or open water surfaces
- Do not apply if air temp is above 25°C in cereals or 20°C in maize as some necrosis may occur

Crop-specific information
- Latest use: before 10 fully expanded leaf stage of crop for maize, sweetcorn; before crop 20 cm tall and before flower buds visible for linseed; before 2nd node detectable (GS 32) for cereals

SEE SECTION 3 FOR PRODUCTS ALSO REGISTERED

- Foliar scorch, which rapidly disappears without affecting growth, will occur if treatment made in hot weather or during rapid growth

Following crops guidance
- There are no restrictions on the establishment of succeeding or replacement crops

Environmental safety
- Very toxic to aquatic organisms
- Keep livestock out of treated areas for at least 6 wk after treatment
- High risk to bees. Do not apply to crops in flower or to those in which bees are actively foraging
- Do not contaminate surface waters or ditches with chemical or used container
- LERAP Category B

Hazard classification and safety precautions
Hazard Harmful, Dangerous for the environment, Harmful if swallowed, Harmful if inhaled
Transport code 9
Packaging group III
UN Number 3082
Risk phrases H317, H361
Operator protection A, H, K; U05a, U14, U15, U23a
Environmental protection E15a, E16a, E34, E38, H410
Storage and disposal D01, D02, D05, D09a, D10c, D12a
Medical advice M03

61 bromoxynil + diflufenican

A contact and residual herbicide mixture for weed control in cereals
HRAC mode of action code: C3 + F1

See also diflufenican

Products

Cyclops	Nufarm UK	160:26.7 g/l	EC	16676

Uses
- Chickweed in *grass seed crops* (off-label), *winter barley*, *winter wheat*
- Fat hen in *grass seed crops* (off-label), *winter barley* (at 1.9 l/ha only)
- Field pansy in *grass seed crops* (off-label), *winter barley*, *winter wheat*
- Groundsel in *grass seed crops* (off-label), *winter barley* (at 1.9 l/ha only)
- Ivy-leaved speedwell in *grass seed crops* (off-label), *winter barley* (at 1.9 l/ha only)
- Volunteer beans in *grass seed crops* (off-label), *winter barley*, *winter wheat*
- Volunteer oilseed rape in *grass seed crops* (off-label), *winter barley*, *winter wheat*

Extension of Authorisation for Minor Use (EAMUs)
- *grass seed crops* 20161452

Approval information
- Bromoxynil and diflufenican included in Annex I under EC Regulation 1107/2009

Restrictions
- Do not allow spray to drift on to non-target plants.
- Do not treat stressed crops.
- Do not overlap spray swathes.

Following crops guidance
- In the event of crop failure, winter wheat can be drilled immediately after cultivation but winter barley can only be sown after ploughing. Before spring crops of wheat, barley, oilseed rape, peas, sugar beet, field beans, potatoes, carrots, edible brassicas or onions are sown, 12 weeks should elapse since application and the land must be ploughed. Do not plant any crop other than those listed here before the subsequent autumn.

Environmental safety
- Buffer zone requirement 7 m [1]
- To protect non target insects/arthropods respect an unsprayed buffer zone of 5 m to non-crop land [1]
- LERAP Category B

Hazard classification and safety precautions

Hazard Corrosive, Dangerous for the environment, Harmful if swallowed
Transport code 9
Packaging group III
UN Number 3082
Risk phrases H317, H361, R70
Operator protection A, C, H; U04b, U05a, U09a, U11, U19a, U20a
Environmental protection E07a, E15b, E16a, E34, E38, H410
Storage and disposal D01, D02, D05, D09a, D10b, D12a, D12b
Medical advice M03, M05a

62   bromuconazole

A systemic triazole fungicide for cereals only available in mixtures
FRAC mode of action code: 3

63   bromuconazole + tebuconazole

A triazole mixture for disease control in wheat, rye and triticale
FRAC mode of action code: 3

See also tebuconazole

Products

Soleil	Sumitomo	167:107 g/l	EC	16869

Uses
- Alternaria in *rye*, *spring wheat*, *triticale*, *winter wheat*
- Brown rust in *rye*, *spring wheat*, *triticale*, *winter wheat*
- Cladosporium in *rye*, *spring wheat*, *triticale*, *winter wheat*
- Fusarium in *rye*, *spring wheat*, *triticale*, *winter wheat*
- Powdery mildew in *rye* (moderate control), *spring wheat* (moderate control), *triticale* (moderate control), *winter wheat* (moderate control)
- Septoria leaf blotch in *rye* (moderate control), *spring wheat* (moderate control), *triticale* (moderate control), *winter wheat* (moderate control)

Approval information
- Bromuconazole and tebuconazole included in Annex 1 under EC Directive 1107/2009

Restrictions
- Newer authorisations for tebuconazole products require application to cereals only after GS 30 and applications to oilseed rape and linseed after GS20 - check label

Following crops guidance
- Only sugar beet, cereals, oilseed rape, field beans, peas, potatoes, linseed and Italian rye-grass may be sown as the following crop. The effect on other crops has not been assessed.
- Some effects may be seen on sugar beet crops grown following a cereal crop but the sugar beet will usually recover completely.

Environmental safety
- LERAP Category B

Hazard classification and safety precautions

Hazard Harmful, Dangerous for the environment, Very toxic to aquatic organisms
Transport code 9

Packaging group III
UN Number 3082
Risk phrases H304, H318, H336, H361
Operator protection A, C, H; U11, U19a, U19f, U20b
Environmental protection E16a, E34, E38, H410
Storage and disposal D01, D02, D05, D09a, D11a, D12a, D12b
Medical advice M03

64 bupirimate

A systemic aminopyrimidinol fungicide active against powdery mildew
FRAC mode of action code: 8

Products

Nimrod	Adama	250 g/l	EC	18522

Uses

- Powdery mildew in *apples*, *begonias*, *blackcurrants*, *chrysanthemums*, *gooseberries*, *ornamental plant production*, *pears*, *protected chrysanthemums*, *protected cucumbers*, *protected ornamentals*, *protected raspberries*, *protected roses*, *protected strawberries*, *protected tomatoes* (off-label), *raspberries*, *redcurrants*, *roses*, *strawberries*

Extension of Authorisation for Minor Use (EAMUs)

- *protected tomatoes* 20192504

Approval information

- Bupirimate included in Annex 1 under EC Regulation 1107/2009
- Accepted by BBPA for use on hops

Efficacy guidance

- On apples during periods that favour disease development lower doses applied weekly give better results than higher rates fortnightly
- Not effective in protected crops against strains of mildew resistant to bupirimate

Restrictions

- Maximum number of treatments or maximum total dose depends on crop and dose (see label for details)
- Reasonable precautions must be taken to prevent access of birds, wild mammals and honey bees to treated crops

Crop-specific information

- HI: 1 d for apples, pears, strawberries; 2 d for cucurbits; 7 d for blackcurrants; 8 d for raspberries; 14 d for gooseberries, hops
- Apply before or at first signs of disease and repeat at 7-14 d intervals. Timing and maximum dose vary with crop. See label for details
- With apples, hops and ornamentals cultivars may vary in sensitivity to spray. See label for details
- If necessary to spray cucurbits in winter or early spring spray a few plants 10-14 d before spraying whole crop to test for likelihood of leaf spotting problem
- On roses some leaf puckering may occur on young soft growth in early spring or under low light intensity. Avoid use of high rates or wetter on such growth
- Never spray flowering begonias (or buds showing colour) as this can scorch petals
- Do not mix with other chemicals for application to begonias, cucumbers or gerberas

Environmental safety

- Dangerous for the environment
- Toxic to aquatic organisms
- Flammable
- Product has negligible effect on *Phytoseiulus* and *Encarsia* and may be used in conjunction with biological control of red spider mite

Hazard classification and safety precautions
> **Hazard** Irritant, Flammable, Dangerous for the environment, Flammable liquid and vapour
> **Transport code** 3
> **Packaging group** III
> **UN Number** 1993
> **Risk phrases** H304, H319, H335, H351
> **Operator protection** A, C; U05a, U20b
> **Environmental protection** E15a, H410
> **Storage and disposal** D01, D02, D09a, D10c
> **Medical advice** M05b

65   buprofezin

A moulting inhibitor, thiadiazine insecticide for whitefly control
IRAC mode of action code: 16

Products

Applaud 25 SC	Certis	250 g/l	SL	17196

Uses
- Glasshouse whitefly in **protected aubergines, protected chilli peppers, protected cucumbers, protected melons, protected ornamentals, protected peppers, protected tomatoes**
- Tobacco whitefly in **protected aubergines, protected chilli peppers, protected cucumbers, protected melons, protected ornamentals, protected peppers, protected tomatoes**

Approval information
- Buprofezin included in Annex 1 under EC Regulation 1107/2009

Efficacy guidance
- Product has contact, residual and some vapour activity
- Whitefly most susceptible at larval stages but residual effect can also kill nymphs emerging from treated eggs and application to pupae reduces emergence
- Adult whitefly not directly affected. Resistant strains of tobacco whitefly are known and where present control likely to be reduced or ineffective

Restrictions
- Maximum number of treatments 8 per crop for tomatoes and cucumbers; 4 per crop on protected ornamentals; 2 per crop for aubergines and peppers
- Do not apply more than 2 sprays within a 65 d period on tomatoes, or within a 45 d period on cucumbers
- Do not treat *Dieffenbachia* or *Closmoplictrum*
- Do not apply to crops under stress
- Do not leave spray liquid in sprayer for long periods
- Do not apply as fog or mist

Crop-specific information
- HI for all crops 3 d
- In IPM programme apply as single application and allow at least 60 d before re-applying
- In All Chemical programme apply twice at 7-14 d interval and allow at least 60 d before re-applying
- See label for list of ornamentals successfully treated but small scale test advised to check varietal tolerance. This is especially important if spraying flowering ornamentals with buds showing colour

Environmental safety
- Product may be used either in IPM programme in association with *Encarsia formosa* or in All Chemical programme

Hazard classification and safety precautions
> **Transport code** 9

Packaging group III
UN Number 3082
Operator protection A; U02a, U09a, U09c, U19a, U20a
Environmental protection E15b, H411
Storage and disposal D05, D09a, D10c, D12b
Medical advice M03a

66   captan

A protectant phthalimide fungicide with horticultural uses
FRAC mode of action code: M4

Products

1 Captan 80 WDG	Adama	80% w/w	WG	16293
2 Clayton Core	Clayton	80% w/w	WG	16934
3 Multicap	Belcrop	80% w/w	WG	17652
4 PP Captan 80 WG	Arysta	80% w/w	WG	16294

Uses

- Botrytis in **ornamental plant production** *(off-label)* [1]
- Disease control in **hops in propagation** *(off-label)*, **nursery fruit trees** *(off-label)*, **quinces** *(off-label)* [1]
- Downy mildew in **ornamental plant production** *(off-label)* [1]
- Gloeosporium rot in **apples** [1-4]; **pears** [1]
- Scab in **apples** [1-4]; **ornamental plant production** *(off-label)* [1]; **pears** [1, 3, 4]
- Shot-hole in **ornamental plant production** *(off-label)* [1]

Extension of Authorisation for Minor Use (EAMUs)

- **hops in propagation** *20142510* [1]
- **nursery fruit trees** *20142510* [1]
- **ornamental plant production** *20151919* [1], *20151920* [1]
- **quinces** *20141276* [1]

Approval information

- Captan included in Annex I under EC Regulation 1107/2009

Restrictions

- Maximum number of treatments 12 per yr on apples and pears as pre-harvest sprays
- Do not use on apple cultivars Bramley, Monarch, Winston, King Edward, Spartan, Kidd's Orange or Red Delicious or on pear cultivar D'Anjou
- Do not mix with alkaline materials or oils
- Do not use on fruit for processing
- Powered visor respirator with hood and neck cape must be used when handling concentrate

Crop-specific information

- HI apples, pears 14 d; strawberries 7 d
- For control of scab apply at bud burst and repeat at 10-14 d intervals until danger of scab infection ceased
- For suppression of fruit storage rots apply from late Jul and repeat at 2-3 wk intervals
- For black spot control in roses apply after pruning with 3 further applications at 14 d intervals or spray when spots appear and repeat at 7-10 d intervals
- For grey mould in strawberries spray at first open flower and repeat every 7-10 d
- Do not leave diluted material for more than 2 h. Agitate well before and during spraying

Environmental safety

- Dangerous for the environment
- Very toxic to aquatic organisms
- LERAP Category B [3, 4]

Hazard classification and safety precautions

Hazard Harmful, Dangerous for the environment [1-4]; Very toxic to aquatic organisms [1-3]

FOR FULL CONDITIONS OF USE ALWAYS READ THE PRODUCT LABEL

Transport code 9
Packaging group III
UN Number 3077
Risk phrases H317, H351 [1-4]; H318, H320 [4]; H319 [1-3]
Operator protection A, D, E, H; U05a, U09a, U11, U19a, U20c
Environmental protection E15a [1-4]; E16a [3, 4]; H410 [4]; H412 [1-3]
Storage and disposal D01, D02, D09a, D11a, D12a

SECTION 2

67 carbetamide

A residual pre- and post-emergence carbamate herbicide for a range of field crops
HRAC mode of action code: K2

Products

Crawler	Adama	60% w/w	WG	16856

Uses

- Annual dicotyledons in *corn gromwell* (off-label), *linseed* (off-label), *winter field beans*, *winter oilseed rape*
- Annual grasses in *winter field beans*, *winter oilseed rape*
- Annual meadow grass in *corn gromwell* (off-label), *linseed* (off-label)
- Blackgrass in *corn gromwell* (off-label), *linseed* (off-label)
- Chickweed in *corn gromwell* (off-label), *linseed* (off-label)
- Cleavers in *corn gromwell* (off-label), *linseed* (off-label)
- Couch in *corn gromwell* (off-label), *linseed* (off-label)
- Fat hen in *corn gromwell* (off-label), *linseed* (off-label)
- Fumitory in *corn gromwell* (off-label), *linseed* (off-label)
- Knotgrass in *corn gromwell* (off-label), *linseed* (off-label)
- Ryegrass in *corn gromwell* (off-label), *linseed* (off-label)
- Small nettle in *corn gromwell* (off-label), *linseed* (off-label)
- Speedwells in *corn gromwell* (off-label), *linseed* (off-label)
- Volunteer cereals in *corn gromwell* (off-label), *linseed* (off-label), *winter field beans*, *winter oilseed rape*
- Wild oats in *corn gromwell* (off-label), *linseed* (off-label)
- Yorkshire fog in *corn gromwell* (off-label), *linseed* (off-label)

Extension of Authorisation for Minor Use (EAMUs)

- *corn gromwell* 20151828
- *linseed* 20152544

Approval information

- Carbetamide included in Annex I under EC Regulation 1107/2009

Efficacy guidance

- Best results pre- or early post-emergence of weeds under cool, moist conditions. Adequate soil moisture is essential
- May be applied pre or post emergence up to end of Feb in year of harvest
- Maintain an interval of at least 7 days when using split applications
- Dicotyledons controlled include chickweed, cleavers and speedwell
- Weed growth stops rapidly after treatment but full effects may take 6-8 wk to develop
- Various tank mixes are recommended to broaden the weed spectrum. See label for details
- Always follow WRAG guidelines for preventing and managing herbicide resistant weeds. See Section 5 for more information

Restrictions

- Maximum number of treatments 1 per crop for all crops
- Do not treat any crop on waterlogged soil
- Do not use on soils with more than 10% organic matter as residual activity is impaired
- Do not apply during prolonged periods of cold weather when weeds are dormant
- Do not tank-mix with prothioconazole or liquid manganese

SEE SECTION 3 FOR PRODUCTS ALSO REGISTERED

Crop-specific information
- HI 6 wk for all crops
- Apply to brassicas from mid-Oct to end-Feb provided crop has at least 4 true leaves (spring cabbage, spring greens), 3-4 true leaves (seed crops, oilseed rape)
- Apply to established lucerne or sainfoin from Nov to end-Feb
- Apply to established red or white clover from Feb to mid-Mar

Following crops guidance
- Succeeding crops may be sown 2 wk after treatment for brassicas, field beans, 8 wk after treatment for peas, runner beans, 16 wk after treatment for cereals, maize
- Ploughing is not necessary before sowing subsequent crops

Environmental safety
- Do not graze crops for at least 6 wk after treatment
- Some pesticides pose a greater threat of contamination of water than others and carbetamide is one of these pesticides. Take special care when applying carbetamide near water and do not apply if heavy rain is forecast

Hazard classification and safety precautions
Hazard Dangerous for the environment
Transport code 9
Packaging group III
UN Number 3077
Risk phrases H319, H351, H360
Operator protection A, C, H; U20c
Environmental protection E15a, E38, H411
Storage and disposal D01, D02, D09a, D11a

68   carbon dioxide (commodity substance)

A gas for the control of trapped rodents and other vertebrates. Approval valid until 31/8/2019

Products
carbon dioxide	various	99.9%	GA	-

Uses
- Birds in *traps*
- Mice in *traps*
- Rats in *traps*

Approval information
- Carbon dioxide included in Annex 1 under EC Regulation 1107/2009
- Approval expiry 31 Aug 2020 [1]

Efficacy guidance
- Use to destroy trapped rodent pests
- Use to control birds covered by general licences issued by the Agriculture and Environment Departments under Section 16(1) of the Wildlife and Countryside Act (1981) for the control of opportunistic bird species, where birds have been trapped or stupefied with alphachloralose/seconal

Restrictions
- Operators must wear self-contained breathing apparatus when carbon dioxide levels are greater than 0.5% v/v
- Operators must be suitably trained and competent

Environmental safety
- Unprotected persons and non-target animals must be excluded from the treatment enclosures and surrounding areas unless the carbon dioxide levels are below 0.5% v/v

Hazard classification and safety precautions
Operator protection G

FOR FULL CONDITIONS OF USE ALWAYS READ THE PRODUCT LABEL

69   carfentrazone-ethyl

A triazolinone contact herbicide
HRAC mode of action code: E

Products

1	Aurora 40 WG	FMC Agro	40% w/w	WG	18703
2	Headlite	AgChem Access	60 g/l	ME	13624
3	Shark	FMC Agro	60 g/l	ME	18700
4	Shylock	AgChem Access	60 g/l	ME	15405
5	Shylock	Nurture	60 g/l	ME	18928
6	Spotlight Plus	FMC Agro	60 g/l	ME	18698

Uses

- Annual and perennial weeds in *all edible crops (outdoor)*, *all non-edible crops (outdoor)*, *potatoes* [6]; *bilberries* (off-label), *blackberries* (off-label), *blackcurrants* (off-label), *blueberries* (off-label), *cranberries* (off-label), *forestry plantations* (off-label), *gooseberries* (off-label), *loganberries* (off-label), *ornamental plant production* (off-label), *protected bilberries* (off-label), *protected blackberries* (off-label), *protected blackcurrants* (off-label), *protected blueberry* (off-label), *protected cranberries* (off-label), *protected forest* (off-label), *protected gooseberries* (off-label), *protected loganberries* (off-label), *protected ornamentals* (off-label), *protected raspberries* (off-label), *protected redcurrants* (off-label), *protected rubus hybrids* (off-label), *protected strawberries* (off-label), *raspberries* (off-label), *redcurrants* (off-label), *ribes species* (off-label), *rubus hybrids* (off-label), *strawberries* (off-label), *wine grapes* (off-label) [3]
- Annual dicotyledons in *all edible crops (outdoor)* (before planting), *all non-edible crops (outdoor)* (before planting), *potatoes* [3-5]; *rye* (off-label), *spring barley*, *spring oats*, *spring wheat*, *triticale*, *winter barley*, *winter oats*, *winter wheat* [1]
- Black bindweed in *asparagus* (off-label) [3]; *potatoes* [3-5]
- Black nightshade in *strawberries* (off-label) [3]
- Cleavers in *asparagus* (off-label), *strawberries* (off-label) [3]; *potatoes* [3-5]
- Desiccation in *blackberries* (off-label), *loganberries* (off-label), *protected blackberries* (off-label), *protected loganberries* (off-label), *protected raspberries* (off-label), *protected rubus hybrids* (off-label), *raspberries* (off-label), *rubus hybrids* (off-label) [3]; *narcissi* (off-label), *potatoes* [6]
- Fat hen in *asparagus* (off-label) [3]; *potatoes* [3-5]
- Field bindweed in *asparagus* (off-label) [3]
- Field pansy in *asparagus* (off-label) [3]
- Groundsel in *asparagus* (off-label) [3]
- Hairy bittercress in *strawberries* (off-label) [3]
- Haulm destruction in *seed potatoes*, *ware potatoes* [2]
- Ivy-leaved speedwell in *potatoes* [3-5]
- Knotgrass in *asparagus* (off-label), *strawberries* (off-label) [3]; *potatoes* [3-5]
- Polygonums in *blackcurrants* (off-label), *blueberries* (off-label), *cranberries* (off-label), *gooseberries* (off-label), *redcurrants* (off-label), *ribes species* (off-label) [3]
- Redshank in *asparagus* (off-label), *strawberries* (off-label) [3]; *potatoes* [3-5]
- Shepherd's purse in *asparagus* (off-label) [3]
- Small nettle in *asparagus* (off-label), *strawberries* (off-label) [3]
- Sowthistle in *asparagus* (off-label) [3]
- Speedwells in *asparagus* (off-label), *strawberries* (off-label) [3]
- Thistles in *blackcurrants* (off-label), *blueberries* (off-label), *cranberries* (off-label), *gooseberries* (off-label), *redcurrants* (off-label), *ribes species* (off-label) [3]
- Volunteer oilseed rape in *all edible crops (outdoor)* (before planting), *all non-edible crops (outdoor)* (before planting), *potatoes* [3-5]; *asparagus* (off-label) [3]
- Willowherb in *strawberries* (off-label) [3]

Extension of Authorisation for Minor Use (EAMUs)

- *asparagus* 20190634 [3]
- *bilberries* 20190633 [3]
- *blackberries* 20190622 [3], 20190633 [3]

SEE SECTION 3 FOR PRODUCTS ALSO REGISTERED

- ***blackcurrants*** *20190627* [3], *20190633* [3]
- ***blueberries*** *20190627* [3], *20190633* [3]
- ***cranberries*** *20190627* [3], *20190633* [3]
- ***forestry plantations*** *20190630* [3]
- ***gooseberries*** *20190627* [3], *20190633* [3]
- ***loganberries*** *20190622* [3], *20190633* [3]
- ***narcissi*** *20190896* [6]
- ***ornamental plant production*** *20190630* [3], *20190633* [3]
- ***protected bilberries*** *20190633* [3]
- ***protected blackberries*** *20190622* [3], *20190633* [3]
- ***protected blackcurrants*** *20190633* [3]
- ***protected blueberry*** *20190633* [3]
- ***protected cranberries*** *20190633* [3]
- ***protected forest*** *20190630* [3]
- ***protected gooseberries*** *20190633* [3]
- ***protected loganberries*** *20190622* [3], *20190633* [3]
- ***protected ornamentals*** *20190630* [3], *20190633* [3]
- ***protected raspberries*** *20190622* [3], *20190633* [3]
- ***protected redcurrants*** *20190633* [3]
- ***protected rubus hybrids*** *20190622* [3], *20190633* [3]
- ***protected strawberries*** *20190633* [3]
- ***raspberries*** *20190622* [3], *20190633* [3]
- ***redcurrants*** *20190627* [3], *20190633* [3]
- ***ribes species*** *20190627* [3]
- ***rubus hybrids*** *20190622* [3], *20190633* [3]
- ***rye*** *20190752* [1]
- ***strawberries*** *20190623* [3], *20190633* [3]
- ***wine grapes*** *20190624* [3]

Approval information
- Carfentrazone-ethyl included in Annex I under EC Regulation 1107/2009
- Accepted by BBPA for use on malting barley and hops

Efficacy guidance
- Best weed control results achieved from good spray cover applied to small actively growing weeds
- Carfentrazone-ethyl acts by contact only; see label for optimum timing on specified weeds. Weeds emerging after application will not be controlled
- For weed control in cereals use as two spray programme with one application in autumn and one in spring
- Efficacy of haulm destruction will be reduced where flailed haulm covers the stems at application
- For potato crops with very dense vigorous haulm or where regrowth occurs following a single application a second application may be necessary to achieve satisfactory desiccation. A minimum interval between applications of 7 d should be observed to achieve optimum performance

Restrictions
- Maximum number of treatments 2 per crop for cereals (1 in Autumn and 1 in Spring); 1 per crop for weed control in potatoes; 1 per yr for pre-planting treatments
- Maximum total dose for potato haulm destruction equivalent to 1.6 full dose treatments
- Do not treat cereal crops under stress from drought, waterlogging, cold, pests, diseases, nutrient or lime deficiency or any factors reducing plant growth
- Do not treat cereals undersown with clover or other legumes
- Allow at least 2-3 wk between application and lifting potatoes to allow skins to set if potatoes are to be stored
- Follow label instructions for sprayer cleaning
- Do not apply through knapsack sprayers
- Contact processor before using a split dose on potatoes for processing
- Do not apply within 10 days of an application of iodosulfuron-methyl-sodium + mesosulfuron-methyl formulations, e.g. MAPP 12478

FOR FULL CONDITIONS OF USE ALWAYS READ THE PRODUCT LABEL

Crop-specific information

- Latest use: 1 mth before planting edible and non-edible crops; before 3rd node detectable (GS 33) on cereals
- HI 7 d for potatoes
- For weed control in cereals treat from 2 leaf stage
- When used as a treatment prior to planting a subsequent crop apply before weeds exceed maximum sizes indicated in the label
- If treated potato tubers are to be stored then allow at least 2-3 wk between the final application and lifting to allow skins to set

Following crops guidance

- No restrictions apply on the planting of succeeding crops 1 mth after application to potatoes for haulm destruction or as a pre-planting treatment, or 3 mth after application to cereals for weed control
- In the event of failure of a treated cereal crop, all cereals, ryegrass, maize, oilseed rape, peas, sunflowers, *Phacelia*, vetches, carrots or onions may be planted within 1 mth of treatment

Environmental safety

- Dangerous for the environment
- Very toxic to aquatic organisms
- Some non-target crops are sensitive. Avoid drift onto broad-leaved plants outside the treated area, or onto ponds waterways or ditches

Hazard classification and safety precautions

Hazard Irritant, Dangerous for the environment, Very toxic to aquatic organisms
Transport code 9
Packaging group III
UN Number 3077 [1]; 3082 [2-6]
Risk phrases H317 [2-6]
Operator protection A, H; U05a, U14 [1-6]; U08, U13 [1]; U20a [2-6]
Environmental protection E15a [1, 3-5]; E15b [2, 6]; E34 [1]; E38, H410 [1-6]
Storage and disposal D01, D02, D09a, D12a [1-6]; D10b [1]; D10c [2-6]
Medical advice M05a [1]

70 carfentrazone-ethyl + mecoprop-P

A foliar applied herbicide for cereals
HRAC mode of action code: E + O

See also mecoprop-P

Products

1 Jewel	ICL (Everris) Ltd	1.5:60% w/w	WG	16212
2 Pan Glory	Pan Amenity	1.5:60% w/w	WG	14487
3 Platform S	FMC Agro	1.5:60% w/w	WG	18701

Uses

- Annual and perennial weeds in *amenity grassland, managed amenity turf* [1, 2]
- Charlock in *spring barley, spring oats, spring wheat, winter barley, winter oats, winter wheat* [3]
- Chickweed in *spring barley, spring oats, spring wheat, winter barley, winter oats, winter wheat* [3]
- Cleavers in *spring barley, spring oats, spring wheat, winter barley, winter oats, winter wheat* [3]
- Field speedwell in *spring barley, spring oats, spring wheat, winter barley, winter oats, winter wheat* [3]
- Ivy-leaved speedwell in *spring barley, spring oats, spring wheat, winter barley, winter oats, winter wheat* [3]
- Moss in *amenity grassland, managed amenity turf* [1, 2]
- Red dead-nettle in *spring barley, spring oats, spring wheat, winter barley, winter oats, winter wheat* [3]

SEE SECTION 3 FOR PRODUCTS ALSO REGISTERED

Approval information
- Carfentrazone-ethyl and mecoprop-P included in Annex I under EC Regulation 1107/2009
- Accepted by BBPA for use on malting barley

Efficacy guidance
- Best results obtained when weeds have germinated and growing vigorously in warm moist conditions
- Treatment of large weeds and poor spray coverage may result in reduced weed control

Restrictions
- Maximum number of treatments 2 per crop. The total amount of mecoprop-P applied in a single yr must not exceed the maximum total dose approved for any single product for the crop per situation
- Do not treat crops suffering from stress from any cause
- Do not treat crops undersown or to be undersown
- Do not apply between 1st October and 1st March

Crop-specific information
- Latest use: before 3rd node detectable (GS 33)
- Can be used on all varieties of wheat and barley in autumn or spring from the beginning of tillering

Following crops guidance
- In the event of crop failure, any cereal, maize, oilseed rape, peas, vetches or sunflowers may be sown 1 mth after a spring treatment. Any crop may be planted 3 mth after treatment

Environmental safety
- Dangerous for the environment
- Very toxic to aquatic organisms
- Keep livestock out of treated areas for at least two weeks following treatment and until poisonous weeds, such as ragwort, have died down and become unpalatable
- Some pesticides pose a greater threat of contamination of water than others and mecoprop-P is one of these pesticides. Take special care when applying mecoprop-P near water and do not apply if heavy rain is forecast

Hazard classification and safety precautions
Hazard Harmful, Dangerous for the environment, Harmful if swallowed [1-3]; Very toxic to aquatic organisms [3]
Transport code 9
Packaging group III
UN Number 3077
Risk phrases H317, H318
Operator protection A, C, H [1-3]; M [3]; U05a, U11, U13, U14, U20b [1-3]; U08 [3]; U09a [1, 2]
Environmental protection E07a, E34, E38 [1-3]; E15a, H411 [3]; E15b, H410 [1, 2]
Storage and disposal D01, D02, D09a, D10b [1-3]; D12a [3]
Medical advice M05a

71   cerevisane

A yeast derived from micro-organisms for downy mildew control in some crops

Products
Romeo	Fargro	94.1% w/w	PO	19170

Uses
- Downy mildew in *protected aubergines, protected courgettes, protected cucumbers, protected gherkins, protected lettuce, protected melons, protected pumpkins, protected strawberries, protected summer squash, protected tomatoes, protected watermelon, protected winter squash*
- Late blight in *protected tomatoes*

FOR FULL CONDITIONS OF USE ALWAYS READ THE PRODUCT LABEL

- Powdery mildew in ***protected aubergines, protected courgettes, protected cucumbers, protected gherkins, protected lettuce, protected melons, protected pumpkins, protected strawberries, protected summer squash, protected watermelon, protected winter squash***

Approval information
- Cerevisane included in Annex I under EC Regulation 1107/2009

Restrictions
- Treatment must only be made under 'permanent protection' situations which provide full enclosure (including continuous top and side barriers down to below ground level) and which are present and maintained over a number of years.
- Reasonable precautions must be taken to prevent access of birds, wild mammals and honey bees to treated crops.
- To minimise airborne environmental exposure, vents, doors and other openings must be closed during and after application until the applied product has fully settled.

Hazard classification and safety precautions
    UN Number N/C
    Operator protection A, D, H; U19a
    Environmental protection E15b, E34
    Storage and disposal D01, D05, D09a, D10b, D12b

72   chlorantraniliprole

An ingested and contact insecticide for insect pest control in apples and pears and, in Ireland only, for colorado beetle control in potatoes.
IRAC mode of action code: 28

Products

1	Clayton Courage	Clayton	200 g/l	SC	18215
2	Coragen	FMC Agro	200 g/l	SC	18804
3	Pan Genie	Pan Agriculture	200 g/l	SC	18491

Uses
- Carnation tortrix moth in ***bilberries*** *(off-label),* ***blackcurrants*** *(off-label),* ***blueberries*** *(off-label),* ***cranberries*** *(off-label),* ***gooseberries*** *(off-label),* ***redcurrants*** *(off-label)* [2]
- Carrot fly in ***carrots*** *(off-label),* ***celeriac*** *(off-label),* ***horseradish*** *(off-label),* ***parsley root*** *(off-label),* ***parsnips*** *(off-label),* ***protected swedes*** *(off-label),* ***protected turnips*** *(off-label),* ***red beet*** *(off-label),* ***swedes*** *(off-label),* ***turnips*** *(off-label)* [2]
- Caterpillars in ***baby leaf crops*** *(off-label),* ***celery (outdoor)*** *(off-label),* ***herbs (see appendix 6)*** *(off-label),* ***hops*** *(off-label),* ***horseradish*** *(off-label),* ***lettuce*** *(off-label),* ***protected swedes*** *(off-label),* ***protected turnips*** *(off-label),* ***red beet*** *(off-label),* ***swedes*** *(off-label),* ***turnips*** *(off-label)* [2]
- Codling moth in ***apples, pears*** [1-3]
- Currant pug moth in ***hops*** *(off-label)* [2]
- Diamond-back moth in ***baby leaf crops*** *(off-label),* ***celery (outdoor)*** *(off-label),* ***herbs (see appendix 6)*** *(off-label),* ***lettuce*** *(off-label)* [2]
- European corn borer in ***hops*** *(off-label)* [2]
- Hylobius abietes in ***ornamental plant production*** *(off-label)* [2]
- Insect pests in ***ornamental plant production*** *(off-label)* [2]
- Light brown apple moth in ***bilberries*** *(off-label),* ***blackcurrants*** *(off-label),* ***blueberries*** *(off-label),* ***cranberries*** *(off-label),* ***gooseberries*** *(off-label),* ***redcurrants*** *(off-label)* [2]
- Rosy rustic moth in ***hops*** *(off-label)* [2]
- Silver Y moth in ***baby leaf crops*** *(off-label),* ***celery (outdoor)*** *(off-label),* ***herbs (see appendix 6)*** *(off-label),* ***hops*** *(off-label),* ***horseradish*** *(off-label),* ***lettuce*** *(off-label),* ***protected swedes*** *(off-label),* ***protected turnips*** *(off-label),* ***red beet*** *(off-label),* ***swedes*** *(off-label),* ***turnips*** *(off-label)* [2]
- South American Tomato Moth in ***protected tomatoes*** *(off-label)* [2]

Extension of Authorisation for Minor Use (EAMUs)
- ***baby leaf crops*** *20191343* [2]
- ***bilberries*** *20190809 expires 26 Jan 2020* [2]

SEE SECTION 3 FOR PRODUCTS ALSO REGISTERED

- **blackcurrants** 20190809 expires 26 Jan 2020 [2]
- **blueberries** 20190809 expires 26 Jan 2020 [2]
- **carrots** 20190830 expires 26 Jan 2020 [2]
- **celeriac** 20190830 expires 26 Jan 2020 [2]
- **celery (outdoor)** 20191341 [2]
- **cranberries** 20190809 expires 26 Jan 2020 [2]
- **gooseberries** 20190809 expires 26 Jan 2020 [2]
- **herbs (see appendix 6)** 20191343 [2]
- **hops** 20191344 expires 31 Dec 2020 [2]
- **horseradish** 20190827 expires 26 Jan 2020 [2]
- **lettuce** 20191343 [2]
- **ornamental plant production** 20190828 expires 26 Jan 2020 [2]
- **parsley root** 20190830 expires 26 Jan 2020 [2]
- **parsnips** 20190830 expires 26 Jan 2020 [2]
- **protected swedes** 20190827 expires 26 Jan 2020 [2]
- **protected tomatoes** 20190829 expires 26 Jan 2020 [2]
- **protected turnips** 20190827 expires 26 Jan 2020 [2]
- **red beet** 20190827 expires 26 Jan 2020 [2]
- **redcurrants** 20190809 expires 26 Jan 2020 [2]
- **swedes** 20190827 expires 26 Jan 2020 [2]
- **turnips** 20190827 expires 26 Jan 2020 [2]

Approval information
- Chlorantraniliprole is included in Annex 1 under EC Regulation 1107/2009

Efficacy guidance
- Can be used as part of an Integrated Pest Management programme.
- For best fruit protection in apples and pears, apply before egg hatch.
- Best applied early morning or late evening to avoid applications when bees may be present.

Restrictions
- Maximum number of applications is two per year

Crop-specific information
- Latest time of application is 14 days before harvest

Environmental safety
- To protect bees and pollinating insects, do not apply to crops when in flower. Do not apply when bees are actively foraging or when flowering plants are present.
- Broadcast air-assisted LERAP (10 m); LERAP Category B

Hazard classification and safety precautions
Hazard Dangerous for the environment [1-3]; Very toxic to aquatic organisms [2]
Transport code 9
Packaging group III
UN Number 3082
Risk phrases R50, R53a [1, 3]
Environmental protection E12e, E15b, E16a, E22b, E38 [1-3]; E17b [1-3] (10 m); H410 [2]
Storage and disposal D01, D02, D05, D09a, D12a

73   chlormequat

A plant-growth regulator for reducing stem growth and lodging

Products

1	3C Chlormequat 750	BASF	750 g/l	SL	16690
2	Agrovista 3 See 750	Agrovista	750 g/l	SL	15975
3	BEC CCC 750	Becesane	750 g/l	SL	17930
4	CCC 720	Nurture	720 g/l	SL	19008
5	CCC 750	Harvest	750 g/l	SL	17753
6	CleanCrop Transist	Agrii	720 g/l	SL	16684

FOR FULL CONDITIONS OF USE ALWAYS READ THE PRODUCT LABEL

SECTION 2

Products – continued

7	Jadex Plus	Clayton	720 g/l	SL	17524
8	Jadex-o-720	SFP Europe	720 g/l	SL	16284
9	Palermo	SFP Europe	720 g/l	SL	17900
10	Stabilan 750	Nufarm UK	750 g/l	SL	09303
11	Stefes CCC 720	Stefes	720 g/l	SL	17731

Uses

- Growth regulation in *canary seed* (off-label), *ornamental specimens* (off-label) [10]; *rye*, *spring barley* [11]; *spring oats*, *spring wheat*, *winter barley*, *winter oats* [4, 6, 7, 9, 11]; *triticale* [3, 5, 10, 11]; *winter wheat* [4, 6-9, 11]
- Lodging control in *oats* [1]; *rye* [1, 10, 11]; *spring barley* [1, 11]; *spring oats*, *winter oats* [2-7, 9-11]; *spring rye* [2]; *spring wheat*, *winter barley* [1-7, 9-11]; *triticale* [1-3, 5, 10, 11]; *winter rye* [2, 3, 5]; *winter wheat* [1-11]

Extension of Authorisation for Minor Use (EAMUs)

- *canary seed* 20180371 expires 31 May 2020 [10]
- *ornamental specimens* 20171416 expires 31 May 2020 [10]

Approval information

- Chlormequat included in Annex 1 under EC Regulation 1107/2009
- Accepted by BBPA for use on malting barley

Efficacy guidance

- Most effective results on cereals normally achieved from application from Apr onwards, on wheat and rye from leaf sheath erect to first node detectable (GS 30-31), on oats at second node detectable (GS 32), on winter barley from mid-tillering to leaf sheath erect (GS 25-30). However, recommendations vary with product. See label for details
- Influence on growth varies with crop and growth stage. Risk of lodging reduced by application at early stem extension. Root development and yield can be improved by earlier treatment
- Results on barley can be variable
- In tank mixes with other pesticides on cereals optimum timing for herbicide action may differ from that for growth reduction. See label for details of tank mix recommendations
- Most products recommended for use on oats require addition of approved non-ionic wetter. Check label
- Some products are formulated with trace elements to help compensate for increased demand during rapid growth

Restrictions

- Maximum number of treatments or maximum total dose varies with crop and product and whether split dose treatments are recommended. Check labels
- Do not use on very late sown spring wheat or oats or on crops under stress
- Mixtures with liquid nitrogen fertilizers may cause scorch and are specifically excluded on some labels
- Do not use on soils of low fertility unless such crops regularly receive adequate dressings of nitrogen
- At least 6 h, preferably 24 h, required before rain for maximum effectiveness. Do not apply to wet crops
- Check labels for tank mixtures known to be incompatible
- Must only be applied between growth stages BBCH 31 to BBCH 39
- A maximum dose 1.33 L per hectare (750 g/l formulations) or 1.38 l/ha (720 g/l formulations) must not be exceeded when applied to winter wheat and spring barley before stem elongation (GS 30)
- No applications to be made before 1st February in the year of harvest [2, 10]

Crop-specific information

- Latest use varies with crop and product. See label for details
- May be used on cereals undersown with grass or clovers
- Ornamentals to be treated must be well established and growing vigorously. Do not treat in strong sunlight or when temperatures are likely to fall below 10°C

SEE SECTION 3 FOR PRODUCTS ALSO REGISTERED

- Temporary yellow spotting may occur on poinsettias. It can be minimised by use of a non-ionic wetting agent - see label

Environmental safety
- Wash equipment thoroughly with water and wetting agent immediately after use and spray out. Spray out again before storing or using for another product. Traces can cause harm to susceptible crops sprayed later
- Do not use straw from treated cereals as horticultural growth medium or mulch

Hazard classification and safety precautions
Hazard Harmful [1-3, 5, 6, 10, 11]; Harmful if swallowed [1-11]
Transport code 8
Packaging group III
UN Number 1760
Risk phrases H290 [1, 2, 4, 5, 7, 9-11]; H312 [1-11]
Operator protection A [1-11]; C [4, 7-9]; H [4, 6-9, 11]; U05a, U20b [1-11]; U08 [1, 2, 6, 11]; U09a [3, 5, 10]; U13 [4, 7-9]; U19a [1-3, 5, 6, 10, 11]
Environmental protection E15a [1-3, 5, 6, 10, 11]; E15b, E40e [4, 7-9]; E34 [1-11]; H411 [4, 7, 8, 11]; H412 [2, 3, 5, 10]
Storage and disposal D01, D02 [1-11]; D05 [1-3, 5, 10]; D09a [3-11]; D10a [3, 5, 10]; D10b [1, 2, 4, 6-9, 11]
Medical advice M03 [1-11]; M05a [3, 5, 6, 10, 11]

74 chlormequat + 2-chloroethylphosphonic acid

A plant growth regulator for use in cereals

See also ethephon

Products
1 BOGOTA UK	SFP Europe	305:155 g/l	SL	16318
2 Chlormephon UK	Clayton	305:155 g/l	SL	16215
3 Ormet Plus	SFP Europe	300:150 g/l	SL	17816
4 Socom	Clayton	305:155 g/l	SL	17155
5 Spatial Plus	Sumitomo	300:150 g/l	SL	16665
6 Vivax	Sumitomo	300:150 g/l	SL	16682

Uses
- Lodging control in **spring barley**, **winter barley**, **winter wheat**

Approval information
- Chlormequat and 2-chloroethylphosphonic acid (ethephon) included in Annex I under EC Regulation 1107/2009
- Accepted by BBPA for use on malting barley
- All products containing 2-chloroethylphosphonic acid carry the warning: '2-chloroethylphosphonic acid is an anticholinesterase organophosphate. Handle with care'.

Efficacy guidance
- Best results obtained when crops growing vigorously
- Recommended dose varies with growth stage. See labels for details and recommendations for use of sequential treatments

Restrictions
- 2-chloroethylphosphonic acid is an anticholinesterase organophosphorus compound. Do not use if under medical advice not to work with such compounds
- Maximum number of treatments 1 per crop; maximum total dose equivalent to one full dose treatment
- Product must always be used with specified non-ionic wetter - see labels
- Do not use on any crop in sequence with any other product containing 2-chloroethylphosphonic acid
- Do not spray when crop wet or rain imminent

- Do not spray during cold weather or periods of night frost, when soil is very dry, when crop diseased or suffering pest damage, nutrient deficiency or herbicide stress
- If used on seed crops grown for certification inform seed merchant beforehand
- Do not use on wheat variety Moulin or on any winter varieties sown in spring
- Do not use on spring barley variety Triumph
- Do not treat barley on soils with more than 10% organic matter
- Only crops growing under conditions of high fertility should be treated

Crop-specific information
- Latest use: before flag leaf ligule/collar just visible (GS 39) or 1st spikelet visible (GS 51) for wheat or barley at top dose; or before flag leaf sheath opening (GS 47) for winter wheat at reduced dose
- Apply before lodging has started

Environmental safety
- Harmful to fish or other aquatic life. Do not contaminate surface waters or ditches with chemical or used container
- Do not use straw from treated cereals as a horticultural growth medium or as a mulch

Hazard classification and safety precautions
    Hazard Harmful, Harmful if swallowed
    Transport code 8
    Packaging group III
    UN Number 3265
    Risk phrases H335
    Operator protection A; U05a, U08, U19a, U20b [1-6]; U14 [1-4]
    Environmental protection E15a, E34, E38 [1-6]; H411 [3]; H412 [5, 6]
    Storage and disposal D01, D02, D09a, D10b [1-6]; D12a [1-4]
    Medical advice M01, M03, M05a

75   chlorotoluron

A contact and residual urea herbicide for cereals only available in mixtures
HRAC mode of action code: C2

76   chlorotoluron + diflufenican + pendimethalin

A herbicide mixture for weed control in cereals
HRAC mode of action code: C2 + F1 + K1

Products

1	Tower	Adama	250:40:300 g/l	SC	16586
2	Tribal	Adama	250:40:300 g/l	SC	17075

Uses
- Annual meadow grass in *spring barley, spring wheat, triticale, winter barley, winter rye, winter wheat*
- Charlock in *spring barley, spring wheat, triticale, winter barley, winter rye, winter wheat*
- Chickweed in *spring barley, spring wheat, triticale, winter barley, winter rye, winter wheat*
- Cleavers in *spring barley, spring wheat, triticale, winter barley, winter rye, winter wheat*
- Crane's-bill in *spring barley, spring wheat, triticale, winter barley, winter rye, winter wheat*
- Dead nettle in *spring barley, spring wheat, triticale, winter barley, winter rye, winter wheat*
- Field pansy in *spring barley, spring wheat, triticale, winter barley, winter rye, winter wheat*
- Forget-me-not in *spring barley, spring wheat, triticale, winter barley, winter rye, winter wheat*
- Fumitory in *spring barley, spring wheat, triticale, winter barley, winter rye, winter wheat*
- Loose silky bent in *spring barley, spring wheat, triticale, winter barley, winter rye, winter wheat*
- Mayweeds in *spring barley, spring wheat, triticale, winter barley, winter rye, winter wheat*

SEE SECTION 3 FOR PRODUCTS ALSO REGISTERED

- Penny cress in *spring barley, spring wheat, triticale, winter barley, winter rye, winter wheat*
- Poppies in *spring barley, spring wheat, triticale, winter barley, winter rye, winter wheat*
- Runch in *spring barley, spring wheat, triticale, winter barley, winter rye, winter wheat*
- Shepherd's purse in *spring barley, spring wheat, triticale, winter barley, winter rye, winter wheat*
- Speedwells in *spring barley, spring wheat, triticale, winter barley, winter rye, winter wheat*
- Volunteer oilseed rape in *spring barley, spring wheat, triticale, winter barley, winter rye, winter wheat*

Approval information
- Chlorotoluron, diflufenican and pendimethalin included in Annex I under EC Regulation 1107/2009

Efficacy guidance
- Minimum water volume for full dose (2.0 l/ha) is 200 l/ha while reducing the dose to 1.0 l/ha allows a lowest water volume of 100 l/ha.

Restrictions
- Maximum number of treatments 1 per crop
- Use only on listed crop varieties when applying post-emergence to winter wheat. Do not apply to undersown crops
- Substances containing the chlorotoluron active agent may not be applied more than once per year on the same area
- Application must not be made after 31 October in the year of drilling
- Do not use if it is frosty
- DO NOT apply to soils with greater than 10% organic matter

Crop-specific information
- Following normal harvest, there are no restrictions. In the event of crop failure, plough before drilling the following crop in the spring. Only winter wheat can be re-drilled in the same autumn following crop failure

Environmental safety
- Buffer zone requirement 6 m [1, 2]
- LERAP Category B

Hazard classification and safety precautions
Transport code 9
Packaging group III
UN Number 3082
Risk phrases H351, H361
Operator protection A; U02a, U04a, U20b
Environmental protection E15b, E16a, E34, E38, H410
Storage and disposal D01, D09a, D10b, D12a
Medical advice M03

77   chlorpropham

A residual carbamate herbicide and potato sprout suppressant
HRAC mode of action code: K2

Products

1	Aceto Chlorpropham 50M	Aceto	500 g/l	HN	14134
2	Aceto Sprout Nip Pellet	Aceto	100% w/w	HN	15397
3	Aceto Sprout Nip Ultra	Aceto	500 g/l	HN	15456
4	Aliacine 400 EC	Arysta	400 g/l	EC	16283
5	CIPC Fog 300	AgChem Access	300 g/l	HN	15904
6	Cleancrop Amigo 2	UPL Europe	400 g/l	EC	15419
7	Gro-Stop 100	Certis	300 g/l	HN	14182

FOR FULL CONDITIONS OF USE ALWAYS READ THE PRODUCT LABEL

Products – continued

8	Gro-Stop Fog	Certis	300 g/l	HN	14183
9	Gro-Stop Ready	Certis	120 g/l	EW	15109
10	Intruder	UPL Europe	400 g/l	EC	15076
11	Pro-Long	UPL Europe	500 g/l	HN	14389

SECTION 2

Uses

- Annual dicotyledons in *baby leaf crops* [10]; *bulb onions, chard* (off-label), *endives, herbs (see appendix 6)* (off-label), *lettuce, ornamental plant production, salad onions, shallots, spinach* (off-label) [6, 10]; *forest nurseries* (off-label), *leaf brassicas* (off-label), *leeks* (off-label) [6]
- Annual grasses in *chard* (off-label), *herbs (see appendix 6)* (off-label), *spinach* (off-label) [10]
- Annual meadow grass in *baby leaf crops* [10]; *bulb onions, endives, lettuce, ornamental plant production, salad onions, shallots* [6, 10]; *chard* (off-label), *forest nurseries* (off-label), *herbs (see appendix 6)* (off-label), *leaf brassicas* (off-label), *spinach* (off-label) [6]
- Chickweed in *baby leaf crops* (off-label), *bulb onions, endives, lettuce, ornamental plant production, salad onions, shallots, spinach* (off-label) [4]; *chard* (off-label), *forest nurseries* (off-label), *herbs (see appendix 6)* (off-label), *leeks* (off-label) [4, 6]; *leaf brassicas* (off-label) [6]
- Polygonums in *baby leaf crops* (off-label), *bulb onions, endives, lettuce, ornamental plant production, salad onions, shallots, spinach* (off-label) [4]; *chard* (off-label), *forest nurseries* (off-label), *herbs (see appendix 6)* (off-label), *leeks* (off-label) [4, 6]; *leaf brassicas* (off-label) [6]
- Small nettle in *baby leaf crops* (off-label), *bulb onions, endives, lettuce, ornamental plant production, salad onions, shallots, spinach* (off-label) [4]; *chard* (off-label), *forest nurseries* (off-label), *herbs (see appendix 6)* (off-label), *leeks* (off-label) [4, 6]; *leaf brassicas* (off-label) [6]
- Sprout suppression in *potatoes* [1-3, 9]; *ware potatoes* [7]; *ware potatoes* (thermal fog) [5, 8, 11]
- Wild oats in *baby leaf crops* [10]; *bulb onions, endives, lettuce, ornamental plant production, salad onions, shallots* [6, 10]; *chard* (off-label), *forest nurseries* (off-label), *herbs (see appendix 6)* (off-label), *leaf brassicas* (off-label), *spinach* (off-label) [6]

Extension of Authorisation for Minor Use (EAMUs)

- *baby leaf crops 20160788* [4]
- *chard 20160788* [4], *20120728* [6], *20110596* [10]
- *forest nurseries 20160787* [4], *20120727* [6]
- *herbs (see appendix 6) 20160788* [4], *20120728* [6], *20110596* [10]
- *leaf brassicas 20120728* [6]
- *leeks 20160789* [4], *20121841* [6]
- *spinach 20160788* [4], *20120728* [6], *20110596* [10]

Approval information

- Chlorpropham included in Annex I under EC Regulation 1107/2009
- Some products are formulated for application by thermal fogging. See labels for details

Efficacy guidance

- For sprout suppression apply with suitable fogging or rotary atomiser equipment over dry tubers before sprouting commences. Repeat applications may be needed. See labels for details
- Best results on potatoes obtained in purpose-built box stores with suitable forced draft ventilation. Potatoes in bulk stores should not be stacked more than 3 m high. Positive ventilation systems are a requirement for use since 2017
- Blockage of air spaces between tubers prevents circulation of vapour and consequent loss of efficacy
- It is important to treat potatoes before the eyes open to obtain best results
- Effectiveness of fogging reduced in non-dedicated stores without proper insulation and temperature controls. Best results obtained at 5-10°C and 75-80% humidity
- When treating potatoes in store with CIPC best practice guidelines are available on-line at www. BeCIPCcompliant.co.uk - some potatoes have recently been found to exceed the MRL and if more are found then the registration of CIPC is in jeopardy.

SEE SECTION 3 FOR PRODUCTS ALSO REGISTERED

Restrictions

- Only clean, mature, disease-free potatoes should be treated for sprout suppression. Use of chlorpropham can inhibit tuber wound healing and the severity of skin spot infection in store may be increased if damaged tubers are treated
- Do not fog potatoes with a high level of skin spot
- Do not use on potatoes for seed. Do not handle, dry or store seed potatoes or any other seed or bulbs in boxes or buildings in which potatoes are being or have been treated
- Do not remove treated potatoes from store for sale or processing for at least 21 d after application

Crop-specific information

- Cure potatoes according to label instructions before treatment and allow 2 days - 4 wks between completion of loading into store and first treatment - check labels.

Following crops guidance

- There is a risk of damage to seed potatoes which are handled or stored in boxes or buildings previously treated with chlorpropham

Environmental safety

- Dangerous for the environment
- Toxic to aquatic organisms
- Keep unprotected persons out of treated stores for at least 24 h after application
- Keep in original container, tightly closed, in a safe place, under lock and key

Hazard classification and safety precautions

Hazard Harmful [4-6, 9, 10]; Dangerous for the environment [1-11]; Highly flammable liquid and vapour [1, 11]; Toxic if swallowed, Toxic in contact with skin, Toxic if inhaled [1]; Harmful if swallowed [4, 6, 9, 10]; Very toxic to aquatic organisms [9]

Transport code 6.1 [1, 5, 7-9, 11]; 9 [3, 4, 6, 10]

Packaging group III [1, 3-11]

UN Number 2902 [1, 5, 7-9, 11]; 3077 [3]; 3082 [4, 6, 10]

Risk phrases H314, H318 [9]; H315 [4-10]; H317 [4, 6, 9, 10]; H319 [4-8]; H320, H335 [10]; H336 [4, 6, 10]; H351, H373 [1-11]; H370 [11]

Operator protection A, H [1-11]; C [1, 3, 4, 6, 10, 11]; D, M [1, 3, 5, 7, 8, 11]; E [3, 5, 7, 8, 11]; J [3, 11]; U02a [4-8, 10]; U04a, U15, U20a [5, 7, 8]; U05a [1, 3-8, 10, 11]; U08 [1, 3, 5, 7, 8, 11]; U09a, U11, U16b [4, 6, 10]; U14 [5, 7-9]; U19a [1-8, 10, 11]; U20b [1, 3, 4, 6, 10, 11]

Environmental protection E02a [1, 3, 11] (24 h); E15a [1, 3, 5, 7, 8, 11]; E15b [4, 6, 9, 10]; E22c [4, 6, 10]; E34 [1, 3-11]; E38 [1-4, 6, 9, 10]; H410 [9, 11]; H411 [1-4, 6, 10]; H412 [5, 7, 8]

Consumer protection C02b, C12 [5, 7, 8]; C02c [1, 3, 11]

Storage and disposal D01, D02 [1, 3-11]; D05 [1, 3, 5, 7-9, 11]; D07 [4-8, 10]; D09a, D10c [4, 6, 10]; D09b [1, 3, 5, 7, 8, 11]; D10a, D12b [1, 3, 11]; D11a [5, 7-9]; D12a [2, 4-10]

Medical advice M03, M05b [4, 6, 10]; M04a [1, 3, 5, 7, 8, 11]

78   citronella oil

A natural plant extract herbicide

Products

Barrier H	Barrier	22.9% w/w	OD	17145

Uses

- Ragwort in *amenity grassland, land temporarily removed from production*

Approval information

- Plant oils such as citronella oil are included in Annex 1 under EC Regulation 1107/2009

Efficacy guidance

- Best results obtained from spot treatment of ragwort in the rosette stage, during dry still conditions
- Aerial growth of ragwort is rapidly destroyed. Longer term control depends on overall management strategy
- Check for regrowth after 28 d and re-apply as necessary

FOR FULL CONDITIONS OF USE ALWAYS READ THE PRODUCT LABEL

Crop-specific information
- Contact with grasses will result in transient scorch which is outgrown in good growing conditions

Environmental safety
- Apply away from bees
- Harmful to fish or other aquatic life. Do not contaminate surface waters or ditches with chemical or used container
- Keep livestock out of treated areas for at least 2 wk and until foliage of any poisonous weeds such as ragwort has died and become unpalatable

Hazard classification and safety precautions
Hazard Irritant
UN Number N/C
Risk phrases H317
Operator protection A, C; U05a, U20c
Environmental protection E07b (2 wk); E12g, E13c
Storage and disposal D01, D02, D05, D09a, D11a

79   clethodim

A post-emergence grass herbicide for use in listed broad-leaved crops
HRAC mode of action code: A

Products

1	Balistik	Arysta	120 g/l	EC	18129
2	Centurion Max	Arysta	120 g/l	EC	17911
3	Clayton Gatso	Clayton	120 g/l	EC	18611
4	Cleancrop Signifier	Clayton	120 g/l	EC	18998
5	Knight	Harvest	120 g/l	EC	18218
6	Select Prime	Arysta	120 g/l	EC	16304

Uses
- Annual grasses in **balm** (off-label), **broad beans** (off-label), **brussels sprouts** (off-label), **bulb onions** (off-label), **cabbages** (off-label), **carrots** (off-label), **celeriac** (off-label), **celery leaves** (off-label), **chives** (off-label), **dill** (off-label), **dwarf beans** (off-label), **fennel leaves** (off-label), **fodder beet** (off-label), **french beans** (off-label), **garlic** (off-label), **gold-of-pleasure** (off-label), **horseradish** (off-label), **mint** (off-label), **mustard** (off-label), **parsley** (off-label), **parsley root** (off-label), **parsnips** (off-label), **poppies** (off-label), **red beet** (off-label), **runner beans** (off-label), **salsify** (off-label), **shallots** (off-label), **strawberries** (off-label), **sunflowers** (off-label), **swedes** (off-label), **turnips** (off-label), **valerian root** (off-label), **vining peas** (off-label) [1, 2, 6]; **blackberries, blackcurrants, blueberries, cranberries, gooseberries, raspberries, redcurrants, strawberries** [2]; **blackberries** (off-label), **blackcurrants** (off-label), **blueberries** (off-label), **cranberries** (off-label), **gooseberries** (off-label), **leeks** (off-label), **raspberries** (off-label), **redcurrants** (off-label), **salad onions** (off-label) [1, 6]; **corn gromwell** (off-label), **crambe** (off-label), **edible podded peas** (off-label), **linseed** (off-label), **lucerne** (off-label) [6]; **sugar beet, winter oilseed rape** [1-6]
- Annual meadow grass in **balm** (off-label), **borage** (off-label), **broad beans** (off-label), **brussels sprouts** (off-label), **bulb onions** (off-label), **cabbages** (off-label), **carrots** (off-label), **celeriac** (off-label), **celery leaves** (off-label), **chives** (off-label), **corn gromwell** (off-label), **crambe** (off-label), **dill** (off-label), **dwarf beans** (off-label), **edible podded peas** (off-label), **fennel leaves** (off-label), **fodder beet** (off-label), **french beans** (off-label), **garlic** (off-label), **gold-of-pleasure** (off-label), **horseradish** (off-label), **leeks** (off-label), **linseed** (off-label), **mint** (off-label), **mustard** (off-label), **parsley** (off-label), **parsley root** (off-label), **parsnips** (off-label), **poppies** (off-label), **red beet** (off-label), **runner beans** (off-label), **salad onions** (off-label), **salsify** (off-label), **shallots** (off-label), **strawberries** (off-label), **sunflowers** (off-label), **swedes** (off-label), **turnips** (off-label), **valerian root** (off-label), **vining peas** (off-label) [1, 2, 6]; **blackberries, blackcurrants, blueberries, cranberries, gooseberries, raspberries, redcurrants, strawberries** [2]; **blackberries** (off-label), **blackcurrants** (off-label), **blueberries** (off-label), **cranberries** (off-label), **gooseberries** (off-label), **raspberries** (off-label), **redcurrants** (off-label) [1, 6]; **lucerne** (off-label) [6]; **sugar beet, winter oilseed rape** [1-6]

SEE SECTION 3 FOR PRODUCTS ALSO REGISTERED

- Blackgrass in **balm** (off-label), **borage** (off-label), **broad beans** (off-label), **brussels sprouts** (off-label), **bulb onions** (off-label), **cabbages** (off-label), **carrots** (off-label), **celeriac** (off-label), **celery leaves** (off-label), **chives** (off-label), **corn gromwell** (off-label), **crambe** (off-label), **dill** (off-label), **dwarf beans** (off-label), **edible podded peas** (off-label), **fennel leaves** (off-label), **fodder beet** (off-label), **french beans** (off-label), **garlic** (off-label), **gold-of-pleasure** (off-label), **horseradish** (off-label), **leeks** (off-label), **linseed** (off-label), **mint** (off-label), **mustard** (off-label), **parsley** (off-label), **parsley root** (off-label), **parsnips** (off-label), **poppies** (off-label), **red beet** (off-label), **runner beans** (off-label), **salad onions** (off-label), **salsify** (off-label), **shallots** (off-label), **strawberries** (off-label), **sunflowers** (off-label), **swedes** (off-label), **turnips** (off-label), **valerian root** (off-label), **vining peas** (off-label) [1, 2, 6]; **blackberries, blackcurrants, blueberries, cranberries, gooseberries, raspberries, redcurrants, strawberries** [2]; **blackberries** (off-label), **blackcurrants** (off-label), **blueberries** (off-label), **cranberries** (off-label), **gooseberries** (off-label), **raspberries** (off-label), **redcurrants** (off-label) [1, 6]; **lucerne** (off-label) [6]; **sugar beet, winter oilseed rape** [1-6]
- Couch in **balm** (off-label), **brussels sprouts** (off-label), **bulb onions** (off-label), **cabbages** (off-label), **carrots** (off-label), **celeriac** (off-label), **celery leaves** (off-label), **chives** (off-label), **dill** (off-label), **fennel leaves** (off-label), **fodder beet** (off-label), **garlic** (off-label), **horseradish** (off-label), **mint** (off-label), **parsley** (off-label), **parsley root** (off-label), **parsnips** (off-label), **red beet** (off-label), **salsify** (off-label), **shallots** (off-label), **strawberries** (off-label), **swedes** (off-label), **turnips** (off-label), **valerian root** (off-label) [1, 2, 6]; **blackberries, blackcurrants, blueberries, cranberries, gooseberries, raspberries, redcurrants, strawberries** [2]; **blackberries** (off-label), **blackcurrants** (off-label), **blueberries** (off-label), **broad beans** (off-label), **cranberries** (off-label), **dwarf beans** (off-label), **french beans** (off-label), **gooseberries** (off-label), **raspberries** (off-label), **redcurrants** (off-label), **runner beans** (off-label), **vining peas** (off-label) [1, 6]; **leeks** (off-label), **lucerne** (off-label), **salad onions** (off-label) [6]
- Volunteer barley in **borage** (off-label), **corn gromwell** (off-label), **crambe** (off-label), **linseed** (off-label) [1, 2, 6]; **sugar beet, winter oilseed rape** [1-6]
- Volunteer cereals in **broad beans** (off-label), **brussels sprouts** (off-label), **bulb onions** (off-label), **cabbages** (off-label), **carrots** (off-label), **dwarf beans** (off-label), **french beans** (off-label), **garlic** (off-label), **runner beans** (off-label), **shallots** (off-label), **vining peas** (off-label) [2, 6]; **leeks** (off-label), **salad onions** (off-label) [6]
- Volunteer wheat in **borage** (off-label), **corn gromwell** (off-label), **crambe** (off-label), **linseed** (off-label) [1, 2, 6]; **sugar beet, winter oilseed rape** [1-6]
- Wild oats in **balm** (off-label), **borage** (off-label), **celeriac** (off-label), **celery leaves** (off-label), **chives** (off-label), **corn gromwell** (off-label), **crambe** (off-label), **dill** (off-label), **fennel leaves** (off-label), **fodder beet** (off-label), **gold-of-pleasure** (off-label), **horseradish** (off-label), **linseed** (off-label), **mint** (off-label), **mustard** (off-label), **parsley** (off-label), **parsley root** (off-label), **parsnips** (off-label), **poppies** (off-label), **red beet** (off-label), **salsify** (off-label), **strawberries** (off-label), **sunflowers** (off-label), **swedes** (off-label), **turnips** (off-label), **valerian root** (off-label) [1, 2, 6]; **blackberries, blackcurrants, blueberries, cranberries, gooseberries, raspberries, redcurrants, strawberries** [2]; **blackberries** (off-label), **blackcurrants** (off-label), **blueberries** (off-label), **cranberries** (off-label), **gooseberries** (off-label), **raspberries** (off-label), **redcurrants** (off-label) [1, 6]; **broad beans** (off-label), **brussels sprouts** (off-label), **bulb onions** (off-label), **cabbages** (off-label), **carrots** (off-label), **dwarf beans** (off-label), **french beans** (off-label), **garlic** (off-label), **runner beans** (off-label), **shallots** (off-label), **vining peas** (off-label) [2, 6]; **leeks** (off-label), **lucerne** (off-label), **salad onions** (off-label) [6]

Extension of Authorisation for Minor Use (EAMUs)

- **balm** 20192137 expires 09 Nov 2020 [1], 20193642 expires 09 Nov 2020 [2], 20192127 expires 09 Nov 2020 [6]
- **blackberries** 20192136 expires 09 Nov 2020 [1], 20192128 expires 09 Nov 2020 [6]
- **blackcurrants** 20192136 expires 09 Nov 2020 [1], 20192128 expires 09 Nov 2020 [6]
- **blueberries** 20192136 expires 09 Nov 2020 [1], 20192128 expires 09 Nov 2020 [6]
- **borage** 20193402 expires 09 Nov 2020 [1], 20193183 expires 09 Nov 2020 [2], 20193403 expires 09 Nov 2020 [6]
- **broad beans** 20192141 expires 09 Nov 2020 [1], 20170460 expires 09 Nov 2020 [2], 20192134 expires 09 Nov 2020 [6]
- **brussels sprouts** 20192141 expires 09 Nov 2020 [1], 20170460 expires 09 Nov 2020 [2], 20192134 expires 09 Nov 2020 [6]

- **bulb onions** *20192141 expires 09 Nov 2020* [1], *20170460 expires 09 Nov 2020* [2], *20192134 expires 09 Nov 2020* [6]
- **cabbages** *20192141 expires 09 Nov 2020* [1], *20170460 expires 09 Nov 2020* [2], *20192134 expires 09 Nov 2020* [6]
- **carrots** *20192141 expires 09 Nov 2020* [1], *20170460 expires 09 Nov 2020* [2], *20192134 expires 09 Nov 2020* [6]
- **celeriac** *20192138 expires 09 Nov 2020* [1], *20193641 expires 09 Nov 2020* [2], *20192126 expires 09 Nov 2020* [6]
- **celery leaves** *20192137 expires 09 Nov 2020* [1], *20193642 expires 09 Nov 2020* [2], *20192127 expires 09 Nov 2020* [6]
- **chives** *20192137 expires 09 Nov 2020* [1], *20193642 expires 09 Nov 2020* [2], *20192127 expires 09 Nov 2020* [6]
- **corn gromwell** *20193402 expires 09 Nov 2020* [1], *20193183 expires 09 Nov 2020* [2], *20192132 expires 09 Nov 2020* [6], *20193403 expires 09 Nov 2020* [6]
- **crambe** *20192143 expires 09 Nov 2020* [1], *20170461 expires 09 Nov 2020* [2], *20192130 expires 09 Nov 2020* [6]
- **cranberries** *20192136 expires 09 Nov 2020* [1], *20192128 expires 09 Nov 2020* [6]
- **dill** *20192137 expires 09 Nov 2020* [1], *20193642 expires 09 Nov 2020* [2], *20192127 expires 09 Nov 2020* [6]
- **dwarf beans** *20192141 expires 09 Nov 2020* [1], *20170460 expires 09 Nov 2020* [2], *20192134 expires 09 Nov 2020* [6]
- **edible podded peas** *20192142 expires 09 Nov 2020* [1], *20193633 expires 09 Nov 2020* [2], *20192131 expires 09 Nov 2020* [6]
- **fennel leaves** *20192137 expires 09 Nov 2020* [1], *20193642 expires 09 Nov 2020* [2], *20192127 expires 09 Nov 2020* [6]
- **fodder beet** *20192138 expires 09 Nov 2020* [1], *20193641 expires 09 Nov 2020* [2], *20192126 expires 09 Nov 2020* [6]
- **french beans** *20192141 expires 09 Nov 2020* [1], *20170460 expires 09 Nov 2020* [2], *20192134 expires 09 Nov 2020* [6]
- **garlic** *20192141 expires 09 Nov 2020* [1], *20170460 expires 09 Nov 2020* [2], *20192134 expires 09 Nov 2020* [6]
- **gold-of-pleasure** *20192135 expires 09 Nov 2020* [1], *20193632* [2], *20192125 expires 09 Nov 2020* [6]
- **gooseberries** *20192136 expires 09 Nov 2020* [1], *20192128 expires 09 Nov 2020* [6]
- **horseradish** *20192138 expires 09 Nov 2020* [1], *20193641 expires 09 Nov 2020* [2], *20192126 expires 09 Nov 2020* [6]
- **leeks** *20192140 expires 09 Nov 2020* [1], *20190792 expires 09 Nov 2020* [2], *20192133 expires 09 Nov 2020* [6]
- **linseed** *20192144 expires 09 Nov 2020* [1], *20170458 expires 09 Nov 2020* [2], *20192129 expires 09 Nov 2020* [6]
- **lucerne** *20192126 expires 09 Nov 2020* [6]
- **mint** *20192137 expires 09 Nov 2020* [1], *20193642 expires 09 Nov 2020* [2], *20192127 expires 09 Nov 2020* [6]
- **mustard** *20192135 expires 09 Nov 2020* [1], *20193632* [2], *20192125 expires 09 Nov 2020* [6]
- **parsley** *20192137 expires 09 Nov 2020* [1], *20193642 expires 09 Nov 2020* [2], *20192127 expires 09 Nov 2020* [6]
- **parsley root** *20192138 expires 09 Nov 2020* [1], *20193641 expires 09 Nov 2020* [2], *20192126 expires 09 Nov 2020* [6]
- **parsnips** *20192138 expires 09 Nov 2020* [1], *20193641 expires 09 Nov 2020* [2], *20192126 expires 09 Nov 2020* [6]
- **poppies** *20192135 expires 09 Nov 2020* [1], *20193632* [2], *20192125 expires 09 Nov 2020* [6]
- **raspberries** *20192136 expires 09 Nov 2020* [1], *20192128 expires 09 Nov 2020* [6]
- **red beet** *20192138 expires 09 Nov 2020* [1], *20193641 expires 09 Nov 2020* [2], *20192126 expires 09 Nov 2020* [6]
- **redcurrants** *20192136 expires 09 Nov 2020* [1], *20192128 expires 09 Nov 2020* [6]
- **runner beans** *20192141 expires 09 Nov 2020* [1], *20170460 expires 09 Nov 2020* [2], *20192134 expires 09 Nov 2020* [6]
- **salad onions** *20192140 expires 09 Nov 2020* [1], *20190792 expires 09 Nov 2020* [2], *20192133 expires 09 Nov 2020* [6]

SEE SECTION 3 FOR PRODUCTS ALSO REGISTERED

- **salsify** *20192138 expires 09 Nov 2020* [1], *20193641 expires 09 Nov 2020* [2], *20192126 expires 09 Nov 2020* [6]
- **shallots** *20192141 expires 09 Nov 2020* [1], *20170460 expires 09 Nov 2020* [2], *20192134 expires 09 Nov 2020* [6]
- **strawberries** *20192136 expires 09 Nov 2020* [1], *20192137 expires 09 Nov 2020* [1], *20193642 expires 09 Nov 2020* [2], *20192127 expires 09 Nov 2020* [6], *20192128 expires 09 Nov 2020* [6]
- **sunflowers** *20192135 expires 09 Nov 2020* [1], *20193632* [2], *20192125 expires 09 Nov 2020* [6]
- **swedes** *20192138 expires 09 Nov 2020* [1], *20193641 expires 09 Nov 2020* [2], *20192126 expires 09 Nov 2020* [6]
- **turnips** *20192138 expires 09 Nov 2020* [1], *20193641 expires 09 Nov 2020* [2], *20192126 expires 09 Nov 2020* [6]
- **valerian root** *20192137 expires 09 Nov 2020* [1], *20193642 expires 09 Nov 2020* [2], *20192127 expires 09 Nov 2020* [6]
- **vining peas** *20192141 expires 09 Nov 2020* [1], *20170460 expires 09 Nov 2020* [2], *20192134 expires 09 Nov 2020* [6]

Approval information
- Clethodim included in Annex I under EC Regulation 1107/2009
- Approval expiry 09 Nov 2020 [5]

Efficacy guidance
- Optimum efficacy is achieved from applications made when grasses are actively growing and have 3 lvs up to beginning of tillering
- Avoid late applications to winter oilseed rape in mixture with other products or adjuvants due to risk of damage to terminal shoot

Restrictions
- To avoid the build-up of resistance, do not apply products containing an ACCase inhibitor more than twice to any crop.
- Label requires a 14 day interval before and after application before other products are applied, particularly in oilseed rape.
- Do not apply in tank mix with other products.
- Do not apply to winter oilseed rape after end of October in year of sowing

Crop-specific information
- Consult processors before use on crops grown for processing.
- Do not apply to crops suffering from stress for any reason.

Following crops guidance
- Broad-leaved crops can be sown at any time after application but when planting cereals, maize or grasses, it is advisable to wait at least 4 weeks after application and to cultivate to at least 20 cm deep before sowing.

Hazard classification and safety precautions
Hazard Harmful, Dangerous for the environment
Transport code 9
Packaging group III
UN Number 3082
Risk phrases H304, H336
Operator protection A, H; U02a, U04a, U05a, U08, U19a, U20a
Environmental protection E15b, E38, H411
Storage and disposal D01, D02, D05, D09a, D10b, D12a
Medical advice M05b

80   clodinafop-propargyl

A foliar acting herbicide for annual grass weed control in wheat, triticale and rye
HRAC mode of action code: A

Products
1	Epee	Harvest	240 g/l	EC	17601

Products – continued

2	Kipota	Life Scientific	240 g/l	EC	17993
3	Ravena	FMC Agro	240 g/l	EC	18782
4	Sword	Nufarm UK	240 g/l	EC	18592
5	Topik	Syngenta	240 g/l	EC	15123
6	Topik	Adama	240 g/l	EC	18568
7	Tuli	AgChem Access	240 g/l	EC	16401
8	Viscount	Adama	240 g/l	EC	18567

Uses

- Blackgrass in *durum wheat, spring rye, spring wheat, triticale, winter rye, winter wheat* [1-8]; *grass seed crops* (off-label) [5]
- Rough-stalked meadow grass in *durum wheat, spring rye, spring wheat, triticale, winter rye, winter wheat* [1-8]; *grass seed crops* (off-label) [5]
- Wild oats in *durum wheat, spring rye, spring wheat, triticale, winter rye, winter wheat* [1-8]; *grass seed crops* (off-label) [5]

Extension of Authorisation for Minor Use (EAMUs)

- *grass seed crops* 20132310 expires 31 Oct 2020 [5]

Approval information

- Clodinafop-propargyl included in Annex I under EC Regulation 1107/2009

Efficacy guidance

- Spray when majority of weeds have germinated but before competition reduces yield
- Products contain a herbicide safener (cloquintocet-mexyl) that improves crop tolerance to clodinafop-propargyl
- Optimum control achieved when all grass weeds emerged. Wait for delayed germination on dry or cloddy seedbed
- A mineral oil additive is recommended to give more consistent control of very high blackgrass populations or for late season treatments. See label for details
- Weed control not affected by soil type, organic matter or straw residues
- Control may be reduced if rain falls within 1 h of treatment
- Clodinafop-propargyl is an ACCase inhibitor herbicide. To avoid the build up of resistance do not apply products containing an ACCase inhibitor more than twice to any crop. In addition do not use any product containing clodinafop-propargyl in mixture or sequence with any other product containing the same ingredient
- Use these products as part of a resistance management strategy that includes cultural methods of control and does not use ACCase inhibitors as the sole chemical method of grass weed control
- Applying a second product containing an ACCase inhibitor to a crop will increase the risk of resistance development; only use a second ACCase inhibitor to control different weeds at a different timing
- Always follow WRAG guidelines for preventing and managing herbicide resistant weeds. See Section 5 for more information

Restrictions

- Maximum number of treatments 1 per crop
- Do not use on barley or oats
- Do not treat crops under stress or suffering from waterlogging, pest attack, disease or frost
- Do not treat crops undersown with grass mixtures
- Do not mix with products containing MCPA, mecoprop-P, 2,4-D or 2,4-DB
- MCPA, mecoprop, 2,4-D or 2,4-DB should not be applied within 21 d before, or 7 d after, treatment

Crop-specific information

- Latest use: before second node detectable stage (GS 32) for durum wheat, triticale, rye; before flag leaf sheath extending (GS 41) for wheat
- Spray in autumn, winter or spring from 1 true leaf stage (GS 11) to before second node detectable (GS 32) on durum, rye, triticale; before flag leaf sheath extends (GS 41) on wheat

SEE SECTION 3 FOR PRODUCTS ALSO REGISTERED

Following crops guidance
- Any broad leaved crop or cereal (except oats) may be sown after failure of a treated crop provided that at least 3 wk have elapsed between application and drilling a cereal
- After normal harvest of a treated crop any broad leaved crop or wheat, durum wheat, rye, triticale or barley should be sown. Oats and grass should not be sown until the following spring

Environmental safety
- Dangerous for the environment
- Very toxic to aquatic organisms

Hazard classification and safety precautions
  Hazard Dangerous for the environment, Very toxic to aquatic organisms [1-8]; Harmful if swallowed [1, 4]
  Transport code 9
  Packaging group III
  UN Number 3082
  Risk phrases H304, H373
  Operator protection A, H [1-8]; C, K [1, 2, 4-8]; U02a, U05a, U09a, U20a [1-8]; U14 [1, 3, 4]
  Environmental protection E15a, E38, H410
  Storage and disposal D01, D02, D05, D09a, D10c, D12a

81   clodinafop-propargyl + cloquintocet-mexyl

A grass herbicide for use in winter wheat and durum wheat
HRAC mode of action code: A

Products

Buguis	UPL Europe	100:25 g/l	EC	17151

Uses
- Annual grasses in *durum wheat*, *winter wheat*
- Blackgrass in *durum wheat*, *winter wheat*

Approval information
- Clodinafop-propargyl included in Annex I under EC Regulation 1107/2009 but cloquintocet-mexyl not yet listed

Efficacy guidance
- Reduced doses require addition of an adjuvant - see label

Restrictions
- Do not use on barley or oats.
- Do not spray crops under stress or crops suffering from waterlogging, pest attack, disease or frost
- Do not spray crops undersown with grass mixtures
- Rain within one hour after application may reduce grass weed control
- Avoid the use of hormone-containing herbicides in mixture or sequence

Following crops guidance
- Activity is not affected by soil type, organic matter or straw residues. In the event of crop failure, any broad-leaved crop may be sown or after an interval of 3 weeks any cereal may be sown. Following normal harvest of a treated crop, any broad-leaved or cereal crop maybe sown.

Hazard classification and safety precautions
  Transport code 9
  Packaging group III
  UN Number 3082
  Risk phrases H304, H317, H319, H373
  Operator protection A, H
  Environmental protection H411

FOR FULL CONDITIONS OF USE ALWAYS READ THE PRODUCT LABEL

82   clofentezine

A selective ovicidal tetrazine acaricide for use in top fruit
IRAC mode of action code: 10A

Products

1	Apollo 50 SC	Adama	500 g/l	SC	17187
2	Clayton Sputnik	Clayton	500 g/l	SC	17344

Uses
- Conifer spinning mite in **forest nurseries** *(off-label)*, **ornamental plant production** *(off-label)* [1]
- Red spider mites in **apples**, **pears** [1, 2]
- Rust mite in **forest nurseries** *(off-label)*, **ornamental plant production** *(off-label)* [1]
- Spider mites in **blackberries** *(off-label)*, **forest nurseries** *(off-label)*, **ornamental plant production** *(off-label)*, **protected strawberries** *(off-label)*, **raspberries** *(off-label)*, **strawberries** *(off-label)* [1]

Extension of Authorisation for Minor Use (EAMUs)
- **blackberries** *20180620* [1]
- **forest nurseries** *20162082* [1]
- **ornamental plant production** *20162082* [1]
- **protected strawberries** *20180620* [1]
- **raspberries** *20180620* [1]
- **strawberries** *20180620* [1]

Approval information
- Clofentezine included in Annex I under EC Regulation 1107/2009

Efficacy guidance
- Acts on eggs and early motile stages of mites. For effective control total cover of plants is essential, particular care being needed to cover undersides of leaves

Restrictions
- Maximum number of treatments 1 per yr for apples, pears, cherries, plums

Crop-specific information
- HI apples, pears 28 d; cherries, plums 8 wk
- For red spider mite control spray apples and pears between bud burst and pink bud, plums and cherries between white bud and first flower. Rust mite is also suppressed
- On established infestations apply in conjunction with an adult acaricide

Environmental safety
- Harmful to aquatic organisms
- Product safe on predatory mites, bees and other predatory insects

Hazard classification and safety precautions
  UN Number N/C
  Operator protection U05a, U08, U20b
  Environmental protection E15a, H412
  Storage and disposal D01, D02, D05, D09a, D11a

83   clomazone

An isoxazolidinone residual herbicide for oilseed rape, field beans, combining and vining peas
HRAC mode of action code: F3

Products

1	Angelus	Rotam	360 g/l	CS	18451
2	Blanco	Adama	360 g/l	CS	16704
3	Centium 360 CS	FMC Agro	360 g/l	CS	18719
4	Cirrus CS	FMC Agro	360 g/l	CS	18721

Products – continued

5	Cleancrop Chicane	FMC Agro	360 g/l	CS	18720
6	Cleancrop Covert	FMC Agro	360 g/l	CS	18722
7	Clomate	Albaugh UK	360 g/l	CS	15565
8	Clozone	UPL Europe	360 g/l	CS	18979
9	Gamit 36 CS	FMC Agro	360 g/l	CS	18718
10	Mohawk CS	Sipcam	360 g/l	CS	18505
11	Notion	AgChem Access	360 g/l	CS	16773
12	Sirtaki CS	Sipcam	360 g/l	CS	18032
13	Throne	Harvest	360 g/l	CS	17826

Uses

- Annual dicotyledons in *asparagus* (off-label), *courgettes* (off-label), *protected courgettes* (off-label), *protected summer squash* (off-label), *summer squash* (off-label) [6]; *baby leaf crops* (off-label), *borage* (off-label), *broad beans* (off-label), *broccoli* (off-label), *brussels sprouts* (off-label), *cabbages* (off-label), *calabrese* (off-label), *cauliflowers* (off-label), *celeriac* (off-label), *celery (outdoor)* (off-label), *choi sum* (off-label), *collards* (off-label), *fennel* (off-label), *french beans* (off-label), *herbs (see appendix 6)* (off-label), *kale* (off-label), *lupins* (off-label), *oriental cabbage* (off-label), *poppies for morphine production* (off-label), *rhubarb* (off-label), *runner beans* (off-label), *soya beans* (off-label), *spinach* (off-label), *spring cabbage* (off-label), *swedes* (off-label), *sweet potato* (off-label) [6, 9]; *beans without pods (fresh)* (off-label), *edible podded peas* (off-label) [9]; *carrots* [7, 11]; *combining peas, spring field beans, vining peas, winter field beans* [2, 7, 11]; *forest nurseries* (off-label), *game cover* (off-label), *hops* (off-label), *ornamental plant production* (off-label) [3, 4, 6, 9]; *potatoes, winter oilseed rape* [2, 7, 8, 11]; *spring oilseed rape* [2]
- Chickweed in *carrots* [6, 9, 10, 12, 13]; *combining peas, spring field beans, spring oilseed rape, vining peas* [3-5, 10, 12]; *courgettes* (off-label), *cucumbers* (off-label), *gherkins* (off-label), *melons* (off-label), *okra* (off-label), *pumpkins* (off-label), *summer squash* (off-label), *watermelons* (off-label), *winter squash* (off-label) [9]; *potatoes* [1, 6, 9, 10, 12, 13]; *winter field beans* [3-5, 12]; *winter oilseed rape* [1, 3-5, 10, 12]
- Cleavers in *carrots* [6, 7, 9-13]; *combining peas, spring field beans, vining peas* [2-5, 7, 10-12]; *courgettes* (off-label), *cucumbers* (off-label), *gherkins* (off-label), *melons* (off-label), *okra* (off-label), *pumpkins* (off-label), *summer squash* (off-label), *watermelons* (off-label), *winter squash* (off-label) [9]; *potatoes* [1, 2, 6-13]; *spring oilseed rape* [2-5, 10, 12]; *winter field beans* [2-5, 7, 11, 12]; *winter oilseed rape* [1-5, 7, 8, 10-12]
- Fool's parsley in *carrots* [6, 9, 10, 12, 13]; *combining peas, spring field beans, spring oilseed rape, vining peas* [3-5, 10, 12]; *courgettes* (off-label), *cucumbers* (off-label), *gherkins* (off-label), *melons* (off-label), *okra* (off-label), *pumpkins* (off-label), *summer squash* (off-label), *watermelons* (off-label), *winter squash* (off-label) [9]; *potatoes* [1, 6, 9, 10, 12, 13]
- Ivy-leaved speedwell in *carrots, potatoes* [6]; *courgettes* (off-label), *cucumbers* (off-label), *gherkins* (off-label), *melons* (off-label), *okra* (off-label), *pumpkins* (off-label), *summer squash* (off-label), *watermelons* (off-label), *winter squash* (off-label) [9]
- Red dead-nettle in *carrots* [6, 9, 10, 12, 13]; *combining peas, spring field beans, spring oilseed rape, vining peas* [3-5, 10, 12]; *courgettes* (off-label), *cucumbers* (off-label), *gherkins* (off-label), *melons* (off-label), *okra* (off-label), *pumpkins* (off-label), *summer squash* (off-label), *watermelons* (off-label), *winter squash* (off-label) [9]; *potatoes* [1, 6, 9, 10, 12, 13]; *winter field beans* [3-5, 12]; *winter oilseed rape* [1, 3-5, 10, 12]
- Shepherd's purse in *carrots* [6, 9, 10, 12, 13]; *combining peas, spring field beans, spring oilseed rape, vining peas* [3-5, 10, 12]; *courgettes* (off-label), *cucumbers* (off-label), *gherkins* (off-label), *melons* (off-label), *okra* (off-label), *pumpkins* (off-label), *summer squash* (off-label), *watermelons* (off-label), *winter squash* (off-label) [9]; *potatoes* [1, 6, 9, 10, 12, 13]; *winter field beans* [3-5, 12]; *winter oilseed rape* [1, 3-5, 10, 12]
- Speedwells in *carrots* [7, 11]; *combining peas, spring field beans, vining peas, winter field beans* [2, 7, 11]; *potatoes, winter oilseed rape* [2, 7, 8, 11]; *spring oilseed rape* [2]
- Spring-germinating perennial weeds in *asparagus* (off-label) [9]

Extension of Authorisation for Minor Use (EAMUs)

- *asparagus* 20190601 [6], 20190802 [9]
- *baby leaf crops* 20190613 [6], 20190778 [9]
- *beans without pods (fresh)* 20193047 [9]

FOR FULL CONDITIONS OF USE ALWAYS READ THE PRODUCT LABEL

- **borage** *20190603* [6], *20190794* [9]
- **broad beans** *20190606* [6], *20190776* [9]
- **broccoli** *20190604* [6], *20190799* [9]
- **brussels sprouts** *20190604* [6], *20190799* [9]
- **cabbages** *20190604* [6], *20190799* [9]
- **calabrese** *20190604* [6], *20190799* [9]
- **cauliflowers** *20190604* [6], *20190799* [9]
- **celeriac** *20190602* [6], *20190803* [9]
- **celery (outdoor)** *20190611* [6], *20190772* [9]
- **choi sum** *20190604* [6], *20190799* [9]
- **collards** *20190604* [6], *20190799* [9]
- **courgettes** *20190599* [6], *20190793* [9]
- **cucumbers** *20190793* [9]
- **edible podded peas** *20193047* [9]
- **fennel** *20190609* [6], *20190771* [9]
- **forest nurseries** *20190743* [3], *20190744* [4], *20190612* [6], *20190774* [9]
- **french beans** *20190600* [6], *20190798* [9]
- **game cover** *20190743* [3], *20190744* [4], *20190612* [6], *20190774* [9]
- **gherkins** *20190793* [9]
- **herbs (see appendix 6)** *20190610* [6], *20190777* [9]
- **hops** *20190743* [3], *20190744* [4], *20190612* [6], *20190774* [9]
- **kale** *20190604* [6], *20190799* [9]
- **lupins** *20190605* [6], *20190795* [9]
- **melons** *20190793* [9]
- **okra** *20190793* [9]
- **oriental cabbage** *20190604* [6], *20190799* [9]
- **ornamental plant production** *20190743* [3], *20190744* [4], *20190612* [6], *20190774* [9]
- **poppies for morphine production** *20190607* [6], *20190796* [9]
- **protected courgettes** *20190599* [6]
- **protected summer squash** *20190599* [6]
- **pumpkins** *20190793* [9]
- **rhubarb** *20190614* [6], *20190797* [9]
- **runner beans** *20183494* [6], *20190800* [9]
- **soya beans** *20190597* [6], *20190801* [9]
- **spinach** *20190613* [6], *20190778* [9]
- **spring cabbage** *20190604* [6], *20190799* [9]
- **summer squash** *20190599* [6], *20190793* [9]
- **swedes** *20190596* [6], *20190775* [9]
- **sweet potato** *20190608* [6], *20190773* [9]
- **watermelons** *20190793* [9]
- **winter squash** *20190793* [9]

Approval information
- Clomazone included in Annex I under EC Regulation 1107/2009

Efficacy guidance
- Best results obtained from application as soon as possible after sowing crop and before emergence of crop or weeds
- Uptake is via roots and shoots. Seedbeds should be firm, level and free from clods. Loose puffy seedbeds should be consolidated before spraying
- Efficacy is reduced on organic soils, on dry cloddy seedbeds and if prolonged dry weather follows application
- Clomazone acts by inhibiting synthesis of chlorophyll pigments. Susceptible weeds emerge but are chlorotic and die shortly afterwards
- Season-long control of weeds may not be achieved
- Always follow WRAG guidelines for preventing and managing herbicide resistant weeds. See Section 5 for more information

Restrictions
- Maximum number of treatments one per crop

SEE SECTION 3 FOR PRODUCTS ALSO REGISTERED

- Crops must be covered by a minimum of 20 mm settled soil. Do not apply to broadcast crops. Direct-drilled crops should be harrowed across the slits to cover seed before spraying
- Do not use on compacted soils or soils of poor structure that may be liable to waterlogging
- Do not use on Sands or Very Light soils or those with more than 10% organic matter
- Do not treat two consecutive crops of carrots with clomazone in one calendar yr
- Consult manufacturer or your advisor before use on potato seed crops
- Do not overlap spray swaths. Crop plants emerged at time of treatment may be severely damaged
- Application must be made using a coarse spray quality.

Crop-specific information
- Latest use: pre-emergence of crop
- Severe, but normally transient, crop damage may occur in overlaps on field beans
- Some transient crop bleaching may occur under certain climatic conditions and can be severe where heavy rain follows application. This is normally rapidly outgrown and has no effect on final crop yield.

Following crops guidance
- Following normal harvest of a spring or autumn treated crop, cereals, oilseed rape, field beans, combining peas, potatoes, maize, turnips, linseed or sugar beet may be sown
- In the event of failure of an autumn treated crop, winter cereals or winter beans may be sown in the autumn if 6 wk have elapsed since treatment. In the spring following crop failure combining peas, field beans or potatoes may be sown if 6 wk have elapsed since treatment, and spring cereals, maize, turnips, onions, carrots or linseed may be sown if 7 mth have elapsed since treatment
- In the event of a failure of a spring treated crop a wide range of crops may be sown provided intervals of 6-9 wk have elapsed since treatment. See label for details
- Prior to resowing any listed replacement crop the soil should be ploughed and cultivated to 15 cm

Environmental safety
- Take extreme care to avoid drift outside the target area, or on to ponds, waterways or ditches as considerable damage may occur. Apply using a coarse quality spray
- To protect non-target plants respect an untreated buffer zone of 10 meters to non-crop land for [1, 2, 7, 8, 11, 12] or 5 meters for [2]

Hazard classification and safety precautions
Transport code 9 [1, 2, 8]
Packaging group III [1, 2, 8]
UN Number 3082 [1]; N/C [2-13]
Risk phrases H315, H319 [1, 8]
Operator protection A [1-10, 12, 13]; C [1]; H [1, 2, 8]; U05a, U14, U20b [1, 3-7, 9, 10, 12, 13]; U05b [2, 8]
Environmental protection E15b [1, 3-7, 9, 10, 12, 13]; E39 [7]; H410 [1, 8, 10, 12]; H412 [2]; H413 [3-6, 9, 13]
Storage and disposal D01, D02 [1-10, 12, 13]; D09a [1, 3-7, 9, 10, 12, 13]; D10b [1, 3-6, 9, 10, 12, 13]; D10c [7]; D12a [2, 8]

84   clomazone + metribuzin

A mixture of isoxazolidinone and triazinone residual herbicides for use in potatoes
HRAC mode of action code: F3 + C1

See also diflufenican + flufenacet + metribuzin
metribuzin

Products
Metric	Belchim	60:233 g/l	CC	16720

Uses
- Annual dicotyledons in **potatoes**
- Annual meadow grass in **potatoes**

Approval information
- Clomazone and metribuzin included in Annex I under EC Regulation 1107/2009

Restrictions
- For use on specified varieties of potato only
- Safety to daughter tubers has not been tested; consult manufacturer before treating seed crops
- Application must be made using a coarse spray quality.

Following crops guidance
- Before drilling or planting any succeeding crop, soil MUST be mouldboard ploughed to a depth of at least 15 cm (6") taking care to ensure that the furrow slice is inverted. Ploughing should be carried out as soon as possible (preferably within 3-4 weeks) after lifting the potato crop, but certainly no later than the end of December.
- In the same year: Provided at least 16 weeks have elapsed after the application of the recommended rate cereals and winter beans may be grown as following crops.In the following year: Do not grow any vegetable brassica crop (including cauliflower, calabrese, Brussels sprout and cabbage), lettuce or radish on land treated in the previous year.
- Cereals, oilseed rape, field beans, combining peas, potatoes, maize, turnip, linseed and sugar beet may be sown from spring onwards in the year following use.

Environmental safety
- Buffer zone requirement 10 m [1]
- To protect non-target plants respect an untreated buffer zone of 5 meters to non-crop land
- LERAP Category B

Hazard classification and safety precautions
Hazard Dangerous for the environment, Very toxic to aquatic organisms
Transport code 9
Packaging group III
UN Number 3082
Operator protection A, H; U05a, U19a, U20b
Environmental protection E15b, E16a, E16b, E34, E38, H410
Storage and disposal D01, D02, D09a, D10b, D12a
Medical advice M05a

85   clomazone + napropamide

A herbicide mixture for use in winter oilseed rape
HRAC mode of action code: F3 + K3

See also clomazone
napropamide

Products

Altiplano DAMtec	FMC Agro	3.5:40% w/w	WG	18717

Uses
- Annual dicotyledons in **winter oilseed rape**
- Annual grasses in **winter oilseed rape**
- Chickweed in **winter oilseed rape**
- Cleavers in **winter oilseed rape**
- Crane's-bill in **winter oilseed rape**
- Dead nettle in **winter oilseed rape**
- Mayweeds in **winter oilseed rape**
- Poppies in **winter oilseed rape**

Approval information
- Clomazone and napropamide included in Annex I under EC Regulation 1107/2009

SEE SECTION 3 FOR PRODUCTS ALSO REGISTERED

Restrictions
- Do not use on sands, very light soils or on soils with more than 10% organic matter.
- Do not roll after application.
- Do not irrigate within 3 weeks of application.

Crop-specific information
- Ensure seed is covered by a minimum of 2 cm settled soil before application

Environmental safety
- LERAP Category B

Hazard classification and safety precautions
    Hazard Dangerous for the environment, Very toxic to aquatic organisms
    Transport code 9
    Packaging group III
    UN Number 3077
    Operator protection A, H
    Environmental protection E16a, E38, H410
    Storage and disposal D12a

86   clomazone + pendimethalin

A residual herbicide mixture for weed control in combining peas and field beans
HRAC mode of action code: F3 + K1

Products

Stallion Sync Tec	FMC Agro	30:333 g/l	CS	18714

Uses
- Annual dicotyledons in *carrots, combining peas, potatoes, spring field beans, vining peas, winter field beans*
- Cleavers in *carrots, combining peas, potatoes, spring field beans, vining peas, winter field beans*

Approval information
- Clomazone and pendimethalin included in Annex I under EC Regulation 1107/2009

Restrictions
- Application must be made using a coarse spray quality.

Following crops guidance
- In the event of crop failure plough the soil to at least 25cm and see label for required elapsed time before planting. Note that sugar beet should not be sown until 12 months after treatment.

Environmental safety
- LERAP Category B

Hazard classification and safety precautions
    Hazard Dangerous for the environment, Very toxic to aquatic organisms
    Transport code 9
    Packaging group III
    UN Number 3082
    Risk phrases H317
    Operator protection A; U02a, U05a, U08, U13, U14, U19a, U20b
    Environmental protection E15b, E16a, E38, H410
    Storage and disposal D01, D02, D05, D09a, D10b, D12a

87   clopyralid

A foliar translocated picolinic herbicide for a wide range of crops
HRAC mode of action code: O

See also 2,4-D + clopyralid + MCPA
bromoxynil + clopyralid

Products

1	Cliophar 400	Arysta	400 g/l	SL	15008
2	Dow Shield 400	Dow (Corteva)	400 g/l	SL	14984
3	Leash	Life Scientific	200 g/l	SL	17969
4	Vivendi 200	UPL Europe	200 g/l	SL	16966

Uses

- Annual dicotyledons in *brussels sprouts, forage maize, linseed* [1, 2]; *fodder beet, mangels, red beet, spring oilseed rape, sugar beet, swedes, turnips, winter oilseed rape* [1-4]; *grassland, ornamental plant production, spring barley, spring oats, spring wheat, winter barley, winter oats, winter wheat* [1, 2, 4]
- Black bindweed in *forest* (off-label) [2]
- Clovers in *grass seed crops* (off-label) [2]
- Corn marigold in *brussels sprouts, fodder beet, forage maize, grassland, linseed, mangels, ornamental plant production, red beet, spring barley, spring oats, spring oilseed rape, spring wheat, sugar beet, swedes, turnips, winter barley, winter oats, winter oilseed rape, winter wheat* [1, 2]; *forest* (off-label) [2]
- Creeping thistle in *brussels sprouts, fodder beet, forage maize, grassland, linseed, mangels, ornamental plant production, red beet, spring barley, spring oats, spring oilseed rape, spring wheat, sugar beet, swedes, turnips, winter barley, winter oats, winter oilseed rape, winter wheat* [1, 2]; *forest* (off-label), *sweetcorn* (off-label) [2]
- Dandelions in *forest* (off-label) [2]
- Groundsel in *all edible seed crops grown outdoors* (off-label), *all non-edible seed crops grown outdoors* (off-label), *bilberries* (off-label), *blackcurrants* (off-label), *blueberries* (off-label), *chard* (off-label), *corn gromwell* (off-label), *cranberries* (off-label), *elderberries* (off-label), *farm forestry* (off-label), *forest* (off-label), *garlic* (off-label), *gooseberries* (off-label), *grass seed crops* (off-label), *hemp grown for fibre production* (off-label), *hops in propagation* (off-label), *miscanthus* (off-label), *mulberries* (off-label), *nursery fruit trees* (off-label), *outdoor leaf herbs* (off-label), *redcurrants* (off-label), *rhubarb* (off-label), *rose hips* (off-label), *shallots* (off-label), *spinach* (off-label), *spinach beet* (off-label), *strawberries* (off-label), *sweetcorn* (off-label) [2]; *forest nurseries* (off-label) [2, 4]; *game cover* (off-label) [1, 2, 4]
- Mayweeds in *all edible seed crops grown outdoors* (off-label), *all non-edible seed crops grown outdoors* (off-label), *apples* (off-label), *bilberries* (off-label), *blackcurrants* (off-label), *blueberries* (off-label), *borage for oilseed production* (off-label), *canary flower (echium spp.)* (off-label), *chard* (off-label), *corn gromwell* (off-label), *crab apples* (off-label), *cranberries* (off-label), *elderberries* (off-label), *evening primrose* (off-label), *farm forestry* (off-label), *forest* (off-label), *garlic* (off-label), *gooseberries* (off-label), *grass seed crops* (off-label), *hemp grown for fibre production* (off-label), *honesty* (off-label), *hops in propagation* (off-label), *leeks* (off-label), *miscanthus* (off-label), *mulberries* (off-label), *mustard* (off-label), *nursery fruit trees* (off-label), *outdoor leaf herbs* (off-label), *pears* (off-label), *poppies for morphine production* (off-label), *redcurrants* (off-label), *rhubarb* (off-label), *rose hips* (off-label), *rye* (off-label), *salad onions* (off-label), *shallots* (off-label), *spinach* (off-label), *spinach beet* (off-label), *strawberries* (off-label), *triticale* (off-label) [2]; *brussels sprouts, forage maize, linseed* [1, 2]; *fodder beet, mangels, red beet, spring oilseed rape, sugar beet, swedes, turnips, winter oilseed rape* [1-4]; *forest nurseries* (off-label) [2, 4]; *game cover* (off-label), *grassland, ornamental plant production, spring barley, spring oats, spring wheat, winter barley, winter oats, winter wheat* [1, 2, 4]
- Perennial dicotyledons in *leeks* (off-label), *salad onions* (off-label) [2]
- Ragwort in *forest* (off-label) [2]
- Sowthistle in *bilberries* (off-label), *blackcurrants* (off-label), *blueberries* (off-label), *cranberries* (off-label), *elderberries* (off-label), *forest* (off-label), *garlic* (off-label), *gooseberries* (off-label), *mulberries* (off-label), *poppies for morphine production* (off-label), *redcurrants* (off-label),

rose hips (off-label), *rye* (off-label), *shallots* (off-label), *triticale* (off-label) [2]; *forest nurseries* (off-label), *game cover* (off-label) [4]
- Thistles in *apples* (off-label), *asparagus* (off-label), *borage for oilseed production* (off-label), *canary flower (echium spp.)* (off-label), *chinese cabbage* (off-label), *choi sum* (off-label), *collards* (off-label), *corn gromwell* (off-label), *crab apples* (off-label), *evening primrose* (off-label), *forest* (off-label), *garlic* (off-label), *honesty* (off-label), *kale* (off-label), *leeks* (off-label), *mustard* (off-label), *pak choi* (off-label), *pears* (off-label), *poppies for morphine production* (off-label), *rhubarb* (off-label), *salad onions* (off-label), *shallots* (off-label), *spring greens* (off-label), *strawberries* (off-label), *tatsoi* (off-label), *trees* (off-label) [2]; *fodder beet, mangels, red beet, spring oilseed rape, sugar beet, swedes, turnips, winter oilseed rape* [3, 4]; *forest nurseries* (off-label) [2, 4]; *game cover* (off-label) [1, 4]; *grassland, ornamental plant production, spring barley, spring oats, spring wheat, winter barley, winter oats, winter wheat* [4]
- Volunteer potatoes in *chinese cabbage* (off-label), *choi sum* (off-label), *collards* (off-label), *forest nurseries* (off-label), *garlic* (off-label), *kale* (off-label), *leeks* (off-label), *pak choi* (off-label), *salad onions* (off-label), *shallots* (off-label), *spring greens* (off-label), *sweetcorn* (off-label), *tatsoi* (off-label) [2]

Extension of Authorisation for Minor Use (EAMUs)
- *all edible seed crops grown outdoors* 20150007 [2]
- *all non-edible seed crops grown outdoors* 20150007 [2]
- *apples* 20102080 [2]
- *asparagus* 20102079 [2]
- *bilberries* 20161629 [2]
- *blackcurrants* 20161629 [2]
- *blueberries* 20161629 [2]
- *borage for oilseed production* 20102086 [2]
- *canary flower (echium spp.)* 20102086 [2]
- *chard* 20102081 [2]
- *chinese cabbage* 20131710 [2]
- *choi sum* 20131710 [2]
- *collards* 20131710 [2]
- *corn gromwell* 20121046 [2]
- *crab apples* 20102080 [2]
- *cranberries* 20161629 [2]
- *elderberries* 20161629 [2]
- *evening primrose* 20102086 [2]
- *farm forestry* 20150007 [2]
- *forest* 20152633 [2]
- *forest nurseries* 20130514 [2], 20150007 [2], 20151529 [4]
- *game cover* 20160786 [1], 20150007 [2], 20151529 [4]
- *garlic* 20130292 [2]
- *gooseberries* 20161629 [2]
- *grass seed crops* 20180041 [2]
- *hemp grown for fibre production* 20150007 [2]
- *honesty* 20102086 [2]
- *hops in propagation* 20150007 [2]
- *kale* 20131710 [2]
- *leeks* 20140400 [2]
- *miscanthus* 20150007 [2]
- *mulberries* 20161629 [2]
- *mustard* 20102086 [2]
- *nursery fruit trees* 20150007 [2]
- *outdoor leaf herbs* 20113236 [2]
- *pak choi* 20131710 [2]
- *pears* 20102080 [2]
- *poppies for morphine production* 20102082 [2]
- *redcurrants* 20161629 [2]
- *rhubarb* 20192582 [2]

- **rose hips** *20161629* [2]
- **rye** *20193254* [2]
- **salad onions** *20140400* [2]
- **shallots** *20130292* [2]
- **spinach** *20102081* [2]
- **spinach beet** *20102081* [2]
- **spring greens** *20131710* [2]
- **strawberries** *20131822* [2]
- **sweetcorn** *20152626* [2]
- **tatsoi** *20131710* [2]
- **trees** *20102083* [2]
- **triticale** *20193254* [2]

Approval information
- Clopyralid included in Annex I under EC Regulation 1107/2009
- Accepted by BBPA for use on malting barley

Efficacy guidance
- Best results achieved by application to young actively growing weed seedlings. Treat creeping thistle at rosette stage and repeat 3-4 wk later as directed
- High activity on weeds of Compositae family. For most crops recommended for use in tank mixes. See label for details

Restrictions
- Maximum total dose varies between the equivalent of one and two full dose treatments, depending on the crop treated. See labels for details
- Do not apply to cereals later than the second node detectable stage (GS 32)
- Do not apply when crop damp or when rain expected within 6 h
- Do not use straw from treated cereals in compost or any other form for glasshouse crops. Straw may be used for strawing down strawberries
- Straw from treated grass seed crops or linseed should be baled and carted away. If incorporated do not plant winter beans in same year
- Do not use on onions at temperatures above 20°C or when under stress
- Do not treat maiden strawberries or runner beds or apply to early leaf growth during blossom period or within 4 wk of picking. Aug or early Sep sprays may reduce yield
- Applications must not be made earlier than 1 March in the year of harvest [3]

Crop-specific information
- Latest use: 7 d before cutting grass for hay or silage; before 3rd node detectable (GS 33) for cereals; before flower buds visible from above for oilseed rape, linseed
- HI grassland 7 d; apples, pears, strawberries 4 wk; maize, sweetcorn, onions, Brussels sprouts, broccoli, cabbage, cauliflowers, calabrese, kale, fodder rape, oilseed rape, swedes, turnips, sugar beet, red beet, fodder beet, mangels, sage, honesty 6 wk
- Timing of application varies with weed problem, crop and other ingredients of tank mixes. See labels for details
- Apply as directed spray in woody ornamentals, avoiding leaves, buds and green stems. Do not apply in root zone of families Compositae or Leguminosae

Following crops guidance
- Do not plant susceptible autumn-sown crops in same year as treatment. Do not apply later than Jul where susceptible crops are to be planted in spring. See label for details

Environmental safety
- Harmful to aquatic organisms [2]
- Wash spray equipment thoroughly with water and detergent immediately after use. Traces of product can damage susceptible plants sprayed later
- Keep livestock out of treated areas for at least 7 d and until foliage of any poisonous weeds such as ragwort has died and become unpalatable
- Some pesticides pose a greater threat of contamination of water than others and clopyralid is one of these pesticides. Take special care when applying clopyralid near water and do not apply if heavy rain is forecast

SEE SECTION 3 FOR PRODUCTS ALSO REGISTERED

Hazard classification and safety precautions
  UN Number N/C
  Risk phrases R52, R53a [1]
  Operator protection A, C [1-4]; H [3]; U05a [3, 4]; U08, U19a, U20b [1-4]
  Environmental protection E07a [1, 2]; E15a [1-4]; E34 [3, 4]
  Storage and disposal D01, D05, D09a, D10b [1-4]; D02 [3, 4]; D12a [1-3]

88   clopyralid + 2,4-D + MCPA

A translocated herbicide mixture for managed amenity turf
HRAC mode of action code: O + O + O

See also 2,4-D
      MCPA

Products

Redeem	Headland Amenity	35:150:175 g/l	SL	17096

Uses
  • Annual dicotyledons in *managed amenity turf*
  • Buttercups in *managed amenity turf*
  • Clover in *managed amenity turf*
  • Daisies in *managed amenity turf*
  • Dandelions in *managed amenity turf*
  • Perennial dicotyledons in *managed amenity turf*
  • Plantains in *managed amenity turf*

Approval information
  • Clopyralid, 2,4-D and MCPA included in Annex I under EC Regulation 1107/2009

Efficacy guidance
  • Apply when soil is moist and weeds growing actively, normally between Apr and Oct
  • Ensure sufficient leaf area for uptake and select appropriate water volume to achieve good spray coverage
  • To allow maximum translocation do not cut grass for 3 d after treatment

Restrictions
  • Maximum number of treatments on managed amenity turf 1 per yr
  • Do not spray in periods of drought unless irrigation is applied
  • Do not apply if night temrperatures are low, if ground frost is imminent or in periods of prolonged cold weather
  • Do not treat turf less than 1 yr old
  • Do not use any treated plant materials for composting or mulching
  • Do not mow for 3 d before or after treatment

Crop-specific information
  • Consult manufacturer or carry out small scale safety test on turf grass cultivars not previously treated

Following crops guidance
  • If reseeding of treated turf is required allow at least 6 wk after treatment before drilling grasses into the sward
  • Where treated land is subsequently to be sown with broad-leaved plants allow an interval of at least 6 mth

Environmental safety
  • Dangerous for the environment
  • Harmful to aquatic organisms
  • Wash spray equipment thoroughly with water and detergent immediately after use. Traces of product can damage susceptible plants sprayed later
  • Do not spray outside the target area and avoid drift onto non-target plants

FOR FULL CONDITIONS OF USE ALWAYS READ THE PRODUCT LABEL

- Some pesticides pose a greater threat of contamination of water than others and clopyralid is one of these pesticides. Take special care when applying clopyralid near water and do not apply if heavy rain is forecast

Hazard classification and safety precautions

Hazard Harmful, Dangerous for the environment, Harmful if swallowed
Transport code 9
Packaging group III
UN Number 3082
Risk phrases H318
Operator protection A, C, H; U05a, U11, U15
Environmental protection E15a, E34, H411
Storage and disposal D01, D02, D12a

89   clopyralid + florasulam

A translocated herbicide mixture for weed control in cereals
HRAC mode of action code: O + B

Products

Gartrel	Dow (Corteva)	300:25 g/l	SC	16828

Uses

- Black bindweed in *durum wheat*, *spelt*, *spring barley*, *spring oats*, *spring wheat*, *triticale*, *winter barley*, *winter rye*, *winter wheat*
- Charlock in *durum wheat*, *spelt*, *spring barley*, *spring oats*, *spring wheat*, *triticale*, *winter barley*, *winter rye*, *winter wheat*
- Chickweed in *durum wheat*, *spelt*, *spring barley*, *spring oats*, *spring wheat*, *triticale*, *winter barley*, *winter rye*, *winter wheat*
- Cleavers in *durum wheat*, *spelt*, *spring barley*, *spring oats*, *spring wheat*, *triticale*, *winter barley*, *winter rye*, *winter wheat*
- Mayweeds in *durum wheat*, *spelt*, *spring barley*, *spring oats*, *spring wheat*, *triticale*, *winter barley*, *winter rye*, *winter wheat*
- Runch in *durum wheat*, *spelt*, *spring barley*, *spring oats*, *spring wheat*, *triticale*, *winter barley*, *winter rye*, *winter wheat*
- Shepherd's purse in *durum wheat*, *spelt*, *spring barley*, *spring oats*, *spring wheat*, *triticale*, *winter barley*, *winter rye*, *winter wheat*
- Volunteer oilseed rape in *durum wheat*, *spelt*, *spring barley*, *spring oats*, *spring wheat*, *triticale*, *winter barley*, *winter rye*, *winter wheat*
- Wild radish in *durum wheat*, *spelt*, *spring barley*, *spring oats*, *spring wheat*, *triticale*, *winter barley*, *winter rye*, *winter wheat*

Approval information

- Clopyralid and florasulam included in Annex I under EC Regulation 1107/2009

Efficacy guidance

- Best results obtained when weeds are small and actively growing
- Effectiveness may be reduced when soil is very dry
- Use adequate water volume to achieve complete spray coverage of the weeds
- Florasulam is a member of the ALS-inhibitor group of herbicides

Restrictions

- Do not roll or harrow for 7 d before or after application
- Do not use any treated plant material for composting or mulching
- Do not use manure from animals fed on treated crops for composting
- Specific restrictions apply to use in sequence or tank mixture with other sulfonylurea or ALS-inhibiting herbicides. See label for details

Following crops guidance

- Where residues of a treated crop have not completely decayed by the time of planting a succeeding crop, avoid planting peas, beans and other legumes, carrots and other Umbelliferae, potatoes, lettuce and other Compositae, glasshouse and protected crops
- Where the product has been used in mixture with certain named products (see label) only cereals or grass may be sown in the autumn following harvest. Otherwise cereals, oilseed rape, grass or vegetable brassicas as transplants may be sown as a following crop in the same calendar yr as treatment. Oilseed rape may show some temporary reduction of vigour after a dry summer, but yields are not affected
- In addition to the above, field beans, linseed, peas, sugar beet, potatoes, maize, clover (for use in grass/clover mixtures) or carrots may be sown in the calendar yr following treatment
- In the event of failure of a treated crop in spring, only spring wheat, spring barley, spring oats, maize or ryegrass may be sown

Environmental safety

- LERAP Category B

Hazard classification and safety precautions

Hazard Very toxic to aquatic organisms
Transport code 9
Packaging group III
UN Number 3082
Operator protection U05a, U19a, U20a
Environmental protection E15a, E16a, E34, H410
Storage and disposal D05, D09a, D10b, D12a
Medical advice M03

90   clopyralid + florasulam + fluroxypyr

A translocated herbicide mixture for cereals
HRAC mode of action code: O + B + O

See also florasulam
fluroxypyr

Products

1 Dakota	Dow (Corteva)	80:2.5:100 g/l	EC	19179
2 Dingo	Dow (Corteva)	80:2.5:100 g/l	EC	18412
3 Galaxy	Dow (Corteva)	80:2.5:100 g/l	EC	18952
4 Leystar	Dow (Corteva)	80:2.5:100 g/l	EC	17921
5 Praxys	Dow (Corteva)	80:2.5;100 g/l	EC	18953

Uses

- Annual dicotyledons in *amenity grassland, lawns, managed amenity turf* [5]; *forage maize, grass seed crops, grassland, newly sown grass leys* [2, 4]; *spelt, spring barley, spring oats, spring rye, spring wheat, triticale, winter barley, winter oats, winter rye, winter wheat* [1, 3]; *spring triticale, undersown barley, undersown oats, undersown rye, undersown spelt, undersown wheat* [3]
- Black medick in *amenity grassland* [5]
- Bristly oxtongue in *amenity grassland* [5]; *grass seed crops, grassland, newly sown grass leys* [2, 4]
- Buttercups in *amenity grassland* [5]; *grass seed crops, grassland, newly sown grass leys* [2, 4]
- Chickweed in *forage maize* [2, 4]; *spelt, spring barley, spring oats, spring rye, spring wheat, triticale, winter barley, winter oats, winter rye, winter wheat* [1, 3]; *spring triticale, undersown barley, undersown oats, undersown rye, undersown spelt, undersown wheat* [3]
- Cleavers in *forage maize* [2, 4]; *spelt, spring barley, spring oats, spring rye, spring wheat, triticale, winter barley, winter oats, winter rye, winter wheat* [1, 3]; *spring triticale,*

FOR FULL CONDITIONS OF USE ALWAYS READ THE PRODUCT LABEL

undersown barley, undersown oats, undersown rye, undersown spelt, undersown wheat [3]
- Common mouse-ear in *amenity grassland* [5]; *grass seed crops, grassland, newly sown grass leys* [2, 4]
- Creeping thistle in *forage maize* [2, 4]; *spelt, spring barley, spring oats, spring rye, spring wheat, triticale, winter barley, winter oats, winter rye, winter wheat* [1]
- Daisies in *amenity grassland* [5]; *grass seed crops, grassland, newly sown grass leys* [2, 4]
- Dandelions in *amenity grassland* [5]; *grass seed crops, grassland, newly sown grass leys* [2, 4]
- Mayweeds in *forage maize* [2, 4]; *spelt, spring barley, spring oats, spring rye, spring wheat, triticale, winter barley, winter oats, winter rye, winter wheat* [1]
- Plantains in *amenity grassland* [5]; *grass seed crops, grassland, newly sown grass leys* [2, 4]
- Self-heal in *amenity grassland* [5]
- Slender speedwell in *amenity grassland* [5]

Approval information
- Clopyralid, florasulam and fluroxypyr are all included in Annex I under EC Regulation 1107/2009

Efficacy guidance
- Best results obtained when weeds are small and actively growing
- Effectiveness may be reduced when soil is very dry
- Use adequate water volume to achieve complete spray coverage of the weeds
- Florasulam is a member of the ALS-inhibitor group of herbicides

Restrictions
- Maximum number of treatments 1 per yr for all crops
- Do not spray when crops are under stress from any cause
- Do not apply through CDA applicators
- Do not roll or harrow for 7 d before or after application
- Do not use any treated plant material for composting or mulching
- Do not use manure from animals fed on treated crops for composting
- Specific restrictions apply to use in sequence or tank mixture with other sulfonylurea or ALS-inhibiting herbicides. See label for details
- Applications to cereals must only be made between 1st February and 30th June
- Application to any grass crop must not be made before 1st February [2]
- To protect surface water, application to maize must not take place before 10th April [2]
- Application to any grass crop must not be made before 1st February [2]
- Following treatment a period of at least 125 days must be observed prior to planting of succeeding crops [2]
- Extreme care must be taken to avoid spray drift onto non-crop plants outside the target area
- For autumn planted crops a maximum total dose of 3.75 g florasulam must be observed for applications made between crop emergence in the year of planting and 1st Feb in the year of harvest. The total amount of florasulam applied to a cereal crop must not exceed 7.5 g/ha.

Crop-specific information
- Latest use: before second node detectable for oats; before third node detectable for spring barley and spring wheat; before flag leaf detectable in winter barley and winter wheat, before 30th Sept in established grassland, before 31st Aug in newly sown leys and before 31st May in maize.

Following crops guidance
- Where residues of a treated crop have not completely decayed by the time of planting a succeeding crop, avoid planting peas, beans and other legumes, carrots and other Umbelliferae, potatoes, lettuce and other Compositae, glasshouse and protected crops
- Where the product has been used in mixture with certain named products (see label) only cereals or grass may be sown in the autumn following harvest. Otherwise cereals, oilseed rape, grass or vegetable brassicas as transplants may be sown as a following crop in the same calendar yr as treatment. Oilseed rape may show some temporary reduction of vigour after a dry summer, but yields are not affected
- In addition to the above, field beans, linseed, peas, sugar beet, potatoes, maize, clover (for use in grass/clover mixtures) or carrots may be sown in the calendar yr following treatment

SEE SECTION 3 FOR PRODUCTS ALSO REGISTERED

- In the event of failure of a treated crop in spring, only spring wheat, spring barley, spring oats, maize or ryegrass may be sown
- A period of at least 125 days must be observed prior to planting of succeeding crops [2, 4]

Environmental safety
- Dangerous for the environment
- Very toxic to aquatic organisms
- Take extreme care to avoid drift outside the target area
- Some pesticides pose a greater threat of contamination of water than others and clopyralid is one of these pesticides. Take special care when applying clopyralid near water and do not apply if heavy rain is forecast
- LERAP Category B

Hazard classification and safety precautions
Hazard Harmful, Dangerous for the environment, Harmful if inhaled
Transport code 9
Packaging group III
UN Number 3082
Risk phrases H304, H315, H317, H319
Operator protection A, C, H; U05a, U11, U19a, U20a
Environmental protection E15b, E16a, E34, E38, H410
Storage and disposal D01, D02, D09a, D10b, D12a [1-5]; D05 [5]
Medical advice M03 [1-5]; M05a [1-4]

---

91 clopyralid + fluroxypyr + MCPA

A translocated herbicide mixture for use in sports and amenity turf
HRAC mode of action code: O + O + O

See also fluroxypyr
MCPA

Products

1 Columbus	Headland Amenity	20:40:200 g/l	ME	18462
2 Greenor	Rigby Taylor	20:40:200 g/l	ME	15204
3 Kinvara	Barclay	28:72:233 g/l	ME	18436

Uses
- Annual dicotyledons in *managed amenity turf* [1, 2]; *rye*, *spring barley*, *spring oats*, *spring wheat*, *triticale*, *winter barley*, *winter oats*, *winter wheat* [3]

Approval information
- Clopyralid, fluroxypyr and MCPA included in Annex I under EC Regulation 1107/2009

Efficacy guidance
- Best results achieved when weeds actively growing and turf grass competitive
- Treatment should normally be between Apr-Sep when the soil is moist
- Do not apply during drought unless irrigation is applied
- Allow 3 d before or after mowing established turf to ensure sufficient weed leaf surface present to allow uptake and movement

Restrictions
- Maximum number of treatments 2 per yr [2]
- Do not treat grass under stress from frost, drought, waterlogging, trace element deficiency, disease or pest attack [2]
- Do not treat if night temperatures are low, when frost is imminent or during prolonged cold weather [2]
- At least 125 days must pass prior to planting of succeeding crops. [3]
- To protect groundwater applications must not be made between 31st August and 1st March [3]

FOR FULL CONDITIONS OF USE ALWAYS READ THE PRODUCT LABEL

Crop-specific information

- Treat young turf only in spring when at least 2 mth have elapsed since sowing
- Allow 5 d after mowing young turf before treatment
- Product selective on a number of turf grass species (see label) but consultation or testing recommended before treatment of any cultivar
- A period of at least 125 days must be observed prior to planting of succeeding crops [3]

Environmental safety

- Dangerous for the environment
- Very toxic to aquatic organisms
- Wash spray equipment thoroughly with water and detergent immediately after use. Traces of product can damage susceptible plants sprayed later
- Some pesticides pose a greater threat of contamination of water than others and clopyralid is one of these pesticides. Take special care when applying clopyralid near water and do not apply if heavy rain is forecast
- LERAP Category B

Hazard classification and safety precautions

Hazard Irritant, Dangerous for the environment [1-3]; Very toxic to aquatic organisms [3]
Transport code 9
Packaging group III
UN Number 3082
Risk phrases H317 [1-3]; H319 [3]; H320 [1, 2]
Operator protection A, C; U05a, U08, U14, U19a, U20b
Environmental protection E15a, E16a, E38, H410
Storage and disposal D01, D02, D05, D09a, D10b, D12a

92   clopyralid + picloram

A post-emergence herbicide mixture for oilseed rape
HRAC mode of action code: O + O

See also picloram

Products

1	Clopic	UPL Europe	267:67 g/l	SL	17387
2	Galera	Dow (Corteva)	267:67 g/l	SL	16413
3	Legara	AgChem Access	267:67 g/l	SL	16789

Uses

- Annual dicotyledons in *corn gromwell* (off-label), *crambe* (off-label), *forest nurseries* (off-label), *game cover* (off-label), *mustard* (off-label), *ornamental plant production* (off-label) [2]
- Cleavers in *corn gromwell* (off-label), *crambe* (off-label), *forest nurseries* (off-label), *game cover* (off-label), *mustard* (off-label), *ornamental plant production* (off-label), *spring oilseed rape* (off-label) [2]; *winter oilseed rape* [1-3]
- Mayweeds in *corn gromwell* (off-label), *crambe* (off-label), *forest nurseries* (off-label), *game cover* (off-label), *mustard* (off-label), *ornamental plant production* (off-label) [2]; *winter oilseed rape* [1-3]

Extension of Authorisation for Minor Use (EAMUs)

- *corn gromwell* 20160559 [2]
- *crambe* 20161622 [2]
- *forest nurseries* 20142720 [2]
- *game cover* 20142720 [2]
- *mustard* 20152939 [2]
- *ornamental plant production* 20150968 [2]
- *spring oilseed rape* 20140827 [2]

Approval information

- Clopyralid and picloram included in Annex I under EC Regulation 1107/2009

Efficacy guidance
- Best results obtained from treatment when weeds are small and actively growing
- Cleavers that germinate after treatment will not be controlled

Restrictions
- Maximum total dose equivalent to one full dose treatment
- Do not treat crops under stress from cold, drought, pest damage, nutrient deficiency or any other cause
- Do not roll or harrow for 7 d before or after spraying
- Do not apply through CDA applicators
- Extreme care must be taken to avoid spray drift onto non-crop plants outside of the target area
- Do not use any treated plant material for composting or mulching
- Do not use manure from animals fed on treated crops for composting
- Chop and incorporate all treated plant remains in early autumn, or as soon as possible after harvest, to release any residues into the soil. Ensure that all treated plant remains have completely decayed before planting susceptible crops

Crop-specific information
- Latest use: before flower buds visible above crop canopy for winter oilseed rape

Following crops guidance
- Wheat, barley, oats, maize, or oilseed rape may be sown 120 days after application, all other crops should only be sown 12 months after application [2]
- Ploughing or thorough cultivation should be carried out before planting leguminous crops
- Do not attempt to plant peas, beans, other legumes, carrots, other umbelliferous crops, potatoes, lettuce, other Compositae, or any glasshouse or protected crops if treated crop remains have not fully decayed by the time of planting
- In the event of failure of an autumn treated crop only oilseed rape, wheat, barley, oats, maize or ryegrass may be sown in the spring and only after ploughing or thorough cultivation

Environmental safety
- Dangerous for the environment
- Toxic to aquatic organisms
- Take extreme care to avoid drift onto crops and non-target plants outside the target area
- Some pesticides pose a greater threat of contamination of water than others and clopyralid is one of these pesticides. Take special care when applying clopyralid near water and do not apply if heavy rain is forecast

Hazard classification and safety precautions
Hazard Dangerous for the environment
UN Number N/C
Operator protection A, C; U05a
Environmental protection E15a, E34 [1-3]; E38 [1, 2]; H411 [2]; H413 [1]
Storage and disposal D01, D02, D05, D07, D09a [1-3]; D12a [1, 2]

93   clopyralid + triclopyr

A perennial and woody weed herbicide for use in grassland
HRAC mode of action code: O + O

See also triclopyr

Products

1	Blaster Pro	Headland Amenity	60:240 g/l	EC	18074
2	Grazon Pro	Dow (Corteva)	60:240 g/l	EC	15785
3	Prevail	Dow (Corteva)	200:200 g/l	SL	17395
4	Thistlex	Dow (Corteva)	200:200 g/l	SL	16123

Uses
- Annual and perennial weeds in **grassland** [2]
- Annual dicotyledons in **game cover** *(off-label)* [4]

- Brambles in *amenity grassland* [1]; *grassland* [1, 2]
- Broom in *amenity grassland* [1]; *grassland* [1, 2]
- Creeping thistle in *grassland* [2-4]; *rotational grass* [3]
- Docks in *amenity grassland* [1]; *grassland* [1, 2]
- Gorse in *amenity grassland* [1]; *grassland* [1, 2]
- Mayweeds in *game cover* (off-label) [4]
- Perennial dicotyledons in *amenity grassland*, *grassland* [1]
- Stinging nettle in *amenity grassland*, *grassland* [1]
- Thistles in *amenity grassland*, *grassland* [1]; *game cover* (off-label) [4]

Extension of Authorisation for Minor Use (EAMUs)
- *game cover* 20142512 [4]

Approval information
- Clopyralid and triclopyr included in Annex I under EC Regulation 1107/2009

Efficacy guidance
- Must be applied to actively growing weeds
- Correct timing crucial for good control. Spray stinging nettle before flowering, docks in rosette stage in spring, creeping thistle before flower stems 15-20 cm high, brambles, broom and gorse in Jun-Aug
- Allow 2-3 wk regrowth after grazing or mowing before spraying perennial weeds
- Where there is a large reservoir of weed seed in the soil further treatment in the following yr may be needed
- [4] available as a twin pack with Doxstar as Pas.Tor for control of docks, nettles and thistles in grassland

Restrictions
- Maximum number of treatments 1 per yr
- Only use on permanent pasture or rotational grassland established for at least 1 yr
- Do not apply where clover is an important constituent of sward
- Do not roll or harrow within 10 d before or 7 d after spraying
- Do not cut grass for 21 d before or 28 d after spraying
- Do not use any treated plant material for composting or mulching, and do not use manure for composting from animals fed on treated crops
- Do not apply by hand-held rotary atomiser equipment
- Do not allow drift onto other crops, amenity plantings or gardens, ponds, lakes or water courses. All conifers, especially pine and larch, are very sensitive
- Maximum concentration must not exceed 60 mls product per 10 litres of water (6 ml product per litre of water) [1, 2]

Crop-specific information
- Latest use: 7 d before grazing or cutting grass
- Some transient yellowing of treated swards may occur but is quickly outgrown

Following crops guidance
- Residues in plant tissues which have not completely decayed may affect succeeding susceptible crops such as peas, beans, other legumes, carrots, parsnips, potatoes, tomatoes, lettuce, glasshouse and protected crops
- Do not plant susceptible autumn-sown crops (eg winter beans) in same year as treatment and allow at least 9 mth from treatment before planting a susceptible crop in the following yr
- Do not direct drill kale, swedes, turnips, grass or grass mixtures within 6 wk of spraying
- Do not spray after end Jul where susceptible crops are to be planted in the next spring

Environmental safety
- Dangerous for the environment
- Very toxic to aquatic organisms
- Keep livestock out of treated areas for at least 7 d after spraying and until foliage of any poisonous weeds such as ragwort or buttercup has died down and become unpalatable

---

SEE SECTION 3 FOR PRODUCTS ALSO REGISTERED

- Some pesticides pose a greater threat of contamination of water than others and clopyralid is one of these pesticides. Take special care when applying clopyralid near water and do not apply if heavy rain is forecast
- LERAP Category B

Hazard classification and safety precautions

Hazard Harmful, Flammable liquid and vapour [1, 2]; Irritant [1, 3, 4]; Dangerous for the environment [1-4]

Transport code 3 [1, 2]

Packaging group III [1, 2]

UN Number 1993 [1, 2]; N/C [3, 4]

Risk phrases H304, H315, H317, H319, H336 [1, 2]; H318, H373 [3, 4]; H335 [1-4]

Operator protection A, C [1-4]; H, M [1, 2]; U02a, U05a, U11, U20b [1-4]; U08, U14, U19a, U23b [1, 2]; U15 [1, 3, 4]; U23a [3, 4]

Environmental protection E07a, E15a, E23, E34, E38, H411 [1, 2]; E07c [3, 4] (7 days); E16a [1-4]; H410 [3, 4]

Consumer protection C01 [1, 2]

Storage and disposal D01, D02, D05, D09a, D10b, D12a

Medical advice M03 [3, 4]; M05b [1]

94   Coniothyrium minitans

A fungal parasite of sclerotia in soil
FRAC mode of action code: 44

Products

| Contans WG | Bayer CropScience | 1000 IU/mg | WG | 17985 |

Uses

- Sclerotinia in *all edible crops (outdoor)*, *all non-edible crops (outdoor)*

Approval information

- Coniothyrium minitans included in Annex 1 under EC Regulation 1107/2009

Efficacy guidance

- *C.minitans* is a soil acting biological fungicide with specific action against the resting bodies (sclerotia) of *Sclerotinia sclerotiorum* and *S.minor*
- Treat 3 mth before disease protection is required to allow time for the infective sclerotia in the soil to be reduced
- A post-harvest treatment of soil and debris prevents further contamination of the soil with sclerotia produced by the previous crop
- For best results the soil should be moist and the temperature between 12-20°C
- Application should be followed by soil incorporation into the surface layer to 10 cm with a rotovator or rotary harrow. Thorough spray coverage of the soil is essential to ensure uniform distribution
- If soil temperature drops below 0°C or rises above 27°C fungicidal activity is suspended but restarts when soil temperature returns within this range
- In glasshouses untreated areas at the margins should be covered by a film to avoid spread of spores from untreated sclerotia

Restrictions

- Maximum number of treatments 1 per crop or situation
- Do not plough or cultivate after treatment
- Do not mix with other pesticides, acids, alkalines or any product that attacks organic material
- Store product under cool dry conditions away from direct sunlight and away from heat sources
- Must only be applied to Sclerotinia susceptible crops

Crop-specific information

- Latest use: pre-planting
- HI: zero

Following crops guidance
- There are no restrictions on following crops and no waiting interval is specified

Hazard classification and safety precautions
   UN Number N/C
   Operator protection A, C, D; U05a, U19a
   Environmental protection E15b, E34
   Storage and disposal D01, D02, D12a
   Medical advice M03

95   copper oxychloride

A protectant copper fungicide and bactericide
FRAC mode of action code: M1

Products

Cuprokylt	Certis	87% w/w (copper)	WP	17079

Uses
- Dothistroma needle blight in **forest nurseries** (off-label)
- Downy mildew in **table grapes**, **wine grapes**

Extension of Authorisation for Minor Use (EAMUs)
- **forest nurseries** 20152416

Approval information
- Copper oxychloride included in Annex 1 under EC Regulation 1107/2009

Efficacy guidance
- Spray crops at high volume when foliage dry but avoid run off. Do not spray if rain expected soon
- Spray interval commonly 10-14 d but varies with crop, see label for details

Environmental safety
- Dangerous for the environment
- Very toxic to aquatic organisms
- Keep all livestock out of treated areas for at least 3 wks
- Buffer zone requirement 18 m in potatoes and ornamentals when using a horizontal sprayer, 50m when using broadcast air-assisted sprayer on crops up to 1.2m height
- LERAP Category B

Hazard classification and safety precautions
   Hazard Dangerous for the environment, Harmful if swallowed, Harmful if inhaled, Very toxic to aquatic organisms
   Transport code 9
   Packaging group III
   UN Number 3077
   Operator protection A, D, H; U20a
   Environmental protection E06a (3 wk); E13c, E16a, E34, E38, H410, H411; E17a (50 m)
   Storage and disposal D01, D09a, D10b

96   cos-oga

An oligosaccharidic elicitor for powdery mildew control in protected crops
FRAC mode of action code: Unclassified

Products

Fytosave	Gowan	12.5 g/l	SC	18433

Uses
- Downy mildew in **baby leaf crops** (off-label), **herbs (see appendix 6)** (off-label), **lettuce** (off-label), **ornamental plant production** (off-label), **protected baby leaf crops** (off-label),

---

protected herbs (see appendix 6) (off-label), *protected lettuce* (off-label), *protected ornamentals* (off-label), *table grapes* (off-label), *wine grapes* (off-label)

- Powdery mildew in *baby leaf crops* (off-label), *blackberries* (off-label), *herbs (see appendix 6)* (off-label), *hops* (off-label), *lettuce* (off-label), *loganberries* (off-label), *ornamental plant production* (off-label), *protected aubergines*, *protected baby leaf crops* (off-label), *protected blackberries* (off-label), *protected chilli peppers*, *protected courgettes*, *protected cucumbers*, *protected herbs (see appendix 6)* (off-label), *protected lettuce* (off-label), *protected loganberries* (off-label), *protected melons*, *protected ornamentals* (off-label), *protected peppers*, *protected pumpkins*, *protected raspberries* (off-label), *protected rubus hybrids* (off-label), *protected strawberries* (off-label), *protected summer squash*, *protected tomatoes*, *protected winter squash*, *raspberries* (off-label), *rubus hybrids* (off-label), *strawberries* (off-label), *table grapes* (off-label), *wine grapes* (off-label)

Extension of Authorisation for Minor Use (EAMUs)
- *baby leaf crops* 20191911
- *blackberries* 20191910
- *herbs (see appendix 6)* 20191911
- *hops* 20191910
- *lettuce* 20191911
- *loganberries* 20191910
- *ornamental plant production* 20191911
- *protected baby leaf crops* 20191911
- *protected blackberries* 20191910
- *protected herbs (see appendix 6)* 20191911
- *protected lettuce* 20191911
- *protected loganberries* 20191910
- *protected ornamentals* 20191911
- *protected raspberries* 20191910
- *protected rubus hybrids* 20191910
- *protected strawberries* 20191910
- *raspberries* 20191910
- *rubus hybrids* 20191910
- *strawberries* 20191910
- *table grapes* 20191911
- *wine grapes* 20191911

Approval information
- Cos-oga included in Annex I under EC Regulation 1107/2009

Efficacy guidance
- A minimum interval of 7 days must be observed between applications
- To minimise airborne environmental exposure, vents, doors and other openings must be closed during and after application until the applied product has fully settled

Restrictions
- Reasonable precautions must be taken to prevent access of birds, wild mammals and honey bees to treated crops

Hazard classification and safety precautions
UN Number N/C
Operator protection U05a
Storage and disposal D01, D12a

97   cyantraniliprole

An ingested and contact insecticide for insect pest control in brassica crops
IRAC mode of action code: 28

Products
Verimark 20 SC	FMC Agro	200 g/l	SC	18756

FOR FULL CONDITIONS OF USE ALWAYS READ THE PRODUCT LABEL

Uses
- Cabbage root fly in **broccoli**, **brussels sprouts**, **cabbages**, **cauliflowers**
- Strawberry blossom weevil in **protected strawberries** *(off-label)*, **strawberries** *(off-label)*
- Whitefly in **protected aubergines** *(off-label)*, **protected tomatoes** *(off-label)*

Extension of Authorisation for Minor Use (EAMUs)
- **protected aubergines** *20193129*
- **protected strawberries** *20192766*
- **protected tomatoes** *20193129*
- **strawberries** *20192766*

Approval information
- Cyantraniliprole is included in Annex 1 under EC Regulation 1107/2009

Efficacy guidance
- May be used as part of an Integrated Pest Management (IPM) programme

Restrictions
- To protect bees and pollinating insects avoid application to crops when in flower.
- Consult processor when the product is to be applied to crops grown for processing
- The maximum total dose of cyantraniliprole per crop must not be exceeded in any calendar year. Any land treated with cyantraniliprole at the maximum total dose must not be treated with any other cyantraniliprole containing products in the same calendar year, including either foliar applications in the growing crop or drench treatments to transplants applied pre-planting.
- Must not be applied in each calendar year after 31 August in brassica crops
- Should only be applied to crops grown in artificial media such as rockwool or perlite. Not to be used on crops grown in soil or organic media

Environmental safety
- Aquatic buffer zone 5 m

Hazard classification and safety precautions
Hazard Very toxic to aquatic organisms
Transport code 9
Packaging group III
UN Number 3082
Operator protection U05a, U09a, U19a, U20a
Environmental protection E12e, E15b, H410
Storage and disposal D01, D02, D09a, D12a, D22
Medical advice M03

98 cyazofamid

A cyanoimidazole sulfonamide protectant fungicide for potatoes
FRAC mode of action code: 21

Products

1	Linford	AgChem Access	400 g/l	SC	13824
2	Ranman Top	Belchim	160 g/l	SC	14753
3	Swallow	AgChem Access	400 g/l	SC	14078

Uses
- Blight in **potatoes** [1-3]
- Downy mildew in **aubergines** *(off-label)*, **courgettes** *(off-label)*, **cucumbers** *(off-label)*, **gherkins** *(off-label)*, **pumpkins** *(off-label)*, **summer squash** *(off-label)*, **tomatoes** *(off-label)*, **winter squash** *(off-label)* [2]
- Late blight in **aubergines** *(off-label)*, **courgettes** *(off-label)*, **cucumbers** *(off-label)*, **gherkins** *(off-label)*, **pumpkins** *(off-label)*, **summer squash** *(off-label)*, **tomatoes** *(off-label)*, **winter squash** *(off-label)* [2]

Extension of Authorisation for Minor Use (EAMUs)
- **aubergines** *20180489* [2]

SEE SECTION 3 FOR PRODUCTS ALSO REGISTERED

SECTION 2

- *courgettes* 20180489 [2]
- *cucumbers* 20180489 [2]
- *gherkins* 20180489 [2]
- *pumpkins* 20180489 [2]
- *summer squash* 20180489 [2]
- *tomatoes* 20180489 [2]
- *winter squash* 20180489 [2]

Approval information
- Cyazofamid included in Annex I under EC Regulation 1107/2009

Efficacy guidance
- Apply as a protectant treatment before blight enters the crop and repeat every 5 -10 d depending on severity of disease pressure
- Commence spray programme immediately the risk of blight in the locality occurs, usually when the crop meets along the rows
- Product must always be used with organosilicone adjuvant provided in the twin pack
- To minimise the chance of development of resistance no more than three applications should be made consecutively (out of a permissible total of six) in the blight control programme. For more information on Resistance Management see Section 5

Restrictions
- Maximum number of treatments 6 per crop (no more than three of which should be consecutive)
- Mixed product must not be allowed to stand overnight
- Consult processor before using on crops intended for processing

Crop-specific information
- HI 7 d

Environmental safety
- Dangerous for the environment
- Very toxic to aquatic organisms
- Do not empty into drains

Hazard classification and safety precautions
Hazard Harmful [1]; Harmful [3] (adjuvant); Irritant [2]; Dangerous for the environment [1-3]
Transport code 9
Packaging group III
UN Number 3082
Risk phrases H315 [2]; H319 [1]; R20, R36, R48 [3] (adjuvant); R41, R50, R53a [3]
Operator protection A, C; U02a, U05a, U11, U15 [1-3]; U19a [1]; U19a [3] (adjuvant); U20a [1, 3]; U20b [2]
Environmental protection E15a, E34 [1, 3]; E15b, H410 [2]; E19b, E38 [1-3]; H412 [1]
Storage and disposal D01, D02, D09a, D10c, D12a [1-3]; D05, D12b [2]
Medical advice M03

99   cycloxydim

A translocated post-emergence cyclohexanedione oxime herbicide for grass weed control
HRAC mode of action code: A

Products

Laser	BASF	200 g/l	EC	17339

Uses
- Agrostis spp. in *managed amenity turf* (off-label)
- Annual grasses in *angelica* (off-label), *baby leaf crops* (off-label), *beans without pods (fresh)* (off-label), *broad beans* (off-label), *broccoli* (off-label), *calabrese* (off-label), *caraway* (off-label), *carrots*, *cauliflowers*, *celeriac* (off-label), *celery leaves* (off-label), *christmas trees*, *combining peas*, *coriander* (off-label), *cress* (off-label), *dill* (off-label), *dwarf beans*, *early potatoes*, *edible podded peas* (off-label), *endives* (off-label), *fennel leaves* (off-label), *fodder beet*, *forest*,

SECTION 2

forest nurseries, herbs (see appendix 6) (off-label), horseradish (off-label), jerusalem artichokes (off-label), lamb's lettuce (off-label), land cress (off-label), leeks, lentils (off-label), lettuce (off-label), linseed, lovage (off-label), maincrop potatoes, mangels, ornamental plant production, ornamental plant production (off-label), purslane (off-label), red beet (off-label), rocket (off-label), runner beans (off-label), salad burnet (off-label), salad onions, salsify (off-label), soya beans (off-label), spinach (off-label), spinach beet (off-label), spring field beans, spring oilseed rape, strawberries, sugar beet, swedes, sweet ciceley (off-label), turnips (off-label), vining peas, watercress (off-label), winter field beans, winter oilseed rape

- Bent grasses in **managed amenity turf** *(off-label)*
- Black bent in **carrots, cauliflowers, combining peas, dwarf beans, early potatoes, fodder beet, leeks, linseed, maincrop potatoes, mangels, salad onions, spring field beans, spring oilseed rape, strawberries, sugar beet, swedes, vining peas, winter field beans, winter oilseed rape**
- Blackgrass in **carrots, cauliflowers, combining peas, dwarf beans, early potatoes, fodder beet, leeks, linseed, maincrop potatoes, mangels, salad onions, spring field beans, spring oilseed rape, strawberries, sugar beet, swedes, vining peas, winter field beans, winter oilseed rape**
- Couch in **carrots, cauliflowers, combining peas, dwarf beans, early potatoes, fodder beet, leeks, linseed, maincrop potatoes, mangels, salad onions, spring field beans, spring oilseed rape, strawberries, sugar beet, swedes, vining peas, winter field beans, winter oilseed rape**
- Creeping bent in **carrots, cauliflowers, combining peas, dwarf beans, early potatoes, fodder beet, leeks, linseed, maincrop potatoes, mangels, salad onions, spring field beans, spring oilseed rape, strawberries, sugar beet, swedes, vining peas, winter field beans, winter oilseed rape**
- Green cover in **land temporarily removed from production**
- Onion couch in **carrots, cauliflowers, combining peas, dwarf beans, early potatoes, fodder beet, leeks, linseed, maincrop potatoes, mangels, salad onions, spring field beans, spring oilseed rape, strawberries, sugar beet, swedes, vining peas, winter field beans, winter oilseed rape**
- Perennial grasses in **angelica** *(off-label)*, **beans without pods (fresh)** *(off-label)*, **broad beans** *(off-label)*, **caraway** *(off-label)*, **celeriac** *(off-label)*, **celery leaves** *(off-label)*, **christmas trees**, **coriander** *(off-label)*, **cress** *(off-label)*, **dill** *(off-label)*, **edible podded peas** *(off-label)*, **endives** *(off-label)*, **fennel leaves** *(off-label)*, **forest, forest nurseries, horseradish** *(off-label)*, **jerusalem artichokes** *(off-label)*, **lamb's lettuce** *(off-label)*, **land cress** *(off-label)*, **lentils** *(off-label)*, **lettuce** *(off-label)*, **lovage** *(off-label)*, **ornamental plant production, ornamental plant production** *(off-label)*, **purslane** *(off-label)*, **red beet** *(off-label)*, **rocket** *(off-label)*, **runner beans** *(off-label)*, **salad burnet** *(off-label)*, **salsify** *(off-label)*, **soya beans** *(off-label)*, **spinach** *(off-label)*, **spinach beet** *(off-label)*, **sweet ciceley** *(off-label)*, **turnips** *(off-label)*
- Volunteer cereals in **angelica** *(off-label)*, **baby leaf crops** *(off-label)*, **beans without pods (fresh)** *(off-label)*, **broad beans** *(off-label)*, **broccoli** *(off-label)*, **calabrese** *(off-label)*, **caraway** *(off-label)*, **carrots, cauliflowers, celery leaves** *(off-label)*, **combining peas, coriander** *(off-label)*, **cress** *(off-label)*, **dill** *(off-label)*, **dwarf beans, early potatoes, edible podded peas** *(off-label)*, **endives** *(off-label)*, **fennel leaves** *(off-label)*, **fodder beet, herbs (see appendix 6)** *(off-label)*, **lamb's lettuce** *(off-label)*, **land cress** *(off-label)*, **leeks, lentils** *(off-label)*, **lettuce** *(off-label)*, **linseed, lovage** *(off-label)*, **maincrop potatoes, mangels, purslane** *(off-label)*, **rocket** *(off-label)*, **runner beans** *(off-label)*, **salad burnet** *(off-label)*, **salad onions, soya beans** *(off-label)*, **spinach** *(off-label)*, **spinach beet** *(off-label)*, **spring field beans, spring oilseed rape, strawberries, sugar beet, swedes, sweet ciceley** *(off-label)*, **vining peas, watercress** *(off-label)*, **winter field beans, winter oilseed rape**
- Wild oats in **carrots, cauliflowers, combining peas, dwarf beans, early potatoes, fodder beet, leeks, linseed, maincrop potatoes, mangels, salad onions, spring field beans, spring oilseed rape, strawberries, sugar beet, swedes, vining peas, winter field beans, winter oilseed rape**

Extension of Authorisation for Minor Use (EAMUs)
- **angelica** *20171266*
- **baby leaf crops** *20171266*
- **beans without pods (fresh)** *20171265*

SEE SECTION 3 FOR PRODUCTS ALSO REGISTERED

- **broad beans** *20171265*
- **broccoli** *20171267*
- **calabrese** *20171267*
- **caraway** *20171266*
- **celeriac** *20170438*
- **celery leaves** *20171266*
- **coriander** *20171266*
- **cress** *20171266*
- **dill** *20171266*
- **edible podded peas** *20171265*
- **endives** *20171266*
- **fennel leaves** *20171266*
- **herbs (see appendix 6)** *20171266*
- **horseradish** *20170438*
- **jerusalem artichokes** *20170438*
- **lamb's lettuce** *20171266*
- **land cress** *20171266*
- **lentils** *20171265*
- **lettuce** *20171266*
- **lovage** *20171266*
- **managed amenity turf** *20171715*
- **ornamental plant production** *20170437*
- **purslane** *20171266*
- **red beet** *20170438*
- **rocket** *20171266*
- **runner beans** *20171265*
- **salad burnet** *20171266*
- **salsify** *20170438*
- **soya beans** *20171265*
- **spinach** *20171266*
- **spinach beet** *20171266*
- **sweet ciceley** *20171266*
- **turnips** *20170438*
- **watercress** *20171266*

Approval information
- Cycloxydim included in Annex 1 under EC Regulation 1107/2009

Efficacy guidance
- Best results achieved when weeds small and have not begun to compete with crop. Effectiveness reduced by drought, cool conditions or stress. Weeds emerging after application are not controlled
- Foliage death usually complete after 3-4 wk but longer under cool conditions, especially late treatments to winter oilseed rape
- Perennial grasses should have sufficient foliage to absorb spray and should not be cultivated for at least 14 d after treatment
- On established couch pre-planting cultivation recommended to fragment rhizomes and encourage uniform emergence
- Split applications to volunteer wheat and barley at GS 12-14 will often give adequate control in winter oilseed rape. See label for details
- Apply to dry foliage when rain not expected for at least 2 h
- Cycloxydim is an ACCase inhibitor herbicide. To avoid the build up of resistance do not apply products containing an ACCase inhibitor herbicide more than twice to any crop. In addition do not use any product containing cycloxydim in mixture or sequence with any other product containing the same ingredient
- Use these products as part of a resistance management strategy that includes cultural methods of control and does not use ACCase inhibitors as the sole chemical method of grass weed control

- Applying a second product containing an ACCase inhibitor to a crop will increase the risk of resistance development; only use a second ACCase inhibitor to control different weeds at a different timing
- Always follow WRAG guidelines for preventing and managing herbicide resistant weeds. See Section 5 for more information

Restrictions
- Maximum number of treatments 1 per crop or yr in most situations. See label
- Must be used with authorised adjuvant oil. See label
- Do not apply to crops damaged or stressed by adverse weather, pest or disease attack or other pesticide treatment
- Prevent drift onto other crops, especially cereals and grass

Crop-specific information
- HI cabbage, cauliflower, calabrese, salad onions 4 wk; peas, dwarf beans 5 wk; bulb onions, carrots, parsnips, strawberries 6 wk; sugar and fodder beet, leeks, mangels, potatoes, field beans, swedes, Brussels sprouts, winter field beans 8 wk; oilseed rape, soya beans, linseed 12 wk
- Recommended time of application varies with crop. See label for details
- On peas a crystal violet wax test should be done if leaf wax likely to have been affected by weather conditions or other chemical treatment. The wax test is essential if other products are to be sprayed before or after treatment
- May be used on ornamental bulbs when crop 5-10 cm tall. Product has been used on tulips, narcissi, hyacinths and irises but some subjects may be more sensitive and growers advised to check tolerance on small number of plants before treating the rest of the crop
- May be applied to land temporarily removed from production where the green cover is made up predominantly of tolerant crops listed on label. Use on industrial crops of linseed and oilseed rape on land temporarily removed from production also permitted.

Following crops guidance
- Guideline intervals for sowing succeeding crops after failed treated crop: field beans, peas, sugar beet, rape, kale, swedes, radish, white clover, lucerne 1 wk; dwarf French beans 4 wk; wheat, barley, maize 8 wk
- Oats should not be sown after failure of a treated crop

Environmental safety
- Dangerous for the environment
- Toxic to aquatic organisms
- Harmful to fish or other aquatic life. Do not contaminate surface waters or ditches with chemical or used container

Hazard classification and safety precautions
> **Hazard** Harmful, Dangerous for the environment
> **Transport code** 9
> **Packaging group** III
> **UN Number** 3082
> **Risk phrases** H304, H315, H318, H336, H361
> **Operator protection** A, C, H; U05a, U08, U20b
> **Environmental protection** E15a, E38, H411
> **Storage and disposal** D01, D02, D09a, D10c, D12a
> **Medical advice** M05b

100   Cydia pomonella GV

An baculovirus insecticide for codling moth control

Products

1	Carpovirusine	Arysta	1.0 x10^13 GV/l	SC	15243
2	Carpovirusine EVO 2	Arysta	1.0 x10^13 GV/l	SC	17565

SEE SECTION 3 FOR PRODUCTS ALSO REGISTERED

Uses
- Codling moth in *apples*, *pears*

Approval information
- Cydia pomonella included in Annex I under EC Regulation 1107/2009

Efficacy guidance
- Optimum results require a precise spraying schedule commencing soon after egg laying and before first hatch
- First generation of codling moth normally occurs in first 2 wk of Jun in UK
- Apply in sufficient water to achieve good coverage of whole tree
- Normally 3 applications at intervals of 8 sunny days are needed for each codling moth generation
- Fruit damage is reduced significantly in the first yr and the population of codling moth in the following year will be reduced

Restrictions
- Maximum number of treatments 3 per season
- Do not tank mix with copper or any products with a pH lower than 5 or higher than 8
- Do not use as the exclusive measure for codling moth control. Adopt an anti-resistance strategy. See Section 5 for more information
- Consult processor before use on crops for processing

Crop-specific information
- HI for apples and pears 1 day [1]
- Evidence of efficacy and crop safety on pears is limited

Environmental safety
- Product has no adverse effect on predatory insects and is fully compatible with IPM programmes

Hazard classification and safety precautions
UN Number N/C
Risk phrases H317 [2]
Operator protection A, D, H; U05a, U14, U16b, U19a
Environmental protection E15b, E34
Storage and disposal D01, D02, D09a, D10b, D12a
Medical advice M05a

101   cyflufenamid

An amidoxime fungicide for cereals
FRAC mode of action code: U6

Products

1 CF-50	Pan Agriculture	50 g/l	EW	18487
2 Clayton Midas	Clayton	50 g/l	EW	18362
3 Clayton Roulette	Clayton	50 g/l	EW	18017
4 Cosine	Certis	50 g/l	EW	16404
5 Cyflamid	Certis	50 g/l	EW	12403
6 Flinch	Pan Agriculture	50 g/l	EW	18464
7 Laquna	Pan Agriculture	50 g/l	EW	19232
8 Takumi SC	Certis	100 g/l	SC	16000
9 Vegas	Certis	50 g/l	EW	15238

Uses
- Disease control in *courgettes*, *cucumbers*, *melons*, *summer squash* [8]
- Powdery mildew in *apples*, *pears* [1, 4]; *courgettes*, *cucumbers*, *melons*, *ornamental plant production* (off-label), *protected tomatoes* (off-label), *pumpkins* (off-label), *strawberries* (off-label), *summer squash*, *winter squash* (off-label) [8]; *durum wheat*, *spring barley*, *spring oats*, *spring wheat*, *triticale*, *winter barley*, *winter oats*, *winter rye*, *winter wheat* [2, 3, 5-7, 9]; *grass seed crops* (off-label), *rye* (off-label) [5]; *wine grapes* (off-label) [4]

FOR FULL CONDITIONS OF USE ALWAYS READ THE PRODUCT LABEL

SECTION 2

Extension of Authorisation for Minor Use (EAMUs)
- *grass seed crops* 20060602 [5]
- *ornamental plant production* 20131294 [8]
- *protected tomatoes* 20140800 [8]
- *pumpkins* 20162915 [8]
- *rye* 20060602 [5]
- *strawberries* 20162055 [8]
- *wine grapes* 20170846 [4]
- *winter squash* 20162915 [8]

Approval information
- Cyflufenamid included in Annex 1 under EC Regulation 1107/2009
- Accepted by BBPA on malting barley

Efficacy guidance
- Best results obtained from treatment at the first visible signs of infection by powdery mildew
- Sustained disease pressure may require a second treatment
- Disease spectrum may be broadened by appropriate tank mixtures. See label
- Product is rainfast within 1 h
- Must be used as part of an Integrated Crop Management programme that includes alternating use, or mixture, with fungicides with a different mode of action effective against powdery mildew

Restrictions
- Maximum number of treatments 2 per crop on all recommended cereals
- Apply only in the spring
- Do not apply to crops under stress from drought, waterlogging, cold, pests or diseases, lime or nutrient deficiency or other factors affecting crop growth

Crop-specific information
- Latest use: before start of flowering of cereal crops

Following crops guidance
- No restrictions

Environmental safety
- Dangerous for the environment
- Very toxic to aquatic organisms
- Avoid drift onto ponds, waterways or ditches

Hazard classification and safety precautions
Hazard Harmful, Harmful if swallowed [1, 4]; Dangerous for the environment [1-9]
Transport code 9 [1-7, 9]
Packaging group III [1-7, 9]
UN Number 3082 [1-7, 9]; N/C [8]
Operator protection A; U05a
Environmental protection E15b, E34, H411 [1-9]; E17a [1, 4] (5 m); E38 [1, 4]
Storage and disposal D01, D02, D05, D10c, D12b
Medical advice M05b [1, 4]

102   cymoxanil

A cyanoacetamide oxime fungicide for potatoes
FRAC mode of action code: 27

See also cyazofamid + cymoxanil

Products

1	Chime	Harvest	45% w/w	WG	18557
2	Cymbal 45	Belchim	45% w/w	WG	16647
3	Danso Flow	Belchim	225 g/l	SC	17963
4	Dauphin 45	SFP Europe	45% w/w	WG	16434

SEE SECTION 3 FOR PRODUCTS ALSO REGISTERED

Products – continued

5 Drum	Belchim	45% w/w	WG	16750
6 Option	DuPont (Corteva)	60% w/w	WG	16959
7 Sacron WG	UPL Europe	45% w/w	WG	16433
8 Sipcam C50 WG	Sipcam	50% w/w	WG	16743

Uses

- Blight in *early potatoes* [8]; *potatoes* [1-8]
- Downy mildew in *hops* *(off-label)*, *wine grapes* *(off-label)* [6]
- Late blight in *potatoes* [4]

Extension of Authorisation for Minor Use (EAMUs)

- *hops* *20161101* [6]
- *wine grapes* *20160542* [6]

Approval information

- Cymoxanil included in Annex I under EC Regulation 1107/2009
- Accepted by BBPA for use on hops

Efficacy guidance

- Product to be used in mixture with specified mixture partners in order to combine systemic and protective activity. See labels for details
- Commence spray programme as soon as weather conditions favourable for disease development occur and before infection appears. At latest the first treatment should be made as the foliage meets along the rows
- Repeat treatments at 7-14 day intervals according to disease incidence and weather conditions
- Treat irrigated crops as soon as possible after irrigation and repeat at 10 d intervals

Restrictions

- Minimum spray interval 7 d
- Do not allow packs to become wet during storage and do not keep opened packs from one season to another
- Cymoxanil must be used in tank mixture with other fungicides
- The maximum concentration must not exceed 220 g of product per 300 litre water [7]

Crop-specific information

- HI for potatoes 1 d [6], 7 d [2, 5, 8], 14 d [7]
- Consult processor before treating potatoes grown for processing

Environmental safety

- Dangerous for the environment
- Very toxic to aquatic organisms
- LERAP Category B [4, 7]

Hazard classification and safety precautions

Hazard Harmful [1-4, 6-8]; Irritant [5]; Dangerous for the environment [1-7]; Harmful if swallowed [1, 2, 6-8]; Very toxic to aquatic organisms [6]

Transport code 9 [1-7]

Packaging group III [1-7]

UN Number 3077 [1, 2, 4-7]; 3082 [3]; N/C [8]

Risk phrases H314 [3]; H317 [1-3, 5-8]; H319 [5]; H361, H373 [1-3, 5, 7]; R22a, R43, R48, R51, R53a, R62, R63 [4]

Operator protection A [1-8]; C [3]; D [4, 7]; H [1-5, 7]; U05a [1-8]; U08 [6, 8]; U09a [1-5, 7]; U14 [1-5, 7, 8]; U19a [6]; U20a [1-4]; U20b [5-8]

Environmental protection E13a, E34 [6]; E13c [8]; E15a [1-5, 7]; E16a [4, 7]; E38 [1-7]; H410 [5, 6]; H411 [1-3, 7, 8]

Storage and disposal D01, D02, D09a, D11a [1-8]; D12a [1-7]

Medical advice M03 [6, 8]; M05a [1-4, 7]

FOR FULL CONDITIONS OF USE ALWAYS READ THE PRODUCT LABEL

103 cymoxanil + fluazinam

A fungicide mixture for blight control in potatoes
FRAC mode of action code: 27 + 29

Products

1	Grecale	Sipcam	200:300 g/l	SC	17002
2	Kunshi	Belchim	25:37.5% w/w	WG	16991
3	Tezuma	Belchim	25:37.5% w/w	WG	17396

Uses
- Blight in *potatoes*
- Late blight in *potatoes*
- Phytophthora in *potatoes*

Approval information
- Cymoxanil and fluazinam included in Annex 1 under EC Regulation 1107/2009

Restrictions
- Use as part of a programme with fungicides acting via a different mode of action to reduce the risk of resistance
- Use low drift spraying equipment up to 30m from the top of the bank of any surface water bodies [1-3]

Following crops guidance
- No restrictions on following crops

Environmental safety
- Buffer zone requirement 6 m [1-3]
- To protect non target insects/arthropods respect an unsprayed buffer zone of 5 m to non-crop land [1]
- Horizontal boom sprayers must be fitted with three star drift reduction technology for all uses

Hazard classification and safety precautions
Hazard Dangerous for the environment, Harmful if swallowed [1]; Very toxic to aquatic organisms [1-3]
Transport code 9
Packaging group III
UN Number 3077 [2, 3]; 3082 [1]
Risk phrases H317, H361, H373 [1-3]; H319 [2, 3]
Operator protection A, C, H [1]
Environmental protection H410

104 cymoxanil + fludioxonil + metalaxyl-M

A fungicide seed dressing for peas
FRAC mode of action code: 27 + 12 + 4

See also fludioxonil
    metalaxyl-M

Products

Wakil XL	Syngenta	10:5:17.5% w/w	WS	17217

Uses
- Damping off in *carrots* (off-label), *parsnips* (off-label)
- Downy mildew in *combining peas*, *vining peas*
- Pythium in *carrots* (off-label), *parsnips* (off-label)

SEE SECTION 3 FOR PRODUCTS ALSO REGISTERED

SECTION 2

Extension of Authorisation for Minor Use (EAMUs)
- *carrots* 20170896
- *parsnips* 20170896

Approval information
- Cymoxanil, fludioxonil and metalaxyl-M included in Annex I under EC Regulation 1107/2009

Efficacy guidance
- Apply through continuous flow seed treaters which should be calibrated before use

Restrictions
- Max number of treatments 1 per seed batch
- Ensure moisture content of treated seed satisfactory and store in a dry place
- Check calibration of seed drill with treated seed before drilling and sow as soon as possible after treatment
- Consult before using on crops for processing
- Treated seeds must not be sown between 29 September and 1 April

Crop-specific information
- Latest use: pre-drilling

Environmental safety
- Harmful to aquatic organisms
- Do not use treated seed as food or feed

Hazard classification and safety precautions
UN Number N/C
Risk phrases H361, H373
Operator protection A, D, H; U02a, U05a, U20b
Environmental protection E03, E15a, H411
Storage and disposal D01, D02, D05, D07, D09a, D11a, D12a
Treated seed S02, S04b, S05, S07

105   cymoxanil + mancozeb

A protectant and systemic fungicide for potato blight control
FRAC mode of action code: 27 + M3

See also mancozeb

Products

1	Curzate M WG	DuPont (Corteva)	4.5:68% w/w	WG	11901
2	Cymax	Belchim	4.5:68% w/w	WG	19045
3	Cymozeb	Belchim	4.5:68% w/w	WG	18085
4	Nautile DG	UPL Europe	5:68% w/w	WG	16653
5	Profilux	Belchim	4.5:68% w/w	WG	16125
6	Rhapsody	DuPont (Corteva)	4.5:68% w/w	WG	11958
7	Video	UPL Europe	5:68% w/w	WG	16685

Uses
- Blight in *potatoes* [1, 2, 4-7]
- Disease control in *hops in propagation* (off-label) [1]
- Late blight in *potatoes* [2, 3, 5]

Extension of Authorisation for Minor Use (EAMUs)
- *hops in propagation* 20162919 [1]

Approval information
- Cymoxanil and mancozeb included in Annex I under EC Regulation 1107/2009

Efficacy guidance
- Apply immediately after blight warning or as soon as local conditions dictate and repeat at 7-14 d intervals until haulm dies down or is burnt off

FOR FULL CONDITIONS OF USE ALWAYS READ THE PRODUCT LABEL

- Spray interval should not be more than 7-10 d in irrigated crops (see product label). Apply treatment after irrigation
- To minimise the likelihood of development of resistance these products should be used in a planned Resistance Management strategy. See Section 5 for more information

Restrictions
- Not specified for most products
- Do not apply at less than 7 d intervals
- Do not allow packs to become wet during storage
- Use low drift spraying equipment up to 30m from the top of the bank of any surface water bodies [4, 7]
- Horizontal boom sprayers must be fitted with three star drift reduction technology for all uses [4, 7]
- The maximum concentration must not exceed 2 kg of product per 300 litres water [7]

Crop-specific information
- Destroy and remove any haulm that remains after harvest of early varieties to reduce blight pressure on neighbouring maincrop potatoes

Following crops guidance
- HI 7 days [5]

Environmental safety
- Dangerous for the environment
- Very toxic to aquatic organisms
- Keep product away from fire or sparks
- Buffer zone requirement 6m [4, 7]
- LERAP Category B [1-6]

Hazard classification and safety precautions
Hazard Harmful [1-3, 5, 6]; Irritant [4, 7]; Dangerous for the environment, Very toxic to aquatic organisms [1-7]
Transport code 9
Packaging group III
UN Number 3077
Risk phrases H317, H361
Operator protection A [1-7]; C [1-3, 5, 6]; D [6]; H [4, 7]; U05a, U08, U14, U19a, U20b [1-7]; U11 [4, 7]
Environmental protection E13c [1-3, 5, 6]; E15a [1-3, 5-7]; E16a [1-6]; E38 [1-7]; H410 [1, 4, 6, 7]; H411 [2, 3, 5]
Storage and disposal D01, D02, D09a [1-7]; D06b, D12a [4, 7]; D11a [1-3, 5, 6]
Medical advice M05a [7]

106   cymoxanil + mandipropamid

A fungicide mixture for use in potatoes
FRAC mode of action code: 27 + 40

See also cymoxanil

Products

Carial Flex	Syngenta	18:25% w/w	WG	16629

Uses
- Blight in **potatoes**
- Late blight in **potatoes**

Approval information
- Cymoxanil and mandipropamid are included in Annex 1 under EC Regulation 1107/2009

Efficacy guidance
- Where possible, use an alternating strategy using fungicides from different mode of action groups.
- Where CAA fungicides are applied as a mixture (co-formulated or as a tank mix) up to six applications (or max. of 50% of the total number of applications) may be made per crop or season.
- No more than 3 applications of any CAA fungicide should be made consecutively

Restrictions
- Application may only be made between 1 May and 31 August

Hazard classification and safety precautions
   Hazard Harmful if swallowed
   Transport code 9
   Packaging group III
   UN Number 3077
   Risk phrases H361, H373
   Operator protection A, H; U05a, U08, U20b
   Environmental protection E15b, E34, E38, H410
   Storage and disposal D01, D02, D09a, D10c, D12a, D19
   Medical advice M03, M05a

107   cymoxanil + propamocarb

A cyanoacetamide oxime fungicide + a carbamate fungicide for potatoes
FRAC mode of action code: 27 + 28

Products

1	Axidor	Arysta	50:400 g/l	SC	16830
2	Proxanil	Arysta	50:400 g/l	SC	16664

Uses
- Blight in **potatoes**

Approval information
- Cymoxanil and propamocarb hydrochloride included in Annex I under EC Regulation 1107/2009

Restrictions
- Must be applied using 300-350 l water/ha

Hazard classification and safety precautions
   Hazard Irritant
   Transport code 8
   Packaging group III
   UN Number 3265
   Risk phrases H290, H317, H361, H411
   Operator protection A, H; U05a, U08, U14, U20b
   Environmental protection E15a, E38, H411
   Storage and disposal D01, D02, D09a, D10a

108   cymoxanil + zoxamide

A protectant and systemic fungicide for blight control in potatoes
FRAC mode of action code: 27 + 22

See also cymoxanil
        dimethomorph + zoxamide

Products

	Lieto	Sipcam	33:33% w/w	WG	16703

Uses
- Blight in **potatoes**

Approval information
- Cymoxanil and zoxamide included in Annex I under EC Regulation 1107/2009

Environmental safety
- Buffer zone requirement 6 m [1]

Hazard classification and safety precautions
Hazard Harmful, Dangerous for the environment, Harmful if swallowed
Transport code 9
Packaging group III
UN Number 3077
Risk phrases H317
Operator protection A, C, H; U05a, U11, U14, U15, U19a
Environmental protection E38, H410
Storage and disposal D01, D02, D12a, D12b
Medical advice M05a

109　cypermethrin

A contact and stomach acting pyrethroid insecticide
IRAC mode of action code: 3

Products

1	Cyper 500	Nurture	500 g/l	EC	18670
2	Cythrin 500 EC	Arysta	500 g/l	EC	16993
3	Cythrin Max	Arysta	500 g/l	EC	17138
4	Forester	Arysta	100 g/l	EW	13164
5	Signal 300 ES	Arysta	300 g/l	ES	15949
6	Talisma EC	Arysta	80 g/l	EC	16541
7	Talisma UL	Arysta	20 g/l	UL	16542

Uses
- Aphids in **asparagus, broccoli, brussels sprouts, cabbages, calabrese, carrots, cauliflowers, celeriac, dwarf beans, edible podded peas, forage maize, french beans, grain maize, horseradish, linseed, lupins, mustard, ornamental plant production, parsley root, parsnips, radishes, runner beans, salsify, spelt, spring barley** (autumn sown), **spring oats, spring rye, spring wheat** (autumn sown), **triticale, vining peas, winter barley, winter oats, winter rye, winter wheat** [1-3]
- Asparagus beetle in **asparagus** (off-label) [3]
- Barley yellow dwarf virus vectors in **spring barley** (autumn sown), **spring oats, spring rye, spring wheat** (autumn sown), **triticale, winter barley, winter oats, winter rye, winter wheat** [1-3]
- Beetles in **cut logs** [4]
- Bladder pod midge in **spring oilseed rape, winter oilseed rape** [1-3]
- Cabbage stem flea beetle in **winter oilseed rape** [1-3]
- Carrot fly in **carrots, celeriac, horseradish, parsley root, parsnips** [1-3]
- Caterpillars in **broccoli, brussels sprouts, cabbages, calabrese, cauliflowers** [1-3]
- Cutworms in **fodder beet, mangels, potatoes, red beet, sugar beet, swedes, turnips** [1-3]
- Frit fly in **spring barley** (qualified recommendation), **spring wheat** (qualified recommendation), **winter barley** (qualified recommendation), **winter wheat** (qualified recommendation) [5]
- Insect pests in **asparagus, carrots, celeriac, dwarf beans, edible podded peas, french beans, grain maize, horseradish, linseed, lupins, mustard, ornamental plant production, parsley root, parsnips, radishes, runner beans, salsify** [1-3]; **crop handling & storage structures, spring rye, winter oats, winter rye** [6]; **rice, spring barley, spring oats, spring wheat, triticale, winter barley, winter wheat** [6, 7]; **rye** [7]; **spelt** [1-3, 6, 7]
- Large pine weevil in **forest** [4]

SEE SECTION 3 FOR PRODUCTS ALSO REGISTERED

- Pea and bean weevil in **combining peas**, **spring field beans**, **vining peas**, **winter field beans** [1-3]
- Pea moth in **vining peas** [1-3]
- Pollen beetle in **spring oilseed rape**, **winter oilseed rape** [1-3]
- Rape winter stem weevil in **winter oilseed rape** [1-3]
- Seed weevil in **spring oilseed rape**, **winter oilseed rape** [1-3]
- Wheat bulb fly in **spring barley** (reduction), **spring wheat** (reduction), **winter barley** (reduction), **winter wheat** (reduction) [5]
- Whitefly in **broccoli**, **brussels sprouts**, **cabbages**, **calabrese**, **cauliflowers** [1-3]
- Yellow cereal fly in **spring barley** (autumn sown), **spring oats**, **spring rye**, **spring wheat** (autumn sown), **triticale**, **winter barley**, **winter oats**, **winter rye**, **winter wheat** [1-3]

Extension of Authorisation for Minor Use (EAMUs)
- **asparagus** 20170903 [3]

Approval information
- Cypermethrin included in Annex I under EC Regulation 1107/2009
- Accepted by BBPA for use on malting barley and hops

Efficacy guidance
- Products combine rapid action, good persistence, and high activity on Lepidoptera.
- As effect is mainly via contact, good coverage is essential for effective action. Spray volume should be increased on dense crops
- A repeat spray after 10-14 d is needed for some pests of outdoor crops
- Rates and timing of sprays vary with crop and pest. See label for details

Restrictions
- Maximum number of treatments varies with crop and product. See label or approval notice for details [4]
- Horizontal boom sprayers must be fitted with three star drift reduction technology for all uses
- Apply by knapsack or hand-held sprayer [4]
- Some new products advise using low drift spraying equipment up to 30m from the top of the bank of any surface water bodies - check label before use.
- Must only be applied to cereal seed sown in the autumn/winter [5]

Crop-specific information
- HI vining peas 7 d; other crops 0 d
- Test spray sample of new or unusual ornamentals or trees before committing whole batches [4]
- Post-planting forestry applications should be made before damage is seen or at the onset of damage during the first 2 yrs after transplanting [4]

Environmental safety
- Dangerous for the environment
- Very toxic to aquatic organisms
- High risk to bees. Do not apply to crops in flower, or to those in which bees are actively foraging, except as directed. Do not apply when flowering weeds are present
- Give local beekeepers warning when using in flowering crops
- Do not spray cereals after 31 Mar within 6 m of the edge of the growing crop [4]
- Buffer zone requirement 18 m [2, 3]
- To protect non target insects/arthropods respect an unsprayed buffer zone of 5 m to non-crop land [2]
- Horizontal boom sprayers must be fitted with three star drift reduction technology for all uses [2]
- Use low drift spraying equipment up to 30m from the top of the bank of any surface water bodies [2]
- Broadcast air-assisted LERAP [3] (18 m); LERAP Category A [1-3]

Hazard classification and safety precautions
Hazard Harmful, Dangerous for the environment [1-5]; Flammable [1-3]; Flammable liquid and vapour, Harmful if inhaled [3]; Harmful if swallowed [1, 2, 4-6]; Very toxic to aquatic organisms [6, 7]; Toxic to aquatic organisms [5]

Transport code 3 [1-3]; 9 [4-7]

Packaging group III

FOR FULL CONDITIONS OF USE ALWAYS READ THE PRODUCT LABEL

UN Number 1993 [1-3]; 3082 [4-7]
Risk phrases H304, H318 [3]; H315 [1-4]; H317 [1, 2, 4, 6]; H320, H373 [1, 2]; H335 [1-3, 5]; H336 [1-3]
Operator protection A, H [1-7]; C [6, 7]; D [5]; M [4]; U02a, U04a, U10, U16b [4]; U05a [1-7]; U08, U20b [1-3, 5]; U14 [4, 5]; U19a [1-4, 6, 7]
Environmental protection E12c, E12f, H410 [1-7]; E13a, E16c, E16d, E16h [1-3]; E15a [4]; E15b [5]; E17b [3] (18 m); E34 [1-5]; E38 [4, 5]
Storage and disposal D01, D02 [1-7]; D05, D12a [1-4, 6]; D09a [1-6]; D09b [7]; D10a [5]; D10b [1-4]
Treated seed S01, S02, S03, S04b, S05, S06a, S07 [5]
Medical advice M03 [4, 5]; M05b [1-3]

110   cypermethrin + tetramethrin

A pyrethroid insecticide mixture for control of crawling and flying insects in a variety of situations
IRAC mode of action code: 3

See also cypermethrin

Products

Vazor Cypermax Plus	Killgerm	10% + 5% w/w	ME	H9988

Uses
* Ants in *agricultural premises*, *manure heaps*, *poultry houses*, *refuse tips*
* Bedbugs in *agricultural premises*, *manure heaps*, *poultry houses*, *refuse tips*
* Cockroaches in *agricultural premises*, *manure heaps*, *poultry houses*, *refuse tips*
* Mosquitoes in *agricultural premises*, *manure heaps*, *poultry houses*, *refuse tips*
* Moths in *agricultural premises*, *manure heaps*, *poultry houses*, *refuse tips*
* Silverfish in *agricultural premises*, *manure heaps*, *poultry houses*, *refuse tips*
* Wasps in *agricultural premises*, *manure heaps*, *poultry houses*, *refuse tips*

Approval information
* Cypermethrin is included but tetramethrin not included in Annex 1 under EC Regulation 1107/2009

Hazard classification and safety precautions
Transport code 9
Packaging group III
UN Number 3082
Operator protection A, C, D, H; U20b
Environmental protection E05c, E12b, H410
Storage and disposal D09a, D12a

111   cyproconazole + trifloxystrobin

A conazole and strobilurin fungicide mixture for disease control in sugar beet
FRAC mode of action code: 3 + 11

See also trifloxystrobin

Products

1	Clayton Cast	Clayton	160:375 g/l	SC	17727
2	Escolta	Bayer CropScience	160:375 g/l	SC	17397

Uses
* Cercospora leaf spot in *fodder beet*, *sugar beet* [1, 2]; *red beet* (off-label) [2]
* Powdery mildew in *fodder beet*, *sugar beet* [1, 2]; *red beet* (off-label) [2]
* Ramularia leaf spots in *fodder beet*, *sugar beet* [1, 2]; *red beet* (off-label) [2]
* Rust in *fodder beet*, *sugar beet* [1, 2]; *red beet* (off-label) [2]

SEE SECTION 3 FOR PRODUCTS ALSO REGISTERED

SECTION 2

Extension of Authorisation for Minor Use (EAMUs)
- *red beet* 20161089 [2]

Approval information
- Cyproconazole and trifloxystrobin included in Annex I under EC Regulation 1107/2009

Efficacy guidance
- Best results obtained from treatment at early stages of disease development. Further treatment may be needed if disease attack is prolonged
- Trifloxystrobin is a member of the QoI cross resistance group. Product should be used preventatively and not relied on for its curative potential
- Use product as part of an Integrated Crop Management strategy incorporating other methods of control, including where appropriate other fungicides with a different mode of action.
- Strains of barley powdery mildew resistant to QoIs are common in the UK

Restrictions
- Maximum total dose per crop equivalent to two full dose treatments
- First application not before BBCH 31 (10% ground cover); second application not before BBCH 40 (full canopy & start of harvestable root development); maintain at least 21 day interval between sprays.

Crop-specific information
- HI 35 d

Environmental safety
- Dangerous for the environment
- Very toxic to aquatic organisms
- LERAP Category B

Hazard classification and safety precautions
Hazard Harmful, Dangerous for the environment, Very toxic to aquatic organisms
Transport code 9
Packaging group III
UN Number 3082
Risk phrases H361
Operator protection A; U02a, U05a, U09a, U19a, U20a
Environmental protection E15b, E16a, E38, H410
Storage and disposal D01, D02, D05, D09a, D10b, D12a
Medical advice M03

112   cyprodinil

An anilinopyrimidine systemic broad spectrum fungicide for cereals
FRAC mode of action code: 9

See also cyproconazole + cyprodinil

Products
1	Coracle	AgChem Access	300 g/l	EC	15866
2	Kayak	Syngenta	300 g/l	EC	14847

Uses
- Disease control in **forest nurseries** *(off-label)* [2]
- Eyespot in **spring barley**, **winter barley** [1, 2]
- Net blotch in **spring barley**, **winter barley** [1, 2]
- Powdery mildew in **spring barley**, **winter barley** [1, 2]
- Rhynchosporium in **spring barley**, **winter barley** [1, 2]

Extension of Authorisation for Minor Use (EAMUs)
- **forest nurseries** 20122056 [2]

FOR FULL CONDITIONS OF USE ALWAYS READ THE PRODUCT LABEL

Approval information
- Cyprodinil included in Annex I under EC Regulation 1107/2009
- Accepted by BBPA for use on malting barley

Efficacy guidance
- Best results obtained from treatment at early stages of disease development
- For best control of eyespot spray before or during the period of stem extension in spring. Control may be reduced if very dry conditions follow treatment

Restrictions
- Maximum number of treatments 2 per crop on wheat and barley [2]

Crop-specific information
- Latest use: before first spikelet of inflorescence just visible stage for barley, before early milk stage for wheat; up to and including first awns visible stage (GS49) in barley [2]

Environmental safety
- Dangerous for the environment
- Very toxic to aquatic organisms
- LERAP Category B

Hazard classification and safety precautions
Hazard Irritant, Dangerous for the environment, Very toxic to aquatic organisms
Transport code 9
Packaging group III
UN Number 3082
Risk phrases H317
Operator protection A, H; U05a, U09a, U20a
Environmental protection E15b, E16a, E34, E38, H410
Storage and disposal D01, D02, D05, D07, D09a, D10c, D12a

113 cyprodinil + fludioxonil

A broad-spectrum fungicide mixture for fruit crops, legumes, some vegetables, forest nurseries and ornamentals
FRAC mode of action code: 9 + 12

See also fludioxonil

Products

1	Clayton Gear	Clayton	37.5:25% w/w	WG	17169
2	Ludo	Pan Agriculture	37.5:25 % w/w	WG	18314
3	Modif	Life Scientific	37.5:25 % w/w	WG	18583
4	Shift	Clayton	37.5:25 % w/w	WG	18881
5	Switch	Syngenta	37.5:25 % w/w	WG	15129

Uses
- Alternaria in *apples*, *pears*, *quinces* [1-5]; *brassica leaves and sprouts* (off-label), *broccoli* (off-label), *brussels sprouts* (off-label), *cabbages* (off-label), *calabrese* (off-label), *cauliflowers* (off-label), *celery (outdoor)* (off-label), *collards* (off-label), *cress* (off-label), *endives* (off-label), *herbs (see appendix 6)* (off-label), *kale* (off-label), *lamb's lettuce* (off-label), *lettuce* (off-label), *protected baby leaf crops* (off-label), *protected cress* (off-label), *protected endives* (off-label), *protected herbs (see appendix 6)* (off-label), *protected lamb's lettuce* (off-label), *protected lettuce* (off-label), *protected radishes* (off-label), *protected red mustard* (off-label), *protected rocket* (off-label), *protected sorrel* (off-label), *protected spinach* (off-label), *protected spinach beet* (off-label), *red mustard* (off-label), *rocket* (off-label), *sorrel* (off-label), *spinach* (off-label), *spinach beet* (off-label), *spring cabbage* (off-label) [5]; *crab apples* [1, 2, 4, 5]
- Alternaria blight in *carrots (moderate control)* [1, 2, 4, 5]
- Anthracnose in *protected blueberry* (off-label) [5]
- Ascochyta in *edible podded peas* [3]; *mange-tout peas, sugar snap peas, vining peas* [1-5]
- Basal stem rot in *ornamental bulbs* (off-label) [5]

- Black spot in **protected strawberries** *(qualified minor use)* [1, 2, 4, 5]; **strawberries** *(qualified minor use)* [1-5]
- Blossom wilt in **apricots** *(off-label)*, **cherries** *(off-label)*, **peaches** *(off-label)*, **plums** *(off-label)* [5]
- Botrytis in **apricots** *(off-label)*, **asparagus** *(off-label)*, **brassica leaves and sprouts** *(off-label)*, **broccoli** *(off-label)*, **brussels sprouts** *(off-label)*, **bulb onions** *(off-label)*, **cabbages** *(off-label)*, **calabrese** *(off-label)*, **cauliflowers** *(off-label)*, **cherries** *(off-label)*, **chicory** *(off-label)*, **collards** *(off-label)*, **cress** *(off-label)*, **endives** *(off-label)*, **garlic** *(off-label)*, **grapevines** *(off-label)*, **herbs (see appendix 6)** *(off-label)*, **kale** *(off-label)*, **lamb's lettuce** *(off-label)*, **lettuce** *(off-label)*, **lupins** *(off-label)*, **peaches** *(off-label)*, **plums** *(off-label)*, **protected asparagus** *(off-label)*, **protected aubergines** *(off-label)*, **protected baby leaf crops** *(off-label)*, **protected blueberry** *(off-label)*, **protected courgettes** *(off-label)*, **protected cress** *(off-label)*, **protected cucumbers** *(off-label)*, **protected endives** *(off-label)*, **protected gherkins** *(off-label)*, **protected herbs (see appendix 6)** *(off-label)*, **protected lamb's lettuce** *(off-label)*, **protected lettuce** *(off-label)*, **protected peppers** *(off-label)*, **protected pumpkins** *(off-label)*, **protected red mustard** *(off-label)*, **protected rocket** *(off-label)*, **protected sorrel** *(off-label)*, **protected spinach** *(off-label)*, **protected spinach beet** *(off-label)*, **protected summer squash** *(off-label)*, **protected tomatoes** *(off-label)*, **pumpkins** *(off-label)*, **red mustard** *(off-label)*, **rocket** *(off-label)*, **salad onions** *(off-label)*, **shallots** *(off-label)*, **sorrel** *(off-label)*, **spinach** *(off-label)*, **spinach beet** *(off-label)*, **spring cabbage** *(off-label)* [5]; **broad beans** *(moderate control)*, **combining peas** *(moderate control)*, **french beans (dry-harvested)** *(moderate control)*, **mange-tout peas** *(moderate control)*, **ornamental plant production**, **protected ornamentals**, **runner beans** *(moderate control)*, **strawberries**, **sugar snap peas** *(moderate control)*, **vining peas** *(moderate control)* [1-5]; **broad beans (dry-harvested)** *(moderate control)*, **carrots** *(moderate control)*, **celeriac** *(moderate control)*, **forest nurseries**, **green beans** *(moderate control)*, **protected forest nurseries**, **protected strawberries** [1, 2, 4, 5]; **dwarf beans** *(moderate control)*, **edible podded peas** *(moderate control)*, **spring field beans** *(moderate control)*, **winter field beans** *(moderate control)* [3]
- Botrytis fruit rot in **apples**, **blackberries**, **pears**, **quinces**, **raspberries** [1-5]; **crab apples** [1, 2, 4, 5]
- Botrytis root rot in **bilberries** *(qualified minor use recommendation)*, **blackcurrants** *(qualified minor use recommendation)*, **blueberries** *(qualified minor use recommendation)*, **cranberries** *(qualified minor use recommendation)*, **gooseberries** *(qualified minor use recommendation)*, **redcurrants** *(qualified minor use recommendation)* [1-5]; **whitecurrants** *(qualified minor use recommendation)* [1, 2, 4, 5]
- Fusarium in **protected cucumbers** *(off-label)* [5]
- Fusarium bulb rot in **ornamental bulbs** *(off-label)* [5]
- Fusarium diseases in **apples**, **pears**, **quinces** [1-5]; **crab apples** [1, 2, 4, 5]
- Gloeosporium rot in **apples**, **pears**, **quinces** [1-5]; **crab apples** [1, 2, 4, 5]
- Leaf spot in **celery (outdoor)** *(off-label)*, **protected radishes** *(off-label)* [5]
- Mycosphaerella in **edible podded peas** [3]; **mange-tout peas**, **sugar snap peas**, **vining peas** [1-5]
- Penicillium rot in **apples**, **pears**, **quinces** [1-5]; **crab apples** [1, 2, 4, 5]
- Phoma in **brassica leaves and sprouts** *(off-label)*, **broccoli** *(off-label)*, **brussels sprouts** *(off-label)*, **cabbages** *(off-label)*, **calabrese** *(off-label)*, **cauliflowers** *(off-label)*, **collards** *(off-label)*, **cress** *(off-label)*, **endives** *(off-label)*, **herbs (see appendix 6)** *(off-label)*, **kale** *(off-label)*, **lamb's lettuce** *(off-label)*, **lettuce** *(off-label)*, **protected baby leaf crops** *(off-label)*, **protected cress** *(off-label)*, **protected endives** *(off-label)*, **protected herbs (see appendix 6)** *(off-label)*, **protected lamb's lettuce** *(off-label)*, **protected lettuce** *(off-label)*, **protected red mustard** *(off-label)*, **protected rocket** *(off-label)*, **protected sorrel** *(off-label)*, **protected spinach** *(off-label)*, **protected spinach beet** *(off-label)*, **red mustard** *(off-label)*, **rocket** *(off-label)*, **sorrel** *(off-label)*, **spinach** *(off-label)*, **spinach beet** *(off-label)*, **spring cabbage** *(off-label)* [5]
- Rhizoctonia in **brassica leaves and sprouts** *(off-label)*, **broccoli** *(off-label)*, **brussels sprouts** *(off-label)*, **cabbages** *(off-label)*, **calabrese** *(off-label)*, **cauliflowers** *(off-label)*, **collards** *(off-label)*, **cress** *(off-label)*, **endives** *(off-label)*, **herbs (see appendix 6)** *(off-label)*, **kale** *(off-label)*, **lamb's lettuce** *(off-label)*, **lettuce** *(off-label)*, **protected baby leaf crops** *(off-label)*, **protected cress** *(off-label)*, **protected endives** *(off-label)*, **protected herbs (see appendix 6)** *(off-label)*, **protected lamb's lettuce** *(off-label)*, **protected lettuce** *(off-label)*, **protected red mustard** *(off-label)*, **protected rocket** *(off-label)*, **protected sorrel** *(off-label)*, **protected spinach** *(off-label)*,

protected spinach beet *(off-label)*, *red mustard* *(off-label)*, *rocket* *(off-label)*, *sorrel* *(off-label)*, *spinach* *(off-label)*, *spinach beet* *(off-label)*, *spring cabbage* *(off-label)* [5]

- Sclerotinia in *beetroot* *(off-label)*, *celery (outdoor)* *(off-label)*, *ginger* *(off-label)*, *herbs for medicinal uses (see appendix 6)* *(off-label)*, *horseradish* *(off-label)*, *parsley root* *(off-label)*, *parsnips* *(off-label)*, *protected radishes* *(off-label)*, *salsify* *(off-label)*, *spice roots* *(off-label)*, *turmeric* *(off-label)* [5]; *broad beans*, *combining peas*, *french beans (dry-harvested)*, *mange-tout peas*, *runner beans*, *sugar snap peas*, *vining peas* [1-5]; *broad beans (dry-harvested)*, *carrots* *(moderate control)*, *green beans* [1, 2, 4, 5]; *dwarf beans*, *edible podded peas*, *spring field beans*, *winter field beans* [3]

Extension of Authorisation for Minor Use (EAMUs)

- *apricots* *20103092* [5]
- *asparagus* *20103173* [5]
- *beetroot* *20103087* [5]
- *brassica leaves and sprouts* *20150085* [5]
- *broccoli* *20150085* [5]
- *brussels sprouts* *20150085* [5]
- *bulb onions* *20122285* [5]
- *cabbages* *20150085* [5]
- *calabrese* *20150085* [5]
- *cauliflowers* *20150085* [5]
- *celery (outdoor)* *20150865* [5]
- *cherries* *20103092* [5]
- *chicory* *20122285* [5]
- *collards* *20150085* [5]
- *cress* *20150085* [5]
- *endives* *20150085* [5]
- *garlic* *20122285* [5]
- *ginger* *20103087* [5]
- *grapevines* *20112098* [5]
- *herbs (see appendix 6)* *20150085* [5]
- *herbs for medicinal uses (see appendix 6)* *20103087* [5]
- *horseradish* *20103087* [5]
- *kale* *20150085* [5]
- *lamb's lettuce* *20150085* [5]
- *lettuce* *20150085* [5]
- *lupins* *20103170* [5]
- *ornamental bulbs* *20152274* [5]
- *parsley root* *20103087* [5]
- *parsnips* *20103087* [5]
- *peaches* *20103092* [5]
- *plums* *20103092* [5]
- *protected asparagus* *20170202* [5]
- *protected aubergines* *20103172* [5]
- *protected baby leaf crops* *20150085* [5]
- *protected blueberry* *20160648* [5]
- *protected courgettes* *20170202* [5]
- *protected cress* *20150085* [5]
- *protected cucumbers* *20103171* [5]
- *protected endives* *20150085* [5]
- *protected gherkins* *20170202* [5]
- *protected herbs (see appendix 6)* *20150085* [5]
- *protected lamb's lettuce* *20150085* [5]
- *protected lettuce* *20150085* [5]
- *protected peppers* *20103172* [5]
- *protected pumpkins* *20170202* [5]
- *protected radishes* *20150865* [5]
- *protected red mustard* *20150085* [5]
- *protected rocket* *20150085* [5]

SEE SECTION 3 FOR PRODUCTS ALSO REGISTERED

SECTION 2

- **protected sorrel** *20150085* [5]
- **protected spinach** *20150085* [5]
- **protected spinach beet** *20150085* [5]
- **protected summer squash** *20170202* [5]
- **protected tomatoes** *20110302* [5]
- **pumpkins** *20170202* [5]
- **red mustard** *20150085* [5]
- **rocket** *20150085* [5]
- **salad onions** *20103174* [5]
- **salsify** *20103087* [5]
- **shallots** *20122285* [5]
- **sorrel** *20150085* [5]
- **spice roots** *20103087* [5]
- **spinach** *20150085* [5]
- **spinach beet** *20150085* [5]
- **spring cabbage** *20150085* [5]
- **turmeric** *20103087* [5]

Approval information
- Cyprodinil and fludioxonil included in Annex I under EC Regulation 1107/2009

Efficacy guidance
- Best results obtained from application at the earliest stage of disease development or as a protective treatment following disease risk assessment
- Subsequent treatments may follow after a minimum of 10 d if disease pressure remains high
- To minimise the likelihood of development of resistance product should be used in a planned Resistance Management strategy. See Section 5 for more information

Restrictions
- Maximum number of treatments 2 per crop
- Consult processor before treating any crop for processing
- Spray equipment must only be used where the operator's normal working position is within a closed cab on a tractor or on a self-propelled sprayer when making air-assisted applications to top fruit.

Crop-specific information
- HI broad beans, green beans, mange-tout peas, runner beans, sugar snap peas, vining peas 14 d; protected strawberries, strawberries 3 d
- First signs of disease infection likely to be seen from early flowering for all crops
- Ensure peas are free from any stress before treatment. If necessary check wax with a crystal violet test

Environmental safety
- Dangerous for the environment
- Very toxic to aquatic organisms
- Do not allow direct spray from broadcast air-assisted sprayers to fall within 10 m of the top of the bank of any static or flowing waterbody when treating blackberry, raspberry, blueberry, bilberry, cranberry, redcurrant , whitecurrant, blackcurrant, gooseberry, ornamental plant production and forest nursery; increase this to 30 m when treating apple, crab apple, pear and quince and not within 5 m of a ditch which is dry at the time of application
- LERAP Category B

Hazard classification and safety precautions
Hazard Dangerous for the environment [1-5]; Very toxic to aquatic organisms [1-3, 5]
Transport code 9
Packaging group III
UN Number 3077
Risk phrases H317
Operator protection A, H; U05a, U20a
Environmental protection E15b, E16a, E16b, E38, H410 [1-5]; E17a [3] (30 m)
Storage and disposal D01, D02, D05, D07, D09a, D10c, D12a

FOR FULL CONDITIONS OF USE ALWAYS READ THE PRODUCT LABEL

114   cyprodinil + isopyrazam

A fungicide mixture for disease control in barley
FRAC mode of action code: 7 + 9

See also cyprodinil

Products

1	Bontima	Syngenta	187.5:62.5 g/l	EC	17188
2	Bontima	Adama	187.5:62.5 g/l	EC	18566
3	Cebara	Syngenta	187.5:62.5 g/l	EC	17286
4	Cebara	Adama	187.5:62.5 g/l	EC	18569
5	Concorde	Syngenta	150:62.5 g/l	EC	17248
6	Concorde	Adama	150:62.5 g/l	EC	18570

Uses

- Brown rust in *spring barley*, *winter barley*
- Net blotch in *spring barley*, *winter barley*
- Powdery mildew in *spring barley*, *winter barley*
- Ramularia leaf spots in *spring barley*, *winter barley*
- Rhynchosporium in *spring barley*, *winter barley*

Approval information

- Cyprodinil and isopyrazam are included in Annex 1 under EC Regulation 1107/2009
- Accepted by BBPA for use on malting barley before ear emergence only

Restrictions

- Isopyrazam is an SDH respiration inhibitor; Do not apply more than two foliar applications of products containing an SDH inhibitor to any cereal crop

Crop-specific information

- There are no restrictions on succeeding crops in a normal rotation

Environmental safety

- LERAP Category B

Hazard classification and safety precautions

Hazard Harmful, Dangerous for the environment, Harmful if inhaled [1-6]; Very toxic to aquatic organisms [5, 6]
Transport code 9
Packaging group III
UN Number 3082
Risk phrases H304, H361 [1-6]; H351 [5, 6]
Operator protection A, C, H; U05a, U09a, U20b [1-6]; U11 [1-4]
Environmental protection E15b, E16a, E16b, E38, H410
Storage and disposal D01, D02, D05, D09a, D10c, D11a, D12a
Medical advice M05b [5, 6]

115   2,4-D

A translocated phenoxycarboxylic acid herbicide for cereals, grass and amenity use
HRAC mode of action code: O

See also amitrole + 2,4-D + diuron
        clopyralid + 2,4-D + MCPA

Products

1	Depitone Ultra	Nufarm UK	600 g/l	EC	17615
2	Depitox	Nufarm UK	500 g/l	SL	13258
3	Dioweed 50	UPL Europe	500 g/l	SL	13197
4	Headland Staff 500	FMC Agro	500 g/l	SL	13196

SEE SECTION 3 FOR PRODUCTS ALSO REGISTERED

Products – continued

5 Herboxone	FMC Agro	500 g/l	SL	13958
6 HY-D Super	Agrichem	500 g/l	SL	13198

Uses

- Annual dicotyledons in *amenity grassland, managed amenity turf, spring barley, spring wheat, winter barley, winter oats, winter rye, winter wheat* [1-6]; *apple orchards, farm forestry* (off-label), *forest nurseries* (off-label), *game cover* (off-label), *ornamental plant production* (off-label), *pear orchards* [2]; *forage maize, grain maize* [1]; *grassland* [1-3]; *miscanthus* (off-label), *undersown oats* [6]; *permanent grassland* [4-6]; *spring rye* [1, 2]; *undersown barley, undersown rye, undersown wheat* [2, 6]; *undersown spring cereals, undersown winter cereals* [3]

- Perennial dicotyledons in *amenity grassland, managed amenity turf, spring barley, spring wheat, winter barley, winter oats, winter rye, winter wheat* [1, 2, 4-6]; *apple orchards, farm forestry* (off-label), *forest nurseries* (off-label), *game cover* (off-label), *ornamental plant production* (off-label), *pear orchards* [2]; *forage maize, grain maize* [1]; *grassland, spring rye* [1, 2]; *permanent grassland* [4-6]; *undersown barley, undersown rye, undersown wheat* [2, 6]; *undersown oats* [6]

Extension of Authorisation for Minor Use (EAMUs)

- *farm forestry* 20082843 [2]
- *forest nurseries* 20082843 [2]
- *game cover* 20082843 [2]
- *miscanthus* 20082914 [6]
- *ornamental plant production* 20082843 [2]

Approval information

- 2,4-D included in Annex I under EC Regulation 1107/2009
- Approved for aquatic weed control [2]. See Section 5 for information on use of herbicides in or near water
- Accepted by BBPA for use on malting barley

Efficacy guidance

- Best results achieved by spraying weeds in seedling to young plant stage when growing actively in a strongly competing crop
- Most effective stage for spraying perennials varies with species. See label for details

Restrictions

- Maximum number of treatments normally 1 per crop and in forestry, 1 per yr in grassland and 2 or 3 per yr in amenity turf. Check individual labels
- Do not use on newly sown leys containing clover
- Do not spray grass seed crops after ear emergence
- Do not spray within 6 mth of laying turf or sowing fine grass
- Do not dump surplus herbicide in water or ditch bottoms
- Do not plant conifers until at least 1 mth after treatment
- Do not spray crops stressed by cold weather or drought or if frost expected
- Do not roll or harrow within 7 d before or after spraying
- Do not spray if rain falling or imminent
- Do not mow or roll turf or amenity grass 4 d before or after spraying. The first 4 mowings after treatment must be composted for at least 6 mth before use
- Do not mow grassland or graze for at least 10 d after spraying

Crop-specific information

- Latest use: before 1st node detectable in cereals; end Aug for conifer plantations; before established grassland 25 cm high
- Spray winter cereals in spring when leaf-sheath erect but before first node detectable (GS 31), spring cereals from 5-leaf stage to before first node detectable (GS 15-31)
- Cereals undersown with grass and/or clover, but not with lucerne, may be treated
- Selective treatment of resistant conifers can be made in Aug when growth ceased and plants hardened off, spray must be directed if applied earlier. See label for details

FOR FULL CONDITIONS OF USE ALWAYS READ THE PRODUCT LABEL

Following crops guidance
- Do not use shortly before or after sowing any crop
- Do not plant succeeding crops within 3 mth [4]
- Do not direct drill brassicas or grass/clover mixtures within 3 wk of application

Environmental safety
- Dangerous for the environment
- Very toxic to aquatic organisms [4]
- Dangerous to aquatic higher plants. Do not contaminate surface waters or ditches with chemical or used container
- 2,4-D is active at low concentrations. Take extreme care to avoid drift onto neighbouring crops, especially beet crops, brassicas, most market garden crops including lettuce and tomatoes under glass, pears and vines
- May be used to control aquatic weeds in presence of fish if used in strict accordance with directions for waterweed control and precautions needed for aquatic use [2]
- Keep livestock out of treated areas for at least 2 wk following treatment and until poisonous weeds, such as ragwort, have died down and become unpalatable
- Water containing the herbicide must not be used for irrigation purposes within 3 wk of treatment or until the concentration in water is below 0.05 ppm

Hazard classification and safety precautions
Hazard Harmful, Dangerous for the environment [3-6]; Irritant [1, 2]; Harmful if swallowed [2-6]
Transport code 9 [1-5]
Packaging group III [1-5]
UN Number 3082 [1-5]; N/C [6]
Risk phrases H315 [1]; H317 [1, 6]; H318 [2-6]
Operator protection A, C, H [1-6]; M [1, 2]; U05a, U11 [1-6]; U08, U20b [1, 2, 4-6]; U09a, U20a [3]; U14, U15 [6]
Environmental protection E07a, E34 [1-6]; E15a [1-3]; E15b, E38 [3-6]; H410 [1, 2, 4, 5]; H411 [6]
Storage and disposal D01, D02, D05 [1-6]; D09a [1, 2, 4-6]; D10a [4-6]; D10b [1, 2]; D10c [3]; D12a [3-6]
Medical advice M03 [1-6]; M05a [3, 6]

116   2,4-D + dicamba

A translocated herbicide for use on turf
HRAC mode of action code: O + O

See also dicamba

Products

1 Magneto	Nufarm UK	344:120 g/l	SL	17444
2 Thrust	Nufarm UK	344:120 g/l	SL	15408

Uses
- Annual dicotyledons in *amenity grassland, grassland*
- Perennial dicotyledons in *amenity grassland, grassland*

Approval information
- 2,4-D and dicamba included in Annex I under EC Regulation 1107/2009
- Accepted by BBPA for use on malting barley

Efficacy guidance
- Best results achieved by application when weeds growing actively in spring or early summer (later with irrigation and feeding)
- More resistant weeds may need repeat treatment after 3 wk
- Improved control of some weeds can be obtained by use of specifed oil adjuvant [1, 2]
- Do not use during drought conditions

SEE SECTION 3 FOR PRODUCTS ALSO REGISTERED

Restrictions

- Maximum number of treatments 2 per yr on amenity grass and established grassland [1, 2]
- Do not treat newly sown or turfed areas or grass less than 1 yr old
- Do not treat grass crops intended for seed production [1, 2]
- Do not treat grass suffering from drought, disease or other adverse factors [1, 2]
- Do not roll or harrow for 7 d before or after treatment [1, 2]
- Do not apply when grassland is flowering [1, 2]
- Avoid spray drift onto cultivated crops or ornamentals
- Do not graze grass for at least seven days after spraying [1, 2]
- Do not mow or roll four days before or after application. The first four mowings after treatment must be composted for at least six months before use [1, 2]
- Must only be applied between 1st May and 1st August

Crop-specific information

- Latest use: before grass 25 cm high for amenity grass, established grassland [2]
- The first four mowings after treatment must be composted for at least 6 mth before use

Environmental safety

- Dangerous for the environment [2]
- Toxic to aquatic organisms
- Keep livestock out of treated areas for at least two weeks following treatment and until poisonous weeds, such as ragwort, have died down and become unpalatable [1, 2]

Hazard classification and safety precautions

Hazard Harmful, Dangerous for the environment, Harmful if swallowed
Transport code 9
Packaging group III
UN Number 3082
Risk phrases H317, H318
Operator protection A, C, H, M; U05a, U11
Environmental protection E07a, E15a, E34, E38, H411 [1, 2]; E22c [1]
Storage and disposal D01, D02, D05, D09a, D10a, D12a
Medical advice M03

---

117   2,4-D + dicamba + fluroxypyr

A translocated and contact herbicide mixture for amenity turf
HRAC mode of action code: O + O + O

See also dicamba
     fluroxypyr

Products

1 Enstar	Headland Amenity	285:52.5:105 g/l	EC	18911
2 Holster XL	Barclay	285:52.5:105 g/l	EC	16407
3 Mascot Crossbar	Rigby Taylor	285:52.5:105 g/l	EC	16454
4 Speedster	Pan Amenity	285:52.5:105 g/l	EC	17025

Uses

- Annual dicotyledons in *amenity grassland, managed amenity turf* [1-4]
- Buttercups in *amenity grassland, managed amenity turf* [1, 2]
- Clover in *amenity grassland, managed amenity turf* [1, 2]
- Daisies in *amenity grassland, managed amenity turf* [1, 2]
- Dandelions in *amenity grassland, managed amenity turf* [1, 2]
- Yarrow in *amenity grassland, managed amenity turf* [1, 2]

Approval information

- 2,4-D, dicamba and fluroxypyr included in Annex I under EC Regulation 1107/2009

Efficacy guidance

- Apply when weeds actively growing (normally between Apr and Sep) and when soil is moist

FOR FULL CONDITIONS OF USE ALWAYS READ THE PRODUCT LABEL

- Best results obtained from treatment in spring or early summer before weeds begin to flower
- Do not apply if turf is wet or if rainfall expected within 4 h of treatment. Both circumstances will reduce weed control

Restrictions
- Maximum number of treatments 2 per yr
- Do not mow for 3 d before or after treatment
- Avoid overlapping or overdosing, especially on newly sown turf
- Do not spray in drought conditions or if turf under stress from frost, waterlogging, trace element deficiency, pest or disease attack

Crop-specific information
- Latest use: normally Sep for managed amenity turf
- New turf may be treated in spring provided at least 2 mth have elapsed since sowing

Environmental safety
- Dangerous for the environment
- Toxic to aquatic organisms
- Keep livestock out of treated areas for at least two weeks following treatment and until poisonous weeds, such as ragwort, have died down and become unpalatable
- LERAP Category B

Hazard classification and safety precautions

Hazard Harmful, Dangerous for the environment, Harmful if swallowed [1-4]; Very toxic to aquatic organisms [1-3]

Transport code 9

Packaging group III

UN Number 3082

Risk phrases H304, H319 [1-3]; H315, H336 [1-4]; H320, H371 [4]

Operator protection A, C, H, M; U05a, U08, U11, U14, U15, U19a, U20b

Environmental protection E07a, E15a, E16a, E34, E38 [1-4]; H410 [1-3]; H411 [4]

Storage and disposal D01, D02, D09a, D10a

Medical advice M03, M05a, M05b

118   2,4-D + dicamba + MCPA + mecoprop-P

A translocated herbicide mixture for grassland
HRAC mode of action code: O + O + O + O

See also dicamba
MCPA
mecoprop-P

Products

1	Enforcer	ICL (Everris) Ltd	70:20:70:42 g/l	SL	17274
2	Longbow	Bayer CropScience	70:20:70:42 g/l	SL	16528

Uses
- Buttercups in *managed amenity turf*
- Clovers in *managed amenity turf*
- Daisies in *managed amenity turf*
- Dandelions in *managed amenity turf*
- Self-heal in *managed amenity turf*

Approval information
- 2,4-D, dicamba, MCPA and mecoprop-P included in Annex I under EC Regulation 1107/2009

Environmental safety
- Some pesticides pose a greater threat of contamination of water than others and mecoprop-P is one of these pesticides. Take special care when applying mecoprop-P near water and do not apply if heavy rain is forecast

SEE SECTION 3 FOR PRODUCTS ALSO REGISTERED

Hazard classification and safety precautions
    UN Number N/C
    Risk phrases H319
    Operator protection A, C, H, M
    Environmental protection E07a, E15a, H410
    Storage and disposal D05, D09a, D10a, D12a

119   2,4-D + florasulam

A post-emergence herbicide mixture for control of broad leaved weeds in managed amenity turf
HRAC mode of action code: O + B

See also florasulam

Products

1	Esteron T	ICL (Everris) Ltd	300:6.25 g/l	ME	17432
2	Junction	Rigby Taylor	452:6.25 g/l	ME	12493

Uses
- Clover in *amenity grassland* [1]; *managed amenity turf* [1, 2]
- Daisies in *amenity grassland* [1]; *managed amenity turf* [1, 2]
- Dandelions in *amenity grassland* [1]; *managed amenity turf* [1, 2]
- Plantains in *amenity grassland* [1]; *managed amenity turf* [1, 2]
- Sticky mouse-ear in *amenity grassland* [1]; *managed amenity turf* [1, 2]

Approval information
- 2,4-D and florasulam included in Annex I under EC Regulation 1107/2009

Efficacy guidance
- Best results obtained from treatment of actively growing weeds between Mar and Oct when soil is moist
- Do not apply when rain is imminent or during periods of drought unless irrigation is applied

Restrictions
- Maximum number of treatments on managed amenity turf: 1 per yr

Crop-specific information
- Avoid mowing turf 3 d before and after spraying
- Ensure newly sown turf has established before treating. Turf sown in late summer or autumn should not be sprayed until growth is resumed in the following spring

Following crops guidance
- An interval of 4 wk must elapse between application and re-seeding turf

Environmental safety
- Dangerous for the environment
- Toxic to aquatic organisms

Hazard classification and safety precautions
    Hazard Harmful, Dangerous for the environment, Harmful if swallowed
    Transport code 9
    Packaging group III
    UN Number 3082
    Risk phrases H317
    Operator protection A, H; U02a, U05a, U08, U14, U20a
    Environmental protection E15a, E34, E38 [1, 2]; H410 [1]; H411 [2]
    Storage and disposal D01, D02, D09a, D10b, D12a

FOR FULL CONDITIONS OF USE ALWAYS READ THE PRODUCT LABEL

120 2,4-D + glyphosate

A total herbicide for weed control in crops and on railway ballast
HRAC mode of action code: O + G

See also glyphosate

Products

1 Diamond	Pan Amenity	160:240 g/l	SL	18832
2 Kyleo	Nufarm UK	160:240 g/l	SL	16567

Uses

- Annual and perennial weeds in *all edible crops (outdoor), all non-edible seed crops grown outdoors, amenity grassland, amenity vegetation, apple orchards, green cover on land temporarily removed from production, pear orchards, railway tracks*

Approval information

- 2,4-D and glyphosate included in Annex I under EC Regulation 1107/2009

Restrictions

- Do not use under polythene or glass
- Application of lime, fertiliser, farmyard manure and pesticides should be delayed until 5 days after application of Kyleo. After application large quantities of decaying foliage, stolons, roots or rhizomes should be dispersed or buried by thorough cultivation before crop drilling
- Do not use in or alongside hedges
- Do not apply by hand-held equipment
- Windfall (fruit fall to ground) must not be used as food or feeding stuff

Environmental safety

- LERAP Category B

Hazard classification and safety precautions

Transport code 9
Packaging group III
UN Number 3082
Risk phrases H317, H319
Operator protection A, C, H; U05a, U19a
Environmental protection E15a, E16a, E34, E38, H410
Storage and disposal D01, D09a, D10a, D12a, D12b

121 2,4-D + MCPA

A translocated herbicide mixture for cereals and grass
HRAC mode of action code: O + O

See also MCPA

Products

1 Cirran 360	Nufarm UK	360:315 g/l	SL	17346
2 Headland Polo	FMC Agro	360:315 g/l	SL	14933
3 Lupo	Nufarm UK	360:315 g/l	SL	14931
4 PastureMaster	Nufarm UK	360:315 g/l	SL	17994

Uses

- Annual dicotyledons in *grassland, spring barley, spring wheat, winter barley, winter oats, winter wheat* [1-4]
- Perennial dicotyledons in *grassland* [1, 3, 4]; *spring barley, spring wheat, winter barley, winter oats, winter wheat* [1-4]

Approval information

- 2,4-D and MCPA included in Annex I under EC Regulation 1107/2009
- Accepted by BBPA for use on malting barley

Efficacy guidance
- Best results achieved by spraying weeds in seedling to young plant stage when growing actively in a strongly competing crop
- Most effective stage for spraying perennials varies with species. See label for details

Restrictions
- Maximum number of treatments 1 per crop in cereals, 2 per year in grassland and 3 per year in managed amenity turf [2]; 1 per crop in cereals and 1 per year in grassland [3]
- Do not spray if rain falling or imminent
- Do not cut grass or graze for at least 10 d after spraying
- Do not use on newly sown leys containing clover or other legumes
- Do not spray crops stressed by cold weather or drought or if frost expected
- Do not use shortly before or after sowing any crop
- Do not roll or harrow within 7 d before or after spraying

Crop-specific information
- Spray winter cereals in spring when leaf-sheath erect but before first node detectable (GS 31), spring cereals from 5-leaf stage to before first node detectable (GS 15-31)
- Latest use: before first node detectable (GS 31) in cereals;

Environmental safety
- Dangerous for the environment
- Toxic to aquatic organisms
- Keep livestock out of treated areas for at least two weeks following treatment and until poisonous weeds, such as ragwort, have died down and become unpalatable
- 2,4-D and MCPA are active at low concentrations. Take extreme care to avoid drift onto neighbouring crops, especially beet crops, brassicas, most market garden crops including lettuce and tomatoes under glass, pears and vines

Hazard classification and safety precautions
Hazard Harmful [1-4]; Dangerous for the environment [2]
Transport code 9
Packaging group III
UN Number 3082
Risk phrases H318
Operator protection A, C, H, M; U05a, U08, U11, U14, U20b [1-4]; U15 [2]
Environmental protection E07a, E15a, E34 [1-4]; E38 [1, 3, 4]; H410 [1, 4]
Storage and disposal D01, D02, D05, D09a, D10a [1-4]; D12a [1, 3, 4]
Medical advice M03 [1-4]; M05a [2]

122   daminozide

A hydrazide plant growth regulator for use in certain ornamentals

Products

1 B-Nine SG	Arysta	85% w/w	SG	14434
2 Dazide Enhance	Fargro	85% w/w	SG	16092
3 Stature	Pure Amenity	85% w/w	SG	18064

Uses
- Growth regulation in **protected ornamentals**

Approval information
- Daminozide included in Annex I under EC Regulation 1107/2009

Efficacy guidance
- Best results obtained by application in late afternoon when glasshouse has cooled down
- Apply a fine spray using compressed air or power sprayers to give good coverage of dry foliage without run off

FOR FULL CONDITIONS OF USE ALWAYS READ THE PRODUCT LABEL

- Response to treatment differs widely depending on variety, stage of growth and physiological condition of the plant. It is recommended that any new variety is first tested on a small scale to observe if adverse effects occur

Restrictions
- Apply only to turgid, well watered plants. Do not water for 24 h after spraying
- Do not mix with other spray chemicals unless specifically recommended
- Do not store product in metal containers

Crop-specific information
- Do not use on chrysanthemum variety Fandango
- Evidence of effectiveness on Azaleas is limited
- See label for guidance on a range of bedding plant species

Hazard classification and safety precautions
UN Number N/C
Operator protection A, H; U02a, U08, U19a, U20b [2, 3]; U05a [1-3]
Environmental protection E15b
Storage and disposal D01, D02, D09a [1-3]; D10b, D12a [2, 3]
Medical advice M05a [2, 3]

123   2,4-DB

A translocated phenoxycarboxylic acid herbicide
HRAC mode of action code: O

Products

1	CloverMaster	Nufarm UK	400 g/l	SL	18251
2	DB Straight	UPL Europe	400 g/l	SL	18256
3	Headland Spruce	FMC Agro	400 g/l	SL	18259

Uses
- Annual dicotyledons in *grassland, lucerne, red clover, spring barley, spring oats, spring wheat, white clover, winter barley, winter oats, winter wheat* [1-3]; *undersown barley, undersown oats, undersown wheat* [1, 2]
- Buttercups in *grassland* [2, 3]
- Fat hen in *grassland* [2, 3]
- Penny cress in *grassland* [2, 3]
- Perennial dicotyledons in *grassland, lucerne, red clover, spring barley, spring oats, spring wheat, undersown barley, undersown oats, undersown wheat, white clover, winter barley, winter oats, winter wheat* [1]
- Shepherd's purse in *grassland* [2, 3]

Approval information
- 2,4-DB included in Annex I under EC Regulation 1107/2009
- Accepted by BBPA for use on malting barley

Efficacy guidance
- Best results achieved on young seedling weeds under good growing conditions. Treatment less effective in cold weather and dry soil conditions
- Rain within 12 h may reduce effectiveness

Restrictions
- Do not allow spray drift onto neighbouring crops
- Livestock must be kept out of treated areas for at least 2 weeks following treatment

Crop-specific information
- Latest use: before first node detectable (GS 31) for undersown cereals; fourth trifoliate leaf for lucerne
- In direct sown lucerne spray when seedlings have reached first trifoliate leaf stage. Optimum time 3-4 trifoliate leaves

SEE SECTION 3 FOR PRODUCTS ALSO REGISTERED

- Do not treat any lucerne after fourth trifoliate leaf
- In spring barley and spring oats undersown with lucerne spray from when cereal has 1 leaf unfolded and lucerne has first trifoliate leaf
- In spring wheat undersown with lucerne spray from when cereal has 3 leaves unfolded and lucerne has first trifoliate leaf

Following crops guidance
- Do not sow any crop into soil treated with 2,4-DB for at least 3 months after application.

Environmental safety
- Dangerous for the environment
- Toxic to aquatic organisms
- Keep livestock out of treated areas for at least two weeks following treatment and until poisonous weeds, such as ragwort, have died down and become unpalatable

Hazard classification and safety precautions
  Hazard Harmful, Dangerous for the environment, Harmful if swallowed
  Transport code 9 [2, 3]
  Packaging group III [2, 3]
  UN Number 3082 [2, 3]; N/C [1]
  Risk phrases H318
  Operator protection A, C, H; U05a, U09a, U11, U20a
  Environmental protection E07a, E15a, E15b, E34, E38
  Storage and disposal D01, D02, D05, D10c, D12a
  Medical advice M03, M05a

124   2,4-DB + MCPA

A translocated herbicide for cereals, clovers and leys
HRAC mode of action code: O + O

See also MCPA

Products

1	Clovermax	Nufarm UK	240:40 g/l	SL	15780
2	Redlegor	UPL Europe	240:40 g/l	SL	15783

Uses
- Annual dicotyledons in *grassland*, *spring wheat*, *winter wheat* [1, 2]; *oats*, *spring barley*, *undersown barley*, *undersown wheat*, *winter barley* [1]; *undersown spring cereals*, *undersown winter cereals* [2]
- Buttercups in *grassland*, *oats*, *spring barley*, *spring wheat*, *undersown barley*, *undersown wheat*, *winter barley*, *winter wheat* [1]
- Fat hen in *grassland*, *oats*, *spring barley*, *spring wheat*, *undersown barley*, *undersown wheat*, *winter barley*, *winter wheat* [1]
- Penny cress in *grassland*, *oats*, *spring barley*, *spring wheat*, *undersown barley*, *undersown wheat*, *winter barley*, *winter wheat* [1]
- Perennial dicotyledons in *grassland*, *spring wheat*, *undersown spring cereals*, *undersown winter cereals*, *winter wheat* [2]
- Plantains in *grassland* [1]
- Polygonums in *grassland*, *spring wheat*, *undersown spring cereals*, *undersown winter cereals*, *winter wheat* [2]
- Shepherd's purse in *grassland*, *oats*, *spring barley*, *spring wheat*, *undersown barley*, *undersown wheat*, *winter barley*, *winter wheat* [1]

Approval information
- 2,4-DB and MCPA included in Annex I under EC Regulation 1107/2009
- Accepted by BBPA for use on malting barley

Efficacy guidance
- Best results achieved on young seedling weeds under good growing conditions

- Spray thistles and other perennials when 10-20 cm high provided clover at correct stage
- Effectiveness may be reduced by rain within 12 h, by very cold conditions or drought

Restrictions
- Maximum number of treatments 1 per crop
- Do not spray established clover crops or lucerne
- Do not roll or harrow within 7 d before or after spraying
- Do not spray immediately before or after sowing any crop
- Do not graze crops within a week before or 2 weeks after applying
- Do not apply during drought, rain or if rain is expected
- Do not apply in very cold conditions as effectiveness may be reduced
- Do not use immediately before or after sowing any crop
- Do not use on cereals undersown with lucerne or in seed mixtures containing lucerne
- Do not spray in windy conditions as the spray drift may cause damage to neighbouring crops. The following crops are particularly susceptible: beet, Brassicae (e.g. turnips, swedes, oilseed rape), and most market garden crops including lettuce and tomatoes under glass, pears and vines

Crop-specific information
- Latest use: before first node detectable (GS 31) for cereals; 4th trifoliate leaf stage of clover for grass re-seeds
- Apply in spring to winter cereals from leaf sheath erect stage, to spring barley or oats from 2-leaf stage (GS 12), to spring wheat from 5-leaf stage (GS 15)
- Spray clovers as soon as possible after first trifoliate leaf, grasses after 2-3 leaf stage
- Red clover may suffer temporary distortion after treatment

Environmental safety
- Harmful to aquatic organisms
- Harmful to fish or other aquatic life. Do not contaminate surface waters or ditches with chemical or used container
- Keep livestock out of treated areas for at least two weeks following treatment and until poisonous weeds, such as ragwort, have died down and become unpalatable

Hazard classification and safety precautions
Hazard Harmful, Dangerous for the environment, Harmful if swallowed
UN Number N/C
Risk phrases H315 [2]; H318 [1, 2]
Operator protection A, C [1]; U05a, U20b [1, 2]; U08, U11, U13, U14, U15 [2]; U12, U19a [1]
Environmental protection E07a [1, 2]; E13c, H412 [2]; E15b, E34, E38, H411 [1]
Storage and disposal D01, D02, D05, D09a [1, 2]; D10b, D12a [1]; D10c [2]
Medical advice M03 [1, 2]; M05a [2]

125　deltamethrin

A pyrethroid insecticide with contact and residual activity
IRAC mode of action code: 3

Products

1 Bandu	FMC Agro	25 g/l	EC	16153
2 Decis Forte	Bayer CropScience	100 g/l	EC	16110
3 Decis Protech	Bayer CropScience	15 g/l	EW	16160
4 Grainstore	Pan Agriculture	25 g/l	EC	14932
5 Grainstore 25EC	Pan Agriculture	25.4 g/l	EC	19218
6 Grainstore ULV	Pan Agriculture	0.69% w/v	UL	19195
7 K-Obiol EC 25	Bayer CropScience	25 g/l	EC	13573
8 K-Obiol ULV 6	Bayer CropScience	0.69% w/v	UL	13572

Uses
- American serpentine leaf miner in **protected aubergines** *(off-label)*, **protected courgettes** *(off-label)*, **protected gherkins** *(off-label)*, **protected squashes** *(off-label)* [1]; **protected ornamentals** *(off-label)* [1-3]

SEE SECTION 3 FOR PRODUCTS ALSO REGISTERED

- Aphids in **amenity vegetation, apples, cress** *(off-label)*, **endives** *(off-label)*, **grass seed crops** *(off-label)*, **lamb's lettuce** *(off-label)*, **land cress** *(off-label)*, **leaves and shoots** *(off-label)*, **lettuce, protected peppers, rye** *(off-label)*, **spring barley, spring oats, spring wheat, triticale** *(off-label)*, **winter barley, winter oats, winter wheat** [1-3]; **brussels sprouts, cabbages, cauliflowers, protected chilli peppers** [2]; **celeriac** *(off-label)*, **evening primrose** *(off-label)*, **herbs (see appendix 6)** *(off-label)*, **jerusalem artichokes** *(off-label)*, **radishes** *(off-label)*, **red beet** *(off-label)*, **salsify** *(off-label)* [3]; **forest nurseries** *(off-label)*, **protected forest nurseries** *(off-label)* [1]; **ornamental plant production, protected cucumbers, protected ornamentals, protected tomatoes** [1, 2]
- Apple sucker in **apples** [1-3]
- Barley yellow dwarf virus vectors in **winter barley, winter wheat** [1-3]
- Beet flea beetle in **red beet** *(off-label)* [3]
- Beet fly in **red beet** *(off-label)* [3]
- Beet virus yellows vectors in **winter oilseed rape** [1-3]
- Black bean aphid in **dwarf beans** *(off-label)*, **edible podded peas** *(off-label)*, **french beans** *(off-label)*, **runner beans** *(off-label)* [3]
- Bramble shoot moth in **blackberries** *(off-label)* [1-3]
- Cabbage seed weevil in **mustard** [1]; **spring mustard, winter mustard** [2, 3]; **spring oilseed rape, winter oilseed rape** [1-3]
- Cabbage stem flea beetle in **winter oilseed rape** [1-3]
- Cabbage stem weevil in **mustard** [1]; **spring mustard, winter mustard** [2, 3]; **spring oilseed rape, winter oilseed rape** [1-3]
- Capsids in **amenity vegetation, apples** [1-3]; **blackcurrants** *(off-label)*, **gooseberries** *(off-label)*, **redcurrants** *(off-label)* [3]; **ornamental plant production** [1, 2]
- Carrot fly in **carrots** *(off-label)*, **horseradish** *(off-label)*, **mallow (althaea spp.)** *(off-label)*, **parsnips** *(off-label)* [1-3]
- Caterpillars in **amenity vegetation, apples, blackberries** *(off-label)*, **brussels sprouts, cabbages, cauliflowers, cress** *(off-label)*, **endives** *(off-label)*, **lamb's lettuce** *(off-label)*, **land cress** *(off-label)*, **leaves and shoots** *(off-label)*, **protected peppers, swedes, turnips** [1-3]; **blackcurrants** *(off-label)*, **celeriac** *(off-label)*, **evening primrose** *(off-label)*, **gooseberries** *(off-label)*, **herbs (see appendix 6)** *(off-label)*, **jerusalem artichokes** *(off-label)*, **lettuce, protected celery** *(off-label)*, **protected rhubarb** *(off-label)*, **radishes** *(off-label)*, **red beet** *(off-label)*, **redcurrants** *(off-label)*, **salsify** *(off-label)*, **strawberries** *(off-label)* [3]; **mustard** [1]; **ornamental plant production, protected cucumbers, protected ornamentals, protected tomatoes** [1, 2]; **protected chilli peppers** [2]
- Cereal flea beetle in **rye** *(off-label)*, **triticale** *(off-label)* [1-3]
- Codling moth in **apples** [1-3]
- Cutworms in **carrots** *(off-label)*, **horseradish** *(off-label)*, **leeks** *(off-label)*, **mallow (althaea spp.)** *(off-label)*, **parsnips** *(off-label)* [1-3]; **herbs (see appendix 6)** *(off-label)* [3]; **lettuce** [1, 3]
- Flea beetle in **brussels sprouts, cabbages, cauliflowers, swedes, turnips** [2]; **cress** *(off-label)*, **endives** *(off-label)*, **lamb's lettuce** *(off-label)*, **land cress** *(off-label)*, **leaves and shoots** *(off-label)*, **sugar beet** [1-3]; **herbs (see appendix 6)** *(off-label)*, **lettuce, radishes** *(off-label)*, **salsify** *(off-label)* [3]; **mustard** [1]
- Gall midge in **dwarf beans** *(off-label)*, **edible podded peas** *(off-label)*, **french beans** *(off-label)*, **runner beans** *(off-label)* [3]
- Gout fly in **rye** *(off-label)*, **triticale** *(off-label)* [1-3]
- Insect pests in **blackberries** *(off-label)*, **carrots** *(off-label)*, **cress** *(off-label)*, **endives** *(off-label)*, **forest nurseries** *(off-label)*, **grass seed crops** *(off-label)*, **horseradish** *(off-label)*, **lamb's lettuce** *(off-label)*, **land cress** *(off-label)*, **leaves and shoots** *(off-label)*, **leeks** *(off-label)*, **mallow (althaea spp.)** *(off-label)*, **parsnips** *(off-label)*, **protected forest nurseries** *(off-label)*, **protected ornamentals** *(off-label)*, **protected spring onions** *(off-label)* [1]; **evening primrose** *(off-label)* [3]; **grain** *(off-label)* [8]; **grain stores** [4, 5, 7]; **lettuce** [2]; **stored grain, stored pulses** [4-8]
- Leaf beetle in **celeriac** *(off-label)*, **jerusalem artichokes** *(off-label)*, **radishes** *(off-label)*, **salsify** *(off-label)* [3]
- Leaf miner in **herbs (see appendix 6)** *(off-label)* [3]; **protected spring onions** *(off-label)* [1-3]
- Leaf moth in **leeks** *(off-label)* [1-3]
- Leafhoppers in **lettuce** [3]

- Mealybugs in *amenity vegetation*, *protected peppers* [1-3]; *ornamental plant production*, *protected cucumbers*, *protected ornamentals*, *protected tomatoes* [1, 2]; *protected chilli peppers* [2]
- Pea and bean weevil in *broad beans*, *combining peas*, *spring field beans*, *vining peas*, *winter field beans* [1-3]; *dwarf beans* (off-label), *edible podded peas* (off-label), *french beans* (off-label), *runner beans* (off-label) [3]
- Pea aphid in *dwarf beans* (off-label), *edible podded peas* (off-label), *french beans* (off-label), *runner beans* (off-label) [3]
- Pea midge in *combining peas*, *vining peas* [2, 3]
- Pea moth in *combining peas*, *vining peas* [1-3]; *dwarf beans* (off-label), *edible podded peas* (off-label), *french beans* (off-label), *runner beans* (off-label) [3]
- Peach-potato aphid in *dwarf beans* (off-label), *edible podded peas* (off-label), *french beans* (off-label), *runner beans* (off-label) [3]
- Pear sucker in *pears* [1-3]
- Pollen beetle in *lettuce* [3]; *mustard* [1]; *spring mustard*, *winter mustard* [2, 3]; *spring oilseed rape*, *winter oilseed rape* [1-3]
- Raspberry beetle in *blackberries* (off-label), *raspberries* [1-3]
- Saddle gall midge in *rye* (off-label), *triticale* (off-label) [1-3]
- Sawflies in *apples* [1-3]; *blackcurrants* (off-label), *gooseberries* (off-label), *redcurrants* (off-label) [3]
- Scale insects in *amenity vegetation*, *protected peppers* [1-3]; *ornamental plant production*, *protected cucumbers*, *protected ornamentals*, *protected tomatoes* [1, 2]; *protected chilli peppers* [2]
- Silver Y moth in *dwarf beans* (off-label), *edible podded peas* (off-label), *french beans* (off-label), *runner beans* (off-label) [3]
- Thrips in *amenity vegetation*, *protected peppers* [1-3]; *blackcurrants* (off-label), *celeriac* (off-label), *dwarf beans* (off-label), *edible podded peas* (off-label), *french beans* (off-label), *gooseberries* (off-label), *jerusalem artichokes* (off-label), *lettuce*, *radishes* (off-label), *red beet* (off-label), *redcurrants* (off-label), *runner beans* (off-label), *salsify* (off-label), *strawberries* (off-label) [3]; *ornamental plant production*, *protected cucumbers*, *protected tomatoes* [1, 2]; *protected chilli peppers* [2]
- Tortrix moths in *apples* [1-3]
- Weevils in *celeriac* (off-label), *jerusalem artichokes* (off-label), *radishes* (off-label), *salsify* (off-label) [3]
- Western flower thrips in *protected aubergines* (off-label), *protected courgettes* (off-label), *protected gherkins* (off-label), *protected squashes* (off-label), *strawberries* (off-label) [1]; *protected chilli peppers* [2]; *protected cucumbers*, *protected tomatoes* [1, 2]; *protected ornamentals* (off-label), *protected peppers* [1-3]
- Whitefly in *amenity vegetation*, *protected peppers* [1-3]; *brussels sprouts*, *cabbages*, *cauliflowers*, *protected chilli peppers* [2]; *herbs (see appendix 6)* (off-label) [3]; *ornamental plant production*, *protected cucumbers*, *protected ornamentals*, *protected tomatoes* [1, 2]
- Wireworm in *herbs (see appendix 6)* (off-label) [3]
- Yellow cereal fly in *rye* (off-label), *triticale* (off-label) [1-3]

Extension of Authorisation for Minor Use (EAMUs)
- *blackberries* 20141106 [1], 20140917 [2], 20131673 [3]
- *blackcurrants* 20190635 [3]
- *carrots* 20141105 [1], 20140916 [2], 20131672 [3]
- *celeriac* 20190679 [3]
- *cress* 20141104 [1], 20140911 [2], 20131670 [3]
- *dwarf beans* 20190678 [3]
- *edible podded peas* 20190678 [3]
- *endives* 20141104 [1], 20140911 [2], 20131670 [3]
- *evening primrose* 20142584 [3]
- *forest nurseries* 20131856 [1]
- *french beans* 20190678 [3]
- *gooseberries* 20190635 [3]
- *grain* 20112491 [8]
- *grass seed crops* 20141107 [1], 20140920 [2], 20131671 [3]

SEE SECTION 3 FOR PRODUCTS ALSO REGISTERED

- **herbs (see appendix 6)** *20191272* [3]
- **horseradish** *20141105* [1], *20140916* [2], *20131672* [3]
- **jerusalem artichokes** *20190679* [3]
- **lamb's lettuce** *20141104* [1], *20140911* [2], *20131670* [3]
- **land cress** *20141104* [1], *20140911* [2], *20131670* [3]
- **leaves and shoots** *20141104* [1], *20140911* [2], *20131670* [3]
- **leeks** *20141103* [1], *20140901 expires 31 Jul 2020* [2], *20131665* [3]
- **mallow (althaea spp.)** *20141105* [1], *20140916* [2], *20131672* [3]
- **parsnips** *20141105* [1], *20140916* [2], *20131672* [3]
- **protected aubergines** *20132523* [1]
- **protected celery** *20142585* [3]
- **protected courgettes** *20132524* [1]
- **protected forest nurseries** *20131856* [1]
- **protected gherkins** *20132524* [1]
- **protected ornamentals** *20141102* [1], *20141012* [2], *20131662* [3]
- **protected rhubarb** *20142585* [3]
- **protected spring onions** *20141101* [1], *20140914* [2], *20131664* [3]
- **protected squashes** *20132524* [1]
- **radishes** *20190679* [3]
- **red beet** *20190679* [3]
- **redcurrants** *20190635* [3]
- **runner beans** *20190678* [3]
- **rye** *20193268* [1], *20193279* [2], *20193280* [3]
- **salsify** *20190679* [3]
- **strawberries** *20132527* [1], *20190636* [3]
- **triticale** *20193268* [1], *20193279* [2], *20193280* [3]

Approval information
- Deltamethrin included in Annex I under EC Regulation 1107/2009
- Accepted by BBPA for use on malting barley and hops

Efficacy guidance
- A contact and stomach poison with 3-4 wk persistence, particularly effective on caterpillars and sucking insects
- Normally applied at first signs of damage with follow-up treatments where necessary at 10-14 d intervals. Rates, timing and recommended combinations with other pesticides vary with crop and pest. See label for details
- Spray is rainfast within 1 h
- May be applied in frosty weather provided foliage not covered in ice
- Temperatures above 35°C may reduce effectiveness or persistence

Restrictions
- Maximum number of treatments varies with crop and pest, 4 per crop for wheat and barley, only 1 application between 1 Apr and 31 Aug. See label for other crops
- Do not apply more than 1 aphicide treatment to cereals in summer
- Do not spray crops suffering from drought or other physical stress
- Consult processer before treating crops for processing
- Do not apply to a cereal crop if any product containing a pyrethroid insecticide or dimethoate has been applied to that crop after the start of ear emergence (GS 51)
- Do not spray cereals after 31 Mar in the year of harvest within 6 m of the outside edge of the crop
- Reduced volume spraying must not be used on cereals after 31 Mar in yr of harvest

Crop-specific information
- Latest use: early dough (GS 83) for barley, oats, wheat; before flowering for mustard, oilseed rape; before 31 Mar for grass seed crops [1, 3]

Environmental safety
- Dangerous for the environment
- Toxic to aquatic organisms
- Flammable [1-3]

FOR FULL CONDITIONS OF USE ALWAYS READ THE PRODUCT LABEL

- Extremely dangerous to fish or other aquatic life. Do not contaminate surface waters or ditches with chemical or used container
- Dangerous to bees. Do not apply to crops in flower or to those in which bees are actively foraging. Do not apply when flowering weeds are present
- Do not apply in tank-mixture with a triazole-containing fungicide when bees are likely to be actively foraging in the crop
- To protect non target insects/arthropods respect an unsprayed buffer zone of 5 m to non-crop land [1-3]
- Buffer zone requirement 7 m [1-3]
- Broadcast air-assisted LERAP [3, 4] (30m); Broadcast air-assisted LERAP (30m raspberry, 50m apple & pear; LERAP category A [1, 3, 4]
- Broadcast air-assisted LERAP [3-5] (18 m); Broadcast air-assisted LERAP [2] (30m (raspberry); 50m (apple & pear)); LERAP Category A [1-5]

Hazard classification and safety precautions

Hazard Harmful, Flammable, Harmful if swallowed, Harmful if inhaled [1, 2, 4, 5]; Dangerous for the environment [1-5]; Flammable liquid and vapour [1, 2]; Very toxic to aquatic organisms [3]
Transport code 3 [1-5]; 9 [6-8]
Packaging group III
UN Number 1993 [1-5]; 3082 [6-8]
Risk phrases H304, H335, H336 [1, 2]; H315 [1, 4, 5]; H318 [1, 2, 4, 5]; H371 [4, 5]
Operator protection A, C, H [1-8]; D, M [4-8]; U04a, U05a, U19a [1, 2, 4-8]; U08 [1, 2, 4, 5]; U09a, U20c [6-8]; U10 [1]; U11, U19c [2]; U14 [3]; U20b [1-5]
Environmental protection E12a [2, 4, 5]; E12c [1]; E12f [1] (cereals, oilseed rape, peas, beans); E12f [2-5] (cereals, oilseed rape, peas, beans - see label for guidance); E15a, E16d [1, 2, 4, 5]; E15b [3, 6-8]; E16c [1, 3-5]; E16j [2]; E17b [2] (30m (raspberry); 50m (apple & pear)); E17b [3-5] (18 m); E22c [3]; E34 [1, 2, 4-8]; E38 [1-5]; H410 [1-3]; H411 [4, 5]
Storage and disposal D01, D02, D09a, D10c [1-8]; D05 [1, 2, 4, 5]; D12a [2-5]
Medical advice M03 [1, 2, 4-8]; M05b [1, 2, 4, 5]

126   desmedipham

A contact phenyl carbamate herbicide available only in mixtures
HRAC mode of action code: C1

127   desmedipham + ethofumesate + lenacil + phenmedipham

A selective contact and residual herbicide mixture for weed control in beet
HRAC mode of action code: C1 + N + C1 + C1

Products

1 Betanal MaxxPro	Bayer CropScience	47:75:27:60 g/l	OD	15086
2 Clayton Fore	Clayton	47:75:27:60 g/l	OD	18481

Uses

- Annual dicotyledons in *fodder beet*, *mangels*, *sugar beet* [1, 2]; *forest nurseries* (off-label) [1]
- Annual meadow grass in *beetroot* (off-label), *forest nurseries* (off-label) [1]; *fodder beet*, *mangels*, *sugar beet* [1, 2]
- Black bindweed in *beetroot* (off-label) [1]
- Black nightshade in *beetroot* (off-label) [1]
- Bugloss in *beetroot* (off-label) [1]
- Charlock in *beetroot* (off-label) [1]
- Chickweed in *beetroot* (off-label) [1]
- Cleavers in *beetroot* (off-label) [1]
- Field speedwell in *beetroot* (off-label) [1]
- Fumitory in *beetroot* (off-label) [1]

Extension of Authorisation for Minor Use (EAMUs)
- *beetroot 20152038* [1]
- *forest nurseries 20162138* [1]

Approval information
- Desmedipham, ethofumesate, lenacil and phenmedipham included in Annex I under EC Regulation 1107/2009
- Approval expires 1/7/2020

Efficacy guidance
- On soils with more than 5% organic matter content, residual activity may be reduced.

Restrictions
- The maximum total dose must not exceed 1.0 kg/ha ethofumesate in any 3 year period.

Following crops guidance
- Beet crops may be sown at any time after the use of Betanal MaxxPro. Any other crop may be sown 3 months after using Betanal MaxxPro. Ploughing (mould board) to a minimum depth of 15 cm should precede preparation of a new seed bed.
- If a crop is suffering from manganese deficiency it may be checked. To avoid crop check, manganese should ideally be applied to the crop first.
- Crops suffering from lime deficiency may also be checked. Growers should ensure that the lime status of the soil is satisfactory before drilling.
- When the temperature is, or is likely to be, above 21°C (70°F) on the day of spraying, application should be made after 5 pm otherwise crop check may occur. If crops are subjected to substantial day to night temperature changes shortly before or after spraying, a check may occur from which the crop may not fully recover.

Hazard classification and safety precautions
Hazard Irritant, Dangerous for the environment, Very toxic to aquatic organisms
Transport code 9
Packaging group III
UN Number 3082
Risk phrases H317, H318
Operator protection A, C, H; U05a, U11, U14, U20b
Environmental protection E15b, E38, E39, H410
Storage and disposal D09a, D10b, D12a
Medical advice M03

128   desmedipham + ethofumesate + phenmedipham

A selective contact and residual herbicide for beet
HRAC mode of action code: C1 + N + C1

See also ethofumesate
*   phenmedipham*

Products

1 Betanal Elite	Bayer CropScience	71:112:91 g/l	EC	17721
2 Betasana Trio SC	UPL Europe	15:115:75 g/l	SC	15551
3 Clayton Trinity	Clayton	15:115:75 g/l	SC	18493
4 Sniper	Adama	50:200:150 g/l	SE	16133
5 Trilogy	UPL Europe	15:115:75 g/l	SC	15644

Uses
- Annual dicotyledons in *fodder beet*, *sugar beet* [1-5]; *mangels* [2, 3, 5]
- Annual meadow grass in *fodder beet*, *sugar beet* [1, 4]
- Barnyard grass in *fodder beet*, *sugar beet* [1]
- Blackgrass in *fodder beet* (moderately susceptible), *sugar beet* (moderately susceptible) [1]
- Bristle grasses in *fodder beet*, *sugar beet* [1]
- Gallant soldier in *fodder beet*, *sugar beet* [1]

FOR FULL CONDITIONS OF USE ALWAYS READ THE PRODUCT LABEL

- Loose silky bent in **fodder beet**, **sugar beet** [1]
- Thorn apple in **fodder beet**, **sugar beet** [1]
- Wild oats in **fodder beet** *(moderately susceptible)*, **sugar beet** *(moderately susceptible)* [1]

Approval information

- Desmedipham, ethofumesate and phenmedipham included in Annex I under EC Regulation 1107/2009

Efficacy guidance

- Product recommended for low-volume overall application in a planned spray programme
- Product acts mainly by contact action. A full programme also gives some residual control but this may be reduced on soils with more than 5% organic matter
- Best results achieved from treatments applied at fully expanded cotyledon stage of largest weeds present. Occasional larger weeds will usually be controlled from a full programme of sprays
- Where a pre-emergence band spray has been applied treatment must be timed according to size of the untreated weeds between the rows
- Susceptible weeds may not all be killed by the first spray. Repeat applications as each flush of weeds reaches cotyledon size normally necessary for season long control
- Sequential treatments should be applied when the previous one is still showing an effect on the weeds
- Various mixtures with other beet herbicides are recommended. See label for details

Restrictions

- Maximum total dose 3.9 l/ha [4] and 7.0 l/ha [2, 5]
- Do not spray crops stressed by nutrient deficiency, wind damage, pest or disease attack, or previous herbicide treatments. Stressed crops treated under conditions of high light intensity may be checked and not recover fully
- If temperature likely to exceed 21°C spray after 5 pm
- Crystallisation may occur if spray volume exceeds that recommended or spray mixture not used within 2 h, especially if the water temperature is below 5°C
- Before use, wash out sprayer to remove all traces of previous products, especially hormone and sulfonyl urea weedkillers.
- The maximum total dose must not exceed 1.0 kg/ha ethofumesate in any three year period

Crop-specific information

- Latest use: before crop meets between rows
- Apply first treatment when majority of crop plants have reached the fully expanded cotyledon stage
- Frost within 7 d of treatment may cause check from which the crop may not recover

Following crops guidance

- Beet crops may be sown at any time after treatment. Any other crop may be sown 3 mth after treatment following mouldboard ploughing to 15 cm minimum

Environmental safety

- Dangerous for the environment
- Toxic to aquatic organisms
- LERAP Category B [1, 4]

Hazard classification and safety precautions

Hazard Irritant [2, 3, 5]; Dangerous for the environment [1-5]; Very toxic to aquatic organisms [1]
Transport code 9
Packaging group III
UN Number 3082
Risk phrases H319 [2, 3, 5]
Operator protection A, H [1, 4]; U08 [1-5]; U20a [1, 4]
Environmental protection E15a, E16a, H410 [1, 4]; E15b, H411 [2, 3, 5]; E38 [1-5]
Storage and disposal D05, D09a, D12a [1-5]; D10b [1, 4]; D10c [2, 3, 5]
Treated seed S06a [2, 3, 5]
Medical advice M03

SEE SECTION 3 FOR PRODUCTS ALSO REGISTERED

129   desmedipham + phenmedipham

A mixture of contact herbicides for use in sugar beet
HRAC mode of action code: C1 + C1

See also phenmedipham

Products

1	Beetup Compact SC	UPL Europe	80:80 g/l	SC	15566
2	Betanal Turbo	Bayer CropScience	160:160 g/l	EC	15505

Uses

- Amaranthus in **fodder beet**, **sugar beet** [1]
- Annual dicotyledons in **fodder beet**, **sugar beet** [1, 2]; **mangels** [2]
- Black nightshade in **fodder beet**, **sugar beet** [1]
- Chickweed in **fodder beet**, **sugar beet** [1]
- Fat hen in **fodder beet**, **sugar beet** [1]
- Field pansy in **fodder beet**, **sugar beet** [1]
- Penny cress in **fodder beet**, **sugar beet** [1]
- Red dead-nettle in **fodder beet**, **sugar beet** [1]
- Shepherd's purse in **fodder beet**, **sugar beet** [1]

Approval information

- Desmedipham and phenmedipham included in Annex I under EC Regulation 1107/2009
- Approval expires 1/7/2020

Efficacy guidance

- Product is recommended for low volume overall spraying in a planned programme involving pre- and/or post-emergence treatments at doses recommended for low volume programmes
- Best results obtained from treatment when earliest germinating weeds have reached cotyledon stage
- Further treatments must be applied as each flush of weeds reaches cotyledon stage but allowing a minimum of 7 d between each spray
- Where a pre-emergence band spray has been applied, the first treatment should be timed according to the size of the weeds in the untreated area between the rows
- Product is absorbed by leaves of emerged weeds which are killed by scorching action in 2-10 d. Apply overall as a fine spray to optimise weed cover and spray retention
- Various tank mixtures and sequences recommended to widen weed spectrum and add residual activity - see label for details

Restrictions

- Maximum total dose equivalent to three full dose treatments
- Do not spray crops stressed by nutrient deficiency, frost, wind damage, pest or disease attack, or previous herbicide treatments. Stressed crops may be checked and not recover fully
- If temperature likely to exceed 21°C spray after 5 pm
- Before use, wash out sprayer to remove all traces of previous products, especially hormone and sulfonyl urea weedkillers
- Crystallisation may occur if spray volume exceeds that recommended or spray mixture not used within 2 h, especially if the water temperature is below 5°C
- Product may cause non-reinforced PVC pipes and hoses to soften and swell. Wherever possible, use reinforced PVC or synthetic rubber hoses

Crop-specific information

- Latest use: before crop leaves meet between rows
- Product safe to use on all soil types

Environmental safety

- Dangerous for the environment
- Toxic to aquatic organisms
- Risk to certain non-target insects or other arthropods. Avoid spraying within 6 m of field boundary
- LERAP Category B

FOR FULL CONDITIONS OF USE ALWAYS READ THE PRODUCT LABEL

Hazard classification and safety precautions
> **Hazard** Irritant [1]; Dangerous for the environment [1, 2]; Very toxic to aquatic organisms [2]
> **Transport code** 9
> **Packaging group** III
> **UN Number** 3082
> **Risk phrases** H315, H317, H319 [1]
> **Operator protection** A; U05a, U19a [1, 2]; U08, U20a [2]; U09a, U11, U14, U20b [1]
> **Environmental protection** E15b, E16a, H410 [1, 2]; E16b, E38, E39 [2]
> **Storage and disposal** D01, D02, D09a [1, 2]; D05, D10c [1]; D10b, D12a [2]

130   dicamba

A translocated benzoic herbicide available in mixtures or formulated alone for weed control in maize
HRAC mode of action code: O

See also 2,4-D + dicamba
> *2,4-D + dicamba + dichlorprop-P*
> *2,4-D + dicamba + fluroxypyr*
> *2,4-D + dicamba + iron sulphate + mecoprop-P*
> *2,4-D + dicamba + MCPA + mecoprop-P*
> *2,4-D + dicamba + triclopyr*

Products

1	Antuco	Ventura	70% w/w	SG	17409
2	Oceal	Rotam	70% w/w	SG	15618

Uses
- Amaranthus in **forage maize**, **grain maize** [1]; **maize** [2]
- Fat hen in **forage maize**, **grain maize** [1]; **maize** [2]
- Field pansy in **forage maize**, **grain maize** [1]; **maize** [2]
- Maple-leaved goosefoot in **forage maize**, **grain maize** [1]; **maize** [2]

Approval information
- Dicamba included in Annex I under EC Regulation 1107/2009

Restrictions
- Do not treat maize during cold or frosty conditions, during periods of high day/night temperature variations or during periods of very high temperatures.
- Avoid drift into greenhouses or onto other agricultural or horticultural crops, amenity plantings or gardens. No to be used in glasshouses. Do not allow spray applications to come into contact with desired broad-leaved trees. Beets, all brassicae (including oilseed rape), lettuce, peas, tomatoes, potatoes, all crops and ornamentals are particularly susceptible to dicamba and may be damaged by spray drift.
- Application may only be made between 1 May and 30 September

Following crops guidance
- In the event of crop failure, only redrill with maize but any crop may be sown after a normal harvest date

Hazard classification and safety precautions
> **Hazard** Dangerous for the environment
> **Transport code** 9
> **Packaging group** III
> **UN Number** 3077
> **Operator protection** U20b
> **Environmental protection** E15b, E38, H411
> **Storage and disposal** D01, D02, D10c, D12b

131   dicamba + MCPA + mecoprop-P

A translocated herbicide for cereals, grassland, amenity grass and orchards
HRAC mode of action code: O + O + O

See also MCPA
      mecoprop-P

Products

1	Hyprone P	UPL Europe	16:101:92 g/l	SL	17418
2	Hysward-P	Agrichem	16:101:92 g/l	SL	17609
3	Mircam Plus	Nufarm UK	19.5:245:43.3 g/l	SL	15868
4	T2 Green Pro	Nufarm UK	19.5:245:43.3 g/l	SL	16366
5	Turfmaster	Nufarm UK	19.5:245:43.3 g/l	SL	16344

Uses

- Annual dicotyledons in *amenity grassland* [2, 4, 5]; *canary seed (off-label)*, *grass seed crops* [3]; *managed amenity turf* [2-5]; *spring barley, spring oats, spring wheat, winter barley, winter oats, winter wheat* [1, 3]
- Black bindweed in *amenity grassland, managed amenity turf* [4, 5]
- Charlock in *amenity grassland, managed amenity turf* [4, 5]
- Chickweed in *amenity grassland, managed amenity turf* [4, 5]
- Clover in *amenity grassland (moderately susceptible), managed amenity turf (moderately susceptible)* [4, 5]
- Common orache in *amenity grassland, managed amenity turf* [4, 5]; *rye (off-label)*, *triticale (off-label)* [3]
- Dandelions in *amenity grassland (moderately susceptible), managed amenity turf (moderately susceptible)* [4, 5]
- Docks in *amenity grassland, managed amenity turf* [2]
- Fat hen in *amenity grassland, managed amenity turf* [4, 5]
- Forget-me-not in *amenity grassland, managed amenity turf* [4, 5]; *rye (off-label)*, *triticale (off-label)* [3]
- Fumitory in *amenity grassland, managed amenity turf* [4, 5]
- Groundsel in *amenity grassland, managed amenity turf* [4, 5]
- Knotgrass in *amenity grassland, managed amenity turf* [4, 5]
- Mayweeds in *amenity grassland, managed amenity turf* [4, 5]
- Pale persicaria in *amenity grassland, managed amenity turf* [4, 5]; *rye (off-label)*, *triticale (off-label)* [3]
- Penny cress in *amenity grassland, managed amenity turf* [4, 5]
- Perennial dicotyledons in *amenity grassland* [2, 4, 5]; *grass seed crops* [3]; *managed amenity turf* [2-5]; *spring barley, spring oats, spring wheat, winter barley, winter oats, winter wheat* [1, 3]
- Plantains in *amenity grassland (moderately susceptible), managed amenity turf (moderately susceptible)* [4, 5]
- Poppies in *amenity grassland, managed amenity turf* [4, 5]; *rye (off-label)*, *triticale (off-label)* [3]
- Redshank in *amenity grassland, managed amenity turf* [4, 5]
- Shepherd's purse in *amenity grassland, managed amenity turf* [4, 5]
- Wild radish in *amenity grassland, managed amenity turf* [4, 5]

Extension of Authorisation for Minor Use (EAMUs)

- *canary seed 20160549* [3]
- *rye 20180355* [3]
- *triticale 20180355* [3]

Approval information

- Dicamba, MCPA and mecoprop-P included in Annex I under EC Regulation 1107/2009
- Accepted by BBPA for use on malting barley

Efficacy guidance
- Treatment should be made when weeds growing actively. Weeds hardened by winter weather may be less susceptible
- For best results apply in fine warm weather, preferably when soil is moist. Do not spray if rain expected within 6 h or in drought
- Application of fertilizer 1-2 wk before spraying aids weed control in turf
- Where a second treatment later in the season is needed in amenity situations and on grass allow 4-6 wk between applications to permit sufficient foliage regrowth for uptake

Restrictions
- Maximum number of treatments (including other mecoprop-P products) or maximum total dose varies with crop and product. See label for details. The total amount of mecoprop-P applied in a single yr must not exceed the maximum total dose approved for any single product for the crop/situation
- Do not apply to cereals after the first node is detectable (GS 31), or to grass under stress from drought or cold weather
- Do not spray cereals undersown with clovers or legumes, to be undersown with grass or legumes or grassland where clovers or other legumes are important
- Do not spray leys established less than 18 mth or orchards established less than 3 yr
- Do not roll or harrow within 7 d before or after treatment, or graze for at least 7 d afterwards (longer if poisonous weeds present)
- Do not use on turf or grass in year of establishment. Allow 6-8 wk after treatment before seeding bare patches
- The first 4 mowings after use should not be used for mulching unless composted for 6 mth
- Turf should not be mown for 24 h before or after treatment (3-4 d for closely mown turf)
- Avoid drift onto all broad-leaved plants outside the target area

Crop-specific information
- Latest use: before first node detectable (GS 31) for cereals; 5-6 wk before head emergence for grass seed crops; mid-Oct for established grass
- HI 7-14 d before cutting or grazing for leys, permanent pasture
- Apply to winter cereals from the leaf sheath erect stage (GS 30), and to spring cereals from the 5 expanded leaf stage (GS 15)
- Spray grass seed crops 4-6 wk before flower heads begin to emerge (Timothy 6 wk)
- Turf containing bulbs may be treated once the foliage has died down completely [2]

Environmental safety
- Harmful to aquatic organisms
- Keep livestock out of treated areas for at least 2 wk following treatment and until poisonous weeds, such as ragwort, have died down and become unpalatable
- Harmful to fish or other aquatic life. Do not contaminate surface waters or ditches with chemical or used container
- Some pesticides pose a greater threat of contamination of water than others and mecoprop-P is one of these pesticides. Take special care when applying mecoprop-P near water and do not apply if heavy rain is forecast

Hazard classification and safety precautions
Hazard Harmful [4, 5]; Irritant [1-3]; Dangerous for the environment [3-5]
Transport code 9 [3-5]
Packaging group III [3-5]
UN Number 3082 [3-5]; N/C [1, 2]
Risk phrases H315 [1, 2]; H318 [1-5]
Operator protection A, C, H [1-5]; M [1-3]; U05a, U08, U11, U20b [1-5]; U15 [3]; U19a [2]
Environmental protection E07a, H411 [1-5]; E13c [1, 2]; E15a, E38 [3]; E15b [4, 5]; E34 [3-5]
Storage and disposal D01, D02, D05, D09a [1-5]; D10b [3-5]; D10c [1, 2]; D12a [3]
Medical advice M03 [3-5]; M05a [1, 2]

SEE SECTION 3 FOR PRODUCTS ALSO REGISTERED

SECTION 2

132   dicamba + mecoprop-P

A translocated post-emergence herbicide for cereals and grassland
HRAC mode of action code: O + O

See also mecoprop-P

Products

1 Foundation	FMC Agro	84:600 g/l	SL	16414
2 Headland Saxon	FMC Agro	84:600 g/l	SL	11947
3 High Load Mircam	Nufarm UK	80:600 g/l	SL	11930
4 Hyban P	Agrichem	18.7:150 g/l	SL	16799
5 Hygrass P	Agrichem	18.7:150 g/l	SL	16802
6 Prompt	FMC Agro	84:600 g/l	SL	11948
7 Quickfire	Headland Amenity	18.7:150 g/l	SL	17245

Uses

* Annual dicotyledons in *amenity grassland* [1-3, 6]; *managed amenity turf* [4, 5, 7]; *rye (off-label)*, *triticale (off-label)* [1, 3, 6]; *spring barley, spring oats, spring wheat, winter barley, winter oats, winter wheat* [1-4, 6]
* Chickweed in *amenity grassland, rye (off-label), triticale (off-label)* [3]; *managed amenity turf* [4]; *spring barley, spring oats, spring wheat, winter barley, winter oats, winter wheat* [1-4, 6]
* Cleavers in *amenity grassland, rye (off-label), triticale (off-label)* [3]; *managed amenity turf* [4]; *spring barley, spring oats, spring wheat, winter barley, winter oats, winter wheat* [1-4, 6]
* Docks in *managed amenity turf* [5, 7]
* Mayweeds in *amenity grassland, rye (off-label), triticale (off-label)* [3]; *managed amenity turf* [4]; *spring barley, spring oats, spring wheat, winter barley, winter oats, winter wheat* [1-4, 6]
* Perennial dicotyledons in *amenity grassland* [1, 2, 6]; *managed amenity turf* [4, 5, 7]; *spring barley, spring oats, spring wheat, winter barley, winter oats, winter wheat* [4]
* Plantains in *managed amenity turf, spring barley, spring oats, spring wheat, winter barley, winter oats, winter wheat* [4]
* Polygonums in *amenity grassland, rye (off-label), triticale (off-label)* [3]; *managed amenity turf* [4]; *spring barley, spring oats, spring wheat, winter barley, winter oats, winter wheat* [1-4, 6]
* Thistles in *managed amenity turf* [5, 7]

Extension of Authorisation for Minor Use (EAMUs)

* *rye* 20193256 [1], 20180470 [3], 20193234 [6]
* *triticale* 20193256 [1], 20180470 [3], 20193234 [6]

Approval information

* Dicamba and mecoprop-P included in Annex I under EC Regulation 1107/2009
* Accepted by BBPA for use on malting barley

Efficacy guidance

* Best results by application in warm, moist weather when weeds are actively growing

Restrictions

* Maximum number of treatments 1 per crop for cereals and 1 or 2 per yr on grass depending on label. The total amount of mecoprop-P applied in a single yr must not exceed the maximum total dose approved for any single product for the crop/situation
* Do not spray in cold or frosty conditions
* Do not spray if rain expected within 6 h
* Do not treat undersown grass until tillering begins
* Do not spray cereals undersown with clover or legume mixtures
* Do not roll or harrow within 7 d before or after spraying
* Do not treat crops suffering from stress from any cause
* Use product immediately following dilution; do not allow diluted product to stand before use [2]

FOR FULL CONDITIONS OF USE ALWAYS READ THE PRODUCT LABEL

- Avoid treatment when drift may damage neighbouring susceptible crops
- Do not use on new grass for at least 6 months after establishment [7]
- Do not cut grass for at least 1 day after treatment
- Do not apply via hand-held equipment [3]

Crop-specific information
- Latest use: before 1st node detectable for cereals; 7 d before cutting or 14 d before grazing grass
- Apply to winter sown crops from 5 expanded leaf stage (GS 15)
- Apply to spring sown cereals from 5 expanded leaf stage but before first node is detectable (GS 15-31)
- Treat grassland just before perennial weeds flower
- Transient crop prostration may occur after spraying but recovery is rapid

Following crops guidance
- The total amount of mecoprop-P applied in a single year must not exceed the maximum total dose approved for that crop for any single product

Environmental safety
- Dangerous for the environment [2, 3, 6]
- Toxic to aquatic organisms
- Harmful to fish or other aquatic life. Do not contaminate surface waters or ditches with chemical or used container
- Keep livestock out of treated areas for at least 2 wk and until foliage of poisonous weeds such as ragwort has died and become unpalatable
- Some pesticides pose a greater threat of contamination of water than others and mecoprop-P is one of these pesticides. Take special care when applying mecoprop-P near water and do not apply if heavy rain is forecast

Hazard classification and safety precautions
Hazard Harmful, Dangerous for the environment, Harmful if swallowed [1-3, 6]; Irritant [4, 5, 7]
Transport code 9 [3]
Packaging group III [3]
UN Number 3082 [3]; N/C [1, 2, 4-7]
Risk phrases H315 [1-3, 6, 7]; H318 [2-7]; H319 [1]
Operator protection A, C [1-7]; H, M [1, 2, 4-7]; U05a, U11, U20b [1-7]; U08 [1, 2, 4, 6]; U09a [3, 5, 7]; U15 [4, 5, 7]; U19a [5, 7]
Environmental protection E07a, E38 [1, 2, 4-7]; E13c [3-5, 7]; E15a [1, 2, 6]; E34 [1-7]; H410 [3]; H411 [1]; H412 [2, 4-7]
Storage and disposal D01, D02, D05, D09a [1-7]; D10b [1, 2, 6]; D10c [4, 5, 7]; D12a [1, 2, 4-7]
Medical advice M03 [1-7]; M05a [4, 5, 7]

133   dicamba + nicosulfuron

A herbicide mixture for weed control in forage and grain maize
HRAC mode of action code: O + B

Products

Kaltor	Rotam	60:15% w/w	SG	18749

Uses
- Annual dicotyledons in *forage maize*, *grain maize*
- Annual meadow grass in *forage maize*
- Chickweed in *forage maize*
- Cockspur grass in *forage maize*
- Docks in *forage maize*, *grain maize*
- Fat hen in *forage maize*
- Redshank in *forage maize*

Approval information
- Dicamba and nicosulfuron included in Annex I under EC Regulation 1107/2009

SEE SECTION 3 FOR PRODUCTS ALSO REGISTERED

Efficacy guidance
- To avoid damage to crops other than maize, immediately after spraying thoroughly clean all spray equipment as advised on the label, including inside and outside the lid

Restrictions
- To avoid the buildup of resistance do not apply this or any other product containing an ALS inhibitor herbicide with claims for control of grass-weeds more than once to any crop
- Application may only be made between 1 May and 30 September
- Do not apply to crops suffering from herbicide damage or stress caused by pests, nutrition defects or weather.
- Do not spray when cold or frosty conditions are prevalent nor during periods of high temperature
- Do not spray maize subjected to substantial day and night temperature fluctuations.
- Avoid overlapping spray swaths as considerable crop damage may occur which may not grow out and could lead to yield reductions.
- Do not mix with foliar or liquid fertilisers.
- In the case of particularly sensitive varieties, e.g. Abraxas, Fjord, Rival and Nancis, some crop damage may be caused from which recovery may not be complete.
- Do not apply in sequence or in tank-mixture with a product containing any other sulfonylurea.
- Do not mix with adjuvants except those specified on the label

Following crops guidance
- Following normal harvest or in the event of crop failure maize or sunflowers may be sown immediately after ploughing, winter wheat may be sown 5 months after harrowing and the following year sunflower, barley, wheat, oilseed rape, maize may be sown after harrowing.

Environmental safety
- LERAP Category B

Hazard classification and safety precautions
Transport code 9
Packaging group III
UN Number 3077
Risk phrases H318
Operator protection A, C, H; U05a, U08, U20b
Environmental protection E15b, E16a, E39, H410
Storage and disposal D01, D02, D09a, D10c, D11a

---

134   dicamba + prosulfuron

A herbicide mixture for weed control in forage and grain maize
HRAC mode of action code: O + B

Products

1 Casper	Syngenta	50:5% w/w	WG	15573
2 Rosan	FMC Agro	50:5% w/w	WG	18778

Uses
- Annual dicotyledons in *forage maize*, *grain maize*
- Bindweeds in *forage maize*, *grain maize*
- Docks in *forage maize* (seedlings only), *grain maize* (seedlings only)

Approval information
- Dicamba and prosulfuron included in Annex I under EC Regulation 1107/2009

Efficacy guidance
- Always apply in mixture with a non-ionic adjuvant.
- Do not apply in mixture with organo-phosphate insecticides

FOR FULL CONDITIONS OF USE ALWAYS READ THE PRODUCT LABEL

Following crops guidance
- Following normal harvest wheat, barley, rye, triticale and perennial ryegrass may be sown in the autumn, wheat, barley, rye, triticale, combining peas, maize, field beans, forage kale, broccoli and cauliflower may be sown the following spring but sugar beet, sunflowers or lucerne are not recommended.

Environmental safety
- LERAP Category B

Hazard classification and safety precautions
Hazard Dangerous for the environment, Very toxic to aquatic organisms
Transport code 9
Packaging group III
UN Number 3077
Operator protection A, C, H; U02a, U05a, U14, U20b
Environmental protection E15b, E16a, E38, H410
Storage and disposal D01, D02, D09a, D10c, D12a

135   dichlorprop-P + MCPA + mecoprop-P

A translocated herbicide mixture for winter and spring cereals
HRAC mode of action code: O + O + O

See also MCPA
*   mecoprop-P*

Products

1	Duplosan Super	Nufarm UK	310:160:130 g/l	SL	18231
2	Isomec Ultra	Nufarm UK	310:160:130 g/l	SL	16033
3	Optica Trio	FMC Agro	310:160:130 g/l	SL	16113

Uses
- Annual dicotyledons in *durum wheat, rye, spelt, spring barley, spring oats, spring wheat, triticale, winter barley, winter oats, winter wheat*
- Chickweed in *durum wheat, rye, spelt, spring barley, spring oats, spring wheat, triticale, winter barley, winter oats, winter wheat*
- Cleavers in *durum wheat, rye, spelt, spring barley, spring oats, spring wheat, triticale, winter barley, winter oats, winter wheat*
- Field pansy in *durum wheat, rye, spelt, spring barley, spring oats, spring wheat, triticale, winter barley, winter oats, winter wheat*
- Mayweeds in *durum wheat, rye, spelt, spring barley, spring oats, spring wheat, triticale, winter barley, winter oats, winter wheat*
- Poppies in *durum wheat, rye, spelt, spring barley, spring oats, spring wheat, triticale, winter barley, winter oats, winter wheat*

Approval information
- Dichlorprop-P, MCPA and mecoprop-P included in Annex I under EC Regulation 1107/2009
- Accepted by BBPA for use on malting barley

Efficacy guidance
- Best results obtained if application is made while majority of weeds are at seedling stage but not if temperatures are too low
- Optimum results achieved by spraying when temperature is above 10°C. If temperatures are lower delay spraying until growth becomes more active

Restrictions
- Maximum number of treatments 1 per crop
- Do not spray in windy conditions where spray drift may cause damage to neighbouring crops, especially sugar beet, oilseed rape, peas, turnips and most horticultural crops including lettuce and tomatoes under glass
- Do not apply before 1st March in year of application.

SEE SECTION 3 FOR PRODUCTS ALSO REGISTERED

Crop-specific information
- Latest use: before second node detectable (GS 32) for all crops

Environmental safety
- Harmful to aquatic organisms
- Harmful to fish or other aquatic life. Do not contaminate surface waters or ditches with chemical or used container
- Some pesticides pose a greater threat of contamination of water than others and mecoprop-P is one of these pesticides. Take special care when applying mecoprop-P near water and do not apply if heavy rain is forecast

Hazard classification and safety precautions
Hazard Harmful, Dangerous for the environment, Harmful if swallowed
Transport code 9 [1, 2]
Packaging group III [1, 2]
UN Number 3082 [1, 2]; N/C [3]
Risk phrases H315 [3]; H317, R51 [1, 2]; H318 [1-3]
Operator protection A, C, H, M; U05a, U08, U11, U20b [1-3]; U14, U15 [1, 2]
Environmental protection E13c, E34, E38 [1-3]; H410 [1, 2]; H411 [3]
Storage and disposal D01, D02, D05, D09a, D10c [1-3]; D12a [1, 2]
Medical advice M03 [3]; M05a [1, 2]

136   difenacoum

An anticoagulant coumarin rodenticide

Products
1	Difenag Wax Bait	Killgerm	0.005% w/w	RB	UK14-0851
2	Difenag Whole Wheat	Killgerm	0.005% w/w	RB	UK12-0497
3	Neokil	BASF	0.005% w/w	RB	UK12-0298
4	Neosorexa Bait Blocks	BASF	0.005% w/w	RB	UK12-0360
5	Neosorexa Gold	BASF	0.005% w/w	RB	UK12-0304
6	Neosorexa Gold Ratpacks	BASF	0.005% w/w	RB	UK12-0304
7	Neosorexa Pasta Bait	BASF	0.005% w/w	RB	UK12-0365
8	Ratak Cut Wheat	Killgerm	0.005% w/w	RB	UK12-0313
9	Ratak Cut Wheat	BASF	0.005% w/w	RB	UK12-0313
10	Sakarat D Pasta Bait	Killgerm	0.005% w/w	RB	UK12-0371
11	Sakarat D Wax Bait	Killgerm	0.005% w/w	RB	UK12-0370
12	Sorexa D	BASF	0.005% w/w	RB	UK12-0319

Uses
- Mice in *farm buildings* [3-5, 7, 8, 10-12]; *farm buildings/yards* [1, 2]; *farmyards* [3-5, 7, 8, 10, 11]; *sewers* [3, 8]
- Rats in *farm buildings* [3-5, 7, 8, 12]; *farm buildings/yards* [1, 2]; *farmyards* [3-5, 7, 8]; *sewers* [3, 8]

Approval information
- Difenacoum included in Annex I under EC Regulation 1107/2009
- Difenacoum approval was to expire 31/3/2015 but the EU has postponed expiry until 30/6/2018 to allow time for a decision to be reached on the applications for renewal.
- Approval expiry 31 Aug 2020 [2]

Efficacy guidance
- Difenacoum is a chronic poison and rodents need to feed several times before accumulating a lethal dose. Effective against rodents resistant to other commonly used anticoagulants
- Best results achieved by placing baits at points between nesting and feeding places, at entry points, in holes and where droppings are seen
- A minimum of five baiting points normally required for a small infestation; more than 40 for a large infestation
- Inspect bait sites frequently and top up as long as there is evidence of feeding

- Maintain a few baiting points to guard against reinfestation after a successful control campaign
- Resistance to difenacoum in rats is now widespread in the UK. Take local professional advice before relying on these products to control infestations.

Restrictions
- Only for use by farmers, horticulturists and other professional users
- When working in rodent infested areas wear synthetic rubber/PVC gloves to protect against rodent-borne diseases

Environmental safety
- Harmful to wildlife
- Cover baits by placing in bait boxes, drain pipes or under boards to prevent access by children, animals or birds
- Products contain human taste deterrent

Hazard classification and safety precautions
  UN Number N/C [1-5, 7, 8, 10-12]
  Operator protection U13, U20b [1-5, 7, 8, 10-12]
  Environmental protection E10b [3-5, 7, 8, 12]
  Storage and disposal D09a, D11a [1-5, 7, 8, 10-12]
  Vertebrate/rodent control products V01b, V02, V04b [3-5, 7, 8, 10-12]; V03b [10, 11]; V04c [3-5, 7, 8, 12]
  Medical advice M03 [1-5, 7, 8, 10-12]

137   difenoconazole

A diphenyl-ether triazole protectant and curative fungicide
FRAC mode of action code: 3

See also azoxystrobin + difenoconazole

Products

1	Alternet	Belcrop	250 g/l	EC	17689
2	Difcon 250	Harvest	250 g/l	EC	17851
3	Difcor 250 EC	Belchim	250 g/l	EC	13917
4	Difenostar	Life Scientific	250 g/l	EC	19118
5	Difference	Belchim	250 g/l	EC	16129
6	Kix	Belchim	250 g/l	EC	17424
7	Narita	Belchim	250 g/l	EC	16210
8	Plover	Syngenta	250 g/l	EC	17288
9	Septuna	Novastar	250 g/l	EC	17390

Uses
- Alternaria in **borage for oilseed production** *(off-label)*, **canary flower (echium spp.)** *(off-label)*, **evening primrose** *(off-label)*, **honesty** *(off-label)*, **linseed** *(off-label)*, **mustard** *(off-label)*, **rye** *(off-label)*, **triticale** *(off-label)* [3]; **broccoli, brussels sprouts, cabbages, calabrese, cauliflowers, spring oilseed rape, winter oilseed rape** [2-4, 8, 9]; **choi sum** *(off-label)* [8]; **collards, kale** [4, 8, 9]
- Blight in **forest nurseries** *(off-label)* [3]
- Brown rust in **rye** *(off-label)*, **triticale** *(off-label)* [3]; **winter wheat** [2, 3]
- Cladosporium in **rye** *(off-label)*, **triticale** *(off-label)* [3]
- Disease control in **potatoes** [1, 6, 7]
- Late blight in **cardoons** *(off-label)*, **celeriac** *(off-label)*, **celery (outdoor)** *(off-label)*, **florence fennel** *(off-label)*, **rhubarb** *(off-label)* [8]
- Leaf blight in **cardoons** *(off-label)*, **celery (outdoor)** *(off-label)*, **florence fennel** *(off-label)*, **rhubarb** *(off-label)* [8]
- Leaf spot in **cardoons** *(off-label)*, **celery (outdoor)** *(off-label)*, **choi sum** *(off-label)*, **florence fennel** *(off-label)*, **rhubarb** *(off-label)* [8]; **forest nurseries** *(off-label)* [3]
- Light leaf spot in **spring oilseed rape, winter oilseed rape** [2-4, 8, 9]

SEE SECTION 3 FOR PRODUCTS ALSO REGISTERED

- Powdery mildew in **borage for oilseed production** *(off-label)*, **canary flower (echium spp.)** *(off-label)*, **evening primrose** *(off-label)*, **honesty** *(off-label)*, **linseed** *(off-label)*, **mustard** *(off-label)* [3]
- Ring spot in **broccoli**, **brussels sprouts**, **cabbages**, **calabrese**, **cauliflowers** [2-4, 8, 9]; **choi sum** *(off-label)* [8]; **collards**, **kale** [4, 8, 9]
- Rust in **asparagus** *(off-label)* [8]; **forest nurseries** *(off-label)* [3]
- Scab in **apples**, **pears** [5]
- Sclerotinia in **borage for oilseed production** *(off-label)*, **canary flower (echium spp.)** *(off-label)*, **evening primrose** *(off-label)*, **honesty** *(off-label)*, **linseed** *(off-label)*, **mustard** *(off-label)* [3]
- Septoria leaf blotch in **rye** *(off-label)*, **triticale** *(off-label)* [3]; **winter wheat** [2, 3]
- Stem canker in **spring oilseed rape**, **winter oilseed rape** [2-4, 8, 9]

Extension of Authorisation for Minor Use (EAMUs)
- **asparagus** *20171713* [8]
- **borage for oilseed production** *20140211* [3]
- **canary flower (echium spp.)** *20140211* [3]
- **cardoons** *20192713* [8]
- **celeriac** *20182953* [8]
- **celery (outdoor)** *20192713* [8]
- **choi sum** *20171714* [8]
- **evening primrose** *20140211* [3]
- **florence fennel** *20192713* [8]
- **forest nurseries** *20140205* [3]
- **honesty** *20140211* [3]
- **linseed** *20140211* [3]
- **mustard** *20140211* [3]
- **rhubarb** *20192713* [8]
- **rye** *20193253* [3]
- **triticale** *20193253* [3]

Approval information
- Difenoconazole included in Annex I under EC Regulation 1107/2009

Efficacy guidance
- In brassicas a 3-spray programme should be used starting at the first sign of disease and repeated at 14-21 d intervals
- Product is fully rainfast 2 h after application
- For most effective control of Septoria, apply as part of a programme of sprays which includes a suitable flag leaf treatment
- Difenoconazole is a DMI fungicide. Resistance to some DMI fungicides has been identified in Septoria leaf blotch which may seriously affect performance of some products. For further advice contact a specialist advisor and visit the Fungicide Resistance Action Group (FRAG)-UK website

Restrictions
- Maximum number of treatments 3 per crop for brassicas; 2 per crop for oilseed rape; 1 per crop for wheat [3, 5], 4 for potatoes [6, 7]
- Maximum total dose equivalent to 3 full dose treatments on brassicas; 2 full dose treatments on oilseed rape; 1 full dose treatment on wheat [8]
- Apply to wheat any time from ear fully emerged stage but before early milk-ripe stage (GS 59-73)
- Maintain an interval of at least 14 days between applications to Brassica crops, 10 - 14 days in potatoes [6, 7]

Crop-specific information
- Latest use: before grain early milk-ripe stage (GS 73) for cereals; end of flowering for oilseed rape
- HI brassicas 14 or 21 d
- Treat oilseed rape in autumn from 4 expanded true leaf stage (GS 1,4). A repeat spray may be made in spring at the beginning of stem extension (GS 2,0) if visible symptoms develop

Environmental safety
- Dangerous for the environment

FOR FULL CONDITIONS OF USE ALWAYS READ THE PRODUCT LABEL

- Very toxic to aquatic organisms
- Buffer zone requirement 6 m [1]
- Broadcast air-assisted LERAP [5] (20m); LERAP Category B [1-4, 6-9]

Hazard classification and safety precautions

Hazard Harmful [5]; Irritant [2, 3]; Dangerous for the environment [1-9]; Harmful if swallowed [2, 3, 5-7]; Very toxic to aquatic organisms [4]
Transport code 9
Packaging group III
UN Number 3082
Risk phrases H304 [1-4, 6-9]; H315, H336 [1, 5]; H318 [1]; H319 [2-4, 6, 7, 9]; H320 [5]
Operator protection A, C, H; U05a, U20b [1-9]; U09a [1-4, 6-9]; U11 [2, 3, 5]
Environmental protection E15a [1-4, 6-8]; E15b [5]; E16a [1-4, 6-9]; E17b [5] (20m); E38 [1-9]; H410 [1, 4, 8, 9]; H411 [2, 3, 5-7]
Storage and disposal D01, D02 [1-9]; D05, D07, D10c [1, 4, 6-9]; D09a, D12a [1-4, 6-9]; D10b [2, 3]
Medical advice M05b [5]

138   difenoconazole + fludioxonil

A triazole + phenylpyrrole seed treatment for use in cereals
FRAC mode of action code: 3 + 12

See also fludioxonil

Products

1 Difend Extra	Certis	25:25 g/l	FS	17739
2 Instrata Elite	ICL (Everris) Ltd	80.3:80.3 g/l	SC	17976

Uses

- Disease control in **managed amenity turf** [2]
- Seed-borne diseases in **winter wheat** [1]

Approval information

- Difenoconazole and fludioxonil included in Annex I under EC Regulation 1107/2009

Efficacy guidance

- Effective against benzimidazole-resistant and benzimidazole-sensitive strains of *Microdochium nivale*

Crop-specific information

- Under adverse environmental or soil conditions, seed rates should be increased to compensate for a slight drop in germination capacity. Flow rates of treated seed should be checked before drilling commences.

Environmental safety

- Buffer zone requirement 10 m [2]
- LERAP Category B [2]

Hazard classification and safety precautions

Hazard Dangerous for the environment [1]
Transport code 9
Packaging group III
UN Number 3082
Operator protection A, H; U05a [1]
Environmental protection E16a [2]; E38 [1]; H410 [1, 2]
Storage and disposal D01, D02, D05 [1, 2]; D12a [1]
Treated seed S01, S02, S03, S04a, S05, S06a, S07, S08 [1]

SEE SECTION 3 FOR PRODUCTS ALSO REGISTERED

139   difenoconazole + fluxapyroxad

A fungicide mixture for disease control in a range of horticultural crops
FRAC mode of action code: 3 + 7

Products

1	Charm	BASF	50:75 g/l	SC	18396
2	Perseus	BASF	50:75 g/l	SC	18397

Uses

- Alternaria in **broccoli** *(reduction)*, **calabrese** *(reduction)*, **celeriac** *(off-label)*, **horseradish** *(off-label)*, **jerusalem artichokes** *(off-label)*, **parsnips** *(off-label)*, **potatoes** *(moderate control)*, **radishes** *(off-label)*, **red beet** *(off-label)*, **swedes** *(off-label)*, **turnips** *(off-label)* [2]
- Blight in **beans with pods** *(off-label)*, **dwarf beans** *(off-label)*, **edible podded peas** *(off-label)*, **french beans** *(off-label)*, **runner beans** *(off-label)* [2]
- Botrytis in **beans with pods** *(off-label)*, **dwarf beans** *(off-label)*, **edible podded peas** *(off-label)*, **french beans** *(off-label)*, **runner beans** *(off-label)* [2]
- Disease control in **protected chilli peppers**, **protected cucumbers**, **protected peppers**, **protected strawberries**, **protected tomatoes**, **strawberries** [1]
- Leaf and pod spot in **vining peas** *(moderate control)* [2]
- Leaf spot in **beans with pods** *(off-label)*, **dwarf beans** *(off-label)*, **edible podded peas** *(off-label)*, **french beans** *(off-label)*, **runner beans** *(off-label)* [2]
- Powdery mildew in **celeriac** *(off-label)*, **horseradish** *(off-label)*, **jerusalem artichokes** *(off-label)*, **parsnips** *(off-label)*, **radishes** *(off-label)*, **red beet** *(off-label)*, **swedes** *(off-label)*, **turnips** *(off-label)* [2]
- Rhizoctonia in **celeriac** *(off-label)*, **horseradish** *(off-label)*, **jerusalem artichokes** *(off-label)*, **parsnips** *(off-label)*, **radishes** *(off-label)*, **red beet** *(off-label)*, **swedes** *(off-label)*, **turnips** *(off-label)* [2]
- Ring spot in **broccoli** *(reduction)*, **calabrese** *(reduction)* [2]
- Sclerotinia in **celeriac** *(off-label)*, **horseradish** *(off-label)*, **jerusalem artichokes** *(off-label)*, **lettuce**, **parsnips** *(off-label)*, **protected lettuce** *(moderate control)*, **radishes** *(off-label)*, **red beet** *(off-label)*, **swedes** *(off-label)*, **turnips** *(off-label)* [2]

Extension of Authorisation for Minor Use (EAMUs)

- **beans with pods** *20193285* [2]
- **celeriac** *20193426* [2]
- **dwarf beans** *20193285* [2]
- **edible podded peas** *20193285* [2]
- **french beans** *20193285* [2]
- **horseradish** *20193426* [2]
- **jerusalem artichokes** *20193426* [2]
- **parsnips** *20193426* [2]
- **radishes** *20193426* [2]
- **red beet** *20193426* [2]
- **runner beans** *20193285* [2]
- **swedes** *20193426* [2]
- **turnips** *20193426* [2]

Approval information

- Difenoconazole and fluxapyroxad included in Annex I under EC Regulation 1107/2009

Restrictions

- Do not use in protected tomato and pepper crops between 1st October and 1st March [1]
- Consult processors before using Perseus on vining peas or potatoes.
- Do not use on crops intended for fermentation.
- Do not use on protected crops intended for production of seed or propagation material.
- Avoid overlapping sprays.
- The total number of applications of SDHI containing products should make up no more than 50% of the total number of fungicides applied per season.
- No more than two consecutive applications of SHDI fungicides should be applied per crop.

SECTION 2

Environmental safety
- LERAP Category B

Hazard classification and safety precautions
    Transport code 9
    Packaging group III
    UN Number 3082
    Risk phrases H319, H351
    Operator protection A, C; U11 [2]; U20b [1, 2]
    Environmental protection E15b [1]; E16a, E34, H410 [1, 2]
    Storage and disposal D01, D02, D09a, D10c

140 difenoconazole + mandipropamid

A triazole/mandelamide fungicide mixture for blight control in potatoes
FRAC mode of action code: 3 + 40

Products

1	Amphore Plus	Syngenta	250:250 g/l	SL	16327
2	Carial Star	Syngenta	250:250 g/l	SL	16323

Uses
- Alternaria in **protected aubergines** *(off-label)*, **protected tomatoes** *(off-label)* [2]
- Alternaria blight in **potatoes** [1, 2]
- Blight in **potatoes** [1, 2]
- Early blight in **potatoes** [1, 2]
- Late blight in **protected aubergines** *(off-label)*, **protected tomatoes** *(off-label)* [2]
- Phytophthora in **protected aubergines** *(off-label)*, **protected tomatoes** *(off-label)* [2]

Extension of Authorisation for Minor Use (EAMUs)
- **protected aubergines** *20183162* [2]
- **protected tomatoes** *20183162* [2]

Approval information
- Difenoconazole and mandipropamid included in Annex I under EC Regulation 1107/2009

Efficacy guidance
- Rainfast within 15 minutes
- Spray programme must start before blight enters the crop

Crop-specific information
- Can be used on all varieties of potato, including seed potatoes

Environmental safety
- LERAP Category B

Hazard classification and safety precautions
    Transport code 9
    Packaging group III
    UN Number 3082
    Operator protection A, H; U05a, U20b
    Environmental protection E15b, E16a, E22c, E34, E38, H410
    Storage and disposal D01, D09a, D10c
    Medical advice M03

141 difenoconazole + paclobutrazol

A triazole fungicide mixture for growth regulation and disease control in oilseed rape
FRAC mode of action code: 3

Products

Toprex	Syngenta	250:125 g/l	SC	16456

SEE SECTION 3 FOR PRODUCTS ALSO REGISTERED

Uses

- Disease control in *winter oilseed rape*
- Kabatiella lini in *linseed (off-label)*
- Light leaf spot in *winter oilseed rape*
- Mildew in *linseed (off-label)*
- Phoma in *winter oilseed rape*
- Septoria leaf blotch in *linseed (off-label)*

Extension of Authorisation for Minor Use (EAMUs)

- *linseed* 20151633

Approval information

- Difenoconazole and paclobutrazol included in Annex I under EC Regulation 1107/2009

Efficacy guidance

- Contains a growth regulator for height reduction and lodging control and a fungicide for disease control. It should only be used when both growth regulation and disease control are required. If this is not the case, use appropriate alternative products at the required timing

Hazard classification and safety precautions

    Hazard Dangerous for the environment, Very toxic to aquatic organisms
    Transport code 9
    Packaging group III
    UN Number 3082
    Risk phrases H361
    Operator protection A, H; U05a
    Environmental protection E38, H410
    Storage and disposal D01, D02, D12a
    Medical advice M05a

142   diflubenzuron

A selective, persistent, contact and stomach acting insecticide
IRAC mode of action code: 15

Products

Dimilin Flo	Arysta	480 g/l	SC	08769

Uses

- Insect pests in *amenity vegetation*, *forest*, *hedges*, *ornamental plant production*

Approval information

- Approved for aerial application in forestry when average wind velocity does not exceed 18 knots and gusts do not exceed 20 knots. See Section 5 for more information
- Diflubenzuron included in Annex I under EC Regulation 1107/2009

Efficacy guidance

- Most active on young caterpillars and most effective control achieved by spraying as eggs start to hatch
- Dose and timing of spray treatments vary with pest and crop. See label for details
- Addition of wetter recommended for use on brassicas and for pear sucker control in pears
- No person may carry out aerial spraying or cause or permit another person to carry out aerial spraying unless such spraying is authorised by an aerial spraying permit issued by the Chemicals Regulation Directorate. The spray droplet spectra produced must be of a minimum volume median diameter (VMD) of 200 microns.

Restrictions

- Maximum number of treatments 3 per yr for apples, pears; 2 per yr for plums, blackcurrants; 2 per crop for brassicas; 1 per yr for forest
- Before treating ornamentals check varietal tolerance on a small sample
- Do not use as a compost drench or incorporated treatment on ornamental crops

- Do not spray protected plants in flower or with flower buds showing colour
- For use only on the food crops specified on the label

Crop-specific information
- HI apples, pears, plums, blackcurrants, brassicas 14 d

Environmental safety
- Dangerous for the environment
- Very toxic to aquatic organisms

Hazard classification and safety precautions
Hazard Dangerous for the environment
Transport code 9
Packaging group III
UN Number 3082
Operator protection U20c
Environmental protection E15a, E38, H410; E17a (20m)
Storage and disposal D09a, D11a

143   diflufenican

A shoot absorbed pyridinecarboxamide herbicide for winter cereals
HRAC mode of action code: F1

See also bromoxynil + diflufenican
bromoxynil + diflufenican + ioxynil
chlorotoluron + diflufenican
chlorotoluron + diflufenican + pendimethalin
clodinafop-propargyl + diflufenican
clopyralid + diflufenican + MCPA

Products

1	Cachet	Agform	500 g/l	SC	18584
2	Clayton El Nino	Clayton	500 g/l	SC	17337
3	Dican	Albaugh UK	50% w/w	WG	16922
4	Diflanil 500 SC	Belchim	500 g/l	SC	17024
5	Hurricane SC	Adama	500 g/l	SC	16027
6	Ossetia	Rotam	50% w/w	WG	17741
7	Sempra XL	UPL Europe	500 g/l	SC	17447
8	Solo 500 SC	Sipcam	500 g/l	SC	17622
9	Twister 500	Harvest	500 g/l	SC	17627

Uses

- Annual dicotyledons in *game cover* (off-label), *grass seed crops* (off-label), *spring oats* (off-label), *winter oats* (off-label) [5]; *spring barley*, *triticale*, *winter barley*, *winter rye*, *winter wheat* [4]
- Annual meadow grass in *game cover* (off-label), *spring oats* (off-label), *winter oats* (off-label) [5]; *grass seed crops* (off-label) [5, 7]; *oat seed crops* (off-label) [7]
- Black bindweed in *grass seed crops* (off-label), *oat seed crops* (off-label) [7]
- Brassica spp. in *durum wheat*, *spring barley*, *spring wheat*, *triticale*, *winter barley*, *winter rye*, *winter wheat* [6]
- Charlock in *forest nurseries* (off-label), *game cover* (off-label), *grass seed crops* (off-label), *oat seed crops* (off-label), *ornamental plant production* (off-label) [7]
- Chickweed in *carrots* (off-label) [5]; *forest nurseries* (off-label), *game cover* (off-label), *grass seed crops* (off-label), *oat seed crops* (off-label), *ornamental plant production* (off-label) [7]; *spring barley*, *triticale*, *winter barley*, *winter rye*, *winter wheat* [4]
- Cleavers in *durum wheat*, *spring wheat* [1, 2, 5-7, 9]; *rye* [8]; *spring barley*, *triticale* [1, 2, 4-9]; *winter barley*, *winter wheat* [1-9]; *winter rye* [1, 2, 4-7, 9]
- Corn spurrey in *forest nurseries* (off-label), *game cover* (off-label), *grass seed crops* (off-label), *oat seed crops* (off-label), *ornamental plant production* (off-label) [7]

SEE SECTION 3 FOR PRODUCTS ALSO REGISTERED

- Dead nettle in *forest nurseries* (off-label), *game cover* (off-label), *grass seed crops* (off-label), *oat seed crops* (off-label), *ornamental plant production* (off-label) [7]
- Field pansy in *carrots* (off-label) [5]; *durum wheat*, *spring wheat*, *winter rye* [1, 2, 5-7, 9]; *forest nurseries* (off-label), *game cover* (off-label), *grass seed crops* (off-label), *oat seed crops* (off-label), *ornamental plant production* (off-label) [7]; *rye* [8]; *spring barley*, *triticale* [1, 2, 5-9]; *winter barley*, *winter wheat* [1-3, 5-9]
- Field speedwell in *durum wheat*, *spring wheat* [1, 2, 5-7, 9]; *rye* [8]; *spring barley*, *triticale* [1, 2, 4-9]; *winter barley*, *winter wheat* [1-9]; *winter rye* [1, 2, 4-7, 9]
- Flixweed in *forest nurseries* (off-label), *game cover* (off-label), *grass seed crops* (off-label), *oat seed crops* (off-label), *ornamental plant production* (off-label) [7]
- Forget-me-not in *forest nurseries* (off-label), *game cover* (off-label), *ornamental plant production* (off-label) [7]
- Ivy-leaved speedwell in *durum wheat*, *spring wheat* [1, 2, 5-7, 9]; *rye* [8]; *spring barley*, *triticale* [1, 2, 4-9]; *winter barley*, *winter wheat* [1-9]; *winter rye* [1, 2, 4-7, 9]
- Knotgrass in *grass seed crops* (off-label), *oat seed crops* (off-label) [7]
- Mayweeds in *durum wheat*, *spring wheat*, *winter rye* [1, 2, 5-7, 9]; *rye* [8]; *spring barley*, *triticale* [1, 2, 5-9]; *winter barley*, *winter wheat* [1-3, 5-9]
- Mouse-ear chickweed in *forest nurseries* (off-label), *game cover* (off-label), *grass seed crops* (off-label), *oat seed crops* (off-label), *ornamental plant production* (off-label) [7]
- Nipplewort in *grass seed crops* (off-label), *oat seed crops* (off-label) [7]
- Parsley-piert in *forest nurseries* (off-label), *game cover* (off-label), *grass seed crops* (off-label), *oat seed crops* (off-label), *ornamental plant production* (off-label) [7]
- Poppies in *durum wheat*, *spring wheat*, *winter rye* [1, 2, 5-7, 9]; *forest nurseries* (off-label), *game cover* (off-label), *grass seed crops* (off-label), *oat seed crops* (off-label), *ornamental plant production* (off-label) [7]; *rye* [8]; *spring barley*, *triticale* [1, 2, 5-9]; *winter barley*, *winter wheat* [1-3, 5-9]
- Red dead-nettle in *durum wheat*, *spring wheat*, *winter rye* [1, 2, 5, 7, 9]; *rye* [8]; *spring barley*, *triticale* [1, 2, 5, 7-9]; *winter barley*, *winter wheat* [1-3, 5, 7-9]
- Runch in *forest nurseries* (off-label), *game cover* (off-label), *grass seed crops* (off-label), *oat seed crops* (off-label), *ornamental plant production* (off-label) [7]
- Shepherd's purse in *carrots* (off-label) [5]; *forest nurseries* (off-label), *game cover* (off-label), *ornamental plant production* (off-label) [7]
- Sowthistle in *grass seed crops* (off-label), *oat seed crops* (off-label) [7]
- Speedwells in *carrots* (off-label) [5]; *forest nurseries* (off-label), *game cover* (off-label), *grass seed crops* (off-label), *oat seed crops* (off-label), *ornamental plant production* (off-label) [7]
- Treacle mustard in *forest nurseries* (off-label), *game cover* (off-label), *grass seed crops* (off-label), *oat seed crops* (off-label), *ornamental plant production* (off-label) [7]
- Volunteer oilseed rape in *carrots* (off-label) [5]; *forest nurseries* (off-label), *game cover* (off-label), *grass seed crops* (off-label), *oat seed crops* (off-label), *ornamental plant production* (off-label) [7]; *spring barley*, *triticale*, *winter barley*, *winter rye*, *winter wheat* [4]
- Wild radish in *forest nurseries* (off-label), *game cover* (off-label), *grass seed crops* (off-label), *oat seed crops* (off-label), *ornamental plant production* (off-label) [7]

Extension of Authorisation for Minor Use (EAMUs)
- *carrots* 20190180 [5]
- *forest nurseries* 20170374 [7]
- *game cover* 20183440 [5], 20170374 [7]
- *grass seed crops* 20141324 [5], 20162935 [7]
- *oat seed crops* 20162935 [7]
- *ornamental plant production* 20170374 [7]
- *spring oats* 20141269 [5]
- *winter oats* 20141269 [5]

Approval information
- Diflufenican included in Annex I under EC Regulation 1107/2009
- Accepted by BBPA for use on malting barley

Efficacy guidance
- Best results achieved from treatment of small actively growing weeds in early autumn or spring

FOR FULL CONDITIONS OF USE ALWAYS READ THE PRODUCT LABEL

- Good weed control depends on efficient burial of trash or straw before or during seedbed preparation
- Loose or fluffy seedbeds should be rolled before application
- The final seedbed should be moist, fine and firm with clods no bigger than fist size
- Ensure good even spray coverage and increase spray volume for post-emergence treatments where the crop or weed foliage is dense
- Activity may be slow under cool conditions and final level of weed control may take some time to appear
- Where cleavers are a particular problem a separate specific herbicide treatment may be required
- Efficacy may be impaired on soils with a Kd factor greater than 6
- Always follow WRAG guidelines for preventing and managing herbicide resistant weeds. See Section 5 for more information

Restrictions
- Do not treat broadcast crops [4, 5]
- Do not roll treated crops or harrow at any time after treatment
- Do not apply to soils with more than 10% organic matter or on Sands, or very stony or gravelly soils
- Do not treat after a period of cold frosty weather
- Horizontal boom sprayers must be fitted with three star drift reduction technology for all uses [1-3, 5, 7-9]
- Must not be applied via hand-held equipment
- Low drift spraying equipment must be used for 30 m from the top of the bank of any surface water bodies [2, 3, 5-9]

Crop-specific information
- Latest use: before end of tillering (GS 29) for wheat and barley [4, 5], pre crop emergence for triticale and winter rye [4, 5]
- Treat only named varieties of rye or triticale [4, 5]

Following crops guidance
- Labels vary slightly but in general ploughing to 150 mm and thoroughly mixing the soil is recommended before drilling or planting any succeeding crops either after crop failure or after normal harvest
- In the event of crop failure only winter wheat or winter barley may be re-drilled immediately after ploughing. Spring crops of wheat, barley, oilseed rape, peas, field beans, sugar beet [4, 5], potatoes, carrots, edible brassicas or onions may be sown provided an interval of 12 wk has elapsed after ploughing
- After normal harvest of a treated crop winter cereals, oilseed rape, field beans, leaf brassicas, sugar beet seed crops and winter onions may be drilled in the following autumn. Other crops listed above for crop failure may be sown in the spring after normal harvest
- Successive treatments with any products containing diflufenican can lead to soil build-up and inversion ploughing to 150 mm must precede sowing any following non-cereal crop. Even where ploughing occurs some crops may be damaged

Environmental safety
- Dangerous for the environment
- Very toxic to aquatic organisms
- Buffer zone requirement 6m [1-3, 5-8]
- Buffer zone requirement 7 m for spring barley, 8m for winter barley, winter rye, triticale and winter wheat [4]
- Buffer zone requirement 12 m on crops grown for game cover or ornamental plant production [5]
- LERAP Category B

Hazard classification and safety precautions
Hazard Dangerous for the environment [1-9]; Very toxic to aquatic organisms [1, 3, 5, 6, 8, 9]
Transport code 9
Packaging group III
UN Number 3077 [3, 6]; 3082 [1, 2, 4, 5, 7-9]
Operator protection A [1-9]; C [1, 2, 4, 5, 7-9]; H [3]; U05a [8, 9]; U08, U13, U19a, U20a [1-3, 5-9]; U15 [1-3, 5-7]

SEE SECTION 3 FOR PRODUCTS ALSO REGISTERED

Environmental protection E13c [8, 9]; E15a [4]; E16a, H410 [1-9]; E16i [1, 2, 5]; E34, E38 [1-3, 5-9]
Storage and disposal D01, D02 [1-3, 5-9]; D05, D09a, D10b [1-9]; D12a [1-7]
Medical advice M05a [1-3, 5-7]

144   diflufenican + florasulam

A herbicide mixture for weed control in cereals
HRAC mode of action code: B + F1

See also diflufenican
florasulam

Products

1 Bow	Nufarm UK	500:50 g/l	SC	19054
2 Lector Delta	Nufarm UK	500:50 g/l	SC	19055

Uses

- Annual dicotyledons in *rye* [2]; *spring barley*, *triticale*, *winter barley*, *winter wheat* [1, 2]; *winter rye* [1]
- Black bindweed in *rye* [2]; *spring barley*, *triticale*, *winter barley*, *winter wheat* [1, 2]; *winter rye* [1]
- Charlock in *rye* [2]; *spring barley*, *triticale*, *winter barley*, *winter wheat* [1, 2]; *winter rye* [1]
- Chickweed in *rye* [2]; *spring barley*, *triticale*, *winter barley*, *winter wheat* [1, 2]; *winter rye* [1]
- Cleavers in *rye* [2]; *spring barley*, *triticale*, *winter barley*, *winter wheat* [1, 2]; *winter rye* [1]
- Field pansy in *rye* [2]; *spring barley*, *triticale*, *winter barley*, *winter wheat* [1, 2]; *winter rye* [1]
- Forget-me-not in *rye* [2]; *spring barley*, *triticale*, *winter barley*, *winter wheat* [1, 2]; *winter rye* [1]
- Hemp-nettle in *rye* [2]; *spring barley*, *triticale*, *winter barley*, *winter wheat* [1, 2]; *winter rye* [1]
- Mayweeds in *rye* [2]; *spring barley*, *triticale*, *winter barley*, *winter wheat* [1, 2]; *winter rye* [1]
- Penny cress in *rye* [2]; *spring barley*, *triticale*, *winter barley*, *winter wheat* [1, 2]; *winter rye* [1]
- Shepherd's purse in *rye* [2]; *spring barley*, *triticale*, *winter barley*, *winter wheat* [1, 2]; *winter rye* [1]
- Volunteer oilseed rape in *rye* [2]; *spring barley*, *triticale*, *winter barley*, *winter wheat* [1, 2]; *winter rye* [1]

Approval information

- Diflufenican and florasulam included in Annex I under EC Regulation 1107/2009

Restrictions

- Following application to grass seed crops, the treated grass must not be fed to livestock.
- The total amount of florasulam applied to a cereal crop must not exceed 7.5 g.a.i/ha
- Only one other product with an ALS inhibitor mode of action may be applied to a treated cereal crop. However, a further application or another product containing florasulam may also be made providing the maximum total dose of florasulam is not exceeded. A joint application with one of the listed ALS products can be made to the same cereal crop (see label).
- To avoid subsequent injury to crops other than cereals all spraying equipment must be thoroughly cleaned both inside and outside using All Clear Extra.

Following crops guidance

- In the event of crop failure in the spring, cultivate to 20 cms and then only plant spring wheat, spring barley or spring oats.
- Crops that can be sown in the same calendar year following treatment are cereals, oilseed rape, field beans, grass, peas, sugar beet, potatoes, maize and vegetable brassicas as transplants.

FOR FULL CONDITIONS OF USE ALWAYS READ THE PRODUCT LABEL

Environmental safety
- Buffer zone requirement 6 m for spring barley, 5 m for winter barley, winter wheat, winter rye and triticale
- LERAP Category B

Hazard classification and safety precautions
Hazard Dangerous for the environment
Transport code 9
Packaging group III
UN Number 3082
Operator protection A; U05a [1]
Environmental protection E15b, E16a, E38, H410
Storage and disposal D01, D02, D05, D06a, D09a, D10c, D12a [1, 2]; D03 [2]

145 diflufenican + flufenacet

A contact and residual herbicide mixture for cereals
HRAC mode of action code: F1 + K3

See also flufenacet

Products

1	Ambush	Agform	100:400 g/l	SC	18136
2	Ascent	Agform	100:400 g/l	SC	17902
3	Clayton Aspect	Clayton	100:400 g/l	SC	18550
4	Clayton Facet	Clayton	100:400 g/l	SC	18265
5	Clayton Sabre	Clayton	100:400 g/l	SC	18535
6	Clayton Vista	Clayton	200:400 g/l	SC	16705
7	Dephend	FMC Agro	100:400 g/l	SC	18739
8	Firestarter	Life Scientific	100:400 g/l	SC	18422
9	Firestorm	Certis	100:400 g/l	SC	17631
10	Flunican	Becesane	200:400 g/l	SC	18305
11	Giddo	Bayer CropScience	100:400 g/l	SC	19128
12	Golding	Rotam	100:400 g/l	SC	18654
13	Herold	Adama	200:400 g/l	SC	16195
14	Liberator	Bayer CropScience	100:400 g/l	SC	15206
15	Mertil	UPL Europe	200:400 g/l	SC	18504
16	Naceto	Belchim	200:400 g/l	SC	18063
17	Nucleus	FMC Agro	100:400 g/l	SC	19026
18	Pincer	Agform	100:400 g/l	SC	17130
19	Regatta	Bayer CropScience	100:400 g/l	SC	15353
20	Reliance	UPL Europe	200:400 g/l	SC	18372
21	Terrane	FMC Agro	100:400 g/l	SC	18738
22	Trumpet	Harvest	200:400 g/l	SC	18183

Uses
- Annual dicotyledons in *grass seed crops* (off-label) [14]; *rye* (off-label), *triticale* (off-label) [11, 13, 14]; *spring barley* (off-label) [13, 14, 19]; *spring wheat* (off-label) [13]; *winter oats* (off-label) [13, 19]
- Annual grasses in *winter oats* (off-label) [14]
- Annual meadow grass in *grass seed crops* (off-label) [14]; *rye* (off-label), *triticale* (off-label) [11, 13, 14]; *spring barley* (off-label) [13, 14, 19]; *spring wheat* [8, 11, 14]; *spring wheat* (off-label) [13]; *triticale, winter rye* [7, 15-17, 20, 21]; *winter barley, winter wheat* [1-22]; *winter oats* (off-label) [13, 19]
- Blackgrass in *grass seed crops* (off-label) [14]; *rye* (off-label), *triticale* (off-label) [11, 14]; *spring barley* (off-label), *winter oats* (off-label) [14, 19]; *spring wheat* [8, 11, 14]; *triticale, winter rye* [7, 15-17, 20, 21]; *winter barley, winter wheat* [1-22]
- Chickweed in *spring wheat* [8, 11, 14]; *triticale, winter rye* [7, 15-17, 20, 21]; *winter barley, winter wheat* [1-22]

- Field pansy in **spring wheat** [8, 11, 14]; **triticale, winter rye** [7, 15-17, 20, 21]; **winter barley, winter wheat** [1-22]
- Field speedwell in **spring wheat** [8, 11, 14]; **triticale, winter rye** [7, 15-17, 20, 21]; **winter barley, winter wheat** [1-22]
- Ivy-leaved speedwell in **spring wheat** [8, 11, 14]; **triticale, winter rye** [7, 15-17, 20, 21]; **winter barley, winter wheat** [1-22]
- Mayweeds in **spring wheat** [8, 11, 14]; **triticale, winter rye** [7, 15-17, 20, 21]; **winter barley, winter wheat** [1-22]
- Red dead-nettle in **spring wheat** [8, 11, 14]; **winter barley, winter wheat** [1-5, 8, 9, 11, 12, 14, 18, 19]
- Shepherd's purse in **triticale, winter rye** [7, 15-17, 20, 21]; **winter barley, winter wheat** [6, 7, 10, 13, 15-17, 20-22]
- Volunteer oilseed rape in **triticale, winter rye** [7, 15-17, 20, 21]; **winter barley, winter wheat** [6, 7, 10, 13, 15-17, 20-22]
- Wild oats in **winter oats** *(off-label)* [14]

Extension of Authorisation for Minor Use (EAMUs)
- **grass seed crops** *20193241* [14]
- **rye** *20193349* [11], *20193259* [13], *20193241* [14]
- **spring barley** *20141698* [13], *20121010* [14], *20121905* [19]
- **spring wheat** *20141698* [13]
- **triticale** *20193349* [11], *20193259* [13], *20193241* [14]
- **winter oats** *20141693* [13], *20111550* [14], *20121904* [19]

Approval information
- Diflufenican and flufenacet included in Annex I under EC Regulation 1107/2009
- Accepted by BBPA for use on malting barley

Efficacy guidance
- Best results obtained when there is moist soil at and after application and rain falls within 7 d
- Residual control may be reduced under prolonged dry conditions
- Activity may be slow under cool conditions and final level of weed control may take some time to appear
- Good weed control depends on burying any trash or straw before or during seedbed preparation
- Established perennial grasses and broad-leaved weeds will not be controlled
- Do not use as a stand-alone treatment for blackgrass control. Always follow WRAG guidelines for preventing and managing herbicide resistant weeds. Section 5 for more information
- Application can only be made between BBCH growth stages 10-13 [16]

Restrictions
- Maximum number of treatments 1 per crop
- Do not treat undersown cereals or those to be undersown
- Do not use on waterlogged soils or soils prone to waterlogging
- Do not use on Sands or Very Light soils, or very stony or gravelly soils, or on soils containing more than 10% organic matter
- Do not treat broadcast crops and treat shallow-drilled crops post-emergence only
- Do not incorporate into the soil or disturb the soil after application by rolling or harrowing
- Avoid treating crops under stress from whatever cause and avoid treating during periods of prolonged or severe frosts
- Where the second application of a sequence is made after 31st December, or where the total dose exceeds 0.6 L/ha, the first application must be made before GS13 of the crop and a minimum interval of 6 weeks must be observed between applications [11]
- Horizontal boom sprayers must be fitted with three star drift reduction technology for all uses [15, 16, 20]
- Must not be applied via hand-held equipment [15, 20]
- Low drift spraying equipment must be operated according to the specific conditions stated in the official three star rating for that equipment as published on HSE website. These operating conditions must be maintained until the operator is 30m from the top of the bank of any surface water bodies [15, 16, 20]

FOR FULL CONDITIONS OF USE ALWAYS READ THE PRODUCT LABEL

Crop-specific information
- Latest use: before 31 Dec in yr of sowing and before 3rd tiller stage (GS 23) for wheat or 4th tiller stage (GS 24) for barley
- For pre-emergence treatments the seed should be covered with a minimum of 32 mm settled soil

Following crops guidance
- In the event of crop failure wheat, barley or potatoes may be sown provided the soil is ploughed to 15 cm, and a minimum of 12 weeks elapse between treatment and sowing spring wheat or spring barley
- After normal harvest wheat, barley or potatoes my be sown without special cultivations. Soil must be ploughed or cultivated to 15 cm before sowing oilseed rape, field beans, peas, sugar beet, carrots, onions or edible brassicae
- Successive treatments with any products containing diflufenican can lead to soil build-up and inversion ploughing to 150 mm must precede sowing any following non-cereal crop. Even where ploughing occurs some crops may be damaged

Environmental safety
- Dangerous for the environment
- Very toxic to aquatic organisms
- Risk to non-target insects or other arthropods. Avoid spraying within 6 m of the field boundary to reduce the effects on non-target insects or other arthropods
- Buffer zone requirement 6 m [16, 17]
- Buffer zone requirement 10 m
- Buffer zone requirement 12 m [20]
- LERAP Category B

Hazard classification and safety precautions
Hazard Harmful, Dangerous for the environment [1-22]; Harmful if swallowed, Very toxic to aquatic organisms [1-6, 8, 9, 11, 12, 14-16, 18-20]
Transport code 9
Packaging group III
UN Number 3082
Risk phrases H317 [3-6, 15, 16, 20]; H373 [1-9, 11, 12, 14-21]; R22a, R48, R50, R53a [10, 13, 22]; R43 [10, 13, 21, 22]
Operator protection A, H; U05a, U14 [1-22]; U23a [7, 17]
Environmental protection E15a [6, 12, 18, 19]; E15b [7, 10, 13, 15-17, 20-22]; E16a, E34, E38 [1-22]; E22c [1-11, 13-17, 20-22]; H410 [1-9, 11, 12, 14-21]
Storage and disposal D01, D02, D09a, D10b, D12a
Medical advice M03, M05a

146   diflufenican + flufenacet + metribuzin

A residual herbicide mixture for weed control in winter cereals
HRAC mode of action code: F1 + K3 + C1

See also diflufenican + flufenacet
        diflufenican + metribuzin
        flufenacet + metribuzin

Products

1	Alternator Met	Bayer CropScience	60:240:70 g/l	SC	18267
2	Octavian Met	Bayer CropScience	90:240:70 g/l	SC	18266

Uses
- Annual dicotyledons in *winter barley, winter wheat* [1, 2]
- Annual meadow grass in *winter barley, winter wheat* [1, 2]
- Chickweed in *winter barley, winter wheat* [1]
- Common field speedwell in *winter barley, winter wheat* [1]
- Field pansy in *winter barley, winter wheat* [1]
- Red dead-nettle in *winter barley, winter wheat* [1]

SEE SECTION 3 FOR PRODUCTS ALSO REGISTERED

SECTION 2

- Scented mayweed in **winter barley**, **winter wheat** [1]
- Wild radish in **winter barley**, **winter wheat** [1]

Approval information
- Diflufenican, flufenacet and metribuzin included in Annex I under EC Regulation 1107/2009

Following crops guidance
- Following normal harvest wheat and barley can be sown without cultivation but thorough mixing of the top 15 cm of soil is necessary before sowing oilseed rape, mustard or edible Brassicae. The following spring after normal harvest wheat, barley, oilseed rape, sugarbeet, mustard, peas, sunflowers or maize can be sown without special cultivations.
- In the event of crop failure, allow at least 5 months from application before sowing wheat or barley and at least 6 months before sowing maize or sunflowers.
- Be aware of the potential for diflufenican to build up in the soil from successive applications over the crop rotation.

Environmental safety
- Buffer zone requirement 6 m
- Horizontal boom sprayers must be fitted with three star drift reduction technology for all uses
- Low drift spraying equipment must be used until the operator is 30m from the top of the bank of any surface water bodies
- LERAP Category B

Hazard classification and safety precautions
  Hazard Harmful, Dangerous for the environment, Very toxic to aquatic organisms
  Transport code 9
  Packaging group III
  UN Number 3082
  Risk phrases H373
  Operator protection A, H; U05a, U08, U13, U14, U19a, U20b
  Environmental protection E15a, E16a, E38, H410
  Storage and disposal D01, D02, D09a, D11a, D12a [1, 2]; D05 [1]
  Medical advice M03

147   diflufenican + glyphosate

A foliar non-selective herbicide mixture
HRAC mode of action code: F1 + G

See also glyphosate

Products

1	Pistol	Bayer CropScience	40:250 g/l	SC	17451
2	Pistol	BASF	40:250 g/l	SC	18854
3	Pistol Rail	BASF	40:250 g/l	SC	18853
4	Proshield	ICL (Everris) Ltd	40:250 g/l	SC	17525

Uses
- Annual and perennial weeds in **hard surfaces** [1-3]; **natural surfaces not intended to bear vegetation**, **permeable surfaces overlying soil** [1, 2, 4]

Approval information
- Diflufenican and glyphosate included in Annex I under EC Regulation 1107/2009
- Approval expiry 14 Aug 2020 [1]

Efficacy guidance
- Treat when weeds actively growing from Mar to end Sep and before they begin to senesce
- Performance may be reduced if application is made to plants growing under stress, such as drought or water-logging
- Pre-emergence activity is reduced on soils containing more than 10% organic matter or where organic debris has collected

FOR FULL CONDITIONS OF USE ALWAYS READ THE PRODUCT LABEL

- Perennial weeds such as docks, perennial sowthistle and willowherb are best treated before flowering or setting seed
- Perennial weeds emerging from established rootstocks after treatment will not be controlled
- A rainfree period of at least 6 h (preferably 24 h) should follow spraying for optimum control
- For optimum control do not cultivate or rake after treatment

Restrictions
- Maximum number of treatments 1 per yr
- May only be used on porous surfaces overlying soil. Must not be used if an impermeable membrane lies between the porous surface and the soil, and must not be used on any non-porous man-made surfaces
- Do not add any wetting agent or adjuvant oil
- Do not spray in windy weather
- Use a knapsack sprayer giving a coarse spray via anti-drift nozzles for application [1]
- Low drift spraying equipment must be operated according to the specific conditions stated in the official three star rating for that equipment as published on HSE Chemicals Regulation Division's website. These operating conditions must be maintained until the operator is 30m from the top of the bank of any surface water bodies [2]

Following crops guidance
- A period of at least 6 mth must be allowed after treatment of sites that are to be cleared and grubbed before sowing and planting. Soil should be ploughed or dug first to ensure thorough mixing and dilution of any herbicide residues

Environmental safety
- Dangerous for the environment
- Very toxic to aquatic organisms
- Avoid drift onto non-target plants
- Heavy rain after application may wash product onto sensitive areas such as newly sown grass or areas about to be planted
- Buffer zone requirement 2 m [1, 2, 4] for hand-held sprayers, 6 m for horizontal boom sprayers
- LERAP Category B

Hazard classification and safety precautions

Hazard Dangerous for the environment, Very toxic to aquatic organisms
Transport code 9
Packaging group III
UN Number 3082
Operator protection A, C [1-4]; H [4]; U20b
Environmental protection E15a, E16a, E16b, E38, H410
Storage and disposal D01, D02, D09a, D11a, D12a

148 diflufenican + iodosulfuron-methyl-sodium + mesosulfuron-methyl

A contact and residual herbicide mixture containing sulfonyl ureas for winter wheat
HRAC mode of action code: F1 + B + B

See also iodosulfuron-methyl-sodium
mesosulfuron-methyl

Products

1	Hamlet	Bayer CropScience	50:2.5:7.5 g/l	OD	17370
2	Othello	Bayer CropScience	50:2.5:7.5 g/l	OD	16149

Uses
- Annual dicotyledons in *winter wheat*
- Annual meadow grass in *winter wheat*
- Cleavers in *winter wheat*
- Mayweeds in *winter wheat*

SEE SECTION 3 FOR PRODUCTS ALSO REGISTERED

- Rough-stalked meadow grass in **winter wheat**
- Volunteer oilseed rape in **winter wheat**

Approval information
- Diflufenican, iodosulfuron-methyl-sodium and mesosulfuron-methyl included in Annex I under EC Regulation 1107/2009

Efficacy guidance
- Optimum control obtained when all weeds are emerged at spraying. Activity is primarily via foliar uptake and good spray coverage of the target weeds is essential
- Translocation occurs readily within the target weeds and growth is inhibited within hours of treatment but symptoms may not be apparent for up to 4 wk, depending on weed species, timing of treatment and weather conditions
- Iodosulfuron-methyl and mesosulfuron-methyl are both members of the ALS-inhibitor group of herbicides. To avoid the build up of resistance do not use any product containing an ALS-inhibitor herbicide with claims for control of grass weeds more than once on any crop
- Use this product as part of a Resistance Management Strategy that includes cultural methods of control and does not use ALS inhibitors as the sole chemical method of weed control in successive crops. See Section 5 for more information

Restrictions
- Maximum number of treatments 1 per crop
- Do not use on crops undersown with grasses, clover or other legumes or any other broad-leaved crop
- Do not use where annual grass weeds other than annual meadow grass and rough meadow grass are present
- Do not use as a stand-alone treatment for control of common chickweed or common poppy. Only use mixtures with non ALS-inhibitor herbicides for these weeds
- Do not use as the sole means of weed control in successive crops
- Specific restrictions apply to use in sequence or tank mixture with other sulfonylurea or ALS-inhibiting herbicides. See label for details
- Do not apply to crops under stress from any cause
- Do not apply when rain is imminent or during periods of frosty weather
- Specified adjuvant must be used. See label

Crop-specific information
- Latest use: before 2nd node detectable for winter wheat
- Transitory crop effects may occur, particularly on overlaps and after late season/spring applications. Recovery is normally complete and yield not affected

Following crops guidance
- In the event of crop failure winter or spring wheat may be drilled after normal cultivation and an interval of 6 wk
- Winter wheat, winter barley or winter oilseed rape may be drilled in the autumn following normal harvest of a treated crop. Spring wheat, spring barley, spring oilseed rape or sugar beet may be drilled in the following spring
- Where the product has been applied in sequence with a permitted ALS-inhibitor herbicide (see label) only winter or spring wheat or barley, or sugar beet may be sown as following crops
- Successive treatments with any products containing diflufenican can lead to soil build-up and inversion ploughing to 150 mm must precede sowing any following non-cereal crop. Even where ploughing occurs some crops may be damaged

Environmental safety
- Dangerous for the environment
- Very toxic to aquatic organisms
- Take extreme care to avoid drift outside the target area
- LERAP Category B

Hazard classification and safety precautions
Hazard Irritant, Dangerous for the environment, Very toxic to aquatic organisms
Transport code 9
Packaging group III

FOR FULL CONDITIONS OF USE ALWAYS READ THE PRODUCT LABEL

UN Number 3082
Risk phrases H319
Operator protection A, C; U05a, U11, U20b
Environmental protection E15b, E16a, E38, H410
Storage and disposal D01, D02, D05, D09a, D10a, D12a

149 diflufenican + metsulfuron-methyl

A contact and residual herbicide mixture containing sulfonyl ureas for cereals
HRAC mode of action code: F1 + B

Products

Pelican Delta	Nufarm UK	60:6% w/w	WG	18513

Uses

- Annual dicotyledons in *spring barley, winter barley, winter wheat*
- Charlock in *spring barley, winter barley, winter wheat*
- Corn spurrey in *spring barley, winter barley, winter wheat*
- Docks in *spring barley, winter barley, winter wheat*
- Hemp-nettle in *spring barley, winter barley, winter wheat*
- Mayweeds in *spring barley, winter barley, winter wheat*
- Pale persicaria in *spring barley, winter barley, winter wheat*
- Parsley-piert in *spring barley, winter barley, winter wheat*
- Redshank in *spring barley, winter barley, winter wheat*
- Scarlet pimpernel in *spring barley, winter barley, winter wheat*
- Shepherd's purse in *spring barley, winter barley, winter wheat*

Approval information

- Diflufenican and metsulfuron-methyl included in Annex I under EC Regulation 1107/2009

Efficacy guidance

- Do not apply to frosted crops or scorch may occur
- Can only be applied after 1st Feb up to latest growth stage stipulated

Restrictions

- Do not use on crops grown for seed
- Do not apply in mixture or sequence with any other ALS herbicide
- For winter cereals no application before 15 March in the year of harvest, for spring barley no application before 1st April in the year of harvest

Crop-specific information

- Do not use on soils containing more than 10% organic matter

Following crops guidance

- In the event of crop failure, only sow wheat within 3 months of application

Environmental safety

- Buffer zone requirement 7m [1]
- LERAP Category B

Hazard classification and safety precautions

Hazard Irritant, Dangerous for the environment
Transport code 9
Packaging group III
UN Number 3077
Risk phrases H315, H318
Operator protection A, C; U05a, U08, U11, U19a, U20b
Environmental protection E15a, E16a, E34, E38, H410
Storage and disposal D01, D02, D05, D09a, D10b, D12a

SECTION 2

150   diflufenican + pendimethalin

A mixture of pyridinecarboxamide and dinitroanaline herbicides for grass and broad-leaved weed control in cereals
HRAC mode of action code: F1 + K1

See also pendimethalin

Products

1 Bulldog	Sipcam	15.6:313 g/l	SC	17538
2 Omaha 2	Adama	40:400 g/l	SC	16846

Uses
- Annual dicotyledons in *spring barley*, *spring wheat* [2]; *triticale*, *winter barley*, *winter rye*, *winter wheat* [1, 2]
- Annual grasses in *spring barley*, *spring wheat*, *triticale*, *winter barley*, *winter rye*, *winter wheat* [2]
- Annual meadow grass in *triticale*, *winter barley*, *winter rye*, *winter wheat* [1]

Approval information
- Diflufenican and pendimethalin included in Annex I under EC Regulation 1107/2009

Efficacy guidance
- Do not apply pre-emergence to winter crops drilled after 20th November
- Do not treat broadcast crops

Following crops guidance
- Plough to 150 mm and thoroughly mix the soil before planting any following crop

Environmental safety
- Buffer zone requirement 6 m [1, 2]

Hazard classification and safety precautions
    Hazard Dangerous for the environment [1, 2]; Very toxic to aquatic organisms [1]
    Transport code 9
    Packaging group III
    UN Number 3082
    Risk phrases H334 [1]; R50, R53a [2]
    Operator protection A; U05a, U08, U20a [1]
    Environmental protection E38 [1, 2]; H410 [1]
    Storage and disposal D01, D02, D09a, D10c [1]; D12a [1, 2]

151   dimethachlor

A residual anilide herbicide for use in oilseed rape
HRAC mode of action code: K3

Products

Teridox	Syngenta	500 g/l	EC	15876

Uses
- Annual dicotyledons in *winter oilseed rape*
- Chickweed in *winter oilseed rape*
- Docks in *winter oilseed rape*
- Groundsel in *winter oilseed rape*
- Loose silky bent in *winter oilseed rape*
- Mayweeds in *winter oilseed rape*
- Red dead-nettle in *winter oilseed rape*
- Scarlet pimpernel in *winter oilseed rape*

Approval information
- Dimethachlor included in Annex 1 under EC Regulation 1107/2009

FOR FULL CONDITIONS OF USE ALWAYS READ THE PRODUCT LABEL

Efficacy guidance
- Optimum weed control requires moist soils.
- Heavy rain within a few days of application may lead to poor weed control

Restrictions
- Do not use on sands or very light soils, stony soils, organic soil, broadcast crops or late-drilled crops.
- Apply after drilling but before crop germination which can occur within 48 hours of drilling.

Crop-specific information
- High transpiration rates or persistent wet weather during the first weeks after treatment may reduce crop vigour and plant stand.
- Crops must be covered by at least 20mm of settled soil before application.
- Do not cultivate after application.

Following crops guidance
- In the event of autumn crop failure, winter oilseed rape may be re-drilled after cultivating the soil to at least 15 cms.
- If re-drilling the following spring, spring oilseed rape, field beans, peas, maize or potatoes may be sown. If planting spring wheat, barley, oats, spring linseed, sunflowers, sugar beet or grass crops the land should be ploughed before sowing.

Environmental safety
- Buffer zone requirement 10m [1]
- LERAP Category B

Hazard classification and safety precautions
Hazard Irritant, Dangerous for the environment, Very toxic to aquatic organisms
Transport code 9
Packaging group III
UN Number 3082
Risk phrases H304, H315, H317
Operator protection A, C, H; U05a, U09a, U19a, U20c
Environmental protection E15b, E16a, E40b, H410
Storage and disposal D01, D02, D05, D09a, D10c, D12a
Medical advice M05a

152   dimethenamid-p

A chloroacetamide herbicide available only in mixtures
HRAC mode of action code: K3

See also clomazone + dimethenamid-p + metazachlor

153   dimethenamid-p + metazachlor

A soil acting herbicide mixture for oilseed rape
HRAC mode of action code: K3 + K3

See also metazachlor

Products

Springbok	BASF	200:200 g/l	EC	16786

Uses
- Annual dicotyledons in **broccoli** (off-label), **brussels sprouts** (off-label), **cabbages** (off-label), **calabrese** (off-label), **cauliflowers** (off-label), **chinese leaf** (off-label), **chives** (off-label), **choi sum** (off-label), **collards** (off-label), **forest nurseries** (off-label), **game cover** (off-label), **kale** (off-label), **kohlrabi** (off-label), **leeks** (off-label), **oriental cabbage** (off-label), **ornamental plant production** (off-label), **salad onions** (off-label), **swedes** (off-label), **tatsoi** (off-label), **turnips** (off-label), **winter oilseed rape**

SEE SECTION 3 FOR PRODUCTS ALSO REGISTERED

- Annual grasses in **broccoli** *(off-label)*, **brussels sprouts** *(off-label)*, **cabbages** *(off-label)*, **calabrese** *(off-label)*, **cauliflowers** *(off-label)*, **chinese leaf** *(off-label)*, **chives** *(off-label)*, **choi sum** *(off-label)*, **collards** *(off-label)*, **forest nurseries** *(off-label)*, **game cover** *(off-label)*, **kale** *(off-label)*, **kohlrabi** *(off-label)*, **leeks** *(off-label)*, **oriental cabbage** *(off-label)*, **ornamental plant production** *(off-label)*, **salad onions** *(off-label)*, **swedes** *(off-label)*, **tatsoi** *(off-label)*, **turnips** *(off-label)*
- Annual meadow grass in **broccoli** *(off-label)*, **brussels sprouts** *(off-label)*, **cabbages** *(off-label)*, **calabrese** *(off-label)*, **cauliflowers** *(off-label)*, **chinese leaf** *(off-label)*, **chives** *(off-label)*, **choi sum** *(off-label)*, **collards** *(off-label)*, **kale** *(off-label)*, **kohlrabi** *(off-label)*, **leeks** *(off-label)*, **oriental cabbage** *(off-label)*, **ornamental plant production** *(off-label)*, **salad onions** *(off-label)*, **swedes** *(off-label)*, **tatsoi** *(off-label)*, **turnips** *(off-label)*
- Chickweed in **ornamental plant production** *(off-label)*, **winter oilseed rape**
- Cleavers in **ornamental plant production** *(off-label)*, **winter oilseed rape**
- Common storksbill in **winter oilseed rape**
- Crane's-bill in **chives** *(off-label)*, **forest nurseries** *(off-label)*, **game cover** *(off-label)*, **leeks** *(off-label)*, **ornamental plant production** *(off-label)*, **salad onions** *(off-label)*, **swedes** *(off-label)*, **turnips** *(off-label)*
- Fat hen in **broccoli** *(off-label)*, **brussels sprouts** *(off-label)*, **cabbages** *(off-label)*, **calabrese** *(off-label)*, **cauliflowers** *(off-label)*, **chinese leaf** *(off-label)*, **choi sum** *(off-label)*, **collards** *(off-label)*, **kale** *(off-label)*, **kohlrabi** *(off-label)*, **oriental cabbage** *(off-label)*, **tatsoi** *(off-label)*
- Field speedwell in **winter oilseed rape**
- Groundsel in **broccoli** *(off-label)*, **brussels sprouts** *(off-label)*, **cabbages** *(off-label)*, **calabrese** *(off-label)*, **cauliflowers** *(off-label)*, **chinese leaf** *(off-label)*, **choi sum** *(off-label)*, **collards** *(off-label)*, **kale** *(off-label)*, **kohlrabi** *(off-label)*, **oriental cabbage** *(off-label)*, **ornamental plant production** *(off-label)*, **tatsoi** *(off-label)*
- Mayweeds in **broccoli** *(off-label)*, **brussels sprouts** *(off-label)*, **cabbages** *(off-label)*, **calabrese** *(off-label)*, **cauliflowers** *(off-label)*, **chinese leaf** *(off-label)*, **choi sum** *(off-label)*, **collards** *(off-label)*, **kale** *(off-label)*, **kohlrabi** *(off-label)*, **oriental cabbage** *(off-label)*, **ornamental plant production** *(off-label)*, **tatsoi** *(off-label)*
- Poppies in **winter oilseed rape**
- Redshank in **broccoli** *(off-label)*, **brussels sprouts** *(off-label)*, **cabbages** *(off-label)*, **calabrese** *(off-label)*, **cauliflowers** *(off-label)*, **chinese leaf** *(off-label)*, **choi sum** *(off-label)*, **collards** *(off-label)*, **kale** *(off-label)*, **kohlrabi** *(off-label)*, **oriental cabbage** *(off-label)*, **tatsoi** *(off-label)*
- Scented mayweed in **winter oilseed rape**
- Shepherd's purse in **broccoli** *(off-label)*, **brussels sprouts** *(off-label)*, **cabbages** *(off-label)*, **calabrese** *(off-label)*, **cauliflowers** *(off-label)*, **chinese leaf** *(off-label)*, **chives** *(off-label)*, **choi sum** *(off-label)*, **collards** *(off-label)*, **forest nurseries** *(off-label)*, **game cover** *(off-label)*, **kale** *(off-label)*, **kohlrabi** *(off-label)*, **leeks** *(off-label)*, **oriental cabbage** *(off-label)*, **ornamental plant production** *(off-label)*, **salad onions** *(off-label)*, **swedes** *(off-label)*, **tatsoi** *(off-label)*, **turnips** *(off-label)*, **winter oilseed rape**
- Small nettle in **broccoli** *(off-label)*, **brussels sprouts** *(off-label)*, **cabbages** *(off-label)*, **calabrese** *(off-label)*, **cauliflowers** *(off-label)*, **chinese leaf** *(off-label)*, **choi sum** *(off-label)*, **collards** *(off-label)*, **kale** *(off-label)*, **kohlrabi** *(off-label)*, **oriental cabbage** *(off-label)*, **ornamental plant production** *(off-label)*, **tatsoi** *(off-label)*
- Sowthistle in **ornamental plant production** *(off-label)*

Extension of Authorisation for Minor Use (EAMUs)
- **broccoli** *20151540*
- **brussels sprouts** *20151540*
- **cabbages** *20151540*
- **calabrese** *20151540*
- **cauliflowers** *20151540*
- **chinese leaf** *20151540*
- **chives** *20143008*
- **choi sum** *20151540*
- **collards** *20151540*
- **forest nurseries** *20143006*
- **game cover** *20143006*
- **kale** *20151540*

- *kohlrabi 20151540*
- *leeks 20143008*
- *oriental cabbage 20151540*
- *ornamental plant production 20143006, 20152108*
- *salad onions 20143008*
- *swedes 20143007*
- *tatsoi 20151540*
- *turnips 20143007*

Approval information
- Dimethenamid-p and metazachlor included in Annex I under EC Regulation 1107/2009

Efficacy guidance
- Best results obtained from treatments to fine, firm and moist seedbeds
- Apply pre- or post-emergence of the crop and ideally before weed emergence
- Residual weed control may be reduced under prolonged dry conditions
- Weeds germinating from depth may not be controlled
- Dimethenamid-p is more active than metazachlor under dry soil conditions

Restrictions
- Maximum total dose on winter oilseed rape equivalent to one full dose treatment
- Do not disturb soil after application
- Do not treat broadcast crops until they have attained two fully expanded cotyledons
- Do not use on Sands, Very Light soils, or soils containing 10% organic matter
- Do not apply when heavy rain forecast or on soils waterlogged or prone to waterlogging
- Do not treat crops suffering from stress from any cause
- Do not use pre-emergence when crop seed has started to germinate or if not covered with 15 mm of soil
- Applications shall be limited to a total dose of not more than 1.0 kg metazachlor/ha in a three year period on the same field
- Do not apply in mixture with phosphate liquid fertilisers
- Metazachlor stewardship requires that all autumn applications should be made before the end of September to reduce the risk to water

Crop-specific information
- Latest use: before 7th true leaf for winter oilseed rape
- All varieties of winter oilseed rape may be treated

Following crops guidance
- Any crop may follow a normally harvested treated winter oilseed rape crop. Ploughing is not essential before a following cereal crop but is required for all other crops
- In the event of failure of a treated crop winter wheat (excluding durum) or winter barley may be drilled in the same autumn, and any cereal (excluding durum wheat), spring oilseed rape, peas or field beans may be sown in the following spring. Ploughing to at least 150 mm should precede planting in all cases

Environmental safety
- Dangerous for the environment
- Very toxic to aquatic organisms
- Take extreme care to avoid spray drift onto non-crop plants outside the target area
- Some pesticides pose a greater threat of contamination of water than others and metazachlor is one of these pesticides. Take special care when applying metazachlor near water and do not apply if heavy rain is forecast
- Metazachlor stewardship guidelines advise a maximum dose of 750 g.a.i/ha/annum. Applications to drained fields should be complete by 15th Oct but, if drains are flowing, complete applications by 1st Oct.
- LERAP Category B

Hazard classification and safety precautions
Hazard Harmful, Dangerous for the environment, Harmful if swallowed, Harmful if inhaled, Very toxic to aquatic organisms
Transport code 9

SEE SECTION 3 FOR PRODUCTS ALSO REGISTERED

Packaging group III
UN Number 3082
Risk phrases H304, H317, H319, H351
Operator protection A, C, H; U05a, U08, U20c
Environmental protection E15a, E16a, E34, E38
Storage and disposal D01, D02, D09a, D10c, D12a
Medical advice M03, M05a

154   dimethenamid-p + metazachlor + quinmerac

A soil-acting herbicide mixture for use in oilseed rape
HRAC mode of action code: K3 + K3 + O

See also metazachlor
quinmerac

Products

1	Banastar	BASF	100:300:100 g/l	SE	16834
2	Elk	BASF	200:200:100 g/l	SE	16920
3	Katamaran Turbo	BASF	200:200:100 g/l	SE	16921
4	Shadow	BASF	200:200:100 g/l	SE	16804

Uses

- Annual dicotyledons in *winter oilseed rape* [1-4]
- Annual grasses in *winter oilseed rape* [2-4]
- Annual meadow grass in *winter oilseed rape* [1]
- Blackgrass in *winter corn gromwell* (off-label) [3]
- Chickweed in *winter corn gromwell* (off-label) [3]
- Cleavers in *winter corn gromwell* (off-label) [3]; *winter oilseed rape* [1-4]
- Crane's-bill in *winter corn gromwell* (off-label) [3]; *winter oilseed rape* [1]
- Fat hen in *winter corn gromwell* (off-label) [3]
- Field pansy in *winter corn gromwell* (off-label) [3]
- Forget-me-not in *winter corn gromwell* (off-label) [3]
- Groundsel in *winter corn gromwell* (off-label) [3]
- Ivy-leaved speedwell in *winter corn gromwell* (off-label) [3]
- Mayweeds in *winter corn gromwell* (off-label) [3]
- Poppies in *winter corn gromwell* (off-label) [3]; *winter oilseed rape* [1]
- Red dead-nettle in *winter corn gromwell* (off-label) [3]
- Shepherd's purse in *winter corn gromwell* (off-label) [3]
- Sowthistle in *winter corn gromwell* (off-label) [3]
- Speedwells in *winter oilseed rape* [1]

Extension of Authorisation for Minor Use (EAMUs)

- *winter corn gromwell* 20152318 [3]

Approval information

- Dimethenamid-p, metazachlor and quinmerac included in Annex I under EC Regulation 1107/2009

Efficacy guidance

- Dimethenamid-p is more active than metazachlor under dry soil conditions

Restrictions

- Applications shall be limited to a total dose of not more than 1.0 kg metazachlor/ha in a three year period on the same field
- Do not apply in mixture with phosphate liquid fertilisers
- Metazachlor stewardship requires that all autumn applications should be made before the end of September to reduce the risk to water

FOR FULL CONDITIONS OF USE ALWAYS READ THE PRODUCT LABEL

Following crops guidance
- For all situations following or rotational crops must not be planted until four months after application

Environmental safety
- Some pesticides pose a greater threat of contamination of water than others and metazachlor is one of these pesticides. Take special care when applying metazachlor near water and do not apply if heavy rain is forecast.
- Metazachlor stewardship guidelines advise a maximum dose of 750 g.a.i/ha/annum. Applications to drained fields should be complete by 15th Oct but, if drains are flowing, complete applications by 1st Oct.
- LERAP Category B

Hazard classification and safety precautions
Hazard Harmful [3, 4]; Irritant [2]; Dangerous for the environment, Very toxic to aquatic organisms [2-4]
Transport code 9
Packaging group III
UN Number 3082
Risk phrases H317, H351
Operator protection A, H; U05a, U08 [1-4]; U14, U20c [2-4]; U20a [1]
Environmental protection E07a, E38 [2-4]; E07d [1]; E15b, E16a, E34, H410 [1-4]
Storage and disposal D01, D02, D09a, D10c [1-4]; D05 [1]; D12a [2-4]
Medical advice M03, M05a [2-4]

155   dimethenamid-p + pendimethalin

A chloracetamide and dinitroaniline herbicide mixture for weed control in maize crops
HRAC mode of action code: K1 + K3

Products

1 Clayton Launch	Clayton	212.5:250 g/l	EC	16277
2 Dime	Pan Agriculture	212.5:250 g/l	EC	18661
3 Dime	Pan Agriculture	212.5:250 g/l	EC	18825
4 Wing - P	BASF	212.5:250 g/l	EC	15425

Uses
- Annual dicotyledons in *baby leaf crops (off-label)*, *broccoli (off-label)*, *brussels sprouts (off-label)*, *bulb onion sets (off-label)*, *bulb onions (off-label)*, *cabbages (off-label)*, *calabrese (off-label)*, *cauliflowers (off-label)*, *chives (off-label)*, *courgettes (off-label)*, *forage maize (under plastic mulches) (off-label)*, *garlic (off-label)*, *grain maize (under plastic mulches) (off-label)*, *herbs (see appendix 6) (off-label)*, *leeks (off-label)*, *lettuce (off-label)*, *ornamental plant production (off-label)*, *pumpkins (off-label)*, *salad onions (off-label)*, *shallots (off-label)*, *strawberries (off-label)*, *summer squash (off-label)*, *sweetcorn (off-label)*, *sweetcorn under plastic mulches (off-label)*, *winter squash (off-label)* [4]; *forage maize, grain maize* [1-4]
- Annual grasses in *baby leaf crops (off-label)*, *broccoli (off-label)*, *brussels sprouts (off-label)*, *bulb onion sets (off-label)*, *bulb onions (off-label)*, *cabbages (off-label)*, *calabrese (off-label)*, *cauliflowers (off-label)*, *chives (off-label)*, *forage maize (under plastic mulches) (off-label)*, *garlic (off-label)*, *grain maize (under plastic mulches) (off-label)*, *herbs (see appendix 6) (off-label)*, *leeks (off-label)*, *lettuce (off-label)*, *ornamental plant production (off-label)*, *salad onions (off-label)*, *shallots (off-label)*, *strawberries (off-label)* [4]
- Annual meadow grass in *broccoli (off-label)*, *brussels sprouts (off-label)*, *calabrese (off-label)*, *cauliflowers (off-label)*, *courgettes (off-label)*, *pumpkins (off-label)*, *summer squash (off-label)*, *sweetcorn (off-label)*, *sweetcorn under plastic mulches (off-label)*, *winter squash (off-label)* [4]; *forage maize, grain maize* [1-4]
- Wild oats in *ornamental plant production (off-label)* [4]

Extension of Authorisation for Minor Use (EAMUs)
- *baby leaf crops 20170810* [4]
- *broccoli 20131656* [4]

SEE SECTION 3 FOR PRODUCTS ALSO REGISTERED

- *brussels sprouts* *20131656* [4]
- *bulb onion sets* *20122250* [4]
- *bulb onions* *20122250* [4]
- *cabbages* *20122251* [4]
- *calabrese* *20131656* [4]
- *cauliflowers* *20131656* [4]
- *chives* *20122249* [4]
- *courgettes* *20180619* [4]
- *forage maize (under plastic mulches)* *20150101* [4]
- *garlic* *20122250* [4]
- *grain maize (under plastic mulches)* *20111841* [4]
- *herbs (see appendix 6)* *20170810* [4]
- *leeks* *20122248* [4]
- *lettuce* *20170810* [4]
- *ornamental plant production* *20130253* [4]
- *pumpkins* *20180619* [4]
- *salad onions* *20122249* [4]
- *shallots* *20122250* [4]
- *strawberries* *20160933* [4]
- *summer squash* *20180619* [4]
- *sweetcorn* *20180917* [4]
- *sweetcorn under plastic mulches* *20180917* [4]
- *winter squash* *20180619* [4]

Approval information
- Dimethenamid-p and pendimethalin are included in Annex 1 under EC Regulation 1107/2009

Efficacy guidance
- Do not use on soil types with more than 10% organic matter
- Best results obtained when rain falls within 7 days of application

Crop-specific information
- Use on maize crops grown under plastic mulches is by EAMU only; do not use in greenhouses, covers or other forms of protection
- Risk of crop damage if heavy rain falls soon after application to stoney or gravelly soils
- Seed should be covered with at least 5 cm of settled soil

Environmental safety
- LERAP Category B

Hazard classification and safety precautions
Hazard Harmful, Dangerous for the environment [1-4]; Harmful if swallowed, Very toxic to aquatic organisms [2-4]
Transport code 9
Packaging group III
UN Number 3082
Risk phrases H304, H315, H317 [2-4]; R22a, R22b, R38, R43, R50, R53a [1]
Operator protection A, H; U05a, U14
Environmental protection E15b, E16a, E34, E38 [1-4]; H410 [2-4]
Storage and disposal D01, D02, D09a, D10c, D12a
Medical advice M03, M05a

FOR FULL CONDITIONS OF USE ALWAYS READ THE PRODUCT LABEL

156   dimethenamid-p + quinmerac

A herbicide mixture for weed control in oilseed rape
HRAC mode of action code: K3 + O

See also dimethenamid-p + metazachlor + quinmerac
quinmerac

Products

1	Tanaris	BASF	333 + 167 g/l	SE	17173
2	Topkat	BASF	333 + 167 g/l	SE	17356

Uses

- Annual dicotyledons in *fodder beet, sugar beet, winter oilseed rape*
- Cleavers in *fodder beet, sugar beet, winter oilseed rape*
- Crane's-bill in *fodder beet, sugar beet, winter oilseed rape*
- Dead nettle in *fodder beet, sugar beet, winter oilseed rape*
- Mayweeds in *fodder beet, sugar beet, winter oilseed rape*
- Poppies in *fodder beet, sugar beet, winter oilseed rape*
- Shepherd's purse in *fodder beet, sugar beet, winter oilseed rape*
- Sowthistle in *fodder beet, sugar beet, winter oilseed rape*
- Speedwells in *fodder beet, sugar beet, winter oilseed rape*

Approval information

- Dimethenamid-p and quinmerac included in Annex I under EC Regulation 1107/2009

Environmental safety

- LERAP Category B

Hazard classification and safety precautions

Hazard Very toxic to aquatic organisms
Transport code 9
Packaging group III
UN Number 3082
Risk phrases H317, H319
Operator protection A, C, H; U05a, U08, U20a
Environmental protection E15b, E16a, E34, H410
Storage and disposal D01, D02, D09a, D10c

157   dimethoate

A contact and systemic organophosphorus insecticide and acaricide
IRAC mode of action code: 1B

Products

Danadim Progress	FMC Agro	400 g/l	EC	18745

Uses

- Aphids in *durum wheat, ornamental plant production, rye, spring wheat, triticale, winter wheat*
- Leaf miner in *ornamental plant production*
- Red spider mites in *ornamental plant production*

Approval information

- Dimethoate included in Annex I under EC Regulation 1107/2009
- In 2006 CRD required that all products containing this active ingredient should carry the following warning in the main area of the container label: "Dimethoate is an anticholinesterase organophosphate. Handle with care"

Efficacy guidance

- Chemical has quick knock-down effect and systemic activity lasts for up to 14 d

SEE SECTION 3 FOR PRODUCTS ALSO REGISTERED

SECTION 2

- With some crops, products differ in range of pests listed as controlled. Uses section above provides summary. See labels for details
- For most pests apply when pest first seen and repeat 2-3 wk later or as necessary. Timing and number of sprays varies with crop and pest. See labels for details
- Best results achieved when crop growing vigorously. Systemic activity reduced when crops suffering from drought or other stress
- In hot weather apply in early morning or late evening
- Where aphids or spider mites resistant to organophosphorus compounds occur control is unlikely to be satisfactory and repeat treatments may result in lower levels of control

Restrictions
- Contains an anticholinesterase organophosphorus compound. Do not use if under medical advice not to work with such compounds
- Maximum number of treatments 6 per crop for Brussels sprouts, broccoli, cauliflower, calabrese, lettuce; 4 per crop for grass seed crops, cereals; 2 per crop for beet crops, triticale, rye; 1 per crop for mangel and sugar beet seed crops
- On beet crops only one treatment per crop may be made for control of leaf miners (max 84 g a.i./ha) and black bean aphid (max 400 g a.i./ha)
- In beet crops resistant strains of peach-potato aphid (*Myzus persicae*) are common and dimethoate products must not be used to control this pest
- Test for varietal susceptibility on all unusual plants or new cultivars
- Do not tank mix with alkaline materials. See label for recommended tank-mixes
- Consult processor before spraying crops grown for processing
- Must not be applied to cereals if any product containing a pyrethroid insecticide or dimethoate has been sprayed after the start of ear emergence (GS 51)

Crop-specific information
- Latest use: flowering just complete stage (GS 30) for wheat, rye, triticale
- HI cereals 14 d; ornamental plant production 7 d

Environmental safety
- Dangerous for the environment
- Harmful to aquatic organisms
- Flammable
- Harmful to game, wild birds and animals
- Harmful to livestock. Keep all livestock out of treated areas for at least 7 d
- Likely to cause adverse effects on beneficial arthropods
- To protect non target insects/arthropods respect an unsprayed buffer zone of 5 m to non-crop land
- Dangerous to bees. Do not apply to crops in flower or to those in which bees are actively foraging. Do not apply when flowering weeds are present
- Surface residues may also cause bee mortality following spraying
- Dangerous to fish or other aquatic life. Do not contaminate surface waters or ditches with chemical or used container
- Keep in original container, tightly closed, in a safe place, under lock and key
- LERAP Category A

Hazard classification and safety precautions
Hazard Harmful, Flammable, Flammable liquid and vapour, Harmful if swallowed, Harmful if inhaled
Transport code 6.1
Packaging group III
UN Number 3017
Risk phrases H304, H317
Operator protection A, H, M; U02a, U04a, U08, U14, U15, U19a, U20b
Environmental protection E06c (7 d); E10b, E12a, E12e, E13b, E16c, E16d, E22a, E34, H410
Storage and disposal D01, D02, D05, D09b, D12b
Medical advice M01, M03

---

158   dimethomorph

A cinnamic acid fungicide with translaminar activity
FRAC mode of action code: 40

See also ametoctradin + dimethomorph

Products

1 Coronam	Pan Agriculture	50% w/w	WP	18366
2 Dimix 500 SC	Arysta	500 g/l	SC	18459
3 Paraat	BASF	50% w/w	WP	15445

Uses

- Blight in **potatoes** [2]
- Crown rot in **protected strawberries** *(moderate control)*, **strawberries** *(moderate control)* [1, 3]
- Downy mildew in **brassica leaves and sprouts** *(off-label)*, **chard** *(off-label)*, **herbs (see appendix 6)** *(off-label)*, **lamb's lettuce** *(off-label)*, **lettuce** *(off-label)*, **protected baby leaf crops** *(off-label)*, **protected herbs (see appendix 6)** *(off-label)*, **protected lamb's lettuce** *(off-label)*, **protected lettuce** *(off-label)*, **protected ornamentals** *(off-label)*, **radishes** *(off-label)*, **rocket** *(off-label)*, **seedling brassicas** *(off-label)*, **spinach** *(off-label)* [3]
- Root rot in **blackberries**, **protected blackberries** *(moderate control)*, **protected raspberries** *(moderate control)*, **raspberries** *(moderate control)* [1, 3]

Extension of Authorisation for Minor Use (EAMUs)

- **brassica leaves and sprouts** *20122244* [3]
- **chard** *20122244* [3]
- **herbs (see appendix 6)** *20122244* [3]
- **lamb's lettuce** *20122244* [3]
- **lettuce** *20122244* [3]
- **protected baby leaf crops** *20112584* [3]
- **protected herbs (see appendix 6)** *20112584* [3]
- **protected lamb's lettuce** *20112584* [3]
- **protected lettuce** *20112584* [3]
- **protected ornamentals** *20112585* [3]
- **radishes** *20160190* [3]
- **rocket** *20122244* [3]
- **seedling brassicas** *20130274* [3]
- **spinach** *20122244* [3]

Approval information

- Dimethomorph included in Annex I under EC Regulation 1107/2009

Restrictions

- Do not sow clover until the spring following application.
- Dimethomorph is a carboxylic acid amide fungicide. No more than three consecutive applications of CAA fungicides should be made.

Crop-specific information

- After application, allow 10 months before sowing clover [3]

Following crops guidance

- Cereal crops may be planted as part of a normal rotation following use. 120 days must elapse after the application before any other crops are planted.

Environmental safety

- LERAP Category B [1, 3]

Hazard classification and safety precautions

Hazard Harmful if swallowed [2]
Transport code 9
Packaging group III
UN Number 3077 [1, 3]; 3082 [2]

Operator protection A, D, H [1, 3]; U05a [1-3]; U09a, U20b [2]
Environmental protection E15b, E38 [1-3]; E16a [1, 3]; H410 [2]; H411 [3]
Storage and disposal D01, D02, D09a, D12a [1-3]; D10a [1, 3]; D10c [2]

159   dimethomorph + fluazinam

A mixture of cinnamic acid and dinitroaniline fungicides for blight control in potatoes
FRAC mode of action code: 29 + 40

See also fluazinam

Products

Hubble	Adama	200:200 g/l	SC	16089

Uses
- Blight in **potatoes**
- Late blight in **potatoes**

Approval information
- Dimethomorph and fluazinam included in Annex I under EC Regulation 1107/2009

Environmental safety
- Buffer zone requirement 6m [1]
- LERAP Category B

Hazard classification and safety precautions
Hazard Harmful, Dangerous for the environment
Transport code 9
Packaging group III
UN Number 3082
Risk phrases H361
Operator protection A, H; U02a, U04a, U05a, U08, U14, U15, U20a
Environmental protection E15b, E16a, E34, E38, H410
Storage and disposal D01, D02, D05, D09a, D12a
Medical advice M03

160   dimethomorph + mancozeb

A systemic and protectant fungicide for potato blight control
FRAC mode of action code: 40 + M3

See also mancozeb

Products

Invader	BASF	7.5:66.7% w/w	WG	15223

Uses
- Blight in **potatoes**
- Disease control in **chives** (off-label), **cress** (off-label), **forest nurseries** (off-label), **frise** (off-label), **herbs (see appendix 6)** (off-label), **lamb's lettuce** (off-label), **leaf brassicas** (off-label), **leeks**, **lettuce** (off-label), **ornamental plant production** (off-label), **poppies for morphine production** (off-label), **radicchio** (off-label), **salad onions** (off-label), **scarole** (off-label), **shallots**
- Downy mildew in **bulb onions**, **chives** (off-label), **cress** (off-label), **forest nurseries** (off-label), **frise** (off-label), **garlic**, **herbs (see appendix 6)** (off-label), **lamb's lettuce** (off-label), **leaf brassicas** (off-label), **lettuce** (off-label), **ornamental plant production** (off-label), **poppies for morphine production** (off-label), **radicchio** (off-label), **salad onions** (off-label), **scarole** (off-label)
- White tip in **leeks** (reduction)

Extension of Authorisation for Minor Use (EAMUs)
- **chives** 20120111

- **cress** *20120110*
- **forest nurseries** *20120109*
- **frise** *20120110*
- **herbs (see appendix 6)** *20120110*
- **lamb's lettuce** *20120110*
- **leaf brassicas** *20120110*
- **lettuce** *20120110*
- **ornamental plant production** *20120109*
- **poppies for morphine production** *20120112*
- **radicchio** *20120110*
- **salad onions** *20120111*
- **scarole** *20120110*

Approval information
- Dimethomorph and mancozeb included in Annex I under EC Regulation 1107/2009

Efficacy guidance
- Commence treatment as soon as there is a risk of blight infection
- In the absence of a warning treatment should start before the crop meets along the rows
- Repeat treatments every 7-14 d depending in the degree of infection risk
- Irrigated crops should be regarded as at high risk and treated every 7 d
- For best results good spray coverage of the foliage is essential
- To minimise the likelihood of development of resistance these products should be used in a planned Resistance Management strategy. See Section 5 for more information

Restrictions
- Maximum total dose equivalent to eight full dose treatments on potatoes

Crop-specific information
- HI 7 d for all crops

Environmental safety
- Dangerous for the environment
- Very toxic to aquatic organisms
- LERAP Category B

Hazard classification and safety precautions
Hazard Irritant, Dangerous for the environment, Very toxic to aquatic organisms
Transport code 9
Packaging group III
UN Number 3077
Risk phrases H317, H319, H335, H361
Operator protection A, C, D, H; U02a, U04a, U05a, U08, U13, U19a, U20b
Environmental protection E16a, E16b, E38, H410
Storage and disposal D01, D02, D05, D09a, D12a
Medical advice M05a

161   dimethomorph + zoxamide

A fungicide mixture for blight control in potatoes
FRAC mode of action code: 40 + 22

See also dimethomorph
zoxamide

Products

Presidium	Gowan	180:180 g/l	SC	18119

Uses
- Blight in **potatoes**
- Late blight in **potatoes**

SEE SECTION 3 FOR PRODUCTS ALSO REGISTERED

Approval information
- Dimethomorph and zoxamide included in Annex I under EC Regulation 1107/2009

Restrictions
- Horizontal boom sprayers must be fitted with three star drift reduction technology for all uses
- Low drift spraying equipment must be operated according to the specific conditions stated in the official three star rating for that equipment as published on HSE Chemicals Regulation Division's website. These operating conditions must be maintained until the operator is 30m from the top of the bank of any surface water bodies

Crop-specific information
- Cereal crops may be planted as part of a normal rotation following the use of Presidium. 120 days must elapse after the application of Presidium before any other crops are planted

Environmental safety
- Buffer zone requirement 6 m.

Hazard classification and safety precautions
   Hazard Very toxic to aquatic organisms
   Transport code 9
   Packaging group III
   UN Number 3082
   Risk phrases H317
   Operator protection A, H
   Environmental protection H410

162   dimoxystrobin

A protectant strobilurin fungicide for cereals available only in mixtures
FRAC mode of action code: 11

See also boscalid + dimoxystrobin

163   dimoxystrobin + epoxiconazole

A protectant and curative fungicide mixture for cereals
FRAC mode of action code: 11 + 3

See also epoxiconazole

Products

Swing Gold	BASF	133:50 g/l	SC	16081

Uses
- Brown rust in *durum wheat, spring wheat, winter wheat*
- Fusarium ear blight in *durum wheat, spring wheat, winter wheat*
- Glume blotch in *durum wheat, spring wheat, winter wheat*
- Septoria leaf blotch in *durum wheat, spring wheat, winter wheat*
- Tan spot in *durum wheat, spring wheat, winter wheat*

Approval information
- Dimoxystrobin and epoxiconazole included in Annex I under EC Regulation 1107/2009

Efficacy guidance
- For best results apply from the start of ear emergence
- Dimoxystrobin is a member of the QoI cross resistance group. Product should be used preventatively and not relied on for its curative potential
- Use product as part of an Integrated Crop Management strategy incorporating other methods of control, including where appropriate other fungicides with a different mode of action. Do not apply more than two foliar applications of QoI containing products to any cereal crop

FOR FULL CONDITIONS OF USE ALWAYS READ THE PRODUCT LABEL

- There is a significant risk of widespread resistance occurring in *Septoria tritici* populations in UK. Failure to follow resistance management action may result in reduced levels of disease control
- Epoxiconazole is a DMI fungicide. Resistance to some DMI fungicides has been identified in Septoria leaf blotch which may seriously affect performance of some products. For further advice contact a specialist advisor and visit the Fungicide Resistance Action Group (FRAG)-UK website

Restrictions
- Maximum number of treatments 1 per crop
- Do not apply before the start of ear emergence

Crop-specific information
- Latest use: up to and including flowering (GS 69)

Environmental safety
- Dangerous for the environment
- Very toxic to aquatic organisms
- Avoid drift onto neighbouring crops
- LERAP Category B

Hazard classification and safety precautions
Hazard Harmful, Dangerous for the environment, Harmful if swallowed, Harmful if inhaled
Transport code 9
Packaging group III
UN Number 3082
Risk phrases H317, H351, H360
Operator protection A; U05a, U20b
Environmental protection E15a, E16a, E34, E38, H410
Storage and disposal D01, D02, D05, D09a, D10c, D12a
Medical advice M05a

164 dithianon

A protectant and eradicant dicarbonitrile fungicide for scab control
FRAC mode of action code: M9

See also dithianon + potassium phosphonates

Products
1 Alcoban	Belchim	70% w/w	WG	18151
2 Dithianon WG	BASF	70% w/w	WG	17018

Uses
- Scab in *apples*, *pears*

Approval information
- Dithianon included in Annex I under EC Regulation 1107/2009

Efficacy guidance
- Apply at bud-burst and repeat every 7-14 d until danger of scab infection ceases
- Application at high rate within 48 h of a Mills period prevents new infection
- Spray programme also reduces summer infection with apple canker

Restrictions
- Maximum number of treatments on apples and pears 6 per crop
- Do not use on Golden Delicious apples after green cluster
- Do not mix with lime sulphur or highly alkaline products

Crop-specific information
- HI 4 wk for apples, pears

SEE SECTION 3 FOR PRODUCTS ALSO REGISTERED

Environmental safety
- Dangerous for the environment
- Very toxic to aquatic organisms

Hazard classification and safety precautions
   Hazard Dangerous for the environment, Toxic if swallowed, Very toxic to aquatic organisms
   Transport code 9
   Packaging group III
   UN Number 3077
   Risk phrases H317, H318 [1, 2]; H351 [1]
   Operator protection A, D, H; U05a, U11
   Environmental protection E15a, E34, E38, H410 [1, 2]; E17a [1] (30 m)
   Storage and disposal D01, D02, D05, D12a

165   dithianon + potassium phosphonates

A fungicide mixture for use in apple and pears
FRAC mode of action code: M9 + P07

See also dithianon

Products

Delan Pro	BASF	125:561 g/l	SC	17374

Uses
- Scab in **apples**, **pears**

Approval information
- Dithianon and potassium phosphonate included in Annex I under EC Regulation 1107/2009

Efficacy guidance
- Apply at bud-burst and repeat every 7-14 d until danger of scab infection ceases
- Application at high rate within 48 h of a Mills period prevents new infection
- Spray programme also reduces summer infection with apple canker

Restrictions
- When mixed with products containing carbonate or bicarbonate carbon dioxide may be released and foaming may occur.

Environmental safety
- Dangerous for the environment
- Very toxic to aquatic organisms
- LERAP Category B

Hazard classification and safety precautions
   Hazard Very toxic to aquatic organisms
   Transport code 9
   Packaging group III
   UN Number 3082
   Risk phrases H317, H319, H351
   Operator protection A, C, H; U19a, U20a
   Environmental protection E16a, H410; E17a (30 m)
   Storage and disposal D01, D02, D05

166   dodecadienol + tetradecenyl acetate + tetradecylacetate

An insect pheromone mixture

Products

RAK 3+4	BASF	3.82 : 4.1 : 1.9 % w/ w	VP	17824

FOR FULL CONDITIONS OF USE ALWAYS READ THE PRODUCT LABEL

SECTION 2

Uses
- Codling moth in *apples*, *pears*
- Fruit tree tortrix moth in *apples*, *cherries*, *pears*

Approval information
- Dodecadienol, tetradecenyl acetate and tetradecylacetate are included in Annex 1 under EC Regulation 1107/2009

Efficacy guidance
- Use in conjunction with an insecticide if damage threshold is exceeded during treatment.

Restrictions
- Do not use in orchards less than 1 ha in area
- Efficacy impaired if high density of Tortrix moths in surrounding untreated area

Hazard classification and safety precautions
   UN Number N/C
   Risk phrases H315, H317
   Operator protection A; U05a, U20b
   Environmental protection E15b, E34, H411
   Storage and disposal D01, D02, D09a

167   dodine

A protectant and eradicant guanidine fungicide
FRAC mode of action code: M7

Products

Syllit 400 SC	Arysta	400 g/l	SC	13363

Uses
- Leaf spot in *blackcurrants*
- Scab in *apples*, *pears*

Approval information
- Dodine included in Annex I under EC Regulation 1107/2009

Efficacy guidance
- Apply protective spray on apples and pears at bud-burst and at 10-14 d intervals until late Jun to early Jul
- Apply post-infection spray within 36 h of rain responsible for initiating infection. Where scab already present spray prevents production of spores
- On blackcurrants commence spraying at early grape stage and repeat at 2-3 wk intervals, and at least once after picking

Restrictions
- Do not apply in very cold weather (under 5°C) or under slow drying conditions to pears or dessert apples during bloom or immediately after petal fall
- Do not mix with lime sulphur or tetradifon
- Consult processors before use on crops grown for processing

Crop-specific information
- Latest use: early Jul for culinary apples; pre-blossom for dessert apples and pears

Environmental safety
- Dangerous for the environment
- Very toxic to aquatic organisms

Hazard classification and safety precautions
   Hazard Toxic if inhaled
   Transport code 6.1
   Packaging group III
   UN Number 2902

SEE SECTION 3 FOR PRODUCTS ALSO REGISTERED

Risk phrases H315, H318
Operator protection A, C, H; U09c, U11
Environmental protection H410
Storage and disposal D01, D02, D09a, D12a

168   epoxiconazole

A systemic, protectant and curative triazole fungicide for use in cereals
FRAC mode of action code: 3

See also boscalid + epoxiconazole
boscalid + epoxiconazole + pyraclostrobin
dimoxystrobin + epoxiconazole
epoxiconazole + fluxapyroxad

Products

1	Bassoon EC	BASF	83 g/l	EC	15219
2	Bowman	Adama	125 g/l	SC	16286
3	Cortez	Adama	125 g/l	SC	16280
4	Cube	Becesane	125 g/l	SC	16876
5	Epic	BASF	125 g/l	SC	16798
6	Mendoza	Adama	125 g/l	SC	17845
7	Rubric	FMC Agro	125 g/l	SC	18697
8	Ruby	AgChem Access	125 g/l	SC	16809
9	Scholar	Becesane	125 g/l	SC	16402
10	Spike 125 SC	Belchim	125 g/l	SC	18377
11	Strand	FMC Agro	125 g/l	SC	18811

Uses

- Brown rust in *durum wheat* [1, 9]; *rye* [1]; *spring barley, spring wheat, triticale, winter barley, winter wheat* [1-11]; *spring rye, winter rye* [2-11]
- Cercospora leaf spot in *red beet* *(off-label)* [7]
- Crown rust in *oats* [1]
- Disease control in *forest nurseries* *(off-label)* [3]; *grass seed crops* *(off-label)* [3, 7]
- Eyespot in *winter barley* *(reduction)* [3-11]; *winter wheat* *(reduction)* [2-11]
- Fusarium ear blight in *durum wheat* *(good reduction)*, *winter wheat* *(good reduction)* [1]; *spring wheat* [5, 10]; *spring wheat* *(good reduction)* [1-3, 6]; *winter wheat* [2, 3, 5, 6, 10]
- Glume blotch in *durum wheat* [1, 9]; *spring wheat, triticale, winter wheat* [1-11]
- Net blotch in *spring barley* [4, 5, 7-11]; *spring barley* *(moderate control)* [1-3, 6]; *winter barley* [3-11]; *winter barley* *(moderate control)* [1, 2]
- Powdery mildew in *durum wheat* [9]; *fodder beet, sugar beet* [4, 7, 8, 11]; *grass seed crops* *(off-label)* [7]; *oats, rye* [1]; *spring barley, winter barley* [1, 4, 5, 7-11]; *spring oats, winter oats* [2-9, 11]; *spring rye, winter rye, winter wheat* [2-11]; *spring wheat, triticale* [4, 5, 7-11]
- Ramularia leaf spots in *spring barley* *(reduction)*, *winter barley* *(reduction)* [1]
- Rhynchosporium in *rye* [1]; *spring barley, winter barley* [1-11]; *spring rye, winter rye* [2-11]
- Rust in *fodder beet, sugar beet* [4, 7, 8, 11]; *grass seed crops* *(off-label)* [3, 7]
- Septoria leaf blotch in *durum wheat* [1, 9]; *spring wheat, triticale, winter wheat* [1-11]
- Sooty moulds in *durum wheat* *(reduction)* [9]; *spring wheat* *(reduction)*, *winter wheat* *(reduction)* [2-11]
- Yellow rust in *durum wheat* [1, 9]; *rye* [1]; *spring barley, spring wheat, triticale, winter barley, winter wheat* [1-11]; *spring rye, winter rye* [2-11]

Extension of Authorisation for Minor Use (EAMUs)

- *forest nurseries* 20141278 [3]
- *grass seed crops* 20193263 [3], 20190893 [7]
- *red beet* 20190894 [7]

Approval information

- Accepted by BBPA for use on malting barley (before ear emergence only)
- Epoxiconazole included in Annex I under EC Regulation 1107/2009

FOR FULL CONDITIONS OF USE ALWAYS READ THE PRODUCT LABEL

Efficacy guidance
- Apply at the start of foliar disease attack
- Optimum effect against eyespot achieved by spraying between leaf-sheath erect and second node detectable stages (GS 30-32)
- Best control of ear diseases of wheat obtained by treatment during ear emergence
- Mildew control improved by use of tank mixtures. See label for details
- For Septoria spray after third node detectable stage (GS 33) when weather favouring disease development has occurred
- Epoxiconazole is a DMI fungicide. Resistance to some DMI fungicides has been identified in Septoria leaf blotch which may seriously affect performance of some products. For further advice contact a specialist advisor and visit the Fungicide Resistance Action Group (FRAG)-UK website

Restrictions
- Maximum total dose equivalent to two full dose treatments
- Product may cause damage to broad-leaved plant species
- Avoid spray drift onto neighbouring crops

Crop-specific information
- Latest use: up to and including flowering just complete (GS 69) in wheat, rye, triticale; up to and including emergence of ear just complete (GS 59) in barley, oats

Environmental safety
- Dangerous for the environment
- Very toxic to aquatic organisms
- LERAP Category B

Hazard classification and safety precautions
Hazard Harmful, Dangerous for the environment [1-11]; Harmful if inhaled [5-7, 10]; Very toxic to aquatic organisms [5, 6, 10]
Transport code 9
Packaging group III
UN Number 3082
Risk phrases H315 [5, 10, 11]; H351 [5-7, 10, 11]; H360 [5, 6, 10]; H361 [7, 11]; R20 [1]; R38 [1, 4, 8, 9]; R40, R50, R53a [1-4, 8, 9]; R62, R63 [1-4, 7-9]
Operator protection A [1-11]; H [1, 5]; U05a, U20b
Environmental protection E15a, E16a, E38 [1-11]; H410 [5, 7, 10, 11]; H411 [6]
Storage and disposal D01, D02, D05, D09a, D10c, D12a [1-11]; D19 [5]
Medical advice M05a

169   epoxiconazole + fluxapyroxad

An SDHI and triazole fungicide mixture for disease control in cereals
FRAC mode of action code: 3 + 7

See also epoxiconazole
      fluxapyroxad

Products

1	Adexar	BASF	62.5:62.5 g/l	EC	17109
2	Clayton Index	Clayton	62.5:62.5 g/l	EC	17266
3	Pexan	BASF	62.5:59.4 g/l	EC	17110

Uses
- Brown rust in **durum wheat**, **rye**, **spring barley**, **spring wheat**, **triticale**, **winter barley**, **winter wheat**
- Crown rust in **oats**
- Eyespot in **durum wheat** (moderate control), **oats** (moderate control), **rye** (moderate control), **triticale** (moderate control), **winter wheat** (moderate control)
- Fusarium ear blight in **durum wheat** (good reduction), **spring wheat** (good reduction), **winter wheat** (good reduction)

SEE SECTION 3 FOR PRODUCTS ALSO REGISTERED

- Glume blotch in *durum wheat*, *spring wheat*, *triticale*, *winter wheat*
- Net blotch in *spring barley*, *winter barley*
- Powdery mildew in *durum wheat* (*moderate control*), *oats* (*moderate control*), *rye* (*moderate control*), *spring barley* (*moderate control*), *spring wheat* (*moderate control*), *triticale* (*moderate control*), *winter barley* (*moderate control*), *winter wheat* (*moderate control*)
- Ramularia leaf spots in *spring barley*, *winter barley*
- Rhynchosporium in *spring barley*, *winter barley*
- Septoria leaf blotch in *durum wheat*, *spring wheat*, *triticale*, *winter wheat*
- Sooty moulds in *durum wheat* (*reduction*), *spring wheat* (*reduction*), *winter wheat* (*reduction*)
- Tan spot in *durum wheat* (*moderate control*), *spring wheat* (*moderate control*), *winter wheat* (*moderate control*)
- Yellow rust in *durum wheat*, *rye*, *spring barley*, *spring wheat*, *triticale*, *winter barley*, *winter wheat*

Approval information
- Epoxiconazole and fluxapyroxad included in Annex I under EC Regulation 1107/2009
- Accepted by BBPA for use on malting barley up to GS 45 only (max. dose, 1 litre/ha)

Restrictions
- Avoid drift on to neighbouring crops since it may damage broad-leaved species.
- Do not apply more that two foliar applications of products containing SDHI fungicides to any cereal crop.
- Must not be applied before GS30 (pseudo-stem erect) [1]

Crop-specific information
- After treating a cereal crop cabbage, carrot, clover, dwarf beans, field beans, lettuce, maize, oats, oilseed rape, onions, peas, potatoes, ryegrass, sugar beet, sunflower, winter barley and winter wheat may be sown as the following crop

Environmental safety
- LERAP Category B

Hazard classification and safety precautions
Hazard Harmful, Dangerous for the environment, Harmful if swallowed, Very toxic to aquatic organisms
Transport code 9
Packaging group III
UN Number 3082
Risk phrases H317, H319, H351, H360
Operator protection A [1-3]; C [1]; H [1, 3]; U05a, U14, U20b
Environmental protection E15b, E16a, E38, H410
Storage and disposal D01, D02, D05, D10c, D12a, D19
Medical advice M05a

170   epoxiconazole + fluxapyroxad + pyraclostrobin

A triazole + SDHI + strobilurin fungicide mixture for disease control in cereals
FRAC mode of action code: 3 + 7 + 11

Products

1 Ceriax	BASF	41.6:41.6:66.6 g/l	EC	17111
2 Vortex	BASF	41.6:41.6:61.0 g/l	EC	17112

Uses
- Brown rust in *durum wheat*, *rye*, *spring barley*, *spring wheat*, *triticale*, *winter barley*, *winter wheat*
- Crown rust in *oats*
- Eyespot in *durum wheat* (*moderate control*), *oats* (*moderate control*), *rye* (*moderate control*), *triticale* (*moderate control*), *winter wheat* (*moderate control*)
- Fusarium ear blight in *durum wheat* (*good reduction*), *spring wheat* (*good reduction*), *winter wheat* (*good reduction*)

FOR FULL CONDITIONS OF USE ALWAYS READ THE PRODUCT LABEL

- Glume blotch in *durum wheat, spring wheat, triticale, winter wheat*
- Net blotch in *spring barley, winter barley*
- Powdery mildew in *durum wheat* (moderate control), *oats* (moderate control), *rye* (moderate control), *spring barley* (moderate control), *spring wheat* (moderate control), *triticale* (moderate control), *winter barley* (moderate control), *winter wheat* (moderate control)
- Ramularia leaf spots in *spring barley, winter barley*
- Rhynchosporium in *rye, spring barley, winter barley*
- Septoria leaf blotch in *durum wheat, spring wheat, triticale, winter wheat*
- Sooty moulds in *durum wheat* (reduction), *spring wheat* (reduction), *winter wheat* (reduction)
- Tan spot in *durum wheat* (moderate control), *spring wheat* (moderate control), *winter wheat* (moderate control)
- Yellow rust in *durum wheat, rye, spring barley, spring wheat, triticale, winter barley, winter wheat*

Approval information
- Epoxiconazole, fluxapyroxad and pyraclostrobin included in Annex I under EC Regulation 1107/2009

Restrictions
- No more than two applications of any QoI fungicide may be applied to wheat, barley, oats, rye or triticale.
- Do not apply more that two foliar applications of products containing SDHI fungicides to any cereal crop.
- May cause damage to broad-leaved plant species.

Following crops guidance
- Cabbage, carrot, clover, dwarf French beans, lettuce, maize, oats, oilseed rape, onions, peas, potatoes, ryegrass, sugarbeet, sunflower, winter barley and winter wheat may be sown as a following crop but the effect on other crops has not been assessed.

Environmental safety
- LERAP Category B

Hazard classification and safety precautions
Hazard Harmful, Dangerous for the environment, Harmful if swallowed, Harmful if inhaled, Very toxic to aquatic organisms
Transport code 9
Packaging group III
UN Number 3082
Risk phrases H315, H317, H318, H351, H360
Operator protection A, H; U05a, U14, U20b
Environmental protection E15b, E16a, E38, H410
Storage and disposal D01, D02, D05, D07, D10c, D12a
Medical advice M05a

171   epoxiconazole + folpet

A systemic and protectant fungicide mixture for disease control in cereals
FRAC mode of action code: 3 + M4

See also epoxiconazole
          folpet

Products
1	Manitoba	Adama	50:375 g/l	SC	16539
2	Scotia	Adama	50:375 g/l	SC	17415

Uses
- Brown rust in *spring wheat, triticale, winter wheat*
- Disease control in *spring barley, triticale, winter barley*
- Glume blotch in *spring wheat, triticale, winter wheat*

- Rhynchosporium in *spring barley*, *triticale*, *winter barley*
- Septoria leaf blotch in *spring wheat*, *triticale*, *winter wheat*
- Yellow rust in *spring wheat*, *triticale*, *winter wheat*

Approval information
- Epoxiconazole and folpet included in Annex I under EC Regulation 1107/2009

Environmental safety
- Dangerous for the environment
- Very toxic to aquatic organisms
- Buffer zone requirement 6m

Hazard classification and safety precautions
Hazard Harmful, Dangerous for the environment, Harmful if inhaled, Very toxic to aquatic organisms
Transport code 9
Packaging group III
UN Number 3082
Risk phrases H319, H351, H360
Operator protection A, H; U05a, U14
Environmental protection E38, H412
Storage and disposal D01, D12a

172   epoxiconazole + isopyrazam

A protectant and eradicant fungicide mixture for use in cereals
FRAC mode of action code: 3 + 7

Products

1 Keystone	Syngenta	99:125 g/l	SC	15493
2 Keystone	Adama	99:125 g/l	SC	18565
3 Micaraz	Syngenta	90:125 g/l	SC	16106
4 Seguris	Syngenta	90:125 g/l	SC	15246

Uses
- Brown rust in *spring barley*, *triticale*, *winter barley*, *winter rye*, *winter wheat*
- Glume blotch in *winter wheat*
- Net blotch in *spring barley*, *winter barley*
- Ramularia leaf spots in *spring barley*, *winter barley*
- Rhynchosporium in *spring barley*, *winter barley*, *winter rye*
- Septoria leaf blotch in *triticale*, *winter wheat*
- Yellow rust in *winter wheat*

Approval information
- Epoxiconazole and isopyrazam are included in Annex I under EC Regulation 1107/2009
- Approved by BBPA for use on malting barley before ear emergence only

Restrictions
- Isopyrazam is an SDH respiration inhibitor; Do not apply more than two foliar applications of products containing an SDH inhibitor to any cereal crop

Environmental safety
- Buffer zone requirement 6 m [3, 4]
- LERAP Category B

Hazard classification and safety precautions
Hazard Harmful, Dangerous for the environment, Harmful if inhaled, Very toxic to aquatic organisms
Transport code 9
Packaging group III
UN Number 3082

Risk phrases H317, H319, H351, H361 [1-4]; H360 [1, 2]
Operator protection A, H; U09a, U14, U20b
Environmental protection E15b, E16a, E16b, E38, H410
Storage and disposal D01, D02, D05, D09a, D10c, D11a, D12a
Medical advice M05a

173   epoxiconazole + metconazole

A triazole mixture for use in cereals
FRAC mode of action code: 3 + 3

See also metconazole

Products

1	Brutus	BASF	37.5:27.5 g/l	EC	14353
2	Osiris P	BASF	56.25:41.25 g/l	EC	15627

Uses

- Brown rust in *durum wheat*, *spring barley*, *spring wheat*, *triticale*, *winter barley*, *winter wheat* [1, 2]; *rye* [2]; *spring rye*, *winter rye* [1]
- Eyespot in *durum wheat* (reduction), *triticale* (reduction), *winter wheat* (reduction) [1, 2]; *rye* (reduction) [2]; *spring rye* (reduction), *winter barley* (reduction), *winter rye* (reduction) [1]
- Fusarium ear blight in *durum wheat* (good reduction), *spring wheat* (good reduction), *triticale* (good reduction), *winter wheat* (good reduction) [1]; *durum wheat* (moderate control), *spring wheat* (moderate control), *triticale* (moderate control), *winter wheat* (moderate control) [2]
- Glume blotch in *durum wheat*, *spring wheat*, *triticale*, *winter wheat* [1, 2]
- Net blotch in *spring barley*, *winter barley* [1, 2]
- Powdery mildew in *durum wheat* (moderate control), *spring barley* (moderate control), *spring wheat* (moderate control), *triticale* (moderate control), *winter barley* (moderate control), *winter wheat* (moderate control) [1, 2]; *rye* (moderate control) [2]; *spring rye* (moderate control), *winter rye* (moderate control) [1]
- Ramularia leaf spots in *spring barley* (moderate control), *winter barley* (moderate control) [1, 2]
- Rhynchosporium in *rye* (moderate control) [2]; *spring barley*, *winter barley* [1, 2]
- Septoria leaf blotch in *durum wheat*, *spring wheat*, *triticale*, *winter wheat* [1, 2]
- Sooty moulds in *durum wheat* (good reduction), *spring wheat* (good reduction), *winter wheat* (good reduction) [1]; *durum wheat* (moderate control), *spring wheat* (moderate control), *winter wheat* (moderate control) [2]
- Tan spot in *durum wheat* (moderate control), *spring wheat* (moderate control), *winter wheat* (moderate control) [1, 2]
- Yellow rust in *durum wheat*, *spring barley*, *spring wheat*, *triticale*, *winter barley*, *winter wheat* [1, 2]; *rye* [2]; *spring rye*, *winter rye* [1]

Approval information

- Epoxiconazole and metconazole included in Annex I under EC Regulation 1107/2009

Following crops guidance

- After treating a cereal crop, oilseed rape, cereals, sugar beet, linseed, maize, clover, beans, peas, carrots, potatoes, lettuce, cabbage, sunflower, ryegrass or onions may be sown as a following crop. The effect on other crops has not been evaluated.

Environmental safety

- LERAP Category B

Hazard classification and safety precautions

Hazard Harmful, Dangerous for the environment, Very toxic to aquatic organisms
Transport code 9
Packaging group III
UN Number 3082
Risk phrases H317, H351, H360 [1, 2]; H319 [2]
Operator protection A, H; U05a, U14 [1]; U20b [1, 2]
Environmental protection E15b, E16a, E34, H410 [1, 2]; E38 [2]

SEE SECTION 3 FOR PRODUCTS ALSO REGISTERED

Storage and disposal D01, D02, D09a, D10c [1, 2]; D05, D12a, D19 [2]
Medical advice M05a

174   epoxiconazole + pyraclostrobin

A protectant, systemic and curative fungicide mixture for cereals
FRAC mode of action code: 3 + 11

See also pyraclostrobin

Products

1	Envoy	BASF	62.5:85 g/l	SE	16297
2	Gemstone	BASF	62.5:80 g/l	SE	16298
3	Opera	BASF	50:133 g/l	SE	16420

Uses

* Brown rust in *durum wheat*, *rye*, *spring barley*, *spring wheat*, *triticale*, *winter barley*, *winter wheat* [1-3]
* Cercospora leaf spot in *sugar beet* [3]
* Crown rust in *spring oats*, *winter oats* [1-3]
* Disease control in *grass seed crops* (off-label) [3]
* Eyespot in *forage maize*, *forage maize* (off-label), *grain maize*, *grain maize* (off-label) [3]
* Fusarium ear blight in *durum wheat* (good reduction), *spring wheat* (good reduction), *winter wheat* (good reduction) [1-3]; *triticale* (good reduction) [3]
* Glume blotch in *durum wheat*, *spring wheat*, *triticale*, *winter wheat* [1-3]
* Leaf blight in *forage maize* (off-label), *grain maize* (off-label) [3]
* Net blotch in *spring barley*, *winter barley* [1-3]
* Northern leaf blight in *forage maize*, *grain maize* [3]
* Powdery mildew in *fodder beet* (off-label), *sugar beet* [3]; *rye*, *spring barley*, *spring oats*, *winter barley*, *winter oats* [1, 2]
* Ramularia leaf spots in *rye* (moderate control), *spring barley* (moderate control), *sugar beet*, *triticale* (moderate control), *winter barley* (moderate control) [3]; *spring barley* (reduction), *winter barley* (reduction) [1, 2]
* Rhynchosporium in *rye*, *spring barley*, *winter barley* [1-3]
* Rust in *fodder beet* (off-label), *forage maize* (off-label), *grain maize* (off-label), *sugar beet* [3]
* Septoria leaf blotch in *durum wheat*, *spring wheat*, *triticale*, *winter wheat* [1, 2]; *durum wheat* (moderate control), *spring wheat* (moderate control), *triticale* (moderate control), *winter wheat* (moderate control) [3]
* Sooty moulds in *spring wheat* (reduction), *winter wheat* (reduction) [1, 2]
* Tan spot in *durum wheat*, *spring wheat*, *winter wheat* [1, 2]; *durum wheat* (moderate control), *spring wheat* (moderate control), *triticale* (moderate control), *winter wheat* (moderate control) [3]
* Yellow rust in *durum wheat*, *rye*, *spring barley*, *spring wheat*, *triticale*, *winter barley*, *winter wheat* [1-3]

Extension of Authorisation for Minor Use (EAMUs)

* *fodder beet* 20140392 [3]
* *forage maize* 20141592 [3]
* *grain maize* 20141592 [3]
* *grass seed crops* 20140522 [3]

Approval information

* Epoxiconazole and pyraclostrobin included in Annex I under EC Regulation 1107/2009
* Accepted by BBPA for use on malting barley (before ear emergence only)

Efficacy guidance

* For best results apply at the start of foliar disease attack
* Good reduction of Fusarium ear blight can be obtained from treatment during flowering
* Yield response may be obtained in the absence of visual disease symptoms

FOR FULL CONDITIONS OF USE ALWAYS READ THE PRODUCT LABEL

- Epoxiconazole is a DMI fungicide. Resistance to some DMI fungicides has been identified in Septoria leaf blotch which may seriously affect performance of some products. For further advice contact a specialist advisor and visit the Fungicide Resistance Action Group (FRAG)-UK website
- Pyraclostrobin is a member of the QoI cross resistance group. Product should be used preventatively and not relied on for its curative potential
- Use product as part of an Integrated Crop Management strategy incorporating other methods of control, including where appropriate other fungicides with a different mode of action. Do not apply more than two foliar applications of QoI containing products to any cereal crop
- There is a significant risk of widespread resistance occurring in *Septoria tritici* populations in UK. Failure to follow resistance management action may result in reduced levels of disease control

Restrictions
- Maximum number of treatments 2 per crop
- Do not use on sugar beet crops grown for seed [3]

Crop-specific information
- Latest use: before grain watery ripe (GS 71) for wheat; up to and including emergence of ear just complete (GS 59) for barley and oats
- HI 6 wk for sugar beet [3]

Environmental safety
- Dangerous for the environment
- Very toxic to aquatic organisms
- LERAP Category B

Hazard classification and safety precautions
Hazard Harmful, Dangerous for the environment, Harmful if inhaled [1-3]; Toxic if swallowed [3]; Harmful if swallowed, Very toxic to aquatic organisms [1, 2]
Transport code 9
Packaging group III
UN Number 3082
Risk phrases H315, H317, H351 [1-3]; H360 [1, 3]
Operator protection A; U05a, U14, U20b
Environmental protection E15a, E16a, E34, E38, H410 [1-3]; E16b [3]
Storage and disposal D01, D02, D09a, D10c, D12a [1-3]; D05 [3]
Medical advice M03, M05a

175   esfenvalerate

A contact and ingested pyrethroid insecticide
IRAC mode of action code: 3

Products

1	Sumi-Alpha	Sumitomo	25 g/l	EC	14023
2	Sven	Sumitomo	25 g/l	EC	14859

Uses
- Aphids in *broccoli, brussels sprouts, cabbages, calabrese, cauliflowers, chinese cabbage, combining peas, edible podded peas, kale, kohlrabi, ornamental plant production, potatoes, spring barley, spring field beans, spring wheat, vining peas, winter barley, winter field beans, winter wheat* [1, 2]; *grass seed crops (off-label), grassland, protected ornamentals* [1]
- Bibionids in *grass seed crops, grassland, managed amenity turf* [2]
- Caterpillars in *broccoli, brussels sprouts, cabbages, calabrese, cauliflowers, chinese cabbage, kale, kohlrabi, ornamental plant production* [2]
- Insect pests in *protected ornamentals* [2]
- Leaf miner in *ornamental plant production* [2]
- Pea and bean weevil in *combining peas, edible podded peas, spring field beans, vining peas, winter field beans* [2]

SEE SECTION 3 FOR PRODUCTS ALSO REGISTERED

- Thrips in **combining peas**, **edible podded peas**, **ornamental plant production**, **vining peas** [2]
- Whitefly in **ornamental plant production** [2]

Extension of Authorisation for Minor Use (EAMUs)
- **grass seed crops** *20081266* [1]

Approval information
- Esfenvalerate included in Annex I under EC Regulation 1107/2009
- Accepted by BBPA for use on malting barley

Efficacy guidance
- For best reduction of spread of barley yellow dwarf virus winter sown crops at high risk (e.g. after grass or in areas with history of BYDV) should be treated when aphids first seen or by mid-Oct. Otherwise treat in late Oct-early Nov
- High risk winter sown crops will need a second treatment
- Spring sown crops should be treated from the 2-3 leaf stage if aphids are found colonising in the crop and a further application may be needed before the first node stage if they reinfest
- Product also recommended between onset of flowering and milky ripe stages (GS 61-73) for control of summer cereal aphids

Restrictions
- Maximum number of treatments 3 per crop of which 2 may be in autumn
- Do not use if another pyrethroid or dimethoate has been applied to crop after start of ear emergence (GS 51)
- Do not leave spray solution standing in spray tank

Crop-specific information
- Latest use: 31 Mar in yr of harvest (winter use dose); early milk stage for barley (summer use); late milk stage for wheat (summer use)

Environmental safety
- Dangerous for the environment
- Very toxic to aquatic organisms
- High risk to non-target insects or other arthropods. Do not spray within 5 m of the field boundary
- Give local beekeepers warning when using in flowering crops
- Flammable
- Store product in dark away from direct sunlight
- LERAP Category A

Hazard classification and safety precautions
Hazard Harmful, Flammable, Dangerous for the environment, Flammable liquid and vapour, Harmful if swallowed, Harmful if inhaled, Very toxic to aquatic organisms
Transport code 3
Packaging group III
UN Number 1993
Risk phrases H304, H317, H318, H373
Operator protection A, C, H; U04a, U05a, U08, U11, U14, U19a, U20b
Environmental protection E15a, E16c, E16d, E22a, E34, E38, H410
Storage and disposal D01, D02, D09a, D10c, D12a
Medical advice M03

176 ethephon

A plant growth regulator for cereals and various horticultural crops

See also chlormequat + 2-chloroethylphosphonic acid
chlormequat + 2-chloroethylphosphonic acid + imazaquin
chlormequat + 2-chloroethylphosphonic acid + mepiquat chloride

Products

1	Cerone	Nufarm UK	480 g/l	SL	15944
2	Chrysal Plus	Chrysal	480 g/l	SL	17847

FOR FULL CONDITIONS OF USE ALWAYS READ THE PRODUCT LABEL

Products – continued

3	Coryx	SFP Europe	480 g/l	SL	17620
4	Ephon Top	Nufarm UK	660 g/l	SL	18872
5	Ethe 480	Agroquimicos	480 g/l	SL	16887
6	Floralife Tulipa	SFP Europe	480 g/l	SL	17996
7	Ipanema	Belchim	480 g/l	SL	15961
8	Padawan	SFP Europe	480 g/l	SL	15577
9	Telsee	Clayton	480 g/l	SL	16325

Uses

- Growth regulation in **apples** *(off-label)*, **ornamental plant production** *(off-label)*, **protected tomatoes** *(off-label)* [1]; **durum wheat, rye, spelt, spring barley, spring wheat, triticale, winter barley, winter wheat** [4]; **ornamental plant production** [2]; **tulips** [6]
- Lodging control in **spring barley, winter barley, winter wheat** [1, 3, 5, 7-9]; **triticale, winter rye** [1, 5]

Extension of Authorisation for Minor Use (EAMUs)

- **apples** *20122367* [1]
- **ornamental plant production** *20122366* [1]
- **protected tomatoes** *20122369* [1]

Approval information

- Ethephon (2-chloroethylphosphonic acid) included in Annex I under EC Regulation 1107/2009
- All products containing ethephon carry the warning: 'ethephon is an anticholinesterase organophosphate. Handle with care'
- Accepted by BBPA for use on malting barley

Efficacy guidance

- Best results achieved on crops growing vigorously under conditions of high fertility
- Optimum timing varies between crops and products. See labels for details
- Do not spray crops when wet or if rain imminent

Restrictions

- Ethephon (2-chloroethylphosphonic acid) is an anticholinesterase organophosphorus compound. Do not use if under medical advice not to work with such compounds
- Maximum number of treatments 1 per crop or yr
- Do not spray crops suffering from stress caused by any factor, during cold weather or period of night frost nor when soil very dry
- Do not apply to cereals within 10 d of herbicide or liquid fertilizer application
- Do not spray wheat or triticale where the leaf sheaths have split and the ear is visible
- When used as a cut stem treatment on tulips the maximum treatment period must not exceed 10 days [6]

Crop-specific information

- Latest use: before 1st spikelet visible (GS 51) for spring barley, winter barley, winter rye; before flag leaf sheath opening (GS 47) for triticale, winter wheat
- HI cider apples, tomatoes 5 d

Environmental safety

- Harmful to aquatic organisms
- Avoid accidental deposits on painted objects such as cars, trucks, aircraft

Hazard classification and safety precautions

Hazard Harmful [1, 5]; Irritant [3, 6-9]; Harmful if inhaled [1]
Transport code 8
Packaging group III
UN Number 3265
Risk phrases H290 [4-9]; H312, H314, H335 [4]; H315 [5-9]; H318 [1, 3, 5-9]
Operator protection A, C [1-9]; H [3, 4]; U05a, U08, U11, U13, U19c, U20b [1, 3, 5-9]
Environmental protection E13c, E38 [1, 3, 5-9]; H411 [2, 4]; H412 [1]
Storage and disposal D01, D02 [1-9]; D05 [2, 4]; D09a, D10b, D12a [1, 3, 5-9]
Medical advice M01 [1, 3, 5-9]

177   ethephon + mepiquat chloride

A plant growth regulator for use in cereals

See also mepiquat chloride

Products

1 Clayton Proud	Clayton	155:305 g/l	SL	17067
2 Terpal	BASF	155:305 g/l	SL	16463

Uses
- Growth regulation in **forest nurseries** *(off-label)*, **ornamental plant production** *(off-label)* [2]
- Increasing yield in **winter barley** *(low lodging situations)*, **winter wheat** *(low lodging situations)* [1, 2]
- Lodging control in **spring barley**, **triticale**, **winter barley**, **winter rye**, **winter wheat** [1, 2]

Extension of Authorisation for Minor Use (EAMUs)
- **forest nurseries** *20142725* [2]
- **ornamental plant production** *20180151* [2]

Approval information
- Ethephon and mepiquat chloride included in Annex I under EC Regulation 1107/2009
- All products containing ethephon (2-chloroethylphosphonic acid) carry the warning: '2-chloroethylphosphonic acid is an anticholinesterase organophosphate. Handle with care'
- Accepted by BBPA for use on malting barley

Efficacy guidance
- Best results achieved on crops growing vigorously under conditions of high fertility
- Recommended dose and timing vary with crop, cultivar, growing conditions, previous treatment and desired degree of lodging control. See label for details
- May be applied to crops undersown with grass or clovers
- Do not apply to crops if wet or rain expected as efficacy will be impaired

Restrictions
- Ethephon is an anticholinesterase organophosphorus compound. Do not use if under medical advice not to work with such compounds
- Maximum number of treatments 2 per crop
- Add an authorised non-ionic wetter to spray solution. See label for recommended product and rate
- Do not treat crops damaged by herbicides or stressed by drought, waterlogging etc
- Do not treat crops on soils of low fertility unless adequately fertilized
- Do not apply to winter cultivars sown in spring or treat winter barley, triticale or winter rye on soils with more than 10% organic matter (winter wheat may be treated)
- Do not apply at temperatures above 21°C

Crop-specific information
- Latest use: before ear visible (GS 49) for winter barley, spring barley, winter wheat and triticale; flag leaf just visible (GS 37) for winter rye
- Late tillering may be increased with crops subject to moisture stress and may reduce quality of malting barley

Environmental safety
- Do not use straw from treated cereals as a mulch or growing medium

Hazard classification and safety precautions
> **Hazard** Harmful [1, 2]; Harmful if swallowed [2]
> **Transport code** 8
> **Packaging group** III
> **UN Number** 3265
> **Risk phrases** H290 [2]
> **Operator protection** A, C; U20b
> **Environmental protection** E15a [1, 2]; H413 [2]
> **Storage and disposal** D01, D02, D08, D09a, D10c
> **Medical advice** M01, M05a

FOR FULL CONDITIONS OF USE ALWAYS READ THE PRODUCT LABEL

178   ethofumesate

A benzofuran herbicide for grass weed control in various crops
HRAC mode of action code: N

See also desmedipham + ethofumesate + lenacil + phenmedipham
desmedipham + ethofumesate + metamitron + phenmedipham
desmedipham + ethofumesate + phenmedipham

Products

1	Efeckt	UPL Europe	500 g/l	SC	17177
2	Ethofol	UPL Europe	500 g/l	SC	18224
3	Oblix 500	UPL Europe	500 g/l	SC	16671
4	Xerton	UPL Europe	417 g/l	SC	17335

Uses

- Annual dicotyledons in *fodder beet*, *grass seed crops*, *mangels*, *red beet*, *sugar beet* [1-3]; *winter wheat* [4]
- Annual grasses in *fodder beet*, *grass seed crops*, *mangels*, *red beet*, *sugar beet* [1, 3]; *winter wheat* [4]
- Blackgrass in *fodder beet*, *grass seed crops*, *mangels*, *red beet*, *sugar beet* [2]; *winter wheat* [4]
- Cleavers in *fodder beet*, *grass seed crops*, *mangels*, *red beet*, *sugar beet* [2]

Approval information

- Ethofumesate included in Annex I under EC Regulation 1107/2009

Efficacy guidance

- Most products may be applied pre- or post-emergence of crop or weeds but some restricted to pre-emergence or post-emergence use only. Check label
- Some products recommended for use only in mixtures. Check label
- Volunteer cereals not well controlled pre-emergence, weed grasses should be sprayed before fully tillered
- Grass crops may be sprayed during rain or when wet. Not recommended in very dry conditions or prolonged frost

Restrictions

- The maximum total dose must not exceed 1.0 kg/ha ethofumesate in any three year period
- Do not use on Sands or Heavy soils, Very Light soils containing a high percentage of stones, or soils with more than 5-10% organic matter (percentage varies according to label)
- Do not use on swards reseeded without ploughing
- Clovers will be killed or severely checked
- Do not graze or cut grass for 14 d after, or roll less than 7 d before or after spraying

Crop-specific information

- Latest use: before crops meet across rows for beet crops and mangels; not specified for other crops
- Apply in beet crops in tank mixes with other pre- or post-emergence herbicides. Recommendations vary for different mixtures. See label for details
- Safe timing on beet crops varies with other ingredient of tank mix. See label for details
- In grass crops apply to moist soil as soon as possible after sowing or post-emergence when crop in active growth, normally mid-Oct to mid-Dec. See label for details
- May be used in Italian, hybrid and perennial ryegrass, timothy, cocksfoot, meadow fescue and tall fescue. Apply pre-emergence to autumn-sown leys, post-emergence after 2-3 leaf stage. See label for details

Following crops guidance

- In the event of failure of a treated crop only sugar beet, red beet, fodder beet and mangels may be redrilled within 3 mth of application
- Any crop may be sown 3 mth after application of mixtures in beet crops following ploughing, 5 mth after application in grass crops

SEE SECTION 3 FOR PRODUCTS ALSO REGISTERED

Environmental safety
- Dangerous for the environment
- Toxic to aquatic organisms
- Do not empty into drains

Hazard classification and safety precautions
    Hazard Irritant [2]
    Transport code 9 [1, 2]
    Packaging group III [1, 2]
    UN Number 3082 [1, 2]; N/C [3, 4]
    Risk phrases H317 [4]
    Operator protection A [2]; H [1-4]; U05a [1, 3, 4]; U09a, U14, U19a, U20b [2]
    Environmental protection E15b, H411 [1-4]; E23, E34 [1, 3, 4]; E38 [2-4]
    Storage and disposal D01, D02, D09a [1-4]; D05, D10c [2]; D10a, D12a [1, 3, 4]
    Medical advice M05a [2]

179   ethofumesate + metamitron

A contact and residual herbicide mixture for beet crops
HRAC mode of action code: N + C1

See also metamitron

Products

1	Oblix MT	UPL Europe	150:350 g/l	SC	18857
2	Torero	Adama	150:350 g/l	SC	18625
3	Volcano	UPL Europe	150:350 g/l	SC	18926

Uses
- Annual dicotyledons in *fodder beet*, *mangels*, *sugar beet*
- Annual meadow grass in *fodder beet*, *mangels*, *sugar beet*

Approval information
- Ethofumesate and metamitron included in Annex I under EC Regulation 1107/2009

Efficacy guidance
- Best results obtained from a series of treatments applied as an overall fine spray commencing when earliest germinating weeds are no larger than fully expanded cotyledon and the majority of the crop at fully expanded cotyledon
- Apply subsequent sprays as each new flush of weeds reaches early cotyledon and continue until weed emergence ceases
- Product may be used on all soil types but residual activity may be reduced on those with more than 5% organic matter

Restrictions
- Maximum total dose equivalent to three full dose treatments
- The maximum total dose must not exceed 1.0 kg/ha ethofumesate in any three year period

Crop-specific information
- Latest use: before crop leaves meet between rows
- Crop tolerance may be reduced by stress caused by growing conditions, effects of pests, disease or other pesticides, nutrient deficiency etc

Following crops guidance
- Beet crops may be sown at any time after treatment. Any other crop may be sown after mouldboard ploughing to 15 cm and a minimum interval of 3 mth after treatment

Environmental safety
- Dangerous for the environment
- Very toxic to aquatic organisms
- Do not empty into drains

Hazard classification and safety precautions

> **Hazard** Harmful [1]; Dangerous for the environment [1-3]; Harmful if swallowed [1, 3]
> **Transport code** 9
> **Packaging group** III
> **UN Number** 3082
> **Operator protection** A, H [1, 3]; U05a, U19a, U20a [1-3]; U08, U14, U15 [2, 3]
> **Environmental protection** E13c, E34 [2, 3]; E15a [1]; E19b, E38 [1-3]; H410 [1, 3]; H411 [2]
> **Storage and disposal** D01, D02, D09a, D10b, D12a [1-3]; D05 [2, 3]
> **Medical advice** M05a [1]

180 ethofumesate + phenmedipham

A contact and residual herbicide for use in beet crops
HRAC mode of action code: N + C1

See also phenmedipham

Products

Betanal Tandem	Bayer CropScience	190:200 g/l	SL	19257

Uses

- Annual dicotyledons in *fodder beet*, *mangels*, *sugar beet*
- Annual meadow grass in *fodder beet*, *mangels*, *sugar beet*
- Blackgrass in *fodder beet*, *mangels*, *sugar beet*

Approval information

- Ethofumesate and phenmedipham included in Annex I under EC Regulation 1107/2009

Efficacy guidance

- Best results achieved by repeat applications to cotyledon stage weeds. Larger susceptible weeds not killed by first treatment usually checked and controlled by second application
- Apply on all soil types at 5-10 d intervals
- On soils with more than 5-10% organic matter residual activity may be reduced

Restrictions

- Maximum number of treatments normally 3 per crop or maximum total dose equivalent to three full dose treatments, or less - see labels for details
- Do not spray wet foliage or if rain imminent
- Spray in evening if daytime temperatures above 21°C expected
- Avoid or delay treatment if frost expected within 7 days
- Avoid or delay treating crops under stress from wind damage, manganese or lime deficiency, pest or disease attack etc.
- The maximum total dose must not exceed 1.0 kg/ha ethofumesate in any three year period

Crop-specific information

- Latest use: before crop foliage meets in the rows
- Check from which recovery may not be complete may occur if treatment made during conditions of sharp diurnal temperature fluctuation

Following crops guidance

- Beet crops may be sown at any time after treatment. Any other crop may be sown after mouldboard ploughing to 15 cm and a minimum interval of 3 mth after treatment

Environmental safety

- Dangerous for the environment
- Toxic to aquatic organisms
- Do not empty into drains
- Extra care necessary to avoid drift because product is recommended for use as a fine spray

Hazard classification and safety precautions

> **Hazard** Irritant, Dangerous for the environment
> **Transport code** 9

SEE SECTION 3 FOR PRODUCTS ALSO REGISTERED

Packaging group III
UN Number 3082
Risk phrases H319
Operator protection A, H; U05a, U08, U14, U19a, U20a
Environmental protection E15a, E19b, E34, H410
Storage and disposal D01, D02, D05, D09a, D10b, D12a
Medical advice M05a

181   etoxazole

A mite growth inhibitor
IRAC mode of action code: 10B

Products

1 Borneo	Sumitomo	110 g/l	SC	13919
2 Clayton Java	Clayton	110 g/l	SC	15155

Uses

- Mites in **protected aubergines**, **protected tomatoes**
- Spider mites in **protected ornamentals** *(off-label)*, **protected strawberries** *(off-label)*
- Two-spotted spider mite in **protected ornamentals** *(off-label)*, **protected strawberries** *(off-label)*

Extension of Authorisation for Minor Use (EAMUs)

- **protected ornamentals** *20081216* [1], *20111544* [2]
- **protected strawberries** *20092400* [1], *20111545* [2]

Approval information

- Etoxazole included in Annex I under EC Regulation 1107/2009

Restrictions

- A maximum individual dose of 500 ml/ha must not be exceeded

Hazard classification and safety precautions

Hazard Dangerous for the environment
Transport code 9
Packaging group III
UN Number 3082
Operator protection A; U05a, U20b
Environmental protection E15a, E34, H410
Storage and disposal D01, D02, D09b, D11a, D12b

182   fenazaquin

A mitochondrial electron transport inhibitor
IRAC mode of action code: 21

Products

Matador 200 SC	Gowan	200 g/l	SC	17870

Uses

- Red spider mites in **protected ornamentals**
- Two-spotted spider mite in **protected ornamentals**

Approval information

- Fenazaquin included in Annex I under EC Regulation 1107/2009

Efficacy guidance

- Acts by contact to give rapid knockdown
- Product should be used as part of a pest control programme. A further acaricide treatment may be necessary after application
- Control may be reduced where water volumes are reduced

FOR FULL CONDITIONS OF USE ALWAYS READ THE PRODUCT LABEL

Restrictions
- Maximum number of treatments 1 per yr
- Other mitochondrial electron transport inhibitor (METI) acaricides should not be applied to the same crop in the same calendar yr either separately or in mixture
- Do not apply to roses. Do not treat new ornamental species or varieties without first testing a few plants on a small scale
- Do not treat ornamentals when in blossom or under stress

Environmental safety
- Extremely dangerous to fish or other aquatic life. Do not contaminate surface waters or ditches with chemical or used container
- Risk to certain non-target insects or other arthropods. For advice on risk management and use in Integrated Pest Management (IPM) see directions for use. Limited evidence shows some adverse effects on predators and parasites such as lacewings and ladybirds
- Broadcast air-assisted LERAP (15 m); LERAP Category B

Hazard classification and safety precautions
> **Hazard** Harmful, Harmful if swallowed, Harmful if inhaled, Very toxic to aquatic organisms
> **Transport code** 6.1
> **Packaging group** III
> **UN Number** 2902
> **Operator protection** A; U05a, U19a
> **Environmental protection** E13a, E16a, E20, E22b, E34, H410; E17b (15 m)
> **Consumer protection** C02a (30 d)
> **Storage and disposal** D01, D02, D09a, D10b
> **Medical advice** M03

183   fenhexamid

A protectant hydroxyanilide fungicide for soft fruit and a range of horticultural crops
FRAC mode of action code: 17

Products

Teldor	Bayer CropScience	50% w/w	WG	11229

Uses
- Botrytis in *bilberries* (off-label), *blackberries*, *blackcurrants*, *blueberries* (off-label), *cherries* (off-label), *frise* (off-label), *gooseberries*, *herbs (see appendix 6)* (off-label), *hops* (off-label), *lamb's lettuce* (off-label), *leaf brassicas* (off-label - baby leaf production), *loganberries*, *plums* (off-label), *protected aubergines* (off-label), *protected blueberry* (off-label), *protected chilli peppers* (off-label), *protected courgettes* (off-label), *protected cucumbers* (off-label), *protected forest nurseries* (off-label), *protected gherkins* (off-label), *protected peppers* (off-label), *protected squashes* (off-label), *protected tomatoes* (off-label), *raspberries*, *redcurrants*, *rocket* (off-label), *rubus hybrids*, *scarole* (off-label), *soft fruit* (off-label), *strawberries*, *tomatoes (outdoor)* (off-label), *whitecurrants*, *wine grapes*
- Grey mould in *protected aubergines* (off-label), *protected chilli peppers* (off-label), *protected courgettes* (off-label), *protected cucumbers* (off-label), *protected gherkins* (off-label), *protected peppers* (off-label), *protected squashes* (off-label), *protected tomatoes* (off-label)

Extension of Authorisation for Minor Use (EAMUs)
- *bilberries* 20161214
- *blueberries* 20161214
- *cherries* 20031866
- *frise* 20082062
- *herbs (see appendix 6)* 20082062
- *hops* 20082926
- *lamb's lettuce* 20082062
- *leaf brassicas* (baby leaf production) 20082062
- *plums* 20031866
- *protected aubergines* 20042087

SEE SECTION 3 FOR PRODUCTS ALSO REGISTERED

- **protected blueberry** *20161214*
- **protected chilli peppers** *20042086*
- **protected courgettes** *20042085*
- **protected cucumbers** *20042085*
- **protected forest nurseries** *20082926*
- **protected gherkins** *20042085*
- **protected peppers** *20042086*
- **protected squashes** *20042085*
- **protected tomatoes** *20042087*
- **rocket** *20082062*
- **scarole** *20082062*
- **soft fruit** *20082926*
- **tomatoes (outdoor)** *20042399*

Approval information
- Fenhexamid included in Annex I under EC Regulation 1107/2009

Efficacy guidance
- Use as part of a programme of sprays throughout the flowering period to achieve effective control of Botrytis
- To minimise possibility of development of resistance, no more than two sprays of the product may be applied consecutively. Other fungicides from a different chemical group should then be used for at least two consecutive sprays. If only two applications are made on grapevines, only one may include fenhexamid
- Complete spray cover of all flowers and fruitlets throughout the blossom period is essential for successful control of Botrytis
- Spray programmes should normally start at the start of flowering

Restrictions
- Maximum number of treatments 2 per yr on grapevines; 4 per yr on other listed crops but no more than 2 sprays may be applied consecutively

Crop-specific information
- HI 1 d for strawberries, raspberries, loganberries, blackberries, Rubus hybrids; 3 d for cherries; 7 d for blackcurrants, redcurrants, whitecurrants, gooseberries; 21d for outdoor grapes

Environmental safety
- Dangerous for the environment
- Harmful to fish or other aquatic life. Do not contaminate surface waters or ditches with chemical or used container

Hazard classification and safety precautions
Hazard Dangerous for the environment
Transport code 9
Packaging group III
UN Number 3077
Operator protection U08, U19a, U20b
Environmental protection E13c, E38, H411
Storage and disposal D05, D09a, D11a, D12a

184   fenoxaprop-P-ethyl

An aryloxyphenoxypropionate herbicide for use in wheat
HRAC mode of action code: A

See also diclofop-methyl + fenoxaprop-P-ethyl

Products

1	Foxtrot	FMC Agro	69 g/l	EW	18808
2	Oskar	FMC Agro	69 g/l	EW	18696
3	Polecat	Sumitomo	69 g/l	EW	16112

FOR FULL CONDITIONS OF USE ALWAYS READ THE PRODUCT LABEL

Uses

- Awned canary grass in *spring wheat*, *winter wheat* [3]
- Blackgrass in *grass seed crops* (off-label), *spring barley*, *winter barley* [1, 2]; *spring wheat*, *winter wheat* [1-3]
- Canary grass in *spring barley*, *spring wheat*, *winter barley*, *winter wheat* [1, 2]
- Rough-stalked meadow grass in *grass seed crops* (off-label), *spring barley*, *winter barley* [1, 2]; *spring wheat*, *winter wheat* [1-3]
- Wild oats in *spring barley*, *winter barley* [1, 2]; *spring wheat*, *winter wheat* [1-3]

Extension of Authorisation for Minor Use (EAMUs)

- *grass seed crops* 20190835 [1], 20190880 [2]

Approval information

- Fenoxaprop-P-ethyl included in Annex I under EC Regulation 1107/2009

Efficacy guidance

- Treat weeds from 2 fully expanded leaves up to flag leaf ligule just visible; for awned canary-grass and rough meadow-grass from 2 leaves to the end of tillering
- A second application may be made in spring where susceptible weeds emerge after an autumn application
- Spray is rainfast 1 h after application
- Dry conditions resulting in moisture stress may reduce effectiveness
- Fenoxaprop-P-ethyl is an ACCase inhibitor herbicide. To avoid the build up of resistance do not apply products containing an ACCase inhibitor herbicide more than twice to any crop. In addition do not use any product containing fenoxaprop-P-ethyl in mixture or sequence with any other product containing the same ingredient
- Use these products as part of a resistance management strategy that includes cultural methods of control and does not use ACCase inhibitors as the sole chemical method of grass weed control
- Applying a second product containing an ACCase inhibitor to a crop will increase the risk of resistance development; only use a second ACCase inhibitor to control different weeds at a different timing
- Always follow WRAG guidelines for preventing and managing herbicide resistant weeds. See Section 5 for more information

Restrictions

- Maximum total dose equivalent to one or two full dose treatments depending on product used
- Do not apply to barley, durum wheat, undersown crops or crops to be undersown
- Do not roll or harrow within 1 wk of spraying
- Do not spray crops under stress, suffering from drought, waterlogging or nutrient deficiency or those grazed or if soil compacted
- Avoid spraying immediately before or after a sudden drop in temperature or a period of warm days/cold nights
- Do not mix with hormone weedkillers

Crop-specific information

- Latest use: before flag leaf sheath extending (GS 41)
- Treat from crop emergence to flag leaf fully emerged (GS 41).
- Product may be sprayed in frosty weather provided crop hardened off but do not spray wet foliage or leaves covered with ice
- Broadcast crops should be sprayed post-emergence after plants have developed well-established root system

Environmental safety

- Dangerous for the environment
- Very toxic to aquatic organisms

Hazard classification and safety precautions

Hazard Irritant, Dangerous for the environment
Transport code 9
Packaging group III
UN Number 3082

SEE SECTION 3 FOR PRODUCTS ALSO REGISTERED

Risk phrases H315 [1, 2]; H317 [1, 3]
Operator protection A, C, H; U05a, U08, U14, U20b [1-3]; U11 [3]; U19a [1, 2]
Environmental protection E15a, E38, H411 [1-3]; E34 [1, 2]
Storage and disposal D01, D02, D05, D09a, D10b [1-3]; D12a [3]

185   fenpyrazamine

A botriticide for protected crops
FRAC mode of action code: 17

Products

1	Empire	Pan Agriculture	50% w/w	WG	17357
2	Prolectus	Sumitomo	50% w/w	WG	16607

Uses

- Botrytis in **protected aubergines**, **protected courgettes**, **protected cucumbers**, **protected peppers**, **protected strawberries**, **protected tomatoes**, **strawberries**, **wine grapes**

Approval information

- Fenpyrazamine is included in Annex I under EC Regulation 1107/2009

Restrictions

- Consult processors before use on crops grown for processing
- No more than a third of the intended botryticide applications made per crop, per year, should contain 3-keto reductase (FRAC code 17) fungicides

Environmental safety

- Buffer zone requirement for ornamental plant production 5 m

Hazard classification and safety precautions

Hazard Dangerous for the environment
Transport code 9
Packaging group III
UN Number 3077
Operator protection U05a, U11, U13, U14, U15, U20b
Environmental protection E15b, H410
Storage and disposal D01, D02, D09a, D11a, D12a, D12b
Medical advice M03

186   ferric phosphate

A molluscicide bait for controlling slugs and snails

Products

1	Fe-est	De Sangosse	3% w/w	RB	18517
2	Fe-Lyn	De Sangosse	3% w/w	RB	18518
3	Ferrex	Cropco Ltd	3.1% w/w	GB	19138
4	Ferrimax Pro	De Sangosse	3% w/w	RB	18519
5	Ironclad	Gemini	3.7% w/w	RB	18970
6	Ironmax Pro	De Sangosse	2.42% w/w	RB	17122
7	Iroxx	Certis	2.97% w/w	GB	16640
8	Menorexx	Certis	2.97% w/w	GB	18842
9	Sigon	De Sangosse	2.4% w/w	RB	19014
10	Sluxx HP	Certis	2.97% w/w	GB	16571
11	X-Ecute	De Sangosse	2.42% w/w	RB	18536

Uses

- Slugs and snails in **all edible crops (outdoor and protected)**, **all non-edible crops (outdoor)** [1, 2, 4, 6-11]; **all protected non-edible crops** [6, 7, 10]; **amenity vegetation** [7, 8, 10]; **asparagus**, **aubergines**, **basil**, **beans without pods (dry)**, **cardoons**, **chickpeas**, **chives**, **choi** *sum, collards, coriander, courgettes, edible flowers, edible podded peas, florence fennel, gherkins, kohlrabi, lupins, melons, mint, newly sown grass leys, oriental cabbage, ornamental plant production, parsley, pumpkins, rhubarb, rosemary, seakale, shallots, spinach, strawberries, summer squash, thyme, winter squash* [3]; *baby leaf crops, borage, chicory, endives, fennel leaves, land cress, mustard, potatoes, purslane, witloof* [5]; *beans without pods (fresh), broad beans, broccoli, brussels sprouts, bulb onions, cabbages, calabrese, cauliflowers, celery (outdoor), combining peas, durum wheat, dwarf beans, forage maize, french beans, garlic, globe artichoke, grain maize, kale, lamb's lettuce, leeks, lettuce, linseed, poppies, red mustard, rocket, runner beans, rye, salad onions, soya beans, spinach beet, spring barley, spring field beans, spring oats, spring oilseed rape, spring wheat, sunflowers, sweetcorn, triticale, vining peas, winter barley, winter field beans, winter oats, winter oilseed rape, winter wheat* [3, 5]; *carrots, celeriac, radishes, red beet, turnips* [1, 2, 4-6, 9, 11]; *fodder beet, sugar beet* [1-6, 9, 11]; *grassland, managed amenity turf* [1, 2, 4, 6, 9, 11]; *swedes* [5, 9]

Approval information
- Ferric phosphate included in Annex 1 under EC Regulation 1107/2009
- Accepted by BBPA for use on malting barley

Efficacy guidance
- Treat as soon as damage first seen preferably in early evening. Repeat as necessary to maintain control
- Best results obtained from moist soaked granules. This will occur naturally on moist soils or in humid conditions
- Ferric phosphate does not cause excessive slime secretion and has no requirement to collect moribund slugs from the soil surface
- Active ingredient is degraded by micro-organisms to beneficial plant nutrients

Restrictions
- Maximum total dose equivalent to four full dose treatments
- The product must not be broadcast near water [5]

Crop-specific information
- Latest use not specified for any crop

Hazard classification and safety precautions
UN Number N/C
Risk phrases H319 [3]
Operator protection A, H [1, 2, 4-6, 9, 11]; U05a, U20a
Environmental protection E15a [3, 7, 8]; E34 [1-11]
Storage and disposal D01, D09a

187   ferrous sulphate

A herbicide/fertilizer combination for moss control in turf

See also dicamba + dichlorprop-P + ferrous sulphate + MCPA
    dichlorprop-P + ferrous sulphate + MCPA

Products

1	Ferromex Mosskiller Concentrate	Omex	381.15 g/l	SL	16896
2	Greenmaster Autumn	ICL (Everris) Ltd	16.3% w/w	GR	16762
3	Greentec Mosskiller Pro	Headland Amenity	16.78% w/w	GR	16728

Uses
- Moss in *amenity grassland* [1, 3]; *managed amenity turf* [1-3]

Approval information
- Ferrous sulphate included in Annex 1 under EC Regulation 1107/2009

Efficacy guidance
- For best results apply when turf is actively growing and the soil is moist
- Fertilizer component of most products encourages strong root growth and tillering
- Mow 3 d before treatment and do not mow for 3-4 d afterwards
- Water after 2 d if no rain
- Rake out dead moss thoroughly 7-14 d after treatment. Re-treatment may be necessary for heavy infestations

Restrictions
- Maximum number of treatments - see labels
- Do not apply during drought or when heavy rain expected
- Do not apply in frosty weather or when the ground is frozen
- Do not walk on treated areas until well watered
- Use only between September and mid-November

Crop-specific information
- If spilt on paving, concrete, clothes etc brush off immediately to avoid discolouration
- Observe label restrictions for interval before cutting after treatment

Environmental safety
- Harmful to fish or other aquatic life. Do not contaminate surface waters or ditches with chemical or used container

Hazard classification and safety precautions
Hazard Harmful [3]
UN Number N/C
Risk phrases H315, H318 [3]; H319 [2]
Operator protection A [1]; U20b [1, 3]; U20c [2]
Environmental protection E13c [2]; E15a [3]; E15b [1]
Storage and disposal D01 [1, 3]; D09a [1-3]; D11a [2, 3]; D12a [1]

188   ferrous sulphate + MCPA + mecoprop-P

A translocated herbicide and moss killer mixture
HRAC mode of action code: O + O

See also MCPA
        mecoprop-P

Products

1	Landscaper Pro Feed Weed Mosskiller	ICL (Everris) Ltd	16.3:0.49:0.29% w/ w	GR	17549
2	Renovator Pro	ICL (Everris) Ltd	16.3:0.49:0.29 % w/ w	GR	16803

Uses
- Annual dicotyledons in *managed amenity turf*
- Moss in *managed amenity turf*
- Perennial dicotyledons in *managed amenity turf*

Approval information
- Ferrous sulphate, MCPA and mecoprop-P included in Annex I under EC Regulation 1107/2009

Efficacy guidance
- Apply from Apr to Sep when weeds are growing
- For best results apply when light showers or heavy dews are expected
- Apply with a suitable calibrated fertilizer distributor
- Retreatment may be necessary after 6 wk if weeds or moss persist
- For best control of moss scarify vigorously after 2 wk to remove dead moss
- Where regrowth of moss or weeds occurs a repeat treatment may be made after 6 wk
- Avoid treatment of wet grass or during drought. If no rain falls within 48 h water in thoroughly

Restrictions

- Maximum number of treatments 3 per yr
- Do not treat new turf until established for 6 mth
- The first 4 mowings after treatment should not be used to mulch cultivated plants unless composted at least 6 mth
- Avoid walking on treated areas until it has rained or they have been watered
- Do not re-seed or turf within 8 wk of last treatment
- Do not cut grass for at least 3 d before and at least 4 d after treatment
- Do not apply during freezing conditions or when rain imminent

Environmental safety

- Keep livestock out of treated areas for up to two weeks following treatment and until poisonous weeds, such as ragwort, have died down and become unpalatable
- Harmful to fish or other aquatic life. Do not contaminate surface waters or ditches with chemical or used container
- Do not empty into drains
- Some pesticides pose a greater threat of contamination of water than others and mecoprop-P is one of these pesticides. Take special care when applying mecoprop-P near water and do not apply if heavy rain is forecast

Hazard classification and safety precautions

UN Number N/C
Risk phrases H319
Operator protection A, H [1]; U20b
Environmental protection E07a, E13c, E19b
Storage and disposal D01, D09a, D12a
Medical advice M05a

189   flazasulfuron

A sulfonylurea herbicide for non-crop use
HRAC mode of action code: B

Products

1 Chikara Weed Control	Belchim	25% w/w	WG	14189
2 Clayton Apt	Clayton	25% w/w	WG	15157
3 Katana	Belchim	25% w/w	WG	16162
4 Pacaya	ProKlass	25% w/w	WG	16797
5 Paradise	Pan Agriculture	25% w/w	WG	18851

Uses

- Annual and perennial weeds in *amenity vegetation*, *hard surfaces* [1, 3, 5]; *natural surfaces not intended to bear vegetation*, *permeable surfaces overlying soil* [1-5]; *railway tracks* [2, 4]

Approval information

- Flazasulfuron included in Annex 1 under EC Regulation 1107/2009

Restrictions

- Extreme care must be taken to avoid spray drift onto non-crop plants outside of the target area
- Direct spray from the train sprayer must not be allowed to fall within 5m of the top of the bank of a static or flowing water body. Do not allow direct overspray of static or flowing surface waters
- Must not be applied to any non-porous man made surfaces

Environmental safety

- Buffer zone requirement 6 m
- LERAP Category B

Hazard classification and safety precautions

Hazard Dangerous for the environment, Very toxic to aquatic organisms
Transport code 9

SEE SECTION 3 FOR PRODUCTS ALSO REGISTERED

Packaging group III
UN Number 3077
Operator protection A, H, M; U02a, U05a, U20b [1-5]; U08 [1, 3-5]; U09a [2]
Environmental protection E15a, E16a, E16b, E38, H410
Storage and disposal D01, D02, D09a, D11a, D12a

190   flocoumafen

A second generation anti-coagulant rodenticide

Products

1	Storm Secure	BASF	0.005% w/w	RB	UK15-0850
2	Storm Ultra	BASF	0.0025% w/w	RB	UK18-1164
3	Storm Ultra Bait Blocks	BASF	0.0025% w/w	RB	UK18-1164
4	Storm Ultra Secure	BASF	0.0025% w/w	RB	UK18-1164

Uses
- Mice in *farm buildings*, *sewers* [1, 2]
- Rodents in *farm buildings*, *sewers* [1, 2]

Approval information
- Flocoumafen is not included in Annex 1 under EC Regulation 1107/2009

Restrictions
- Not approved for outdoor use except in sewers and covered drains

Hazard classification and safety precautions
Hazard Harmful [1, 2]
UN Number N/C [1, 2]
Operator protection A [1, 2]; U05a, U20a [1, 2]
Environmental protection E38 [1, 2]
Storage and disposal D01, D02, D09b, D12b [1, 2]
Medical advice M03 [1, 2]

191   flonicamid

A selective feeding blocker aphicide
IRAC mode of action code: 9C

Products

1	Mainman	Belchim	50% w/w	WG	13123
2	Teppeki	Belchim	50% w/w	WG	12402

Uses
- Aphids in *apples*, *herbs (see appendix 6)* (off-label), *pears*, *protected chilli peppers* (off-label), *protected peppers* (off-label) [1]; *cabbages* (off-label), *canary seed* (off-label), *dwarf beans* (off-label), *edible podded peas* (off-label), *fodder beet*, *french beans* (off-label), *potatoes*, *runner beans* (off-label), *sugar beet*, *vining peas* (off-label), *winter oilseed rape*, *winter wheat* [2]
- Black bean aphid in *dwarf beans* (off-label), *edible podded peas* (off-label), *french beans* (off-label), *runner beans* (off-label), *vining peas* (off-label) [2]
- Damson-hop aphid in *hops* (off-label) [1]
- Glasshouse whitefly in *courgettes* (off-label), *cucumbers* (off-label), *protected courgettes* (off-label), *protected cucumbers* (off-label), *protected summer squash* (off-label), *summer squash* (off-label) [1]
- Mealybugs in *protected aubergines* (off-label), *protected tomatoes* (off-label) [1]
- Pea aphid in *dwarf beans* (off-label), *edible podded peas* (off-label), *french beans* (off-label), *runner beans* (off-label), *vining peas* (off-label) [2]
- Peach-potato aphid in *cabbages* (off-label), *dwarf beans* (off-label), *edible podded peas* (off-label), *french beans* (off-label), *runner beans* (off-label), *vining peas* [2]; *courgettes*

(off-label), **cucumbers** *(off-label)*, **protected courgettes** *(off-label)*, **protected cucumbers** *(off-label)*, **protected summer squash** *(off-label)*, **summer squash** *(off-label)* [1]
- Tobacco whitefly in **ornamental plant production** *(off-label)* [1]
- Whitefly in **ornamental plant production** *(off-label)*, **protected aubergines** *(off-label)*, **protected tomatoes** *(off-label)* [1]

Extension of Authorisation for Minor Use (EAMUs)
- *cabbages 20183740* [2]
- *canary seed 20170673* [2]
- *courgettes 20142923* [1]
- *cucumbers 20142923* [1]
- *dwarf beans 20183739* [2]
- *edible podded peas 20183739* [2]
- *french beans 20183739* [2]
- *herbs (see appendix 6) 20172096* [1]
- *hops 20142225* [1]
- *ornamental plant production 20130045* [1]
- *protected aubergines 20160191* [1]
- *protected chilli peppers 20191139* [1]
- *protected courgettes 20142923* [1]
- *protected cucumbers 20142923* [1]
- *protected peppers 20191139* [1]
- *protected summer squash 20142923* [1]
- *protected tomatoes 20160191* [1]
- *runner beans 20183739* [2]
- *summer squash 20142923* [1]
- *vining peas 20183739* [2]

Approval information
- Flonicamid included in Annex 1 under EC Regulation 1107/2009
- Accepted by BBPA for use on hops

Efficacy guidance
- Apply when warning systems forecast significant aphid infestations
- Persistence of action is 21 d
- Currently has activity against pyrethroid-resistant aphids with no cross-resistance to carbamates or neo-nics.

Restrictions
- Maximum number of treatments 2 per crop for potatoes, winter wheat [2], 3 per year on apples and pears [1]
- Must not be applied to winter wheat before 50% ear emerged stage (GS 53)
- Use a maximum of two consecutive applications of [1] in apples and pears. If further treatments are required, use an insecticide with a different mode of action before applying the final application of [1]
- Fodder beet must not be grazed by livestock or harvested for animal consumption until at least 60 days following the last application

Crop-specific information
- HI: potatoes 14 d; winter wheat 28 d;
- HI: apples and pears 21 days [1]

Environmental safety
- Dangerous for the environment
- Harmful to aquatic organisms
- Dangerous to bees. To protect bees and pollinating insects do not apply to crop plants when in flower. Do not use where bees are actively foraging. Do not apply when flowering weeds are present

Hazard classification and safety precautions
Hazard Dangerous for the environment [2]
UN Number N/C

SEE SECTION 3 FOR PRODUCTS ALSO REGISTERED

Operator protection A, H, M [1, 2]; C [2]; E [1]; U05a [1, 2]; U13, U14, U15 [1]
Environmental protection E12c [2]; E15a, H412 [1, 2]
Storage and disposal D01, D02, D12b [1, 2]; D03, D08 [1]; D07 [2]

192   florasulam

A triazolopyrimidine herbicide for cereals
HRAC mode of action code: B

See also 2,4-D + florasulam
clopyralid + florasulam
clopyralid + florasulam + fluroxypyr
diflufenican + florasulam

Products

1	Boxer	Dow (Corteva)	50 g/l	SC	09819
2	Lector	Nufarm UK	50 g/l	SC	18728
3	Omasa	Nurture	50 g/l	SC	19043
4	Paramount	Nufarm UK	50 g/l	SC	18727
5	Solstice	Nufarm UK	50 g/l	SC	18752
6	Sumir	Life Scientific	50 g/l	SC	18473

Uses

* Annual dicotyledons in **farm forestry** *(off-label)*, **forest nurseries** *(off-label)*, **game cover** *(off-label)*, **ornamental plant production** *(off-label)* [1]; **grass seed crops** *(off-label)* [1, 2, 4]; **spring barley, spring oats, spring wheat, winter barley, winter oats, winter wheat** [1, 4-6]; **triticale, winter rye** [4, 5]
* Chickweed in **grass seed crops** *(off-label)* [1]; **spring barley, spring oats, spring wheat, winter barley, winter oats, winter wheat** [1-6]; **triticale, winter rye** [2-5]
* Cleavers in **grass seed crops** *(off-label)* [1]; **spring barley, spring oats, spring wheat, winter barley, winter oats, winter wheat** [1-6]; **triticale, winter rye** [2-5]
* Hedge mustard in **spring barley, spring oats, spring wheat, triticale, winter barley, winter oats, winter rye, winter wheat** [2, 3]
* Mayweeds in **spring barley, spring oats, spring wheat, winter barley, winter oats, winter wheat** [1, 4-6]; **triticale, winter rye** [4, 5]
* Runch in **spring barley, spring oats, spring wheat, triticale, winter barley, winter oats, winter rye, winter wheat** [2, 3]
* Scented mayweed in **spring barley, spring oats, spring wheat, triticale, winter barley, winter oats, winter rye, winter wheat** [2, 3]
* Scentless mayweed in **spring barley, spring oats, spring wheat, triticale, winter barley, winter oats, winter rye, winter wheat** [2, 3]
* Shepherd's purse in **spring barley, spring oats, spring wheat, triticale, winter barley, winter oats, winter rye, winter wheat** [2, 3]
* Volunteer oilseed rape in **spring barley, spring oats, spring wheat, winter barley, winter oats, winter wheat** [1-6]; **triticale, winter rye** [2-5]
* Wild radish in **spring barley, spring oats, spring wheat, triticale, winter barley, winter oats, winter rye, winter wheat** [2, 3]

Extension of Authorisation for Minor Use (EAMUs)

* **farm forestry** *20082826* [1]
* **forest nurseries** *20082826* [1]
* **game cover** *20082826* [1]
* **grass seed crops** *20150314* [1], *20183488* [2], *20183493* [4]
* **ornamental plant production** *20082826* [1]

Approval information

* Florasulam included in Annex I under EC Regulation 1107/2009
* Accepted by BBPA for use on malting barley

FOR FULL CONDITIONS OF USE ALWAYS READ THE PRODUCT LABEL

Efficacy guidance
- Best results obtained from treatment of small actively growing weeds in good conditions
- Apply in autumn or spring once crop has 3 leaves
- Product is mainly absorbed by leaves of weeds and is effective on all soil types
- Florasulam is a member of the ALS-inhibitor group of herbicides

Restrictions
- The total amount of florasulam applied to a cereal crop must not exceed 7.5 g.a.i/ha
- For autumn planted crops a maximum total dose of 3.75 g of florasulam must be observed for applications made between crop emergence in the year of planting and 1st January in the year of harvest [6]
- Do not roll or harrow within 7 d before or after application
- Do not spray when crops under stress from cold, drought, pest damage, nutrient deficiency or any other cause
- Specific restrictions apply to use in sequence or tank mixture with other sulfonylurea or ALS-inhibiting herbicides. See label for details

Crop-specific information
- Latest use: up to and including flag leaf just visible stage for all crops

Following crops guidance
- Where the product has been used in mixture with certain named products (see label) only cereals or grass may be sown in the autumn following harvest
- Unless otherwise restricted cereals, oilseed rape, field beans, grass or vegetable brassicas as transplants may be sown as a following crop in the same calendar yr as treatment. Oilseed rape may show some temporary reduction of vigour after a dry summer, but yields are not affected
- In addition to the above, linseed, peas, sugar beet, potatoes, maize, clover (for use in grass/clover mixtures) or carrots may be sown in the calendar yr following treatment
- In the event of failure of a treated crop in spring only spring wheat, spring barley, spring oats, maize or ryegrass may be sown

Environmental safety
- Dangerous for the environment
- Very toxic to aquatic organisms
- See label for detailed instructions on tank cleaning
- LERAP Category B [2-5]

Hazard classification and safety precautions

Hazard Dangerous for the environment [1-6]; Very toxic to aquatic organisms [6]
Transport code 9
Packaging group III
UN Number 3082
Risk phrases R50, R53a, R70 [5]
Operator protection U05a
Environmental protection E15a, E34 [1, 6]; E15b, E16a [2-5]; E38 [1-6]; H410 [1-4, 6]
Storage and disposal D01, D02, D05, D09a [1-6]; D06a, D10c, D12b [2-5]; D10b, D12a [1, 6]

193 florasulam + fluroxypyr

A post-emergence herbicide mixture for cereals or grass
HRAC mode of action code: B + O

See also fluroxypyr

Products

1	Cabadex	Headland Amenity	2.5:100 g/l	SE	13948
2	Cleave	Adama	2.5:100 g/l	SE	16774
3	Envy	Dow (Corteva)	2.5:100 g/l	SE	17901
4	GF 184	Dow (Corteva)	2.5:100 g/l	SE	10878
5	Hunter	Dow (Corteva)	2.5:100 g/l	SE	12836
6	Nevada	Dow (Corteva)	5:100 g/l	SE	17349

SEE SECTION 3 FOR PRODUCTS ALSO REGISTERED

SECTION 2

Products – continued

7	Sickle	Dow (Corteva)	2.5:100 g/l	SE	17923
8	Slalom	Dow (Corteva)	2.5:100 g/l	SE	13772
9	Spitfire	Dow (Corteva)	5:100 g/l	SE	15101
10	Starane XL	Dow (Corteva)	2.5:100 g/l	SE	10921
11	Trafalgar	Pan Amenity	2.5:100 g/l	SE	14888

Uses

- Annual and perennial weeds in *amenity grassland, lawns, managed amenity turf* [1]; *grass seed crops, grassland, newly sown grass leys* [7]
- Annual dicotyledons in *amenity grassland, lawns, managed amenity turf* [11]; *game cover (off-label)* [4, 10]; *grass seed crops (off-label)* [4, 9, 10]; *oats, rye* [6, 9]; *ornamental plant production (off-label), spring rye (off-label), spring triticale (off-label), winter rye (off-label), winter triticale (off-label)* [10]; *spring barley, spring wheat, winter barley, winter wheat* [2, 4-6, 8-10]; *spring oats, winter oats* [2, 4, 5, 8, 10]; *triticale* [2, 6, 9]; *undersown barley, undersown oats, undersown rye, undersown spring cereals, undersown triticale, undersown wheat* [9]; *winter rye* [2]
- Black bindweed in *oats, rye, spring barley, spring wheat, triticale, winter barley, winter wheat* [6, 9]; *undersown barley, undersown oats, undersown rye, undersown spring cereals, undersown triticale, undersown wheat* [9]
- Black nightshade in *oats, rye, spring barley, spring wheat, triticale, winter barley, winter wheat* [6, 9]; *undersown barley, undersown oats, undersown rye, undersown spring cereals, undersown triticale, undersown wheat* [9]
- Buttercups in *amenity grassland, lawns, managed amenity turf* [1]; *grass seed crops, grassland, newly sown grass leys* [7]
- Charlock in *grass seed crops, grassland, newly sown grass leys* [3]; *oats, rye, spring barley, spring wheat, triticale, winter barley, winter wheat* [6, 9]; *undersown barley, undersown oats, undersown rye, undersown spring cereals, undersown triticale, undersown wheat* [9]
- Chickweed in *amenity grassland, lawns, managed amenity turf* [11]; *grass seed crops, grassland, newly sown grass leys* [3]; *oats, rye* [6, 9]; *spring barley, spring wheat, winter barley, winter wheat* [2, 4-6, 8-10]; *spring oats, winter oats* [2, 4, 5, 8, 10]; *triticale* [2, 6, 9]; *undersown barley, undersown oats, undersown rye, undersown spring cereals, undersown triticale, undersown wheat* [9]; *winter rye* [2]
- Cleavers in *grass seed crops, grassland, newly sown grass leys* [3]; *grass seed crops (off-label), undersown barley, undersown oats, undersown rye, undersown spring cereals, undersown triticale, undersown wheat* [9]; *oats, rye* [6, 9]; *spring barley, spring wheat, winter barley, winter wheat* [2, 4-6, 8-10]; *spring oats, winter oats* [2, 4, 5, 8, 10]; *triticale* [2, 6, 9]; *winter rye* [2]
- Clover in *amenity grassland, lawns, managed amenity turf* [1]; *grass seed crops, grassland, newly sown grass leys* [7]
- Common mouse-ear in *grassland* [3]
- Creeping buttercup in *grassland* [3]
- Daisies in *amenity grassland, lawns, managed amenity turf* [1]; *grass seed crops, newly sown grass leys* [7]; *grassland* [3, 7]
- Dandelions in *amenity grassland, lawns, managed amenity turf* [1]; *grass seed crops, newly sown grass leys* [7]; *grassland* [3, 7]
- Forget-me-not in *grass seed crops, grassland, newly sown grass leys* [3]; *oats, rye, spring barley, spring wheat, triticale, winter barley, winter wheat* [6, 9]; *undersown barley, undersown oats, undersown rye, undersown spring cereals, undersown triticale, undersown wheat* [9]
- Groundsel in *grass seed crops (off-label)* [9]
- Knotgrass in *grass seed crops, grassland, newly sown grass leys* [3]; *oats, rye, spring barley, spring wheat, triticale, winter barley, winter wheat* [6, 9]; *undersown barley, undersown oats, undersown rye, undersown spring cereals, undersown triticale, undersown wheat* [9]
- Mayweeds in *grass seed crops, grassland, newly sown grass leys* [3]; *grass seed crops (off-label), undersown barley, undersown oats, undersown rye, undersown spring cereals, undersown triticale, undersown wheat* [9]; *oats, rye* [6, 9]; *spring barley, spring wheat,*

SECTION 2

winter barley, winter wheat [2, 4-6, 8-10]; *spring oats, winter oats* [2, 4, 5, 8, 10]; *triticale* [2, 6, 9]; *winter rye* [2]

- Plantains in *amenity grassland, lawns, managed amenity turf* [1]; *grass seed crops, grassland, newly sown grass leys* [7]
- Poppies in *grass seed crops, grassland, newly sown grass leys* [3]; *oats, rye, spring barley, spring wheat, triticale, winter barley, winter wheat* [6, 9]; *undersown barley, undersown oats, undersown rye, undersown spring cereals, undersown triticale, undersown wheat* [9]
- Redshank in *oats, rye, spring barley, spring wheat, triticale, winter barley, winter wheat* [6, 9]; *undersown barley, undersown oats, undersown rye, undersown spring cereals, undersown triticale, undersown wheat* [9]
- Shepherd's purse in *grass seed crops, grassland, newly sown grass leys* [3]
- Sowthistle in *grass seed crops (off-label)* [9]
- Volunteer beans in *oats, rye, spring barley, spring wheat, triticale, winter barley, winter wheat* [6, 9]; *undersown barley, undersown oats, undersown rye, undersown spring cereals, undersown triticale, undersown wheat* [9]
- Volunteer oilseed rape in *grass seed crops, grassland, newly sown grass leys* [3]; *oats, rye, spring barley, spring wheat, triticale, winter barley, winter wheat* [6, 9]; *undersown barley, undersown oats, undersown rye, undersown spring cereals, undersown triticale, undersown wheat* [9]

Extension of Authorisation for Minor Use (EAMUs)
- *game cover* 20082858 [4], 20082904 [10]
- *grass seed crops* 20072809 [4], 20132321 [9], 20193248 [10]
- *ornamental plant production* 20082904 [10]
- *spring rye* 20193248 [10]
- *spring triticale* 20193248 [10]
- *winter rye* 20193248 [10]
- *winter triticale* 20193248 [10]

Approval information
- Florasulam and fluroxypyr included in Annex I under EC Regulation 1107/2009
- Accepted by BBPA for use on malting barley

Efficacy guidance
- Best results obtained when weeds are small and growing actively
- Products are mainly absorbed through weed foliage. Cleavers emerging after application will not be controlled
- Florasulam is a member of the ALS-inhibitor group of herbicides

Restrictions
- Maximum total dose equivalent to one full dose treatment
- Do not roll or harrow 7 d before or after application
- Do not spray when crops are under stress from cold, drought, pest damage or nutrient deficiency
- Do not apply through CDA applicators
- Specific restrictions apply to use in sequence or tank mixture with other sulfonylurea or ALS-inhibiting herbicides. See label for details
- The total amount of florasulam applied to a cereal crop must not exceed 7.5 g.a.i/ha
- For autumn planted crops a maximum total dose of 3.75 g of florasulam must be observed for applications made between crop emergence in the year of planting and February 1st in the year of harvest [2, 6, 8, 9]

Crop-specific information
- Latest use: before flag leaf sheath extended (before GS 41) for spring barley and spring wheat; before flag leaf sheath opening (before GS 47) for winter barley and winter wheat; before second node detectable (before GS 32) for winter oats

Following crops guidance
- Cereals, oilseed rape, field beans or grass may follow treated crops in the same yr. Oilseed rape may suffer temporary vigour reduction after a dry summer

SEE SECTION 3 FOR PRODUCTS ALSO REGISTERED

- In addition to the above, linseed, peas, sugar beet, potatoes, maize or clover may be sown in the calendar yr following treatment
- In the event of failure of a treated crop in the spring, only spring cereals, maize or ryegrass may be planted
- May also be used on crops undersown with grass [6]

Environmental safety
- Dangerous for the environment
- Toxic to aquatic organisms
- Take extreme care to avoid drift onto non-target crops or plants
- Aquatic buffer zone of 5m [7]
- LERAP Category B [2, 3, 6, 7, 9]

Hazard classification and safety precautions
Hazard Irritant, Dangerous for the environment [1-11]; Flammable, Flammable liquid and vapour [6, 9]; Very toxic to aquatic organisms [1, 4, 5]
Transport code 3 [6, 9]; 9 [1-5, 7, 8, 10, 11]
Packaging group III
UN Number 1993 [6, 9]; 3082 [1-5, 7, 8, 10, 11]
Risk phrases H304 [3, 6, 7, 9]; H315, H336 [1-11]; H317, H319 [1-10]; H320 [11]; H335 [1, 3-10]
Operator protection A [1-11]; C [1-10]; H [3, 6, 9]; U05a, U11, U19a [1-11]; U08 [1, 3]; U09a [6, 9]; U14, U20b [1, 3, 6, 9]
Environmental protection E07c [3] (7 days); E15a [2, 4-11]; E15b [1]; E16a [2, 3, 6, 7, 9]; E34, E38 [1-11]; H410 [1-10]; H411 [11]
Storage and disposal D01, D02, D09a, D10c, D12a [1-11]; D05 [3, 11]
Medical advice M05a [6, 9, 11]

194   florasulam + halauxifen-methyl

A herbicide mixture for control of broad-leaved weeds in cereals
HRAC mode of action code: B + O

See also florasulam
florasulam + fluroxypyr

Products

Zypar	Dow (Corteva)	5:6.25 g/l	OD	17938

Uses
- Annual dicotyledons in *durum wheat*, *rye*, *spelt*, *spring barley*, *spring wheat*, *triticale*, *winter barley*, *winter wheat*
- Black bindweed in *grass seed crops* (off-label)
- Chickweed in *durum wheat*, *grass seed crops* (off-label), *rye*, *spelt*, *spring barley*, *spring wheat*, *triticale*, *winter barley*, *winter wheat*
- Cleavers in *durum wheat*, *grass seed crops* (off-label), *rye*, *spelt*, *spring barley*, *spring wheat*, *triticale*, *winter barley*, *winter wheat*
- Crane's-bill in *durum wheat*, *grass seed crops* (off-label), *rye*, *spelt*, *spring barley*, *spring wheat*, *triticale*, *winter barley*, *winter wheat*
- Fat hen in *durum wheat*, *grass seed crops* (off-label), *rye*, *spelt*, *spring barley*, *spring wheat*, *triticale*, *winter barley*, *winter wheat*
- Fumitory in *durum wheat*, *grass seed crops* (off-label), *rye*, *spelt*, *spring barley*, *spring wheat*, *triticale*, *winter barley*, *winter wheat*
- Groundsel in *durum wheat*, *rye*, *spelt*, *spring barley*, *spring wheat*, *triticale*, *winter barley*, *winter wheat*
- Hemp-nettle in *grass seed crops* (off-label)
- Ivy-leaved speedwell in *grass seed crops* (off-label)
- Mayweeds in *durum wheat*, *grass seed crops* (off-label), *rye*, *spelt*, *spring barley*, *spring wheat*, *triticale*, *winter barley*, *winter wheat*
- Poppies in *durum wheat*, *grass seed crops* (off-label), *rye*, *spelt*, *spring barley*, *spring wheat*, *triticale*, *winter barley*, *winter wheat*

FOR FULL CONDITIONS OF USE ALWAYS READ THE PRODUCT LABEL

- Red dead-nettle in **grass seed crops** *(off-label)*
- Shepherd's purse in **grass seed crops** *(off-label)*
- Volunteer oilseed rape in **durum wheat**, **rye**, **spelt**, **spring barley**, **spring wheat**, **triticale**, **winter barley**, **winter wheat**

Extension of Authorisation for Minor Use (EAMUs)
- **grass seed crops** *20181557*

Approval information
- Florasulam and halauxifen-methyl included in Annex I under EC Regulation 1107/2009

Efficacy guidance
- Rainfast 1 hour after application

Restrictions
- One application per season
- The total amount of halauxifen-methyl applied to a winter cereal must not exceed 13.5 g a.e/ha per season: 7.5 g a.e/ha in the autumn followed by 6 g a.e/ha in the spring, with a minimal interval of 3 months between both applications of products which contain halauxifen-methyl.
- Do not apply more than 0.75 litre [1] per hectare to any crop before the 15th February
- For autumn planted crops a maximum total dose of 3.75 g of florasulam must be observed for applications made between crop emergence in the year of planting and February 15th in the year of harvest. The total amount of florasulam applied to a cereal crop must not exceed 7.5 g
- To avoid subsequent injury to crops other than cereals (wheat, durum wheat, spelt, barley, rye, triticale), all spraying equipment must be thoroughly cleaned both inside and out using All Clear Extra spray cleaner at 0.5 % v/v or bleach containing 5 % hypochlorite.

Following crops guidance
- There are no restrictions for sowing any succeeding crop after the cereal harvest but for sensitive species such as clover, lentils or sunflower ploughing is recommended prior to drilling.
- Where crop failure after an autumn application occurs, sowing the following spring crops is possible: 1 month after application (no cultivation restrictions): spring wheat, spring barley, maize, ryegrass; 3 months after application (after ploughing): spring oilseed rape, field beans, peas, sunflower
- Where crop failure after a spring application occurs, it is possible to sow spring wheat and spring barley 1 month after application with no need to cultivate while maize can be sown 2 months after application after ploughing.

Environmental safety
- LERAP Category B

Hazard classification and safety precautions
Transport code 9
Packaging group III
UN Number 3082
Risk phrases H315, H317, H319
Operator protection A, C, H; U05a, U08, U19a, U20c
Environmental protection E15b, E16a, E34, H410
Storage and disposal D09a, D10c

195 florasulam + pyroxsulam

A mixture of two triazolopirimidine sulfonamides for winter wheat
HRAC mode of action code: B

See also pyroxsulam

Products
1	Broadway Star	Dow (Corteva)	1.4:7.1% w/w	WG	18273
2	Palio	Dow (Corteva)	1.4:7.1% w/w	WG	18349

Uses

- Charlock in *spring wheat, triticale, winter rye, winter wheat*
- Chickweed in *spring wheat, triticale, winter rye, winter wheat*
- Cleavers in *spring wheat, triticale, winter rye, winter wheat*
- Field pansy in *spring wheat, triticale, winter rye, winter wheat*
- Field speedwell in *spring wheat, triticale, winter rye, winter wheat*
- Geranium species in *spring wheat, triticale, winter rye, winter wheat*
- Ivy-leaved speedwell in *spring wheat, triticale, winter rye, winter wheat*
- Mayweeds in *spring wheat, triticale, winter rye, winter wheat*
- Poppies in *spring wheat, triticale, winter rye, winter wheat*
- Ryegrass in *spring wheat, triticale, winter rye, winter wheat*
- Sterile brome in *spring wheat, triticale, winter rye, winter wheat*
- Volunteer oilseed rape in *spring wheat, triticale, winter rye, winter wheat*
- Wild oats in *spring wheat, triticale, winter rye, winter wheat*

Approval information

- Florasulam and pyroxsulam included in Annex I under EC Regulation 1107/2009.

Efficacy guidance

- Rainfast within 1 hour of application
- Requires an authorised adjuvant at application and recommended for use in a programme with herbicides employing a different mode of action.

Restrictions

- For autumn planted crops a maximum total dose of 3.75 g.a.i/ha of florasulam must be observed for applications made between crop emergence in the year of planting and February 1st in the year of harvest.
- The total amount of florasulam applied to a cereal crop must not exceed 7.5 g.a.i/ha
- Do not apply this or any other ALS inhibitor herbicide with claims for grass weed control more than once to any crop.

Crop-specific information

- Crop injury may occur if applied in tank mixture with plant growth regulators - allow a minimum interval of 7 days.
- Crop injury may occur if applied in tank mixture with OP insecticides, MCPB or dicamba - allow a minimum interval of 14 days

Following crops guidance

- Crop failure before 1st Feb - plough and allow 6 weeks to elapse and then drill spring wheat, spring barley, grass or maize.
- Crop failure after 1st Feb - plough and allow 6 weeks to elapse before drilling grass or maize.

Environmental safety

- Take extreme care to avoid drift on to susceptible crops, non-target plants or waterways
- LERAP Category B

Hazard classification and safety precautions

Hazard Dangerous for the environment
Transport code 9
Packaging group III
UN Number 3077
Operator protection U05a, U14, U15, U20a
Environmental protection E15b, E16a, E34, E38, E39, H410
Storage and disposal D01, D02, D09a, D12a

196 florasulam + tribenuron-methyl

A herbicide mixture for weed control in winter & spring cereals
HRAC mode of action code: B + B

See also tribenuron-methyl

Products

1	Bolt	Nufarm UK	20:60% w/w	WG	18753
2	Paramount Max	Nufarm UK	20:60% w/w	WG	18751

Uses

- Annual dicotyledons in **spring barley, spring oats, triticale, winter barley, winter oats, winter rye, winter wheat**

Approval information

- Florasulam and tribenuron-methyl included in Annex I under EC Regulation 1107/2009

Restrictions

- For autumn planted crops a maximum total dose of 3.75 g.a.i/ha of florasulam must be observed for applications made between crop emergence in the year of planting and February 1st in the year of harvest.
- The total amount of florasulam applied to a cereal crop must not exceed 7.5 g.a.i/ha
- Only one other product with an ALS inhibitor mode of action may be applied to a cereal crop treated with [1] or [2]. However, a further application of [1], [2] or another product containing florasulam may also be made providing the maximum total dose of florasulam is not exceeded. [1] or [2] may be applied in joint application to the same cereal crop with one of the listed ALS products (see label).
- To avoid subsequent injury to crops other than cereals all spraying equipment must be thoroughly cleaned both inside and outside using a cleaner such as All Clear Extra.
- Applications must not be made to winter cereals before 1 March in the year of harvest, and before 5 tillers (GS25); Applications must not be made to spring barley before 15 March, and before 3 tillers (GS23).

Following crops guidance

- In the event of crop failure in the spring after application only cereals may be planted.
- Only cereals, oilseed rape, field beans and grass can be sown in the same year as a treated crop is harvested. Vigour reductions may be seen in the following crops of oilseed rape after a dry summer. This will be outgrown and will not result in yield loss.
- Only cereals, oilseed rape, field beans, grass, linseed, peas, sugar beet, potatoes, maize, clover (for use in grass/clover mixtures), carrots and vegetable brassicas as transplants can be sown in the calendar year after treatment.

Environmental safety

- LERAP Category B

Hazard classification and safety precautions

Transport code 9
Packaging group III
UN Number 3077
Operator protection A; U05a
Environmental protection E15b [1]; E16a, E34, H410 [1, 2]
Storage and disposal D01, D02, D05, D07, D10c [1, 2]; D06e [1]

197 fluazifop-P-butyl

A phenoxypropionic acid grass herbicide for broadleaved crops
HRAC mode of action code: A

Products

1	Clayton Maximus	Clayton	125 g/l	EC	12543

SEE SECTION 3 FOR PRODUCTS ALSO REGISTERED

Products – continued

2 Fusilade Forte	Nufarm UK	150 g/l	EC	19019
3 Fusilade Max	Nufarm UK	125 g/l	EC	19013

Uses

*   Annual grasses in *almonds, apples, apricots, asparagus, baby leaf crops, beans without pods (dry), bilberries, blackberries, blueberries, broad beans, cardoons, celeriac, celery (outdoor), chestnuts, chicory root, cranberries, edible podded peas, elderberries, endives, florence fennel, garlic, ginger, ginseng root, globe artichoke, hazel nuts, herbs (see appendix 6), horseradish, lamb's lettuce, land cress, lettuce, liquorice, loganberries, mallow (althaea spp.), mangels, mulberries, parsley root, parsnips, pears, plums, potatoes, purslane, quinces, radishes, red beet, red mustard, redcurrants, rhubarb, rocket, rose hips, rubus hybrids, safflower, salsify, shallots, spinach, spinach beet, sunflowers, table grapes, turmeric, valerian root, walnuts, wine grapes* [3]; *blackcurrants, bulb onions, carrots, farm forestry, gooseberries, hops, raspberries, swedes* (stockfeed only), *turnips* (stockfeed only), *vining peas* [1, 3]; *combining peas, fodder beet, linseed, spring field beans, spring oilseed rape, sugar beet, winter field beans, winter oilseed rape* [1-3]; *fodder rape, lupins, poppies* [2, 3]; *kale* (stockfeed only), *spring oilseed rape for industrial use, strawberries, winter oilseed rape for industrial use* [1]; *mustard* [2]
*   Blackgrass in *hops* [1]; *spring field beans, winter field beans* [1-3]; *sunflowers* [3]
*   Green cover in *land temporarily removed from production* [1]
*   Perennial grasses in *almonds, apples, apricots, asparagus, baby leaf crops, beans without pods (dry), bilberries, blackberries, blueberries, broad beans, cardoons, celeriac, celery (outdoor), chestnuts, chicory root, cranberries, edible podded peas, elderberries, endives, florence fennel, garlic, ginger, ginseng root, globe artichoke, hazel nuts, herbs (see appendix 6), horseradish, lamb's lettuce, land cress, lettuce, liquorice, loganberries, mallow (althaea spp.), mangels, mulberries, parsley root, parsnips, pears, plums, potatoes, purslane, quinces, radishes, red beet, red mustard, redcurrants, rhubarb, rocket, rose hips, rubus hybrids, safflower, salsify, shallots, spinach, spinach beet, sunflowers, table grapes, turmeric, valerian root, walnuts, wine grapes* [3]; *blackcurrants, bulb onions, carrots, farm forestry, gooseberries, hops, raspberries, swedes* (stockfeed only), *turnips* (stockfeed only), *vining peas* [1, 3]; *combining peas, fodder beet, linseed, spring field beans, spring oilseed rape, sugar beet, winter field beans, winter oilseed rape* [1-3]; *fodder rape, lupins, poppies* [2, 3]; *kale* (stockfeed only), *spring oilseed rape for industrial use, strawberries, winter oilseed rape for industrial use* [1]; *mustard* [2]
*   Volunteer cereals in *almonds, apples, apricots, asparagus, baby leaf crops, beans without pods (dry), bilberries, blackberries, blueberries, broad beans, cardoons, celeriac, celery (outdoor), chestnuts, chicory root, cranberries, edible podded peas, elderberries, endives, florence fennel, garlic, ginger, ginseng root, globe artichoke, hazel nuts, herbs (see appendix 6), horseradish, lamb's lettuce, land cress, lettuce, liquorice, loganberries, mallow (althaea spp.), mangels, mulberries, parsley root, parsnips, pears, plums, potatoes, purslane, quinces, radishes, red beet, red mustard, redcurrants, rhubarb, rocket, rose hips, rubus hybrids, safflower, salsify, shallots, spinach, spinach beet, sunflowers, table grapes, turmeric, valerian root, walnuts, wine grapes* [3]; *blackcurrants, bulb onions, carrots, farm forestry, gooseberries, hops, raspberries, swedes* (stockfeed only), *turnips* (stockfeed only), *vining peas* [1, 3]; *combining peas, fodder beet, linseed, spring field beans, spring oilseed rape, sugar beet, winter field beans, winter oilseed rape* [1-3]; *fodder rape, lupins, poppies* [2, 3]; *kale* (stockfeed only), *spring oilseed rape for industrial use, strawberries, winter oilseed rape for industrial use* [1]; *mustard* [2]
*   Wild oats in *almonds, apples, apricots, asparagus, baby leaf crops, beans without pods (dry), bilberries, blackberries, blueberries, broad beans, cardoons, celeriac, celery (outdoor), chestnuts, chicory root, cranberries, edible podded peas, elderberries, endives, florence fennel, garlic, ginger, ginseng root, globe artichoke, hazel nuts, herbs (see appendix 6), horseradish, lamb's lettuce, land cress, lettuce, liquorice, loganberries, mallow (althaea spp.), mangels, mulberries, parsley root, parsnips, pears, plums, potatoes, purslane, quinces, radishes, red beet, red mustard, redcurrants, rhubarb, rocket, rose hips, rubus hybrids, safflower, salsify, shallots, spinach, spinach beet, sunflowers, table grapes, turmeric, valerian root, walnuts, wine grapes* [3]; *blackcurrants, bulb onions, carrots, farm forestry, gooseberries, hops, raspberries, swedes* (stockfeed only), *turnips*

FOR FULL CONDITIONS OF USE ALWAYS READ THE PRODUCT LABEL

(stockfeed only), **vining peas** [1, 3]; **combining peas, fodder beet, linseed, spring field beans, spring oilseed rape, sugar beet, winter field beans, winter oilseed rape** [1-3]; **fodder rape, lupins, poppies** [2, 3]; **kale** *(stockfeed only)*, **spring oilseed rape for industrial use, strawberries, winter oilseed rape for industrial use** [1]; **mustard** [2]

Approval information
- Fluazifop-P-butyl included in Annex I under EC Regulation 1107/2009
- Accepted by BBPA for use on hops

Efficacy guidance
- Best results achieved by application when weed growth active under warm conditions with adequate soil moisture.
- Spray weeds from 2-expanded leaf stage to fully tillered, couch from 4 leaves when majority of shoots have emerged, with a second application if necessary
- Control may be reduced under dry conditions. Do not cultivate for 2 wk after spraying couch
- Annual meadow grass is not controlled
- May also be used to remove grass cover crops
- Fluazifop-P-butyl is an ACCase inhibitor herbicide. To avoid the build up of resistance do not apply products containing an ACCase inhibitor herbicide more than twice to any crop. In addition do not use any product containing fluazifop-P-butyl in mixture or sequence with any other product containing the same ingredient
- Use these products as part of a resistance management strategy that includes cultural methods of control and does not use ACCase inhibitors as the sole chemical method of grass weed control

Restrictions
- Maximum number of treatments 1 per crop or yr for all crops
- Do not sow cereals or grass crops for at least 8 wk after application of high rate or 2 wk after low rate
- Do not apply through CDA sprayer, with hand-held equipment or from air
- Avoid treatment before spring growth has hardened or when buds opening
- Do not treat bush and cane fruit or hops between flowering and harvest
- Consult processors before treating crops intended for processing
- Oilseed rape, linseed and flax for industrial use must not be harvested for human or animal consumption nor grazed
- Do not use for forestry establishment on land not previously under arable cultivation or improved grassland
- Treated vegetation in field margins, land temporarily removed from production etc, must not be grazed or harvested for human or animal consumption and unprotected persons must be kept out of treated areas for at least 24 h

Crop-specific information
- Latest use: before 50% ground cover for swedes, turnips; before 5 leaf stage for spring oilseed rape; before flowering for blackcurrants, gooseberries, hops, raspberries, strawberries; before flower buds visible for field beans, peas, linseed, flax, winter oilseed rape; 2 wk before sowing cereals or grass for field margins, land temporarily removed from production
- HI beet crops, kale, carrots 8 wk; onions 4 wk; oilseed rape for industrial use 2 wk
- Apply to sugar and fodder beet from 1-true leaf to 50% ground cover
- Apply to winter oilseed rape from 1-true leaf to established plant stage
- Apply to spring oilseed rape from 1-true leaf but before 5-true leaves
- Apply in fruit crops after harvest. See label for timing details on other crops
- Before using on onions or peas use crystal violet test to check that leaf wax is sufficient

Environmental safety
- Dangerous for the environment
- Very toxic to aquatic organisms

Hazard classification and safety precautions
Hazard Harmful, Dangerous for the environment [1-3]; Very toxic to aquatic organisms [3]
Transport code 9
Packaging group III

SEE SECTION 3 FOR PRODUCTS ALSO REGISTERED

UN Number 3082
Risk phrases H315 [1]; H317 [2]; H361 [1-3]
Operator protection A, C, H, M; U05a, U08, U20b
Environmental protection E15a, H410 [1-3]; E38 [2, 3]
Storage and disposal D01, D02, D05, D09a, D12a [1-3]; D10b [1]; D10c [2, 3]
Medical advice M05b [1]

198   fluazinam

A dinitroaniline fungicide for use in potatoes
FRAC mode of action code: 29

See also azoxystrobin + fluazinam
cymoxanil + fluazinam
dimethomorph + fluazinam

Products

1 Fluazinova	Barclay	500 g/l	SC	17625
2 Gando	Harvest	500 g/l	SC	18001
3 Nando 500SC	Nufarm UK	500 g/l	SC	16388
4 Shirlan	Belchim	500 g/l	SC	18406
5 Tizca	FMC Agro	500 g/l	SC	18813
6 Volley	Adama	500 g/l	SC	16451

Uses

- Blight in *potatoes* [1-6]
- Late blight in *potatoes* [1]
- Powdery scab in *potatoes grown for seed* (off-label) [3, 4]; *seed potatoes* (off-label) [5]
- Raspberry root rot in *blackberries* (off-label), *loganberries* (off-label), *protected blackberries* (off-label), *protected loganberries* (off-label), *protected raspberries* (off-label), *protected ribes hybrids* (off-label), *raspberries* (off-label), *rubus hybrids* (off-label) [5]

Extension of Authorisation for Minor Use (EAMUs)

- *blackberries* 20192066 [5]
- *loganberries* 20192066 [5]
- *potatoes grown for seed* 20180200 [3], 20192051 [4]
- *protected blackberries* 20192066 [5]
- *protected loganberries* 20192066 [5]
- *protected raspberries* 20192066 [5]
- *protected ribes hybrids* 20192066 [5]
- *raspberries* 20192066 [5]
- *rubus hybrids* 20192066 [5]
- *seed potatoes* 20190877 [5]

Approval information

- Fluazinam included in Annex I under EC Regulation 1107/2009

Efficacy guidance

- Commence treatment at the first blight risk warning (before blight enters the crop). Products are rainfast within 1 h
- In the absence of a warning, treatment should start before foliage of adjacent plants meets in the rows
- Spray at 5-14 d intervals depending on severity of risk (see label)
- Ensure complete coverage of the foliage and stems, increasing volume as haulm growth progresses, in dense crops and if blight risk increases

Restrictions

- Horizontal boom sprayers must be fitted with three star drift reduction technology for all uses
- Do not use with hand-held sprayers
- Low drift spraying equipment must be operated according to the specific conditions stated in the official three star rating for that equipment as published on HSE Chemicals Regulation Division's

FOR FULL CONDITIONS OF USE ALWAYS READ THE PRODUCT LABEL

website. These operating conditions must be maintained until the operator is 30m from the top of the bank of any surface water bodies

Crop-specific information
* HI 0 - 10 d for potatoes. Check label
* Ensure complete kill of potato haulm before lifting and do not lift crops for storage while there is any green tissue left on the leaves or stem bases

Environmental safety
* Dangerous for the environment
* Very toxic to aquatic organisms
* Buffer zone requirement 6m [1, 6]
* Buffer zone requirement 7m [4, 5]
* Buffer zone requirement 8m [2, 3]
* LERAP Category B

Hazard classification and safety precautions
Hazard Harmful [2, 3]; Irritant [1, 4, 5]; Dangerous for the environment [1-6]; Very toxic to aquatic organisms [4, 6]
Transport code 9
Packaging group III
UN Number 3082
Risk phrases H315 [2, 3]; H317 [1-5]; H318 [5]; H361 [1-4, 6]
Operator protection A, C, H [1, 4-6]; U02a, U04a, U08 [1-4, 6]; U05a, U14 [1-5]; U11 [5]; U15, U20a [1-4]; U20b [5, 6]
Environmental protection E15a, E16a, E38, H410 [1-6]; E16b, E34 [1-4, 6]
Storage and disposal D01, D02, D05, D10c [1-5]; D09a, D12a [1-6]; D10a [6]
Medical advice M03 [1-4, 6]; M05a [1-4]

199 fludioxonil

A phenylpyrrole fungicide seed treatment for wheat and barley
FRAC mode of action code: 12

See also cymoxanil + fludioxonil + metalaxyl-M
cyprodinil + fludioxonil
difenoconazole + fludioxonil
difenoconazole + fludioxonil + tebuconazole

Products
1	Beret Gold	Syngenta	25 g/l	FS	16430
2	Emblem	Pure Amenity	125 g/l	SC	18066
3	Geoxe	Syngenta	50% w/w	WG	16596
4	Maxim 100FS	Syngenta	100 g/l	FS	15683
5	Medallion TL	Syngenta	125 g/l	SC	15287

Uses
* Alternaria in **apples** *(reduction)*, **crab apples** *(reduction)*, **pears** *(reduction)*, **quinces** *(reduction)* [3]
* Anthracnose in **amenity grassland** *(reduction)*, **managed amenity turf** *(reduction)* [2, 5]
* Black dot in **potatoes** *(some reduction)* [4]
* Black scurf in **potatoes** [4]
* Botrytis in **apples** *(reduction)*, **crab apples** *(reduction)*, **pears** *(reduction)*, **quinces** *(reduction)* [3]
* Bunt in **spring wheat** *(seed treatment)*, **winter wheat** *(seed treatment)* [1]
* Covered smut in **spring barley** *(seed treatment)*, **winter barley** *(seed treatment)* [1]
* Disease control in **apples**, **crab apples**, **pears**, **quinces** [3]
* Drechslera leaf spot in **amenity grassland** *(useful levels of control)*, **managed amenity turf** *(useful levels of control)* [2, 5]

SEE SECTION 3 FOR PRODUCTS ALSO REGISTERED

- Fusarium foot rot and seedling blight in *rye* *(seed treatment)*, *spring barley* *(seed treatment)*, *spring oats* *(seed treatment)*, *spring wheat* *(seed treatment)*, *triticale* *(seed treatment)*, *winter barley* *(seed treatment)*, *winter oats* *(seed treatment)*, *winter wheat* *(seed treatment)* [1]
- Fusarium patch in *amenity grassland*, *managed amenity turf* [2, 5]
- Leaf stripe in *spring barley* *(seed treatment - reduction)*, *winter barley* *(seed treatment - reduction)* [1]
- Microdochium nivale in *amenity grassland*, *managed amenity turf* [2, 5]; *rye* *(seed treatment)*, *triticale* [1]
- Monilinia spp. in *apples* *(reduction)*, *crab apples* *(reduction)*, *pears* *(reduction)*, *quinces* *(reduction)* [3]
- Nectria spp in *apples* *(reduction)*, *crab apples* *(reduction)*, *pears* *(reduction)*, *quinces* *(reduction)* [3]
- Penicillium rot in *apples* *(reduction)*, *crab apples* *(reduction)*, *pears* *(reduction)*, *quinces* *(reduction)* [3]
- Phlyctema vagabunda in *apples* *(reduction)*, *crab apples* *(reduction)*, *pears* *(reduction)*, *quinces* *(reduction)* [3]
- Pyrenophora leaf spot in *spring oats* *(seed treatment)*, *winter oats* *(seed treatment)* [1]
- Scab in *seed potatoes* *(off-label)* [4]
- Septoria seedling blight in *spring wheat* *(seed treatment)*, *winter wheat* *(seed treatment)* [1]
- Silver scurf in *potatoes* *(reduction)* [4]
- Snow mould in *spring barley* *(seed treatment)*, *spring oats* *(seed treatment)*, *spring wheat* *(seed treatment)*, *winter barley* *(seed treatment)*, *winter oats* *(seed treatment)*, *winter wheat* *(seed treatment)* [1]
- Stripe smut in *rye* [1]

Extension of Authorisation for Minor Use (EAMUs)
- *seed potatoes* *20160736* [4]

Approval information
- Fludioxonil included in Annex I under EC Regulation 1107/2009
- Accepted by BBPA for use on malting barley

Efficacy guidance
- Apply direct to seed using conventional seed treatment equipment. Continuous flow treaters should be calibrated using product before use
- Effective against benzimidazole-resistant strains of *Microdochium nivale*

Restrictions
- Maximum number of treatments 1 per seed batch
- Do not apply to cracked, split or sprouted seed
- Sow treated seed within 6 mth

Crop-specific information
- Latest use: before drilling
- Product may reduce flow rate of seed through drill. Recalibrate with treated seed before drilling

Environmental safety
- Dangerous for the environment
- Toxic to aquatic organisms
- Do not use treated seed as food or feed
- Treated seed harmful to game and wildlife
- LERAP Category B [2, 3, 5]

Hazard classification and safety precautions
Hazard Irritant, Very toxic to aquatic organisms [3]; Dangerous for the environment [1-5]
Transport code 9
Packaging group III
UN Number 3077 [3]; 3082 [1, 2, 4, 5]
Risk phrases H317 [3]
Operator protection A [1, 3, 4]; H [1, 3]; U05a [1-5]; U14, U19a, U20a [3]; U20b [1, 2, 4, 5]
Environmental protection E03, E15a [1]; E15b [2-5]; E16a [2, 3, 5]; E16b [2, 5]; E34 [2, 4, 5]; E38 [1-5]; H410 [3]; H411 [1, 2, 4, 5]

FOR FULL CONDITIONS OF USE ALWAYS READ THE PRODUCT LABEL

Storage and disposal D01, D02, D09a, D12a [1-5]; D05, D11a [1, 4]; D10c [2-5]
Treated seed S01, S03 [4]; S02, S05 [1, 4]; S07 [1]

200 fludioxonil + metalaxyl-M

A seed dressing for use in forage maize
FRAC mode of action code: 12 + 4

Products

Maxim XL	Syngenta	25:9.69 g/l	FS	16599

Uses
- Damping off in *forage maize*
- Fusarium in *forage maize*
- Pythium in *forage maize*

Approval information
- Fludioxonil and metalaxyl-M included in Annex I under EC Regulation 1107/2009

Restrictions
- For advice on resistance management, refer to the latest Fungicide Resistance Action Group (FRAG) guidelines

Hazard classification and safety precautions
UN Number N/C
Operator protection A, H; U05a
Environmental protection E34, E38, H412
Storage and disposal D01, D02, D05, D09a, D10c, D11a, D12a
Treated seed S01, S02, S03, S04a, S05, S07

201 fludioxonil + metalaxyl-M + sedaxane

A fungicide seed treatment for use on sugarbeet
FRAC mode of action code: 12 + 4 + 7

Products

Vibrance SB	Syngenta	22.5:14.4:15 g/l	FS	18588

Uses
- Phoma in *fodder beet*, *sugar beet*
- Pythium in *fodder beet*, *sugar beet*
- Rhizoctonia in *fodder beet*, *sugar beet*

Approval information
- Fludioxonil, metalaxyl-M and sedaxane included in Annex 1 under EC Regulation 1107/2009

Restrictions
- Treated seed must not be used for food or feed
- Sacks containing treated seed must not be re-used for food or feed
- Treated seed must not be applied from the air

Hazard classification and safety precautions
Transport code 9
Packaging group III
UN Number 3082
Risk phrases H351
Operator protection A, H
Environmental protection H411

SEE SECTION 3 FOR PRODUCTS ALSO REGISTERED

202   fludioxonil + pyrimethanil

A fungicide mixture for use in apples and pears
FRAC mode of action code: 9 + 12

Products

Pomax	Belchim	133:336 g/l	SC	18244

Uses
- Disease control in **apples**, **pears**

Restrictions
- Do not apply by hand-held equipment

Environmental safety
- Broadcast air-assisted LERAP (20 m)

Hazard classification and safety precautions
    Hazard Very toxic to aquatic organisms
    Transport code 9
    Packaging group III
    UN Number 3082
    Operator protection A, H
    Environmental protection E17b (20 m); H410

203   fludioxonil + sedaxane

A fungicide seed treatment for cereals
FRAC mode of action code: 12 + 7

See also sedaxane

Products

Vibrance Duo	Syngenta	25:25 g/l	FS	17838

Uses
- Bunt in **winter wheat**
- Fusarium diseases in **winter wheat**
- Fusarium ear blight in **winter wheat** *(moderate control)*
- Loose smut in **spring oats**, **winter wheat**
- Septoria leaf blotch in **winter wheat**
- Snow mould in **rye**, **triticale**, **winter wheat**
- Stripe smut in **rye**

Approval information
- Fludioxonil and sedaxane included in Annex 1 under EC Regulation 1107/2009

Hazard classification and safety precautions
    Hazard Harmful if inhaled
    Transport code 9
    Packaging group III
    UN Number 3082
    Risk phrases H317
    Operator protection A, H; U05a, U19a
    Environmental protection E15b, E34, H410
    Storage and disposal D01, D02, D05, D09a, D10c, D11a, D12a
    Medical advice M03

FOR FULL CONDITIONS OF USE ALWAYS READ THE PRODUCT LABEL

204 fluidioxonil + tebuconazole

A seed treatment for use in cereals
FRAC mode of action code: 3 + 12

See also fludioxonil
tebuconazole

Products

Fountain	Certis	50:10 g/l	FS	17708

Uses
- Seed-borne diseases in **triticale**, **winter barley**, **winter oats**, **winter rye**, **winter wheat**

Approval information
- Fludioxonil and tebuconazole included in Annex I under EC Regulation 1107/2009

Efficacy guidance
- Do not broadcast treated seed. Ensure that it is covered by at least 40mm settled soil.

Hazard classification and safety precautions
    Transport code 9
    Packaging group III
    UN Number 3082
    Operator protection A, H
    Environmental protection H410

205 fluidioxonil + tefluthrin

A fungicide and insecticide seed treatment mixture for cereals
FRAC mode of action code: 12 + IRAC 3

See also tefluthrin

Products

Austral Plus	Syngenta	10:40 g/l	FS	13314

Uses
- Bunt in **spring wheat**, **winter wheat**
- Covered smut in **spring barley**, **winter barley**
- Fusarium foot rot and seedling blight in **spring barley**, **spring oats**, **spring wheat**, **winter barley**, **winter oats**, **winter wheat**
- Leaf stripe in **spring barley** (partial control), **winter barley** (partial control)
- Pyrenophora leaf spot in **spring oats**, **winter oats**
- Seed-borne diseases in **triticale** (off-label)
- Septoria seedling blight in **spring wheat**, **winter wheat**
- Snow mould in **spring barley**, **spring wheat**, **winter barley**, **winter wheat**
- Wheat bulb fly in **spring barley**, **spring wheat**, **winter barley**, **winter wheat**
- Wireworm in **spring barley**, **spring oats**, **spring wheat**, **winter barley**, **winter oats**, **winter wheat**

Extension of Authorisation for Minor Use (EAMUs)
- **triticale** 20191387 expires 31 Aug 2020

Approval information
- Fludioxonil and tefluthrin included in Annex 1 under EC Regulation 1107/2009
- Accepted by BBPA for use on malting barley
- Approval expiry 31 Aug 2020 [1]

Efficacy guidance
- Apply direct to seed using conventional seed treatment equipment. Continuous flow treaters should be calibrated using product before use

SEE SECTION 3 FOR PRODUCTS ALSO REGISTERED

- Best results obtained from seed drilled into a firm even seedbed
- Tefluthrin is released into soil after drilling and repels or kills larvae of wheat bulb fly and wireworm attacking below ground. Pest control may be reduced by deep or shallow drilling
- Where egg counts of wheat bulb fly or population counts of wireworm indicate a high risk of severe attack follow up spray treatments may be needed
- Control of leaf stripe in barley may not be sufficient in crops grown for seed certification
- Effective against benzimidazole-resistant strains of *Microdochium nivale*
- Under adverse soil or environmental conditions seed rates should be increased to compensate for possible reduced germination capacity

Restrictions
- Maximum number of treatments 1 per batch
- Store treated seed in cool dry conditions and drill within 3 mth
- Do not use on seed above 16% moisture content or on sprouted, cracked or damaged seed

Crop-specific information
- Latest use: before drilling

Environmental safety
- Dangerous for the environment
- Very toxic to aquatic organisms

Hazard classification and safety precautions
Hazard Dangerous for the environment, Very toxic to aquatic organisms
Transport code 9
Packaging group III
UN Number 3082
Operator protection A, C, D, H; U02a, U04a, U05a, U07, U08, U14, U20b
Environmental protection E03, E15b, E34, E38, H410
Storage and disposal D01, D02, D05, D09b, D10c, D11a, D12a
Treated seed S02, S04b, S05, S07, S08

206   flufenacet

A broad spectrum oxyacetamide herbicide for weed control in winter cereals
HRAC mode of action code: K3

See also diflufenican + flufenacet
diflufenican + flufenacet + flurtamone
diflufenican + flufenacet + metribuzin
flufenacet + isoxaflutole
flufenacet + metribuzin
flufenacet + pendimethalin

Products

1	Clayton Tacit	Clayton	480 g/l	SL	19073
2	Firecloud	Certis	500 g/l	SL	18175
3	Gorgon	Sipcam	480 g/l	SL	17685
4	Iconic	Adama	500 g/l	SC	19094
5	Preto	Nurture	480 g/l	SL	18594
6	Staket	Top Crop	480 g/l	SL	18662
7	Starfire	Certis	500 g/l	SL	18179
8	Sunfire	Certis	500 g/l	SL	16745
9	System 50	Certis	500 g/l	SL	16612

Uses
- Annual dicotyledons in **ornamental plant production** *(off-label)* [8]; **winter barley, winter wheat** [1-9]
- Annual meadow grass in **ornamental plant production** *(off-label)*, **protected forest nurseries** *(off-label)* [8]; **rye** *(off-label)*, **triticale** *(off-label)* [2, 7-9]

- Blackgrass in *ornamental plant production* *(off-label)* [8]; *rye (off-label)*, *triticale (off-label)* [2, 7-9]; *winter barley*, *winter wheat* [1-9]
- Chickweed in *protected forest nurseries* *(off-label)* [8]
- Field pansy in *protected forest nurseries* *(off-label)* [8]
- Penny cress in *protected forest nurseries* *(off-label)* [8]
- Scented mayweed in *protected forest nurseries* *(off-label)* [8]
- Shepherd's purse in *protected forest nurseries* *(off-label)* [8]
- Sowthistle in *protected forest nurseries* *(off-label)* [8]
- Speedwells in *protected forest nurseries* *(off-label)* [8]

Extension of Authorisation for Minor Use (EAMUs)
- *ornamental plant production 20171065* [8]
- *protected forest nurseries 20170951* [8]
- *rye 20193068* [2], *20193070* [7], *20170952* [8], *20162019* [9]
- *triticale 20193068* [2], *20193070* [7], *20170952* [8], *20162019* [9]

Approval information
- Flufenacet included in Annex I under EC Regulation 1107/2009

Efficacy guidance
- Reports of reduced sensitivity in blackgrass and isolated instances of Enhanced Metabolic Resistance to flufenacet have been confirmed in a paper presented in Pest Management Science, March 2019

Restrictions
- Do not apply by hand-held equipment [3]
- Do not handle treated crops for at least 2 days after treatment

Environmental safety
- Buffer zone requirement 8 m
- LERAP Category B

Hazard classification and safety precautions
Hazard Harmful, Dangerous for the environment, Harmful if swallowed [1-9]; Very toxic to aquatic organisms [2, 4, 7-9]
Transport code 9
Packaging group III
UN Number 3082
Risk phrases H317 [2, 4, 7-9]; H373 [1-9]
Operator protection A, H; U05a, U09a, U19a [4]
Environmental protection E16a, H410
Storage and disposal D01, D02 [1-7]; D05, D09a [2, 7]; D10a, D11a [7]; D10b, D12a [2]; D12b [4]
Medical advice M05a [4]

207   flufenacet + isoxaflutole

A residual herbicide mixture for maize
HRAC mode of action code: K3 + F2

See also isoxaflutole

Products

| Amethyst | AgChem Access | 48:10% w/w | WG | 14079 |

Uses
- Annual dicotyledons in *forage maize*, *grain maize*
- Grass weeds in *forage maize*, *grain maize*

Approval information
- Flufenacet and isoxaflutole included in Annex I under EC Regulation 1107/2009

SEE SECTION 3 FOR PRODUCTS ALSO REGISTERED

Efficacy guidance
- Best results obtained from applications to a fine, firm seedbed in the presence of some soil moisture
- Efficacy may be reduced on cloddy seedbeds or under prolonged dry conditions
- Established perennial grasses and broad-leaved weeds growing from rootstocks will not be controlled
- Always follow WRAG guidelines for preventing and managing herbicide resistant weeds. See Section 5 for more information

Restrictions
- Maximum number of treatments 1 per crop
- Do not use on soils both above 70% sand and less than 2% organic matter
- Do not use on maize crops intended for seed production

Crop-specific information
- Latest use: before crop emergence
- Ideally treatment should be made within 4 d of sowing, before the maize seeds have germinated

Following crops guidance
- In the event of failure of a treated crop maize, sweet corn or potatoes may be grown after surface cultivation or ploughing
- After normal harvest of a treated crop oats or barley may be sown after ploughing and wheat, rye, triticale, sugar beet, field beans, peas, soya, sorghum or sunflowers may be sown after surface cultivation or ploughing.

Environmental safety
- Dangerous for the environment
- Very toxic to aquatic organisms
- Take care to avoid drift over other crops. Beet crops and sunflowers are particularly sensitive
- LERAP Category B

Hazard classification and safety precautions
  Hazard Harmful, Dangerous for the environment
  Transport code 9
  Packaging group III
  UN Number 3077
  Risk phrases R22a, R36, R43, R48, R50, R53a, R63
  Operator protection A, C, H; U05a, U11, U20b
  Environmental protection E15b, E16a, E34, E38
  Storage and disposal D01, D02, D05, D09a, D10a, D12a
  Medical advice M03

208   flufenacet + metribuzin

A herbicide mixture for potatoes
HRAC mode of action code: K3 + C1

See also metribuzin

Products

Artist	Bayer CropScience	24:17.5% w/w	WP	17049

Uses
- Annual dicotyledons in *bilberries* (off-label), *blackcurrants* (off-label), *cranberries* (off-label), *early potatoes*, *gooseberries* (off-label), *maincrop potatoes*, *redcurrants* (off-label), *ribes species* (off-label), *soya beans* (off-label)
- Annual grasses in *bilberries* (off-label), *blackcurrants* (off-label), *cranberries* (off-label), *gooseberries* (off-label), *redcurrants* (off-label), *ribes species* (off-label), *soya beans* (off-label)
- Annual meadow grass in *early potatoes*, *maincrop potatoes*
- Black bindweed in *soya beans* (off-label)

FOR FULL CONDITIONS OF USE ALWAYS READ THE PRODUCT LABEL

Extension of Authorisation for Minor Use (EAMUs)
- *bilberries 20152968*
- *blackcurrants 20152968*
- *cranberries 20152968*
- *gooseberries 20152968*
- *redcurrants 20152968*
- *ribes species 20152968*
- *soya beans 20171098*

Approval information
- Flufenacet and metribuzin included in Annex I under EC Regulation 1107/2009

Efficacy guidance
- Product acts through root uptake and needs sufficient soil moisture at and shortly after application
- Effectiveness is reduced under dry soil conditions
- Residual activity is reduced on mineral soils with a high organic matter content and on peaty or organic soils
- Ensure application is made evenly to both sides of potato ridges
- Perennial weeds are not controlled

Restrictions
- Maximum total dose equivalent to one full dose treatment
- Potatoes must be sprayed before emergence of crop and weeds
- See label for list of tolerant varieties. Do not treat Maris Piper grown on Sands or Very Light soils
- Do not use on Sands
- On stony or gravelly soils there is risk of crop damage especially if heavy rain falls soon after application

Crop-specific information
- Latest use: before potato crop emergence
- Consult processor before use on crops for processing

Following crops guidance
- Before drilling or planting any succeeding crop soil must be mouldboard ploughed to at least 15 cm as soon as possible after lifting and no later than end Dec
- In W Cornwall on soils with more than 5% organic matter treated early potatoes may be followed by summer planted brassica crops 14 wk after treatment and after mouldboard ploughing. Elsewhere cereals or winter beans may be grown in the same year if at least 16 wk have elapsed since treatment
- In the yr following treatment any crop may be grown except lettuce or radish, or vegetable brassica crops on silt soils in Lincs

Environmental safety
- Dangerous for the environment
- Very toxic to aquatic organisms
- Take care to avoid spray drift onto neighbouring crops, especially lettuce or brassicas
- LERAP Category B

Hazard classification and safety precautions
Hazard Harmful, Dangerous for the environment, Harmful if swallowed
Transport code 9
Packaging group III
UN Number 3077
Risk phrases H317, H373
Operator protection A, C, D, H; U05a, U08, U13, U14, U19a, U20b
Environmental protection E15a, E16a, E38, H410
Storage and disposal D01, D02, D09a, D11a, D12a
Medical advice M03

SEE SECTION 3 FOR PRODUCTS ALSO REGISTERED

209 flufenacet + pendimethalin

A broad spectrum residual and contact herbicide mixture for winter cereals
HRAC mode of action code: K3 + K1

See also pendimethalin

Products

1 Clayton Glacier	Clayton	60:300 g/l	EC	15920
2 Confluence	Pan Agriculture	60:300 g/l	EC	17199
3 Crystal	BASF	60:300 g/l	EC	13914
4 Shooter	BASF	60:300 g/l	EC	14106
5 Trooper	BASF	60:300 g/l	EC	13924

Uses

- Annual dicotyledons in *game cover* (off-label), *spring barley* (off-label) [3]; *winter barley, winter wheat* [1-5]
- Annual grasses in *winter barley, winter wheat* [1-5]
- Annual meadow grass in *game cover* (off-label), *spring barley* (off-label) [3]; *winter barley, winter wheat* [1-5]
- Blackgrass in *game cover* (off-label), *spring barley* (off-label) [3]; *winter barley, winter wheat* [1-5]
- Chickweed in *winter barley, winter wheat* [1-5]
- Corn marigold in *winter barley, winter wheat* [1-5]
- Field speedwell in *winter barley, winter wheat* [1-5]
- Ivy-leaved speedwell in *winter barley, winter wheat* [1-5]
- Wild oats in *spring barley* (off-label) [3]

Extension of Authorisation for Minor Use (EAMUs)

- *game cover* 20090450 [3]
- *spring barley* 20130951 [3]

Approval information

- Flufenacet and pendimethalin included in Annex I under EC Regulation 1107/2009
- Accepted by BBPA for use on malting barley

Efficacy guidance

- Best results achieved when applied from pre-emergence of weeds up to 2 leaf stage but post emergence treatment is not recommended on clay soils
- Product requires some soil moisture to be activated ideally from rain within 7 d of application. Prolonged dry conditions may reduce residual control
- Product is slow acting and final level of weed control may take some time to appear
- For effective weed control seed bed preparations should ensure even incorporation of any trash, straw and ash to 15 cm
- Efficacy may be reduced on soils with more than 6% organic matter
- Always follow WRAG guidelines for preventing and managing herbicide resistant weeds. See Section 5 for more information

Restrictions

- Maximum total dose equivalent to one full dose treatment
- For pre-emergence treatments seed should be covered with at least 32 mm settled soil. Shallow drilled crops should be treated post-emergence only
- Do not treat undersown crops
- Avoid spraying during periods of prolonged or severe frosts
- Do not use on stony or gravelly soils or those with more than 10% organic matter
- Pre-emergence treatment may only be used on crops drilled before 30 Nov. All crops must be treated before 31 Dec in yr of planting.
- Concentrated or diluted product may stain clothing or skin

Crop-specific information

- Latest use: before third tiller stage (GS 23) and before 31 Dec in yr of planting

FOR FULL CONDITIONS OF USE ALWAYS READ THE PRODUCT LABEL

- Very wet weather before and after treatment may result in loss of crop vigour and reduced yield, particularly where soils become waterlogged

Following crops guidance
- Any crop may follow a failed or normally harvested treated crop provided ploughing to at least 15 cm is carried out beforehand

Environmental safety
- Dangerous for the environment
- Very toxic to aquatic organisms
- Risk to certain non-target insects or other arthropods - avoid spraying within 6 m of field boundary
- Some products supplied in small volume returnable packs. Follow instructions for use
- LERAP Category B

Hazard classification and safety precautions
Hazard Harmful, Dangerous for the environment, Harmful if swallowed [1-5]; Very toxic to aquatic organisms [1, 3-5]
Transport code 3
Packaging group III
UN Number 1993 [3-5]; 3082 [1, 2]
Risk phrases H304 [1, 3-5]; H315 [1-5]; H351, H370 [2]
Operator protection A, C, H; U02a, U05a [1-5]; U14, U20b [4]; U20c [1-3, 5]
Environmental protection E15a [1-3, 5]; E15b [4]; E16a, E16b, E22b, E34, E38, H410 [1-5]
Storage and disposal D01, D02, D09a, D12a [1-5]; D08, D10c [4]; D10a [1-3, 5]
Medical advice M03 [1-5]; M05b [1-3, 5]

210   flufenacet + picolinafen

A herbicide mixture for weed control in winter cereals
HRAC mode of action code: K3 + F1

Products

1	Lantern	BASF	240:95 g/l	SC	19085
2	Pontos	BASF	240:100 g/l	SC	17811
3	Quirinus	BASF	240:50 g/l	SC	17711

Uses
- Annual dicotyledons in *rye, triticale, winter barley, winter wheat*
- Annual grasses in *rye, triticale, winter barley, winter wheat*
- Annual meadow grass in *rye, triticale, winter barley, winter wheat*
- Charlock in *rye, triticale, winter barley, winter wheat*
- Chickweed in *rye, triticale, winter barley, winter wheat*
- Field pansy in *rye, triticale, winter barley, winter wheat*
- Loose silky bent in *rye, triticale, winter barley, winter wheat*
- Mayweeds in *rye, triticale, winter barley, winter wheat*
- Shepherd's purse in *rye, triticale, winter barley, winter wheat*
- Speedwells in *rye, triticale, winter barley, winter wheat*
- Volunteer oilseed rape in *rye, triticale, winter barley, winter wheat*

Approval information
- Flufenacet and picolinafen are included in Annex I under EC Regulation 1107/2009

Efficacy guidance
- Can be used on all varieties of winter crops of wheat, barley, rye and triticale

Restrictions
- Always follow WRAG guidelines for preventing and managing herbicide resistant weeds
- Use low drift spraying equipment up to 30m from the top of the bank of any surface water bodies
- Horizontal boom sprayers must be fitted with three star drift reduction technology for all uses

SEE SECTION 3 FOR PRODUCTS ALSO REGISTERED

Following crops guidance

- There are no restrictions on following crops after the normal harvest.
- In the event of crop failure, winter wheat can be re-sown in the same autumn provided soil is cultivated to a minimum depth of 15cm. Any of the following crops may be sown provided there has been a minimum of 60 days after the application and the soil is cultivated to a minimum depth of 15cm; legumes, maize, sugar beet and sunflower. Oilseed rape can be re-sown after 90 days following a pre-emergence application or 60 days following a post emergence application and the soil is cultivated to a minimum depth of 15cm. Spring barley can be re-drilled 120 days following application and the soil is cultivated to a minimum depth of 15cm.

Environmental safety

- Buffer zone requirement 6 m [1-3]
- LERAP Category B

Hazard classification and safety precautions

Transport code 9
Packaging group III
UN Number 3082
Risk phrases H373
Operator protection A, H; U05a, U19a
Environmental protection E16a, E34, H410
Storage and disposal D01, D02, D05, D09a, D10c, D12b
Medical advice M03

211   flumioxazin

A phenyphthalimide herbicide for winter wheat
HRAC mode of action code: E

Products

Sumimax	Sumitomo	300 g/l	SC	13548

Uses

- Annual dicotyledons in *bulb onions* (off-label), *carrots* (off-label), *hops* (off-label), *ornamental plant production* (off-label), *parsnips* (off-label), *soft fruit* (off-label), *top fruit* (off-label), *vining peas* (off-label), *winter oats* (off-label), *winter wheat*
- Annual grasses in *bulb onions* (off-label), *carrots* (off-label), *hops* (off-label), *ornamental plant production* (off-label), *parsnips* (off-label), *soft fruit* (off-label), *top fruit* (off-label), *vining peas* (off-label), *winter oats* (off-label)
- Annual meadow grass in *winter wheat*
- Blackgrass in *winter oats* (off-label)
- Chickweed in *winter wheat*
- Cleavers in *winter wheat*
- Groundsel in *bulb onions* (off-label), *carrots* (off-label), *hops* (off-label), *ornamental plant production* (off-label), *parsnips* (off-label), *soft fruit* (off-label), *top fruit* (off-label), *vining peas* (off-label), *winter oats* (off-label)
- Loose silky bent in *winter wheat*
- Mayweeds in *winter wheat*
- Ryegrass in *winter oats* (off-label)
- Speedwells in *winter wheat*
- Volunteer oilseed rape in *bulb onions* (off-label), *carrots* (off-label), *hops* (off-label), *ornamental plant production* (off-label), *parsnips* (off-label), *soft fruit* (off-label), *top fruit* (off-label), *vining peas* (off-label), *winter oats* (off-label), *winter wheat*
- Volunteer potatoes in *bulb onions* (off-label), *carrots* (off-label), *hops* (off-label), *ornamental plant production* (off-label), *parsnips* (off-label), *soft fruit* (off-label), *top fruit* (off-label), *vining peas* (off-label), *winter oats* (off-label)

Extension of Authorisation for Minor Use (EAMUs)

- *bulb onions* 20091108
- *carrots* 20091111
- *hops* 20082881

FOR FULL CONDITIONS OF USE ALWAYS READ THE PRODUCT LABEL

- **ornamental plant production** *20082881*
- **parsnips** *20091111*
- **soft fruit** *20082881*
- **top fruit** *20082881*
- **vining peas** *20091109*
- **winter oats** *20093122*

Approval information
- Flumioxazine is included in Annex 1 under EC Regulation 1107/2009

Efficacy guidance
- Best results obtained from applications to moist soil early post-emergence of the crop when weeds are germinating up to 1 leaf stage
- Flumioxazin is contact acting and relies on weeds germinating in moist soil and coming into sufficient contact with the herbicide
- Flumioxazin remains in the surface layer of the soil and does not migrate to lower layers
- Contact activity is long lasting during the autumn
- Seed beds should be firm, fine and free from clods. Crops should be drilled to 25 mm and well covered by soil
- Efficacy is reduced on soils with more than 10% organic matter
- Flumioxazin is a member of the protoporphyrin oxidase (PPO) inhibitor herbicides. To avoid the build up of resistance do not use PPO-inhibitor herbicides more than once on any crop
- Use as part of a resistance management strategy that includes cultural methods of control and does not use PPO-inhibitors as the sole chemical method of grass weed control

Restrictions
- Maximum number of treatments 1 per crop for winter wheat
- Only apply to crops that have been hardened by cool weather. Do not treat crops with lush or soft growth
- Do not apply to waterlogged soils or to soils with more than 10% organic matter
- Do not follow an application with another pesticide for 14 d
- Do not treat undersown crops
- Do not treat broadcast crops until they are past the 3 leaf stage
- Do not treat crops under stress for any reason, or during prolonged frosty weather
- Do not roll or harrow for 2 wk before treatment or at any time afterwards

Crop-specific information
- Latest use: before 5th true leaf stage (GS 15) for winter wheat
- Light discolouration of leaf margins can occur after treatment and treatment of soft crops will cause transient leaf bleaching

Following crops guidance
- Any crop may be sown following normal harvest of a treated crop
- In the event of failure of a treated crop spring cereals, spring oilseed rape, sugar beet, maize or potatoes may be redrilled after ploughing

Environmental safety
- Dangerous for the environment
- Very toxic to aquatic organisms
- Take extreme care to avoid drift onto plants outside the target area
- LERAP Category B

Hazard classification and safety precautions
Hazard Toxic, Dangerous for the environment
Transport code 9
Packaging group III
UN Number 3082
Risk phrases H360
Operator protection A, H; U05a, U19a
Environmental protection E15b, E16a, E34, E38, H410
Storage and disposal D01, D02, D05, D07, D09a, D12b
Medical advice M04a

SEE SECTION 3 FOR PRODUCTS ALSO REGISTERED

212   fluopicolide

An benzamide fungicide available only in mixtures
FRAC mode of action code: 43

213   fluopicolide + propamocarb hydrochloride

A protectant and systemic fungicide mixture for potato blight
FRAC mode of action code: 43 + 28

See also propamocarb hydrochloride

Products

Infinito	Bayer CropScience	62.5:625 g/l	SC	16335

Uses

* Blight in **potatoes**
* Downy mildew in **baby leaf crops** *(off-label)*, **broccoli** *(off-label)*, **brussels sprouts** *(off-label)*, **bulb onions** *(off-label)*, **cabbages** *(off-label)*, **calabrese** *(off-label)*, **cauliflowers** *(off-label)*, **collards** *(off-label)*, **cress** *(off-label)*, **garlic** *(off-label)*, **herbs (see appendix 6)** *(off-label)*, **kale** *(off-label)*, **lamb's lettuce** *(off-label)*, **land cress** *(off-label)*, **leeks** *(off-label)*, **lettuce** *(off-label)*, **ornamental plant production** *(off-label)*, **protected broccoli** *(off-label)*, **protected brussels sprouts** *(off-label)*, **protected cabbages** *(off-label)*, **protected calabrese** *(off-label)*, **protected cauliflowers** *(off-label)*, **protected collards** *(off-label)*, **protected kale** *(off-label)*, **radishes** *(off-label)*, **red mustard** *(off-label)*, **rocket** *(off-label)*, **salad onions** *(off-label)*, **shallots** *(off-label)*, **spinach** *(off-label)*

Extension of Authorisation for Minor Use (EAMUs)

* **baby leaf crops** *20172431*
* **broccoli** *20152557*
* **brussels sprouts** *20152557*
* **bulb onions** *20161552*
* **cabbages** *20152557*
* **calabrese** *20152557*
* **cauliflowers** *20152557*
* **collards** *20152557*
* **cress** *20172431*
* **garlic** *20161552*
* **herbs (see appendix 6)** *20172431*
* **kale** *20152557*
* **lamb's lettuce** *20172431*
* **land cress** *20172431*
* **leeks** *20161552*
* **lettuce** *20172431*
* **ornamental plant production** *20142251*
* **protected broccoli** *20152557*
* **protected brussels sprouts** *20152557*
* **protected cabbages** *20152557*
* **protected calabrese** *20152557*
* **protected cauliflowers** *20152557*
* **protected collards** *20152557*
* **protected kale** *20152557*
* **radishes** *20172432*
* **red mustard** *20172431*
* **rocket** *20172431*
* **salad onions** *20161552*
* **shallots** *20161552*
* **spinach** *20172431*

FOR FULL CONDITIONS OF USE ALWAYS READ THE PRODUCT LABEL

Approval information
- Fluopicolide and propamocarb hydrochloride included in Annex I under EC Regulation 1107/2009

Efficacy guidance
- Commence spray programme before infection appears as soon as weather conditions favourable for disease development occur. At latest the first treatment should be made as the foliage meets along the rows
- Repeat treatments at 7-10 day intervals according to disease incidence and weather conditions
- Reduce spray interval if conditions are conducive to the spread of blight
- Spray as soon as possible after irrigation
- Increase water volume in dense crops
- When used from full canopy development to haulm desiccation as part of a full blight protection programme tubers will be protected from late blight after harvest and tuber blight incidence will be reduced
- To reduce the development of resistance product should be used in single or block applications with fungicides from a different cross-resistance group

Restrictions
- Maximum total dose on potatoes equivalent to four full dose treatments
- Do not apply if rainfall or irrigation is imminent. Product is rainfast in 1 h provided spray has dried on leaf
- Do not apply as a curative treatment when blight is present in the crop
- Do not apply more than 3 consecutive treatments of the product
- To protect groundwater do not apply more than 400 g/ha fluopicolide in any three year period.

Crop-specific information
- HI 7 d for potatoes
- All varieties of potatoes, including seed crops, may be treated

Environmental safety
- Dangerous for the environment
- Toxic to aquatic organisms

Hazard classification and safety precautions
Hazard Irritant, Dangerous for the environment, Very toxic to aquatic organisms
Transport code 9
Packaging group III
UN Number 3082
Risk phrases H317
Operator protection A, H; U05a, U08, U11, U19a, U20a
Environmental protection E15a, E38, H410
Storage and disposal D01, D02, D09a, D10b, D12a

214   fluopyram

An SDHI fungicide for disease control in fruit trees
FRAC mode of action code: 7

See also bixafen + fluopyram + prothioconazole

Products
1	Luna Privilege	Bayer CropScience	500 g/l	SC	18393
2	Velum Prime	Bayer CropScience	400 g/l	SC	18880

Uses
- Disease control in **apples**, **pears** [1]
- Globodera species in **potatoes** [2]
- Potato cyst nematode in **potatoes** [2]
- Powdery mildew in **hops** *(off-label)* [1]

Extension of Authorisation for Minor Use (EAMUs)
- **hops** *20192567* [1]

SEE SECTION 3 FOR PRODUCTS ALSO REGISTERED

Approval information
- Fluopyram is included in Annex 1 under EC Regulation 1107/2009

Efficacy guidance
- Apply as a medium or coarse spray [2]

Restrictions
- Apply as an in-furrow application but note that it is important to direct spray into the planting furrow and not onto the seed tuber
- Can be used on all potato varieties but crops grown under cover are treated at growers risk only.
- If the crop is intended for processing, consult the processor before the use
- Where [2] has been applied, the first foliar fungicide application made to the crop must not be a member of the SDHI group

Crop-specific information
- Cardoons, celeries, Florence fennels and crops belonging to the category 'other stem vegetables' cannot be allowed as succeeding crops [2]

Environmental safety
- Broadcast air-assisted LERAP [1] (10m)

Hazard classification and safety precautions
Hazard Harmful if swallowed [1]
Transport code 9
Packaging group III
UN Number 3082
Risk phrases H317 [2]
Operator protection A, C, H [2]; U05a [1, 2]; U09a, U20a [1]; U20b [2]
Environmental protection E15a [1]; E15b [2]; E17b [1] (10m); E34, H411 [1, 2]
Storage and disposal D01, D02, D09a [1, 2]; D05, D10b, D12b, D22 [2]; D10a [1]
Medical advice M03

215   fluopyram + prothioconazole

A foliar fungicide for use in oilseed rape
FRAC mode of action code: 3 + 7

Products

1	Propulse	Bayer CropScience	125:125 g/l	SE	17837
2	Recital	Bayer CropScience	125:125 g/l	SE	17909

Uses
- Disease control in **spring oilseed rape**, **winter oilseed rape** [1, 2]
- Light leaf spot in **winter oilseed rape** [1]
- Phoma in **winter oilseed rape** [1]
- Powdery mildew in **winter oilseed rape** [1]
- Sclerotinia in **winter oilseed rape** [1]
- Stem canker in **winter oilseed rape** [1]

Approval information
- Fluopyram and prothioconazole included in Annex 1 under EC Regulation 1107/2009.

Environmental safety
- LERAP Category B

Hazard classification and safety precautions
Hazard Harmful [2]; Dangerous for the environment, Very toxic to aquatic organisms [1, 2]
Transport code 9
Packaging group III
UN Number 3082
Operator protection A, H; U05a, U09b, U19a, U20c

FOR FULL CONDITIONS OF USE ALWAYS READ THE PRODUCT LABEL

Environmental protection E15b, E16a, E34, E38, H410
Storage and disposal D01, D02, D05, D09a, D10b, D12a
Medical advice M03

216   fluopyram + prothioconazole + tebuconazole

An SDHI and triazole mixture for seed treatment in winter barley
FRAC mode of action code: 3 + 7

Products

Raxil Star	Bayer CropScience	20:100:60 g/l	FS	17805

Uses

- Covered smut in **winter barley**
- Fusarium in **winter barley**
- Leaf stripe in **winter barley**
- Loose smut in **winter barley**
- Microdochium nivale in **winter barley**
- Net blotch in **winter barley** *(seed borne)*

Approval information

- Fluopyram, prothioconazole and tebuconazole included in Annex 1 under EC Regulation 1107/2009
- Accepted by BBPA for use on malting barley

Efficacy guidance

- Seed treatment products must be applied by manufacturer's recommended treatment application equipment
- Treated cereal seed should preferably be drilled in the same season
- Follow-up treatments will be needed later in the season to give protection against air-borne and splash-borne diseases
- Seed should be drilled to a depth of 40 mm into a well prepared and firm seedbed. If seed is present on the soil surface, or spills have occurred, then, if conditions are appropriate, the field should be harrowed then rolled to ensure good incorporation
- Do not use on seed with more than 16% moisture content or on sprouted, cracked, skinned or otherwise damaged seed

Hazard classification and safety precautions

Hazard Harmful, Dangerous for the environment
Transport code 9
Packaging group III
UN Number 3082
Risk phrases H361
Operator protection A, C, D, H; U05a, U12b, U20b, U20e
Environmental protection E15b, E34, E38, H410
Storage and disposal D01, D02, D05, D09a, D10d, D12a, D14
Treated seed S01, S02, S04b, S05, S07, S08

217   fluopyram + trifloxystrobin

A fungicide mixture for disease control in protected strawberries
FRAC mode of action code: 7 + 11

See also fluopyram
*   trifloxystrobin*

Products

1	Clayton Stellar	Clayton	250:250 g/l	SC	18417
2	Exteris Stressgard	Bayer CropScience	12.5:12.5 g/l	SC	17825
3	Luna Sensation	Bayer CropScience	250:250 g/l	SC	15793

SEE SECTION 3 FOR PRODUCTS ALSO REGISTERED

Products – continued

4 Robin	Pan Agriculture	250:250 g/l	SC	18915
5 Robin	Pan Agriculture	250:250 g/l	SC	19209

Uses

* Botrytis in *lettuce* (off-label), *protected baby leaf crops* (off-label), *protected lamb's lettuce* (off-label), *protected land cress* (off-label), *protected lettuce* (off-label), *protected purslane* (off-label), *protected red mustard* (off-label), *protected rocket* (off-label) [3]
* Dollar spot in *managed amenity turf* [2]
* Grey mould in *protected chilli peppers*, *protected peppers*, *protected strawberries* (moderate control) [1, 3-5]
* Microdochium nivale in *managed amenity turf* [2]
* Powdery mildew in *lettuce* (off-label), *protected baby leaf crops* (off-label), *protected lamb's lettuce* (off-label), *protected land cress* (off-label), *protected lettuce* (off-label), *protected purslane* (off-label), *protected red mustard* (off-label), *protected rocket* (off-label) [3]; *protected chilli peppers*, *protected peppers*, *protected strawberries* [1, 3-5]
* Sclerotinia in *lettuce* (off-label), *protected baby leaf crops* (off-label), *protected lamb's lettuce* (off-label), *protected land cress* (off-label), *protected lettuce* (off-label), *protected purslane* (off-label), *protected red mustard* (off-label), *protected rocket* (off-label) [3]

Extension of Authorisation for Minor Use (EAMUs)

* *lettuce* 20171179 [3]
* *protected baby leaf crops* 20182998 [3]
* *protected lamb's lettuce* 20182998 [3]
* *protected land cress* 20182998 [3]
* *protected lettuce* 20171179 [3]
* *protected purslane* 20182998 [3]
* *protected red mustard* 20182998 [3]
* *protected rocket* 20182998 [3]

Approval information

* Fluopyram and trifloxystrobin are included in Annex 1 under EC Regulation 1107/2009

Restrictions

* If more than two Qol products are to be used on the crop then the FRAC advice on treatment must be followed.
* Cardoons, celeries, Florence fennels and crops belonging to the category "other stem vegetables" cannot be allowed as succeeding crops

Environmental safety

* Buffer zone requirement 14 m on protected strawberries
* LERAP Category B

Hazard classification and safety precautions

Hazard Harmful if swallowed [1, 3-5]; Very toxic to aquatic organisms [1-5]
Transport code 9
Packaging group III
UN Number 3082
Risk phrases H317 [2]
Operator protection A, H; U05a, U09a, U20a [1, 3-5]
Environmental protection E15a [1, 3-5]; E15b [2]; E16a, E34, H410 [1-5]
Storage and disposal D01, D02, D09a, D10a
Medical advice M03 [1, 3-5]

218   fluoxastrobin

A protectant stobilurin fungicide available in mixtures
FRAC mode of action code: 11

See also bixafen + fluoxastrobin + prothioconazole

219   fluoxastrobin + prothioconazole

A strobilurin and triazole fungicide mixture for cereals
FRAC mode of action code: 11 + 3

See also prothioconazole

Products

1 Fandango	Bayer CropScience	100:100 g/l	EC	17318
2 Firefly 155	Bayer CropScience	45:110 g/l	EC	14818
3 Kurdi	AgChem Access	100:100 g/l	EC	15917
4 Maestro	Bayer CropScience	100:100 g/l	EC	18300
5 Unicur	Bayer CropScience	100:100 g/l	EC	17402

Uses

- Botrytis in **bulb onions** *(useful reduction)*, **shallots** *(useful reduction)* [5]
- Botrytis squamosa in **bulb onions** *(useful reduction)*, **shallots** *(useful reduction)* [5]
- Brown rust in **spring barley**, **winter barley** [1, 3, 4]; **spring wheat**, **winter rye**, **winter wheat** [1-4]
- Crown rust in **spring oats**, **winter oats** [1, 3, 4]
- Downy mildew in **bulb onions** *(control)*, **shallots** *(control)* [5]
- Eyespot in **spring barley** *(reduction)*, **spring oats**, **winter barley** *(reduction)*, **winter oats** [1, 3, 4]; **spring oats** *(reduction of incidence and severity)*, **winter oats** *(reduction of incidence and severity)* [2]; **spring wheat** *(reduction)*, **winter rye** *(reduction)*, **winter wheat** *(reduction)* [1-4]
- Fusarium root rot in **spring wheat** *(reduction)*, **winter wheat** *(reduction)* [1, 3, 4]
- Glume blotch in **spring wheat**, **winter wheat** [1-4]
- Late ear diseases in **spring barley**, **spring wheat**, **winter barley**, **winter wheat** [1, 3, 4]
- Net blotch in **spring barley**, **winter barley** [1, 3, 4]
- Powdery mildew in **spring barley**, **spring oats**, **winter barley**, **winter oats** [1, 3, 4]; **spring wheat**, **winter rye**, **winter wheat** [1-4]
- Rhynchosporium in **spring barley**, **winter barley** [1, 3, 4]; **winter rye** [1-4]
- Septoria leaf blotch in **spring wheat**, **winter wheat** [1-4]
- Sharp eyespot in **spring wheat** *(reduction)*, **winter wheat** [1, 3, 4]
- Sooty moulds in **spring wheat** *(reduction)*, **winter wheat** *(reduction)* [1, 3, 4]
- Take-all in **spring wheat** *(reduction)*, **winter barley** *(reduction)*, **winter wheat** *(reduction)* [1, 3, 4]
- Tan spot in **spring wheat**, **winter wheat** [1, 3, 4]
- Yellow rust in **spring wheat**, **winter wheat** [1-4]

Approval information

- Fluoxastrobin and prothioconazole included in Annex I under EC Regulation 1107/2009
- Accepted by BBPA for use on malting barley

Efficacy guidance

- Best results on foliar diseases obtained from treatment at early stages of disease development. Further treatment may be needed if disease attack is prolonged
- Foliar applications to established infections of any disease are likely to be less effective
- Best control of cereal ear diseases obtained by treatment during ear emergence
- Fluoxastrobin is a member of the QoI cross resistance group. Foliar product should be used preventatively and not relied on for its curative potential
- Use product as part of an Integrated Crop Management strategy incorporating other methods of control, including where appropriate other fungicides with a different mode of action. Do not apply more than two foliar applications of QoI containing products to any cereal crop
- There is a significant risk of widespread resistance occurring in *Septoria tritici* populations in UK. Failure to follow resistance management action may result in reduced levels of disease control
- Strains of wheat and barley powdery mildew resistant to QoIs are common in the UK. Control of wheat mildew can only be relied on from the triazole component
- Prothioconazole is a DMI fungicide. Resistance to some DMI fungicides has been identified in Septoria leaf blotch which may seriously affect performance of some products. For further

advice contact a specialist advisor and visit the Fungicide Resistance Action Group (FRAG)-UK website
- Where specific control of wheat mildew is required this should be achieved through a programme of measures including products recommended for the control of mildew that contain a fungicide from a different cross-resistance group and applied at a dose that will give robust control

Restrictions
- Maximum total dose of foliar sprays equivalent to two full dose treatments for the crop
- Do not apply by hand-held equipment, e.g Knapsack Sprayer [5]

Crop-specific information
- Latest use: before grain milky ripe for spray treatments on wheat and rye; beginning of flowering for barley, oats
- Some transient leaf chlorosis may occur after treatment of wheat or barley but this has not been found to affect yield

Environmental safety
- Dangerous for the environment
- Toxic to aquatic organisms
- Risk to non-target insects or other arthropods. Avoid spraying within 6 m of the field boundary to reduce the effects on non-target insects or other arthropods
- LERAP Category B

Hazard classification and safety precautions
Hazard Dangerous for the environment, Very toxic to aquatic organisms
Transport code 9
Packaging group III
UN Number 3082
Risk phrases H318 [1, 3, 4]
Operator protection A [1-5]; H [1, 2, 4, 5]; U05a [1-5]; U09a, U20b [2]; U09b, U20a [1, 3-5]
Environmental protection E15a, E16a, E34, E38, H410 [1-5]; E22c [1, 3-5]
Storage and disposal D01, D05, D09a, D10b, D12a
Medical advice M03

220   fluoxastrobin + prothioconazole + trifloxystrobin

A triazole and strobilurin fungicide mixture for cereals
FRAC mode of action code: 11 + 3 + 11

See also prothioconazole
trifloxystrobin

Products
Jaunt	Bayer CropScience	75:150:75 g/l	EC	12350

Uses
- Brown rust in **durum wheat, rye, spring barley, spring wheat, triticale, winter barley, winter wheat**
- Eyespot in **durum wheat** (reduction), **rye** (reduction), **spring barley** (reduction), **spring wheat** (reduction), **triticale** (reduction), **winter barley** (reduction), **winter wheat** (reduction)
- Glume blotch in **durum wheat, rye, spring wheat, triticale, winter wheat**
- Late ear diseases in **durum wheat, rye, spring barley, spring wheat, triticale, winter barley, winter wheat**
- Net blotch in **spring barley, winter barley**
- Powdery mildew in **durum wheat, rye, spring barley, spring wheat, triticale, winter barley, winter wheat**
- Rhynchosporium in **spring barley, winter barley**
- Septoria leaf blotch in **durum wheat, rye, spring wheat, triticale, winter wheat**
- Tan spot in **durum wheat, rye, spring wheat, triticale, winter wheat**
- Yellow rust in **durum wheat, rye, spring wheat, triticale, winter wheat**

FOR FULL CONDITIONS OF USE ALWAYS READ THE PRODUCT LABEL

Approval information
- Fluoxastrobin, prothioconazole and trifloxystrobin included in Annex I under EC Regulation 1107/2009
- Accepted by BBPA for use on malting barley

Efficacy guidance
- Best results obtained from treatment at early stages of disease development. Further treatment may be needed if disease attack is prolonged
- Applications to established infections of any disease are likely to be less effective
- Best control of cereal ear diseases obtained by treatment during ear emergence
- Fluoxastrobin and trifloxystrobin are members of the QoI cross resistance group. Product should be used preventatively and not relied on for its curative potential
- Use product as part of an Integrated Crop Management strategy incorporating other methods of control, including where appropriate other fungicides with a different mode of action. Do not apply more than two foliar applications of QoI containing products to any cereal crop
- There is a significant risk of widespread resistance occurring in *Septoria tritici* populations in UK. Failure to follow resistance management action may result in reduced levels of disease control
- Strains of wheat and barley powdery mildew resistant to QoIs are common in the UK. Control of wheat mildew can only be relied on from the triazole component
- Where specific control of wheat mildew is required this should be achieved through a programme of measures including products recommended for the control of mildew that contain a fungicide from a different cross-resistance group and applied at a dose that will give robust control
- Prothioconazole is a DMI fungicide. Resistance to some DMI fungicides has been identified in Septoria leaf blotch which may seriously affect performance of some products. For further advice contact a specialist advisor and visit the Fungicide Resistance Action Group (FRAG)-UK website

Restrictions
- Maximum total dose equivalent to two full dose treatments

Crop-specific information
- Latest use: before grain milky ripe for winter wheat; up to beginning of anthesis (GS 61) for barley

Environmental safety
- Dangerous for the environment
- Very toxic to aquatic organisms
- Risk to non-target insects or other arthropods. Avoid spraying within 6 m of the field boundary to reduce the effects on non-target insects or other arthropods
- LERAP Category B

Hazard classification and safety precautions
Hazard Irritant, Dangerous for the environment, Very toxic to aquatic organisms
Transport code 9
Packaging group III
UN Number 3082
Operator protection A, C, H; U05a, U09b, U19a, U20b
Environmental protection E15a, E16a, E22c, E34, E38, H410
Storage and disposal D01, D02, D05, D09a, D10b, D12a
Medical advice M03

221   fluoxastrobin + tebuconazole

A fungicide mixture for disease control in oilseed rape
FRAC mode of action code: 11 + 3

See also fluoxastrobin + prothioconazole
fluoxastrobin + prothioconazole + tebuconazole

Products

Evito T	Arysta	180:250 g/l	SC	18671

Uses
• Sclerotinia in **spring oilseed rape**, **winter oilseed rape**

Approval information
• Fluoxastrobin and tebuconazole included in Annex I under EC Regulation 1107/2009

Restrictions
• Do not apply before 10% of flowers on main raceme open, main raceme elongating (GS61) or 1 May if this occurs after GS61 in the year of harvest

Environmental safety
• LERAP Category B

Hazard classification and safety precautions
Hazard Very toxic to aquatic organisms
Transport code 9
Packaging group III
UN Number 3082
Risk phrases H319, H361
Operator protection A, H; U05a, U11, U20b
Environmental protection E16a, H410
Medical advice M04a

222   fluroxypyr

A post-emergence pyridinecarboxylic acid herbicide
HRAC mode of action code: O

See also 2,4-D + dicamba + fluroxypyr
aminopyralid + fluroxypyr
bromoxynil + fluroxypyr
bromoxynil + fluroxypyr + ioxynil
clopyralid + florasulam + fluroxypyr
clopyralid + fluroxypyr + MCPA
clopyralid + fluroxypyr + triclopyr
florasulam + fluroxypyr

Products

1	Cleancrop Gallifrey 3	Dow (Corteva)	333 g/l	EC	17399
2	Crescent	Certis	200 g/l	EC	17589
3	Flurostar 200	Globachem	200 g/l	EC	17438
4	Gal-gone	Belchim	200 g/l	EC	17505
5	Hatchet Xtra	Certis	200 g/l	EC	17593
6	Hudson 200	Barclay	200 g/l	EC	17749
7	Hurler	Barclay	200 g/l	EC	17715
8	Minstrel	UPL Europe	200 g/l	EC	13745
9	Starane Hi-Load HL	Dow (Corteva)	333 g/l	EC	16557
10	Tandus	Nufarm UK	200 g/l	EC	18071
11	Tomahawk 2	Nufarm UK	200 g/l	EC	17468

Uses

- Annual dicotyledons in ***durum wheat*** [1-7, 9-11]; ***forage maize*** [1-8, 10, 11]; ***grass seed crops***, ***spelt*** [1, 9]; ***grassland*** [1-10]; ***permanent grassland, rotational grass*** [11]; ***rye*** [10]; ***spring barley*** [2-8, 10, 11]; ***spring oats, spring wheat, triticale, winter barley, winter oats, winter wheat*** [1-11]; ***spring rye*** [1-7, 9, 11]; ***winter rye*** [1-9, 11]
- Black bindweed in ***bulb onions*** *(off-label)*, ***garlic*** *(off-label)*, ***grass seed crops***, ***leeks*** *(off-label)*, ***salad onions*** *(off-label)*, ***shallots*** *(off-label)*, ***spelt*** [1, 9]; ***durum wheat*** [1-7, 9-11]; ***forage maize*** [1-8, 11]; ***grassland*** [1-9]; ***newly sown grass leys*** *(off-label)*, ***poppies*** *(off-label)*, ***poppies grown for seed production*** *(off-label)* [9]; ***permanent grassland, rotational grass*** [11]; ***rye*** [10]; ***spring barley, spring oats, spring wheat, triticale, winter barley, winter oats, winter wheat*** [1-11]; ***spring rye*** [1-7, 9, 11]; ***winter rye*** [1-9, 11]
- Black nightshade in ***forage maize*** [1]
- Chickweed in ***almonds*** *(off-label)*, ***apples*** *(off-label)*, ***bulb onions*** *(off-label)*, ***chestnuts*** *(off-label)*, ***garlic*** *(off-label)*, ***grass seed crops***, ***hazel nuts*** *(off-label)*, ***leeks*** *(off-label)*, ***pears*** *(off-label)*, ***poppies for morphine production*** *(off-label)*, ***salad onions*** *(off-label)*, ***shallots*** *(off-label)*, ***spelt***, ***sweetcorn*** *(off-label)*, ***walnuts*** *(off-label)* [1, 9]; ***canary seed*** *(off-label)*, ***farm forestry*** *(off-label)*, ***forest nurseries*** *(off-label)*, ***game cover*** *(off-label)*, ***miscanthus*** *(off-label)*, ***newly sown grass leys*** *(off-label)*, ***ornamental plant production*** *(off-label)*, ***poppies*** *(off-label)*, ***poppies grown for seed production*** *(off-label)* [9]; ***durum wheat*** [1-7, 9-11]; ***forage maize*** [1-8, 10, 11]; ***grassland*** [1-10]; ***permanent grassland, rotational grass*** [11]; ***rye*** [10]; ***spring barley, spring oats, spring wheat, triticale, winter barley, winter oats, winter wheat*** [1-11]; ***spring rye*** [1-7, 9, 11]; ***winter rye*** [1-9, 11]
- Cleavers in ***almonds*** *(off-label)*, ***apples*** *(off-label)*, ***bulb onions*** *(off-label)*, ***chestnuts*** *(off-label)*, ***garlic*** *(off-label)*, ***grass seed crops***, ***hazel nuts*** *(off-label)*, ***leeks*** *(off-label)*, ***pears*** *(off-label)*, ***poppies for morphine production*** *(off-label)*, ***salad onions*** *(off-label)*, ***shallots*** *(off-label)*, ***spelt***, ***sweetcorn*** *(off-label)*, ***walnuts*** *(off-label)* [1, 9]; ***canary seed*** *(off-label)*, ***farm forestry*** *(off-label)*, ***forest nurseries*** *(off-label)*, ***game cover*** *(off-label)*, ***miscanthus*** *(off-label)*, ***newly sown grass leys*** *(off-label)*, ***ornamental plant production*** *(off-label)*, ***poppies*** *(off-label)*, ***poppies grown for seed production*** *(off-label)* [9]; ***durum wheat*** [1-7, 9-11]; ***forage maize*** [1-8, 11]; ***grassland*** [1-9]; ***permanent grassland, rotational grass*** [11]; ***rye*** [10]; ***spring barley, spring oats, spring wheat, triticale, winter barley, winter oats, winter wheat*** [1-11]; ***spring rye*** [1-7, 9, 11]; ***winter rye*** [1-9, 11]
- Corn spurrey in ***bulb onions*** *(off-label)*, ***garlic*** *(off-label)*, ***leeks*** *(off-label)*, ***salad onions*** *(off-label)*, ***shallots*** *(off-label)* [1, 9]; ***newly sown grass leys*** *(off-label)* [9]
- Dead nettle in ***poppies*** *(off-label)*, ***poppies grown for seed production*** *(off-label)* [9]
- Docks in ***durum wheat*** [2-7, 10, 11]; ***forage maize, winter rye*** [2-8, 11]; ***grassland*** [2-7]; ***permanent grassland, rotational grass*** [11]; ***rye*** [10]; ***spring barley, spring oats, spring wheat, triticale, winter barley, winter oats, winter wheat*** [2-8, 10, 11]; ***spring rye*** [2-7, 11]
- Forget-me-not in ***bulb onions*** *(off-label)*, ***garlic*** *(off-label)*, ***grass seed crops***, ***leeks*** *(off-label)*, ***salad onions*** *(off-label)*, ***shallots*** *(off-label)*, ***spelt*** [1, 9]; ***durum wheat*** [1-7, 9-11]; ***forage maize*** [1-8, 11]; ***grassland*** [1-9]; ***newly sown grass leys*** *(off-label)*, ***poppies*** *(off-label)*, ***poppies grown for seed production*** *(off-label)* [9]; ***permanent grassland, rotational grass*** [11]; ***rye*** [10]; ***spring barley, spring oats, spring wheat, triticale, winter barley, winter oats, winter wheat*** [1-11]; ***spring rye*** [1-7, 9, 11]; ***winter rye*** [1-9, 11]
- Fumitory in ***bulb onions*** *(off-label)*, ***garlic*** *(off-label)*, ***leeks*** *(off-label)*, ***newly sown grass leys*** *(off-label)*, ***poppies*** *(off-label)*, ***poppies grown for seed production*** *(off-label)*, ***salad onions*** *(off-label)*, ***shallots*** *(off-label)* [9]
- Groundsel in ***bulb onions*** *(off-label)*, ***garlic*** *(off-label)*, ***leeks*** *(off-label)*, ***newly sown grass leys*** *(off-label)*, ***poppies*** *(off-label)*, ***poppies grown for seed production*** *(off-label)*, ***salad onions*** *(off-label)*, ***shallots*** *(off-label)* [9]
- Hemp-nettle in ***almonds*** *(off-label)*, ***apples*** *(off-label)*, ***bulb onions*** *(off-label)*, ***chestnuts*** *(off-label)*, ***garlic*** *(off-label)*, ***grass seed crops***, ***hazel nuts*** *(off-label)*, ***leeks*** *(off-label)*, ***pears*** *(off-label)*, ***salad onions*** *(off-label)*, ***shallots*** *(off-label)*, ***spelt***, ***sweetcorn*** *(off-label)*, ***walnuts*** *(off-label)* [1, 9]; ***durum wheat*** [1-7, 9-11]; ***farm forestry*** *(off-label)*, ***forest nurseries*** *(off-label)*, ***game cover*** *(off-label)*, ***miscanthus*** *(off-label)*, ***newly sown grass leys*** *(off-label)*, ***ornamental plant production*** *(off-label)*, ***poppies*** *(off-label)*, ***poppies grown for seed production*** *(off-label)* [9]; ***forage maize*** [1-8, 11]; ***grassland*** [1-9]; ***permanent grassland, rotational grass*** [11]; ***rye*** [10]; ***spring barley, spring oats, spring wheat, triticale, winter barley, winter oats, winter wheat*** [1-11]; ***spring rye*** [1-7, 9, 11]; ***winter rye*** [1-9, 11]

SECTION 2

SEE SECTION 3 FOR PRODUCTS ALSO REGISTERED

- Ivy-leaved speedwell in *poppies* (off-label), **poppies grown for seed production** (off-label) [9]
- Knotgrass in **bulb onions** (off-label), **garlic** (off-label), **leeks** (off-label), **newly sown grass leys** (off-label), **poppies** (off-label), **poppies grown for seed production** (off-label), **salad onions** (off-label), **shallots** (off-label) [9]
- Mayweeds in *poppies* (off-label), **poppies grown for seed production** (off-label) [9]
- Pale persicaria in **bulb onions** (off-label), **garlic** (off-label), **leeks** (off-label), **salad onions** (off-label), **shallots** (off-label) [1, 9]; **newly sown grass leys** (off-label), **poppies** (off-label), **poppies grown for seed production** (off-label) [9]
- Poppies in *poppies* (off-label), **poppies grown for seed production** (off-label) [9]; **poppies for morphine production** (off-label) [1, 9]
- Red dead-nettle in **bulb onions** (off-label), **garlic** (off-label), **leeks** (off-label), **salad onions** (off-label), **shallots** (off-label) [1, 9]; **newly sown grass leys** (off-label) [9]
- Redshank in **bulb onions** (off-label), **garlic** (off-label), **leeks** (off-label), **salad onions** (off-label), **shallots** (off-label) [1, 9]; **newly sown grass leys** (off-label), **poppies** (off-label), **poppies grown for seed production** (off-label) [9]
- Speedwells in **bulb onions** (off-label), **garlic** (off-label), **leeks** (off-label), **newly sown grass leys** (off-label), **salad onions** (off-label), **shallots** (off-label) [9]
- Volunteer potatoes in **bulb onions** (off-label), **farm forestry** (off-label), **forest nurseries** (off-label), **game cover** (off-label), **garlic** (off-label), **leeks** (off-label), **miscanthus** (off-label), **newly sown grass leys** (off-label), **ornamental plant production** (off-label), **poppies** (off-label), **poppies grown for seed production** (off-label), **salad onions** (off-label), **shallots** (off-label) [9]; **poppies for morphine production** (off-label) [1]; **spring barley, spring oats, spring wheat** [10]; **sweetcorn** (off-label) [1, 9]; **winter barley, winter wheat** [2-8, 10, 11]

Extension of Authorisation for Minor Use (EAMUs)
- *almonds* 20171043 [1], 20170227 [9]
- *apples* 20171043 [1], 20170227 [9]
- *bulb onions* 20171042 [1], 20142851 [9]
- *canary seed* 20171133 [9]
- *chestnuts* 20171043 [1], 20170227 [9]
- *farm forestry* 20171268 [9]
- *forest nurseries* 20171268 [9]
- *game cover* 20171268 [9]
- *garlic* 20171042 [1], 20142851 [9]
- *hazel nuts* 20171043 [1], 20170227 [9]
- *leeks* 20171042 [1], 20142851 [9]
- *miscanthus* 20171268 [9]
- *newly sown grass leys* 20142851 [9]
- *ornamental plant production* 20171268 [9]
- *pears* 20171043 [1], 20170227 [9]
- *poppies* 20180145 [9]
- *poppies for morphine production* 20171040 [1], 20161040 [9]
- *poppies grown for seed production* 20180145 [9]
- *salad onions* 20171042 [1], 20142851 [9]
- *shallots* 20171042 [1], 20142851 [9]
- *sweetcorn* 20171041 [1], 20170226 [9]
- *walnuts* 20171043 [1], 20170227 [9]

Approval information
- Fluroxypyr included in Annex I under EC Regulation 1107/2009
- Accepted by BBPA for use on malting barley

Efficacy guidance
- Best results achieved under good growing conditions in a strongly competing crop
- A number of tank mixtures with other herbicides are recommended for use in autumn and spring to extend range of species controlled. See label for details
- Spray is rainfast in 1 h

Restrictions

- Maximum number of treatments 1 per crop or yr or maximum total dose equivalent to one full dose treatment
- Do not apply in any tank-mix on triticale or forage maize
- Do not use on crops undersown with clovers or other legumes
- Do not treat crops suffering stress caused by any factor
- Do not roll or harrow for 7 d before or after treatment
- Do not spray if frost imminent
- Straw from treated crops must not be returned directly to the soil but must be removed and used only for livestock bedding
- A maximum total dose of 0.75 l/ha must be observed for applications made to cereals between crop emergence in the year of planting and 1st February in the year of harvest [8]

Crop-specific information

- Latest use: before flag leaf sheath opening (GS 47) for winter wheat and barley; before flag leaf sheath extending (GS 41) for spring wheat and barley; before second node detectable (GS 32) for oats, rye, triticale and durum wheat; before 7 leaves unfolded and before buttress roots appear for maize
- Apply to new leys from 3 expanded leaf stage
- Timing varies in tank mixtures. See label for details
- Crops undersown with grass may be sprayed provided grasses are tillering

Following crops guidance

- Clovers, peas, beans and other legumes must not be sown for 12 mth following treatment at the highest dose
- In the event of crop failure, spring cereals, spring oilseed rape, maize, onions, poppies and new leys may be sown 5 weeks after application with no requirements for soil cultivation [9]

Environmental safety

- Dangerous for the environment
- Very toxic to aquatic organisms
- Flammable
- Keep livestock out of treated areas for at least 3 d following treatment and until poisonous weeds, such as ragwort, have died down and become unpalatable
- Wash spray equipment thoroughly with water and detergent immediately after use. Traces of product can damage susceptible plants sprayed later
- LERAP Category B [1-5, 8, 9]

Hazard classification and safety precautions

Hazard Harmful [2-5, 7, 8, 10, 11]; Flammable [2-5, 7, 8, 11]; Dangerous for the environment [1-5, 7-11]; Flammable liquid and vapour [2, 5-7, 10]; Very toxic to aquatic organisms [3, 4, 8]

Transport code 3 [1-3, 5, 7-9, 11]; 9 [4, 6, 10]

Packaging group III

UN Number 1993 [1-5, 7-9, 11]; 3082 [6, 10]

Risk phrases H304, H336 [2-8, 10, 11]; H315 [2-7, 10]; H317 [1, 2, 5-7, 9, 10]; H318 [2, 5, 11]; H319 [1, 3, 4, 6-10]; H335 [6-8, 10]

Operator protection A, H [1-10]; C [1-6, 8, 9]; U04c, U12 [1, 9]; U05a [1-5, 7, 9]; U08, U19a [1-11]; U11 [1-5, 9]; U14 [2-5, 7, 10]; U20b [1-9, 11]

Environmental protection E06a [2] (3 d); E07a [3-6, 8, 10, 11]; E07c [7] (14 d); E15a [6, 8, 11]; E15b [1-5, 7, 9, 10]; E16a [1-5, 8, 9]; E34 [1-5, 7, 9, 11]; E38 [2-8, 10, 11]; H410 [2-8, 10]; H411 [1, 9]; H412 [11]

Storage and disposal D01, D02 [1-5, 7, 9]; D05 [1-5, 7, 9, 11]; D09a [1-11]; D10a [7]; D10b [1-6, 8-11]; D12a [2-8, 10, 11]; D12b [1, 9]

Treated seed S06a [10]

Medical advice M03 [2-5, 7]; M05b [2-8, 10, 11]

SEE SECTION 3 FOR PRODUCTS ALSO REGISTERED

223   fluroxypyr + halauxifen-methyl

A herbicide mixture for use in cereals
HRAC mode of action code: O

Products

1	Pixxaro EC	Dow (Corteva)	280:12 g/l	EC	17545
2	Whorl	Dow (Corteva)	280:12 g/l	EC	17819

Uses

- Chickweed in *durum wheat, rye, spelt, spring barley, spring wheat, triticale, winter barley, winter wheat*
- Cleavers in *durum wheat, rye, spelt, spring barley, spring wheat, triticale, winter barley, winter wheat*
- Crane's-bill in *durum wheat, rye, spelt, spring barley, spring wheat, triticale, winter barley, winter wheat*
- Fat hen in *durum wheat, rye, spelt, spring barley, spring wheat, triticale, winter barley, winter wheat*
- Fumitory in *durum wheat, rye, spelt, spring barley, spring wheat, triticale, winter barley, winter wheat*
- Poppies in *durum wheat, rye, spelt, spring barley, spring wheat, triticale, winter barley, winter wheat*

Approval information

- Fluroxypyr and halauxifen-methyl included in Annex I under EC Regulation 1107/2009

Efficacy guidance

- Rainfast one hour after application
- Addition of an adjuvant gives improved reliability against poppy, chickweed and volunteer potatoes

Restrictions

- The total amount of halauxifen-methyl applied to a winter cereal must not exceed 13.5 g.a.e/ha per season: 7.5 g.a.e/ha in the autumn followed by 6 g.a.e/ha in the spring, with a minimal interval of 3 months between both applications of products which contain halauxifen-methyl

Following crops guidance

- After application of 0.5 l/ha up to BBCH 45, white mustard, pea, winter oilseed rape, phacelia, ryegrass, winter wheat and winter barley may be sown in the autumn of the same year after the cereal harvest; spring wheat, spring barley, spring oat, ryegrass, maize, sorghum, spring oilseed rape, sugar beet, potato, field bean, tomato, sunflower, onion, clover, and pea may be sown in the spring of the year after the cereal harvest. Ploughing is recommended prior to drilling lucerne, field bean, soybean, clover, maize or broad bean.
- In the event of crop failure spring wheat, spring barley, spring oats and ryegrass may be sown one month after application with no cultivation. Maize, spring oilseed rape, peas, broad beans and field beans may be sown two months after application and ploughing. Three months after application sorghum may be sown.

Environmental safety

- LERAP Category B

Hazard classification and safety precautions

Transport code 9
Packaging group III
UN Number 3082
Risk phrases H317, H319, H335
Operator protection A, C, H; U05a, U08, U19a, U20c
Environmental protection E15b, E16a, E34, H410
Storage and disposal D09a, D10c

FOR FULL CONDITIONS OF USE ALWAYS READ THE PRODUCT LABEL

224   fluroxypyr + metsulfuron-methyl

A foliar-applied herbicide mixture for use in wheat and barley
HRAC mode of action code: O + B

Products

Croupier OD	Certis	225:9 g/l	OD	18457

Uses

- Annual dicotyledons in **durum wheat**, **spring barley**, **spring wheat**, **winter barley**, **winter wheat**

Approval information

- Fluroxypyr and metsulfuron-methyl included in Annex I under EC Regulation 1107/2009

Efficacy guidance

- See label for ALS herbicides allowed in mixture
- Weed control may be reduced in dry soil conditions

Restrictions

- Do not apply to any cereal crop in tank mixture or sequence with any product containing ALS inhibiting or sulfonylurea herbicides, except as directed for specified products.
- Apply to dry foliage, when rain is not imminent.
- Transient chlorosis may occur following treatment during periods of rapid crop growth.
- Do not apply within 7 days of rolling the crop.
- Consult contract agents before using on crops grown for seed.
- For winter cereals, do not apply before March 15 in the year of harvest

Following crops guidance

- Only cereals, oilseed rape, field beans or grass may be sown in the same calendar year following a treated cereal crop.
- If a crop fails for any reason, only sow wheat within 3 months of the date of treatment.
- Only wheat, barley, rye and triticale can follow a cereal crop treated with a tank mixture of and amidosulfuron e.g. MAPP 16490 or 16491.

Environmental safety

- LERAP Category B

Hazard classification and safety precautions

  Hazard Irritant, Dangerous for the environment
  Transport code 9
  Packaging group III
  UN Number 3077
  Operator protection A, C, H; U05a, U09a, U19a, U20b
  Environmental protection E15b, E16a, E16b, E34, E38, H410
  Storage and disposal D01, D02, D09a, D11a, D12a
  Medical advice M05a

225   fluroxypyr + metsulfuron-methyl + thifensulfuron-methyl

A herbicide mixture for cereals
HRAC mode of action code: O + B + B

See also fluroxypyr + metsulfuron-methyl
*   thifensulfuron-methyl*

Products

Provalia LQM	FMC Agro	135:5:30 g/l	OD	18798

Uses

- Annual dicotyledons in **winter barley**, **winter wheat**
- Charlock in **winter barley**, **winter wheat**

SEE SECTION 3 FOR PRODUCTS ALSO REGISTERED

SECTION 2

- Chickweed in *winter barley, winter wheat*
- Cleavers in *winter barley, winter wheat*
- Field pansy in *winter barley, winter wheat*
- Fumitory in *winter barley, winter wheat*
- Mayweeds in *winter barley, winter wheat*
- Poppies in *winter barley, winter wheat*
- Red dead-nettle in *winter barley, winter wheat*
- Shepherd's purse in *winter barley, winter wheat*

Approval information
- Fluroxypyr, metsulfuron-methyl and thifensulfuron-methyl included in Annex I under EC Regulation 1107/2009

Efficacy guidance
- Apply in a minimum of 150 litres of water per hectare.

Restrictions
- Do not apply to any crop suffering from stress as a result of drought, waterlogging, low temperatures, pest or disease attack, nutrient or lime deficiency or other factors reducing crop growth.
- Do not use on cereal crops undersown with grasses, clover or other legumes or any other broad-leaved crop.
- Take special care to avoid damage by drift onto broad-leaved plants outside the target area, or onto ponds, waterways or ditches. Thorough cleansing of equipment is also very important.
- Do not apply within 7 days of rolling the crop.

Following crops guidance
- Cereals, oilseed rape, field beans or grass may be sown in the same calendar year as harvest of a cereal crop. In case of crop failure for any reason, sow only wheat within three months of application. Before sowing, soil should be ploughed and cultivated to a depth of at least 15cm.

Environmental safety
- LERAP Category B

Hazard classification and safety precautions
Transport code 9
Packaging group III
UN Number 3082
Risk phrases H317
Operator protection A, H; U05b, U09a, U11, U20a
Environmental protection E15b, E16a, H410
Storage and disposal D01, D02, D09a, D10c, D12a

226   fluroxypyr + triclopyr

A foliar acting herbicide for docks in grassland
HRAC mode of action code: O + O

See also triclopyr

Products

1 Doxstar Pro	Dow (Corteva)	150:150 g/l	EC	15664
2 Pivotal	Dow (Corteva)	150:150 g/l	EC	16943

Uses
- Docks in *amenity grassland, grassland* [2]; *established grassland* [1]

Approval information
- Fluroxypyr and triclopyr included in Annex I under EC Regulation 1107/2009

Efficacy guidance
- Seedling docks in established grass only are controlled up to 50 mm diameter. Apply in spring or autumn or, at lower dose, in spring and autumn on docks up to 200 mm. A second application in the subsequent yr may be needed
- Allow 2-3 wk after cutting or grazing to allow sufficient regrowth of docks to occur before spraying
- Control may be reduced if rain falls within 2 h of application
- To allow maximum translocation to the roots of docks do not cut grass for 28 d after spraying
- [1] available as a twin pack with Thistlex as Pas.Tor for control of docks, nettles and thistles in grassland

Restrictions
- Maximum total dose equivalent to one full dose treatment
- Do not roll or harrow for 10 d before or 7 d after spraying
- Do not spray in drought, very hot or very cold weather

Crop-specific information
- Latest use: 7 d before grazing or harvest of grass
- Grass less than one yr old may be treated at half dose from the third leaf visible stage
- Clover will be killed or severely checked by treatment

Following crops guidance
- Do not sow kale, turnips, swedes or grass mixtures containing clover by direct drilling or minimum cultivation techniques within 6 wk of application

Environmental safety
- Dangerous for the environment
- Toxic to aquatic organisms
- Keep livestock out of treated areas for at least 7 d following treatment and until poisonous weeds, such as ragwort, have died down and become unpalatable
- Do not allow drift to come into contact with crops, amenity plantings, gardens, ponds, lakes or watercourses
- Wash spray equipment thoroughly with water and detergent immediately after use. Traces of product can damage susceptible plants sprayed later

Hazard classification and safety precautions
Hazard Irritant, Dangerous for the environment
Transport code 3
Packaging group III
UN Number 1993
Risk phrases H317
Operator protection A; U14
Environmental protection E38, H410
Storage and disposal D01, D02, D05, D12a
Medical advice M05a

227 flutolanil

An carboxamide fungicide for treatment of potato seed tubers
FRAC mode of action code: 7

Products
1 Rhino	Certis	460 g/l	SC	14311
2 Rhino DS	Certis	6% w/w	DS	12763

Uses
- Black scurf in **potatoes** (tuber treatment)
- Stem canker in **potatoes** (tuber treatment)

Approval information
- Flutolanil included in Annex I under EC Regulation 1107/2009

SEE SECTION 3 FOR PRODUCTS ALSO REGISTERED

Efficacy guidance

- Apply to clean tubers before chitting, prior to planting, or at planting
- Apply flowable concentrate through canopied, hydraulic or spinning disc equipment (with or without electrostatics) mounted on a rolling conveyor or table [1]
- Flowable concentrate may be diluted with water up to 2.0 l per tonne to improve tuber coverage. Disease in areas not covered by spray will not be controlled [1]
- Dry powder may be applied via an on-planter applicator [2]

Restrictions

- Maximum number of treatments 1 per batch of seed tubers
- Check with processor before use on crops for processing

Crop-specific information

- Latest use: at planting
- Seed tubers should be of good quality and free from bacterial rots, physical damage or virus infection, and should not be sprouted to such an extent that mechanical damage to the shoots will occur during treatment or planting

Environmental safety

- Harmful to aquatic organisms

Hazard classification and safety precautions

Hazard Irritant
Transport code 9 [1]
Packaging group III [1]
UN Number 3082 [1]; N/C [2]
Risk phrases H317 [1]; H319, R53a [2]
Operator protection A, H [1, 2]; C [2]; U04a, U05a, U19a, U20a [1, 2]; U14 [1]
Environmental protection E15a, E34, E38 [1, 2]; H411 [1]; H412 [2]
Storage and disposal D01, D02, D09a, D12a [1, 2]; D05, D10a [1]; D11a [2]
Treated seed S01, S03, S04a, S05 [1, 2]; S02 [2]
Medical advice M03

228 flutriafol

A broad-spectrum triazole fungicide for cereals
FRAC mode of action code: 3

See also fludioxonil + flutriafol

Products

1	Consul	FMC Agro	125 g/l	SC	18805
2	Pointer	FMC Agro	125 g/l	SC	12975

Uses

- Brown rust in *spring barley, winter barley, winter wheat*
- Powdery mildew in *spring barley, winter barley, winter wheat*
- Rhynchosporium in *spring barley, winter barley*
- Septoria leaf blotch in *winter wheat*
- Yellow rust in *spring barley, winter barley, winter wheat*

Approval information

- Flutriafol included in Annex I under EC Regulation 1107/2009
- Accepted by BBPA for use on malting barley

Efficacy guidance

- Best results obtained from treatment in early stages of disease development. See label for detailed guidance on spray timing for specific diseases and the need for repeat treatments
- Tank mix options available on the label to broaden activity spectrum
- Good spray coverage essential for optimum performance

FOR FULL CONDITIONS OF USE ALWAYS READ THE PRODUCT LABEL

- Flutriafol is a DMI fungicide. Resistance to some DMI fungicides has been identified in Septoria leaf blotch which may seriously affect performance of some products. For further advice contact a specialist advisor and visit the Fungicide Resistance Action Group (FRAG)-UK website

Restrictions
- Maximum number of treatments 2 per crop (including other products containing flutriafol)

Crop-specific information
- Latest use: before early grain milky ripe stage (GS 73)
- Flag leaf tip scorch on wheat caused by stress may be increased by fungicide treatment

Environmental safety
- Harmful to aquatic organisms
- Harmful to fish or other aquatic life. Do not contaminate surface waters or ditches with chemical or used container

Hazard classification and safety precautions
Hazard Harmful
UN Number N/C
Risk phrases H317
Operator protection A, H; U05a, U09a, U19a, U20b
Environmental protection E13c, E38, H411
Storage and disposal D01, D02, D09a, D10c

229 fluxapyroxad

Also known as Xemium, it is an SDHI fungicide for use in cereals.
FRAC mode of action code: 7

See also difenoconazole + fluxapyroxad
epoxiconazole + fluxapyroxad
epoxiconazole + fluxapyroxad + pyraclostrobin
fluxapyroxad + metconazole

Products

1	Allstar	BASF	300 g/l	SC	18138
2	Bugle	BASF	59.4 g/l	EC	17821
3	Imtrex	BASF	62.5 g/l	EC	17108
4	Sercadis	BASF	300 g/l	SC	17776

Uses
- Brown rust in **durum wheat**, **rye**, **spring barley**, **spring wheat**, **triticale**, **winter barley**, **winter wheat** [2, 3]
- Crown rust in **spring oats** [2]; **winter oats** [2, 3]
- Disease control in **potatoes** [1]
- Net blotch in **spring barley**, **winter barley** [2, 3]
- Powdery mildew in **apples**, **nectarines** (off-label), **peaches** (off-label), **pears**, **wine grapes** (off-label) [4]; **durum wheat** (reduction), **rye** (reduction), **spring wheat** (reduction), **triticale** (reduction), **winter barley** (reduction), **winter oats** (reduction), **winter wheat** (reduction) [3]
- Rhynchosporium in **rye**, **spring barley**, **winter barley** [2, 3]
- Scab in **apples**, **nectarines** (off-label), **peaches** (off-label), **pears** [4]
- Septoria leaf blotch in **durum wheat**, **spring wheat**, **triticale**, **winter wheat** [2, 3]
- Tan spot in **durum wheat** (reduction), **spring wheat** (reduction), **winter wheat** (reduction) [3]
- Yellow rust in **durum wheat** (moderate control), **rye** (moderate control), **spring barley** (moderate control), **spring wheat** (moderate control), **triticale** (moderate control), **winter barley** (moderate control), **winter wheat** (moderate control) [2, 3]

Extension of Authorisation for Minor Use (EAMUs)
- **nectarines** 20172430 [4]
- **peaches** 20172430 [4]
- **wine grapes** 20180205 [4]

SEE SECTION 3 FOR PRODUCTS ALSO REGISTERED

Approval information

- Fluxapyroxad included in Annex I under EC Regulation 1107/2009
- Accepted by BBPA for use on malting barley up to GS 45 only (max. dose, 1 litre/ha) [2, 3]

Restrictions

- Do not apply more that two foliar applications of products containing SDHI fungicides to any cereal crop.
- Do not apply by hand-held equipment

Following crops guidance

- Only cereals, cabbages, carrots, chicory, clover, dwarf french beans, field beans, leeks, lettuce, linseed, maize, oats, oilseed rape, onions, peas, potatoes, radishes, ryegrass, soyabean, spinach, sunflowers or sugar beet may be sown as following crops after treatment

Environmental safety

- Buffer zone requirement 15 m [4]
- LERAP Category B [4]

Hazard classification and safety precautions

Hazard Harmful, Dangerous for the environment [2-4]; Harmful if inhaled [2, 3]; Very toxic to aquatic organisms [1, 4]
Transport code 9
Packaging group III
UN Number 3082
Risk phrases H319 [2, 3]; H351 [1-4]
Operator protection A, H; U05a, U20b [2-4]; U23a [1]
Environmental protection E15b, E34, E38 [2-4]; E16a [4]; H410 [1, 4]; H411 [2, 3]
Storage and disposal D01, D02, D05, D09a, D10c, D12a [2-4]
Medical advice M05a [2-4]

230　fluxapyroxad + metconazole

An SDHI and triazole mixture for disease control in cereals
FRAC mode of action code: 3 + 7

Products

1	Clayton Tardis	Clayton	62.5:45 g/l	EC	17459
2	Librax	BASF	62.5:45 g/l	EC	17107
3	Wolverine	Dow (Corteva)	62.5:45 g/l	EC	17729

Uses

- Brown rust in ***durum wheat, rye, spring barley, spring wheat, triticale, winter barley, winter wheat***
- Cladosporium in ***durum wheat*** *(moderate control),* ***spring wheat*** *(moderate control),* ***winter wheat*** *(moderate control)*
- Disease control in ***durum wheat, rye, spring barley, spring wheat, triticale, winter barley, winter wheat***
- Eyespot in ***durum wheat*** *(good reduction),* ***rye*** *(good reduction),* ***spring wheat*** *(good reduction),* ***triticale*** *(good reduction),* ***winter wheat*** *(good reduction)*
- Fusarium ear blight in ***durum wheat*** *(good reduction),* ***spring wheat*** *(good reduction),* ***winter wheat*** *(good reduction)*
- Net blotch in ***spring barley, winter barley***
- Powdery mildew in ***durum wheat*** *(moderate control),* ***rye*** *(moderate control),* ***spring barley*** *(moderate control),* ***spring wheat*** *(moderate control),* ***triticale*** *(moderate control),* ***winter barley*** *(moderate control),* ***winter wheat*** *(moderate control)*
- Ramularia leaf spots in ***spring barley, winter barley***
- Rhynchosporium in ***rye, spring barley, winter barley***
- Septoria leaf blotch in ***durum wheat, spring wheat, triticale, winter wheat***
- Tan spot in ***durum wheat*** *(good reduction),* ***spring wheat*** *(good reduction),* ***winter wheat*** *(good reduction)*

FOR FULL CONDITIONS OF USE ALWAYS READ THE PRODUCT LABEL

- Yellow rust in **durum wheat**, **rye**, **spring barley**, **spring wheat**, **triticale**, **winter barley**, **winter wheat**

Approval information
- Fluxapyroxad and metconazole included in Annex I under EC Regulation 1107/2009

Efficacy guidance
- Metconazole is a DMI fungicide. Resistance to some DMI fungicides has been identified in Septoria leaf blotch which may seriously affect performance of some products. For further advice contact a specialist advisor and visit the Fungicide Resistance Action Group (FRAG)-UK website

Restrictions
- Do not apply more that two foliar applications of products containing SDHI fungicides to any cereal crop.

Following crops guidance
- Only cereals, cabbages, carrots, clover, dwarf french beans, field beans, lettuce, maize, oats, oilseed rape, onions, peas, potatoes, ryegrass, sunflowers or sugar beet may be sown as following crops after treatment

Environmental safety
- LERAP Category B

Hazard classification and safety precautions
Hazard Harmful, Dangerous for the environment, Harmful if inhaled, Very toxic to aquatic organisms
Transport code 9
Packaging group III
UN Number 3082
Risk phrases H317, H319, H351, H361
Operator protection A, C, H; U05a, U20b
Environmental protection E15b, E16a, E34, E38, H410
Storage and disposal D01, D02, D09a, D10c, D12a
Medical advice M03

231   fluxapyroxad + pyraclostrobin

A fungicide mixture for disease control in cereals
FRAC mode of action code: 7 + 11

Products
1	Priaxor EC	BASF	75:150 g/l	EC	17371
2	Serpent	BASF	75:150 g/l	EC	17427

Uses
- Brown rust in **durum wheat**, **rye**, **spring wheat**, **triticale**, **winter wheat** [1]; **spring barley**, **winter barley** [1, 2]
- Crown rust in **spring oats**, **winter oats** [1]
- Glume blotch in **durum wheat** *(moderate control)*, **spring wheat** *(moderate control)*, **winter wheat** *(moderate control)* [1]
- Net blotch in **spring barley**, **winter barley** [1, 2]
- Powdery mildew in **durum wheat** *(moderate control)*, **rye** *(moderate control)*, **spring oats** *(moderate control)*, **spring wheat** *(moderate control)*, **triticale** *(moderate control)*, **winter oats** *(moderate control)*, **winter wheat** *(moderate control)* [1]; **spring barley** *(moderate control)*, **winter barley** *(moderate control)* [1, 2]
- Ramularia leaf spots in **spring barley**, **winter barley** [1, 2]
- Rhynchosporium in **rye** [1]; **spring barley**, **winter barley** [1, 2]
- Septoria leaf blotch in **durum wheat**, **spring wheat**, **triticale**, **winter wheat** [1]
- Tan spot in **durum wheat**, **spring wheat**, **winter wheat** [1]

SEE SECTION 3 FOR PRODUCTS ALSO REGISTERED

- Yellow rust in *durum wheat*, *rye*, *spring wheat*, *triticale*, *winter wheat* [1]; *spring barley*, *winter barley* [1, 2]

Approval information
- Fluxapyroxad and pyraclostrobin included in Annex I under EC Regulation 1107/2009

Environmental safety
- LERAP Category B

Hazard classification and safety precautions
   Hazard Harmful if swallowed, Harmful if inhaled, Very toxic to aquatic organisms
   Transport code 9
   Packaging group III
   UN Number 3082
   Risk phrases H351
   Operator protection A, H; U05a, U20a
   Environmental protection E15b, E16a, H410
   Storage and disposal D01, D02, D10c

232   folpet

A multi-site protectant fungicide for disease control in wheat and barley
FRAC mode of action code: M4

See also epoxiconazole + folpet

Products

1	Arizona	Adama	500 g/l	SC	15318
2	Clayton Canyon	Clayton	500 g/l	SC	18414
3	Phoenix	Adama	500 g/l	SC	15259

Uses
- Rhynchosporium in *spring barley* (reduction), *winter barley* (reduction)
- Septoria leaf blotch in *spring wheat* (reduction), *winter wheat* (reduction)

Approval information
- Folpet included in Annex I under EC Regulation 1107/2009
- Accepted by BBPA for use on cereals

Efficacy guidance
- Folpet is a protectant fungicide and so the first application must be made before the disease becomes established in the crop. A second application should be timed to protect new growth.
- Folpet only has contact activity and so good coverage of the target foliage is essential for good activity.

Environmental safety
- Buffer zone requirement 20 m
- LERAP Category B

Hazard classification and safety precautions
   Hazard Harmful, Dangerous for the environment [1-3]; Very toxic to aquatic organisms [1, 3]
   Transport code 9
   Packaging group III
   UN Number 3082
   Risk phrases H351 [1, 3]; R40, R50, R53a [2]
   Operator protection A, H; U05a, U08, U11, U14, U15, U20a
   Environmental protection E15b, E16a, E34, E38 [1-3]; H411 [1, 3]
   Storage and disposal D01, D02, D09a, D12a
   Medical advice M03

233   foramsulfuron + iodosulfuron-methyl-sodium

A sulfonylurea herbicide mixture for weed control in forage and grain maize.
HRAC mode of action code: B

See also iodosulfuron-methyl-sodium

Products

Maister WG	Bayer CropScience	0.3:0.01% w/w	SG	16116

Uses

- Annual dicotyledons in *forage maize*, *grain maize*
- Annual meadow grass in *forage maize*, *grain maize*
- Black nightshade in *forage maize*, *grain maize*
- Chickweed in *forage maize*, *grain maize*
- Cockspur grass in *forage maize*, *grain maize*
- Couch in *forage maize*, *grain maize*
- Fat hen in *forage maize*, *grain maize*
- Knotgrass in *forage maize*, *grain maize*
- Mayweeds in *forage maize*, *grain maize*
- Shepherd's purse in *forage maize*, *grain maize*

Approval information

- Foramsulfuron and iodosulfuron-methyl-sodium included in Annex 1 under EC Regulation 1107/2009

Efficacy guidance

- Do not spray if rain is imminent or if temperature is above 25°C.
- Do not use in a non-CRD approved tank-mixture or sequence with another ALS or sulfonyl urea herbicide.
- Do not use on crops suffering from stress.
- Do not use on crops undersown with grass or a broad-leaved crop.

Environmental safety

- Buffer zone requirement 11m [1]
- LERAP Category B

Hazard classification and safety precautions

Hazard Irritant, Dangerous for the environment, Very toxic to aquatic organisms
Transport code 9
Packaging group III
UN Number 3077
Risk phrases H317
Operator protection A, D, H; U05a, U14
Environmental protection E15b, E16a, E38, H410
Storage and disposal D01, D02, D10a, D12a

234   foramsulfuron + thiencarbazone-methyl

Herbicide mixture for weed control in ALS-resistant sugar beet varieties
HRAC mode of action code: B

Products

Conviso One	Bayer CropScience	50:30 g/l	OD	19036

Uses

- Annual dicotyledons in *sugar beet*
- Volunteer sugar beet in *sugar beet*

Approval information

- Foramsulfuron and thiencarbazone-methyl included in Annex I under EC Regulation 1107/2009

SEE SECTION 3 FOR PRODUCTS ALSO REGISTERED

SECTION 2

Efficacy guidance
- Application when the temperature is above 25°C under conditions of high light intensity and low water supply may cause phytotoxic symptoms to the crop - wait until the cool of the evening.

Restrictions
- Must only be used on Conviso Smart sugar beet hybrids

Following crops guidance
- After normal harvest plough or cultivate to 20cms before sowing wheat in the same year as application or before sowing wheat, barley, maize, sunflowers, peas or ryegrass the following spring.
- In the event of crop failure, Conviso Smart beet can be re-sown or, after ploughing or cultivating to 20 cms, maize may be sown 1 month after application and wheat may be sown 4 months after application.

Environmental safety
- Buffer zone requirement 10m

Hazard classification and safety precautions
Hazard Harmful if inhaled, Very toxic to aquatic organisms
Risk phrases H304, H315, H317, H318, H351
Operator protection A, C, H; U05a, U12, U14, U20a
Environmental protection H410
Storage and disposal D05, D09a, D10b
Medical advice M03

235   fosetyl-aluminium + propamocarb hydrochloride

A systemic and protectant fungicide mixture for use in horticulture
FRAC mode of action code: 33 + 28

See also propamocarb hydrochloride

Products
1 Pan Cradle	Pan Agriculture	310:530 g/l	SL	15923
2 Previcur Energy	Bayer CropScience	310:530 g/l	SL	15367

Uses
- Damping off in *herbs (see appendix 6)* (off-label), *protected aubergines* (off-label), *protected chilli peppers* (off-label), *protected courgettes* (off-label), *protected forest nurseries* (off-label), *protected gherkins* (off-label), *protected hops* (off-label), *protected marrows* (off-label), *protected pumpkins* (off-label), *protected soft fruit* (off-label), *protected squashes* (off-label), *protected sweet peppers* (off-label), *protected top fruit* (off-label), *protected watermelon* (off-label) [2]; *protected broccoli, protected brussels sprouts, protected cabbages, protected calabrese, protected cauliflowers, protected chinese cabbage, protected collards, protected kale, protected radishes* [1, 2]
- Downy mildew in *herbs (see appendix 6)* (off-label), *ornamental plant production* (off-label), *protected forest nurseries* (off-label), *protected hops* (off-label), *protected ornamentals* (off-label), *protected soft fruit* (off-label), *protected top fruit* (off-label), *spinach* (off-label) [2]; *lettuce, protected broccoli, protected brussels sprouts, protected cabbages, protected calabrese, protected cauliflowers, protected chinese cabbage, protected collards, protected kale, protected lettuce, protected melons, protected radishes* [1, 2]
- Pythium in *lettuce, protected cucumbers, protected lettuce, protected tomatoes* [1, 2]; *protected ornamentals* (off-label) [2]

Extension of Authorisation for Minor Use (EAMUs)
- *herbs (see appendix 6) 20130983* [2]
- *ornamental plant production 20131845* [2]
- *protected aubergines 20111553* [2]
- *protected chilli peppers 20111553* [2]
- *protected courgettes 20111556* [2]

FOR FULL CONDITIONS OF USE ALWAYS READ THE PRODUCT LABEL

- *protected forest nurseries 20122045* [2]
- *protected gherkins 20111556* [2]
- *protected hops 20122045* [2]
- *protected marrows 20111556* [2]
- *protected ornamentals 20111557* [2], *20131845* [2]
- *protected pumpkins 20111556* [2]
- *protected soft fruit 20122045* [2]
- *protected squashes 20111556* [2]
- *protected sweet peppers 20111553* [2]
- *protected top fruit 20122045* [2]
- *protected watermelon 20111556* [2]
- *spinach 20112452* [2]

Approval information
- Fosetyl-aluminium and propamocarb hydrochloride included in Annex I under EC Regulation 1107/2009

Hazard classification and safety precautions
Hazard Irritant
UN Number N/C
Risk phrases H317
Operator protection A, H, M; U04a, U05a, U08, U14, U19a, U20b
Environmental protection E15a
Storage and disposal D09a, D11a
Medical advice M03

236   fosthiazate

An organophosphorus contact nematicide for potatoes
IRAC mode of action code: 1B

Products
Nemathorin 10G	Syngenta	10% w/w	FG	11003

Uses
- Nematodes in *hops* *(off-label)*, *ornamental plant production* *(off-label)*, *soft fruit* *(off-label)*
- Potato cyst nematode in *potatoes*
- Spraing vectors in *potatoes* *(reduction)*
- Wireworm in *potatoes* *(reduction)*

Extension of Authorisation for Minor Use (EAMUs)
- *hops 20082912*
- *ornamental plant production 20082912*
- *soft fruit 20082912*

Approval information
- Fosthiazate included in Annex I under EC Regulation 1107/2009
- In 2006 CRD required that all products containing this active ingredient should carry the following warning in the main area of the container label: "Fosthiazate is an anticholinesterase organophosphate. Handle with care"

Efficacy guidance
- Application best achieved using equipment such as Horstine Farmery Microband Applicator, Matco or Stocks Micrometer applicators together with a rear mounted powered rotary cultivator
- Granules must not become wet or damp before use.
- For optimum efficacy incorporate evenly in to top 20 cm of soil.
- Apply as close as possible to time of planting.

Restrictions
- Contains an anticholinesterase organophosphorus compound. Do not use if under medical advice not to work with such compounds

SEE SECTION 3 FOR PRODUCTS ALSO REGISTERED

- Maximum number of treatments 1 per crop
- Product must only be applied using tractor-mounted/drawn direct placement machinery. Do not use air assisted broadcast machinery other than that referenced on the product label
- Do not allow granules to stand overnight in the application hopper
- Do not apply more than once every four years on the same area of land
- Do not use on crops to be harvested less than 17 wk after treatment
- Consult before using on crops intended for processing

Crop-specific information
- Latest use: at planting
- HI 17 wk for potatoes

Environmental safety
- Dangerous for the environment
- Toxic to aquatic organisms
- Dangerous to game, wild birds and animals
- Dangerous to livestock. Keep all livestock out of treated areas for at least 13 wk
- Incorporation to 10-15 cm and ridging up of treated soil must be carried out immediately after application. Powered rotary cultivators are preferred implements for incorporation but discs, power, spring tine or Dutch harrows may be used provided two passes are made at right angles
- To protect birds and wild mammals remove spillages
- Failure completely to bury granules immediately after application is hazardous to wildlife
- To protect groundwater do not apply any product containing fosthiazate more than once every four yr

Hazard classification and safety precautions
Hazard Harmful, Dangerous for the environment, Toxic if swallowed, Very toxic to aquatic organisms
Transport code 9
Packaging group III
UN Number 3077
Risk phrases H317
Operator protection A, E, G, H, K, M; U02a, U04a, U05a, U09a, U13, U14, U19a, U20a
Environmental protection E06b (13 wk); E15b, E34, E38, H410
Storage and disposal D01, D02, D05, D09a, D11a, D14
Medical advice M01, M03, M05a

237   garlic extract

A naturally occuring nematicide and animal repellent

Products

NEMguard DE	Certis	45% w/w	GR	16749

Uses
- Free-living nematodes in **fodder beet** (off-label), **red beet** (off-label)
- Nematodes in **carrots, parsnips**
- Root-knot nematodes in **bulb onions** (off-label), **garlic** (off-label), **leeks** (off-label), **shallots** (off-label)
- Stem and bulb nematodes in **bulb onions** (off-label), **garlic** (off-label), **leeks** (off-label), **shallots** (off-label)

Extension of Authorisation for Minor Use (EAMUs)
- **bulb onions** 20151838
- **fodder beet** 20151841
- **garlic** 20151838
- **leeks** 20151838
- **red beet** 20151841
- **shallots** 20151838

Approval information
- Garlic extract included in Annex 1 under EC Regulation 1107/2009

Hazard classification and safety precautions

Hazard Irritant, Dangerous for the environment
UN Number N/C
Risk phrases H315, H320
Operator protection A, C, H; U11, U14
Environmental protection E38, H411

238 gibberellins

A plant growth regulator for use in top fruit and grassland

Products

1	Florgib Tablet	Fine	20.4% w/w	ST	18285
2	Gibb 3	Belchim	10% w/w	TB	17013
3	Gibb Plus	Belchim	10 g/l	SL	17251
4	Novagib	Fine	10 g/l	SL	18341
5	Regulex 10 SG	Sumitomo	10% w/w	SG	17158
6	Smartgrass	Sumitomo	40% w/w	SG	15628

Uses
- Fruit retention in *cherries* (off-label) [1]
- Growth regulation in *cherries* (off-label), *protected cherries* (off-label), *protected rhubarb* (off-label) [2]; *grassland* [6]; *ornamental plant production, pears, rhubarb, wine grapes* [1]
- Improved germination in *nothofagus* [5]
- Increasing fruit set in *pears* [2]
- Reducing fruit russeting in *apples* [3, 4]; *pears* [4]; *pears* (off-label), *quinces* (off-label) [3]
- Russet reduction in *apples, pears* [5]

Extension of Authorisation for Minor Use (EAMUs)
- *cherries* 20192566 [1], 20192041 [2]
- *pears* 20171152 [3]
- *protected cherries* 20192041 [2]
- *protected rhubarb* 20162836 [2]
- *quinces* 20171152 [3]

Approval information
- Gibberellins included in Annex 1 under EC Regulation 1107/2009

Efficacy guidance
- For optimum results on apples spray under humid, slow drying conditions and ensure good spray cover
- Treat apples and pears immediately after completion of petal fall and repeat as directed on the label
- Fruit set in pears can be improved when blossom is spare, setting is poor or where frost has killed many flowers [2]
- The maximum concentration must not exceed 6 g of product per 1 litre water [1]
- Tablets must be placed whole into the sprayer. They must not be broken-up or crumbled [1]

Restrictions
- Maximum total dose varies with crop and product. Check label
- Prepared spray solutions are unstable. Do not leave in the sprayer during meal breaks or overnight
- Return bloom may be reduced in yr following treatment
- Consult processor before treating crops grown for processing [2]
- Avoid storage at temperatures above 32°C [2]
- Do not exceed 2.5 g per 5 litres when used as a seed soak in Nothofagus [5]
- Do not apply by hand-held equipment [4]
- Livestock must be kept out of treated areas for at least 14 days after treatment

SEE SECTION 3 FOR PRODUCTS ALSO REGISTERED

Crop-specific information

- HI zero for apples [3, 5]
- Apply to apples at completion of petal fall and repeat 3 or 4 times at 7-10 d intervals. Number of sprays and spray interval depend on weather conditions and dose (see labels)
- Good results achieved on apple varieties Cox's Orange Pippin, Discovery, Golden Delicious and Karmijn. For other cultivars test on a small number of trees
- Split treatment on pears allows second spray to be omitted if pollinating conditions become very favourable [2]
- Pear variety Conference usually responds well to treatment; Doyenne du Comice can be variable [2]

Hazard classification and safety precautions

UN Number N/C
Operator protection A [1]; U05a [5]; U08 [2]; U20b [4]; U20c [2, 3, 5]
Environmental protection E15a [2-5]
Storage and disposal D01 [4, 5]; D02 [5]; D03, D05, D07 [4]; D09a [2-5]; D10b [3-5]

239   Gliocladium catenulatum

A fungus that acts as an antagonist against other fungi
FRAC mode of action code: 44

Products

Prestop	ICL (Everris) Ltd	32% w/w	WP	17223

Uses

- Botrytis in **almonds** *(off-label)*, **apples** *(off-label)*, **apricots** *(off-label)*, **asparagus** *(off-label)*, **baby leaf crops** *(off-label)*, **beans without pods (fresh)** *(off-label)*, **bilberries** *(off-label)*, **blackcurrants** *(off-label)*, **blueberries** *(off-label)*, **broad beans** *(off-label)*, **broccoli** *(off-label)*, **brussels sprouts** *(off-label)*, **bulb vegetables** *(off-label)*, **cabbages** *(off-label)*, **calabrese** *(off-label)*, **cane fruit** *(off-label)*, **cardoons** *(off-label)*, **carrots** *(off-label)*, **cauliflowers** *(off-label)*, **celeriac** *(off-label)*, **celery (outdoor)** *(off-label)*, **celery leaves** *(off-label)*, **cherries** *(off-label)*, **chestnuts** *(off-label)*, **chicory root** *(off-label)*, **chives** *(off-label)*, **choi sum** *(off-label)*, **collards** *(off-label)*, **cranberries** *(off-label)*, **cress** *(off-label)*, **dwarf beans** *(off-label)*, **edible flowers** *(off-label)*, **edible podded peas** *(off-label)*, **endives** *(off-label)*, **florence fennel** *(off-label)*, **fruiting vegetables** *(off-label)*, **globe artichoke** *(off-label)*, **gooseberries** *(off-label)*, **hazel nuts** *(off-label)*, **herbs (see appendix 6)** *(off-label)*, **hops** *(off-label)*, **horseradish** *(off-label)*, **jerusalem artichokes** *(off-label)*, **kale** *(off-label)*, **kohlrabi** *(off-label)*, **lamb's lettuce** *(off-label)*, **leeks** *(off-label)*, **lentils** *(off-label)*, **lettuce** *(off-label)*, **medlar** *(off-label)*, **nectarines** *(off-label)*, **oriental cabbage** *(off-label)*, **ornamental plant production** *(off-label)*, **parsley** *(off-label)*, **parsley root** *(off-label)*, **parsnips** *(off-label)*, **peaches** *(off-label)*, **pears** *(off-label)*, **plums** *(off-label)*, **quinces** *(off-label)*, **radishes** *(off-label)*, **red beet** *(off-label)*, **redcurrants** *(off-label)*, **rhubarb** *(off-label)*, **runner beans** *(off-label)*, **salsify** *(off-label)*, **seakale** *(off-label)*, **soya beans** *(off-label)*, **spinach** *(off-label)*, **spinach beet** *(off-label)*, **swedes** *(off-label)*, **table grapes** *(off-label)*, **turnips** *(off-label)*, **vining peas** *(off-label)*, **walnuts** *(off-label)*, **wine grapes** *(off-label)*
- Didymella in **all protected edible crops** *(moderate control)*, **all protected non-edible crops** *(moderate control)*, **strawberries** *(moderate control)*
- Fusarium in **all protected edible crops** *(moderate control)*, **all protected non-edible crops** *(moderate control)*, **almonds** *(off-label)*, **apples** *(off-label)*, **apricots** *(off-label)*, **asparagus** *(off-label)*, **baby leaf crops** *(off-label)*, **beans without pods (fresh)** *(off-label)*, **bilberries** *(off-label)*, **blackcurrants** *(off-label)*, **blueberries** *(off-label)*, **broad beans** *(off-label)*, **broccoli** *(off-label)*, **brussels sprouts** *(off-label)*, **bulb vegetables** *(off-label)*, **cabbages** *(off-label)*, **calabrese** *(off-label)*, **cane fruit** *(off-label)*, **cardoons** *(off-label)*, **carrots** *(off-label)*, **cauliflowers** *(off-label)*, **celeriac** *(off-label)*, **celery (outdoor)** *(off-label)*, **celery leaves** *(off-label)*, **cherries** *(off-label)*, **chestnuts** *(off-label)*, **chicory root** *(off-label)*, **chives** *(off-label)*, **choi sum** *(off-label)*, **collards** *(off-label)*, **cranberries** *(off-label)*, **cress** *(off-label)*, **dwarf beans** *(off-label)*, **edible flowers** *(off-label)*, **edible podded peas** *(off-label)*, **endives** *(off-label)*, **florence fennel** *(off-label)*, **fruiting vegetables** *(off-label)*, **globe artichoke** *(off-label)*, **gooseberries** *(off-label)*, **hazel nuts** *(off-label)*, **herbs (see appendix 6)** *(off-label)*, **hops** *(off-label)*, **horseradish** *(off-label)*, **jerusalem**

artichokes *(off-label)*, **kale** *(off-label)*, **kohlrabi** *(off-label)*, **lamb's lettuce** *(off-label)*, **leeks** *(off-label)*, **lentils** *(off-label)*, **lettuce** *(off-label)*, **medlar** *(off-label)*, **nectarines** *(off-label)*, **oriental cabbage** *(off-label)*, **ornamental plant production** *(off-label)*, **parsley** *(off-label)*, **parsley root** *(off-label)*, **parsnips** *(off-label)*, **peaches** *(off-label)*, **pears** *(off-label)*, **plums** *(off-label)*, **quinces** *(off-label)*, **radishes** *(off-label)*, **red beet** *(off-label)*, **redcurrants** *(off-label)*, **rhubarb** *(off-label)*, **runner beans** *(off-label)*, **salsify** *(off-label)*, **seakale** *(off-label)*, **soya beans** *(off-label)*, **spinach** *(off-label)*, **spinach beet** *(off-label)*, **strawberries** *(moderate control)*, **swedes** *(off-label)*, **table grapes** *(off-label)*, **turnips** *(off-label)*, **vining peas** *(off-label)*, **walnuts** *(off-label)*, **wine grapes** *(off-label)*

- Phytophthora in **all protected edible crops** *(moderate control)*, **all protected non-edible crops** *(moderate control)*, **almonds** *(off-label)*, **apples** *(off-label)*, **apricots** *(off-label)*, **asparagus** *(off-label)*, **baby leaf crops** *(off-label)*, **beans without pods (fresh)** *(off-label)*, **bilberries** *(off-label)*, **blackcurrants** *(off-label)*, **blueberries** *(off-label)*, **broad beans** *(off-label)*, **broccoli** *(off-label)*, **brussels sprouts** *(off-label)*, **bulb vegetables** *(off-label)*, **cabbages** *(off-label)*, **calabrese** *(off-label)*, **cane fruit** *(off-label)*, **cardoons** *(off-label)*, **carrots** *(off-label)*, **cauliflowers** *(off-label)*, **celeriac** *(off-label)*, **celery (outdoor)** *(off-label)*, **celery leaves** *(off-label)*, **cherries** *(off-label)*, **chestnuts** *(off-label)*, **chicory root** *(off-label)*, **chives** *(off-label)*, **choi sum** *(off-label)*, **collards** *(off-label)*, **cranberries** *(off-label)*, **cress** *(off-label)*, **dwarf beans** *(off-label)*, **edible flowers** *(off-label)*, **edible podded peas** *(off-label)*, **endives** *(off-label)*, **florence fennel** *(off-label)*, **fruiting vegetables** *(off-label)*, **globe artichoke** *(off-label)*, **gooseberries** *(off-label)*, **hazel nuts** *(off-label)*, **herbs (see appendix 6)** *(off-label)*, **hops** *(off-label)*, **horseradish** *(off-label)*, **jerusalem artichokes** *(off-label)*, **kale** *(off-label)*, **kohlrabi** *(off-label)*, **lamb's lettuce** *(off-label)*, **leeks** *(off-label)*, **lentils** *(off-label)*, **lettuce** *(off-label)*, **medlar** *(off-label)*, **nectarines** *(off-label)*, **oriental cabbage** *(off-label)*, **ornamental plant production** *(off-label)*, **parsley** *(off-label)*, **parsley root** *(off-label)*, **parsnips** *(off-label)*, **peaches** *(off-label)*, **pears** *(off-label)*, **plums** *(off-label)*, **quinces** *(off-label)*, **radishes** *(off-label)*, **red beet** *(off-label)*, **redcurrants** *(off-label)*, **rhubarb** *(off-label)*, **runner beans** *(off-label)*, **salsify** *(off-label)*, **seakale** *(off-label)*, **soya beans** *(off-label)*, **spinach** *(off-label)*, **spinach beet** *(off-label)*, **strawberries** *(moderate control)*, **swedes** *(off-label)*, **table grapes** *(off-label)*, **turnips** *(off-label)*, **vining peas** *(off-label)*, **walnuts** *(off-label)*, **wine grapes** *(off-label)*
- Pythium in **all protected edible crops** *(moderate control)*, **all protected non-edible crops** *(moderate control)*, **strawberries** *(moderate control)*
- Rhizoctonia in **all protected edible crops** *(moderate control)*, **all protected non-edible crops** *(moderate control)*, **almonds** *(off-label)*, **apples** *(off-label)*, **apricots** *(off-label)*, **asparagus** *(off-label)*, **baby leaf crops** *(off-label)*, **beans without pods (fresh)** *(off-label)*, **bilberries** *(off-label)*, **blackcurrants** *(off-label)*, **blueberries** *(off-label)*, **broad beans** *(off-label)*, **broccoli** *(off-label)*, **brussels sprouts** *(off-label)*, **bulb vegetables** *(off-label)*, **cabbages** *(off-label)*, **calabrese** *(off-label)*, **cane fruit** *(off-label)*, **cardoons** *(off-label)*, **carrots** *(off-label)*, **cauliflowers** *(off-label)*, **celeriac** *(off-label)*, **celery (outdoor)** *(off-label)*, **celery leaves** *(off-label)*, **cherries** *(off-label)*, **chestnuts** *(off-label)*, **chicory root** *(off-label)*, **chives** *(off-label)*, **choi sum** *(off-label)*, **collards** *(off-label)*, **cranberries** *(off-label)*, **cress** *(off-label)*, **dwarf beans** *(off-label)*, **edible flowers** *(off-label)*, **edible podded peas** *(off-label)*, **endives** *(off-label)*, **florence fennel** *(off-label)*, **fruiting vegetables** *(off-label)*, **globe artichoke** *(off-label)*, **gooseberries** *(off-label)*, **hazel nuts** *(off-label)*, **herbs (see appendix 6)** *(off-label)*, **hops** *(off-label)*, **horseradish** *(off-label)*, **jerusalem artichokes** *(off-label)*, **kale** *(off-label)*, **kohlrabi** *(off-label)*, **lamb's lettuce** *(off-label)*, **leeks** *(off-label)*, **lentils** *(off-label)*, **lettuce** *(off-label)*, **medlar** *(off-label)*, **nectarines** *(off-label)*, **oriental cabbage** *(off-label)*, **ornamental plant production** *(off-label)*, **parsley** *(off-label)*, **parsley root** *(off-label)*, **parsnips** *(off-label)*, **peaches** *(off-label)*, **pears** *(off-label)*, **plums** *(off-label)*, **quinces** *(off-label)*, **radishes** *(off-label)*, **red beet** *(off-label)*, **redcurrants** *(off-label)*, **rhubarb** *(off-label)*, **runner beans** *(off-label)*, **salsify** *(off-label)*, **seakale** *(off-label)*, **soya beans** *(off-label)*, **spinach** *(off-label)*, **spinach beet** *(off-label)*, **strawberries** *(moderate control)*, **swedes** *(off-label)*, **table grapes** *(off-label)*, **turnips** *(off-label)*, **vining peas** *(off-label)*, **walnuts** *(off-label)*, **wine grapes** *(off-label)*

Extension of Authorisation for Minor Use (EAMUs)
- **almonds** *20182843*
- **apples** *20182843*
- **apricots** *20182843*
- **asparagus** *20182843*

SEE SECTION 3 FOR PRODUCTS ALSO REGISTERED

- **baby leaf crops** 20182843
- **beans without pods (fresh)** 20182843
- **bilberries** 20182843
- **blackcurrants** 20182843
- **blueberries** 20182843
- **broad beans** 20182843
- **broccoli** 20182843
- **brussels sprouts** 20182843
- **bulb vegetables** 20182843
- **cabbages** 20182843
- **calabrese** 20182843
- **cane fruit** 20182843
- **cardoons** 20182843
- **carrots** 20182843
- **cauliflowers** 20182843
- **celeriac** 20182843
- **celery (outdoor)** 20182843
- **celery leaves** 20182843
- **cherries** 20182843
- **chestnuts** 20182843
- **chicory root** 20182843
- **chives** 20182843
- **choi sum** 20182843
- **collards** 20182843
- **cranberries** 20182843
- **cress** 20182843
- **dwarf beans** 20182843
- **edible flowers** 20182843
- **edible podded peas** 20182843
- **endives** 20182843
- **florence fennel** 20182843
- **fruiting vegetables** 20182843
- **globe artichoke** 20182843
- **gooseberries** 20182843
- **hazel nuts** 20182843
- **herbs (see appendix 6)** 20182843
- **hops** 20182843
- **horseradish** 20182843
- **jerusalem artichokes** 20182843
- **kale** 20182843
- **kohlrabi** 20182843
- **lamb's lettuce** 20182843
- **leeks** 20182843
- **lentils** 20182843
- **lettuce** 20182843
- **medlar** 20182843
- **nectarines** 20182843
- **oriental cabbage** 20182843
- **ornamental plant production** 20182843
- **parsley** 20182843
- **parsley root** 20182843
- **parsnips** 20182843
- **peaches** 20182843
- **pears** 20182843
- **plums** 20182843
- **quinces** 20182843
- **radishes** 20182843
- **red beet** 20182843
- **redcurrants** 20182843

- **rhubarb** *20182843*
- **runner beans** *20182843*
- **salsify** *20182843*
- **seakale** *20182843*
- **soya beans** *20182843*
- **spinach** *20182843*
- **spinach beet** *20182843*
- **swedes** *20182843*
- **table grapes** *20182843*
- **turnips** *20182843*
- **vining peas** *20182843*
- **walnuts** *20182843*
- **wine grapes** *20182843*

Approval information

- Gliocladium catenulatum included in Annex 1 under EC Regulation 1107/2009

Restrictions

- When used via broadcast air-assisted sprayers, spray equipment must only be used where the operator's normal working position is within a closed cab on a tractor or on a self-propelled sprayer

Hazard classification and safety precautions

UN Number N/C
Operator protection A, D, H; U05a, U10, U14, U15, U20a
Environmental protection E15b, E34; E17a (10 m)
Storage and disposal D01, D02, D09a, D10a
Medical advice M03

240   glyphosate

A translocated non-residual glycine derivative herbicide
HRAC mode of action code: G

See also 2,4-D + glyphosate
diflufenican + glyphosate
diuron + glyphosate
flufenacet + glyphosate + metosulam
glyphosate + pyraflufen-ethyl
glyphosate + sulfosulfuron

Products

1	Amega Duo	Nufarm UK	540 g/l	SL	13358
2	Ardee XL	Barclay	360 g/l	SL	17515
3	Asteroid	FMC Agro	360 g/l	SL	18958
4	Asteroid Pro	FMC Agro	360 g/l	SL	18754
5	Asteroid Pro 450	FMC Agro	450 g/l	SL	18788
6	Azural	Monsanto	360 g/l	SL	16239
7	Barbarian XL	Barclay	360 g/l	SL	17665
8	Barclay Gallup Biograde 360	Barclay	360 g/l	SL	17612
9	Barclay Gallup Biograde Amenity	Barclay	360 g/l	SL	17674
10	Barclay Gallup Hi-Aktiv	Barclay	490 g/l	SL	17646
11	Barclay Glyde 144	Barclay	144 g/l	SL	15362
12	Clinic UP	Nufarm UK	360 g/l	SL	17893
13	Credit	Nufarm UK	540 g/l	SL	16775
14	Crestler	Nufarm UK	540 g/l	SL	16555
15	Discman Biograde	Barclay	216 g/l	SL	12856
16	Ecoplug Max	Monsanto	68 % w/w	GR	17581

SEE SECTION 3 FOR PRODUCTS ALSO REGISTERED

Products – continued

17	Envision	FMC Agro	450 g/l	SL	18770
18	Gallup Hi-Aktiv Amenity	Barclay	360 g/l	SL	17681
19	Gallup XL	Barclay	360 g/l	SL	17663
20	Garryowen XL	Barclay	360 g/l	SL	17508
21	Glyfos Dakar	FMC Agro	68% w/w	SG	13054
22	Hilite	Nomix Enviro	20.1% w/w	RH	18352
23	Landmaster 360 TF	Albaugh UK	360 g/l	SL	17683
24	Liaison	Monsanto	360 g/l	SL	17089
25	Mascot Hi-Aktiv Amenity	Rigby Taylor	490 g/l	SL	17696
26	Master Gly	Ventura	360 g/l	SL	19202
27	Mentor	Monsanto	360 g/l	SL	16508
28	Monsanto Amenity Glyphosate	Monsanto	360 g/l	SL	16382
29	Monsanto Amenity Glyphosate XL	Monsanto	360 g/l	SL	17997
30	Motif	Monsanto	360 g/l	SL	16509
31	Nomix Conqueror Amenity	Nomix Enviro	144 g/l	RC	18369
32	Proliance Quattro	Syngenta	360 g/l	SL	13670
33	Rattler	Nufarm UK	540 g/l	SL	15522
34	Rodeo	Monsanto	360 g/l	SL	16242
35	Rosate 360 TF	Albaugh UK	360 g/l	SL	17682
36	Rosate Green	Albaugh UK	360 g/l	SL	15122
37	Roundup Biactive GL	Monsanto	360 g/l	SL	17348
38	Roundup Energy	Monsanto	450 g/l	SL	12945
39	Roundup Flex	Monsanto	480 g/l	SL	15541
40	Roundup Metro XL	Monsanto	360 g/l	SL	17684
41	Roundup POWERMAX	Monsanto	72% w/w	SG	16373
42	Roundup ProActive	Monsanto	360 g/l	SL	17380
43	Roundup ProVantage	Monsanto	480 g/l	SL	15534
44	Roundup Sonic	Monsanto	450 g/l	SL	17152
45	Roundup Star	Monsanto	450 g/l	SL	15470
46	Roundup Ultimate	Monsanto	450 g/l	SL	12774
47	Roundup Vista Plus	Monsanto	450 g/l	SL	18002
48	Rustler Pro-Green	ChemSource	360 g/l	SL	15798
49	Samurai	Monsanto	360 g/l	SL	16238
50	Scorpion	Monsanto	360 g/l	SL	17516
51	Snapper	Nufarm UK	550 g/l	SL	19037
52	Surrender T/F	Agrovista	360 g/l	SL	19203
53	Tanker	Nufarm UK	540 g/l	SL	15016
54	Trustee Amenity	Barclay	450 g/l	SL	17697
55	Vesuvius Green	Ventura	360 g/l	SL	19201

Uses

- Annual and perennial weeds in *all edible crops (outdoor and protected)* [1, 12-14, 26, 53]; *all edible crops (outdoor)* [3, 12, 26, 33, 37, 39, 42, 43, 51]; *all edible crops (outdoor)* (before planting), *all non-edible crops (outdoor)* (before planting) [2, 7, 8, 10, 19-21, 23, 35, 36, 40, 48, 52, 55]; *all edible crops (stubble)*, *all non-edible crops (stubble)* [2, 7, 8, 10, 37, 42, 55]; *all edible crops (stubble)* (before planting), *all non-edible crops (stubble)* (before planting) [19, 20]; *all non-edible crops (outdoor)*, *spring barley*, *spring field beans*, *spring oats*, *spring wheat*, *winter barley*, *winter field beans*, *winter oats*, *winter wheat* [1, 3, 12-14, 26, 33, 37, 39, 42, 43, 51, 53]; *all non-edible seed crops grown outdoors* [12, 26]; *almonds* (off-label), *apples* (off-label), *cherries* (off-label), *peaches* (off-label), *pears* (off-label), *plums* (off-label), *quinces* (off-label) [3, 37, 38, 41]; *amenity vegetation* [1, 4-6, 9, 11, 13-15, 18, 22, 24, 25, 27-31, 33, 34, 36, 39-45, 47-54]; *apple orchards*, *pear orchards* [1-5, 7, 8, 10, 12-14, 17, 19-21, 23, 26, 33, 35, 36, 38-40, 43-48, 51-53, 55]; *apples*, *managed amenity turf* (pre-establishment only), *pears* [31]; *apricots* (off-label), *chestnuts* (off-label), *hazel nuts* (off-label), *walnuts* (off-label) [37, 38, 41]; *asparagus* [2, 7, 8, 10, 12, 13, 19, 20, 23, 26, 35, 39-43, 48, 52, 55]; *asparagus* (off-

label) [3, 37]; **beetroot** *(off-label)*, **carrots** *(off-label)*, **garlic** *(off-label)*, **horseradish** *(off-label)*, **leeks** *(off-label)*, **lupins** *(off-label)*, **miscanthus** *(off-label)*, **onions** *(off-label)*, **parsley root** *(off-label)*, **parsnips** *(off-label)*, **rye** *(off-label)*, **salad onions** *(off-label)*, **salsify** *(off-label)*, **shallots** *(off-label)*, **swedes** *(off-label)*, **turnips** *(off-label)* [38, 39, 41]; **bilberries** *(off-label)*, **blackcurrants** *(off-label)*, **blueberries** *(off-label)*, **cranberries** *(off-label)*, **gooseberries** *(off-label)*, **nectarines** *(off-label)*, **redcurrants** *(off-label)* [3, 38, 41]; **blackberries** *(off-label)*, **farm forestry** *(off-label)*, **game cover** *(off-label)*, **hemp** *(off-label)*, **hops** *(off-label)*, **loganberries** *(off-label)*, **medlar** *(off-label)*, **quinoa** *(off-label)*, **raspberries** *(off-label)*, **rubus hybrids** *(off-label)*, **strawberries** *(off-label)*, **woad** *(off-label)* [38, 41]; **buckwheat** *(off-label)*, **canary seed** *(off-label)*, **millet** *(off-label)*, **soft fruit** *(off-label)*, **sorghum** *(off-label)*, **top fruit** *(off-label)* [38]; **bulb onions, durum wheat, mustard** [1, 12-14, 26, 33, 39, 42, 43, 51, 53]; **cherries, plums** [1-8, 10, 12-14, 17, 19-21, 23, 24, 26-28, 30, 31, 33-36, 38-53, 55]; **christmas trees** *(off-label)* [42]; **combining peas** [1, 3, 12-14, 26, 33, 37, 39, 42, 43, 47, 51, 53]; **crab apples** *(off-label)*, **rotational grass** [37]; **damsons** [14, 17, 21, 38, 40, 46, 48, 53]; **edible podded peas** *(off-label)*, **vining peas** *(off-label)* [38, 39]; **enclosed waters** [4-6, 9, 18, 24, 25, 27, 30, 31, 34, 36, 41, 44, 45, 47, 49, 50, 54]; **farm forestry** [6, 24, 27, 30, 34, 44, 45, 47, 49, 50]; **forest** [1-15, 17-20, 22-28, 30-36, 39-45, 47-55]; **forest nurseries** [1, 28, 29, 32, 36, 51, 52]; **forestry plantations** [1, 13, 14, 33, 51, 53]; **grapevines** *(off-label)* [3]; **grassland** [1, 6, 12-14, 22, 24, 26-28, 30, 33, 34, 42, 44, 45, 47, 49-51, 53]; **green cover on land temporarily removed from production** [1, 2, 6-8, 10, 12-14, 22, 24, 26-28, 30, 33, 34, 37, 39, 41, 43, 49-51, 53, 55]; **hard surfaces** [1-13, 15, 18-20, 22-37, 39-45, 47-52, 54, 55]; **kentish cobnuts** *(off-label)* [37, 41]; **land immediately adjacent to aquatic areas** [2, 4-11, 15, 18, 23-27, 30, 31, 34-37, 40-42, 44, 45, 47, 49, 50, 54, 55]; **land not intended to bear vegetation** [17]; **leeks, sugar beet, vining peas** [1, 6, 12-14, 24, 26-28, 30, 33, 34, 39, 41-43, 49-51, 53]; **linseed** [1, 3, 6, 12-14, 24, 26-28, 30, 33, 34, 39, 41-43, 49-51, 53]; **managed amenity turf** [22]; **natural surfaces not intended to bear vegetation, permeable surfaces overlying soil** [1-15, 18-20, 22-37, 39-45, 47-55]; **oilseed rape** [6, 13, 24, 27, 28, 30, 34, 41, 49, 50]; **open waters** [6, 9, 18, 24, 25, 27, 30, 31, 34, 36, 44, 45, 47, 49, 50, 54]; **ornamental plant production** [22, 31]; **ornamental plant production** *(off-label - as a directed spray)* [12]; **permanent grassland** [37, 38, 46]; **poppies for morphine production** *(off-label)* [39]; **potatoes** [37, 39, 41-43]; **rhubarb** *(off-label)*, **vaccinium spp.** *(off-label)*, **whitecurrants** *(off-label)* [3, 41]; **salad onions, table grapes, wine grapes** [41]; **spring oilseed rape, winter oilseed rape** [1, 3, 12-14, 26, 33, 39, 42, 43, 51, 53]; **stubbles** [3, 6, 23, 24, 27, 28, 30, 34-36, 38, 40, 41, 44-50, 52]; **swedes, turnips** [1, 6, 12-14, 24, 26-28, 30, 33, 34, 37, 39, 41-43, 49-51, 53]

- Annual dicotyledons in **bulb onions** [2, 6-8, 10, 19-21, 23, 24, 27, 28, 30, 34-36, 38, 40, 41, 44-50, 52, 55]; **combining peas** [2, 7, 8, 10, 19-21, 23, 35, 36, 40, 48, 52, 55]; **durum wheat** [2, 7, 8, 10, 19-21, 23, 28, 35, 36, 38, 40, 44-48, 52, 55]; **grassland** [42, 44, 45, 47]; **leeks, linseed, mustard, spring field beans, spring oilseed rape, sugar beet, swedes, winter field beans, winter oilseed rape** [2, 7, 8, 10, 19-21, 23, 35, 36, 38, 40, 44-48, 52, 55]; **permanent grassland** [38, 46]; **potatoes** [44, 45, 47]; **spring barley, spring oats, spring wheat, winter barley, winter oats, winter wheat** [2, 7, 8, 10, 17, 19-21, 23, 35, 36, 38, 40, 44-48, 52, 55]; **turnips** [2, 7, 8, 10, 19-21, 23, 35, 36, 38, 44-47, 52, 55]; **vining peas** [2, 7, 8, 10, 19, 20, 23, 35, 36, 38, 40, 44-48, 52, 55]

- Annual grasses in **bulb onions** [6, 21, 24, 27, 28, 30, 34, 38, 41, 44-47, 49, 50]; **combining peas** [21]; **cultivated land/soil** [17]; **durum wheat** [21, 28, 38, 44-47]; **grassland** [42, 44, 45, 47]; **leeks, linseed, mustard, spring barley, spring field beans, spring oats, spring oilseed rape, spring wheat, sugar beet, swedes, turnips, winter barley, winter field beans, winter oats, winter oilseed rape, winter wheat** [21, 38, 44-47]; **permanent grassland** [38, 46]; **potatoes** [44, 45, 47]; **vining peas** [38, 44-47]

- Aquatic weeds in **enclosed waters** [6, 18, 24, 27, 30, 34, 36, 37, 39, 41-43, 49, 50, 54]; **land immediately adjacent to aquatic areas** [2, 6-8, 10, 18, 23, 24, 27, 30, 34-37, 39-43, 49, 50, 54, 55]; **open waters** [6, 18, 24, 27, 30, 34, 36, 37, 39, 42, 43, 49, 50, 54]

- Biennial weeds in **beetroot** *(off-label)*, **carrots** *(off-label)*, **garlic** *(off-label)*, **horseradish** *(off-label)*, **leeks** *(off-label)*, **onions** *(off-label)*, **parsley root** *(off-label)*, **parsnips** *(off-label)*, **salad onions** *(off-label)*, **salsify** *(off-label)*, **shallots** *(off-label)*, **swedes** *(off-label)*, **turnips** *(off-label)* [39]

- Black bent in **combining peas, linseed, spring field beans, spring oilseed rape, stubbles, winter field beans, winter oilseed rape** [17]

- Bolters in **sugar beet** *(wiper application)* [21, 38, 46]

SEE SECTION 3 FOR PRODUCTS ALSO REGISTERED

- Bracken in **forest** [2, 7, 8, 10, 17, 19, 20, 23, 35, 36, 39, 40, 43, 48, 52, 55]; **forest nurseries** [36, 52]
- Canary grass in **aquatic areas** [21]
- Chemical thinning in **farm forestry** [6, 24, 27, 30, 34, 44, 45, 47, 49, 50]; **forest** [2-8, 10, 17, 19, 20, 23, 24, 27-30, 32, 34-36, 39, 40, 42-45, 47-50, 52, 55]; **forest nurseries** [28, 32, 36, 52]
- Couch in **combining peas** [3, 6, 17, 21, 24, 27, 28, 30, 34, 38, 41, 44-46, 49, 50]; **durum wheat** [6, 21, 24, 27, 28, 30, 34, 38, 41, 44-47, 49, 50]; **linseed, spring oilseed rape, stubbles, winter oilseed rape** [3, 17, 21, 38, 44-47]; **mustard** [6, 24, 27, 28, 30, 34, 38, 41, 44-47, 49, 50]; **oats** [6, 24, 27, 28, 30, 34, 41, 49, 50]; **spring barley, spring wheat, winter barley, winter wheat** [3, 6, 21, 24, 27, 28, 30, 34, 38, 41, 44-47, 49, 50]; **spring field beans, winter field beans** [3, 6, 17, 21, 24, 27, 28, 30, 34, 38, 41, 44-47, 49, 50]; **spring oats, winter oats** [3, 21, 38, 44-47]
- Creeping bent in **aquatic areas** [21]; **combining peas, linseed, spring field beans, spring oilseed rape, stubbles, winter field beans, winter oilseed rape** [17]
- Desiccation in **amenity vegetation** [9, 36]; **buckwheat** (off-label), **canary seed** (off-label), **hemp** (off-label), **millet** (off-label), **quinoa** (off-label), **sorghum** (off-label) [38]; **combining peas** [37, 39, 43, 47]; **durum wheat, linseed, mustard, spring barley, spring oats, spring oilseed rape, spring wheat, winter barley, winter oats, winter oilseed rape, winter wheat** [39, 43]; **lupins** (off-label), **rye** (off-label) [38, 39, 41]; **poppies for morphine production** (off-label) [38, 39]; **spring field beans, winter field beans** [37, 39, 43]
- Destruction of crops in **all edible crops (outdoor and protected)** [1, 13, 53]; **all edible crops (outdoor)** [6, 12, 24, 26-28, 30, 33, 34, 37, 41, 42, 49-51]; **all edible crops (stubble), all non-edible crops (stubble)** [37, 42]; **all non-edible crops (outdoor)** [6, 13, 24, 27, 28, 30, 33, 34, 37, 41, 42, 49-51]; **all non-edible seed crops grown outdoors** [12, 26]; **amenity vegetation** [15, 42]; **grassland** [1, 13, 33, 39, 43, 51, 53]; **green cover on land temporarily removed from production** [6, 24, 27, 28, 30, 34, 37, 41, 49, 50]; **hard surfaces, natural surfaces not intended to bear vegetation, permeable surfaces overlying soil** [42]; **permanent grassland, rotational grass** [37]
- Grass weeds in **forest, hard surfaces, land immediately adjacent to aquatic areas, natural surfaces not intended to bear vegetation, permeable surfaces overlying soil** [15]
- Green cover in **land immediately adjacent to aquatic areas** [19, 20]; **land temporarily removed from production** [3, 17, 19-21, 23, 35, 36, 38, 40, 42, 44-48, 52]
- Growth suppression in **tree stumps** [12, 33, 51]
- Harvest management/desiccation in **combining peas** [1, 2, 6-8, 10, 12, 13, 19, 20, 23, 24, 26-28, 30, 33-36, 38, 40, 41, 44-46, 48-53, 55]; **durum wheat** [1, 2, 6-8, 10, 12, 13, 19-21, 23, 24, 26-28, 30, 33-36, 38, 40, 41, 44-53, 55]; **grassland** [6, 24, 27, 28, 30, 34, 49, 50]; **linseed, mustard** [1, 2, 6-8, 10, 12, 13, 19, 20, 23, 24, 26-28, 30, 33-36, 38, 40, 41, 44-53, 55]; **oats, oilseed rape** [6, 24, 27, 28, 30, 34, 41, 49, 50]; **spring barley, spring wheat, winter barley, winter wheat** [1, 2, 6-8, 10, 12, 13, 17, 19-21, 23, 24, 26-28, 30, 33-36, 38, 40, 41, 44-53, 55]; **spring field beans, winter field beans** [2, 6-8, 10, 19, 20, 23, 24, 27, 28, 30, 33-36, 38, 40, 41, 44-52, 55]; **spring oats, winter oats** [1, 2, 7, 8, 10, 12, 13, 17, 19-21, 23, 26, 33, 35, 36, 38, 40, 44-48, 51-53, 55]; **spring oilseed rape, winter oilseed rape** [1-3, 7, 8, 10, 12, 13, 17, 19-21, 23, 26, 33, 35, 36, 38, 40, 44-48, 51-53, 55]
- Heather in **forest** [2, 7, 8, 10, 17, 19, 20, 23, 35, 36, 40, 48, 52, 55]; **forest nurseries** [36, 52]
- Perennial dicotyledons in **combining peas, linseed, spring field beans, spring oilseed rape, winter field beans, winter oilseed rape** [17, 21]; **grassland** (wiper application) [44, 45, 47]; **permanent grassland** (wiper application) [38, 46]
- Perennial grasses in **aquatic areas** [4, 5, 17]; **combining peas, linseed, spring field beans, stubbles, winter field beans** [21]; **enclosed waters, land immediately adjacent to aquatic areas** [3]
- Reeds in **aquatic areas** [4, 5, 17, 21]; **enclosed waters, land immediately adjacent to aquatic areas** [3, 39, 43]; **open waters** [39, 43]
- Re-growth suppression in **tree stumps** [3, 42]
- Rhododendrons in **forest** [2, 7, 8, 10, 17, 19, 20, 23, 35, 36, 40, 48, 52, 55]; **forest nurseries** [36, 52]
- Rushes in **aquatic areas** [4, 5, 17, 21]; **enclosed waters, land immediately adjacent to aquatic areas** [3, 39, 43]; **open waters** [39, 43]
- Sedges in **aquatic areas** [4, 5, 17, 21]; **enclosed waters, land immediately adjacent to aquatic areas** [3, 39, 43]; **open waters** [39, 43]
- Sprout suppression in **tree stumps** [18, 39, 43-45, 47]

FOR FULL CONDITIONS OF USE ALWAYS READ THE PRODUCT LABEL

- Sucker control in **apple orchards, cherries, pear orchards, plums** [6, 24, 27, 28, 30, 34, 38, 41, 42, 44-47, 49, 50]; **damsons** [38, 46]
- Sucker inhibition in **enclosed waters, open waters** [16]; **forest** [4, 5]; **tree stumps** [1, 6, 13, 16, 24, 27, 28, 30, 34, 49, 50, 53]
- Sward destruction in **amenity vegetation** [18, 25, 54]; **grassland** [2, 3, 7, 8, 10, 19, 20, 23, 35, 36, 40, 42, 44, 45, 47, 52, 55]; **permanent grassland** [17, 21, 38, 46, 48]; **rotational grass** [48]
- Total vegetation control in **all edible crops (outdoor)** *(pre-sowing/planting),* **all non-edible crops (outdoor)** *(pre-sowing/planting)* [38, 44-47]
- Trees in **amenity vegetation** *(off-label),* **forest** *(off-label)* [16]
- Volunteer cereals in **bulb onions** [6, 21, 24, 27, 28, 30, 34, 38, 41, 44-47, 49, 50]; **combining peas** [21]; **cultivated land/soil** [17]; **durum wheat** [28, 38, 44-47]; **leeks, linseed, mustard, spring field beans, sugar beet, swedes, turnips, winter field beans** [21, 38, 44-47]; **potatoes** [44, 45, 47]; **spring barley, spring oats, spring oilseed rape, spring wheat, vining peas, winter barley, winter oats, winter oilseed rape, winter wheat** [38, 44-47]; **stubbles** [3, 17, 21, 38, 44-47]
- Volunteer potatoes in **stubbles** [3, 21, 38, 44-47]
- Waterlilies in **aquatic areas** [4, 5, 17, 21]; **enclosed waters, land immediately adjacent to aquatic areas** [3, 39, 43]; **open waters** [39, 43]
- Weed beet in **sugar beet** *(wiper application)* [21]
- Woody weeds in **forest** [2, 7, 8, 10, 17, 19, 20, 23, 35, 36, 40, 48, 52, 55]; **forest nurseries** [36, 52]

Extension of Authorisation for Minor Use (EAMUs)
- **almonds** *20190748* [3], *20170234* [37], *20082888* [38], *20141300* [41]
- **amenity vegetation** *20170195* [16]
- **apples** *20190748* [3], *20170234* [37], *20082888* [38], *20141300* [41]
- **apricots** *20170234* [37], *20082888* [38], *20141300* [41]
- **asparagus** *20190751* [3], *20171123* [37]
- **beetroot** *20130354* [38], *20132528* [39], *20141305* [41]
- **bilberries** *20190750* [3], *20082888* [38], *20141300* [41], *20141316* [41]
- **blackberries** *20082888* [38], *20141300* [41]
- **blackcurrants** *20190750* [3], *20082888* [38], *20141300* [41], *20141316* [41]
- **blueberries** *20190750* [3], *20082888* [38], *20141300* [41], *20141316* [41]
- **buckwheat** *20071839* [38]
- **canary seed** *20071839* [38]
- **carrots** *20130354* [38], *20132528* [39], *20141305* [41]
- **cherries** *20190748* [3], *20170234* [37], *20082888* [38], *20141300* [41]
- **chestnuts** *20170234* [37], *20082888* [38], *20141300* [41], *20141317* [41]
- **christmas trees** *20152945* [42]
- **crab apples** *20170234* [37]
- **cranberries** *20190750* [3], *20082888* [38], *20141300* [41], *20141316* [41]
- **edible podded peas** *20141672* [38], *20141671* [39]
- **farm forestry** *20082888* [38], *20141300* [41]
- **forest** *20170195* [16]
- **game cover** *20082888* [38], *20141300* [41]
- **garlic** *20130354* [38], *20132528* [39], *20141305* [41]
- **gooseberries** *20190750* [3], *20082888* [38], *20141300* [41], *20141316* [41]
- **grapevines** *20190745* [3]
- **hazel nuts** *20170234* [37], *20082888* [38], *20141300* [41], *20141317* [41]
- **hemp** *20071840* [38], *20141300* [41], *20141321* [41]
- **hops** *20082888* [38], *20141300* [41]
- **horseradish** *20130354* [38], *20132528* [39], *20141305* [41]
- **kentish cobnuts** *20170234* [37], *20141317* [41]
- **leeks** *20130354* [38], *20132528* [39], *20141305* [41]
- **loganberries** *20082888* [38], *20141300* [41]
- **lupins** *20071279* [38], *20131667* [39], *20141306* [41]
- **medlar** *20082888* [38], *20141300* [41]
- **millet** *20071839* [38]
- **miscanthus** *20082888* [38], *20132131* [39], *20141300* [41]

- **nectarines** *20190748* [3], *20082888* [38], *20141300* [41]
- **onions** *20130354* [38], *20132528* [39], *20141305* [41]
- **ornamental plant production** *(as a directed spray)* *20172202* [12]
- **parsley root** *20130354* [38], *20132528* [39], *20141305* [41]
- **parsnips** *20130354* [38], *20132528* [39], *20141305* [41]
- **peaches** *20190748* [3], *20170234* [37], *20082888* [38], *20141300* [41]
- **pears** *20190748* [3], *20170234* [37], *20082888* [38], *20141300* [41]
- **plums** *20190748* [3], *20170234* [37], *20082888* [38], *20141300* [41]
- **poppies for morphine production** *20081058* [38], *20131668* [39]
- **quinces** *20190748* [3], *20170234* [37], *20082888* [38], *20141300* [41]
- **quinoa** *20071840* [38], *20141321* [41]
- **raspberries** *20082888* [38], *20141300* [41]
- **redcurrants** *20190750* [3], *20082888* [38], *20141300* [41], *20141316* [41]
- **rhubarb** *20190749* [3], *20141320* [41]
- **rubus hybrids** *20082888* [38], *20141300* [41]
- **rye** *20111701* [38], *20131666* [39], *20141307* [41]
- **salad onions** *20130354* [38], *20132528* [39], *20141305* [41]
- **salsify** *20130354* [38], *20132528* [39], *20141305* [41]
- **shallots** *20130354* [38], *20132528* [39], *20141305* [41]
- **soft fruit** *20082888* [38]
- **sorghum** *20071839* [38]
- **strawberries** *20082888* [38], *20141300* [41]
- **swedes** *20130354* [38], *20132528* [39], *20141305* [41]
- **top fruit** *20082888* [38]
- **turnips** *20130354* [38], *20132528* [39], *20141305* [41]
- **vaccinium spp.** *20190750* [3], *20141316* [41]
- **vining peas** *20141672* [38], *20141671* [39]
- **walnuts** *20170234* [37], *20082888* [38], *20141300* [41], *20141317* [41]
- **whitecurrants** *20190750* [3], *20141316* [41]
- **woad** *20082888* [38], *20141300* [41]

Approval information
- Glyphosate included in Annex I under EC Regulation 1107/2009
- Accepted by BBPA for use on malting barley and hops

Efficacy guidance
- For best results apply to actively growing weeds with enough leaf to absorb chemical
- For most products a rainfree period of at least 6 h (preferably 24 h) should follow spraying
- Adjuvants are obligatory for some products and recommended for some uses with others. See labels
- Mixtures with other pesticides or fertilizers may lead to reduced control.
- Products are formulated as isopropylamine, ammonium, potassium, or trimesium salts of glyphosate and may vary in the details of efficacy claims. See individual product labels
- If using to treat hard surfaces, ensure spraying takes place only when weeds are actively growing (normally March to October) and is confined only to visible weeds including those in the 30cm swath covering the kerb edge and road gulley – do not overspray drains.
- For use with hand held rotary atomisers, the spray droplet spectra produced must be of a minimum Volume Median Diameter (VMD) of 200 microns [22, 31]
- With wiper application weeds should be at least 10 cm taller than crop
- Annual weed grasses should have at least 5 cm of leaf and annual broad-leaved weeds at least 2 expanded true leaves
- Perennial grass weeds should have 4-5 new leaves and be at least 10 cm long when treated. Perennial broad-leaved weeds should be treated at or near flowering but before onset of senescence
- Volunteer potatoes and polygonums are not controlled by harvest-aid rates
- Bracken must be treated at full frond expansion
- Fruit tree suckers best treated in late spring
- Chemical thinning treatment can be applied as stump spray or stem injection
- In order to allow translocation, do not cultivate before spraying and do not apply other pesticides, lime, fertilizer or farmyard manure within 5 d of treatment

FOR FULL CONDITIONS OF USE ALWAYS READ THE PRODUCT LABEL

- Recommended intervals after treatment and before cultivation vary. See labels
- When used on managed amenity turf application must only be carried out pre-establishment of the crop

Restrictions
- Maximum total dose per crop or season normally equivalent to one full dose treatment on field and edible crops and no restriction for non-crop uses. However some older labels indicate a maximum number of treatments. Check for details
- Do not treat cereals grown for seed or undersown crops
- Consult grain merchant before treating crops grown on contract or intended for malting
- Do not use treated straw as a mulch or growing medium for horticultural crops
- For use in nursery stock, shrubberies, orchards, grapevines and tree nuts care must be taken to avoid contact with the trees. Do not use in orchards established less than 2 yr and keep off low-lying branches
- Certain conifers may be sprayed overall in dormant season. See label for details
- Use a tree guard when spraying in established forestry plantations
- Do not spray root suckers in orchards in late summer or autumn
- Do not use under glass or polythene as damage to crops may result
- Do not mix, store or apply in galvanised or unlined mild steel containers or spray tanks
- Do not leave diluted chemical in spray tanks for long periods and make sure that tanks are well vented
- The maximum concentration of glyphosate in water must not exceed 0.12 ppm or such lower concentration as the appropriate regulatory body may require

Crop-specific information
- Harvest intervals: 4 wk for blackcurrants, blueberries; 14 d for linseed, oilseed rape; 8 d for mustard; 5-7 d for all other edible crops. Check label for exact details
- Latest use: for most products 2-14 d before cultivating, drilling or planting a crop in treated land; after harvest (post-leaf fall) but before bud formation in the following season for nuts and most fruit and vegetable crops; before fruit set for grapevines. See labels for details

Following crops guidance
- Decaying remains of plants killed by spraying must be dispersed before direct drilling
- Crops may be drilled 48 h after application. Trees and shrubs may be planted 7 d after application.

Environmental safety
- Products differ in their hazard and environmental safety classification. See labels
- Do not dump surplus herbicide in water or ditch bottoms or empty into drains
- Check label for maximum permitted concentration in treated water
- The Environment Agency or Local River Purification Authority must be consulted before use in or near water
- Take extreme care to avoid drift and possible damage to neighbouring crops or plants
- Treated poisonous plants must be removed before grazing or conserving
- Do not use in covered areas such as greenhouses or under polythene
- For field edge treatment direct spray away from hedge bottoms
- Some products require livestock to be excluded from treated areas and do not permit treated forage to be used for hay, silage or bedding. Check label for details

Hazard classification and safety precautions
Hazard Harmful [26, 52]; Irritant [2, 7, 12, 16, 18-20, 23, 28, 29, 33, 35, 36, 40, 51]; Dangerous for the environment [2, 6, 7, 12, 16, 18-20, 22-24, 26-31, 33-36, 39, 40, 43, 49-52]; Harmful if inhaled [40]
Transport code 9 [2, 6, 7, 14, 16, 19, 20, 22-24, 26-30, 33-36, 40, 41, 43, 49-52]
Packaging group III [2, 6, 7, 14, 16, 19, 20, 22-24, 26-30, 33-36, 40, 41, 43, 49-52]
UN Number 3077 [16, 41]; 3082 [2, 6, 7, 14, 19, 20, 22-24, 26-30, 33-36, 40, 43, 49-52]; N/C [1, 3-5, 8-13, 15, 17, 18, 21, 25, 31, 32, 37-39, 42, 44-48, 53-55]
Risk phrases H318 [14, 33, 40, 51]; H319 [2, 6, 20, 24, 27, 29, 30, 34, 44, 45, 47, 50]; R36 [49]; R41 [28]; R51 [25, 28, 49]; R52 [21, 36, 38, 44, 46]; R53a [21, 25, 28, 36, 38, 39, 44, 46, 49]
Operator protection A [1-15, 17-55]; C [2-10, 12, 16, 17, 19, 20, 23-30, 33-35, 38, 40, 42, 44-52, 55]; D [3-5, 17]; H [1-5, 7-11, 13-15, 17-23, 25, 31, 33, 35, 37, 39-43, 48, 51-55]; M [1-11, 13-15,

17-24, 27-31, 33, 35, 37-55]; U02a [1, 8-15, 18, 21, 22, 26, 28, 29, 31, 33, 41, 48, 51, 53-55]; U05a [1, 2, 7, 13, 14, 16, 19, 20, 22, 23, 25, 31, 33, 35, 36, 40, 41, 51-53]; U08 [8-10, 16, 18, 21, 41, 48, 54, 55]; U09a [11, 15, 22, 31]; U11 [2, 6, 7, 12, 16, 19-21, 23-30, 34-36, 40, 49, 50, 52]; U12, U15 [33, 51]; U19a [8-11, 15, 18, 21, 22, 31, 41, 48, 54, 55]; U20a [2, 7, 15, 16, 19, 20, 23, 25, 35-37, 40, 42, 52]; U20b [1, 3-6, 8-14, 17, 18, 21, 22, 24, 26-34, 38, 39, 41, 43-51, 53-55]
Environmental protection E06a [32] (2 weeks); E07d [53]; E13c [10, 22, 31, 54]; E15a [3-5, 17, 37]; E15b [1, 2, 6-9, 11-16, 18-24, 26-36, 38-53, 55]; E19a [3-5, 10, 17, 54]; E34, E36a [13, 14, 53]; E38 [2, 6, 7, 12, 16, 19-21, 23, 24, 26-30, 33-36, 39-41, 43, 49-52]; H410 [22, 33]; H411 [2, 6, 14, 20, 24, 27, 30, 40]; H412 [13, 16, 45, 51]; H413 [1, 12, 43, 53]
Storage and disposal D01 [1-9, 11, 13-16, 18-25, 27, 30, 31, 33-53, 55]; D02 [1, 2, 6-9, 11, 13-16, 18-25, 27, 30, 31, 33-53, 55]; D05 [1, 2, 6-16, 18-20, 22-24, 26-31, 33-40, 43-55]; D09a [2-12, 15-24, 26-52, 54, 55]; D10a [8, 9, 18, 48, 55]; D10b [3-5, 10, 17, 54]; D10c [1, 2, 6, 7, 12-14, 19, 20, 23, 24, 26-30, 32-40, 42-47, 49-53]; D11a [11, 15, 21, 22, 31, 41]; D12a [2, 6, 7, 12, 16, 19-24, 26-31, 34-36, 39, 40, 43, 49, 50, 52]; D14 [13, 14, 53]
Medical advice M05a [6, 12, 24-30, 34, 49, 50]

241 glyphosate + pyraflufen-ethyl

A herbicide mixture for non-selective weed control around amenity plants and on areas not intended to bear vegetation
HRAC mode of action code: G + E

See also glyphosate
pyraflufen-ethyl

Products

Hammer	ICL (Everris) Ltd	240:0.67 g/l	SE	16498

Uses
- Annual and perennial weeds in *amenity vegetation, hard surfaces, land not intended to bear vegetation, natural surfaces not intended to bear vegetation, ornamental plant production, permeable surfaces overlying soil*

Approval information
- Glyphosate and pyraflufen-ethyl included in Annex I under EC Regulation 1107/2009

Hazard classification and safety precautions
Hazard Dangerous for the environment
Transport code 9
Packaging group III
UN Number 3082
Risk phrases R51, R53a
Operator protection A, H, M; U02a, U08, U20b
Environmental protection E15b, E38
Storage and disposal D01, D02, D05, D09a, D10c, D12a

242 glyphosate + sulfosulfuron

A translocated non-residual glycine derivative herbicide + a sulfonyl urea herbicide
HRAC mode of action code: G + B

See also sulfosulfuron

Products

Nomix Dual	Nomix Enviro	120:2.22 g/l	RC	18351

Uses
- Annual and perennial weeds in *amenity vegetation, hard surfaces, natural surfaces not intended to bear vegetation, permeable surfaces overlying soil*

FOR FULL CONDITIONS OF USE ALWAYS READ THE PRODUCT LABEL

Approval information
- Glyphosate and sulfosulfuron included in Annex I under EC Regulation 1107/2009

Efficacy guidance
- For use with hand held rotary atomisers, the spray droplet spectra produced must be of a minimum Volume Median Diameter (VMD) of 200 microns [1]

Hazard classification and safety precautions
Hazard Dangerous for the environment
Transport code 9
Packaging group III
UN Number 3082
Operator protection A, H, M; U02a, U09a, U19a, U20b
Environmental protection E15b, H410
Storage and disposal D01, D05, D09a, D11a

243   halauxifen-methyl

A herbicide for use in cereal crops only available in mixtures
HRAC mode of action code: O

See also aminopyralid + halauxifen-methyl
clopyralid + halauxifen-methyl
florasulam + halauxifen-methyl
fluroxypyr + halauxifen-methyl

244   halauxifen-methyl + picloram

A broad-leaved contact herbicide mixture for weed control in oilseed rape
HRAC mode of action code: O + O

Products
Belkar	Dow (Corteva)	10:48 g/l	EC	18615

Uses
- Annual dicotyledons in **winter oilseed rape**
- Cleavers in **winter oilseed rape**
- Crane's-bill in **winter oilseed rape**
- Fumitory in **winter oilseed rape**
- Poppies in **winter oilseed rape**
- Shepherd's purse in **winter oilseed rape**

Approval information
- Halauxifen-methyl and picloram included in Annex 1 under EC Regulation 1107/2009

Efficacy guidance
- Use in cold or warm (from 2 to 25°C), humid or dry conditions. In severe drought conditions there can be a slight reduction in efficacy.
- Rainfast 1 hour after application

Restrictions
- DO NOT apply 0.25 l/ha before 1st September or BBCH 12 (2 true leaves)
- DO NOT apply 0.5 l/ha before 15th September or BBCH 16 (6 true leaves)

Following crops guidance
- After normal harvest or in the case of crop failure wheat, barley, oats, maize or oilseed rape can be planted 120 days after application. Allow at least 12 months before planting other crops. Ploughing or thorough cultivation is necessary before planting legumes such as peas or beans.

Environmental safety
- LERAP Category B

SEE SECTION 3 FOR PRODUCTS ALSO REGISTERED

Hazard classification and safety precautions
    Hazard Very toxic to aquatic organisms
    Transport code 9
    Packaging group III
    UN Number 3082
    Risk phrases H319, H335
    Operator protection A, C, H; U05a, U08, U10, U19a, U20a
    Environmental protection E15b, E16a, E23, E34, H410
    Storage and disposal D09a, D10c

245   hymexazol

A systemic heteroaromatic fungicide for pelleting sugar beet seed
FRAC mode of action code: 32

Products

Tachigaren 70 WP	Sumi Agro	70% w/w	WP	17977

Uses
- Black leg in *sugar beet* (seed treatment)

Approval information
- Hymexazol included in Annex 1 under EC Regulation 1107/2009

Efficacy guidance
- Incorporate into pelleted seed using suitable seed pelleting machinery

Restrictions
- Maximum number of treatments 1 per batch of seed
- Do not use treated seed as food or feed

Crop-specific information
- Latest use: before planting sugar beet seed

Environmental safety
- Harmful to aquatic organisms
- Harmful to fish or other aquatic life. Do not contaminate surface waters or ditches with chemical or used container
- Treated seed harmful to game and wildlife

Hazard classification and safety precautions
    Hazard Flammable solid
    Transport code 3
    Packaging group III
    UN Number 1325
    Risk phrases H317, H318, H361
    Operator protection A, C, F; U05a, U11, U20b
    Environmental protection E03, E13c, H411
    Storage and disposal D01, D02, D09a, D11a
    Treated seed S01, S02, S03, S04a, S05

246   imazalil

A systemic and protectant imidazole fungicide
FRAC mode of action code: 3

See also guazatine + imazalil

Products

Gavel	Certis	100 g/l	SL	17586

Uses
- Dry rot in *seed potatoes*
- Gangrene in *seed potatoes*
- Silver scurf in *seed potatoes*
- Skin spot in *seed potatoes*

Approval information
- Imazalil included in Annex I under EC Regulation 1107/2009

Efficacy guidance
- For best control of skin and wound diseases of ware potatoes treat as soon as possible after harvest, preferably within 7-10 d, before any wounds have healed

Restrictions
- Maximum number of treatments 1 per batch of ware tubers; 2 per batch of seed tubers
- Consult processor before treating potatoes for processing

Crop-specific information
- Latest use: during storage and before chitting for seed potatoes
- Apply to clean soil-free potatoes post-harvest before putting into store, or at first grading. A further treatment may be applied in early spring before planting
- Apply through canopied hydraulic or spinning disc equipment preferably diluted with up to two litres water per tonne of potatoes to obtain maximum skin cover and penetration
- Use on ware potatoes subject to discharges of imazalil from potato washing plants being within emission limits set by the UK monitoring authority

Environmental safety
- Dangerous for the environment
- Toxic to aquatic organisms
- Do not empty into drains
- Personal protective equipment requirements may vary for each pack size. Check label

Hazard classification and safety precautions
Hazard Irritant
UN Number N/C
Risk phrases H318, H351
Operator protection A, C, H; U04a, U05a, U11, U14, U19a, U20a
Environmental protection E15a, E19b, E34, H410
Storage and disposal D01, D02, D05, D09a, D10c, D12a
Treated seed S01, S02, S03, S04a, S05, S06a

247 imazalil + ipconazole

An imidazole + triazole seed treatment for barley
FRAC mode of action code: 3

Products

Rancona i-MIX	Arysta	50:20 g/l	MS	15574

Uses
- Bunt in *spring wheat*, *winter wheat*
- Fusarium foot rot and seedling blight in *spring barley* (useful protection), *spring wheat* (useful protection), *winter barley* (useful protection), *winter wheat* (useful protection)
- Leaf stripe in *spring barley*
- Loose smut in *spring barley*, *winter barley*
- Microdochium nivale in *spring barley* (useful protection), *spring wheat* (useful protection), *winter barley* (useful protection), *winter wheat* (useful protection)

Approval information
- Imazalil and ipconazole included in Annex I under EC Regulation 1107/2009

SEE SECTION 3 FOR PRODUCTS ALSO REGISTERED

Restrictions
- Treated seed must not be sown between 1 February and 31 August

Hazard classification and safety precautions
Hazard Harmful, Dangerous for the environment, Harmful if swallowed
Transport code 9
Packaging group III
UN Number 3082
Risk phrases H351
Operator protection A, H; U05a, U09b, U20c
Environmental protection E15b, E34, E38, H410
Storage and disposal D01, D02, D09a, D10a
Treated seed S01, S02, S03, S04d, S05, S07
Medical advice M03

248   imazamox

An imidazolinone contact and residual herbicide available only in mixtures
HRAC mode of action code: B

See also imazamox + metazachlor
*   imazamox + pendimethalin*

249   imazamox + metazachlor

A contact and residual herbicide mixture for weed control in oilseed rape
HRAC mode of action code: B + K3

Products

Cleranda	BASF	17.5:375 g/l	SC	15036

Uses
- Annual dicotyledons in **winter oilseed rape**
- Annual grasses in **winter oilseed rape**

Approval information
- Imazamox and metazachlor are included in Annex I under EC Regulation 1107/2009

Efficacy guidance
- Must only be used for weed control in CLEARFIELD oilseed rape hybrids. Treatment of an oilseed rape variety that is not a 'CLEARFIELD Hybrid' will result in complete crop loss.

Restrictions
- Applications shall be limited to a total dose of not more than 1.0 kg metazachlor/ha in a three year period on the same field
- To avoid the buildup of resistance, do not apply this or any other product containing an ALS inhibitor herbicide with claims for grass weed control more than once to any crop.
- Do not apply in mixture with phosphate liquid fertilisers
- Metazachlor stewardship requires that all autumn applications should be made before the end of September to reduce the risk to water

Following crops guidance
- Wheat, barley, oats, oilseed rape, field beans, combining peas and sugar beet can follow normally harvested oilseed rape treated with [1].
- In the event of crop failure CLEARFIELD oilseed rape may be drilled 4 weeks after application after thorough mixing of the soil to distribute residues. If winter field beans are to be planted, allow 10 weeks after application and plough before planting. Wheat and barley also require ploughing before planting but may be drilled 8 weeks after application.

Environmental safety
- Metazachlor stewardship guidelines advise a maximum dose of 750 g.a.i/ha/annum. Applications to drained fields should be complete by 15th Oct but, if drains are flowing, complete applications by 1st Oct.
- LERAP Category B

Hazard classification and safety precautions
Hazard Irritant, Dangerous for the environment
Transport code 9
Packaging group III
UN Number 3082
Operator protection A, H; U05a
Environmental protection E07d, E15b, E16a, E40b
Storage and disposal D01, D02, D09a, D10c

250 imazamox + pendimethalin

A pre-emergence broad-spectrum herbicide mixture for legumes
HRAC mode of action code: B + K1

See also pendimethalin

Products
Nirvana	BASF	16.7:250 g/l	EC	14256

Uses
- Annual dicotyledons in **broad beans** *(off-label)*, **combining peas**, **hops** *(off-label)*, **ornamental plant production** *(off-label)*, **soft fruit** *(off-label)*, **spring field beans**, **top fruit** *(off-label)*, **vining peas**, **winter field beans**
- Black bindweed in **soya beans** *(off-label)*
- Charlock in **soya beans** *(off-label)*
- Chickweed in **soya beans** *(off-label)*
- Common orache in **soya beans** *(off-label)*
- Fat hen in **soya beans** *(off-label)*
- Field speedwell in **soya beans** *(off-label)*
- Fumitory in **soya beans** *(off-label)*
- Poppies in **soya beans** *(off-label)*

Extension of Authorisation for Minor Use (EAMUs)
- *broad beans 20092891*
- *hops 20092894*
- *ornamental plant production 20092894*
- *soft fruit 20092894*
- *soya beans 20152425*
- *top fruit 20092894*

Approval information
- Imazamox and pendimethalin included in Annex I under EC Regulation 1107/2009

Efficacy guidance
- Best results obtained from applications to fine firm seedbeds in the presence of adequate moisture
- Weed control may be reduced on cloddy seedbeds and on soils with over 6% organic matter
- Residual control may be reduced under prolonged dry conditions

Restrictions
- Maximum number of treatments 1 per crop
- Seed must be drilled to at least 2.5 cm of settled soil
- Do not use on soils containing more than 10% organic matter
- Do not apply to soils that are waterlogged or are prone to waterlogging
- Do not apply if heavy rain is forecast

SEE SECTION 3 FOR PRODUCTS ALSO REGISTERED

- Do not soil incorporate the product or disturb the soil after application
- Consult processors before use on crops destined for processing
- To avoid the build-up of resistance do not apply this or any other product containing an ALS inhibitor herbicide with claims for control of grass-weeds more than once to any crop

Crop-specific information
- Latest use: pre-crop emergence for all crops
- Inadequately covered seed may result in cupping of the leaves after application from which recovery is normally complete
- Crop damage may occur on stony or gravelly soils especially if heavy rain follows treatment
- Winter oilseed rape and other brassica crops should not be drilled as the following crop.

Following crops guidance
- Winter wheat or winter barley may be drilled as a following crop provided 3 mth have elapsed since treatment and the land has been at least cultivated by a non-inversion technique such as discing
- Winter oilseed rape or other brassica crops should not be drilled as following crops
- A minimum of 12 mth must elapse between treatment and sowing red beet, sugar beet or spinach

Environmental safety
- Dangerous for the environment
- Very toxic to aquatic organisms
- LERAP Category B

Hazard classification and safety precautions
Hazard Irritant, Dangerous for the environment, Very toxic to aquatic organisms
Transport code 9
Packaging group III
UN Number 3082
Risk phrases H317, R38
Operator protection A, H; U05a, U14, U20b
Environmental protection E15b, E16a, E38, H410
Storage and disposal D01, D02, D09a, D10c, D12a
Medical advice M03

251   imazamox + quinmerac

A herbicide mixture for weed control in CLEARFIELD oilseed rape only
HRAC mode of action code: B + O

Products

Cleravo	BASF	35:250 g/l	SC	16877

Uses
- Annual dicotyledons in *spring oilseed rape, winter oilseed rape*
- Charlock in *spring oilseed rape, winter oilseed rape*
- Cleavers in *spring oilseed rape, winter oilseed rape*
- Crane's-bill in *spring oilseed rape* (cut-leaved), *winter oilseed rape* (cut-leaved)
- Dead nettle in *spring oilseed rape, winter oilseed rape*
- Flixweed in *spring oilseed rape, winter oilseed rape*
- Volunteer oilseed rape in *spring oilseed rape, winter oilseed rape*

Approval information
- Imazamox and quinmerac included in Annex I under EC Regulation 1107/2009

Efficacy guidance
- Transient crop scorch may occur if applied under frosty conditions.
- Soil moisture is required for efficacy via root uptake.
- Maximum activity is from application before or shortly after weed emergence. Uptake is via cotyledons, roots and shoots.

FOR FULL CONDITIONS OF USE ALWAYS READ THE PRODUCT LABEL

Restrictions
- Must only be applied to CLEARFIELD oilseed rape hybrids.
- Do not apply if heavy rain is forecast.
- Do not use on sands, very light soils or soils with more than 10% organic matter.
- Use on stony soils may result in a reduction in plant vigour or crop stand.
- To avoid the build-up of resistance do not apply this or any other product containing an ALS inhibitor herbicide with claims for control of grass-weeds more than once to any crop.

Crop-specific information
- In the event of crop failure, do not re-drill with non CLEARFIELD oilseed rape or with sugar beet. After ploughing Clearfield rape, maize or peas may be sown 4 weeks after application, wheat, barley or oats 8 weeks after application, field beans 10 weeks after application or sunflower 15 weeks after application.

Hazard classification and safety precautions
> **Hazard** Very toxic to aquatic organisms
> **Transport code** 9
> **Packaging group** III
> **UN Number** 3082
> **Operator protection** U05a, U20a
> **Environmental protection** E15b, E38, H410
> **Storage and disposal** D01, D02, D09a, D10c, D12b

252   4-indol-3-ylbutyric acid

A plant growth regulator promoting the rooting of cuttings

Products

1	Chryzoplus Grey 0.8%	Fargro	0.8% w/w	DP	17569
2	Chryzopon Rose 0.1%	Fargro	0.1% w/w	DP	17566
3	Chryzotek Beige 0.4%	Fargro	0.4% w/w	DP	17568
4	Chryzotop Green 0.25%	Fargro	0.25% w/w	DP	17567
5	Rhizopon AA Powder (0.5%)	Fargro	0.5% w/w	DP	17570
6	Rhizopon AA Powder (1%)	Fargro	1% w/w	DP	17571
7	Rhizopon AA Powder (2%)	Fargro	2% w/w	DP	17572
8	Rhizopon AA Tablets	Fargro	50 mg a.i.	WT	17573

Uses
- Rooting of cuttings in *ornamental plant production*

Approval information
- 4-indol-3-ylbutyric acid not included in Annex 1 under EC Regulation 1107/2009

Efficacy guidance
- Dip base of cuttings into powder immediately before planting
- Powders or solutions of different concentration are required for different types of cutting. Lowest concentration for softwood, intermediate for semi-ripe, highest for hardwood
- See label for details of concentration and timing recommended for different species
- Use of planting holes recommended for powder formulations to ensure product is not removed on insertion of cutting. Cuttings should be watered in if necessary

Restrictions
- Maximum number of treatments 1 per situation
- Use of too strong a powder or solution may cause injury to cuttings
- No unused moistened powder should be returned to container

Crop-specific information
- Latest use: before cutting insertion for ornamental specimens

SEE SECTION 3 FOR PRODUCTS ALSO REGISTERED

Hazard classification and safety precautions
UN Number N/C
Operator protection U14, U15 [1-4]; U19a [1-8]; U20a [1-4, 6-8]; U20b [5]
Environmental protection E15a
Storage and disposal D09a, D11a

253   indoxacarb

An oxadiazine insecticide for caterpillar control in a range of crops
IRAC mode of action code: 22A

Products

1	Clayton Purser	Clayton	30% w/w	WG	18283
2	Explicit	FMC Agro	30% w/w	WG	18763
3	Rumo	FMC Agro	30% w/w	WG	18797
4	Steward	FMC Agro	30% w/w	WG	18792

Uses

- Cabbage leafroller in *choi sum* (off-label), *collards* (off-label), *kale* (off-label), *kohlrabi* (off-label), *oriental cabbage* (off-label), *protected baby leaf crops* (off-label), *protected herbs (see appendix 6)* (off-label), *protected kohlrabi* (off-label), *spring greens* (off-label), *tatsoi* (off-label) [2-4]
- Cabbage moth in *choi sum* (off-label), *collards* (off-label), *kale* (off-label), *kohlrabi* (off-label), *oriental cabbage* (off-label), *protected baby leaf crops* (off-label), *protected herbs (see appendix 6)* (off-label), *protected kohlrabi* (off-label), *spring greens* (off-label), *tatsoi* (off-label) [2-4]
- Cabbage white butterfly in *choi sum* (off-label), *collards* (off-label), *kale* (off-label), *kohlrabi* (off-label), *oriental cabbage* (off-label), *protected baby leaf crops* (off-label), *protected herbs (see appendix 6)* (off-label), *protected kohlrabi* (off-label), *spring greens* (off-label), *tatsoi* (off-label) [2-4]
- Capsids in *apricots* (off-label), *nectarines* (off-label), *peaches* (off-label), *strawberries* (off-label) [2, 4]
- Caterpillars in *almonds* (off-label), *apricots* (off-label), *bilberries* (off-label), *blackberries* (off-label), *blackcurrants* (off-label), *blueberries* (off-label), *cherries* (off-label), *chestnuts* (off-label), *cranberries* (off-label), *elderberries* (off-label), *figs* (off-label), *gooseberries* (off-label), *grapevines* (off-label), *hazel nuts* (off-label), *kiwi fruit* (off-label), *loganberries* (off-label), *medlar* (off-label), *mulberries* (off-label), *nectarines* (off-label), *olives* (off-label), *peaches* (off-label), *plums* (off-label), *protected almonds* (off-label), *protected apricots* (off-label), *protected bilberries* (off-label), *protected blackberries* (off-label), *protected blackcurrants* (off-label), *protected blueberry* (off-label), *protected cherries* (off-label), *protected chestnuts* (off-label), *protected cranberries* (off-label), *protected elderberries* (off-label), *protected figs* (off-label), *protected gooseberries* (off-label), *protected hazelnuts* (off-label), *protected kiwi fruit* (off-label), *protected loganberries* (off-label), *protected medlars* (off-label), *protected mulberry* (off-label), *protected nectarines* (off-label), *protected olives* (off-label), *protected peaches* (off-label), *protected plums* (off-label), *protected quince* (off-label), *protected raspberries* (off-label), *protected redcurrants* (off-label), *protected rose hips* (off-label), *protected rubus hybrids* (off-label), *protected strawberries* (off-label), *protected table grapes* (off-label), *protected walnuts* (off-label), *protected wine grapes* (off-label), *quinces* (off-label), *raspberries* (off-label), *redcurrants* (off-label), *rose hips* (off-label), *rubus hybrids* (off-label), *strawberries* (off-label), *table grapes* (off-label), *walnuts* (off-label), *wine grapes* (off-label) [2, 4]; *apples, pears, protected marrows, protected ornamentals, protected squashes* [1, 2, 4]; *broccoli, cabbages, cauliflowers, protected aubergines, protected courgettes, protected cucumbers, protected melons, protected peppers, protected pumpkins, protected tomatoes* [1-4]; *brussels sprouts* (off-label), *choi sum* (off-label), *collards* (off-label), *endives* (off-label), *hops* (off-label), *kale* (off-label), *kohlrabi* (off-label), *lettuce* (off-label), *oriental cabbage* (off-label), *ornamental plant production* (off-label), *protected all edible seed crops* (off-label), *protected all non-edible seed crops* (off-label), *protected baby leaf crops* (off-label), *protected herbs (see appendix 6)* (off-label), *protected hops* (off-label), *protected kohlrabi* (off-label), *protected ornamentals* (off-label), *spring*

greens (off-label), **sweetcorn** (off-label), **tatsoi** (off-label) [2-4]; **calabrese, protected chilli peppers, protected summer squash, protected winter squash** [3]

- Diamond-back moth in **brussels sprouts** (off-label), **choi sum** (off-label), **collards** (off-label), **endives** (off-label), **kale** (off-label), **kohlrabi** (off-label), **lettuce** (off-label), **oriental cabbage** (off-label), **protected baby leaf crops** (off-label), **protected herbs (see appendix 6)** (off-label), **protected kohlrabi** (off-label), **spring greens** (off-label), **sweetcorn** (off-label), **tatsoi** (off-label) [2-4]
- Flax tortrix moth in **endives** (off-label), **lettuce** (off-label) [2-4]
- Garden pebble moth in **choi sum** (off-label), **collards** (off-label), **kale** (off-label), **kohlrabi** (off-label), **oriental cabbage** (off-label), **protected baby leaf crops** (off-label), **protected herbs (see appendix 6)** (off-label), **protected kohlrabi** (off-label), **spring greens** (off-label), **tatsoi** (off-label) [2-4]
- Ghost moth in **endives** (off-label), **lettuce** (off-label) [2-4]
- Grape berry moth in **grapevines** (off-label) [2, 4]
- Insect pests in **almonds** (off-label), **apricots** (off-label), **bilberries** (off-label), **blackberries** (off-label), **blackcurrants** (off-label), **blueberries** (off-label), **cherries** (off-label), **chestnuts** (off-label), **cranberries** (off-label), **elderberries** (off-label), **figs** (off-label), **gooseberries** (off-label), **hazel nuts** (off-label), **kiwi fruit** (off-label), **loganberries** (off-label), **medlar** (off-label), **mulberries** (off-label), **nectarines** (off-label), **olives** (off-label), **peaches** (off-label), **plums** (off-label), **protected almonds** (off-label), **protected apricots** (off-label), **protected bilberries** (off-label), **protected blackberries** (off-label), **protected blackcurrants** (off-label), **protected blueberry** (off-label), **protected cherries** (off-label), **protected chestnuts** (off-label), **protected cranberries** (off-label), **protected elderberries** (off-label), **protected figs** (off-label), **protected gooseberries** (off-label), **protected hazelnuts** (off-label), **protected kiwi fruit** (off-label), **protected loganberries** (off-label), **protected medlars** (off-label), **protected mulberry** (off-label), **protected nectarines** (off-label), **protected olives** (off-label), **protected peaches** (off-label), **protected plums** (off-label), **protected quince** (off-label), **protected raspberries** (off-label), **protected redcurrants** (off-label), **protected rose hips** (off-label), **protected rubus hybrids** (off-label), **protected strawberries** (off-label), **protected table grapes** (off-label), **protected walnuts** (off-label), **protected wine grapes** (off-label), **quinces** (off-label), **raspberries** (off-label), **redcurrants** (off-label), **rose hips** (off-label), **rubus hybrids** (off-label), **strawberries** (off-label), **table grapes** (off-label), **walnuts** (off-label), **wine grapes** (off-label) [2, 4]; **hops** (off-label), **ornamental plant production** (off-label), **protected all edible seed crops** (off-label), **protected all non-edible seed crops** (off-label), **protected hops** (off-label), **protected ornamentals** (off-label) [2-4]
- Light brown apple moth in **cherries** (off-label) [2, 4]
- Peach moth in **apricots** (off-label), **nectarines** (off-label), **peaches** (off-label) [2, 4]; **strawberries** (off-label) [2]
- Pollen beetle in **mustard** (off-label) [2, 4]; **spring oilseed rape, winter oilseed rape** [1-4]
- Silver Y moth in **brussels sprouts** (off-label), **endives** (off-label), **lettuce** (off-label), **protected baby leaf crops** (off-label), **protected herbs (see appendix 6)** (off-label), **protected kohlrabi** (off-label) [2-4]
- Summer-fruit tortrix moth in **cherries** (off-label) [2, 4]
- Swift moth in **endives** (off-label), **lettuce** (off-label) [2-4]
- Turnip moth in **endives** (off-label), **lettuce** (off-label) [2-4]
- Winter moth in **apricots** (off-label), **nectarines** (off-label), **peaches** (off-label) [2, 4]; **strawberries** (off-label) [2]

Extension of Authorisation for Minor Use (EAMUs)
- **almonds** 20190758 [2], 20190588 [4]
- **apricots** 20190758 [2], 20190759 [2], 20190588 [4], 20190590 [4]
- **bilberries** 20190758 [2], 20190588 [4]
- **blackberries** 20190758 [2], 20190767 [2], 20190588 [4], 20190593 [4]
- **blackcurrants** 20190758 [2], 20190588 [4]
- **blueberries** 20190758 [2], 20190767 [2], 20190588 [4], 20190593 [4]
- **brussels sprouts** 20190769 [2], 20190892 [3], 20190587 [4]
- **cherries** 20190758 [2], 20190760 [2], 20190588 [4], 20190591 [4]
- **chestnuts** 20190758 [2], 20190588 [4]
- **choi sum** 20190764 [2], 20190887 [3], 20190594 [4]

SEE SECTION 3 FOR PRODUCTS ALSO REGISTERED

- **collards** *20190764* [2], *20190887* [3], *20190594* [4]
- **cranberries** *20190758* [2], *20190588* [4]
- **elderberries** *20190758* [2], *20190588* [4]
- **endives** *20190770* [2], *20190890* [3], *20190595* [4]
- **figs** *20190758* [2], *20190588* [4]
- **gooseberries** *20190758* [2], *20190588* [4]
- **grapevines** *20190763* [2], *20190586* [4]
- **hazel nuts** *20190758* [2], *20190588* [4]
- **hops** *20190758* [2], *20190888* [3], *20190588* [4]
- **kale** *20190764* [2], *20190887* [3], *20190594* [4]
- **kiwi fruit** *20190758* [2], *20190588* [4]
- **kohlrabi** *20190764* [2], *20190887* [3], *20190594* [4]
- **lettuce** *20190770* [2], *20190890* [3], *20190595* [4]
- **loganberries** *20190758* [2], *20190588* [4]
- **medlar** *20190758* [2], *20190588* [4]
- **mulberries** *20190758* [2], *20190588* [4]
- **mustard** *20190768* [2], *20190592* [4]
- **nectarines** *20190758* [2], *20190759* [2], *20190588* [4], *20190590* [4]
- **olives** *20190758* [2], *20190588* [4]
- **oriental cabbage** *20190764* [2], *20190887* [3], *20190594* [4]
- **ornamental plant production** *20190758* [2], *20190888* [3], *20190588* [4]
- **peaches** *20190758* [2], *20190759* [2], *20190588* [4], *20190590* [4]
- **plums** *20190758* [2], *20190588* [4]
- **protected all edible seed crops** *20190758* [2], *20190888* [3], *20190588* [4]
- **protected all non-edible seed crops** *20190758* [2], *20190888* [3], *20190588* [4]
- **protected almonds** *20190758* [2], *20190588* [4]
- **protected apricots** *20190758* [2], *20190588* [4]
- **protected baby leaf crops** *20190766* [2], *20190889* [3], *20190589* [4]
- **protected bilberries** *20190758* [2], *20190588* [4]
- **protected blackberries** *20190758* [2], *20190767* [2], *20190588* [4], *20190593* [4]
- **protected blackcurrants** *20190758* [2], *20190588* [4]
- **protected blueberry** *20190758* [2], *20190588* [4]
- **protected cherries** *20190758* [2], *20190588* [4]
- **protected chestnuts** *20190758* [2], *20190588* [4]
- **protected cranberries** *20190758* [2], *20190588* [4]
- **protected elderberries** *20190758* [2], *20190588* [4]
- **protected figs** *20190758* [2], *20190588* [4]
- **protected gooseberries** *20190758* [2], *20190588* [4]
- **protected hazelnuts** *20190758* [2], *20190588* [4]
- **protected herbs (see appendix 6)** *20190766* [2], *20190889* [3], *20190589* [4]
- **protected hops** *20190758* [2], *20190888* [3], *20190588* [4]
- **protected kiwi fruit** *20190758* [2], *20190588* [4]
- **protected kohlrabi** *20190766* [2], *20190889* [3], *20190589* [4]
- **protected loganberries** *20190758* [2], *20190588* [4]
- **protected medlars** *20190758* [2], *20190588* [4]
- **protected mulberry** *20190758* [2], *20190588* [4]
- **protected nectarines** *20190758* [2], *20190588* [4]
- **protected olives** *20190758* [2], *20190588* [4]
- **protected ornamentals** *20190758* [2], *20190888* [3], *20190588* [4]
- **protected peaches** *20190758* [2], *20190588* [4]
- **protected plums** *20190758* [2], *20190588* [4]
- **protected quince** *20190758* [2], *20190588* [4]
- **protected raspberries** *20190758* [2], *20190767* [2], *20190588* [4], *20190593* [4]
- **protected redcurrants** *20190758* [2], *20190588* [4]
- **protected rose hips** *20190758* [2], *20190588* [4]
- **protected rubus hybrids** *20190758* [2], *20190588* [4]
- **protected strawberries** *20190758* [2], *20190588* [4]
- **protected table grapes** *20190758* [2], *20190588* [4]
- **protected walnuts** *20190758* [2], *20190588* [4]

FOR FULL CONDITIONS OF USE ALWAYS READ THE PRODUCT LABEL

- **protected wine grapes** *20190758* [2], *20190588* [4]
- **quinces** *20190758* [2], *20190588* [4]
- **raspberries** *20190758* [2], *20190767* [2], *20190588* [4], *20190593* [4]
- **redcurrants** *20190758* [2], *20190588* [4]
- **rose hips** *20190758* [2], *20190588* [4]
- **rubus hybrids** *20190758* [2], *20190588* [4]
- **spring greens** *20190764* [2], *20190887* [3], *20190594* [4]
- **strawberries** *20190758* [2], *20190759* [2], *20190588* [4], *20190590* [4]
- **sweetcorn** *20190757* [2], *20190891* [3], *20190585* [4]
- **table grapes** *20190758* [2], *20190588* [4]
- **tatsoi** *20190764* [2], *20190887* [3], *20190594* [4]
- **walnuts** *20190758* [2], *20190588* [4]
- **wine grapes** *20190758* [2], *20190588* [4]

Approval information
- Indoxacarb included in Annex I under EC Regulation 1107/2009

Efficacy guidance
- Best results in brassica and protected crops obtained from treatment when first caterpillars are detected, or when damage first seen, or 7-10 d after trapping first adults in pheromone traps
- In apples and pears apply at egg-hatch
- Subsequent treatments in all crops may be applied at 8-14 d intervals
- Indoxacarb acts by ingestion and contact. Only larval stages are controlled but there is some ovicidal action against some species

Restrictions
- Maximum number of treatments 3 per crop or yr for apples, pears, brassica crops; 6 per yr for protected crops and 1 per crop for oilseed rape
- Do not apply to any crop suffering from stress from any cause
- When applying to protected crops the maximum concentration must not exceed 12.5 g of product per 100 litre water

Crop-specific information
- HI: brassica crops, protected crops 1 d; apples, pears 7 d

Following crops guidance
- When treating protected crops, observe the maximum concentration permitted.

Environmental safety
- Dangerous for the environment
- Toxic to aquatic organisms
- In accordance with good agricultural practice apply in early morning or late evening when bees are less active
- Broadcast air-assisted LERAP [1, 2, 4] (15m); LERAP Category B [2, 3]

Hazard classification and safety precautions
Hazard Harmful, Dangerous for the environment, Harmful if swallowed
Transport code 9
Packaging group III
UN Number 3077
Risk phrases H371
Operator protection A, D; U04a, U05a
Environmental protection E12f, E15b, E34, E38, H410 [1-4]; E16a [2, 3]; E17b [1, 2, 4] (15m)
Storage and disposal D01, D02, D09a, D12a
Medical advice M03

254   iodosulfuron-methyl-sodium + mesosulfuron-methyl

A sulfonyl urea herbicide mixture for winter wheat
HRAC mode of action code: B + B

See also mesosulfuron-methyl

Products

1	Atlantis OD	Bayer CropScience	2:10 g/l	OD	18100
2	Cintac	Life Scientific	1.0:3.0% w/w	WG	18222
3	Clayton Metropolis	Clayton	2:10 g/l	OD	18401
4	Hatra	Bayer CropScience	2:10 g/l	OD	16190
5	Horus	Bayer CropScience	2:10 g/l	OD	16216
6	Niantic	Life Scientific	0.6:3.0% w/w	WG	18217

Uses

- Annual meadow grass in *winter wheat*
- Blackgrass in *winter wheat*
- Chickweed in *winter wheat*
- Italian ryegrass in *winter wheat*
- Mayweeds in *winter wheat*
- Perennial ryegrass in *winter wheat*
- Rough-stalked meadow grass in *winter wheat*
- Wild oats in *winter wheat*

Approval information

- Iodosulfuron-methyl-sodium and mesosulfuron-methyl included in Annex I under EC Regulation 1107/2009

Efficacy guidance

- Optimum grass weed control obtained when all grass weeds are emerged at spraying. Activity is primarily via foliar uptake and good spray coverage of the target weeds is essential
- Translocation occurs readily within the target weeds and growth is inhibited within hours of treatment but symptoms may not be apparent for up to 4 wk, depending on weed species, timing of treatment and weather conditions
- Residual activity is important for best results and is optimised by treatment on fine moist seedbeds. Avoid application under very dry conditions
- Residual efficacy may be reduced by high soil temperatures and cloddy seedbeds
- Iodosulfuron-methyl and mesosulfuron-methyl are both members of the ALS-inhibitor group of herbicides. To avoid the build up of resistance do not use any product containing an ALS-inhibitor herbicide with claims for control of grass weeds more than once on any crop
- Use these products as part of a Resistance Management Strategy that includes cultural methods of control and does not use ALS inhibitors as the sole chemical method of grass weed control. See Section 5 for more information

Restrictions

- Maximum number of treatments 1 per crop with a maximum total dose equivalent to one full dose treatment
- Do not use on crops undersown with grasses, clover or other legumes or any other broad-leaved crop
- Do not use as a stand-alone treatment for blackgrass, ryegrass or chickweed control
- To avoid the build up of resistance do not apply this or any other product containing an ALS inhibitor herbicide with claims for control of grass-weeds more than once to any crop [6]
- Do not use as the sole means of weed control in successive crops
- Do not use in mixture or in sequence with any other ALS-inhibitor herbicide except those (if any) specified on the label
- Do not apply earlier than 1 Feb in the yr of harvest [2]
- Do not apply to crops under stress from any cause
- Do not apply when rain is imminent or during periods of frosty weather
- Specified adjuvant must be used. See label

FOR FULL CONDITIONS OF USE ALWAYS READ THE PRODUCT LABEL

Crop-specific information
- Latest use: flag leaf just visible (GS 39)
- Winter wheat may be treated from the two-leaf stage of the crop
- Safety to crops grown for seed not established
- Avoid application before a severe frost or transient yellowing of the crop may occur.

Following crops guidance
- In the event of crop failure sow only winter wheat in the same cropping season
- Only winter wheat or winter barley (or winter oilseed rape) may be sown in the year of harvest of a treated crop but ploughing before drilling oilseed rape is advisable, especially in a season where applications are made later than usual.
- Spring wheat, spring barley, sugar beet or spring oilseed rape may be drilled in the following spring. Plough before drilling oilseed rape

Environmental safety
- Dangerous for the environment
- Very toxic to aquatic organisms
- Dangerous to fish or other aquatic life. Do not contaminate surface waters or ditches with chemical or used container
- Take extreme care to avoid drift onto plants outside the target area or on to ponds, waterways or ditches
- LERAP Category B

Hazard classification and safety precautions
Hazard Irritant, Dangerous for the environment [1-6]; Very toxic to aquatic organisms [1-5]
Transport code 9
Packaging group III
UN Number 3077 [2, 6]; 3082 [1, 3-5]
Risk phrases H315, H317 [6]; H318 [2, 6]; H319 [1, 3-5]
Operator protection A, C, H; U05a, U11, U20b
Environmental protection E15a, E16a, E38 [1-6]; H410 [1-5]; H411 [6]
Storage and disposal D01, D02, D05, D09a, D10a, D12a

255 ipconazole

A triazole fungicide for disease control in cereals
FRAC mode of action code: 3

See also imazalil + ipconazole

Products
Rancona 15 ME	Arysta	15.1 g/l	MS	16136

Uses
- Seed-borne diseases in *spring barley, spring wheat, winter barley, winter wheat*
- Soil-borne diseases in *spring barley, spring wheat, winter barley, winter wheat*

Approval information
- Ipconazole included in Annex I under EC Regulation 1107/2009
- Accepted by BBPA for use on malting barley

Hazard classification and safety precautions
Transport code 9
Packaging group III
UN Number 3082
Operator protection A, H; U05a, U09a, U20b
Environmental protection E03, E15b, E34, H410
Storage and disposal D01, D02, D09a, D10a
Treated seed S01, S02, S03, S04d, S05, S07
Medical advice M03

SEE SECTION 3 FOR PRODUCTS ALSO REGISTERED

256   isopyrazam

An SDHI fungicide available for disease control in barley.
FRAC mode of action code: 7

See also azoxystrobin + isopyrazam
cyprodinil + isopyrazam
epoxiconazole + isopyrazam

Products

1 Reflect	Syngenta	125 g/l	EC	17228
2 Reflect	Adama	125 g/l	EC	18573
3 Zulu	Syngenta	125 g/l	EC	16913

Uses

* Alternaria in **celeriac** *(off-label)*, **parsnips** *(off-label)*, **red beet** *(off-label)*, **swedes** *(off-label)*, **turnips** *(off-label)* [1, 2]
* Alternaria blight in **carrots** [1, 2]
* Canker in **celeriac** *(off-label)*, **parsnips** *(off-label)*, **red beet** *(off-label)*, **swedes** *(off-label)*, **turnips** *(off-label)* [1, 2]
* Cercospora leaf spot in **celeriac** *(off-label)*, **parsnips** *(off-label)*, **red beet** *(off-label)*, **swedes** *(off-label)*, **turnips** *(off-label)* [1, 2]
* Disease control in **protected melons**, **protected peppers**, **protected watermelon** [1, 2]; **spring barley**, **winter barley** [3]
* Grey mould in **protected hemp** *(off-label)* [1, 2]
* Powdery mildew in **carrots**, **celeriac** *(off-label)*, **ornamental plant production** *(off-label)*, **parsnips** *(off-label)*, **protected aubergines**, **protected chilli peppers**, **protected courgettes**, **protected cucumbers**, **protected hemp** *(off-label)*, **protected melons**, **protected peppers**, **protected summer squash**, **protected tomatoes**, **protected watermelon**, **red beet** *(off-label)*, **swedes** *(off-label)*, **turnips** *(off-label)* [1, 2]
* Ramularia leaf spots in **celeriac** *(off-label)*, **parsnips** *(off-label)*, **red beet** *(off-label)*, **swedes** *(off-label)*, **turnips** *(off-label)* [1, 2]
* Sclerotinia in **celeriac** *(off-label)*, **parsnips** *(off-label)*, **protected hemp** *(off-label)*, **red beet** *(off-label)*, **swedes** *(off-label)*, **turnips** *(off-label)* [1, 2]

Extension of Authorisation for Minor Use (EAMUs)

* **celeriac** *20182968* [1], *20182976* [2]
* **ornamental plant production** *20182965* [1], *20182975* [2]
* **parsnips** *20182968* [1], *20182976* [2]
* **protected hemp** *20182960* [1], *20182974* [2]
* **red beet** *20182968* [1], *20182976* [2]
* **swedes** *20182968* [1], *20182976* [2]
* **turnips** *20182968* [1], *20182976* [2]

Approval information

* Isopyrazam is included in Appendix 1 under EC Regulation 1107/2009

Efficacy guidance

* Use only in mixture with fungicides with a different mode of action that are effective against the target diseases.

Restrictions

* Isopyrazam is an SDH respiration inhibitor; Do not apply more than two foliar applications of products containing an SDH inhibitor to any cereal crop

Crop-specific information

* There are no restrictions on succeeding crops in a normal rotation

Environmental safety

* LERAP Category B

FOR FULL CONDITIONS OF USE ALWAYS READ THE PRODUCT LABEL

Hazard classification and safety precautions
Hazard Harmful, Irritant, Dangerous for the environment, Harmful if inhaled, Very toxic to aquatic organisms [3]; Harmful if swallowed [1-3]
Transport code 9
Packaging group III
UN Number 3082
Risk phrases H319, H361 [1-3]; H351 [1, 2]
Operator protection A, H [1-3]; C [1, 2]
Environmental protection E16a, H410 [1-3]; E38 [3]
Storage and disposal D01, D02, D12a, D12b [3]

257   isopyrazam + prothioconazole

An SDHI and triazole fungicide mixture for disease control in cereals
FRAC mode of action code: 3 + 7

See also isopyrazam
prothioconazole

Products

Prizm	Adama	125:150 g/l	SC	18940

Uses
- Disease control in *rye*, *spring barley*, *spring wheat*, *triticale*, *winter barley*, *winter wheat*

Approval information
- Isopyrazam and prothioconazole are included in Appendix 1 under EC Regulation 1107/2009

Restrictions
- A minimum interval of 14 days must be observed between applications.
- The earliest application timing is GS31, the second application must be made after GS40

Environmental safety
- Buffer zone requirement 6 m
- LERAP Category B

Hazard classification and safety precautions
Transport code 9
Packaging group III
UN Number 3082
Risk phrases H319, H361
Operator protection A, C, H
Environmental protection E16a, H410

258   isoxaben

A soil-acting benzamide herbicide for use in cereals, grass and fruit
HRAC mode of action code: L

Products

1	Flexidor	Landseer	500 g/l	SC	18042
2	Pan Isoxaben 500	Pan Amenity	500 g/l	SC	18419

Uses
- Annual dicotyledons in *almonds* (off-label), *broad beans* (off-label), *bulb onions* (off-label), *carrots* (off-label), *chestnuts* (off-label), *chicory* (off-label), *courgettes* (off-label), *hazel nuts* (off-label), *horseradish* (off-label), *leeks* (off-label), *parsnips* (off-label), *rhubarb* (off-label), *shallots* (off-label), *summer squash* (off-label), *walnuts* (off-label), *witloof* (off-label) [1]; *amenity vegetation*, *apples*, *asparagus*, *blackberries*, *blackcurrants*, *cherries*, *forest*, *forest nurseries*, *gooseberries*, *hops*, *ornamental plant production*, *pears*, *plums*, *raspberries*,

SEE SECTION 3 FOR PRODUCTS ALSO REGISTERED

SECTION 2

redcurrants, strawberries, triticale, winter barley, winter oats, winter rye, winter wheat [1, 2]
- Chickweed in **rhubarb** *(off-label)* [1]

Extension of Authorisation for Minor Use (EAMUs)
- **almonds** *20180011* [1]
- **broad beans** *20193020* [1]
- **bulb onions** *20181145* [1]
- **carrots** *20180020* [1]
- **chestnuts** *20180011* [1]
- **chicory** *20193019* [1]
- **courgettes** *20180012* [1]
- **hazel nuts** *20180011* [1]
- **horseradish** *20180020* [1]
- **leeks** *20181145* [1]
- **parsnips** *20180020* [1]
- **rhubarb** *20192598* [1]
- **shallots** *20181145* [1]
- **summer squash** *20180012* [1]
- **walnuts** *20180011* [1]
- **witloof** *20193019* [1]

Approval information
- Isoxaben included in Annex 1 under EC Regulation 1107/2009
- Accepted by BBPA for use on hops

Efficacy guidance
- When used alone apply pre-weed emergence
- Effectiveness is reduced in dry conditions. Weed seeds germinating at depth are not controlled
- Activity reduced on soils with more than 10% organic matter. Do not use on peaty soils
- Various tank mixtures are recommended for early post-weed emergence treatment (especially for grass weeds). See label for details

Restrictions
- Maximum number of treatments 1 per crop for all edible crops; 2 per yr on amenity vegetation and non-edible crops

Crop-specific information
- Latest use: before 1 Apr in yr of harvest for edible crops

Following crops guidance
- See label for details of crops which may be sown in the event of failure of a treated crop

Environmental safety
- Keep all livestock out of treated areas for at least 50 d
- Buffer zone requirement 20 m for rhubarb [1]

Hazard classification and safety precautions
Hazard Very toxic to aquatic organisms
Transport code 9
Packaging group III
UN Number 3082
Operator protection A, C, H, M; U05a, U20a
Environmental protection E06a (50 days); E15b, H410
Storage and disposal D01, D02, D09a, D11a

FOR FULL CONDITIONS OF USE ALWAYS READ THE PRODUCT LABEL

259   isoxaflutole

An isoxazole herbicide available only in mixtures
HRAC mode of action code: F2

See also flufenacet + isoxaflutole

260   kresoxim-methyl

A protectant strobilurin fungicide for apples
FRAC mode of action code: 11

See also epoxiconazole + kresoxim-methyl
epoxiconazole + kresoxim-methyl + pyraclostrobin

Products

Stroby WG	BASF	50% w/w	WG	17316

Uses
- American gooseberry mildew in **blueberries** *(off-label)*, **cranberries** *(off-label)*, **gooseberries** *(off-label)*, **whitecurrants** *(off-label)*
- Black rot in **wine grapes** *(off-label)*
- Powdery mildew in **apple for cider making** *(reduction)*, **apples** *(reduction)*, **blackcurrants**, **blueberries** *(off-label)*, **cranberries** *(off-label)*, **gooseberries** *(off-label)*, **ornamental plant production**, **protected strawberries**, **redcurrants**, **strawberries**, **whitecurrants** *(off-label)*, **wine grapes** *(off-label)*
- Scab in **apple for cider making**, **apples**

Extension of Authorisation for Minor Use (EAMUs)
- **blueberries** *20160089*
- **cranberries** *20160089*
- **gooseberries** *20160089*
- **whitecurrants** *20160089*
- **wine grapes** *20160960*

Approval information
- Kresoxim-methyl included in Annex I under EC Regulation 1107/2009
- Approved for use in ULV systems

Efficacy guidance
- Activity is protectant. Best results achieved from treatments prior to disease development. See label for timing details on each crop. Treatments should be repeated at 10-14 d intervals but note limitations below
- To minimise the likelihood of development of resistance to strobilurin fungicides these products should be used in a planned Resistance Management strategy. See Section 5 for more information
- Product may be applied in ultra low volumes (ULV) but disease control may be reduced

Restrictions
- Maximum number of treatments 4 per yr on apples; 3 per yr on other crops. See notes in Efficacy about limitations on consecutive treatments
- Consult before using on crops intended for processing

Crop-specific information
- HI 14 d for blackcurrants, protected strawberries, strawberries; 35 d for apples
- On apples do not spray product more than twice consecutively and separate each block of two consecutive treatments with at least two applications from a different cross-resistance group. For all other crops do not apply consecutively and use a maximum of once in every three fungicide sprays
- Product should not be used as final spray of the season on apples

Environmental safety
- Dangerous for the environment
- Very toxic to aquatic organisms
- Harmless to ladybirds and predatory mites
- Harmless to honey bees and may be applied during flowering. Nevertheless local beekeepers should be notified when treatment of orchards in flower is to occur

Hazard classification and safety precautions

Hazard Harmful, Dangerous for the environment, Very toxic to aquatic organisms
Transport code 9
Packaging group III
UN Number 3077
Risk phrases H351
Operator protection U05a, U20b
Environmental protection E15a, E38, H410
Storage and disposal D01, D02, D08, D09a, D10c, D12a
Medical advice M05a

261 lambda-cyhalothrin

A quick-acting contact and ingested pyrethroid insecticide
IRAC mode of action code: 3

Products

1	Clayton Sparta	Clayton	50 g/l	EC	13457
2	Dalda 5	Globachem	50 g/l	EC	13688
3	Hallmark with Zeon Technology	Syngenta	100 g/l	CS	12629
4	Karate 2.5WG	Syngenta	2.5% w/w	GR	14060
5	Lambda-C 100	Becesane	100 g/l	CS	15987
6	Lambda-C 50	Becesane	50 g/l	EC	15986
7	Lambdastar	Life Scientific	100 g/l	CS	17406
8	Markate 50	Belchim	50 g/l	EC	13529
9	Sceptre	Adama	2.5% w/w	GR	18868
10	Seal Z	AgChem Access	100 g/l	CS	14201
11	Sparviero	Sipcam	100 g/l	CS	15687

Uses
- Aphids in *all edible seed crops grown outdoors* (off-label), *all non-edible seed crops grown outdoors* (off-label), *asparagus* (off-label), *bulb onions* (off-label), *crambe* (off-label), *farm forestry* (off-label), *forest nurseries* (off-label), *garlic* (off-label), *hops* (off-label), *leeks* (off-label), *miscanthus* (off-label), *ornamental plant production* (off-label), *protected asparagus* (off-label), *protected aubergines* (off-label), *protected forest nurseries* (off-label), *protected ornamentals* (off-label), *protected pak choi* (off-label), *protected peppers* (off-label), *protected tatsoi* (off-label), *protected tomatoes* (off-label), *salad onions* (off-label), *shallots* (off-label), *soft fruit* (off-label), *top fruit* (off-label), *willow (short rotation coppice)* (off-label) [3]; *chestnuts* (off-label), *hazel nuts* (off-label), *walnuts* (off-label) [3, 8]; *cob nuts* (off-label), *filberts* (off-label) [8]; *combining peas, vining peas* [1, 2, 6, 8]; *durum wheat* [2-6, 8-10]; *edible podded peas, spring field beans, spring oilseed rape, sugar beet, winter field beans, winter oilseed rape* [2, 6, 8]; *lettuce, pears, rye, triticale* [2]; *potatoes, spring wheat* [1-3, 5-8, 10, 11]; *spring barley, spring oats* [2, 3, 5, 7, 10, 11]; *winter barley, winter oats* [2-5, 7, 9-11]; *winter wheat* [1-11]
- Barley yellow dwarf virus vectors in *durum wheat* [3-5, 9, 10]; *winter barley, winter oats, winter wheat* [3-5, 7, 9-11]
- Bean seed fly in *soya beans* (off-label) [7]
- Beet leaf miner in *sugar beet* [3, 5, 7, 10, 11]
- Beet virus yellows vectors in *spring oilseed rape* [3, 5, 7, 10, 11]; *winter oilseed rape* [3-5, 7, 9-11]

FOR FULL CONDITIONS OF USE ALWAYS READ THE PRODUCT LABEL

- Beetles in **combining peas**, **durum wheat**, **edible podded peas**, **potatoes**, **spring field beans**, **spring oilseed rape**, **spring wheat**, **sugar beet**, **vining peas**, **winter field beans**, **winter oilseed rape**, **winter wheat** [2, 6, 8]; **rye**, **spring barley**, **spring oats**, **triticale**, **winter barley**, **winter oats** [2]
- Black bean aphid in **asparagus** *(off-label)*, **protected asparagus** *(off-label)* [3]
- Brassica pod midge in **spring oilseed rape**, **winter oilseed rape** [1]
- Bruchid beetle in **broad beans** *(off-label)* [8]
- Cabbage seed weevil in **celeriac** *(off-label)*, **radishes** *(off-label)* [8]; **spring oilseed rape**, **winter oilseed rape** [1, 3, 5, 7, 10, 11]
- Cabbage stem flea beetle in **spring oilseed rape** [3, 5, 7, 10, 11]; **winter oilseed rape** [3-5, 7, 9-11]
- Capsids in **blackberries** *(off-label)*, **dewberries** *(off-label)*, **raspberries** *(off-label)*, **rubus hybrids** *(off-label)* [3]; **grapevines** *(off-label)* [3, 8]
- Carrot fly in **carrots** *(off-label)*, **celeriac** *(off-label)*, **parsnips** *(off-label)*, **radishes** *(off-label)* [8]; **celery (outdoor)** *(off-label)*, **horseradish** *(off-label)*, **mallow (althaea spp.)** *(off-label)*, **parsley root** *(off-label)* [3, 8]; **fennel** *(off-label)* [3]
- Caterpillars in **all edible seed crops grown outdoors** *(off-label)*, **all non-edible seed crops grown outdoors** *(off-label)*, **almonds** *(off-label)*, **apples** *(off-label)*, **apricots** *(off-label)*, **bilberries** *(off-label)*, **blackcurrants** *(off-label)*, **blueberries** *(off-label)*, **borage for oilseed production** *(off-label)*, **canary flower (echium spp.)** *(off-label)*, **cherries** *(off-label)*, **chestnuts** *(off-label)*, **cob nuts** *(off-label)*, **cranberries** *(off-label)*, **dwarf beans** *(off-label)*, **evening primrose** *(off-label)*, **farm forestry** *(off-label)*, **filberts** *(off-label)*, **forest nurseries** *(off-label)*, **frise** *(off-label)*, **gooseberries** *(off-label)*, **grapevines** *(off-label)*, **grass seed crops** *(off-label)*, **hazel nuts** *(off-label)*, **herbs (see appendix 6)** *(off-label)*, **hops** *(off-label)*, **lamb's lettuce** *(off-label)*, **land cress** *(off-label)*, **leaf brassicas** *(off-label)*, **loganberries** *(off-label)*, **lupins** *(off-label)*, **medlar** *(off-label)*, **miscanthus** *(off-label)*, **mustard** *(off-label)*, **nectarines** *(off-label)*, **ornamental plant production** *(off-label)*, **peaches** *(off-label)*, **pears** *(off-label)*, **plums** *(off-label)*, **protected all edible seed crops** *(off-label)*, **protected all non-edible seed crops** *(off-label)*, **protected forest nurseries** *(off-label)*, **protected oriental cabbage** *(off-label)*, **protected ornamentals** *(off-label)*, **quinces** *(off-label)*, **raspberries** *(off-label)*, **redcurrants** *(off-label)*, **rubus hybrids** *(off-label)*, **runner beans** *(off-label)*, **rye** *(off-label)*, **scarole** *(off-label)*, **spring linseed** *(off-label)*, **strawberries** *(off-label)*, **triticale** *(off-label)*, **walnuts** *(off-label)*, **winter linseed** *(off-label)* [8]; **beetroot** *(off-label)*, **french beans** *(off-label)*, **swedes** *(off-label)*, **turnips** *(off-label)* [3, 8]; **broccoli**, **brussels sprouts**, **cabbages**, **calabrese**, **cauliflowers** [3, 5, 7, 10]; **celery (outdoor)** *(off-label)*, **navy beans** *(off-label)* [3]; **combining peas**, **durum wheat**, **edible podded peas**, **potatoes**, **spring field beans**, **spring oilseed rape**, **spring wheat**, **sugar beet**, **vining peas**, **winter field beans**, **winter oilseed rape**, **winter wheat** [2, 6, 8]; **pears**, **rye**, **spring barley**, **spring oats**, **triticale**, **winter barley**, **winter oats** [2]
- Clay-coloured weevil in **blackberries** *(off-label)*, **dewberries** *(off-label)*, **raspberries** *(off-label)*, **rubus hybrids** *(off-label)* [3]
- Common green capsid in **hops** *(off-label)* [3]
- Cutworms in **beetroot** *(off-label)* [3, 8]; **carrots**, **lettuce**, **parsnips** [3, 5, 7, 10, 11]; **celeriac** *(off-label)*, **fodder beet** *(off-label)*, **radishes** *(off-label)* [8]; **chicory** *(off-label)*, **fennel** *(off-label)* [3]; **sugar beet** [1, 3, 5, 7, 10, 11]
- Diamond-back moth in **swedes** *(off-label)*, **turnips** *(off-label)* [8]
- Flea beetle in **corn gromwell** *(off-label)*, **hops** *(off-label)*, **poppies** *(off-label)*, **poppies grown for seed production** *(off-label)* [3]; **crambe** *(off-label)*, **fodder beet** *(off-label)*, **protected oriental cabbage** *(off-label)*, **swedes** *(off-label)*, **turnips** *(off-label)* [8]; **poppies for morphine production** *(off-label)* [3, 8]; **spring oilseed rape**, **sugar beet** [1, 3, 5, 7, 10, 11]; **winter oilseed rape** [1, 3-5, 7, 9-11]
- Frit fly in **sweetcorn** *(off-label)* [3, 8]
- Insect pests in **all edible seed crops grown outdoors** *(off-label)*, **all non-edible seed crops grown outdoors** *(off-label)*, **blackcurrants** *(off-label)*, **borage for oilseed production** *(off-label)*, **bulb onions** *(off-label)*, **canary flower (echium spp.)** *(off-label)*, **cherries** *(off-label)*, **chestnuts** *(off-label)*, **evening primrose** *(off-label)*, **farm forestry** *(off-label)*, **forest nurseries** *(off-label)*, **frise** *(off-label)*, **garlic** *(off-label)*, **gooseberries** *(off-label)*, **grass seed crops** *(off-label)*, **hazel nuts** *(off-label)*, **herbs (see appendix 6)** *(off-label)*, **hops** *(off-label)*, **lamb's lettuce** *(off-label)*, **leaf brassicas** *(off-label)*, **leeks** *(off-label)*, **lupins** *(off-label)*, **miscanthus** *(off-label)*, **mustard** *(off-label)*, **ornamental plant production** *(off-label)*, **plums** *(off-label)*, **protected**

forest nurseries (off-label), *protected ornamentals* (off-label), *raspberries* (off-label), *redcurrants* (off-label), *rubus hybrids* (off-label), *salad onions* (off-label), *scarole* (off-label), *shallots* (off-label), *spring linseed* (off-label), *strawberries* (off-label), *triticale* (off-label), *walnuts* (off-label), *winter linseed* (off-label) [3, 8]; *almonds* (off-label), *apples* (off-label), *apricots* (off-label), *bilberries* (off-label), *blueberries* (off-label), *cranberries* (off-label), *land cress* (off-label), *loganberries* (off-label), *medlar* (off-label), *nectarines* (off-label), *peaches* (off-label), *pears* (off-label), *protected all edible seed crops* (off-label), *protected all non-edible seed crops* (off-label), *quinces* (off-label), *rye* (off-label) [8]; *beetroot* (off-label), *blackberries* (off-label), *celeriac* (off-label), *celery (outdoor)* (off-label), *crambe* (off-label), *dewberries* (off-label), *fodder beet* (off-label), *honesty* (off-label), *horseradish* (off-label), *mallow (althaea spp.)* (off-label), *mirabelles* (off-label), *navy beans* (off-label), *parsley root* (off-label), *poppies for morphine production* (off-label), *protected aubergines* (off-label), *protected pak choi* (off-label), *protected peppers* (off-label), *protected strawberries* (off-label), *protected tatsoi* (off-label), *protected tomatoes* (off-label), *radishes* (off-label), *runner beans* (off-label), *soft fruit* (off-label), *spring rye* (off-label), *top fruit* (off-label), *whitecurrants* (off-label), *willow (short rotation coppice)* (off-label), *winter rye* (off-label) [3]
- Leaf curling midge in *blackcurrants* (off-label), *gooseberries* (off-label), *redcurrants* (off-label) [8]
- Leaf midge in *blackcurrants* (off-label), *gooseberries* (off-label), *redcurrants* (off-label), *whitecurrants* (off-label) [3]
- Leaf miner in *fodder beet* (off-label) [8]; *sugar beet* [1]
- Pea and bean weevil in *combining peas, spring field beans, vining peas, winter field beans* [1, 3, 5, 7, 10, 11]; *edible podded peas* [3, 5, 7, 10, 11]
- Pea aphid in *combining peas, edible podded peas, vining peas* [3, 5, 7, 10, 11]
- Pea midge in *combining peas, edible podded peas, vining peas* [3, 5, 7, 10, 11]
- Pea moth in *combining peas, vining peas* [1, 3, 5, 7, 10, 11]; *edible podded peas* [3, 5, 7, 10, 11]
- Peach-potato aphid in *asparagus* (off-label), *protected asparagus* (off-label) [3]
- Pear sucker in *pears* [1, 3, 5, 7, 10, 11]; *winter wheat* [1]
- Pod midge in *poppies* (off-label), *poppies grown for seed production* (off-label) [3]; *spring oilseed rape, winter oilseed rape* [3, 5, 7, 10, 11]
- Pollen beetle in *corn gromwell* (off-label), *poppies* (off-label), *poppies grown for seed production* (off-label) [3]; *crambe* (off-label) [8]; *poppies for morphine production* (off-label) [3, 8]; *spring oilseed rape, winter oilseed rape* [1, 3, 5, 7, 10, 11]
- Rosy rustic moth in *hops* (off-label) [3]
- Sawflies in *blackcurrants* (off-label), *gooseberries* (off-label), *redcurrants* (off-label) [3, 8]; *whitecurrants* (off-label) [3]
- Seed beetle in *broad beans* (off-label) [3]
- Seed weevil in *poppies* (off-label), *poppies grown for seed production* (off-label) [3]
- Silver Y moth in *beetroot* (off-label), *celery (outdoor)* (off-label), *dwarf beans* (off-label) [8]; *french beans* (off-label), *runner beans* (off-label) [3, 8]
- Spotted wing drosophila in *bilberries* (off-label), *blueberries* (off-label), *cranberries* (off-label), *ribes hybrids* (off-label) [3]
- Springtails in *poppies* (off-label), *poppies grown for seed production* (off-label) [3]
- Thrips in *bilberries* (off-label), *blueberries* (off-label), *cranberries* (off-label), *poppies* (off-label), *poppies grown for seed production* (off-label), *ribes hybrids* (off-label) [3]; *broad beans* (off-label), *bulb onions* (off-label), *garlic* (off-label), *leeks* (off-label), *salad onions* (off-label), *shallots* (off-label) [3, 8]
- Wasps in *grapevines* (off-label) [3, 8]
- Weevils in *combining peas, durum wheat, edible podded peas, potatoes, spring field beans, spring oilseed rape, spring wheat, sugar beet, vining peas, winter field beans, winter oilseed rape, winter wheat* [2, 6, 8]; *rye, spring barley, spring oats, triticale, winter barley, winter oats* [2]
- Wheat-blossom midge in *winter wheat* [3, 5, 7, 10, 11]
- Whitefly in *broccoli, brussels sprouts, cabbages, calabrese, cauliflowers* [3, 5, 7, 10]
- Willow aphid in *short rotation coppice willow* (off-label) [8]
- Willow beetle in *short rotation coppice willow* (off-label) [8]
- Willow sawfly in *short rotation coppice willow* (off-label) [8]
- Yellow cereal fly in *winter wheat* [1, 3, 5, 7, 10, 11]

FOR FULL CONDITIONS OF USE ALWAYS READ THE PRODUCT LABEL

Extension of Authorisation for Minor Use (EAMUs)
- *all edible seed crops grown outdoors* 20082944 [3], 20102880 [8]
- *all non-edible seed crops grown outdoors* 20082944 [3], 20102880 [8]
- *almonds* 20102880 [8]
- *apples* 20102880 [8]
- *apricots* 20102880 [8]
- *asparagus* 20182878 [3]
- *beetroot* 20060743 [3], 20073254 [8]
- *bilberries* 20140521 [3], 20102880 [8]
- *blackberries* 20060728 [3]
- *blackcurrants* 20060727 [3], 20073269 [8], 20102880 [8]
- *blueberries* 20140521 [3], 20102880 [8]
- *borage for oilseed production* 20060634 [3], 20073258 [8]
- *broad beans* 20060753 [3], 20073234 [8]
- *bulb onions* 20190109 [3], 20073256 [8]
- *canary flower (echium spp.)* 20060634 [3], 20073258 [8]
- *carrots* 20080201 [8]
- *celeriac* 20060731 [3], 20080204 [8]
- *celery (outdoor)* 20060744 [3], 20073257 [8]
- *cherries* 20131273 [3], 20102880 [8]
- *chestnuts* 20060742 [3], 20080206 [8], 20102880 [8]
- *chicory* 20060740 [3]
- *cob nuts* 20080206 [8]
- *corn gromwell* 20121047 [3]
- *crambe* 20081046 [3], 20102876 [8]
- *cranberries* 20140521 [3], 20102880 [8]
- *dewberries* 20060728 [3]
- *dwarf beans* 20080202 [8]
- *evening primrose* 20060634 [3], 20073258 [8]
- *farm forestry* 20082944 [3], 20102880 [8]
- *fennel* 20060733 [3]
- *filberts* 20080206 [8]
- *fodder beet* 20060637 [3], 20073233 [8]
- *forest nurseries* 20082944 [3], 20102880 [8]
- *french beans* 20060739 [3], 20080202 [8]
- *frise* 20190108 [3], 20073259 [8]
- *garlic* 20190109 [3], 20073256 [8]
- *gooseberries* 20060727 [3], 20073269 [8], 20102880 [8]
- *grapevines* 20060266 [3], 20080205 [8]
- *grass seed crops* 20060624 [3], 20073260 [8]
- *hazel nuts* 20060742 [3], 20080206 [8], 20102880 [8]
- *herbs (see appendix 6)* 20190108 [3], 20073259 [8]
- *honesty* 20060634 [3]
- *hops* 20082944 [3], 20131719 [3], 20102880 [8]
- *horseradish* 20071301 [3], 20080201 [8]
- *lamb's lettuce* 20190108 [3], 20073259 [8]
- *land cress* 20102878 [8]
- *leaf brassicas* 20190108 [3], 20073259 [8]
- *leeks* 20190109 [3], 20073256 [8]
- *loganberries* 20102880 [8]
- *lupins* 20060635 [3], 20102877 [8]
- *mallow (althaea spp.)* 20071301 [3], 20080201 [8]
- *medlar* 20102880 [8]
- *mirabelles* 20131273 [3]
- *miscanthus* 20082944 [3], 20102880 [8]
- *mustard* 20060634 [3], 20073258 [8]
- *navy beans* 20060739 [3]
- *nectarines* 20102880 [8]
- *ornamental plant production* 20082944 [3], 20102880 [8]

SEE SECTION 3 FOR PRODUCTS ALSO REGISTERED

- **parsley root** *20071301* [3], *20080201* [8]
- **parsnips** *20080201* [8]
- **peaches** *20102880* [8]
- **pears** *20102880* [8]
- **plums** *20131273* [3], *20102880* [8]
- **poppies** *20180013* [3]
- **poppies for morphine production** *20060749* [3], *20102875* [8]
- **poppies grown for seed production** *20180013* [3]
- **protected all edible seed crops** *20102880* [8]
- **protected all non-edible seed crops** *20102880* [8]
- **protected asparagus** *20182878* [3]
- **protected aubergines** *20121994* [3]
- **protected forest nurseries** *20082944* [3], *20102880* [8]
- **protected oriental cabbage** *20102879* [8]
- **protected ornamentals** *20082944* [3], *20102880* [8]
- **protected pak choi** *20081263* [3]
- **protected peppers** *20121994* [3]
- **protected strawberries** *20111705* [3]
- **protected tatsoi** *20081263* [3]
- **protected tomatoes** *20121994* [3]
- **quinces** *20102880* [8]
- **radishes** *20060731* [3], *20080204* [8]
- **raspberries** *20060728* [3], *20102880* [8]
- **redcurrants** *20060727* [3], *20073269* [8], *20102880* [8]
- **ribes hybrids** *20140521* [3]
- **rubus hybrids** *20060728* [3], *20102880* [8]
- **runner beans** *20060739* [3], *20080202* [8]
- **rye** *20073260* [8]
- **salad onions** *20190109* [3], *20073256* [8]
- **scarole** *20190108* [3], *20073259* [8]
- **shallots** *20190109* [3], *20073256* [8]
- **short rotation coppice willow** *20073268* [8]
- **soft fruit** *20082944* [3]
- **soya beans** *20180785* [7]
- **spring linseed** *20060634* [3], *20073258* [8]
- **spring rye** *20060624* [3]
- **strawberries** *20111705* [3], *20102880* [8]
- **swedes** *20101856* [3], *20102911* [8]
- **sweetcorn** *20060732* [3], *20080203* [8]
- **top fruit** *20082944* [3]
- **triticale** *20060624* [3], *20073260* [8]
- **turnips** *20101856* [3], *20102911* [8]
- **walnuts** *20060742* [3], *20080206* [8], *20102880* [8]
- **whitecurrants** *20060727* [3]
- **willow (short rotation coppice)** *20060748* [3]
- **winter linseed** *20060634* [3], *20073258* [8]
- **winter rye** *20060624* [3]

Approval information
- Lambda-cyhalothrin included in Annex I under EC Regulation 1107/2009
- Accepted by BBPA for use on malting barley and hops

Efficacy guidance
- Best results normally obtained from treatment when pest attack first seen. See label for detailed recommendations on each crop
- Timing for control of barley yellow dwarf virus vectors depends on specialist assessment of the level of risk in the area
- Repeat applications recommended in some crops where prolonged attack occurs, up to maximum total dose. See label for details
- Where strains of aphids resistant to lambda-cyhalothrin occur control is unlikely to be satisfactory

FOR FULL CONDITIONS OF USE ALWAYS READ THE PRODUCT LABEL

- Addition of wetter recommended for control of certain pests in brassicas and oilseed rape
- Use of sufficient water volume to ensure thorough crop penetration recommended for optimum results
- Use of drop-legged sprayer gives improved results in crops such as Brussels sprouts

Restrictions
- Maximum number of applications or maximum total dose per crop varies - see labels
- Do not apply to a cereal crop if any product containing a pyrethroid insecticide or dimethoate has been applied to the crop after the start of ear emergence (GS 51)
- Do not spray cereals in the spring/summer (i.e. after 1 Apr) within 6 m of edge of crop

Crop-specific information
- Latest use before late milk stage on cereals; before end of flowering for winter oilseed rape
- HI 3 d for radishes, red beet; 7 d for lettuce [3]; 14 d for carrots and parsnips; 25 d for peas, field beans; 6 wk for spring oilseed rape; 8 wk for sugar beet

Environmental safety
- Dangerous for the environment
- Very toxic to aquatic organisms
- Flammable [1]
- To protect non-target arthropods respect an untreated buffer zone of 5 m to non-crop land
- Since there is a high risk to non-target insects or other anthropods, do not spray cereals in spring/summer i.e. after 1st April within 6m of the field boundary [4, 9]
- Dangerous to bees
- Broadcast air-assisted LERAP [2, 3, 5-8, 10] (25 m); Broadcast air-assisted LERAP [1, 11] (38 m); LERAP Category A [1, 4, 11]; LERAP Category B [2, 3, 5-10]

Hazard classification and safety precautions
Hazard Harmful, Dangerous for the environment, Harmful if inhaled [1-11]; Corrosive, Toxic if swallowed [2, 6, 8]; Flammable, Flammable liquid and vapour [1]; Harmful if swallowed [1, 3-5, 7, 10, 11]; Very toxic to aquatic organisms [1, 3-8, 10, 11]
Transport code 3 [1]; 8 [2, 6, 8]; 9 [3-5, 7, 9-11]
Packaging group II [2]; III [1, 3-11]
UN Number 1760 [2, 6, 8]; 1993 [1]; 3072 [9]; 3077 [4]; 3082 [3, 5, 7, 10, 11]
Risk phrases H304, H336 [1, 2, 6, 8]; H314 [2, 6, 8]; H317 [3-5, 10, 11]; H319 [1]; H320 [4]
Operator protection A, H [1-11]; C [1, 2, 4, 6, 8, 9]; J, K, M [4, 9]; U02a [2-11]; U04a [2, 6, 8]; U05a [1-11]; U08 [1-3, 5-8, 10, 11]; U09a [4, 9]; U11 [2, 4, 6, 8, 9]; U14, U20b [3-5, 7, 9-11]; U19a, U20a [1, 2, 6, 8]
Environmental protection E12a [2, 6, 8]; E12c [1, 3-5, 7, 9-11]; E15b [1, 2, 4, 6, 8, 9]; E16a [2, 3, 5-10]; E16c [1, 4, 11]; E17b [1] (38 m); E17b [2, 3, 5-8, 10] (25 m); E17b [11] (38 m); E22b [2-4, 7-9]; E34 [1-3, 5-8, 10, 11]; E38 [2-11]; H410 [1-11]
Storage and disposal D01, D02 [1-11]; D05, D07 [1]; D09a [1-3, 5-8, 10, 11]; D10b [1, 2, 6, 8]; D10c [3, 5, 7, 10, 11]; D12a [1, 3-5, 7, 9-11]
Medical advice M03 [3-5, 7, 9-11]; M04a [2, 6, 8]; M05b [1, 2, 6, 8]

262 laminarin

A fungicide for winter wheat

Products

1	Iodus	Arysta	37 g/l	SL	19163
2	Vacciplant	Arysta	37 g/l	SL	13260

Uses
- Powdery mildew in **winter wheat**
- Septoria leaf blotch in **winter wheat**

Approval information
- Laminarin included in annex 1 under EC Regulation 1107/2009

SEE SECTION 3 FOR PRODUCTS ALSO REGISTERED

Hazard classification and safety precautions
    UN Number N/C
    Risk phrases H290, H312, H314, H335
    Operator protection A, C, H
    Environmental protection E15b, H411
    Storage and disposal D01, D02, D05, D09a, D10b

263   Lecanicillium muscarium

A fungal pathogen of whitefly
FRAC mode of action code: 44

Products

Mycotal	Koppert	4.8% w/w	WP	16644

Uses

- Spider mites in *leaf brassicas* (off-label), *protected bilberries* (off-label), *protected blackberries* (off-label), *protected blackcurrants* (off-label), *protected blueberry* (off-label), *protected cayenne peppers* (off-label), *protected cranberries* (off-label), *protected gooseberries* (off-label), *protected herbs (see appendix 6)* (off-label), *protected hops* (off-label), *protected loganberries* (off-label), *protected nursery fruit trees* (off-label), *protected raspberries* (off-label), *protected redcurrants* (off-label), *protected ribes hybrids* (off-label), *protected rubus hybrids* (off-label), *protected table grapes* (off-label), *protected wine grapes* (off-label)
- Thrips in *leaf brassicas* (off-label), *protected bilberries* (off-label), *protected blackberries* (off-label), *protected blackcurrants* (off-label), *protected blueberry* (off-label), *protected cayenne peppers* (off-label), *protected cranberries* (off-label), *protected gooseberries* (off-label), *protected herbs (see appendix 6)* (off-label), *protected hops* (off-label), *protected loganberries* (off-label), *protected nursery fruit trees* (off-label), *protected raspberries* (off-label), *protected redcurrants* (off-label), *protected ribes hybrids* (off-label), *protected rubus hybrids* (off-label), *protected table grapes* (off-label), *protected wine grapes* (off-label)
- Whitefly in *leaf brassicas* (off-label), *protected bilberries* (off-label), *protected blackberries* (off-label), *protected blackcurrants* (off-label), *protected blueberry* (off-label), *protected cayenne peppers* (off-label), *protected cranberries* (off-label), *protected cucumbers*, *protected gooseberries* (off-label), *protected herbs (see appendix 6)* (off-label), *protected hops* (off-label), *protected loganberries* (off-label), *protected nursery fruit trees* (off-label), *protected ornamentals*, *protected peppers*, *protected raspberries* (off-label), *protected redcurrants* (off-label), *protected ribes hybrids* (off-label), *protected rubus hybrids* (off-label), *protected strawberries*, *protected table grapes* (off-label), *protected tomatoes*, *protected wine grapes* (off-label)

Extension of Authorisation for Minor Use (EAMUs)

- *leaf brassicas* 20142679
- *protected bilberries* 20142685
- *protected blackberries* 20142684
- *protected blackcurrants* 20142685
- *protected blueberry* 20142685
- *protected cayenne peppers* 20142680
- *protected cranberries* 20142685
- *protected gooseberries* 20142685
- *protected herbs (see appendix 6)* 20142679
- *protected hops* 20142681
- *protected loganberries* 20142684
- *protected nursery fruit trees* 20142681
- *protected raspberries* 20142684
- *protected redcurrants* 20142685
- *protected ribes hybrids* 20142685
- *protected rubus hybrids* 20142684
- *protected table grapes* 20142685
- *protected wine grapes* 20142685

FOR FULL CONDITIONS OF USE ALWAYS READ THE PRODUCT LABEL

Approval information
- *Lecanicillium* included in Annex I under EC Regulation 1107/2009

Efficacy guidance
- *Lecanicillium muscarium* is a pathogenic fungus that infects the target pests and destroys them
- Apply spore powder as spray as part of biological control programme keeping the spray liquid well agitated
- Treat before infestations build to high levels and repeat as directed on the label
- Spray during late afternoon and early evening directing spray onto underside of leaves and to growing points
- Best results require minimum 80% relative humidity (70% if applied with the adjuvant Addit) and 18°C within the crop canopy

Restrictions
- Never use in tank mixture
- Do not use a fungicide within 3 d of treatment. Pesticides containing captan, imazalil, or prochloraz may not be used on the same crop
- Keep in a refrigerated store at 2-6°C but do not freeze

Environmental safety
- Products have negligible effects on commercially available natural predators or parasites but consult manufacturer before using with a particular biological control agent for the first time

Hazard classification and safety precautions
UN Number N/C
Operator protection A, D, H; U19a, U20b
Environmental protection E15a
Storage and disposal D09a, D11a

264   lenacil

A residual, soil-acting uracil herbicide for beet crops
HRAC mode of action code: C1

See also desmedipham + ethofumesate + lenacil + phenmedipham

Products

1 Lenazar Flo 500 SC	Belcrop	500 g/l	SC	18124
2 Venzar 500 SC	FMC Agro	500 g/l	SC	18799

Uses
- Annual dicotyledons in **sugar beet** [1, 2]
- Annual meadow grass in **sugar beet** [1, 2]
- Black bindweed in **baby leaf crops** *(off-label)*, **edible flowers** *(off-label)*, **farm woodland** *(off-label)*, **fodder beet** *(off-label)*, **herbs (see appendix 6)** *(off-label)*, **mangels** *(off-label)*, **red beet** *(off-label)*, **red mustard** *(off-label)*, **spinach** *(off-label)* [2]
- Brassica spp. in **baby leaf crops** *(off-label)*, **edible flowers** *(off-label)*, **farm woodland** *(off-label)*, **fodder beet** *(off-label)*, **herbs (see appendix 6)** *(off-label)*, **mangels** *(off-label)*, **red beet** *(off-label)*, **red mustard** *(off-label)*, **spinach** *(off-label)* [2]
- Polygonums in **baby leaf crops** *(off-label)*, **edible flowers** *(off-label)*, **farm woodland** *(off-label)*, **fodder beet** *(off-label)*, **herbs (see appendix 6)** *(off-label)*, **mangels** *(off-label)*, **red beet** *(off-label)*, **red mustard** *(off-label)*, **spinach** *(off-label)* [2]

Extension of Authorisation for Minor Use (EAMUs)
- **baby leaf crops** *20191357* [2]
- **edible flowers** *20191357* [2]
- **farm woodland** *20190731* [2]
- **fodder beet** *20192585* [2]
- **herbs (see appendix 6)** *20191357* [2]
- **mangels** *20192585* [2]

SEE SECTION 3 FOR PRODUCTS ALSO REGISTERED

SECTION 2

- ***red beet*** *20192585* [2]
- ***red mustard*** *20191357* [2]
- ***spinach*** *20191357* [2]

Approval information
- Lenacil included in Annex I under EC Regulation 1107/2009

Efficacy guidance
- Best results, especially from pre-emergence treatments, achieved on fine, even, firm and moist soils free from clods. Continuing presence of moisture from rain or irrigation gives improved residual control of later germinating weeds. Effectiveness may be reduced by dry conditions
- On beet crops may be used pre- or post-emergence, alone or in mixture to broaden weed spectrum
- Apply overall or as band spray to beet crops pre-drilling incorporated, pre- or post-emergence
- All labels have limitations on soil types that may be treated. Residual activity reduced on soils with high OM content
- May only be applied once in every three years

Restrictions
- Do not use any other residual herbicide within 3 mth of the initial application to fruit or ornamental crops
- Do not treat crops under stress from drought, low temperatures, nutrient deficiency, pest or disease attack, or waterlogging
- Must only be applied after BBCH10 growth stage
- May only be applied once in every three years

Crop-specific information
- Latest use: pre-emergence for red beet, fodder beet, spinach, spinach beet, mangels; before leaves meet over rows when used on these crops post-emergence; 24 h after planting new strawberry runners or before flowering for established strawberry crops, blackcurrants, gooseberries, raspberries
- Heavy rain after application to beet crops may cause damage especially if followed by very hot weather
- Reduction in beet stand may occur where crop emergence or vigour is impaired by soil capping or pest attack
- Strawberry runner beds to be treated should be level without depressions around the roots
- New soft fruit cuttings should be planted at least 15 cm deep and firmed before treatment
- Check varietal tolerance of ornamentals before large scale treatment

Following crops guidance
- Succeeding crops should not be planted or sown for at least 4 mth (6 mth on organic soils) after treatment following ploughing to at least 150 mm.
- Only beet crops, mangels or strawberries may be sown within 4 months of treatment

Environmental safety
- Dangerous for the environment
- Very toxic to aquatic organisms
- LERAP Category B

Hazard classification and safety precautions
Hazard Dangerous for the environment
Transport code 9
Packaging group III
UN Number 3082
Risk phrases H351
Operator protection A, C; U05a, U08, U19a, U20a
Environmental protection E13c, E16a, E38, H410
Storage and disposal D01, D02, D05, D09a, D10a, D12a

FOR FULL CONDITIONS OF USE ALWAYS READ THE PRODUCT LABEL

265   lenacil + triflusulfuron-methyl

A foliar and residual herbicide mixture for sugar beet
HRAC mode of action code: C1 + B

See also triflusulfuron-methyl

Products

1 Debut Plus	FMC Agro	71.4:5.4% w/w	WG	18765
2 Safari Lite WSB	FMC Agro	71.4:5.4% w/w	ZZ	18784

Uses
- Annual dicotyledons in *sugar beet* [1, 2]
- Black bindweed in *fodder beet* *(off-label)* [2]
- Charlock in *fodder beet* *(off-label)* [2]
- Chickweed in *fodder beet* *(off-label)* [2]
- Cleavers in *fodder beet* *(off-label)* [2]
- Fat hen in *fodder beet* *(off-label)* [2]
- Field pansy in *fodder beet* *(off-label)* [2]
- Fool's parsley in *fodder beet* *(off-label)* [2]
- Fumitory in *fodder beet* *(off-label)* [2]
- Knotgrass in *fodder beet* *(off-label)* [2]
- Mayweeds in *fodder beet* *(off-label)* [2]
- Red dead-nettle in *fodder beet* *(off-label)* [2]
- Redshank in *fodder beet* *(off-label)* [2]
- Scentless mayweed in *fodder beet* *(off-label)* [2]
- Small nettle in *fodder beet* *(off-label)* [2]
- Volunteer oilseed rape in *fodder beet* *(off-label)* [2]; *sugar beet* [1, 2]

Extension of Authorisation for Minor Use (EAMUs)
- *fodder beet* 20190808 [2]

Approval information
- Lenacil and triflusulfuron-methyl included in Annex I under EC Regulation 1107/2009

Efficacy guidance
- Best results obtained when weeds are small and growing actively
- Product recommended for use in a programme of treatments in tank mixture with a suitable herbicide partner to broaden the weed spectrum
- Ensure good spray cover of weeds. Apply when first weeds have emerged provided crop has reached cotyledon stage
- Susceptible plants cease to grow almost immediately after treatment and symptoms can be seen 5-10 d later
- Weed control may be reduced in very dry soil conditions
- Triflusulfuron-methyl is a member of the ALS-inhibitor group of herbicides

Restrictions
- Maximum number of treatments on sugar beet 3 per crop and do not apply more than 4 applications of any product containing triflusulfuron-methyl
- Do not apply to crops suffering from stress caused by drought, water-logging, low temperatures, pest or disease attack, nutrient deficiency or any other factors affecting crop growth
- Do not use on Sands, stony or gravelly soils or on soils with more than 10% organic matter
- Do not apply when temperature above or likely to exceed 21°C on day of spraying or under conditions of high light intensity

Crop-specific information
- Latest use: before crop leaves meet between rows for sugar beet

Following crops guidance
- Only cereals may be sown in the same calendar yr as a treated sugar beet crop. Any crop may be sown in the following spring
- In the event of crop failure sow only sugar beet within 4 mth of treatment

SECTION 2

Environmental safety
- Dangerous for the environment
- Very toxic to aquatic organisms
- Take extreme care to avoid drift onto broad-leaved plants outside the target area or onto surface waters or ditches, or land intended for cropping
- Spraying equipment should not be drained or flushed onto land planted, or to be planted, with trees or crops other than cereals and should be thoroughly cleansed after use - see label for instructions
- LERAP Category B

Hazard classification and safety precautions
Hazard Irritant, Dangerous for the environment
Transport code 9
Packaging group III
UN Number 3077
Risk phrases H351
Operator protection A, C; U05a, U08, U11, U19a, U20b, U22a
Environmental protection E15a, E16a, E38, H410
Storage and disposal D01, D02, D09a, D12a

266   magnesium phosphide

A phosphine generating compound used to control insect pests in stored commodities

Products

| Degesch Plates | Rentokil | 56% w/w | GE | 17105 |

Uses
- Insect pests in *carob, cocoa, coffee, crop handling & storage structures, herbal infusions, processed consumable products, stored dried spices, stored grain, stored linseed, stored nuts, stored oilseed rape, stored pulses, tea, tobacco*

Approval information
- Magnesium phosphide included in Annex I under EC Regulation 1107/2009
- Accepted by BBPA for use in stores for malting barley

Efficacy guidance
- Product acts as fumigant by releasing poisonous hydrogen phosphide gas on contact with moisture in the air
- Place plates on the floor or wall of the building or on the surface of the commodity. Exposure time varies depending on temperature and pest. See label

Restrictions
- Magnesium phosphide is subject to the Poisons Rules 1982 and the Poisons Act 1972. See Section 5 for more information
- Only to be used by professional operators trained in the use of magnesium phosphide and familiar with the precautionary measures to be observed. See label for full precautions

Environmental safety
- Highly flammable
- Prevent access to buildings under fumigation by livestock, pets and other non-target mammals and birds
- Dangerous to fish or other aquatic life. Do not contaminate surface waters or ditches with chemical or used container
- Keep in original container, tightly closed, in a safe place, under lock and key
- Do not allow plates or their spent residues to come into contact with food other than raw cereal grains
- Remove used plates after treatment. Do not bulk spent plates and residues: spontaneous ignition could result
- Keep livestock out of treated areas

SECTION 2

Hazard classification and safety precautions

Hazard Very toxic, Highly flammable, Dangerous for the environment, In contact with water releases flammable gases which may ignite spontaneously, Fatal if swallowed, Fatal if inhaled, Very toxic to aquatic organisms

Transport code 4.3
Packaging group I
UN Number 2011
Risk phrases H312
Operator protection A, D, H; U01, U05b, U07, U13, U19a, U20a
Environmental protection E02a (4 h min); E02b, E13b, E34
Storage and disposal D01, D02, D05, D07, D09b, D11b
Medical advice M04a

267   maleic hydrazide

A pyridazine plant growth regulator suppressing sprout and bud growth

See also fatty acids + maleic hydrazide
fatty acids + pelargonic acid + maleic hydrazide
maleic hydrazide + pelargonic acid

Products

1	Crown MH	Certis	270 g/l	SL	18018
2	Fazor	Arysta	60% w/w	SG	19074

Uses

- Growth regulation in **bulb onions**, **potatoes**

Approval information

- Maleic hydrazide included in Annex I under EC Regulation 1107/2009

Efficacy guidance

- Uniform coverage and dry weather necessary for effective results
- Accurate timing essential for good results on potatoes but rain or irrigation within 24 h may reduce effectiveness on onions and potatoes
- Mow 2-3 d before and 5-10 d after spraying for best results. Need for mowing reduced for up to 6 wk
- When used for suppression of volunteer potatoes treatment may also give some suppression of sprouting in store but separate treatment will be necessary if sprouting occurs

Restrictions

- Do not apply in drought or when crops are suffering from pest, disease or herbicide damage.
- Do not treat potatoes within 3 wk of applying a haulm desiccant or if temperatures above 26°C
- Consult processor before use on potato crops for processing
- Do not apply to seed potatoes or first year onion sets

Crop-specific information

- Latest use: 3 wk before haulm destruction for potatoes; before 50% necking for onions
- HI onions 1 wk; potatoes 3 wk
- Apply to onions at 10% necking and not later than 50% necking stage when the tops are still green
- Only treat onions in good condition and properly cured, and do not treat more than 2 wk before maturing. Treated onions may be stored until Mar but must then be removed to avoid browning
- Apply to second early or maincrop potatoes at least 3 wk before haulm destruction
- Only treat potatoes of good keeping quality; not on seed, first earlies or crops grown under polythene

Environmental safety

- Do not use treated water for irrigation purposes within 3 wk of treatment or until concentration in water falls below 0.02 ppm

SEE SECTION 3 FOR PRODUCTS ALSO REGISTERED

- Maximum permitted concentration in water 2 ppm
- Do not dump surplus product in water or ditch bottoms
- Avoid drift onto nearby vegetables, flowers or other garden plants

Hazard classification and safety precautions
  Transport code 9
  Packaging group III
  UN Number 3077 [2]; 3082 [1]
  Operator protection A, H [1, 2]; D [2]; U08, U20b
  Environmental protection E15a, H411
  Storage and disposal D09a, D11a

268   maltodextrin

A polysaccharide used as a food additive and with activity against red spider mites
FRAC mode of action code: 12 + IRAC 3

Products

1	Eradicoat	Certis	598 g/l	SL	17121
2	Majestik	Certis	598 g/l	SL	17240

Uses
- Aphids in *all edible crops (outdoor)*, *all non-edible crops (outdoor)*, *all protected edible crops*, *all protected non-edible crops*
- Spider mites in *all edible crops (outdoor)*, *all non-edible crops (outdoor)*, *all protected edible crops*, *all protected non-edible crops*
- Whitefly in *all edible crops (outdoor)*, *all non-edible crops (outdoor)*, *all protected edible crops*, *all protected non-edible crops*

Approval information
- Maltodextrin is included in Annex 1 under EC Regulation 1107/2009

Restrictions
- Reasonable precautions must be taken to prevent access of birds, wild mammals and honey bees to treated crops

Hazard classification and safety precautions
  Hazard Irritant
  UN Number N/C
  Risk phrases H319
  Operator protection A, H; U05a, U09a, U14, U19a, U20b
  Environmental protection E12a [2]; E12e [1]; H412 [1, 2]
  Storage and disposal D01, D02, D09a, D10b, D12a

269   mancozeb

A protective dithiocarbamate fungicide for potatoes and other crops
FRAC mode of action code: M3

See also ametoctradin + mancozeb
        benalaxyl + mancozeb
        benthiavalicarb-isopropyl + mancozeb
        cymoxanil + mancozeb
        dimethomorph + mancozeb
        fenamidone + mancozeb

Products

1	Clayton Zebo	Clayton	75% w/w	WG	18460

Products – continued

2	Dithane 945	Sumitomo	80% w/w	WP	14621
3	Dithane NT Dry Flowable	Sumitomo	75% w/w	WG	14704
4	Laminator 75 WG	Sumitomo	75% w/w	WG	15667
5	Laminator Flo	Sumitomo	455 g/l	SC	19104
6	Malvi	AgChem Access	75% w/w	WG	14981
7	Manzate 75 WG	UPL Europe	75% w/w	WG	15052
8	Mewati	AgChem Access	80% w/w	WP	15763
9	Penncozeb 80 WP	UPL Europe	80% w/w	WP	16953
10	Penncozeb WDG	UPL Europe	75% w/w	WG	16885
11	Quell Flo	Sumitomo	455 g/l	SC	15237
12	Unizeb Gold	UPL Europe	500 g/l	SC	18186

Uses
- Alternaria blight in **carrots**, **parsnips** [3]
- Blight in **potatoes** [1-11]; **tomatoes (outdoor)** [4, 6]
- Botrytis in **flower bulbs** [1, 7]
- Brown rust in **durum wheat** *(useful control)*, **spring wheat** *(useful control)*, **winter wheat** *(useful control)* [2, 3, 8]; **spelt**, **triticale** [12]; **spring wheat**, **winter wheat** [5, 11, 12]
- Disease control in **apples**, **table grapes**, **wine grapes** [4, 6]; **forest nurseries** *(off-label)* [2, 10]; **ornamental plant production** [10]; **ornamental plant production** *(off-label)* [2]
- Downy mildew in **bulb onions** [1, 7, 9, 10]; **forest nurseries** *(off-label)*, **ornamental plant production** [10]
- Early blight in **potatoes** [2, 3, 8]
- Rust in **bulb onions**, **shallots** [3]
- Scab in **apples** [1, 7, 9, 10]
- Septoria leaf blotch in **durum wheat** *(reduction)* [2, 3, 8]; **spelt**, **triticale** [12]; **spring wheat**, **winter wheat** [5, 9-12]; **spring wheat** *(reduction)*, **winter wheat** *(reduction)* [1-3, 7, 8]
- Sooty moulds in **spelt**, **triticale** [12]; **spring wheat**, **winter wheat** [5, 11, 12]
- Yellow rust in **winter wheat** [5, 11]

Extension of Authorisation for Minor Use (EAMUs)
- **forest nurseries** *20152652* [2], *20170900* [10]
- **ornamental plant production** *20152652* [2]

Approval information
- Mancozeb included in Annex I under EC Regulation 1107/2009

Efficacy guidance
- Mancozeb is a protectant fungicide and will give moderate control, suppression or reduction of the cereal diseases listed if treated before they are established but in many cases mixture with carbendazim is essential to achieve satisfactory results. See labels for details
- May be recommended for suppression or control of mildew in cereals depending on product and tank mix. See label for details

Restrictions
- Maximum number of treatments varies with crop and product used - check labels for details
- Check labels for minimum interval that must elapse between treatments
- Avoid treating wet cereal crops or those suffering from drought or other stress
- Keep dry formulations away from fire and sparks
- Use dry formulations immediately. Do not store
- Avoid spraying within 5m of the field boundary to reduce effects on non-target insects or other arthropods [12]
- Spray concentration must not exceed 0.4 kg product per 100 L of water. The rate of product applied must not exceed 2.0 kg product per hectare irrespective of the water volume being used

Crop-specific information
- Latest use: before early milk stage (GS 73) for cereals;
- HI potatoes 7 d; apples, blackcurrants 28 d
- Apply to potatoes before haulm meets across rows (usually mid-Jun) or at earlier blight warning, and repeat every 7-14 d depending on conditions and product used (see label)
- May be used on potatoes up to desiccation of haulm

SEE SECTION 3 FOR PRODUCTS ALSO REGISTERED

- Apply to cereals from 4-leaf stage to before early milk stage (GS 71). Recommendations vary, see labels for details

Environmental safety
- Dangerous for the environment
- Very toxic to aquatic organisms
- Harmful to fish or other aquatic life. Do not contaminate surface waters or ditches with chemical or used container
- Do not empty into drains
- Buffer zone for application via a broadcast sprayer is 10m for grapes, 40m for apples
- LERAP Category B

Hazard classification and safety precautions
Hazard Harmful [4, 9, 10]; Irritant [1-3, 5-8, 11, 12]; Dangerous for the environment [1-4, 6-12]; Very toxic to aquatic organisms [1-3, 8]
Transport code 9
Packaging group III
UN Number 3077 [1-4, 6-10]; 3082 [5, 11, 12]
Risk phrases H317 [1-12]; H319 [1, 3]; H334 [5, 11]; H335 [4, 6, 7]; H360 [10]; H361 [1-5, 8, 9, 11, 12]
Operator protection A [1-8, 10-12]; C [6]; D [1-4, 7, 8, 10]; H [2, 8, 10]; U05a [1-8, 10-12]; U08 [5, 6, 11, 12]; U11 [6]; U14 [1-4, 7-12]; U19a [6, 9, 10]; U20b [1-8, 11, 12]
Environmental protection E13c [5]; E15a [1-4, 6-8, 11, 12]; E15b [9, 10]; E16a [1-12]; E19b [10]; E34 [5, 6, 11, 12]; E38 [1-4, 6-12]; H410 [2-12]; H411 [1]
Storage and disposal D01; D12a [1-4, 6-12]; D02 [1-12]; D05 [1-8, 11, 12]; D07 [5, 6]; D09a [1-8, 10-12]; D10a [6]; D11a [1-5, 7, 8, 11, 12]
Medical advice M05a [9, 10]

270   mancozeb + metalaxyl-M

A systemic and protectant fungicide mixture
FRAC mode of action code: M3 + 4

See also metalaxyl-M

Products

1	Clayton Mohawk	Clayton	64:4% w/w	WG	18133
2	Clayton Mohawk	Clayton	64:4% w/w	WG	18173
3	Ensis	AgChem Access	64:4% w/w	WG	15919
4	Fubol Gold WG	Syngenta	64:4% w/w	WG	14605

Uses
- Blight in **potatoes** [1-4]
- Disease control in **bilberries** *(off-label)*, **blackberries** *(off-label)*, **blackcurrants** *(off-label)*, **blueberries** *(off-label)*, **cranberries** *(off-label)*, **forest nurseries** *(off-label)*, **gooseberries** *(off-label)*, **hops** *(off-label)*, **loganberries** *(off-label)*, **ornamental plant production** *(off-label)*, **protected bilberries** *(off-label)*, **protected blackberries** *(off-label)*, **protected blackcurrants** *(off-label)*, **protected blueberry** *(off-label)*, **protected cranberries** *(off-label)*, **protected forest nurseries** *(off-label)*, **protected gooseberries** *(off-label)*, **protected hops** *(off-label)*, **protected loganberries** *(off-label)*, **protected ornamentals** *(off-label)*, **protected raspberries** *(off-label)*, **protected redcurrants** *(off-label)*, **protected rubus hybrids** *(off-label)*, **protected strawberries** *(off-label)*, **raspberries** *(off-label)*, **redcurrants** *(off-label)*, **rubus hybrids** *(off-label)*, **strawberries** *(off-label)* [4]
- Downy mildew in **baby leaf crops** *(off-label)*, **herbs (see appendix 6)** *(off-label)*, **lettuce** *(off-label)*, **poppies for morphine production** *(off-label)*, **protected herbs (see appendix 6)** *(off-label)*, **rhubarb** *(off-label)*, **salad onions** *(off-label)* [4]; **bulb onions** *(useful control)*, **shallots** *(useful control)* [1-4]
- Fungus diseases in **apple orchards** *(off-label)* [4]
- Pink rot in **potatoes** *(reduction)* [1, 2, 4]
- White blister in **cabbages** *(off-label)* [4]

FOR FULL CONDITIONS OF USE ALWAYS READ THE PRODUCT LABEL

Extension of Authorisation for Minor Use (EAMUs)

- *apple orchards* 20132282 [4]
- *baby leaf crops* 20132285 [4]
- *bilberries* 20132288 [4]
- *blackberries* 20132288 [4]
- *blackcurrants* 20132288 [4]
- *blueberries* 20132288 [4]
- *cabbages* 20132281 [4]
- *cranberries* 20132288 [4]
- *forest nurseries* 20132288 [4]
- *gooseberries* 20132288 [4]
- *herbs (see appendix 6)* 20132285 [4]
- *hops* 20132288 [4]
- *lettuce* 20132285 [4]
- *loganberries* 20132288 [4]
- *ornamental plant production* 20132288 [4]
- *poppies for morphine production* 20132286 [4]
- *protected bilberries* 20132288 [4]
- *protected blackberries* 20132288 [4]
- *protected blackcurrants* 20132288 [4]
- *protected blueberry* 20132288 [4]
- *protected cranberries* 20132288 [4]
- *protected forest nurseries* 20132288 [4]
- *protected gooseberries* 20132288 [4]
- *protected herbs (see appendix 6)* 20132287 [4]
- *protected hops* 20132288 [4]
- *protected loganberries* 20132288 [4]
- *protected ornamentals* 20132288 [4]
- *protected raspberries* 20132288 [4]
- *protected redcurrants* 20132288 [4]
- *protected rubus hybrids* 20132288 [4]
- *protected strawberries* 20132288 [4]
- *raspberries* 20132288 [4]
- *redcurrants* 20132288 [4]
- *rhubarb* 20132283 [4]
- *rubus hybrids* 20132288 [4]
- *salad onions* 20132284 [4]
- *strawberries* 20132288 [4]

Approval information

- Mancozeb and metalaxyl-M included in Annex I under EC Regulation 1107/2009

Efficacy guidance

- Commence potato blight programme before risk of infection occurs as crops begin to meet along the rows and repeat every 7-14 d according to blight risk. Do not exceed a 14 d interval between sprays
- If infection risk conditions occur earlier than the above growth stage commence spraying potatoes immediately
- Complete the potato blight programme using a protectant fungicide starting no later than 10 d after the last phenylamide spray. At least 2 such sprays should be applied
- To minimise the likelihood of development of resistance these products should be used in a planned Resistance Management strategy. See Section 5 for more information

Crop-specific information

- Latest use: before end of active potato haulm growth or before end Aug, whichever is earlier
- HI 7 d for potatoes
- After treating early potatoes destroy and remove any remaining haulm after harvest to minimise blight pressure on neighbouring maincrop potatoes

Environmental safety

- Dangerous for the environment

SEE SECTION 3 FOR PRODUCTS ALSO REGISTERED

- Very toxic to aquatic organisms
- Do not harvest crops for human consumption for at least 7 d after final application
- LERAP Category B

Hazard classification and safety precautions

Hazard Harmful, Dangerous for the environment [1-4]; Very toxic to aquatic organisms [2-4]
Transport code 9
Packaging group III
UN Number 3077
Risk phrases H317, H361 [2-4]; R37, R43, R50, R53a, R63 [1]
Operator protection A; U05a, U08, U20b
Environmental protection E15a, E16a, E34, E38 [1-4]; H410 [2-4]
Consumer protection C02a (7 d)
Storage and disposal D01, D02, D05, D09a, D11a, D12a

271   mancozeb + zoxamide

A protectant fungicide mixture for potatoes
FRAC mode of action code: M3 + 22

See also zoxamide

Products

1	Electis 75WG	Gowan	66.7:8.3% w/w	WG	17871
2	Roxam 75WG	Gowan	66.7:8.3% w/w	WG	17869
3	Unikat 75WG	Gowan	66.7:8.3% w/w	WG	17868

Uses

- Blight in *potatoes* [1-3]
- Disease control in *ornamental plant production* (off-label) [1]; *wine grapes* [1-3]

Extension of Authorisation for Minor Use (EAMUs)

- *ornamental plant production* 20170462 [1]

Approval information

- Mancozeb and zoxamide included in Annex I under EC Regulation 1107/2009

Efficacy guidance

- Apply as protectant spray on potatoes immediately risk of blight in district or as crops begin to meet along the rows and repeat every 7-10 d according to blight risk
- Do not use if potato blight present in crop. Products are not curative
- Spray irrigated potato crops as soon as possible after irrigation once the crop leaves are dry

Restrictions

- Maximum number of treatments 8 per crop for potatoes

Crop-specific information

- HI 7 d for potatoes

Environmental safety

- Dangerous for the environment
- Very toxic to aquatic organisms
- Keep away from fire and sparks
- LERAP Category B

Hazard classification and safety precautions

Hazard Irritant, Dangerous for the environment, Very toxic to aquatic organisms
Transport code 9
Packaging group III
UN Number 3077
Risk phrases H317, H361

Operator protection A, D, H; U05a, U14, U20b [1-3]; U19a [1, 2]
Environmental protection E15a, E16a, E16b, E38, H410
Storage and disposal D01, D02, D05, D09a, D11a, D12a

272   mandipropamid

A mandelamide fungicide for the control of potato blight
FRAC mode of action code: 40

See also cymoxanil + mandipropamid
difenoconazole + mandipropamid

Products

Revus	Syngenta	250 g/l	SC	17443

Uses

- Blight in **potatoes**
- Disease control in **baby leaf crops, celery leaves, chives, cress, endives, herbs (see appendix 6), lamb's lettuce, land cress, lettuce, protected baby leaf crops, protected chives, protected cress, protected endives, protected herbs (see appendix 6), protected lamb's lettuce, protected land cress, protected lettuce, protected red mustard, protected spinach, protected spinach beet, red mustard, spinach, spinach beet**
- Downy mildew in **baby leaf crops, broccoli, calabrese, celery leaves, chives, cress, endives, herbs (see appendix 6), hops** *(off-label)*, **lamb's lettuce, land cress, lettuce, ornamental plant production** *(off-label)*, **protected baby leaf crops, protected chives, protected cress, protected endives, protected herbs (see appendix 6), protected lamb's lettuce, protected land cress, protected lettuce, protected ornamentals** *(off-label)*, **protected red mustard, protected spinach, protected spinach beet, red mustard, spinach, spinach beet**
- Tuber blight in **potatoes** *(protection)*

Extension of Authorisation for Minor Use (EAMUs)

- *hops 20162024*
- *ornamental plant production 20162763*
- *protected ornamentals 20162763*

Approval information

- Mandipropamid is included in Annex 1 under EC Regulation 1107/2009

Efficacy guidance

- Mandipropamid acts preventatively by preventing spore germination and inhibiting mycelial growth during incubation. Apply immediately after blight warning or as soon as local conditions favour disease development but before blight enters the crop
- Spray at 7-10 d intervals reducing the interval as blight risk increases
- Spray programme should include a complete haulm desiccant to prevent tuber infection at harvest
- Eliminate other potential infection sources
- To minimise the likelihood of development of resistance this product should be used in a planned Resistance Management strategy. See Section 5 for more information
- See label for details of tank mixtures that may be used as part of a resistance management strategy
- Rainfast in potatoes within 15 minutes of application

Restrictions

- Maximum number of treatments 4 per crop. Do not apply more than 3 treatments of this, or any other fungicide in the same resistance category, consecutively

Crop-specific information

- HI 3 d for potatoes

Environmental safety

- Buffer zone requirement 10 m in hops [1]

SEE SECTION 3 FOR PRODUCTS ALSO REGISTERED

Hazard classification and safety precautions
 UN Number N/C
 Operator protection A, H; U05a, U20b
 Environmental protection E15b, E34, E38, H411
 Storage and disposal D01, D02, D05, D09a, D10c, D12a
 Medical advice M03, M05a

273 MCPA

A translocated phenoxycarboxylic acid herbicide for cereals and grassland
HRAC mode of action code: O

See also 2,4-D + clopyralid + MCPA
 2,4-D + dicamba + MCPA + mecoprop-P
 2,4-D + dichlorprop-P + MCPA + mecoprop-P
 2,4-D + MCPA
 2,4-DB + linuron + MCPA
 2,4-DB + MCPA
 bentazone + MCPA + MCPB
 bifenox + MCPA + mecoprop-P
 clopyralid + 2,4-D + MCPA
 clopyralid + diflufenican + MCPA
 clopyralid + fluroxypyr + MCPA
 dicamba + dichlorprop-P + ferrous sulphate + MCPA
 dicamba + dichlorprop-P + MCPA
 dicamba + MCPA + mecoprop-P
 dichlorprop-P + ferrous sulphate + MCPA
 dichlorprop-P + MCPA
 dichlorprop-P + MCPA + mecoprop-P
 ferrous sulphate + MCPA + mecoprop-P

Products

1	Agritox	Nufarm UK	500 g/l	SL	14894
2	Agroxone	FMC Agro	500 g/l	SL	14909
3	Clayton Haksar	Clayton	500 g/l	SL	18380
4	Easel	Nufarm UK	750 g/l	SL	15548
5	Headland Spear	FMC Agro	500 g/l	SL	14910
6	HY-MCPA	Agrichem	500 g/l	SL	14927
7	Larke	Nufarm UK	750 g/l	SL	14914

Uses

- Annual and perennial weeds in **grass seed crops**, **grassland** [1, 3, 4, 7]
- Annual dicotyledons in **farm forestry** *(off-label)*, **game cover** *(off-label)* [2]; **grass seed crops**, **grassland** [2, 5, 6]; **miscanthus** *(off-label)* [5]; **spring barley**, **spring oats**, **spring wheat**, **winter barley**, **winter oats**, **winter wheat** [1-7]; **spring rye**, **winter rye** [1, 3, 4, 7]; **undersown barley**, **undersown oats**, **undersown rye**, **undersown wheat** [1, 3, 4, 6, 7]; **undersown barley** *(red clover or grass)*, **undersown wheat** *(red clover or grass)* [2, 5]
- Charlock in **spring barley**, **spring oats**, **spring wheat**, **winter barley**, **winter oats**, **winter wheat** [1-7]; **spring rye**, **winter rye** [1, 3, 4, 7]; **undersown barley** *(red clover or grass)*, **undersown wheat** *(red clover or grass)* [2, 5]
- Fat hen in **spring barley**, **spring oats**, **spring wheat**, **winter barley**, **winter oats**, **winter wheat** [1-7]; **spring rye**, **winter rye** [1, 3, 4, 7]; **undersown barley** *(red clover or grass)*, **undersown wheat** *(red clover or grass)* [2, 5]
- Hemp-nettle in **spring barley**, **spring oats**, **spring wheat**, **winter barley**, **winter oats**, **winter wheat** [1-7]; **spring rye**, **winter rye** [1, 3, 4, 7]; **undersown barley** *(red clover or grass)*, **undersown wheat** *(red clover or grass)* [2, 5]
- Perennial dicotyledons in **farm forestry** *(off-label)*, **game cover** *(off-label)* [2]; **grass seed crops**, **grassland**, **undersown barley** *(red clover or grass)*, **undersown wheat** *(red clover or grass)* [2, 5]; **miscanthus** *(off-label)* [5]; **spring barley**, **spring oats**, **spring wheat**, **winter barley**, **winter oats**, **winter wheat** [1-7]; **spring rye**, **winter rye** [1, 3, 4, 7]

FOR FULL CONDITIONS OF USE ALWAYS READ THE PRODUCT LABEL

- Wild radish in **spring barley**, **spring oats**, **spring wheat**, **winter barley**, **winter oats**, **winter wheat** [1-7]; **spring rye**, **winter rye** [1, 3, 4, 7]; **undersown barley** *(red clover or grass)*, **undersown wheat** *(red clover or grass)* [2, 5]

Extension of Authorisation for Minor Use (EAMUs)
- **farm forestry** *20122061* [2]
- **game cover** *20122061* [2]
- **miscanthus** *20122050* [5]

Approval information
- MCPA included in Annex I under EC Regulation 1107/2009
- Accepted by BBPA for use on malting barley

Efficacy guidance
- Best results achieved by application to weeds in seedling to young plant stage under good growing conditions when crop growing actively
- Spray perennial weeds in grassland before flowering. Most susceptible growth stage varies between species. See label for details
- Do not spray during cold weather, drought, if rain or frost expected or if crop wet

Restrictions
- Maximum number of treatments normally 1 per crop or yr except grass (2 per yr) for some products. See label
- Do not treat grass within 3 mth of germination and preferably not in the first yr of a direct sown ley or after reseeding
- Do not use on cereals before undersowing
- Do not roll, harrow or graze for a few days before or after spraying; see label
- Do not use on grassland where clovers are an important part of the sward
- Do not use on any crop suffering from stress or herbicide damage
- Avoid spray drift onto nearby susceptible crops
- Do not apply by hand-held equipment
- Livestock must be kept out of treated areas until poisonous weeds such as ragwort have died and become unpalatable
- This product must not be applied before the end of February in the year of harvest
- Do not apply in volumes less than 200 litres of water per hectare

Crop-specific information
- Latest use: before 1st node detectable (GS 31) for cereals; 4-6 wk before heading for grass seed crops; before crop 15-25 cm high for linseed
- Apply to winter cereals in spring from fully tillered, leaf sheath erect stage to before first node detectable (GS 31)
- Apply to spring barley and wheat from 5-leaves unfolded (GS 15), to oats from 1-leaf unfolded (GS 11) to before first node detectable (GS 31)
- Apply to cereals undersown with grass after grass has 2-3 leaves unfolded
- Recommendations for crops undersown with legumes vary. Red clover may withstand low doses after 2-trifoliate leaf stage, especially if shielded by taller weeds but white clover is more sensitive. See label for details
- Apply to grass seed crops from 2-3 leaf stage to 5 wk before head emergence
- Temporary wilting may occur on linseed but without long term effects

Following crops guidance
- Do not direct drill brassicas or legumes within 6 wk of spraying grassland

Environmental safety
- Harmful to aquatic organisms
- MCPA is active at low concentrations. Take extreme care to avoid drift onto neighbouring crops, especially beet crops, brassicas, most market garden crops including lettuce and tomatoes under glass, pears and vines
- Keep livestock out of treated areas for at least 2 wk and until foliage of poisonous weeds such as ragwort has died and become unpalatable
- LERAP Category B

SECTION 2

SEE SECTION 3 FOR PRODUCTS ALSO REGISTERED

Hazard classification and safety precautions

Hazard Harmful, Harmful if swallowed [1, 2, 4-7]; Irritant [3]; Dangerous for the environment [2-6]

Transport code 9 [1, 4, 7]

Packaging group III [1, 4, 7]

UN Number 3082 [1, 4, 7]; N/C [2, 3, 5, 6]

Risk phrases H318 [1, 2, 4-7]; H335 [5, 6]; R22a, R37, R41, R50, R53a [3]

Operator protection A, C; U05a, U08, U11, U20b [1-7]; U09a [2]; U12 [4]; U15 [5]

Environmental protection E06a [6] (2 wk); E07a [1-5, 7]; E15a [1-3, 6, 7]; E15b [4, 5]; E16a, E34 [1-7]; E38 [3-5]; E40b [5]; H410 [1, 2, 4-6]

Storage and disposal D01, D02, D05, D09a [1-7]; D10a [1-4, 6, 7]; D10b [5]; D12a [3, 4]

Medical advice M03 [1-7]; M05a [6]

274   MCPA + mecoprop-P

A translocated selective herbicide for amenity grass
HRAC mode of action code: O + O

See also mecoprop-P

Products

Cleanrun Pro	ICL (Everris) Ltd	0.49:0.29% w/w	GR	15828

Uses

* Annual dicotyledons in *managed amenity turf*
* Perennial dicotyledons in *managed amenity turf*

Approval information

* MCPA and mecoprop-P included in Annex I under EC Regulation 1107/2009

Efficacy guidance

* Apply from Apr to Sep, when weeds growing actively and have large leaf area available for chemical absorption

Restrictions

* The total amount of mecoprop-P applied in a single year must not exceed the maximum total dose approved for any single product for use on turf
* Avoid contact with cultivated plants
* Do not use first 4 mowings as compost or mulch unless composted for 6 mth
* Do not treat newly sown or turfed areas for at least 6 mth
* Do not reseed bare patches for 8 wk after treatment
* Do not apply when heavy rain expected or during prolonged drought. Irrigate after 1-2 d unless rain has fallen
* Do not mow within 2-3 d of treatment
* Treat areas planted with bulbs only after the foliage has died down
* Avoid walking on treated areas until it has rained or irrigation has been applied

Crop-specific information

* Granules contain NPK fertilizer to encourage grass growth

Environmental safety

* Take extreme care to avoid drift onto neighbouring crops, especially beet crops, brassicas, most market garden crops including lettuce and tomatoes under glass, pears and vines
* Harmful to fish or other aquatic life. Do not contaminate surface waters or ditches with chemical or used container
* Keep livestock out of treated areas for at least 2 wk and until foliage of any poisonous weeds such as ragwort has died and become unpalatable
* Some pesticides pose a greater threat of contamination of water than others and mecoprop-P is one of these pesticides. Take special care when applying mecoprop-P near water and do not apply if heavy rain is forecast

FOR FULL CONDITIONS OF USE ALWAYS READ THE PRODUCT LABEL

Hazard classification and safety precautions
    UN Number N/C
    Operator protection A, C, H, M; U20b
    Environmental protection E07a, E13c, E19b
    Storage and disposal D01, D09a, D12a
    Medical advice M05a

275   MCPB

A translocated phenoxycarboxylic acid herbicide
HRAC mode of action code: O

See also bentazone + MCPA + MCPB
       bentazone + MCPB
       MCPA + MCPB

Products

1	Butoxone	FMC Agro	400 g/l	SL	18640
2	Tropotox	Nufarm UK	400 g/l	SL	18309

Uses

- Annual dicotyledons in **combining peas**, **vining peas** [1, 2]
- Docks in **combining peas**, **vining peas** [2]
- Perennial dicotyledons in **combining peas**, **vining peas** [1]
- Thistles in **combining peas**, **vining peas** [2]

Approval information

- MCPB included in Annex I under EC Regulation 1107/2009

Efficacy guidance

- Best results achieved by spraying young seedling weeds in good growing conditions
- Best results on perennials by spraying before flowering
- Effectiveness may be reduced by rain within 12 h, by very cold or dry conditions

Restrictions

- Maximum number of treatments 1 per crop or yr.
- Do not roll or harrow for 7-10 d before or after treatment (check label)

Crop-specific information

- Latest use: first node detectable stage (GS 31) for cereals; before flower buds appear in terminal leaf (GS 201) for peas; before flower buds form for clover
- Apply to undersown cereals from 2-leaves unfolded to first node detectable (GS 12-31), and after first trifoliate leaf stage of clover
- Red clover seedlings may be temporarily damaged but later growth is normal
- Apply to white clover seed crops in Mar to early Apr, not after mid-May, and allow 3 wk before cutting and closing up for seed
- Apply to peas from 3-6 leaf stage but before flower bud detectable (GS 103-201). Consult PGRO (see Appendix 2) or label for information on susceptibility of cultivars.
- Do not use on leguminous crops not mentioned on the label
- Apply to cane and bush fruit after harvest and after shoot growth ceased but before weeds are damaged by frost, usually in late Aug or Sep; direct spray onto weeds as far as possible

Environmental safety

- Harmful to aquatic organisms
- Harmful to fish or other aquatic life. Do not contaminate surface waters or ditches with chemical or used container
- Keep livestock out of treated areas until foliage of any poisonous weeds such as ragwort has died and become unpalatable
- Take extreme care to avoid drift onto neighbouring sensitive crops

Hazard classification and safety precautions

    Hazard Harmful, Dangerous for the environment, Harmful if swallowed

SEE SECTION 3 FOR PRODUCTS ALSO REGISTERED

SECTION 2

Transport code 9 [2]
Packaging group III [2]
UN Number 3082 [2]; N/C [1]
Risk phrases H315 [2]; H318 [1, 2]
Operator protection A, C; U05a, U08, U20b [1, 2]; U11, U14, U15 [1]; U19a [2]
Environmental protection E07a, E34, H411 [1, 2]; E13c [1]; E15a, E38 [2]
Storage and disposal D01, D02, D09a, D10b [1, 2]; D05 [1]
Medical advice M03, M05a

276   mecoprop-P

A translocated phenoxycarboxylic acid herbicide for cereals and grassland
HRAC mode of action code: O

See also 2,4-D + dicamba + iron sulphate + mecoprop-P
2,4-D + dicamba + MCPA + mecoprop-P
2,4-D + dichlorprop-P + MCPA + mecoprop-P
2,4-D + mecoprop-P
bifenox + MCPA + mecoprop-P
bromoxynil + ioxynil + mecoprop-P
carfentrazone-ethyl + mecoprop-P
dicamba + MCPA + mecoprop-P
dicamba + mecoprop-P
dichlorprop-P + MCPA + mecoprop-P
diflufenican + mecoprop-P
ferrous sulphate + MCPA + mecoprop-P
fluroxypyr + mecoprop-P
MCPA + mecoprop-P

Products

1	Charge	FMC Agro	600 g/l	SL	18802
2	Compitox Plus	Nufarm UK	600 g/l	SL	18375
3	Duplosan Kv	Nufarm UK	600 g/l	SL	18374
4	Optica	FMC Agro	600 g/l	SL	14373

Uses

- Annual dicotyledons in *amenity grassland* [2, 3]; *game cover* (off-label), *miscanthus* (off-label) [4]; *grass seed crops, managed amenity turf, spring barley, spring oats, spring wheat, winter barley, winter oats, winter wheat* [1-4]
- Chickweed in *amenity grassland* [2, 3]; *grass seed crops, managed amenity turf, spring barley, spring oats, spring wheat, winter barley, winter oats, winter wheat* [1-4]
- Cleavers in *amenity grassland* [2, 3]; *grass seed crops, managed amenity turf, spring barley, spring oats, spring wheat, winter barley, winter oats, winter wheat* [1-4]
- Perennial dicotyledons in *amenity grassland, grass seed crops, managed amenity turf, spring barley, spring oats, spring wheat, winter barley, winter oats, winter wheat* [2, 3]

Extension of Authorisation for Minor Use (EAMUs)

- *game cover 20111125* [4]
- *miscanthus 20111125* [4]

Approval information

- Mecoprop-P included in Annex I under EC Regulation 1107/2009
- Accepted by BBPA for use on malting barley

Efficacy guidance

- Best results achieved by application to seedling weeds which have not been frost hardened, when soil warm and moist and expected to remain so for several days

Restrictions

- Maximum number of treatments normally 1 per crop for spring cereals and 1 per yr for newly sown grass; 2 per crop or yr for winter cereals and grass crops. Check labels for details

FOR FULL CONDITIONS OF USE ALWAYS READ THE PRODUCT LABEL

- The total amount of mecoprop-P applied in a single yr must not exceed the maximum total dose approved for any single product for the crop/situation
- Do not spray cereals undersown with clovers or legumes or to be undersown with legumes or grasses
- Do not spray grass seed crops within 5 wk of seed head emergence
- Do not spray crops suffering from herbicide damage or physical stress
- Do not spray during cold weather, periods of drought, if rain or frost expected or if crop wet
- Do not roll or harrow for 7 d before or after treatment
- Applications to cereals must not be made between 1 October and 1 March

Crop-specific information

- Latest use: generally before 1st node detectable (GS 31) for spring cereals and before 3rd node detectable (GS 33) for winter cereals, but individual labels vary; 5 wk before emergence of seed head for grass seed crops
- Spray winter cereals from 1 leaf stage in autumn up to and including first node detectable in spring (GS 10-31) or up to second node detectable (GS 32) if necessary. Apply to spring cereals from first fully expanded leaf stage (GS 11) but before first node detectable (GS 31)
- Spray cereals undersown with grass after grass starts to tiller
- Spray newly sown grass leys when grasses have at least 3 fully expanded leaves and have begun to tiller. Any clovers will be damaged

Environmental safety

- Harmful to aquatic organisms
- Harmful to fish or other aquatic life. Do not contaminate surface waters or ditches with chemical or used container
- Keep livestock out of treated areas for at least 2 wk and until foliage of any poisonous weeds, such as ragwort, has died and become unpalatable
- Take extreme care to avoid drift onto neighbouring crops, especially beet crops, brassicas, most market garden crops including lettuce and tomatoes under glass, pears and vines
- Some pesticides pose a greater threat of contamination of water than others and mecoprop-P is one of these pesticides. Take special care when applying mecoprop-P near water and do not apply if heavy rain is forecast

Hazard classification and safety precautions

Hazard Harmful, Dangerous for the environment, Harmful if swallowed
Transport code 9
Packaging group III
UN Number 3082
Risk phrases H315, H318
Operator protection A, C [1-4]; H [2, 3]; P [1, 4]; U05a, U08, U11, U20b [1-4]; U15 [1, 4]
Environmental protection E07a, E13c, H411 [1, 4]; E15b, H410 [2, 3]; E34, E38 [1-4]
Storage and disposal D01, D02 [1-4]; D05, D10b [1, 4]; D09a, D10c [2, 3]
Medical advice M03

277   mefentrifluconazole

A triazole fungicide for use in cereals
FRAC mode of action code: 3

See also fluxapyroxad + mefentrifluconazole

Products

Myresa	BASF	97 g/l	EC	19153

Uses

- Brown rust in **durum wheat**, **rye** *(moderate control)*, **spelt**, **spring barley**, **spring wheat**, **triticale**, **winter barley**, **winter wheat**
- Crown rust in **spring oats**, **winter oats**
- Net blotch in **spring barley** *(reduction)*, **winter barley** *(reduction)*

SEE SECTION 3 FOR PRODUCTS ALSO REGISTERED

- Powdery mildew in **durum wheat** *(moderate control)*, **rye** *(moderate control)*, **spelt** *(moderate control)*, **spring oats**, **spring wheat** *(moderate control)*, **triticale** *(moderate control)*, **winter oats**, **winter wheat** *(moderate control)*
- Ramularia leaf spots in **spring barley**, **winter barley**
- Rhynchosporium in **rye** *(moderate control)*, **spring barley**, **winter barley**
- Septoria leaf blotch in **durum wheat**, **rye** *(moderate control)*, **spelt**, **spring wheat**, **triticale**, **winter wheat**
- Yellow rust in **durum wheat**, **rye** *(moderate control)*, **spelt**, **spring barley**, **spring wheat**, **triticale**, **winter barley**, **winter wheat**

Approval information
- Mefentrifluconazole included in Annex I under EC Regulation 1107/2009

Efficacy guidance
- For best results apply as a protectant treatment. Applications can be made from beginning of stem elongation BBCH 30 up to and including flowering anthesis complete (BBCH 69) in wheat, barley, rye, triticale and oats.

Environmental safety
- LERAP Category B

Hazard classification and safety precautions
Hazard Harmful if inhaled, Very toxic to aquatic organisms
Transport code 9
Packaging group III
UN Number 3082
Risk phrases H315, H317, H319, H335
Operator protection A, C, H; U05a, U20c
Environmental protection E15b, E16a, H410, H411
Storage and disposal D01, D02, D05, D08, D09a, D10c
Medical advice M03

278   mepanipyrim

An anilinopyrimidine fungicide for use in horticulture
FRAC mode of action code: 9

Products

Frupica SC	Certis	450 g/l	SC	12067

Uses
- Botrytis in **courgettes** *(off-label)*, **forest nurseries** *(off-label)*, **ornamental plant production** *(off-label)*, **protected aubergines** *(off-label)*, **protected courgettes** *(off-label)*, **protected cucumbers** *(off-label)*, **protected ornamentals** *(off-label)*, **protected tomatoes** *(off-label)*, **strawberries**
- Powdery mildew in **ornamental plant production** *(off-label)*, **protected aubergines** *(off-label)*, **protected courgettes** *(off-label)*, **protected cucumbers** *(off-label)*, **protected ornamentals** *(off-label)*, **protected tomatoes** *(off-label)*

Extension of Authorisation for Minor Use (EAMUs)
- **courgettes** *20093235*
- **forest nurseries** *20082853*
- **ornamental plant production** *20191294*
- **protected aubergines** *20191523*
- **protected courgettes** *20191523*
- **protected cucumbers** *20191523*
- **protected ornamentals** *20191294*
- **protected tomatoes** *20191523*

Approval information
- Mepanipyrim included in Annex I under EC Regulation 1107/2009

FOR FULL CONDITIONS OF USE ALWAYS READ THE PRODUCT LABEL

Efficacy guidance
- Product is protectant and should be applied as a preventative spray when conditions favourable for Botrytis development occur
- To maintain Botrytis control use as part of a programme with other fungicides that control the disease
- To minimise the possibility of development of resistance adopt resistance management procedures by using products from different chemical groups as part of a mixed spray programme

Restrictions
- Maximum number of treatments 2 per crop (including other anilinopyrimidine products)
- Consult processor before use on crops for processing
- Use spray mixture immediately after preparation

Crop-specific information
- HI 3 d

Environmental safety
- Dangerous for the environment
- Very toxic to aquatic organisms
- LERAP Category B

Hazard classification and safety precautions
Hazard Dangerous for the environment
Transport code 9
Packaging group III
UN Number 3082
Risk phrases H351
Operator protection A, H; U20c
Environmental protection E16a, E16b, E34, H410
Storage and disposal D01, D02, D05, D10c, D11a, D12b

279   mepiquat chloride

A quaternary ammonium plant growth regulator available only in mixtures

See also 2-chloroethylphosphonic acid + mepiquat chloride
chlormequat + 2-chloroethylphosphonic acid + mepiquat chloride
chlormequat + mepiquat chloride
ethephon + mepiquat chloride
mepiquat chloride + metconazole
mepiquat chloride + prohexadione-calcium

280   mepiquat chloride + metconazole

A PGR mixture for growth control in winter oilseed rape

Products
Caryx	BASF	210:30 g/l	SL	16100

Uses
- Growth regulation in **winter oilseed rape**
- Lodging control in **winter oilseed rape**

Approval information
- Metconazole and mepiquat chloride included in Annex I under EC Regulation 1107/2009

Efficacy guidance
- Apply in 200 - 400 l/ha water. Use at lower water volumes has not been evaluated.
- Apply from the beginning of stem extension once the crop is actively growing.

SEE SECTION 3 FOR PRODUCTS ALSO REGISTERED

Following crops guidance
- Beans, cabbage, carrots, cereals, clover, lettuce, linseed, maize, oilseed rape, onions, peas, potatoes, ryegrass, sugar beet or sunflowers may be sown as a following crop but the effect on other crops has not been tested.

Environmental safety
- LERAP Category B

Hazard classification and safety precautions
Hazard Harmful, Dangerous for the environment, Harmful if swallowed, Harmful if inhaled
Transport code 9
Packaging group III
UN Number 3082
Risk phrases H317, H318
Operator protection A, C, H; U02a, U05a, U11, U14, U15
Environmental protection E15b, E16a, E38, H411
Storage and disposal D01, D02, D05, D09a, D10c, D19
Medical advice M03, M05a

281   mepiquat chloride + prohexadione-calcium

A growth regulator mixture for cereals

See also prohexadione-calcium

Products

Canopy	BASF	300:50 g/l	SC	16314

Uses
- Increasing yield in *oats*, *spring barley*, *triticale*, *winter barley*, *winter rye*, *winter wheat*
- Lodging control in *oats*, *spring barley*, *triticale*, *winter barley*, *winter rye*, *winter wheat*

Approval information
- Mepiquat chloride and prohexadione-calcium included in Annex I under EC Regulation 1107/2009
- Accepted by BBPA for use on malting barley

Efficacy guidance
- Best results obtained from treatments applied to healthy crops from the beginning of stem extension

Restrictions
- Maximum total dose equivalent to one full dose treatment on all crops
- Do not apply to any crop suffering from physical stress caused by waterlogging, drought or other conditions
- Do not treat on soils with a substantial moisture deficit
- Consult grain merchant or processor before use on crops for bread making or brewing. Effects on these processes have not been tested

Crop-specific information
- Latest use: before flag leaf fully emerged on wheat and barley

Following crops guidance
- Any crop may follow a normally harvested treated crop. Ploughing is not essential.

Environmental safety
- Harmful to aquatic organisms
- Avoid spray drift onto neighbouring crops

Hazard classification and safety precautions
Hazard Harmful, Harmful if swallowed
UN Number N/C
Operator protection A; U05a, U20b

FOR FULL CONDITIONS OF USE ALWAYS READ THE PRODUCT LABEL

Environmental protection E15a, E34, E38, H412
Storage and disposal D01, D02, D05, D09a, D10c
Medical advice M05a

282 mesosulfuron-methyl

A sulfonyl urea herbicide for cereals available only in mixtures
HRAC mode of action code: B

See also amidosulfuron + iodosulfuron-methyl-sodium + mesosulfuron-methyl
diflufenican + iodosulfuron-methyl-sodium + mesosulfuron-methyl
iodosulfuron-methyl-sodium + mesosulfuron-methyl

283 mesosulfuron-methyl + propoxycarbazone-sodium

A herbicide mixture for use in winter cereals
HRAC mode of action code: B + B

Products

Monolith	Bayer CropScience	4.5:6.75% w/w	WG	17687

Uses

- Annual dicotyledons in *durum wheat, spelt, triticale, winter rye, winter wheat*
- Annual meadow grass in *durum wheat, spelt, triticale, winter rye, winter wheat*
- Blackgrass in *durum wheat, spelt, triticale, winter rye, winter wheat*
- Brome grasses in *durum wheat, spelt, triticale, winter rye, winter wheat*
- Chickweed in *durum wheat, spelt, triticale, winter rye, winter wheat*
- Loose silky bent in *durum wheat, spelt, triticale, winter rye, winter wheat*
- Mayweeds in *durum wheat, spelt, triticale, winter rye, winter wheat*
- Ryegrass in *durum wheat, spelt, triticale, winter rye, winter wheat*
- Wild oats in *durum wheat, spelt, triticale, winter rye, winter wheat*

Approval information

- Mesosulfuron-methyl and propoxycarbazone-sodium included in Annex I under EC Regulation 1107/2009

Efficacy guidance

- Apply as early as possible and before GS 29 of grass weeds
- Always use in mixture the with authorised adjuvant Biopower (ADJ: 0617) at a rate of 1 l/ha

Restrictions

- This product must only be applied between 1 February in the year of harvest and the specified latest time of application.
- To avoid the build-up of resistance do not apply this or any other product containing an ALS herbicide with claims of control of grass-weeds more than once to any crop.
- This product must not be applied via hand-held equipment.
- DO NOT use on crops undersown with grasses, clover or other legumes or any other broad-leaved crop.
- Do not use on cereal crops grown for seed as effects on germination have not been established.
- Do not use as the sole means of grass weed or broad-leaved weed control in successive crops

Crop-specific information

- Clean the spray tank with clean water and add a liquid sprayer cleaner specifically formulated for sulfonylurea herbicides after use.

Following crops guidance

- Winter wheat, winter barley, winter oilseed rape, mustard, lupins and phacelia may be sown in the year of harvest to succeed a treated cereal crop. Spray overlaps in the treated cereal crop should be avoided in order to reduce the risk of localised adverse effects on following crops of winter oilseed rape and mustard.

SEE SECTION 3 FOR PRODUCTS ALSO REGISTERED

SECTION 2

- Spring wheat, spring barley, maize, spring oilseed rape, sugar beet, Italian rye-grass, peas and sunflowers may be drilled in the spring following harvest of a treated cereal crop.

Environmental safety
- LERAP Category B

Hazard classification and safety precautions
  Hazard Very toxic to aquatic organisms
  Transport code 9
  Packaging group III
  UN Number 3077
  Risk phrases H319
  Operator protection A, C, H; U05a, U11, U20b
  Environmental protection E15a, E16a, H410
  Storage and disposal D01, D02, D05, D09a, D10a, D12a, D12b

284   mesotrione

A foliar applied triketone herbicide for maize
HRAC mode of action code: F2

Products

1 Barracuda	Albaugh UK	100 g/l	SC	18348
2 Basilico	Life Scientific	100 g/l	SC	18028
3 Callisto	Syngenta	100 g/l	SC	12323
4 Clayton Cob	Clayton	100 g/l	SC	18971
5 Daneva	Rotam	100 g/l	SC	18029
6 Kideka	Nufarm UK	100 g/l	SC	18033
7 Raikiri	Sipcam	100 g/l	SC	17670
8 Temsa SC	Belchim	100 g/l	SC	18261

Uses
- Annual dicotyledons in *asparagus* (off-label), *game cover* (off-label), *linseed* (off-label), *poppies* (off-label), *poppies for morphine production* (off-label), *poppies grown for seed production* (off-label), *sweetcorn* (off-label) [3]; *forage maize* [1-8]; *grain maize* [5]; *sweetcorn* [6]
- Annual grasses in *game cover* (off-label), *poppies* (off-label), *poppies grown for seed production* (off-label), *sweetcorn* (off-label) [3]
- Annual meadow grass in *grain maize* [1-4, 6-8]; *linseed* (off-label) [3]
- Black bindweed in *rhubarb* (off-label) [3]
- Charlock in *rhubarb* (off-label) [3]
- Cleavers in *rhubarb* (off-label) [3]
- Field pansy in *rhubarb* (off-label) [3]
- Groundsel in *rhubarb* (off-label) [3]
- Himalayan balsam in *rhubarb* (off-label) [3]
- Mayweeds in *rhubarb* (off-label) [3]
- Volunteer oilseed rape in *forage maize*, *grain maize* [1-8]; *linseed* (off-label), *poppies* (off-label), *poppies grown for seed production* (off-label) [3]; *sweetcorn* [6]

Extension of Authorisation for Minor Use (EAMUs)
- *asparagus* 20111113 [3]
- *game cover* 20082830 [3]
- *linseed* 20071706 [3]
- *poppies* 20180019 [3]
- *poppies for morphine production* 20113148 [3]
- *poppies grown for seed production* 20180019 [3]
- *rhubarb* 20161761 [3]
- *sweetcorn* 20051893 [3]

Approval information
- Mesotrione included in Annex I under EC Regulation 1107/2009

FOR FULL CONDITIONS OF USE ALWAYS READ THE PRODUCT LABEL

Efficacy guidance
- Best results obtained from treatment of young actively growing weed seedlings in the presence of adequate soil moisture
- Treatment in poor growing conditions or in dry soil may give less reliable control
- Activity is mostly by foliar uptake with some soil uptake
- To minimise the possible development of resistance where continuous maize is grown the product should not be used for more than two consecutive seasons

Restrictions
- Maximum number of treatments 1 per crop of forage or grain maize
- Do not use on seed crops or on sweetcorn varieties
- Do not spray when crop foliage wet or when excessive rainfall is expected to follow application
- Do not treat crops suffering from stress from cold or drought conditions, or when wide temperature fluctuations are anticipated
- Extreme care must be taken to avoid spray drift onto plants outside of the target area
- Must not be applied via hand-held equipment [1]
- Do not handle treated crops for at least 2 days after treatment [1]

Crop-specific information
- Latest use: 8 leaves unfolded stage (GS 18) for forage maize
- Treatment under adverse conditions may cause mild to moderate chlorosis. The effect is transient and does not affect yield

Following crops guidance
- Winter wheat, durum wheat, winter barley or ryegrass may follow a normally harvested treated crop of maize. Oilseed rape may be sown provided it is preceded by deep ploughing to more than 15 cm
- In the spring following application only forage maize, ryegrass, spring wheat or spring barley may be sown
- In the event of crop failure maize may be re-seeded immediately. Some slight crop effects may be seen soon after emergence but these are normally transient

Environmental safety
- Dangerous for the environment
- Very toxic to aquatic organisms
- Take extreme care to avoid drift onto all plants outside the target area
- LERAP Category B [1-4, 6-8]

Hazard classification and safety precautions

Hazard Irritant, Dangerous for the environment [1-8]; Very toxic to aquatic organisms [1-7]
Transport code 9 [3-6]
Packaging group III [3-6]
UN Number 3082 [3-6]; N/C [1, 2, 7, 8]
Risk phrases H317 [1, 7, 8]; H318 [1, 5-8]; H319 [2-4]
Operator protection A, C [1-8]; H [8]; U05a, U09a, U20b
Environmental protection E15b, E38, H410 [1-8]; E16a, E16b [1-4, 6-8]
Storage and disposal D01, D02, D05, D09a, D10c, D11a, D12a

SEE SECTION 3 FOR PRODUCTS ALSO REGISTERED

285   metalaxyl-M

A phenylamide systemic fungicide
FRAC mode of action code: 4

See also cymoxanil + fludioxonil + metalaxyl-M
fluazinam + metalaxyl-M
fludioxonil + metalaxyl-M
fludioxonil + metalaxyl-M + sedaxane
fludioxonil + metalaxyl-M + thiamethoxam
mancozeb + metalaxyl-M

Products

1	Apron XL	Syngenta	339.2 g/l	ES	14654
2	Caveo	Pan Agriculture	465.2 g/l	SL	16361
3	Clayton Tine	Clayton	465 g/l	SL	14072
4	Floreo	Pan Agriculture	465.2 g/l	SL	16358
5	Hobson	AgChem Access	465.2 g/l	SL	16529
6	SL 567A	Syngenta	465.2 g/l	SL	12380
7	Subdue	Fargro	465 g/l	SL	12503

Uses

- Cavity spot in **carrots** [2, 5, 6]; **carrots** *(reduction only)*, **parsnips** *(off-label - reduction)* [3]; **parsnips** *(off-label)* [6]
- Crown rot in **protected water lilies** *(off-label)*, **water lilies** *(off-label)* [6]
- Damping off in **watercress** *(off-label)* [6]
- Downy mildew in **asparagus** *(off-label)*, **baby leaf crops** *(off-label)*, **blackberries** *(off-label)*, **broccoli** *(off-label)*, **calabrese** *(off-label)*, **collards** *(off-label)*, **grapevines** *(off-label)*, **herbs (see appendix 6)** *(off-label)*, **hops** *(off-label)*, **horseradish** *(off-label)*, **kale** *(off-label)*, **protected baby leaf crops** *(off-label)*, **protected cucumbers** *(off-label)*, **protected herbs (see appendix 6)** *(off-label)*, **protected soft fruit** *(off-label)*, **protected spinach** *(off-label)*, **protected spinach beet** *(off-label)*, **protected water lilies** *(off-label)*, **raspberries** *(off-label)*, **rubus hybrids** *(off-label)*, **salad onions** *(off-label)*, **soft fruit** *(off-label)*, **spinach** *(off-label)*, **spinach beet** *(off-label)*, **spring field beans** *(off-label)*, **water lilies** *(off-label)*, **watercress** *(off-label)*, **winter field beans** *(off-label)* [6]
- Phytophthora in **amenity vegetation** *(off-label)*, **forest nurseries** *(off-label)* [7]; **ornamental plant production**, **protected ornamentals** [4]
- Phytophthora root rot in **ornamental plant production**, **protected ornamentals** [7]
- Pythium in **beetroot** *(off-label)*, **brassica leaves and sprouts** *(off-label)*, **broccoli**, **brussels sprouts**, **bulb onions**, **cabbages**, **calabrese**, **cauliflowers**, **chard** *(off-label)*, **chinese cabbage**, **choi sum** *(off-label)*, **collards** *(off-label)*, **herbs (see appendix 6)** *(off-label)*, **kale** *(off-label)*, **kohlrabi**, **ornamental plant production** *(off-label)*, **radishes** *(off-label)*, **shallots** *(off-label)*, **spinach** [1]; **ornamental plant production**, **protected ornamentals** [4, 7]
- Root malformation disorder in **red beet** *(off-label)* [6]
- White blister in **horseradish** *(off-label)* [6]

Extension of Authorisation for Minor Use (EAMUs)

- **amenity vegetation** *20120383* [7]
- **asparagus** *20051502* [6]
- **baby leaf crops** *20051507* [6]
- **beetroot** *20102539* [1]
- **blackberries** *20072195* [6]
- **brassica leaves and sprouts** *20120526* [1]
- **broccoli** *20112502* [6]
- **calabrese** *20112502* [6]
- **chard** *20120526* [1]
- **choi sum** *20192583* [1]
- **collards** *20192583* [1], *20112048* [6]
- **forest nurseries** *20120383* [7]
- **grapevines** *20051504* [6]

- ***herbs (see appendix 6)*** *20102539* [1], *20051507* [6]
- ***hops*** *20051500* [6]
- ***horseradish*** *20051499* [6], *20061040* [6]
- ***kale*** *20192583* [1], *20112048* [6]
- ***ornamental plant production*** *20102539* [1]
- ***parsnips*** *(reduction)* *20111033* [3], *20051508* [6]
- ***protected baby leaf crops*** *20051507* [6]
- ***protected cucumbers*** *20051503* [6]
- ***protected herbs (see appendix 6)*** *20051507* [6]
- ***protected soft fruit*** *20082937* [6]
- ***protected spinach*** *20051507* [6]
- ***protected spinach beet*** *20051507* [6]
- ***protected water lilies*** *20051501* [6]
- ***radishes*** *20102539* [1]
- ***raspberries*** *20072195* [6]
- ***red beet*** *20051307* [6]
- ***rubus hybrids*** *20072195* [6]
- ***salad onions*** *20072194* [6]
- ***shallots*** *20121635* [1]
- ***soft fruit*** *20082937* [6]
- ***spinach*** *20051507* [6]
- ***spinach beet*** *20051507* [6]
- ***spring field beans*** *20130917* [6]
- ***water lilies*** *20051501* [6]
- ***watercress*** *20072193* [6]
- ***winter field beans*** *20130917* [6]

Approval information
- Metalaxyl-M included in Annex I under EC Regulation 1107/2009
- Accepted by BBPA for use on hops

Efficacy guidance
- Best results achieved when applied to damp soil or potting media
- Treatments to ornamentals should be followed immediately by irrigation to wash any residues from the leaves and allow penetration to the rooting zone [7]
- Efficacy may be reduced in prolonged dry weather [6]
- Results may not be satisfactory on soils with high organic matter content [6]
- Control of cavity spot on carrots overwintered in the ground or lifted in winter may be lower than expected [6]
- Use in an integrated pest management strategy and, where appropriate, alternate with products from different chemical groups
- Product should ideally be used preventatively and the number of phenylamide applications should be limited to 1-2 consecutive treatments
- Always follow FRAG guidelines for preventing and managing fungicide resistance. See Section 5 for more information

Restrictions
- Maximum number of treatments on ornamentals 1 per situation for media treatment. See label for details of drench treatment of protected ornamentals [7]
- Maximum total dose on carrots equivalent to one full dose treatment [6]
- Do not use where carrots have been grown on the same site within the previous eight yrs [6]
- Consult before use on crops intended for processing [6]
- Do not re-use potting media from treated plants for subsequent crops [7]
- Disinfect pots thoroughly prior to re-use [7]

Crop-specific information
- Latest use: 6 wk after drilling for carrots [6]
- Because of the large number of species and ornamental cultivars susceptibility should be checked before large scale treatment [7]
- Treatment of *Viburnum* and *Prunus* species not recommended [7]

SEE SECTION 3 FOR PRODUCTS ALSO REGISTERED

Environmental safety
- Harmful to aquatic organisms
- Limited evidence suggests that metalaxyl-M is not harmful to soil dwelling predatory mites

Hazard classification and safety precautions
 Hazard Harmful, Harmful if swallowed
 UN Number N/C
 Risk phrases H319 [2-7]; H335 [7]
 Operator protection A [1-7]; C [7]; D [1]; H [1, 7]; U02a, U05a, U20b [1-7]; U04a, U10, U19a [2-7]
 Environmental protection E15b, E34, E38, H412
 Storage and disposal D01, D02, D05, D09a, D12a [1-7]; D07 [3, 5-7]; D10c [2-7]; D11a [1]
 Treated seed S02, S04d, S05, S07, S08, S09 [1]
 Medical advice M03 [2-7]; M05a [1]

286   metaldehyde

A molluscicide bait for controlling slugs and snails

Products

1	Carakol 3	Adama	3% w/w	PT	14309
2	Certis Metaldehyde 3	Certis	3% w/w	PT	14337
3	Condor 3	Certis	3% w/w	PT	14324
4	Enzo	Adama	3% w/w	PT	14306
5	ESP3	De Sangosse	3% w/w	RB	16088
6	Gusto 3	Adama	3% w/w	PT	14308
7	Lynx H	De Sangosse	3% w/w	RB	14426
8	Osarex W	De Sangosse	3% w/w	RB	14428
9	Prowler	De Sangosse	3% w/w	RB	16087
10	Super 3	Certis	3% w/w	PT	14370
11	TDS Major	De Sangosse	4% w/w	RB	13462
12	Trigger 3	Certis	3% w/w	GB	14304

Uses
- Slugs in *all edible crops except potatoes and cauliflowers, all non-edible crops (outdoor), cauliflowers, natural surfaces not intended to bear vegetation, potatoes* [11]
- Slugs and snails in *all edible crops (outdoor)* [3]; *all edible crops except potatoes and cauliflowers, potatoes* [1, 2, 4-10, 12]; *all non-edible crops (outdoor)* [1, 3-10]; *amenity grassland, managed amenity turf* [2, 12]; *cauliflowers* [1, 4-9]; *natural surfaces not intended to bear vegetation* [2, 3, 5, 7-10, 12]
- Snails in *all edible crops except potatoes and cauliflowers, all non-edible crops (outdoor), cauliflowers, natural surfaces not intended to bear vegetation, potatoes* [11]

Approval information
- Metaldehyde included in Annex I under EC Regulation 1107/2009
- Accepted by BBPA for use on malting barley and hops

Efficacy guidance
- Apply pellets by hand, fiddle drill, fertilizer distributor, by air (check label) or in admixture with seed. See labels for rates and timing.
- Best results achieved from an even spread of granules applied during mild, damp weather when slugs and snails most active. May be applied in standing crops
- To establish the need for pellet application on winter wheat or winter oilseed rape, monitor for slug activity. Where bait traps are used, use a foodstuff attractive to slugs e.g. chicken layer's mash
- Varieties of oilseed rape low in glucosinolates can be more acceptable to slugs than "single low" varieties and control may not be as good
- To prevent slug build up apply at end of season to brassicas and other leafy crops
- To reduce tuber damage in potatoes apply twice in Jul and Aug
- For information on slug trapping and damage risk assessment refer to AHDB Information Sheet 02 - Integrated slug control, available from the AHDB website (www.ahdb.org.uk/slugcontrol)

FOR FULL CONDITIONS OF USE ALWAYS READ THE PRODUCT LABEL

Restrictions

- Do not apply when rain imminent or water glasshouse crops within 4 d of application
- Take care to avoid lodging of pellets in the foliage when making late applications to edible crops.
- The maximum total dose of metaldehyde must not exceed 700 g active substance/ha/year.
- The risk to ground water on drained fields or sloping sites can be avoided by substituting metaldehyde products for ferric phosphate products

Crop-specific information

- Put slug traps out before cultivation, when the soil surface is visibly moist and the weather mild (5-25˚C) (see label for guidance)
- For winter wheat, a catch of 4 or more slugs/trap indicates a possible risk, where soil and weather conditions favour slug activity
- For winter oilseed rape a catch of 4 or more slugs in standing cereals, or 1 or more in cereal stubble, if other conditions were met, would indicate possible risk of damage

Environmental safety

- Dangerous to game, wild birds and animals
- Some products contain proprietary cat and dog deterrent
- Keep poultry out of treated areas for at least 7 d
- Do not use slug pellets in traps in winter wheat or winter oilseed rape since they are a potential hazard to wildlife and pets
- Some pesticides pose a greater threat of contamination of water than others and metaldehyde is one of these pesticides. Take special care when applying metaldehyde near water and do not apply if heavy rain is forecast
- No pellets should fall within 10 metres of a field boundary or watercourse.

Hazard classification and safety precautions

UN Number N/C

Operator protection A, H; U05a [1-9, 11, 12]; U20a [1, 4, 6]; U20b [2, 3, 5, 7-9, 12]; U20c [10, 11]

Environmental protection E05b [1, 2] (7 d); E05b [3] (7 days); E05b [4] (7 d); E05b [5] (7 days); E05b [6] (7 d); E05b [7-9] (7 days); E05b [10-12] (7 d); E07a, E34 [3, 5, 7-9]; E10a [1, 3-11]; E10c [2, 12]; E15a [1-12]

Storage and disposal D01, D09a [1-12]; D02 [1-9, 11, 12]; D05 [2, 12]; D07 [2, 3, 5, 7-12]; D11a [1-10, 12]; D11b [11]

Treated seed S04a [3, 5, 7-9, 11]

Medical advice M04a [2, 12]; M05a [3, 5, 7-9, 11]

287   metamitron

A contact and residual triazinone herbicide for use in beet crops and a fruit thinner for apples and pears.
HRAC mode of action code: C1

See also chlorpropham + metamitron
ethofumesate + metamitron
ethofumesate + metamitron + phenmedipham

Products

1 Beetron 700	AgChem Access	700 g/l	SC	17458
2 Bettix Flo	UPL Europe	700 g/l	SC	16559
3 Bettix Flo SC	UPL Europe	700 g/l	SC	18245
4 Brevis	Adama	15% w/w	SG	17479
5 Clayton Neutron	Clayton	700 g/l	SC	18609
6 Defiant	UPL Europe	700 g/l	SC	18216
7 Defiant SC	UPL Europe	700 g/l	SC	16531
8 Glotron 700 SC	Belchim	700 g/l	SC	17308
9 Goltix 70 SC	Adama	700 g/l	SC	16638
10 Mitron 700 SC	Belcrop	700 g/l	SC	16908
11 Synergy Generics Metamitron	Synergy	700 g/l	SC	18277

SEE SECTION 3 FOR PRODUCTS ALSO REGISTERED

Products – continued

12	Target Flo	UPL Europe	700 g/l	SC	19159
13	Tronix 700 SC	Belchim	700 g/l	SC	18137

Uses

- Annual dicotyledons in *fodder beet, mangels, red beet, sugar beet* [1-3, 5-13]; *horseradish (off-label), parsnips (off-label), strawberries (off-label)* [9]
- Annual grasses in *fodder beet, mangels, red beet, sugar beet* [1-3, 6, 7, 10, 12]
- Annual meadow grass in *fodder beet, mangels, red beet, sugar beet* [1-3, 5-13]; *horseradish (off-label), ornamental plant production (off-label), parsnips (off-label), protected ornamentals (off-label), strawberries (off-label)* [9]
- Fat hen in *fodder beet, mangels, red beet, sugar beet* [1-3, 6, 7, 10, 12]; *ornamental plant production (off-label), protected ornamentals (off-label)* [9]
- Fruit thinning in *apples, pears* [4]
- Groundsel in *ornamental plant production (off-label), protected ornamentals (off-label)* [9]
- Growth regulation in *apples, pears* [4]
- Mayweeds in *ornamental plant production (off-label), protected ornamentals (off-label)* [9]
- Red dead-nettle in *ornamental plant production (off-label), protected ornamentals (off-label)* [9]
- Scarlet pimpernel in *ornamental plant production (off-label), protected ornamentals (off-label)* [9]
- Small nettle in *ornamental plant production (off-label), protected ornamentals (off-label)* [9]

Extension of Authorisation for Minor Use (EAMUs)

- *horseradish* 20142918 [9]
- *ornamental plant production* 20151175 [9]
- *parsnips* 20142918 [9]
- *protected ornamentals* 20151175 [9]
- *strawberries* 20142919 [9]

Approval information

- Metamitron included in Annex I under EC Regulation 1107/2009

Efficacy guidance

- May be used pre-emergence alone or post-emergence in tank mixture or with an authorised adjuvant oil
- Low dose programme (LDP). Apply a series of low-dose post-weed emergence sprays, including adjuvant oil, timing each treatment according to weed emergence and size. See label for details and for recommended tank mixes and sequential treatments. On mineral soils the LDP should be preceded by pre-drilling or pre-emergence treatment
- Traditional application. Apply either pre-drilling before final cultivation with incorporation to 8-10 cm, or pre-crop emergence at or soon after drilling into firm, moist seedbed to emerged weeds from cotyledon to first true leaf stage
- On emerged weeds at or beyond 2-leaf stage addition of adjuvant oil advised
- For control of wild oats and certain other weeds, tank mixes with other herbicides or sequential treatments are recommended. See label for details
- When used for fruit thinning in apples and pears, apply post blossom if fruit set is excessive. Apply in temperatures between 10 and 25°C [4]

Restrictions

- Maximum total dose equivalent to three full dose treatments for most products. Check label
- Using traditional method post-crop emergence on mineral soils do not apply before first true leaves have reached 1 cm long
- Fodder beet and mangels must not be grazed by livestock or harvested for animal consumption until at least 103 days following the last application

Crop-specific information

- Latest use: before crop foliage meets across rows for beet crops
- HI herbs 6 wk

- HI for apples and pears 60 days
- Crop tolerance may be reduced by stress caused by growing conditions, effects of pests, disease or other pesticides, nutrient deficiency etc

Following crops guidance
- Only sugar beet, fodder beet or mangels may be drilled within 4 mth after treatment. Winter cereals may be sown in same season after ploughing, provided 16 wk passed since last treatment

Environmental safety
- Dangerous for the environment
- Very toxic to aquatic organisms
- Dangerous to fish or other aquatic life. Do not contaminate surface waters or ditches with chemical or used container
- Do not empty into drains

Hazard classification and safety precautions
Hazard Harmful [1, 5-11, 13]; Dangerous for the environment [1-13]; Harmful if swallowed [1-6, 8-11, 13]; Very toxic to aquatic organisms [12]
Transport code 9
Packaging group III
UN Number 3077 [4]; 3082 [1-3, 5-13]
Risk phrases H317 [2, 6]; H318 [4]; R22a, R43, R51, R53a [7]
Operator protection A [1-13]; C [4]; H [5, 8, 9]; U02a, U04a, U11 [4]; U05a [4, 10, 11, 13]; U08, U19a [10, 11, 13]; U09a, U20c [2, 3, 6, 7, 12]; U14 [2, 3, 6, 7, 10-13]; U20a [10]; U20b [1, 4, 11, 13]
Environmental protection E13b [2, 3, 6, 7, 12]; E15a [1, 10, 11, 13]; E15b [4]; E19b [1]; E34 [1, 3, 4, 6, 11, 13]; E38 [2, 3, 5-13]; H410 [1, 8, 9, 11-13]; H411 [2-6, 10]
Storage and disposal D01, D02, D12a [1-13]; D05 [2-4, 6, 7, 12]; D09a [1-4, 6, 7, 10-13]; D10c [2, 3, 6, 7, 12]; D11a [1, 4, 11, 13]
Medical advice M03 [1-3, 6, 7, 11-13]; M05a [1-3, 5-10, 12]

288   metamitron + quinmerac

A herbicide mixture for weed control in beet crops
HRAC mode of action code: C1 + O

See also metamitron

Products

Goltix Titan	Adama	525:40 g/l	SC	17301

Uses
- Annual dicotyledons in *fodder beet*, *sugar beet*
- Annual meadow grass in *fodder beet*, *sugar beet*
- Cleavers in *fodder beet*, *sugar beet*

Approval information
- Metamitron and quinmerac included in Annex I under EC Regulation 1107/2009

Efficacy guidance
- Do not handle treated plants for at least 2 days after treatment

Restrictions
- Maintain an interval of at least 5 days between applications

Crop-specific information
- Beet crops may be sown at any time following application. Allow 16 weeks from application before sowing winter cereals in the same season. After normal harvest, any crop can be grown in the following spring. Deep and thorough cultivation is necessary before sowing or planting succeeding crops. Mouldboard ploughing to a minimum depth of 15 cm is recommended before seedbed preparation.

SEE SECTION 3 FOR PRODUCTS ALSO REGISTERED

SECTION 2

Hazard classification and safety precautions
  Transport code 9
  Packaging group III
  UN Number 3082
  Operator protection A, H; U02a, U04a
  Environmental protection E15b, H411
  Storage and disposal D01, D09a, D10b, D12a

289   Metarhizium anisopliae

A naturally occurring insect parasitic fungus, Metarhizium anisopliae var. anisopliae strain F52
IRAC mode of action code: UNF

Products

Met52 granular bioinsecticide	Fargro	2% w/w	GR	15168

Uses

- Cabbage root fly in *vegetable brassicas (off-label)*
- Leatherjackets in *bilberries (off-label)*, *cranberries (off-label)*, *loganberries (off-label)*, *ribes hybrids (off-label)*, *rubus hybrids (off-label)*, *table grapes (off-label)*, *wine grapes (off-label)*
- Lettuce root aphid in *herbs (see appendix 6) (off-label)*, *leafy vegetables (off-label)*
- Midges in *amenity vegetation (off-label)*, *bilberries (off-label)*, *blueberries (off-label)*, *container-grown ornamentals (off-label)*, *cranberries (off-label)*, *herbs (see appendix 6) (off-label)*, *leafy vegetables (off-label)*, *loganberries (off-label)*, *ornamental plant production (off-label)*, *ribes hybrids (off-label)*, *rubus hybrids (off-label)*, *table grapes (off-label)*, *top fruit (off-label)*, *wine grapes (off-label)*
- Sciarid flies in *amenity vegetation (off-label)*, *bilberries (off-label)*, *blueberries (off-label)*, *container-grown ornamentals (off-label)*, *cranberries (off-label)*, *edible fungi (off-label)*, *herbs (see appendix 6) (off-label)*, *leafy vegetables (off-label)*, *loganberries (off-label)*, *ornamental plant production (off-label)*, *ribes hybrids (off-label)*, *rubus hybrids (off-label)*, *table grapes (off-label)*, *top fruit (off-label)*, *wine grapes (off-label)*
- Thrips in *amenity vegetation (off-label)*, *bilberries (off-label)*, *blueberries (off-label)*, *container-grown ornamentals (off-label)*, *cranberries (off-label)*, *herbs (see appendix 6) (off-label)*, *leafy vegetables (off-label)*, *loganberries (off-label)*, *ornamental plant production (off-label)*, *ribes hybrids (off-label)*, *rubus hybrids (off-label)*, *table grapes (off-label)*, *top fruit (off-label)*, *wine grapes (off-label)*
- Vine weevil in *amenity vegetation (off-label)*, *blackberries*, *blackcurrants*, *blueberries*, *gooseberries*, *ornamental plant production*, *raspberries*, *redcurrants*, *strawberries*, *top fruit (off-label)*

Extension of Authorisation for Minor Use (EAMUs)

- *amenity vegetation* 20111568, 20111997
- *bilberries* 20111568, 20111997
- *blueberries* 20111997
- *container-grown ornamentals* 20111997
- *cranberries* 20111568, 20111997
- *edible fungi* 20111568
- *herbs (see appendix 6)* 20111568, 20111997
- *leafy vegetables* 20111568, 20111997
- *loganberries* 20111568, 20111997
- *ornamental plant production* 20111997
- *ribes hybrids* 20111568, 20111997
- *rubus hybrids* 20111568, 20111997
- *table grapes* 20111568, 20111997
- *top fruit* 20111997
- *vegetable brassicas* 20111568
- *wine grapes* 20111568, 20111997

FOR FULL CONDITIONS OF USE ALWAYS READ THE PRODUCT LABEL

Approval information
- Metarhizium anisopliae included in Annex 1 under EC Regulation 1107/2009

Efficacy guidance
- Incorporate into growing media at any growth stage

Restrictions
- Maximum of 24 treatments per glasshouse structure per year

Hazard classification and safety precautions
UN Number N/C
Risk phrases R70
Operator protection A, D, H; U20c
Environmental protection E15b, E34, E38
Storage and disposal D01, D09a, D10a, D12a, D20
Medical advice M03, M05a

290   metazachlor

A residual anilide herbicide for use in brassicas, nurseries and forestry
HRAC mode of action code: K3

See also aminopyralid + metazachlor + picloram
clomazone + dimethenamid-p + metazachlor
clomazone + metazachlor
clomazone + metazachlor + napropamide
dimethenamid-p + metazachlor
dimethenamid-p + metazachlor + quinmerac
imazamox + metazachlor
imazamox + metazachlor + quinmerac

Products

1	Butisan S	BASF	500 g/l	SC	16569
2	Rapsan 500 SC	Belchim	500 g/l	SC	16592
3	Rapsan Solo	Belchim	500 g/l	SC	18516
4	Stalwart	UPL Europe	500 g/l	SC	17405
5	Sultan 50 SC	Adama	500 g/l	SC	16680
6	Taza 500	Becesane	500 g/l	SC	17284

Uses
- Annual dicotyledons in *borage for oilseed production* (off-label), *canary flower (echium spp.)* (off-label), *evening primrose* (off-label), *honesty* (off-label), *mustard* (off-label) [1, 5]; *broccoli, cabbages, calabrese, cauliflowers, nursery fruit trees* [4-6]; *brussels sprouts, ornamental plant production* [1, 3-6]; *chinese cabbage* (off-label), *choi sum* (off-label), *collards* (off-label), *kale* (off-label), *kohlrabi* (off-label), *linseed* (off-label), *pak choi* (off-label), *swedes* (off-label), *tatsoi* (off-label), *turnips* (off-label) [1]; *spring linseed* (off-label), *winter linseed* (off-label) [5]; *spring oilseed rape, winter oilseed rape* [1-6]
- Annual meadow grass in *borage for oilseed production* (off-label), *canary flower (echium spp.)* (off-label), *evening primrose* (off-label), *honesty* (off-label), *mustard* (off-label) [1, 5]; *broccoli, cabbages, calabrese, cauliflowers, nursery fruit trees* [4-6]; *brussels sprouts, ornamental plant production* [1, 3-6]; *chinese cabbage* (off-label), *choi sum* (off-label), *collards* (off-label), *kale* (off-label), *kohlrabi* (off-label), *linseed* (off-label), *pak choi* (off-label), *swedes* (off-label), *tatsoi* (off-label), *turnips* (off-label) [1]; *spring linseed* (off-label), *winter linseed* (off-label) [5]; *spring oilseed rape, winter oilseed rape* [1-6]
- Blackgrass in *broccoli, cabbages, calabrese, cauliflowers, nursery fruit trees* [4-6]; *brussels sprouts, ornamental plant production* [1, 3-6]; *spring oilseed rape, winter oilseed rape* [1-6]
- Chickweed in *choi sum* (off-label), *collards* (off-label), *kale* (off-label), *kohlrabi* (off-label), *oriental cabbage* (off-label), *tatsoi* (off-label) [5]
- Groundsel in *borage for oilseed production* (off-label), *canary flower (echium spp.)* (off-label), *evening primrose* (off-label), *mustard* (off-label), *swedes* (off-label), *turnips* (off-label) [1, 3]; *chinese cabbage* (off-label), *honesty* (off-label), *linseed* (off-label), *pak choi* (off-label) [1]; *choi*

SEE SECTION 3 FOR PRODUCTS ALSO REGISTERED

sum (off-label), **collards** (off-label), **kale** (off-label), **kohlrabi** (off-label), **tatsoi** (off-label) [1, 3, 5]; **oriental cabbage** (off-label) [3, 5]; **winter linseed** (off-label) [3]

- Mayweeds in **borage for oilseed production** (off-label), **canary flower (echium spp.)** (off-label), **evening primrose** (off-label), **mustard** (off-label), **swedes** (off-label), **turnips** (off-label) [1, 3]; **chinese cabbage** (off-label), **honesty** (off-label), **linseed** (off-label), **pak choi** (off-label) [1]; **choi sum** (off-label), **collards** (off-label), **kale** (off-label), **kohlrabi** (off-label), **tatsoi** (off-label) [1, 3, 5]; **oriental cabbage** (off-label) [3, 5]; **winter linseed** (off-label) [3]
- Shepherd's purse in **borage for oilseed production** (off-label), **canary flower (echium spp.)** (off-label), **evening primrose** (off-label), **mustard** (off-label), **swedes** (off-label), **turnips** (off-label) [1, 3]; **chinese cabbage** (off-label), **honesty** (off-label), **linseed** (off-label), **pak choi** (off-label) [1]; **choi sum** (off-label), **collards** (off-label), **kale** (off-label), **kohlrabi** (off-label), **tatsoi** (off-label) [1, 3, 5]; **oriental cabbage** (off-label) [3, 5]; **winter linseed** (off-label) [3]

Extension of Authorisation for Minor Use (EAMUs)

- **borage for oilseed production** 20141053 [1], 20181945 [3], 20142381 [5]
- **canary flower (echium spp.)** 20141053 [1], 20181945 [3], 20142381 [5]
- **chinese cabbage** 20141052 [1]
- **choi sum** 20141052 [1], 20181944 [3], 20161213 [5]
- **collards** 20141052 [1], 20181944 [3], 20161213 [5]
- **evening primrose** 20141053 [1], 20181945 [3], 20142381 [5]
- **honesty** 20141053 [1], 20142381 [5]
- **kale** 20141052 [1], 20181944 [3], 20161213 [5]
- **kohlrabi** 20141052 [1], 20181944 [3], 20161213 [5]
- **linseed** 20141053 [1]
- **mustard** 20141053 [1], 20181945 [3], 20142381 [5]
- **oriental cabbage** 20181944 [3], 20161213 [5]
- **pak choi** 20141052 [1]
- **spring linseed** 20142381 [5]
- **swedes** 20171383 [1], 20181946 [3]
- **tatsoi** 20141052 [1], 20181944 [3], 20161213 [5]
- **turnips** 20171383 [1], 20181946 [3]
- **winter linseed** 20181945 [3], 20142381 [5]

Approval information

- Metazachlor is included in Annex I under EC Regulation 1107/2009

Efficacy guidance

- Activity is dependent on root uptake. For pre-emergence use apply to firm, moist, clod-free seedbed
- Some weeds (chickweed, mayweed, blackgrass etc) susceptible up to 2- or 4-leaf stage. Moderate control of cleavers achieved provided weeds not emerged and adequate soil moisture present
- Split pre- and post-emergence treatments recommended for certain weeds in winter oilseed rape on light and/or stony soils
- Effectiveness is reduced on soils with more than 10% organic matter
- Always follow WRAG guidelines for preventing and managing herbicide resistant weeds. See Section 5 for more information

Restrictions

- Maximum number of treatments 1 per crop for spring oilseed rape, swedes, turnips and brassicas; 2 per crop for winter oilseed rape (split dose treatment); 3 per yr for ornamentals, nursery stock, nursery fruit trees, forestry and farm forestry
- Do not use on sand, very light or poorly drained soils
- Do not treat protected crops or spray overall on ornamentals with soft foliage
- Do not spray crops suffering from wilting, pest or disease
- Do not spray broadcast crops or if a period of heavy rain forecast
- When used on nursery fruit trees any fruit harvested within 1 yr of treatment must be destroyed
- Applications shall be limited to a total dose of not more than 1.0 kg metazachlor/ha in a three year period on the same field
- Do not apply in mixture with phosphate liquid fertilisers

FOR FULL CONDITIONS OF USE ALWAYS READ THE PRODUCT LABEL

- Metazachlor stewardship requires that all autumn applications should be made before the end of September to reduce the risk to water

Crop-specific information
- Latest use: pre-emergence for swedes and turnips; before 10 leaf stage for spring oilseed rape; before end of Jan for winter oilseed rape
- HI brassicas 6 wk
- On winter oilseed rape may be applied pre-emergence from drilling until seed chits, post-emergence after fully expanded cotyledon stage (GS 1,0) or by split dose technique depending on soil and weeds. See label for details
- On spring oilseed rape may also be used pre-weed-emergence from cotyledon to 10-leaf stage of crop (GS 1,0-1,10)
- With pre-emergence treatment ensure seed covered by 15 mm of well consolidated soil. Harrow across slits of direct-drilled crops
- Ensure brassica transplants have roots well covered and are well established. Direct drilled brassicas should not be treated before 3 leaf stage
- In ornamentals and hardy nursery stock apply after plants established and hardened off as a directed spray or, on some subjects, as an overall spray. See label for list of tolerant subjects. Do not treat plants in containers

Following crops guidance
- Any crop can follow normally harvested treated winter oilseed rape. See label for details of crops which may be planted after spring treatment and in event of crop failure

Environmental safety
- Dangerous for the environment
- Very toxic to aquatic organisms
- Keep livestock out of treated areas until foliage of any poisonous weeds such as ragwort has died and become unpalatable
- Keep livestock out of treated areas of swede and turnip for at least 5 wk following treatment
- Some pesticides pose a greater threat of contamination of water than others and metazachlor is one of these pesticides. Take special care when applying metazachlor near water and do not apply if heavy rain is forecast.
- Metazachlor stewardship guidelines advise a maximum dose of 750 g.a.i/ha/annum. Applications to drained fields should be complete by 15th Oct but, if drains are flowing, complete applications by 1st Oct.
- LERAP Category B [1, 3-6]

Hazard classification and safety precautions
Hazard Harmful, Dangerous for the environment [1-6]; Harmful if swallowed [1, 3-6]; Very toxic to aquatic organisms [1, 3, 4]
Transport code 9
Packaging group III
UN Number 3082
Risk phrases H317 [2, 4-6]; H351 [1-6]
Operator protection A, C, H, M; U05a, U19a [1-6]; U08, U14, U20b [1, 3-6]; U09a, U20a [2]; U15 [4-6]
Environmental protection E06a [2, 4, 6] (5 wk for swedes, turnips); E07a [1-4, 6]; E15a, E34, H410 [1-6]; E16a [1, 3-6]; E38 [1, 3]
Consumer protection C02a [2] (5 weeks for swedes and turnips)
Storage and disposal D01, D02, D09a, D10c [1-6]; D05 [2, 4-6]; D12a, D19 [1, 3]; D12b [4-6]
Medical advice M03 [1-6]; M05a [1, 3]

SEE SECTION 3 FOR PRODUCTS ALSO REGISTERED

291   metazachlor + quinmerac

A residual herbicide mixture for oilseed rape
HRAC mode of action code: K3 + O

See also quinmerac

Products

1 Legion	Adama	375:125 g/l	SC	16249
2 Metazerac	Becesane	375:125 g/l	SC	17744
3 Naspar Extra	Belchim	375:125 g/l	SC	17038
4 Parsan Extra	Belchim	375:125 g/l	SC	17050

Uses

- Annual dicotyledons in *spring oilseed rape* [3, 4]; *winter oilseed rape* [1-4]
- Annual meadow grass in *spring oilseed rape* [3, 4]; *winter oilseed rape* [1-4]
- Blackgrass in *corn gromwell* *(off-label)* [1]; *spring oilseed rape* [3, 4]; *winter oilseed rape* [1-4]
- Chickweed in *corn gromwell* *(off-label)* [1]
- Cleavers in *corn gromwell* *(off-label)* [1]; *spring oilseed rape* [3, 4]; *winter oilseed rape* [1-4]
- Mayweeds in *corn gromwell* *(off-label)* [1]; *spring oilseed rape* [3, 4]; *winter oilseed rape* [1-4]
- Poppies in *corn gromwell* *(off-label)* [1]; *spring oilseed rape* [3, 4]; *winter oilseed rape* [1-4]
- Speedwells in *corn gromwell* *(off-label)* [1]

Extension of Authorisation for Minor Use (EAMUs)

- *corn gromwell* 20162047 [1]

Approval information

- Metazachlor and quinmerac included in Annex I under EC Regulation 1107/2009

Efficacy guidance

- Activity is dependent on root uptake. Pre-emergence treatments should be applied to firm moist seedbeds. Applications to dry soil do not become effective until after rain has fallen
- Maximum activity achieved from treatment before weed emergence for some species
- Weed control may be reduced if excessive rain falls shortly after application especially on light soils
- May be used on all soil types except Sands, Very Light Soils, and soils containing more than 10% organic matter. Crop vigour and/or plant stand may be reduced on brashy and stony soils

Restrictions

- Maximum total dose equivalent to one full dose treatment
- Damage may occur in waterlogged conditions. Do not use on poorly drained soils
- Do not treat stressed crops. In frosty conditions transient scorch may occur
- Applications shall be limited to a total dose of not more than 1.0 kg metazachlor/ha in a three year period on the same field
- Do not apply in mixture with phosphate liquid fertilisers
- Metazachlor stewardship requires that all autumn applications should be made before the end of September to reduce the risk to water

Crop-specific information

- Latest use: end Jan in yr of harvest
- To ensure crop safety it is essential that crop seed is well covered with soil to 15 mm. Loose or puffy seedbeds must be consolidated before treatment. Do not use on broadcast crops
- Crop vigour and possibly plant stand may be reduced if excessive rain falls shortly after treatment especially on light soils

Following crops guidance

- In the event of crop failure after use, wheat or barley may be sown in the autumn after ploughing to 15 cm. Spring cereals or brassicas may be planted after ploughing in the spring
- For all situations following or rotational crops must not be planted until four months after application

---

FOR FULL CONDITIONS OF USE ALWAYS READ THE PRODUCT LABEL

Environmental safety

- Dangerous for the environment
- Very toxic to aquatic organisms
- Keep livestock out of treated areas until foliage of any poisonous weeds such as ragwort has died and become unpalatable
- To reduce risk of movement to water do not apply to dry soil or if heavy rain is forecast. On clay soils create a fine consolidated seedbed
- Some pesticides pose a greater threat of contamination of water than others and metazachlor is one of these pesticides. Take special care when applying metazachlor near water and do not apply if heavy rain is forecast.
- Metazachlor stewardship guidelines advise a maximum dose of 750 g.a.i/ha/annum. Applications to drained fields should be complete by 15th Oct but, if drains are flowing, complete applications by 1st Oct.
- LERAP Category B

Hazard classification and safety precautions

Hazard Irritant, Dangerous for the environment [1-4]; Very toxic to aquatic organisms [3, 4]
Transport code 9
Packaging group III
UN Number 3082
Risk phrases H317 [3, 4]; H351 [1-4]
Operator protection A; U05a, U08, U14, U19a, U20b
Environmental protection E07a, E15a, E16a, E38, H410
Storage and disposal D01, D02, D08, D09a, D10c, D12a
Medical advice M05a

292   metconazole

A triazole fungicide for cereals and oilseed rape
FRAC mode of action code: 3

See also boscalid + metconazole
epoxiconazole + metconazole
fluxapyroxad + metconazole
mepiquat chloride + metconazole

Products

1	Ambarac	Life Scientific	60 g/l	EC	17971
2	Caramba 90	BASF	90 g/l	EC	15524
3	Gringo	AgChem Access	60 g/l	EC	15774
4	Juventus	BASF	90 g/l	EC	15528
5	Metcostar	Life Scientific	60 g/l	EC	17446
6	Metfin	Clayton	60 g/l	EC	18660
7	Metfin 90	Clayton	90 g/l	EC	18910
8	Redstar	AgChem Access	90 g/l	SL	15938
9	Sirena	Belchim	60 g/l	EC	17103
10	Sunorg Pro	BASF	90 g/l	EC	15433

Uses

- Alternaria in *spring oilseed rape*, *winter oilseed rape* [1-10]
- Ascochyta in *combining peas* (reduction), *lupins* (qualified minor use), *vining peas* (reduction) [1-10]
- Botrytis in *combining peas* (reduction), *lupins* (qualified minor use), *vining peas* (reduction) [1-10]
- Brown rust in *durum wheat*, *rye*, *spring barley*, *triticale*, *winter barley*, *winter wheat* [1-10]; *spring wheat* [2, 4, 8, 10]
- Disease control in *broad beans* (off-label) [10]; *spring linseed* (off-label), *winter linseed* (off-label) [4, 10]
- Fusarium ear blight in *durum wheat* (reduction) [1, 3-7, 9, 10]; *rye* (reduction) [1-3, 5-10]; *spring wheat* (reduction) [2, 4, 8, 10]; *triticale* (reduction), *winter wheat* (reduction) [1-10]

- Growth regulation in **winter oilseed rape** [10]
- Light leaf spot in **spring oilseed rape**, **winter oilseed rape** [1-3, 5-10]; **spring oilseed rape** *(reduction)*, **winter oilseed rape** *(reduction)* [4]
- Mycosphaerella in **combining peas** *(reduction)*, **vining peas** *(reduction)* [1-10]
- Net blotch in **durum wheat** *(reduction)* [2]; **spring barley** *(reduction)*, **winter barley** *(reduction)* [1-10]
- Phoma in **spring oilseed rape** *(reduction)*, **winter oilseed rape** *(reduction)* [1-10]
- Powdery mildew in **durum wheat** *(moderate control)*, **rye** *(moderate control)*, **spring barley** *(moderate control)*, **triticale** *(moderate control)*, **winter wheat** *(moderate control)* [1-10]; **spring wheat** *(moderate control)* [2, 4, 8, 10]; **winter barley** [1-3, 5-10]; **winter barley** *(moderate control)* [4]
- Rhynchosporium in **rye** *(moderate control)*, **spring barley** *(reduction)*, **winter barley** *(reduction)* [4]; **spring barley**, **winter barley** [1-3, 5-10]
- Rust in **combining peas**, **lupins** *(qualified minor use)*, **spring field beans**, **vining peas**, **winter field beans** [1-10]
- Septoria leaf blotch in **durum wheat** [1, 3-7, 9, 10]; **rye** [1-3, 5-10]; **spring wheat** [2, 4, 8, 10]; **triticale**, **winter wheat** [1-10]
- Yellow rust in **durum wheat** [1, 3-7, 9, 10]; **rye**, **triticale**, **winter wheat** [1-10]; **spring barley**, **winter barley** [4]; **spring wheat** [2, 4, 8, 10]

Extension of Authorisation for Minor Use (EAMUs)
- **broad beans** *20111928* [10]
- **spring linseed** *20130387* [4], *20111929* [10]
- **winter linseed** *20130387* [4], *20111929* [10]

Approval information
- Metconazole included in Annex I under EC Regulation 1107/2009
- Accepted by BBPA for use on malting barley
- Approval expiry 31 Oct 2020 [5]

Efficacy guidance
- Best results from application to healthy, vigorous crops when disease starts to develop
- Good spray cover of the target is essential for best results. Spray volume should be increased to improve spray penetration into dense crops
- Metconazole is a DMI fungicide. Resistance to some DMI fungicides has been identified in Septoria leaf blotch which may seriously affect performance of some products. For further advice contact a specialist advisor and visit the Fungicide Resistance Action Group (FRAG)-UK website

Restrictions
- Maximum total dose equivalent to two full dose treatments on all crops
- Do not apply to oilseed rape crops that are damaged or stressed from previous treatments, adverse weather, nutrient deficiency or pest attack. Spring application may lead to reduction of crop height
- The addition of adjuvants is neither advised nor necessary and can lead to enhanced growth regulatory effects on stressed crops of oilseed rape
- Ensure sprayer is free from residues of previous treatments that may harm the crop, especially oilseed rape. Use of a detergent cleaner is advised before and after use
- Do not apply with pyrethroid insecticides on oilseed rape at flowering
- Maintain an interval of at least 14 days between applications to oilseed rape, peas, beans and lupins [4], 21 days between applications to cereals [4]
- To protect birds, only one application is allowed on cereals before end of tillering (GS29)

Crop-specific information
- Latest use: up to and including milky ripe stage (GS 71) for cereals; 10% pods at final size for oilseed rape
- HI: 14 d for peas, field beans, lupins
- Spring treatments on oilseed rape can reduce the height of the crop
- Treatment for Septoria leaf spot in wheat should be made before second node detectable stage and when weather favouring development of the disease has occurred. If conditions continue to favour disease development a follow-up treatment may be needed

FOR FULL CONDITIONS OF USE ALWAYS READ THE PRODUCT LABEL

- Treat mildew infections in cereals before 3% infection on any green leaf. A specific mildewicide will improve control of established infections
- Treat yellow rust on wheat before 1% infection on any leaf or as preventive treatment after GS 39
- For brown rust in cereals spray susceptible varieties before any of top 3 leaves has more than 2% infection
- Peas should be treated at the start of flowering and repeat 3-4 wk later if required
- Field beans and lupins should be treated at petal fall and repeat 3-4 wk later if required

Following crops guidance
- Only cereals, oilseed rape, sugar beet, linseed, maize, clover, beans, peas, carrots, potatoes or onions may be sown as following crops after treatment

Environmental safety
- Dangerous for the environment
- Very toxic to aquatic organisms
- Avoid treatment close to field boundary, even if permitted by LERAP assessment, to reduce effects on non-target insects or other arthropods
- LERAP Category B

Hazard classification and safety precautions
Hazard Harmful [1, 3-7, 9]; Irritant [2, 8, 10]; Flammable [1, 3, 5-7, 9]; Dangerous for the environment [1-10]; Flammable liquid and vapour [1, 3, 5, 6, 9]; Very toxic to aquatic organisms [1, 3, 5, 6]
Transport code 3 [1, 3, 5-7, 9]; 9 [2, 4, 8, 10]
Packaging group III
UN Number 1993 [1, 3, 5-7, 9]; 3082 [2, 4, 8, 10]
Risk phrases H304, H315, H317, H318, H335 [1, 3, 5, 6, 9]; H319, H373 [2, 4, 7, 8, 10]; H361 [1-10]
Operator protection A, C [1-10]; H [1, 3-7, 9]; M [1]; U02a, U05a, U20c [1-10]; U11, U14 [1, 3, 5-7, 9]
Environmental protection E15a, E16a, E34, E38 [1-10]; E16b [1, 3-7, 9]; H410 [1, 3, 5, 6, 9]; H411 [2, 4, 7, 8, 10]
Storage and disposal D01, D02, D05, D09a, D12a [1-10]; D06c [2, 4, 8, 10]; D10a [1-3, 5-10]; D10c [4]
Medical advice M03, M05a [4]; M05b [1, 3, 5-7, 9]

293   methoxyfenozide

A moulting accelerating diacylhydrazine insecticide
IRAC mode of action code: 18A

Products
1	Agrovista Trotter	Agrovista	240 g/l	SC	14236
2	Runner	Landseer	240 g/l	SC	15629

Uses
- Caterpillars in *aubergines* (off-label), *chillies* (off-label), *peppers* (off-label), *tomatoes* (off-label) [2]
- Codling moth in *apples*, *pears* [1, 2]
- Insect pests in *aubergines* (off-label), *chillies* (off-label), *hops* (off-label), *ornamental plant production* (off-label), *peppers* (off-label), *soft fruit* (off-label), *tomatoes* (off-label) [2]
- Plum fruit moth in *plums* (off-label) [2]
- Tortrix moths in *apples*, *pears* [1, 2]; *table grapes* (off-label), *wine grapes* (off-label) [2]
- Winter moth in *apples*, *pears* [1, 2]

Extension of Authorisation for Minor Use (EAMUs)
- *aubergines* 20170901 [2]
- *chillies* 20170901 [2]
- *hops* 20120448 [2]

SEE SECTION 3 FOR PRODUCTS ALSO REGISTERED

- *ornamental plant production 20120448* [2]
- *peppers 20170901* [2]
- *plums 20161694* [2]
- *soft fruit 20120448* [2]
- *table grapes 20161694* [2]
- *tomatoes 20170901* [2]
- *wine grapes 20161694* [2]

Approval information
- Methoxyfenozide included in Annex 1 under EC Regulation 1107/2009

Efficacy guidance
- To achieve best results uniform coverage of the foliage and full spray penetration of the leaf canopy is important, particularly when spraying post-blossom
- For maximum effectiveness on winter moth and tortrix spray pre-blossom when first signs of active larvae are seen, followed by a further spray in June if larvae of the summer generation are present
- For codling moth spray post-blossom to coincide with early to peak egg deposition. Follow-up treatments will normally be needed
- Methoxyfenozide is a moulting accelerating compound (MAC) and may be used in an anti-resistance strategy with other top fruit insecticides (including chitin biosynthesis inhibitors and juvenile hormones) which have a different mode of action
- To reduce further the likelihood of resistance development use at full recommended dose in sufficient water volume to achieve required spray penetration

Restrictions
- Maximum number of treatments 3 per yr but no more than two should be sprayed consecutively

Crop-specific information
- HI 14 d for apples, pears

Environmental safety
- Risk to non-target insects or other arthropods
- Broadcast air-assisted LERAP (5 m)

Hazard classification and safety precautions
UN Number N/C
Operator protection U20b
Environmental protection E15b, E16b, E22c; E17b (5 m)
Storage and disposal D05, D09a, D11a

294   1-methylcyclopropene

An inhibitor of ethylene production for use in stored apples

Products

1	SmartFresh	Landseer	3.3% w/w	SP	11799
2	SmartFresh ProTabs	Landseer	2% w/w	TB	16546
3	SmartFresh SmartTabs	Landseer	0.63% w/w	TB	12684

Uses
- Botrytis in *plums* (off-label) [3]
- Ethylene inhibition in *apples* (post-harvest use), *pears* (off-label - post harvest only) [1]
- Scald in *apples* (post-harvest use), *pears* (off-label - post harvest only) [1]
- Slow down of ripening in *apples*, *pears*, *plums* [2]; *tomatoes post harvest* [2, 3]
- Storage rots in *broccoli* (off-label), *brussels sprouts* (off-label), *cabbages* (off-label), *calabrese* (off-label), *cauliflowers* (off-label) [1]; *plums* (off-label) [3]

Extension of Authorisation for Minor Use (EAMUs)
- *broccoli 20111502* [1]
- *brussels sprouts 20111503* [1]

- ***cabbages*** *20111503* [1]
- ***calabrese*** *20111502* [1]
- ***cauliflowers*** *20111502* [1]
- ***pears*** *(post harvest only) 20102424* [1]
- ***plums*** *20072234* [3]

Approval information
- Methylcyclopropene included in Annex 1 under EC Regulation 1107/2009

Efficacy guidance
- Best results obtained from treatment of fruit in good condition and of proper quality for long-term storage
- Effects may be reduced in fruit that is in poor condition or ripe prior to storage or harvested late
- Product acts by releasing vapour into store when mixed with water
- Apply as soon as possible after harvest
- Treatment controls superficial scald and maintains fruit firmness and acid content for 3-6 mth in normal air, and 6-9 mth in controlled atmosphere storage
- Ethylene production recommences after removal from storage

Restrictions
- Maximum number of treatments 1 per batch of apples
- Must only be used by suitably trained and competent persons in fumigation operations
- Keep unprotected persons out of treated areas during the 24-hour treatment process
- Consult processors before treatment of fruit destined for processing or cider making
- Do not apply in mixture with other products
- Ventilate all areas thoroughly with all refrigeration fans operating at maximum power for at least 15 min before re-entry

Crop-specific information
- Latest use: 7 d after harvest of apples
- Product tested on Granny Smith, Gala, Jonagold, Bramley and Cox. Consult distributor or supplier before treating other varieties

Environmental safety
- Unprotected persons must be kept out of treated stores during the 24 h treatment period
- Prior to application ensure that the store can be properly and promptly sealed

Hazard classification and safety precautions
Hazard Irritant [2, 3]
UN Number N/C
Risk phrases H319 [2, 3]
Operator protection C [2]; U01, U11, U12 [2, 3]; U05a [1-3]
Environmental protection E02a [1] (24 h); E02a [2, 3] (12 h); E15a [1]; E15b [2, 3]; E34 [1-3]; E38, H412 [2]
Consumer protection C12 [1]
Storage and disposal D01, D02, D14 [1-3]; D07 [1]

295 metobromuron

A pre-emergence herbicide for weed control in potatoes
HRAC mode of action code: C2

Products

1	Inigo	Belchim	500 g/l	SC	16933
2	Praxim	Belchim	500 g/l	SC	16871
3	Soleto	Belchim	500 g/l	SC	16935

Uses
- Annual dicotyledons in ***ornamental bulbs*** *(off-label)*, ***potatoes***
- Annual grasses in ***ornamental bulbs*** *(off-label)*
- Annual meadow grass in ***potatoes***

SEE SECTION 3 FOR PRODUCTS ALSO REGISTERED

- Charlock in *potatoes*
- Chickweed in *potatoes*
- Fat hen in *potatoes*
- Groundsel in *potatoes*
- Mayweeds in *potatoes*
- Shepherd's purse in *potatoes*

Extension of Authorisation for Minor Use (EAMUs)
- *ornamental bulbs* *20183121* [1], *20183073* [2], *20183127* [3]

Approval information
- Metobromuron included in Annex I under EC Regulation 1107/2009

Restrictions
- Heavy rain after application may cause transient discolouration of the crop.

Crop-specific information
- May be used on all varieties of potato

Following crops guidance
- Following normal harvest of a treated crop Brassicas (oilseed rape, turnip, brassica vegetables) and sugar beet may be sown, providing the soil has been ploughed. No restrictions apply when other crops are sown.
- In the event of crop failure maize, beans, peas and carrots may be sown after ploughing. It is not recommended to sow Brassicas (oilseed rape, turnip, brassica vegetables) or sugar beet as a replacement crop after crop failure. Potatoes can be re-planted after minimal cultivation.

Hazard classification and safety precautions
Hazard Very toxic to aquatic organisms [1, 3]
Transport code 9
Packaging group III
UN Number 3082
Risk phrases H351, H373
Operator protection A, H, M; U05a, U09a, U19a, U20a
Environmental protection E15b, E34, E38, H410
Storage and disposal D01, D02, D05, D09a, D12b
Medical advice M05a

296   metrafenone

A benzophenone protectant and curative fungicide for cereals
FRAC mode of action code: U8

See also epoxiconazole + metrafenone

Products

1	Lexi	AgChem Access	300 g/l	SC	14890
2	Vivando	BASF	500 g/l	SC	18026

Uses
- Cobweb in *mushroom boxes* [2]
- Eyespot in *spring wheat* (reduction), *winter wheat* (reduction) [1]
- Powdery mildew in *hops* (off-label), *wine grapes* (off-label) [2]; *spring barley*, *spring oats* (evidence of mildew control on oats is limited), *spring wheat*, *winter barley*, *winter oats* (evidence of mildew control on oats is limited), *winter wheat* [1]

Extension of Authorisation for Minor Use (EAMUs)
- *hops* *20192992* [2]
- *wine grapes* *20192990* [2]

Approval information
- Metrafenone included in Annex I under EC Regulation 1107/2009
- Accepted by BBPA for use on malting barley

Efficacy guidance

- Best results obtained from treatment at the start of foliar disease attack
- Activity against mildew in wheat is mainly protectant with moderate curative control in the latent phase; activity is entirely protectant in barley
- Useful reduction of eyespot in wheat is obtained if treatment applied at GS 30-32
- Should be used as part of a resistance management strategy that includes mixtures or sequences effective against mildew and non-chemical methods

Restrictions

- Maximum number of treatments 2 per crop
- Avoid the use of sequential applications of metrafenone unless it is used in tank mixture with other products active against powdery mildew employing a different mode of action

Crop-specific information

- Latest use: beginning of flowering (GS 31) for wheat, barley, oats

Following crops guidance

- Cereals, oilseed rape, sugar beet, linseed, maize, clover, field beans, peas, turnips, carrots, cauliflowers, onions, lettuce or potatoes may follow a treated cereal crop

Environmental safety

- Dangerous for the environment
- Toxic to aquatic organisms

Hazard classification and safety precautions

Hazard Irritant, Dangerous for the environment
Transport code 9 [2]
Packaging group III [2]
UN Number 3082 [2]; N/C [1]
Risk phrases H317, H319 [1]
Operator protection A, H; U05a, U20b
Environmental protection E15a, E34 [1, 2]; H411 [2]
Storage and disposal D01, D02, D05, D09a, D10c
Medical advice M03

297   metribuzin

A contact and residual triazinone herbicide for use in potatoes
HRAC mode of action code: C1

See also clomazone + metribuzin
diflufenican + flufenacet + metribuzin
diflufenican + metribuzin
flufenacet + metribuzin
mecoprop-P + metribuzin

Products

1	Clayton Tribute	Clayton	70% w/w	WG	18363
2	Cleancrop Frizbee	Agrii	70% w/w	WG	16642
3	Gale	Harvest	70% w/w	WG	18089
4	Matecor New	Clayton	70% w/w	WG	18315
5	Sencorex Flow	Nufarm UK	600 g/l	SL	16167
6	Shotput	Adama	70% w/w	WG	15968

Uses

- Annual dicotyledons in *asparagus (off-label)*, *carrots (off-label)*, *mallow (althaea spp.) (off-label)*, *parsnips (off-label)*, *sweet potato (off-label)* [2, 5, 6]; *celeriac (off-label)*, *ornamental plant production (off-label)* [5]; *early potatoes, maincrop potatoes* [1-6]
- Annual grasses in *asparagus (off-label)*, *carrots (off-label)*, *mallow (althaea spp.) (off-label)*, *parsnips (off-label)*, *sweet potato (off-label)* [5, 6]; *celeriac (off-label)*, *ornamental plant production (off-label)* [5]; *early potatoes, maincrop potatoes* [1-6]

SEE SECTION 3 FOR PRODUCTS ALSO REGISTERED

- Annual meadow grass in *carrots* (off-label), *mallow (althaea spp.)* (off-label), *parsnips* (off-label), *sweet potato* (off-label) [6]; *maincrop potatoes* [1-4, 6]; *ornamental plant production* (off-label), *protected ornamentals* (off-label) [5]
- Charlock in *ornamental plant production* (off-label), *protected ornamentals* (off-label) [5]
- Cleavers in *ornamental plant production* (off-label), *protected ornamentals* (off-label) [5]
- Common orache in *ornamental plant production* (off-label), *protected ornamentals* (off-label) [5]
- Fat hen in *ornamental plant production* (off-label), *protected ornamentals* (off-label) [5]
- Field pansy in *ornamental plant production* (off-label), *protected ornamentals* (off-label) [5]
- Fool's parsley in *asparagus* (off-label), *sweet potato* (off-label) [2, 5]; *carrots* (off-label), *mallow (althaea spp.)* (off-label), *parsnips* (off-label) [2, 5, 6]; *celeriac* (off-label) [5]
- Groundsel in *ornamental plant production* (off-label), *protected ornamentals* (off-label) [5]
- Himalayan balsam in *rhubarb* (off-label) [5]
- Knotgrass in *ornamental plant production* (off-label), *protected ornamentals* (off-label) [5]
- Mayweeds in *ornamental plant production* (off-label), *protected ornamentals* (off-label) [5]
- Pale persicaria in *ornamental plant production* (off-label), *protected ornamentals* (off-label) [5]
- Perennial dicotyledons in *asparagus* (off-label - from seed) [6]
- Perennial weeds in *asparagus* (off-label), *sweet potato* (off-label) [2]
- Redshank in *ornamental plant production* (off-label), *protected ornamentals* (off-label) [5]
- Shepherd's purse in *ornamental plant production* (off-label), *protected ornamentals* (off-label) [5]
- Small nettle in *ornamental plant production* (off-label), *protected ornamentals* (off-label) [5]
- Volunteer oilseed rape in *asparagus* (off-label), *carrots* (off-label), *celeriac* (off-label), *mallow (althaea spp.)* (off-label), *ornamental plant production* (off-label), *parsnips* (off-label), *protected ornamentals* (off-label), *sweet potato* (off-label) [5]; *early potatoes, maincrop potatoes* [1-6]
- Wild mignonette in *asparagus* (off-label), *sweet potato* (off-label) [2, 5]; *carrots* (off-label), *mallow (althaea spp.)* (off-label), *parsnips* (off-label) [2, 5, 6]; *celeriac* (off-label) [5]
- Willowherb in *ornamental plant production* (off-label), *protected ornamentals* (off-label) [5]

Extension of Authorisation for Minor Use (EAMUs)
- *asparagus* 20151018 [2], 20150917 [5], (from seed) 20142321 [6], 20142321 [6]
- *carrots* 20151017 [2], 20150916 [5], 20142320 [6]
- *celeriac* 20150916 [5]
- *mallow (althaea spp.)* 20151017 [2], 20150916 [5], 20142320 [6]
- *ornamental plant production* 20131867 [5], 20171732 [5]
- *parsnips* 20151017 [2], 20150916 [5], 20142320 [6]
- *protected ornamentals* 20171732 [5]
- *rhubarb* 20152218 [5]
- *sweet potato* 20151019 [2], 20150915 [5], 20142252 [6]

Approval information
- Metribuzin included in Annex I under EC Regulation 1107/2009

Efficacy guidance
- Best results achieved on weeds at cotyledon to 1-leaf stage
- May be applied pre- or post-emergence of crop on named maincrop varieties and cv. Marfona; pre-emergence only on named early varieties
- Apply to moist soil with well-rounded ridges and few clods
- Activity reduced by dry conditions and on soils with high organic matter content
- On fen and moss soils pre-planting incorporation to 10-15 cm gives increased activity. Incorporate thoroughly and evenly
- With named maincrop and second early potato varieties on soils with more than 10% organic matter shallow pre- or post-planting incorporation may be used. See label for details
- Effective control using a programme of reduced doses is made possible by using a spray of smaller droplets, thus improving retention

FOR FULL CONDITIONS OF USE ALWAYS READ THE PRODUCT LABEL

Restrictions

- Maximum total dose equivalent to one full dose treatment on early potato varieties and one and a third full doses on maincrop varieties
- Only certain varieties may be treated. Apply pre-emergence only on named first earlies, pre- or post-emergence on named second earlies. On named maincrop varieties apply pre-emergence (except for certain varieties on Sands or Very Light soils) or post-emergence. See label
- All post-emergence treatments must be carried out before longest shoots reach 15 cm
- Do not cultivate after treatment
- Some recommended varieties may be sensitive to post-emergence treatment if crop under stress

Crop-specific information

- Latest use: pre-crop emergence for named early potato varieties; before most advanced shoots have reached 15 cm for post-emergence treatment of potatoes
- On stony or gravelly soils there is risk of crop damage, especially if heavy rain falls soon after application
- When days are hot and sunny delay spraying until evening

Following crops guidance

- Ryegrass, cereals or winter beans may be sown in same season provided at least 16 wk elapsed after treatment and ground ploughed to 15 cm and thoroughly cultivated as soon as possible after harvest and no later than end Dec
- In W Cornwall on soil with more than 5% organic matter early potatoes treated as recommended may be followed by summer planted brassica crops provided the soil has been ploughed, spring rainfall has been normal and at least 14 wk have elapsed since treatment
- Do not grow any vegetable brassicas, lettuces or radishes on land treated the previous yr. Other crops may be sown normally in spring of next yr

Environmental safety

- Dangerous for the environment
- Very toxic to aquatic organisms
- Do not empty into drains
- LERAP Category B

Hazard classification and safety precautions

Hazard Harmful, Dangerous for the environment [1-6]; Very toxic to aquatic organisms [1-4, 6]
Transport code 9
Packaging group III
UN Number 3077 [1-4, 6]; 3082 [5]
Operator protection A, H; U05a, U08, U13, U19a [1-6]; U14, U20b [1-4, 6]; U20a [5]
Environmental protection E15a, E16b, E19b [1-4, 6]; E16a, H410 [1-6]; E38 [5]
Storage and disposal D01, D02, D09a, D11a, D12a
Medical advice M03 [5]; M05a [1-6]

298　metsulfuron-methyl

A contact and residual sulfonylurea herbicide used in cereals, linseed and set-aside
HRAC mode of action code: B

See also carfentrazone-ethyl + metsulfuron-methyl
diflufenican + metsulfuron-methyl
flupyrsulfuron-methyl + metsulfuron-methyl
fluroxypyr + metsulfuron-methyl
fluroxypyr + metsulfuron-methyl + thifensulfuron-methyl
mecoprop-P + metsulfuron-methyl

Products

1	Accurate	Nufarm UK	20% w/w	WG	18614
2	Alias SX	FMC Agro	20% w/w	SG	18668
3	Answer SX	FMC Agro	20% w/w	SG	18684
4	Deft Premium	Rotam	20% w/w	SG	18525

SEE SECTION 3 FOR PRODUCTS ALSO REGISTERED

Products – continued

5	Finy	UPL Europe	20% w/w	WG	18274
6	Gropper SX	FMC Agro	20% w/w	SG	18685
7	Jubilee SX	FMC Agro	20% w/w	SG	18686
8	Laya	Life Scientific	20% w/w	SG	19016
9	Lorate	FMC Agro	20% w/w	SG	18687
10	Savvy	Rotam	20% w/w	WG	14266
11	Savvy Premium	Rotam	20% w/w	SG	18461
12	Simba SX	FMC Agro	20% w/w	SG	18791

Uses

- Amsinkia in *green cover on land temporarily removed from production, linseed, spring barley, spring oats, spring wheat, triticale, winter barley, winter oats, winter wheat* [7]
- Annual dicotyledons in *green cover on land temporarily removed from production* [5, 7]; *linseed* [1-3, 6, 7, 9, 12]; *spring barley, spring oats, spring wheat, triticale, winter barley, winter oats, winter wheat* [1-12]
- Charlock in *green cover on land temporarily removed from production, linseed* [7]; *spring barley, spring oats, spring wheat, triticale, winter barley, winter oats, winter wheat* [4, 7]
- Chickweed in *green cover on land temporarily removed from production* [7]; *linseed* [1-3, 6, 7, 9, 12]; *spring barley, spring oats, spring wheat, triticale, winter barley, winter oats, winter wheat* [1-4, 6-12]
- Docks in *green cover on land temporarily removed from production, linseed, spring barley, spring oats, spring wheat, triticale, winter barley, winter oats, winter wheat* [7]
- Green cover in *land temporarily removed from production* [1-4, 6, 8-12]
- Groundsel in *green cover on land temporarily removed from production, linseed, spring barley, spring oats, spring wheat, triticale, winter barley, winter oats, winter wheat* [7]
- Hemp-nettle in *green cover on land temporarily removed from production, linseed, spring barley, spring oats, spring wheat, triticale, winter barley, winter oats, winter wheat* [7]
- Mayweeds in *green cover on land temporarily removed from production* [7]; *linseed* [1-3, 6, 7, 9, 12]; *spring barley, spring oats, spring wheat, triticale, winter barley, winter oats, winter wheat* [1-4, 6-12]
- Mouse-ear chickweed in *green cover on land temporarily removed from production, linseed, spring barley, spring oats, spring wheat, triticale, winter barley, winter oats, winter wheat* [7]
- Scarlet pimpernel in *green cover on land temporarily removed from production, linseed, spring barley, spring oats, spring wheat, triticale, winter barley, winter oats, winter wheat* [7]
- Venus looking-glass in *green cover on land temporarily removed from production, linseed, spring barley, spring oats, spring wheat, triticale, winter barley, winter oats, winter wheat* [7]
- Volunteer oilseed rape in *green cover on land temporarily removed from production, linseed, spring barley, spring oats, spring wheat, triticale, winter barley, winter oats, winter wheat* [7]
- Volunteer sugar beet in *green cover on land temporarily removed from production, linseed, spring barley, spring oats, spring wheat, triticale, winter barley, winter oats, winter wheat* [7]

Approval information

- Metsulfuron-methyl included in Annex I under EC Regulation 1107/2009
- Accepted by BBPA for use on malting barley

Efficacy guidance

- Best results achieved on small, actively growing weeds up to 6 true leaf stage. Good spray cover is important
- Weed control may be reduced in dry soil conditions
- Commonly used in tank-mixture on wheat and barley with other cereal herbicides to improve control of resistant dicotyledons (cleavers, fumitory, ivy-leaved speedwell), larger weeds and grasses. See label for recommended mixtures
- Metsulfuron-methyl is a member of the ALS-inhibitor group of herbicides and products should be used in a planned Resistance Management strategy. See Section 5 for more information

FOR FULL CONDITIONS OF USE ALWAYS READ THE PRODUCT LABEL

Restrictions
- Maximum number of treatments 1 per crop or per yr (for set-aside)
- Product must only be used after 1 Feb
- Do not apply within 7 d of rolling
- Do not use on cereal crops undersown with grass or legumes
- Do not use in tank mixture on oats, triticale or linseed. On linseed allow at least 7 d before or after other treatments
- Consult contract agents before use on a cereal crop for seed
- Do not use on any crop suffering stress from drought, waterlogging, frost, deficiency, pest or disease attack
- Specific restrictions apply to use in sequence or tank mixture with other sulfonylurea or ALS-inhibiting herbicides. See label for details
- Spraying equipment should not be drained or flushed onto land planted, or to be planted, with trees or crops other than cereals and should be thoroughly cleansed after use - see label for instructions
- Must only be applied from 1 February in the year of harvest until the specified latest time of application and must only be applied to green cover on land not being used for crop production where a full green cover is established [10, 11]
- For winter cereals; application should only be made after 15 March and before specified latest time of application [8]
- For spring cereals and linseed; this product must only be applied from 1 April in the year of harvest until the specified latest time of application; For winter cereals application should only be made after 15 March and before specified latest time of application; For setaside this product must only be used after 1 April and before the specified latest time of application [12]

Crop-specific information
- Latest use: before flag leaf sheath extending stage for cereals (GS 41); before flower buds visible or crop 30 cm tall for linseed; before 1 Aug in yr of treatment for land not being used for crop production
- Apply after 1 Feb to wheat, oats and triticale from 2-leaf (GS 12), and to barley from 3-leaf (GS 13) until flag-leaf sheath extending (GS 41)
- On linseed allow at least 7 d (10 d if crop is growing poorly or under stress) before or after other treatments
- Use in set-aside when a full green cover is established and made up predominantly of grassland, wheat, barley, oats or triticale. Do not use on seedling grasses

Following crops guidance
- Only cereals, oilseed rape, field beans or grass may be sown in same calendar year after treating cereals with the product alone. Other restrictions apply to tank mixtures. See label for details
- Only cereals should be planted within 16 mth of applying to a linseed crop or set-aside
- In the event of failure of a treated crop sow only wheat within 3 mth after treatment
- A minimum period of 120 days must be observed prior to planting of succeeding crops [8]

Environmental safety
- Dangerous for the environment
- Very toxic to aquatic organisms
- Take extreme care to avoid damage by drift onto broad-leaved plants outside the target area, onto surface waters or ditches or onto land intended for cropping
- A range of broad leaved species will be fully or partially controlled when used in land temporarily removed from production, hence product may not be suitable where wild flower borders or other forms of conservation headland are being developed
- Before use on land temporarily removed from production as part of grant-aided scheme, ensure compliance with the management rules
- Green cover on land temporarily removed from production must not be grazed by livestock or harvested for human or animal consumption or used for animal bedding
- LERAP Category B

Hazard classification and safety precautions
Hazard Dangerous for the environment [1-12]; Very toxic to aquatic organisms [1-3, 6-10, 12]
Transport code 9
Packaging group III

SEE SECTION 3 FOR PRODUCTS ALSO REGISTERED

UN Number 3077
Risk phrases H315, H318 [4, 11]; H319 [1]
Operator protection A, C [4, 11]; H [4]; U04b, U11 [4]; U05a [5]; U19a [10]; U20b [4, 10]
Environmental protection E15a [10, 11]; E15b [1-9, 12]; E16a, E38, H410 [1-12]; E16b [3]; E23 [5]
Storage and disposal D01, D02 [5]; D09a [3-5, 10, 11]; D10b [4, 10, 11]; D11a [3, 5]; D12a [1-4, 6-12]; D12b [4, 5]

299   metsulfuron-methyl + thifensulfuron-methyl

A contact residual and translocated sulfonylurea herbicide mixture for use in cereals
HRAC mode of action code: B + B

See also thifensulfuron-methyl

Products

1	Accurate Extra	Nufarm UK	7:68% w/w	WG	18608
2	Assuage	Becesane	6.8:68.2% w/w	SG	15801
3	Avro SX	FMC Agro	2.9:42.9 % w/w	SG	18771
4	Chimera SX	FMC Agro	2.9:42.9% w/w	SG	18823
5	Clayton Paragon	Clayton	6.8:68.2% w/w	WG	19051
6	Concert SX	FMC Agro	4:40% w/w	SG	18806
7	Ergon	Rotam	6.8:68.2% w/w	WG	14382
8	Finish SX	FMC Agro	6.7:33.3% w/w	SG	18762
9	Harmony M SX	FMC Agro	4:40% w/w	SG	18824
10	Mozaic SX	FMC Agro	4:40% w/w	SG	18759
11	Pennant	FMC Agro	4:40% w/w	SG	18779
12	Presite SX	FMC Agro	6.7:33.3% w/w	SG	18776
13	Refine Max SX	FMC Agro	6.7:33.3% w/w	SG	18785

Uses

*   Annual dicotyledons in *miscanthus (off-label)* [4, 9]; *spring barley, spring wheat, winter wheat* [1-13]; *spring oats, triticale, winter rye* [5, 7]; *winter barley* [3, 4, 8, 12, 13]; *winter oats* [8, 12, 13]
*   Black bindweed in *spring barley, spring wheat, winter barley, winter wheat* [3]
*   Charlock in *spring barley, spring wheat, winter wheat* [2, 3, 5, 7]; *spring oats, triticale, winter rye* [5, 7]; *winter barley* [3]
*   Chickweed in *spring barley, spring wheat, winter wheat* [2-13]; *spring oats, triticale, winter rye* [5, 7]; *winter barley* [3, 4, 8, 12, 13]; *winter oats* [8, 12, 13]
*   Creeping thistle in *spring barley, spring wheat, winter barley, winter wheat* [3, 4]
*   Fat hen in *spring barley, spring wheat, winter barley, winter wheat* [3]
*   Field pansy in *spring barley, spring wheat, winter barley, winter wheat* [3, 8, 12, 13]; *winter oats* [8, 12, 13]
*   Field speedwell in *spring barley, spring wheat, winter barley, winter wheat* [3, 8, 12, 13]; *winter oats* [8, 12, 13]
*   Forget-me-not in *spring barley, spring wheat, winter barley, winter wheat* [3]
*   Hemp-nettle in *spring barley, spring wheat, winter wheat* [2, 3, 5, 7]; *spring oats, triticale, winter rye* [5, 7]; *winter barley* [3]
*   Ivy-leaved speedwell in *spring barley, spring wheat, winter barley, winter oats, winter wheat* [8, 12, 13]
*   Knotgrass in *spring barley, spring wheat, winter barley, winter wheat* [3, 8, 12, 13]; *winter oats* [8, 12, 13]
*   Mayweeds in *spring barley, spring wheat, winter wheat* [2-13]; *spring oats, triticale, winter rye* [5, 7]; *winter barley* [3, 4, 8, 12, 13]; *winter oats* [8, 12, 13]
*   Parsley-piert in *spring barley, spring wheat, winter wheat* [2, 5, 7]; *spring oats, triticale, winter rye* [5, 7]
*   Polygonums in *spring barley, spring wheat, winter wheat* [4, 6, 9-11]; *winter barley* [4]
*   Poppies in *spring barley, spring wheat, winter barley, winter wheat* [3]

- Red dead-nettle in *spring barley*, *spring wheat*, *winter wheat* [2-7, 9-11]; *spring oats*, *triticale*, *winter rye* [5, 7]; *winter barley* [3, 4]
- Redshank in *spring barley*, *spring wheat*, *winter wheat* [2, 3, 5, 7]; *spring oats*, *triticale*, *winter rye* [5, 7]; *winter barley* [3]
- Shepherd's purse in *spring barley*, *spring wheat*, *winter wheat* [2-7, 9-11]; *spring oats*, *triticale*, *winter rye* [5, 7]; *winter barley* [3, 4]

Extension of Authorisation for Minor Use (EAMUs)
- *miscanthus* 20183485 [4], 20190866 [9]

Approval information
- Metsulfuron-methyl and thifensulfuron-methyl included in Annex I under EC Regulation 1107/2009
- Accepted by BBPA for use on malting barley

Efficacy guidance
- Best results by application to small, actively growing weeds up to 6-true leaf stage
- Ensure good spray cover
- Susceptible weeds stop growing almost immediately but symptoms may not be visible for about 2 wk
- Effectiveness may be reduced by heavy rain or if soil conditions very dry
- Metsulfuron-methyl and thifensulfuron-methyl are members of the ALS-inhibitor group of herbicides and products should be used in a planned Resistance Management strategy. See Section 5 for more information

Restrictions
- Maximum number of treatments 1 per crop
- Products may only be used after 1 Feb
- Do not use on any crop suffering stress from drought, waterlogging, frost, deficiency, pest or disease attack or any other cause
- Do not use on crops undersown with grasses, clover or legumes, or any other broad leaved crop
- Specific restrictions apply to use in sequence or tank mixture with other sulfonylurea or ALS-inhibiting herbicides. See label for details
- Do not apply within 7 d of rolling
- Consult contract agents before use on a cereal crop grown for seed

Crop-specific information
- Latest use: before flag leaf sheath extending (GS 39) for barley and wheat, before 2nd node (GS32) for winter oats

Following crops guidance
- Only cereals, oilseed rape, field beans or grass may be sown in same calendar year after treatment
- Additional constraints apply after use of certain tank mixtures. See label
- In the event of crop failure sow only winter wheat within 3 mth after treatment and after ploughing and cultivating to a depth of at least 15 cm

Environmental safety
- Dangerous for the environment
- Very toxic to aquatic organisms
- Take extreme care to avoid damage by drift onto broad-leaved plants outside the target area, or onto ponds, waterways or ditches
- Spraying equipment should not be drained or flushed onto land planted, or to be planted, with trees or crops other than cereals and should be thoroughly cleansed after use - see label for instructions
- LERAP Category B

Hazard classification and safety precautions
Hazard Dangerous for the environment [1-13]; Very toxic to aquatic organisms [3, 4, 6, 8-13]
Transport code 9
Packaging group III
UN Number 3077

SEE SECTION 3 FOR PRODUCTS ALSO REGISTERED

Risk phrases R50, R53a [2]
Operator protection U05a [8, 12, 13]; U08, U20b [3, 4, 6, 8-13]; U19a [2-13]; U20c [2, 5, 7]
Environmental protection E15a [8, 12, 13]; E15b [1, 2, 5-7, 9-11]; E16a [1-13]; E38 [1, 3, 4, 6, 8-13]; H410 [1, 3-13]
Storage and disposal D01, D02 [8, 12, 13]; D09a [1-13]; D10b [2, 5, 7]; D11a [1, 6, 8-13]; D12a [1, 3, 4, 6, 8-13]

---

300 metsulfuron-methyl + tribenuron-methyl

A sulfonylurea herbicide mixture for cereals
HRAC mode of action code: B + B

See also tribenuron-methyl

Products

1 Ally Max SX	FMC Agro	14.3:14.3% w/w	SG	18768
2 Biplay SX	FMC Agro	11.1:22.2% w/w	SG	18800
3 Boudha	Rotam	25:25% w/w	WG	15158
4 DP911 SX	FMC Agro	11.1:22.2% w/w	SG	18764
5 Traton SX	FMC Agro	11.1:22.2% w/w	SG	18793

Uses

- Annual dicotyledons in *rye (off-label)* [1]; *spring barley, spring wheat, triticale, winter barley, winter oats, winter wheat* [1-5]; *spring oats* [1, 3]; *winter rye* [3]
- Chickweed in *spring barley, spring wheat, triticale, winter barley, winter oats, winter wheat* [1-3, 5]; *spring oats* [1, 3]; *winter rye* [3]
- Field pansy in *spring barley, spring wheat, triticale, winter barley, winter oats, winter wheat* [1, 2, 5]
- Field speedwell in *spring oats* [1]
- Hemp-nettle in *spring barley, spring wheat, triticale, winter barley, winter oats, winter wheat* [1, 2, 5]; *spring oats* [1]
- Mayweeds in *spring barley, spring wheat, triticale, winter barley, winter oats, winter wheat* [1-3, 5]; *spring oats* [1, 3]; *winter rye* [3]
- Poppies in *spring barley, spring oats, spring wheat, triticale, winter barley, winter oats, winter rye, winter wheat* [3]
- Red dead-nettle in *spring barley, spring wheat, triticale, winter barley, winter oats, winter wheat* [1-3, 5]; *spring oats* [1, 3]; *winter rye* [3]
- Shepherd's purse in *spring barley, spring oats, spring wheat, triticale, winter barley, winter oats, winter rye, winter wheat* [3]
- Volunteer oilseed rape in *rye (off-label)* [1]; *spring barley, spring oats, spring wheat, triticale, winter barley, winter oats, winter wheat* [1, 3]; *winter rye* [3]
- Volunteer sugar beet in *spring barley, spring oats, spring wheat, triticale, winter barley, winter oats, winter wheat* [1]

Extension of Authorisation for Minor Use (EAMUs)

- *rye 20190562* [1]

Approval information

- Metsulfuron-methyl and tribenuron-methyl included in Annex I under EC Regulation 1107/2009
- Accepted by BBPA for use on malting barley

Efficacy guidance

- Best results obtained when applied to small actively growing weeds
- Product acts by foliar and root uptake. Good spray cover essential but performance may be reduced when soil conditions are very dry and residual effects may be reduced by heavy rain
- Weed growth inhibited within hours of treatment and many show marked colour changes as they die back. Full effects may not be apparent for up to 4 wk
- Metsulfuron-methyl and tribenuron-methyl are members of the ALS-inhibitor group of herbicides and products should be used in a planned Resistance Management strategy. See Section 5 for more information

FOR FULL CONDITIONS OF USE ALWAYS READ THE PRODUCT LABEL

Restrictions

- Maximum number of treatments 1 per crop
- Product must only be used after 1 Feb and after crop has three leaves
- Do not apply to a crop suffering from drought, waterlogging, low temperatures, pest or disease attack, nutrient deficiency, soil compaction or any other stress
- Do not use on crops undersown with grasses, clover or other legumes
- Do not apply within 7 d of rolling
- Specific restrictions apply to use in sequence or tank mixture with other sulfonylurea or ALS-inhibiting herbicides. See label for details

Crop-specific information

- Latest use: before flag leaf sheath extending

Following crops guidance

- Only cereals, field beans, grass or oilseed rape may be sown in the same calendar yr as harvest of a treated crop
- In the event of failure of a treated crop only winter wheat may be sown within 3 mth of treatment, and only after ploughing and cultivation to 15 cm minimum

Environmental safety

- Dangerous for the environment
- Very toxic to aquatic organisms
- Some non-target crops are highly sensitive. Take extreme care to avoid drift outside the target area, or onto ponds, waterways or ditches
- Spraying equipment should be thoroughly cleaned in accordance with manufacturer's instructions
- LERAP Category B [2-5]

Hazard classification and safety precautions

Hazard Irritant, Very toxic to aquatic organisms [1, 2, 4, 5]; Dangerous for the environment [1-5]
Transport code 9
Packaging group III
UN Number 3077
Operator protection A, H [1, 2, 5]; U08, U19a [1-5]; U20b [1, 2, 4, 5]
Environmental protection E15a [1, 2, 4, 5]; E15b, E39 [3]; E16a [2-5]; E38 [1-3, 5]; H410 [1-5]
Storage and disposal D01, D02, D09a, D11a, D12a

301   Mild Pepino Mosaic Virus isolate VC1 + VX1

For use in tomatoes grown under permanent protection

Products

V10	Valto	10 - 50 mg/litre	LI	18969

Uses

- Virus in **protected tomatoes**

Approval information

- Pepino mosaic virus isolate VX1 and VC1 included in Annex 1 under EC Regulation 1107/2009

Efficacy guidance

- Plants must be free of Pepino Mosaic Virus at the time of treatment

Hazard classification and safety precautions

UN Number N/C
Operator protection H

SEE SECTION 3 FOR PRODUCTS ALSO REGISTERED

302 1-naphthylacetic acid

A plant growth regulator to promote rooting of cuttings

Products

Fixor	Belcrop	100 g/l	SL	17428

Uses

- Fruit thinning in *apples*
- Growth regulation in *christmas trees* (off-label)

Extension of Authorisation for Minor Use (EAMUs)

- *christmas trees* 20181213

Approval information

- 1-naphthylacetic acid included in Annex I under EC Regulation 1107/2009

Efficacy guidance

- One application should be made when the fruit have a diameter between 8 and 12 mm (BBCH 71)
- Dose rate: Max 150 ml product/ha in 1000 litres water
- Adapt the spray volume to the tree size and canopy density to ensure a thorough fruit and leaf coverage

Hazard classification and safety precautions

UN Number N/C
Risk phrases H318, H361
Operator protection U20c
Environmental protection E15a
Storage and disposal D05, D09a, D11a

303 napropamide

A soil applied alkanamide herbicide for oilseed rape, fruit and woody ornamentals
HRAC mode of action code: K3

See also clomazone + metazachlor + napropamide
clomazone + napropamide

Products

1	AC 650	UPL Europe	450 g/l	SC	17966
2	Devrinol	UPL Europe	450 g/l	SC	17853

Uses

- Amaranthus in *baby leaf crops* (off-label), *celeriac* (off-label), *choi sum* (off-label), *collards* (off-label), *herbs (see appendix 6)* (off-label), *horseradish* (off-label), *lamb's lettuce* (off-label), *oriental cabbage* (off-label), *protected baby leaf crops* (off-label), *protected herbs (see appendix 6)* (off-label), *protected lamb's lettuce* (off-label), *protected red mustard* (off-label), *protected rocket* (off-label), *red mustard* (off-label), *rocket* (off-label), *swedes* (off-label), *turnips* (off-label) [2]
- Annual dicotyledons in *broccoli, brussels sprouts, cabbages, calabrese, cauliflowers, kale* [2]; *winter oilseed rape* [1, 2]
- Annual grasses in *broccoli, brussels sprouts, cabbages, calabrese, cauliflowers, kale* [2]; *winter oilseed rape* [1, 2]
- Annual meadow grass in *baby leaf crops* (off-label), *borage for oilseed production* (off-label), *celeriac* (off-label), *choi sum* (off-label), *collards* (off-label), *evening primrose* (off-label), *game cover* (off-label), *herbs (see appendix 6)* (off-label), *horseradish* (off-label), *lamb's lettuce* (off-label), *linseed* (off-label), *mustard* (off-label), *oriental cabbage* (off-label), *protected baby leaf crops* (off-label), *protected herbs (see appendix 6)* (off-label), *protected lamb's lettuce* (off-label), *protected red mustard* (off-label), *protected rocket* (off-label), *red mustard* (off-label), *rocket* (off-label), *swedes* (off-label), *turnips* (off-label), *winter oilseed rape* [2]

- Barnyard grass in **baby leaf crops** *(off-label)*, **celeriac** *(off-label)*, **choi sum** *(off-label)*, **collards** *(off-label)*, **herbs (see appendix 6)** *(off-label)*, **horseradish** *(off-label)*, **lamb's lettuce** *(off-label)*, **oriental cabbage** *(off-label)*, **protected baby leaf crops** *(off-label)*, **protected herbs (see appendix 6)** *(off-label)*, **protected lamb's lettuce** *(off-label)*, **protected red mustard** *(off-label)*, **protected rocket** *(off-label)*, **red mustard** *(off-label)*, **rocket** *(off-label)*, **swedes** *(off-label)*, **turnips** *(off-label)* [2]
- Black bindweed in **baby leaf crops** *(off-label)*, **celeriac** *(off-label)*, **choi sum** *(off-label)*, **collards** *(off-label)*, **herbs (see appendix 6)** *(off-label)*, **horseradish** *(off-label)*, **lamb's lettuce** *(off-label)*, **oriental cabbage** *(off-label)*, **protected baby leaf crops** *(off-label)*, **protected herbs (see appendix 6)** *(off-label)*, **protected lamb's lettuce** *(off-label)*, **protected red mustard** *(off-label)*, **protected rocket** *(off-label)*, **red mustard** *(off-label)*, **rocket** *(off-label)*, **swedes** *(off-label)*, **turnips** *(off-label)* [2]
- Blackgrass in **borage for oilseed production** *(off-label)*, **evening primrose** *(off-label)*, **game cover** *(off-label)*, **linseed** *(off-label)*, **mustard** *(off-label)*, **winter oilseed rape** [2]
- Chickweed in **baby leaf crops** *(off-label)*, **borage for oilseed production** *(off-label)*, **celeriac** *(off-label)*, **choi sum** *(off-label)*, **collards** *(off-label)*, **evening primrose** *(off-label)*, **game cover** *(off-label)*, **herbs (see appendix 6)** *(off-label)*, **horseradish** *(off-label)*, **lamb's lettuce** *(off-label)*, **linseed** *(off-label)*, **mustard** *(off-label)*, **oriental cabbage** *(off-label)*, **protected baby leaf crops** *(off-label)*, **protected herbs (see appendix 6)** *(off-label)*, **protected lamb's lettuce** *(off-label)*, **protected red mustard** *(off-label)*, **protected rocket** *(off-label)*, **red mustard** *(off-label)*, **rocket** *(off-label)*, **swedes** *(off-label)*, **turnips** *(off-label)* [2]
- Cleavers in **baby leaf crops** *(off-label)*, **broccoli**, **brussels sprouts**, **cabbages**, **calabrese**, **cauliflowers**, **celeriac** *(off-label)*, **choi sum** *(off-label)*, **collards** *(off-label)*, **herbs (see appendix 6)** *(off-label)*, **horseradish** *(off-label)*, **kale**, **lamb's lettuce** *(off-label)*, **oriental cabbage** *(off-label)*, **protected baby leaf crops** *(off-label)*, **protected herbs (see appendix 6)** *(off-label)*, **protected lamb's lettuce** *(off-label)*, **protected red mustard** *(off-label)*, **protected rocket** *(off-label)*, **red mustard** *(off-label)*, **rocket** *(off-label)*, **swedes** *(off-label)*, **turnips** *(off-label)* [2]
- Common purslane in **baby leaf crops** *(off-label)*, **celeriac** *(off-label)*, **choi sum** *(off-label)*, **collards** *(off-label)*, **herbs (see appendix 6)** *(off-label)*, **horseradish** *(off-label)*, **lamb's lettuce** *(off-label)*, **oriental cabbage** *(off-label)*, **protected baby leaf crops** *(off-label)*, **protected herbs (see appendix 6)** *(off-label)*, **protected lamb's lettuce** *(off-label)*, **protected red mustard** *(off-label)*, **protected rocket** *(off-label)*, **red mustard** *(off-label)*, **rocket** *(off-label)*, **swedes** *(off-label)*, **turnips** *(off-label)* [2]
- Fat hen in **baby leaf crops** *(off-label)*, **borage for oilseed production** *(off-label)*, **celeriac** *(off-label)*, **choi sum** *(off-label)*, **collards** *(off-label)*, **evening primrose** *(off-label)*, **herbs (see appendix 6)** *(off-label)*, **horseradish** *(off-label)*, **lamb's lettuce** *(off-label)*, **mustard** *(off-label)*, **oriental cabbage** *(off-label)*, **protected baby leaf crops** *(off-label)*, **protected herbs (see appendix 6)** *(off-label)*, **protected lamb's lettuce** *(off-label)*, **protected red mustard** *(off-label)*, **protected rocket** *(off-label)*, **red mustard** *(off-label)*, **rocket** *(off-label)*, **swedes** *(off-label)*, **turnips** *(off-label)* [2]
- Field pansy in **borage for oilseed production** *(off-label)*, **evening primrose** *(off-label)*, **game cover** *(off-label)*, **linseed** *(off-label)*, **mustard** *(off-label)* [2]
- Field speedwell in **baby leaf crops** *(off-label)*, **borage for oilseed production** *(off-label)*, **celeriac** *(off-label)*, **choi sum** *(off-label)*, **collards** *(off-label)*, **evening primrose** *(off-label)*, **game cover** *(off-label)*, **herbs (see appendix 6)** *(off-label)*, **horseradish** *(off-label)*, **lamb's lettuce** *(off-label)*, **linseed** *(off-label)*, **mustard** *(off-label)*, **oriental cabbage** *(off-label)*, **protected baby leaf crops** *(off-label)*, **protected herbs (see appendix 6)** *(off-label)*, **protected lamb's lettuce** *(off-label)*, **protected red mustard** *(off-label)*, **protected rocket** *(off-label)*, **red mustard** *(off-label)*, **rocket** *(off-label)*, **swedes** *(off-label)*, **turnips** *(off-label)* [2]
- Forget-me-not in **borage for oilseed production** *(off-label)*, **evening primrose** *(off-label)*, **game cover** *(off-label)*, **linseed** *(off-label)*, **mustard** *(off-label)* [2]
- Fumitory in **borage for oilseed production** *(off-label)*, **evening primrose** *(off-label)*, **game cover** *(off-label)*, **linseed** *(off-label)*, **mustard** *(off-label)* [2]
- Groundsel in **baby leaf crops** *(off-label)*, **broccoli**, **brussels sprouts**, **cabbages**, **calabrese**, **cauliflowers**, **celeriac** *(off-label)*, **choi sum** *(off-label)*, **collards** *(off-label)*, **herbs (see appendix 6)** *(off-label)*, **horseradish** *(off-label)*, **kale**, **lamb's lettuce** *(off-label)*, **oriental cabbage** *(off-label)*, **protected baby leaf crops** *(off-label)*, **protected herbs (see appendix 6)** *(off-label)*, **protected lamb's lettuce** *(off-label)*, **protected red mustard** *(off-label)*, **protected rocket** *(off-label)*, **red mustard** *(off-label)*, **rocket** *(off-label)*, **swedes** *(off-label)*, **turnips** *(off-label)* [2]

SEE SECTION 3 FOR PRODUCTS ALSO REGISTERED

- Ivy-leaved speedwell in *borage for oilseed production* (off-label), *evening primrose* (off-label), *game cover* (off-label), *linseed* (off-label), *mustard* (off-label) [2]
- Knotgrass in *baby leaf crops* (off-label), *celeriac* (off-label), *choi sum* (off-label), *collards* (off-label), *herbs (see appendix 6)* (off-label), *horseradish* (off-label), *lamb's lettuce* (off-label), *oriental cabbage* (off-label), *protected baby leaf crops* (off-label), *protected herbs (see appendix 6)* (off-label), *protected lamb's lettuce* (off-label), *protected red mustard* (off-label), *protected rocket* (off-label), *red mustard* (off-label), *rocket* (off-label), *swedes* (off-label), *turnips* (off-label) [2]
- Mayweeds in *baby leaf crops* (off-label), *borage for oilseed production* (off-label), *celeriac* (off-label), *choi sum* (off-label), *collards* (off-label), *evening primrose* (off-label), *game cover* (off-label), *herbs (see appendix 6)* (off-label), *horseradish* (off-label), *lamb's lettuce* (off-label), *linseed* (off-label), *mustard* (off-label), *oriental cabbage* (off-label), *protected baby leaf crops* (off-label), *protected herbs (see appendix 6)* (off-label), *protected lamb's lettuce* (off-label), *protected red mustard* (off-label), *protected rocket* (off-label), *red mustard* (off-label), *rocket* (off-label), *swedes* (off-label), *turnips* (off-label) [2]
- Pale persicaria in *borage for oilseed production* (off-label), *evening primrose* (off-label), *game cover* (off-label), *linseed* (off-label), *mustard* (off-label) [2]
- Poppies in *borage for oilseed production* (off-label), *evening primrose* (off-label), *game cover* (off-label), *linseed* (off-label), *mustard* (off-label) [2]
- Redshank in *baby leaf crops* (off-label), *borage for oilseed production* (off-label), *celeriac* (off-label), *choi sum* (off-label), *collards* (off-label), *evening primrose* (off-label), *game cover* (off-label), *herbs (see appendix 6)* (off-label), *horseradish* (off-label), *lamb's lettuce* (off-label), *linseed* (off-label), *mustard* (off-label), *oriental cabbage* (off-label), *protected baby leaf crops* (off-label), *protected herbs (see appendix 6)* (off-label), *protected lamb's lettuce* (off-label), *protected red mustard* (off-label), *protected rocket* (off-label), *red mustard* (off-label), *rocket* (off-label), *swedes* (off-label), *turnips* (off-label) [2]
- Shepherd's purse in *baby leaf crops* (off-label), *celeriac* (off-label), *choi sum* (off-label), *collards* (off-label), *herbs (see appendix 6)* (off-label), *horseradish* (off-label), *lamb's lettuce* (off-label), *oriental cabbage* (off-label), *protected baby leaf crops* (off-label), *protected herbs (see appendix 6)* (off-label), *protected lamb's lettuce* (off-label), *protected red mustard* (off-label), *protected rocket* (off-label), *red mustard* (off-label), *rocket* (off-label), *swedes* (off-label), *turnips* (off-label) [2]
- Small nettle in *baby leaf crops* (off-label), *celeriac* (off-label), *choi sum* (off-label), *collards* (off-label), *herbs (see appendix 6)* (off-label), *horseradish* (off-label), *lamb's lettuce* (off-label), *oriental cabbage* (off-label), *protected baby leaf crops* (off-label), *protected herbs (see appendix 6)* (off-label), *protected lamb's lettuce* (off-label), *protected red mustard* (off-label), *protected rocket* (off-label), *red mustard* (off-label), *rocket* (off-label), *swedes* (off-label), *turnips* (off-label) [2]

Extension of Authorisation for Minor Use (EAMUs)

- *baby leaf crops* 20171231 [2]
- *borage for oilseed production* 20170447 [2]
- *celeriac* 20171126 [2]
- *choi sum* 20171127 [2]
- *collards* 20171127 [2]
- *evening primrose* 20170447 [2]
- *game cover* 20170446 [2]
- *herbs (see appendix 6)* 20192121 [2]
- *horseradish* 20171126 [2]
- *lamb's lettuce* 20171231 [2]
- *linseed* 20170447 [2]
- *mustard* 20170447 [2]
- *oriental cabbage* 20171127 [2]
- *protected baby leaf crops* 20171231 [2]
- *protected herbs (see appendix 6)* 20192121 [2]
- *protected lamb's lettuce* 20171231 [2]
- *protected red mustard* 20171231 [2]
- *protected rocket* 20171231 [2]
- *red mustard* 20171231 [2]

FOR FULL CONDITIONS OF USE ALWAYS READ THE PRODUCT LABEL

- *rocket* 20171231 [2]
- *swedes* 20171126 [2]
- *turnips* 20171126 [2]

Approval information

- Napropamide included in Annex I under EC Regulation 1107/2009

Efficacy guidance

- Best results obtained from treatment pre-emergence of weeds but product may be used in conjunction with contact herbicides for control of emerged weeds. Otherwise remove existing weeds before application
- Weed control may be reduced where spray is mixed too deeply in the soil
- Manure, crop debris or other organic matter may reduce weed control
- Napropamide broken down by sunlight, so application during such conditions not recommended. Most crops recommended for treatment between Nov and end-Feb
- Post-emergence use of specific grass weedkilller recommended where volunteer cereals are a serious problem

Restrictions

- Maximum number of treatments 1 per crop or yr
- Do not use on Sands
- Do not use on soils with more than 10% organic matter
- Consult processors before use on any crop for processing

Crop-specific information

- Latest use: before transplanting for brassicas; pre-emergence for winter oilseed rape
- Where minimal cultivation used to establish oilseed rape, tank-mixture may be applied directly to stubble and mixed into top 25 mm as part of surface cultivations [1]
- Apply up to 14 d prior to drilling winter oilseed rape [1]

Following crops guidance

- After use in oilseed rape only oilseed rape, swedes, fodder turnips, brassicas or potatoes should be sown within 12 mth of application
- Soil should be mould-board ploughed to a depth of at least 200 mm before drilling or planting any following crop

Environmental safety

- Dangerous for the environment
- Toxic to aquatic organisms
- LERAP Category B

Hazard classification and safety precautions

Hazard Irritant, Dangerous for the environment, Very toxic to aquatic organisms
Transport code 9
Packaging group III
UN Number 3082
Risk phrases H315, H320 [2]
Operator protection A [1, 2]; C [2]; H [1]; U05a, U09a, U11, U20b
Environmental protection E15a [2]; E16a, E34, E38, H410 [1, 2]
Storage and disposal D01, D02, D05, D09a, D10c [1, 2]; D12a [1]; D12b [2]

304   nicosulfuron

A sulfonylurea herbicide for maize
HRAC mode of action code: B

See also dicamba + nicosulfuron
   mesotrione + nicosulfuron

Products

1	Accent	DuPont (Corteva)	75% w/w	WG	15392

SEE SECTION 3 FOR PRODUCTS ALSO REGISTERED

Products – continued

2	Crew	Nufarm UK	40 g/l	OD	15770
3	Entail	FMC Agro	240 g/l	OD	18740
4	Fornet 6 OD	Belchim	60 g/l	OD	15891
5	Ikanos	Nufarm UK	40 g/l	OD	18641
6	Milagro 240 OD	Syngenta	240 g/l	OD	15957
7	Nico Pro 4SC	Belchim	40 g/l	OD	16424
8	Pampa 4 SC	Belchim	40 g/l	OD	17521
9	Primero	Rotam	40.8 g/l	OD	16444
10	Samson Extra 6%	Belchim	60 g/l	OD	16418
11	Templier	Rotam	75% w/w	WG	16866

Uses

- Amaranthus in *forage maize* [2, 5, 7, 8]; *grain maize* [2, 5]
- Annual dicotyledons in *forage maize* [1, 3, 4, 6, 9-11]; *forest nurseries* (off-label), *game cover* (off-label), *ornamental plant production* (off-label), *top fruit* (off-label) [4, 7, 10]; *grain maize* [3, 4, 6, 9-11]; *sweetcorn* (off-label), *sweetcorn under plastic mulches* (off-label) [4]
- Annual grasses in *forest nurseries* (off-label), *game cover* (off-label), *ornamental plant production* (off-label), *top fruit* (off-label) [7]
- Annual meadow grass in *forage maize* [1, 2, 4, 5, 7-9, 11]; *forest nurseries* (off-label), *game cover* (off-label), *ornamental plant production* (off-label), *top fruit* (off-label) [4, 7, 10]; *grain maize* [2, 4, 5, 9, 11]; *sweetcorn* (off-label), *sweetcorn under plastic mulches* (off-label) [4, 7]
- Black nightshade in *sweetcorn* (off-label), *sweetcorn under plastic mulches* (off-label) [4]
- Blackgrass in *forage maize* [1, 11]; *grain maize* [11]
- Cockspur grass in *forage maize* [1, 11]; *grain maize* [11]
- Common amaranth in *sweetcorn* (off-label), *sweetcorn under plastic mulches* (off-label) [7]
- Couch in *forage maize* [1, 11]; *grain maize* [11]
- Grass weeds in *forage maize*, *grain maize* [3, 6, 10]; *forest nurseries* (off-label), *game cover* (off-label), *ornamental plant production* (off-label), *top fruit* (off-label) [10]
- Groundsel in *forage maize* [2, 5, 7, 8]; *grain maize* [2, 5]; *sweetcorn* (off-label), *sweetcorn under plastic mulches* (off-label) [7]
- Mayweeds in *forage maize* [2, 5, 7, 8]; *grain maize* [2, 5]
- Ryegrass in *forage maize* [1, 2, 5, 7, 8, 11]; *forest nurseries* (off-label), *game cover* (off-label), *ornamental plant production* (off-label), *top fruit* (off-label) [4, 10]; *grain maize* [2, 5, 11]; *sweetcorn* (off-label), *sweetcorn under plastic mulches* (off-label) [4, 7]
- Shepherd's purse in *forage maize* [2, 5, 7, 8]; *grain maize* [2, 5]; *sweetcorn* (off-label), *sweetcorn under plastic mulches* (off-label) [7]
- Volunteer cereals in *forage maize* [1, 11]; *grain maize* [11]
- Wild oats in *forage maize* [1, 11]; *grain maize* [11]

Extension of Authorisation for Minor Use (EAMUs)

- *forest nurseries* 20140942 [4], 20140752 [7], 20140490 [10], 20141054 [10]
- *game cover* 20140942 [4], 20140752 [7], 20140490 [10], 20141054 [10]
- *ornamental plant production* 20140942 [4], 20140752 [7], 20140490 [10], 20141054 [10]
- *sweetcorn* 20140895 [4], 20140941 [7]
- *sweetcorn under plastic mulches* 20140895 [4], 20140941 [7]
- *top fruit* 20140942 [4], 20140752 [7], 20140490 [10], 20141054 [10]

Approval information

- Nicosulfuron included in Annex I under EC Regulation 1107/2009

Efficacy guidance

- Product should be applied post-emergence between 2 and 8 crop leaf stage and to emerged weeds from the 2-leaf stage
- Product acts mainly by foliar activity. Ensure good spray cover of the weeds
- Nicosulfuron is a member of the ALS-inhibitor group of herbicides. To avoid the build up of resistance do not use any product containing an ALS-inhibitor herbicide with claims for control of grass weeds more than once on any crop
- Use these products as part of a resistance management strategy that includes cultural methods of control and does not use ALS inhibitors as the sole chemical method of grass weed control

FOR FULL CONDITIONS OF USE ALWAYS READ THE PRODUCT LABEL

Restrictions
- Maximum number of treatments 1 per crop
- Do not use if an organophosphorus soil insecticide has been used on the same crop
- Do not mix with foliar or liquid fertilisers or specified herbicides. See label for details
- Do not apply in mixture, or sequence, with any other sulfonyl-urea containing product
- Do not treat crops under stress
- Do not apply if rainfall is forecast to occur within 6 h of application

Crop-specific information
- Latest use: up to and including 8 true leaves for maize
- Some transient yellowing may be seen from 1-2 wk after treatment
- Only healthy maize crops growing in good field conditions should be treated

Following crops guidance
- Winter wheat (not undersown) may be sown 4 mth after treatment; maize may be sown in the spring following treatment; all other crops may be sown from the next autumn
- In normal crop rotation, after ploughing winter wheat, winter barley, winter rye and triticale can be sown and all other crops can be sown the following spring. In the event of crop failure maize or soybeans can be sown after ploughing [4, 7, 10]

Environmental safety
- Dangerous for the environment
- Very toxic to aquatic organisms
- LERAP Category B

Hazard classification and safety precautions
Hazard Harmful [4, 10]; Irritant [2, 3, 5-8]; Dangerous for the environment [1-11]; Harmful if inhaled [2]; Very toxic to aquatic organisms [1, 4, 6, 9, 10]
Transport code 9
Packaging group III
UN Number 3077 [1, 11]; 3082 [2-10]
Risk phrases H315 [2, 3, 5-8]; H317 [2-6, 10]; H319 [4, 10]
Operator protection A [2-8, 10]; C [3, 10]; H [2, 3, 5-8, 10]; U02a, U20b [3, 6, 10]; U05a [1-8, 10]; U10, U11 [3, 4, 6, 10]; U14, U19a [2-8, 10]; U15 [2, 5, 7, 8]
Environmental protection E15b [1-3, 5-8, 10]; E16a, H410 [1-11]; E16b [3, 6, 10]; E34 [1, 2, 5, 7, 8]; E38 [1-8, 10, 11]; E40b [1]
Storage and disposal D01, D02 [1-10]; D06a [4]; D09a [1, 2, 5, 7, 8]; D09b [9]; D10c [2, 3, 5-8, 10]; D11a [1]; D12a [1, 3, 4, 6, 10, 11]; D12b [2, 4, 5, 7, 8]
Medical advice M03 [2, 5, 7, 8]; M05a [3, 4, 6, 10]

305   oxamyl

A soil-applied, systemic carbamate nematicide and insecticide
IRAC mode of action code: 1A

Products
Vydate 10G	DuPont (Corteva)	10% w/w	GR	16595

Uses
- Docking disorder vectors in *fodder beet*, *sugar beet*
- Free-living nematodes in *fodder beet*, *potatoes*, *sugar beet*
- Potato cyst nematode in *potatoes*
- Stem nematodes in *carrots*, *parsnips*

Approval information
- Oxamyl included in Annex I under EC Regulation 1107/2009
- In 2006 CRD required that all products containing this active ingredient should carry the following warning in the main area of the container label: "Oxamyl is an anticholinesterase carbamate. Handle with care"
- Approval expiry 31 Dec 2020 [1]

SEE SECTION 3 FOR PRODUCTS ALSO REGISTERED

Efficacy guidance
- Apply granules with suitable applicator before drilling or planting. See label for details of recommended machines
- In potatoes incorporate thoroughly to 10 cm and plant within 1-2 d
- In sugar beet apply in seed furrow at drilling

Restrictions
- Oxamyl is subject to the Poisons Rules 1982 and the Poisons Act 1972. See Section 5 for more information
- Contains an anticholinesterase carbamate compound. Do not use if under medical advice not to work with such compounds
- Maximum number of treatments 1 per crop or yr
- Keep in original container, tightly closed, in a safe place, under lock and key
- Wear protective gloves if handling treated compost or soil within 2 wk after treatment

Crop-specific information
- Latest use: at drilling/planting for vegetables; before drilling/planting for potatoes.
- HI: A minimum of 80 days must elapse between application and initial haulm destruction in potatoes

Environmental safety
- Dangerous for the environment
- Toxic to aquatic organisms
- Dangerous to fish or other aquatic life. Do not contaminate surface waters or ditches with chemical or used container
- Dangerous to game, wild birds and animals. Bury spillages

Hazard classification and safety precautions
Hazard Toxic, Fatal if swallowed, Toxic if inhaled
Transport code 6.1
Packaging group II
UN Number 2757
Operator protection A, B, H, K, M; C (or D); U02a, U04a, U05a, U08, U13, U19a, U20a, U23a
Environmental protection E10a, E13b, E15b, H412
Storage and disposal D01, D02, D06e, D09a, D11b
Medical advice M04a, M04c, M05a

306   oxathiapiprolin

A blight fungicide for use in potatoes
FRAC mode of action code: 49

Products
Zorvec Enicade	DuPont (Corteva)	100 g/l	OD	18370

Uses
- Blight in *potatoes*
- Late blight in *potatoes*
- Tuber blight in *potatoes*

Approval information
- Oxathiapiprolin included in Annex I under EC Regulation 1107/2009

Efficacy guidance
- Sold only in twinpacks with either Gachinko (amisulbrom) or Rhapsody (cymoxanil + mancozeb) to combat the risk of resistance

Restrictions
- Do not apply by hand-held equipment.

Hazard classification and safety precautions
Transport code 9

FOR FULL CONDITIONS OF USE ALWAYS READ THE PRODUCT LABEL

Packaging group III
UN Number 3082
Risk phrases H317
Operator protection H
Environmental protection H411

307   paclobutrazol

A triazole plant growth regulator for ornamentals and fruit
FRAC mode of action code: 3

See also difenoconazole + paclobutrazol

Products

1	Bonzi	Syngenta	4 g/l	SC	17576
2	Pirouette	Fargro	4 g/l	SC	17203

Uses

- Growth regulation in *azaleas*, *bedding plants*, *kalanchoes*, *mini roses*, *pelargoniums*, *poinsettias* [2]; *protected ornamentals* [1]
- Increasing flowering in *azaleas*, *bedding plants*, *kalanchoes*, *mini roses*, *ornamental plant production*, *ornamental plant production* (*off-label*), *pelargoniums*, *poinsettias* [2]
- Stem shortening in *ornamental plant production*, *ornamental plant production* (*off-label*) [2]

Extension of Authorisation for Minor Use (EAMUs)

- *ornamental plant production 20171269* [2]

Approval information

- Paclobutrazol included in Annex I under EC Regulation 1107/2009

Efficacy guidance

- Chemical is active via both foliage and root uptake. For best results apply in dull weather when relative humidity not high

Restrictions

- Maximum number of treatments 1 per specimen for some species [2]

Following crops guidance

- Chemical has residual soil activity which can affect growth of following crops. Do not use treated pots or soil for other crops

Environmental safety

- Harmful to aquatic organisms
- Do not use on food crops [2]
- Keep livestock out of treated areas for at least 2 years after treatment

Hazard classification and safety precautions

Transport code 9 [2]
Packaging group III [2]
UN Number 3082 [2]; N/C [1]
Operator protection A [2]; U05a, U08, U20a
Environmental protection E13c [2]; E15a, H411 [1, 2]
Consumer protection C01
Storage and disposal D01, D02, D05, D09a [1, 2]; D10b [2]; D10c [1]

308   pelargonic acid

A naturally occuring 9 carbon acid terminating in a carboxylic acid
HRAC mode of action code: Not classified

See also fatty acids + pelargonic acid + maleic hydrazide
maleic hydrazide + pelargonic acid

Products

1	Finalsan	Certis	186.7 g/l	EC	13102
2	Katoun Gold	Belchim	500 g/l	EC	17879

Uses

* Algae in **amenity grassland, amenity vegetation, beetroot** *(off-label)*, **bulb onions** *(off-label)*, **carrots** *(off-label)*, **garlic** *(off-label)*, **managed amenity turf, natural surfaces not intended to bear vegetation, ornamental plant production, parsnips** *(off-label)*, **permeable surfaces overlying soil, salad onions** *(off-label)*, **shallots** *(off-label)*, **swedes** *(off-label)*, **turnips** *(off-label)* [1]
* Annual and perennial weeds in **amenity grassland, amenity vegetation, managed amenity turf, natural surfaces not intended to bear vegetation, ornamental plant production, permeable surfaces overlying soil** [1]
* Annual dicotyledons in **amenity vegetation, forest nurseries, natural surfaces not intended to bear vegetation, ornamental plant production, permeable surfaces overlying soil** [2]; **beetroot** *(off-label)*, **bulb onions** *(off-label)*, **carrots** *(off-label)*, **garlic** *(off-label)*, **parsnips** *(off-label)*, **salad onions** *(off-label)*, **shallots** *(off-label)*, **swedes** *(off-label)*, **turnips** *(off-label)* [1]
* Annual grasses in **beetroot** *(off-label)*, **bulb onions** *(off-label)*, **carrots** *(off-label)*, **garlic** *(off-label)*, **parsnips** *(off-label)*, **salad onions** *(off-label)*, **shallots** *(off-label)*, **swedes** *(off-label)*, **turnips** *(off-label)* [1]
* Annual meadow grass in **celery (outdoor)** *(off-label)*, **leeks** *(off-label)* [1]
* Black nightshade in **asparagus** *(off-label)*, **rhubarb** *(off-label)* [1]
* Burdock in **asparagus** *(off-label)*, **rhubarb** *(off-label)* [1]
* Buttercups in **asparagus** *(off-label)*, **rhubarb** *(off-label)* [1]
* Canadian fleabane in **asparagus** *(off-label)*, **rhubarb** *(off-label)* [1]
* Chickweed in **celery (outdoor)** *(off-label)*, **leeks** *(off-label)* [1]
* Cleavers in **asparagus** *(off-label)*, **rhubarb** *(off-label)* [1]
* Couch in **asparagus** *(off-label)*, **rhubarb** *(off-label)* [1]
* Creeping thistle in **asparagus** *(off-label)*, **beetroot** *(off-label)*, **bulb onions** *(off-label)*, **carrots** *(off-label)*, **garlic** *(off-label)*, **parsnips** *(off-label)*, **rhubarb** *(off-label)*, **salad onions** *(off-label)*, **shallots** *(off-label)*, **swedes** *(off-label)*, **turnips** *(off-label)* [1]
* Dandelions in **beetroot** *(off-label)*, **bulb onions** *(off-label)*, **carrots** *(off-label)*, **garlic** *(off-label)*, **parsnips** *(off-label)*, **salad onions** *(off-label)*, **shallots** *(off-label)*, **swedes** *(off-label)*, **turnips** *(off-label)* [1]
* Field bindweed in **asparagus** *(off-label)*, **rhubarb** *(off-label)* [1]
* Mayweeds in **celery (outdoor)** *(off-label)*, **leeks** *(off-label)* [1]
* Moss in **amenity grassland, amenity vegetation, beetroot** *(off-label)*, **bulb onions** *(off-label)*, **carrots** *(off-label)*, **celery (outdoor)** *(off-label)*, **garlic** *(off-label)*, **leeks** *(off-label)*, **managed amenity turf, natural surfaces not intended to bear vegetation, ornamental plant production, parsnips** *(off-label)*, **permeable surfaces overlying soil, salad onions** *(off-label)*, **shallots** *(off-label)*, **swedes** *(off-label)*, **turnips** *(off-label)* [1]
* Mugwort in **asparagus** *(off-label)*, **rhubarb** *(off-label)* [1]
* Polygonums in **celery (outdoor)** *(off-label)*, **leeks** *(off-label)* [1]
* Ragwort in **asparagus** *(off-label)*, **rhubarb** *(off-label)* [1]
* Small nettle in **asparagus** *(off-label)*, **rhubarb** *(off-label)* [1]
* Speedwells in **celery (outdoor)** *(off-label)*, **leeks** *(off-label)* [1]

Extension of Authorisation for Minor Use (EAMUs)

* **asparagus** *20191786* [1]
* **beetroot** *20171892* [1]
* **bulb onions** *20171892* [1]
* **carrots** *20171892* [1]

FOR FULL CONDITIONS OF USE ALWAYS READ THE PRODUCT LABEL

- **celery (outdoor)** *20191786* [1]
- **garlic** *20171892* [1]
- **leeks** *20191786* [1]
- **parsnips** *20171892* [1]
- **rhubarb** *20191786* [1]
- **salad onions** *20171892* [1]
- **shallots** *20171892* [1]
- **swedes** *20171892* [1]
- **turnips** *20171892* [1]

SECTION 2

Approval information
- Pelargonic acid included in Annex 1 under EC Regulation 1107/2009

Environmental safety
- LERAP Category B [1]

Hazard classification and safety precautions
Hazard Irritant [1]
UN Number N/C
Risk phrases H319 [2]; R36 [1]
Operator protection A, C [1, 2]; P [1]; U05a, U11, U15, U16b, U19a [1]
Environmental protection E12c, E12e [2]; E15a, E16a, E16b, E19b [1]
Storage and disposal D01, D02 [1, 2]; D05 [1]
Medical advice M05a [1]

309 penconazole

A protectant triazole fungicide with antisporulant activity
FRAC mode of action code: 3

See also captan + penconazole

Products

1	Pan Penco	Pan Amenity	100 g/l	EC	18180
2	Topas	Syngenta	100 g/l	EC	16765

Uses

- Powdery mildew in **apples, blackcurrants, crab apples, pears, protected strawberries, redcurrants, strawberries, table grapes, wine grapes** [1, 2]; **courgettes** *(off-label)*, **cucumbers** *(off-label)*, **gherkins** *(off-label)*, **globe artichoke** *(off-label)*, **hops in propagation** *(off-label)*, **nursery fruit trees** *(off-label)*, **ornamental plant production** *(off-label)*, **protected chilli peppers** *(off-label)*, **protected courgettes** *(off-label)*, **protected cucumbers** *(off-label)*, **protected gherkins** *(off-label)*, **protected hops** *(off-label)*, **protected melons** *(off-label)*, **protected nursery fruit trees** *(off-label)*, **protected ornamentals** *(off-label)*, **protected peppers** *(off-label)*, **protected pumpkins** *(off-label)*, **protected summer squash** *(off-label)*, **protected tomatoes** *(off-label)*, **protected watermelon** *(off-label)*, **protected winter squash** *(off-label)*, **summer squash** *(off-label)* [2]

Extension of Authorisation for Minor Use (EAMUs)

- **courgettes** *20152336* [2]
- **cucumbers** *20152336* [2]
- **gherkins** *20152336* [2]
- **globe artichoke** *20152901* [2]
- **hops in propagation** *20152338* [2]
- **nursery fruit trees** *20152338* [2]
- **ornamental plant production** *20190169* [2]
- **protected chilli peppers** *20152335* [2]
- **protected courgettes** *20152336* [2]
- **protected cucumbers** *20152336* [2]
- **protected gherkins** *20152336* [2]
- **protected hops** *20152338* [2]

SEE SECTION 3 FOR PRODUCTS ALSO REGISTERED

- ***protected melons*** *20152337* [2]
- ***protected nursery fruit trees*** *20152338* [2]
- ***protected ornamentals*** *20190169* [2]
- ***protected peppers*** *20152335* [2]
- ***protected pumpkins*** *20152337* [2]
- ***protected summer squash*** *20152336* [2]
- ***protected tomatoes*** *20152335* [2]
- ***protected watermelon*** *20152337* [2]
- ***protected winter squash*** *20152337* [2]
- ***summer squash*** *20152336* [2]

Approval information
- Accepted by BBPA for use on hops
- Penconazole included in Annex I under EC Regulation 1107/2009

Efficacy guidance
- Use as a protectant fungicide by treating at the earliest signs of disease
- Treat crops every 10-14 d (every 7-10 d in warm, humid weather) at first sign of infection or as a protective spray ensuring complete coverage. See label for details of timing
- Increase dose and volume with growth of hops but do not exceed 2000 l/ha. Little or no activity will be seen on established powdery mildew
- Antisporulant activity reduces development of secondary mildew in apples

Restrictions
- Maximum number of treatments 10 per yr for apples, 6 per yr for hops, 4 per yr for blackcurrants
- Check for varietal susceptibility in roses. Some defoliation may occur after repeat applications on Dearest

Crop-specific information
- HI apples, hops 14 d; currants, bilberries, blueberries, cranberries, gooseberries 4 wk

Environmental safety
- Dangerous for the environment
- Toxic to aquatic organisms

Hazard classification and safety precautions
Hazard Irritant, Dangerous for the environment
Transport code 3
Packaging group III
UN Number 1915
Risk phrases H319, H361
Operator protection A, C; U02a, U05a, U08, U20b
Environmental protection E15a, E38, H411
Storage and disposal D01, D02, D05, D09a, D10c, D12a
Medical advice M05b

310   pencycuron

A non-systemic urea fungicide for use on seed potatoes
FRAC mode of action code: 20

See also imazalil + pencycuron

Products
| Solaren | AgChem Access | 12.5% w/w | DS | 15912 |

Uses
- Black scurf in ***potatoes*** *(tuber treatment)*
- Stem canker in ***potatoes*** *(tuber treatment)*

Approval information
- Pencycuron included in Annex I under EC Regulation 1107/2009

Efficacy guidance
- Provides control of tuber-borne disease and gives some reduction of stem canker
- Seed tubers should be of good quality and vigour, and free from soil deposits when treated
- Dust formulations should be applied immediately before, or during, the planting process by treating seed tubers in chitting trays, in bulk bins immediately before planting or in hopper at planting. Liquid formulations may be applied by misting equipment at any time, into or out of store but treatment over a roller table at the end of grading out, or at planting, is usually most convenient
- Whichever method is chosen, complete and even distribution on the tubers is essential for optimum efficacy
- Apply in accordance with detailed guidelines in manufacturer's literature
- If rain interrupts planting, cover dust treated tubers in hopper

Restrictions
- Maximum number of treatments 1 per batch of seed potatoes
- Treated tubers must be used only as seed and not for human or animal consumption
- Do not use on tubers previously treated with a dry powder seed treatment or hot water
- To prevent rapid absorption through damaged 'eyes' seed tubers should only be treated before the 'eyes' begin to open or immediately before or during planting
- Tubers removed from cold storage must be allowed to attain a temperature of at least 8°C before treatment
- Use of suitable dust mask is mandatory when applying dust, filling the hopper or riding on planter
- Treated tubers in bulk bins must be allowed to dry before being stored, especially when intended for cold storage
- Some internal staining of boxes used to store treated tubers may occur. Ware tubers should not be stored or supplied to packers in such stained boxes

Crop-specific information
- Latest use: at planting

Environmental safety
- Do not use treated seed as food or feed
- Treated seed harmful to game and wildlife

Hazard classification and safety precautions
 UN Number N/C
 Operator protection A, F; U20b
 Environmental protection E03, E15a, H411
 Storage and disposal D09a, D11a
 Treated seed S01, S02, S03, S04a, S05, S06a

311   pendimethalin

A residual dinitroaniline herbicide for cereals and other crops
HRAC mode of action code: K1

See also bentazone + pendimethalin
chlorotoluron + diflufenican + pendimethalin
chlorotoluron + pendimethalin
clomazone + pendimethalin
diflufenican + pendimethalin
dimethenamid-p + pendimethalin
flufenacet + pendimethalin
imazamox + pendimethalin

Products

1	Anthem	Adama	400 g/l	SC	15761
2	Bandit	Harvest	400 g/l	SC	18577
3	Claymore	BASF	400 g/l	SC	13441
4	Cleancrop National	Agrii	400 g/l	SC	17094

SEE SECTION 3 FOR PRODUCTS ALSO REGISTERED

SECTION 2

Products – continued

5	Most Micro	Sipcam	365 g/l	CS	16063
6	Nighthawk 330 EC	AgChem Access	330 g/l	EC	14083
7	Pendifin 400 SC	Clayton	400 g/l	SC	18132
8	Stomp 400 SC	BASF	400 g/l	SC	13405
9	Stomp Aqua	BASF	455 g/l	CS	14664
10	Torres	BASF	400 g/l	SC	19017
11	Trample	AgChem Access	330 g/l	EC	14173
12	Yomp	Becesane	330 g/l	EC	15609

Uses

* Annual dicotyledons in **almonds** *(off-label)*, **asparagus** *(off-label)*, **bilberries** *(off-label)*, **blueberries** *(off-label)*, **celery (outdoor)** *(off-label)*, **chervil** *(off-label)*, **chestnuts** *(off-label)*, **coriander** *(off-label)*, **cranberries** *(off-label)*, **cress** *(off-label)*, **dill** *(off-label)*, **evening primrose** *(off-label)*, **farm forestry** *(off-label)*, **fennel** *(off-label)*, **forest** *(off-label)*, **forest nurseries** *(off-label)*, **frise** *(off-label)*, **grass seed crops** *(off-label)*, **hazel nuts** *(off-label)*, **herbs (see appendix 6)** *(off-label)*, **lamb's lettuce** *(off-label)*, **leeks** *(off-label)*, **lettuce** *(off-label)*, **miscanthus** *(off-label)*, **ornamental plant production** *(off-label)*, **parsley root** *(off-label)*, **protected forest nurseries** *(off-label)*, **quinces** *(off-label)*, **radicchio** *(off-label)*, **redcurrants** *(off-label)*, **rhubarb** *(off-label)*, **salad onions** *(off-label)*, **scarole** *(off-label)*, **top fruit** *(off-label)*, **walnuts** *(off-label)*, **whitecurrants** *(off-label)* [8, 9]; **apple orchards**, **pear orchards** [3, 6, 8, 10]; **apples**, **pears** [1, 2, 4, 7, 9, 12]; **blackberries**, **blackcurrants**, **broccoli**, **brussels sprouts**, **cabbages**, **calabrese**, **carrots**, **cauliflowers**, **cherries**, **gooseberries**, **leeks**, **loganberries**, **parsnips**, **plums**, **raspberries**, **rubus hybrids**, **strawberries** [1-4, 6-10, 12]; **bog myrtle** *(off-label)*, **broccoli** *(off-label)*, **brussels sprouts** *(off-label)*, **cabbages** *(off-label)*, **calabrese** *(off-label)*, **carrots** *(off-label)*, **cauliflowers** *(off-label)*, **collards** *(off-label)*, **french beans**, **hops** *(off-label)*, **horseradish** *(off-label)*, **kale** *(off-label)*, **leaf brassicas** *(off-label)*, **mallow (althaea spp.)** *(off-label)*, **navy beans**, **parsley** *(off-label)*, **parsnips** *(off-label)*, **sweetcorn** *(off-label)*, **sweetcorn under plastic mulches** *(off-label)* [9]; **broad beans** *(off-label)*, **game cover** *(off-label)* [1, 4, 8, 9]; **bulb onion sets** *(off-label)* [1, 8]; **bulb onions** [1-4, 7-10, 12]; **bulb onions** *(off-label)*, **garlic** *(off-label)*, **shallots** *(off-label)*, **spring wheat** *(off-label)* [1, 8, 9]; **bulb onions** *(pre + post emergence treatment)* [8, 10]; **chives** *(off-label)*, **navy beans** *(off-label)*, **salad brassicas** *(off-label - for baby leaf production)*, **sweetcorn** *(off-label - under covers)* [8]; **combining peas**, **forage maize** [1-4, 6-12]; **durum wheat**, **potatoes**, **spring barley**, **sunflowers**, **triticale**, **winter barley**, **winter rye**, **winter wheat** [1-12]; **dwarf beans** *(off-label)*, **french beans** *(off-label)* [5, 8]; **edible podded peas** *(off-label)*, **vining peas** *(off-label)* [3, 5, 6, 9]; **field beans** *(off-label)* [4]; **forage maize (under plastic mulches)**, **grain maize**, **grain maize (under plastic mulches)** [1, 2, 4, 9, 12]; **forage soya bean** *(off-label)*, **soya beans** *(off-label)*, **lupins** *(off-label)*, **runner beans** *(off-label)* [3, 5, 8, 9]; **redcurrants** [4, 7, 8, 10]; **spring field beans** *(off-label)* [1, 3, 5, 8, 9]; **winter field beans** *(off-label)* [1, 3, 5, 9]

* Annual grasses in **apples**, **blackberries**, **blackcurrants**, **broccoli**, **brussels sprouts**, **bulb onions**, **cabbages**, **calabrese**, **carrots**, **cauliflowers**, **cherries**, **gooseberries**, **leeks**, **loganberries**, **parsnips**, **pears**, **plums**, **raspberries**, **strawberries** [1, 2, 4, 7, 9, 12]; **bog myrtle** *(off-label)*, **broccoli** *(off-label)*, **brussels sprouts** *(off-label)*, **calabrese** *(off-label)*, **cauliflowers** *(off-label)*, **celery (outdoor)** *(off-label)*, **collards** *(off-label)*, **fennel** *(off-label)*, **grass seed crops** *(off-label)*, **kale** *(off-label)*, **rhubarb** *(off-label)* [9]; **broad beans** *(off-label)*, **game cover** *(off-label)* [1, 4, 8]; **bulb onion sets** *(off-label)*, **bulb onions** *(off-label)*, **garlic** *(off-label)*, **shallots** *(off-label)* [1, 8]; **chervil** *(off-label)*, **coriander** *(off-label)*, **dill** *(off-label)*, **forest nurseries** *(off-label)*, **leeks** *(off-label)*, **miscanthus** *(off-label)*, **navy beans** *(off-label)*, **ornamental plant production** *(off-label)*, **protected forest nurseries** *(off-label)*, **salad onions** *(off-label)*, **top fruit** *(off-label)* [8]; **combining peas**, **forage maize** [1, 2, 4, 7, 9, 11, 12]; **durum wheat**, **potatoes**, **spring barley**, **sunflowers**, **triticale**, **winter barley**, **winter rye**, **winter wheat** [1, 2, 4, 5, 7, 9, 11, 12]; **dwarf beans** *(off-label)*, **french beans** *(off-label)* [5, 8]; **edible podded peas** *(off-label)*, **vining peas** *(off-label)* [5, 6]; **field beans** *(off-label)* [4]; **forage maize (under plastic mulches)**, **grain maize**, **grain maize (under plastic mulches)** [1, 2, 4, 9, 12]; **forage soya bean** *(off-label)*, **lupins** *(off-label)*, **runner beans** *(off-label)* [5]; **redcurrants** [4, 7]; **rubus hybrids** [1, 2, 4, 6, 7, 9, 12]; **spring field beans** *(off-label)*, **winter field beans** *(off-label)* [1, 5, 6]; **spring wheat** *(off-label)* [1, 8, 9]

* Annual meadow grass in **apple orchards**, **blackberries**, **blackcurrants**, **broccoli**, **brussels sprouts**, **cabbages**, **calabrese**, **carrots**, **cauliflowers**, **cherries**, **combining peas**, **forage**

maize, gooseberries, leeks, loganberries, parsnips, pear orchards, plums, raspberries, strawberries [3, 6, 8, 10]; *broad beans* (off-label), *dwarf beans* (off-label), *french beans* (off-label) [5]; *bulb onion sets* (off-label), *bulb onions* (off-label), *garlic* (off-label), *shallots* (off-label) [1, 8]; *bulb onions, rubus hybrids* [3, 8, 10]; *cabbages* (off-label), *lettuce* (off-label) [9]; *durum wheat, potatoes, spring barley, sunflowers, triticale, winter barley, winter rye, winter wheat* [3, 5, 6, 8, 10]; *edible podded peas* (off-label), *lupins* (off-label), *runner beans* (off-label), *spring field beans* (off-label), *vining peas* (off-label), *winter field beans* (off-label) [3, 5]; *forage soya bean* (off-label), *soya beans* (off-label) [5, 9]; *leeks* (off-label), *salad onions* (off-label) [8]; *redcurrants* [8, 10]

- Blackgrass in *apple orchards, blackberries, blackcurrants, broccoli, brussels sprouts, cabbages, calabrese, carrots, cauliflowers, cherries, combining peas, durum wheat, gooseberries, leeks, loganberries, parsnips, pear orchards, plums, raspberries, spring barley, strawberries, sunflowers, triticale, winter barley, winter rye, winter wheat* [3, 6, 8, 10]; *broad beans* (off-label), *game cover* (off-label), *spring field beans* (off-label), *spring wheat* (off-label), *winter field beans* (off-label) [1]; *bulb onions, rubus hybrids* [3, 8, 10]; *forage soya bean* (off-label) [9]; *redcurrants* [8, 10]; *soya beans* (off-label) [5, 9]
- Chickweed in *broad beans* (off-label) [5]; *celeriac* (off-label), *lettuce* (off-label) [9]
- Cleavers in *broad beans* (off-label), *spring field beans* (off-label) [1]; *winter field beans* (off-label) [1, 6, 8]
- Common orache in *broad beans* (off-label) [5]
- Corn marigold in *broad beans* (off-label) [5]
- Dead nettle in *broad beans* (off-label) [5]
- Fat hen in *broad beans* (off-label) [5]; *celeriac* (off-label) [9]
- Field pansy in *broad beans* (off-label) [5]
- Forget-me-not in *broad beans* (off-label) [5]
- Fumitory in *broad beans* (off-label) [5]
- Groundsel in *celeriac* (off-label) [9]
- Hemp-nettle in *broad beans* (off-label) [5]
- Knotgrass in *edible podded peas* (off-label), *vining peas* (off-label) [6, 8]
- Mayweeds in *celeriac* (off-label) [9]
- Parsley-piert in *broad beans* (off-label) [5]
- Polygonums in *celeriac* (off-label) [9]
- Poppies in *broad beans* (off-label) [5]
- Red dead-nettle in *lettuce* (off-label) [9]
- Rough-stalked meadow grass in *apple orchards, blackberries, blackcurrants, broccoli, brussels sprouts, cabbages, calabrese, carrots, cauliflowers, cherries, combining peas, durum wheat, forage maize, gooseberries, leeks, loganberries, parsnips, pear orchards, plums, potatoes, raspberries, spring barley, strawberries, sunflowers, triticale, winter barley, winter rye, winter wheat* [3, 6, 8, 10]; *broad beans* (off-label) [5]; *bulb onions, rubus hybrids* [3, 8, 10]; *redcurrants* [8, 10]
- Scarlet pimpernel in *broad beans* (off-label) [5]
- Shepherd's purse in *broad beans* (off-label) [5]
- Small nettle in *broad beans* (off-label) [5]; *celeriac* (off-label) [9]
- Sowthistle in *broad beans* (off-label) [5]; *celeriac* (off-label) [9]
- Speedwells in *broad beans* (off-label) [5]; *lettuce* (off-label) [9]
- Volunteer oilseed rape in *apple orchards, pear orchards, spring barley, sunflowers* [3, 6, 8, 10]; *apples, pears* [1, 2, 4, 7, 12]; *blackberries, blackcurrants, broccoli, brussels sprouts, cabbages, calabrese, carrots, cauliflowers, cherries, gooseberries, leeks, loganberries, parsnips, plums, raspberries, rubus hybrids* [1-4, 6-8, 10, 12]; *broad beans* (off-label), *field beans* (off-label), *game cover* (off-label) [4]; *bulb onions* [1-4, 7, 8, 10, 12]; *combining peas, durum wheat, forage maize, potatoes, triticale, winter barley, winter rye, winter wheat* [1-4, 6-8, 10-12]; *forage maize (under plastic mulches), grain maize, grain maize (under plastic mulches)* [1, 2, 4, 12]; *redcurrants* [4, 7, 8, 10]; *strawberries* [3, 6, 8, 10, 12]
- Wild oats in *durum wheat, triticale, winter barley, winter rye, winter wheat* [3, 6, 8, 10]

Extension of Authorisation for Minor Use (EAMUs)

- *almonds* 20071442 [8], 20092916 [9]
- *asparagus* 20071431 [8], 20092920 [9]
- *bilberries* 20071444 [8], 20092907 [9]

SEE SECTION 3 FOR PRODUCTS ALSO REGISTERED

- **blueberries** 20071444 [8], 20092907 [9]
- **bog myrtle** 20120323 [9]
- **broad beans** 20141259 [1], 20152446 [4], 20140230 [5], 20081016 [8], 20092922 [9]
- **broccoli** 20112451 [9]
- **brussels sprouts** 20112451 [9]
- **bulb onion sets** 20171415 [1], 20171389 [8]
- **bulb onions** 20171415 [1], 20171389 [8], 20092914 [9]
- **cabbages** 20100650 [9]
- **calabrese** 20112451 [9]
- **carrots** 20093526 [9]
- **cauliflowers** 20112451 [9]
- **celeriac** 20131625 [9]
- **celery (outdoor)** 20071430 [8], 20111287 [9]
- **chervil** 20081808 [8], 20092913 [9]
- **chestnuts** 20071442 [8], 20092916 [9]
- **chives** 20071432 [8]
- **collards** 20112451 [9]
- **coriander** 20081808 [8], 20092913 [9]
- **cranberries** 20071444 [8], 20092907 [9]
- **cress** 20071432 [8], 20092921 [9]
- **dill** 20081808 [8], 20092913 [9]
- **dwarf beans** 20140232 [5], 20080053 [8]
- **edible podded peas** 20152445 [3], 20140233 [5], 20111007 [6], 20071437 [8], 20092909 [9]
- **evening primrose** 20071434 [8], 20092915 [9]
- **farm forestry** 20071436 [8], 20092918 [9]
- **fennel** 20071430 [8], 20111287 [9]
- **field beans** 20152446 [4]
- **forage soya bean** 20140231 [5], 20121634 [9]
- **forest** 20071436 [8], 20092918 [9]
- **forest nurseries** 20082923 [8], 20092919 [9]
- **french beans** 20140232 [5], 20080053 [8], 20092908 [9]
- **frise** 20071432 [8], 20092921 [9]
- **game cover** 20141260 [1], 20152496 [4], 20082923 [8], 20092919 [9]
- **garlic** 20171415 [1], 20171389 [8], 20092914 [9]
- **grass seed crops** 20071429 [8], 20092923 [9]
- **hazel nuts** 20071442 [8], 20092916 [9]
- **herbs (see appendix 6)** 20071432 [8], 20092921 [9]
- **hops** 20092919 [9]
- **horseradish** 20093526 [9]
- **kale** 20112451 [9]
- **lamb's lettuce** 20071432 [8], 20092921 [9]
- **leaf brassicas** 20092921 [9]
- **leeks** 20171389 [8], 20092914 [9]
- **lettuce** 20071432 [8], 20092921 [9], 20170372 [9], 20170375 [9]
- **lupins** 20152445 [3], 20140233 [5], 20071437 [8], 20092909 [9]
- **mallow (althaea spp.)** 20092921 [9]
- **miscanthus** 20082923 [8], 20092919 [9]
- **navy beans** 20080053 [8], 20092908 [9]
- **ornamental plant production** 20082923 [8], 20092919 [9]
- **parsley** 20092913 [9]
- **parsley root** 20071433 [8], 20092911 [9], 20093526 [9]
- **parsnips** 20093526 [9]
- **protected forest nurseries** 20082923 [8], 20092919 [9]
- **quinces** 20071443 [8], 20092917 [9]
- **radicchio** 20071432 [8], 20092921 [9]
- **redcurrants** 20071444 [8], 20092907 [9]
- **rhubarb** 20071430 [8], 20111287 [9]
- **runner beans** 20152445 [3], 20140233 [5], 20071437 [8], 20092909 [9]
- **salad brassicas** (for baby leaf production) 20071432 [8]

- **salad onions** *20171389* [8], *20092914* [9]
- **scarole** *20071432* [8], *20092921* [9]
- **shallots** *20171415* [1], *20171389* [8], *20092914* [9]
- **soya beans** *20140231* [5], *20121634* [9]
- **spring field beans** *20141259* [1], *20152445* [3], *20140233* [5], *20092551* [6], *20071437* [8], *20092909* [9]
- **spring wheat** *20141261* [1], *20082947* [8], *20092910* [9]
- **sweetcorn** *(under covers)* *20071435* [8], *20092912* [9]
- **sweetcorn under plastic mulches** *20092912* [9]
- **top fruit** *20082923* [8], *20092919* [9]
- **vining peas** *20152445* [3], *20140233* [5], *20111007* [6], *20071437* [8], *20092909* [9]
- **walnuts** *20071442* [8], *20092916* [9]
- **whitecurrants** *20071444* [8], *20092907* [9]
- **winter field beans** *20141259* [1], *20152445* [3], *20140233* [5], *20092552* [6], *20071437* [8], *20092909* [9]

Approval information
- Pendimethalin included in Annex I under EC Regulation 1107/2009
- Accepted by BBPA for use on malting barley and hops

Efficacy guidance
- Apply as soon as possible after drilling. Weeds are controlled as they germinate and emerged weeds will not be controlled by use of the product alone
- For effective blackgrass control apply not more than 2 d after final cultivation and before weed seeds germinate
- Best results by application to fine firm, moist, clod-free seedbeds when rain follows treatment. Effectiveness reduced by prolonged dry weather after treatment
- Effectiveness reduced on soils with more than 6% organic matter. Do not use where organic matter exceeds 10%
- Any trash, ash or straw should be incorporated evenly during seedbed preparation
- Do not disturb soil after treatment
- Apply to potatoes as soon as possible after planting and ridging in tank-mix with metribuzin but note that this will restrict the varieties that may be treated
- Always follow WRAG guidelines for preventing and managing herbicide resistant weeds. See Section 5 for more information

Restrictions
- Maximum number of treatments 1 per crop or yr
- Maximum total dose equivalent to one full dose treatment on most crops; 2 full dose treatments on leeks
- May be applied pre-emergence of cereal crops sown before 30 Nov provided seed covered by at least 32 mm soil, or post-emergence to early tillering stage (GS 23)
- Do not undersow treated crops
- Do not use on crops suffering stress due to disease, drought, waterlogging, poor seedbed conditions or chemical treatment or on soils where water may accumulate
- Do not apply with hand-held equipment [1]

Crop-specific information
- Latest use: pre-emergence for spring barley, carrots, lettuce, fodder maize, parsnips, parsley, combining peas, potatoes, onions and leeks; before transplanting for brassicas; before leaf sheaths erect for winter cereals; before bud burst for blackcurrants, gooseberries, cane fruit, hops; before flower trusses emerge for strawberries; 14 d after transplanting for leaf herbs
- Do not use on spring barley after end Mar (mid-Apr in Scotland on some labels) because dry conditions likely. Do not apply to dry seedbeds in spring unless rain imminent
- Apply to combining peas as soon as possible after sowing. Do not spray if plumule less than 13 mm below soil surface
- Apply to potatoes up to 7 d before first shoot emerges
- Apply to drilled crops as soon as possible after drilling but before crop and weed emergence
- Apply in top fruit, bush fruit and hops from autumn to early spring when crop dormant

- In cane fruit apply to weed free soil from autumn to early spring, immediately after planting new crops and after cutting out canes in established crops
- Apply in strawberries from autumn to early spring (not before Oct on newly planted bed). Do not apply pre-planting or during flower initiation period (post-harvest to mid-Sep)
- Apply pre-emergence in drilled onions or leeks, not on Sands, Very Light, organic or peaty soils or when heavy rain forecast
- Apply to brassicas after final plant-bed cultivation but before transplanting. Avoid unnecessary soil disturbance after application and take care not to introduce treated soil into the root zone when transplanting. Follow transplanting with specified post-planting treatments - see label
- Do not use on protected crops or in greenhouses

Following crops guidance
- Before ryegrass is drilled after a very dry season plough or cultivate to at least 15 cm. If treated spring crops are to be followed by crops other than cereals, plough or cultivate to at least 15 cm
- In the event of crop failure land must be ploughed or thoroughly cultivated to at least 15 cm. See label for minimum intervals that should elapse between treatment and sowing a range of replacement crops

Environmental safety
- Dangerous for the environment
- Very toxic to aquatic organisms
- Dangerous to fish or other aquatic life. Do not contaminate surface waters or ditches with chemical or used container
- LERAP Category B

Hazard classification and safety precautions
Hazard Harmful [11, 12]; Irritant [5, 6]; Dangerous for the environment [1-12]; Harmful if swallowed [6, 11, 12]; Very toxic to aquatic organisms [3, 6, 7, 11, 12]
Transport code 9
Packaging group III
UN Number 3082
Risk phrases H304, H315, H319 [6, 11, 12]; H317 [5]
Operator protection A [1-12]; C [6, 11, 12]; H [1, 2, 4, 5, 7, 9]; U02a [1-4, 7-10]; U05a, U08 [1-12]; U11 [11, 12]; U13 [1-4, 7-12]; U14 [3, 5, 8-10]; U15 [5]; U19a [1-4, 6-12]; U20b [1-5, 7, 8, 10-12]; U23a [6]
Environmental protection E15a [6]; E15b [1-5, 7-12]; E16a, E38 [1-12]; E16b [1, 2, 4, 7]; E34 [6, 11, 12]; E39 [9]; H410 [1-4, 6-12]; H412 [5]
Storage and disposal D01, D02, D09a [1-12]; D05 [1, 2, 4, 6, 7, 9]; D08 [3, 8, 10-12]; D10b, D12a [1-4, 6-12]; D10c [5]
Medical advice M03, M05b [6, 11, 12]; M05a [1-4, 7, 8, 10]

312   pendimethalin + picolinafen

A post-emergence broad-spectrum herbicide mixture for winter cereals
HRAC mode of action code: K1 + F1

See also picolinafen

Products
1	Chronicle	BASF	320:16 g/l	SC	15394
2	Orient	BASF	330:7.5 g/l	SC	12541
3	Parade	BASF	320:16 g/l	SC	17246
4	PicoMax	BASF	320:16 g/l	SC	13456
5	Picona	BASF	320:16 g/l	SC	13428
6	PicoPro	BASF	320:16 g/l	SC	13454
7	PicoStomp	BASF	320:16 g/l	SC	13455

FOR FULL CONDITIONS OF USE ALWAYS READ THE PRODUCT LABEL

Uses

- Annual dicotyledons in *forest nurseries* *(off-label)* [4, 7]; *game cover* *(off-label)* [5]; *rye* [1]; *spring barley, spring wheat, triticale* [1, 3-7]; *spring barley* *(off-label)*, *spring wheat* *(off-label)* [1, 2, 6]; *winter barley, winter wheat* [1-7]; *winter rye* [3-7]
- Annual grasses in *spring barley* *(off-label)*, *spring wheat* *(off-label)* [1]
- Annual meadow grass in *forest nurseries* *(off-label)* [4, 7]; *game cover* *(off-label)* [5]; *rye* [1]; *spring barley, spring wheat, triticale* [1, 3-7]; *spring barley* *(off-label)*, *spring wheat* *(off-label)* [1, 2, 6]; *winter barley, winter wheat* [1-7]; *winter rye* [3-7]
- Chickweed in *rye* [1]; *spring barley, spring wheat, triticale* [1, 3-7]; *winter barley, winter wheat* [1-7]; *winter rye* [3-7]
- Cleavers in *rye* [1]; *spring barley, spring wheat, triticale* [1, 3-7]; *winter barley, winter wheat* [1-7]; *winter rye* [3-7]
- Field speedwell in *rye* [1]; *spring barley, spring wheat, triticale* [1, 3-7]; *winter barley, winter wheat* [1-7]; *winter rye* [3-7]
- Ivy-leaved speedwell in *rye* [1]; *spring barley, spring wheat, triticale* [1, 3-7]; *winter barley, winter wheat* [1-7]; *winter rye* [3-7]
- Loose silky bent in *spring barley* [5]
- Rough-stalked meadow grass in *forest nurseries* *(off-label)* [4, 7]; *game cover* *(off-label)* [5]; *rye* [1]; *spring barley, spring wheat, triticale* [1, 3-7]; *winter barley, winter wheat* [1-7]; *winter rye* [3-7]

Extension of Authorisation for Minor Use (EAMUs)

- *forest nurseries* *20082863* [4], *20082865* [7]
- *game cover* *20082864* [5]
- *spring barley* *20130724* [1], *20130386* [2], *20130725* [6]
- *spring wheat* *20130724* [1], *20130386* [2], *20130725* [6]

Approval information

- Pendimethalin and picolinafen included in Annex I under EC Regulation 1107/2009

Efficacy guidance

- Best results obtained on crops growing in a fine, firm tilth and when rain falls within 7 d of application
- Loose or cloddy seed beds must be consolidated prior to application otherwise reduced weed control may occur
- Residual weed control may be reduced on soils with more than 6% organic matter or under prolonged dry conditions
- Always follow WRAG guidelines for preventing and managing herbicide resistant weeds. See Section 5 for more information

Restrictions

- Maximum number of treatments 1 per crop
- Do not treat undersown cereals or those to be undersown
- Do not roll emerged crops before treatment nor autumn treated crops until the following spring
- Do not use on stony or gravelly soils, on soils that are waterlogged or prone to waterlogging, or on soils with more than 10% organic matter
- Do not treat crops under stress from any cause
- Do not apply pre-emergence to crops drilled after 30 Nov [2]
- Consult processor before treating crops for processing [2]

Crop-specific information

- Latest use: before pseudo-stem erect stage (GS 30)
- Transient bleaching may occur after treatment but it does not lead to yield loss
- Do not use pre-emergence on crops drilled after 30th Nov [4]
- For pre-emergence applications, seed should be covered with a minimum of 3.2 cm settled soil [5, 6]

Following crops guidance

- After normal harvest any crop except ryegrass can be sown. Cultivation or ploughing to at least 15 cm is required before sowing ryegrass.
- In the event of crop failure following autumn application spring wheat, spring barley or winter field beans can be sown after the land is ploughed and at least 8 weeks have elapsed since

SEE SECTION 3 FOR PRODUCTS ALSO REGISTERED

treatment. The following spring oilseed rape, peas or spring field beans can be sown after non-inversion tillage to at least 5 cm but maize should not be planted until at least 5 months have elapsed since application.

- In the event of crop failure following spring application spring wheat, spring barley, peas, spring field beans, maize or spring oilseed rape can be sown after the land is ploughed to at least 15 cm and at least 8 weeks have elapsed since application.

Environmental safety
- Dangerous for the environment
- Very toxic to aquatic organisms
- Product binds strongly to soil minimising likelihood of movement into groundwater
- LERAP Category B

Hazard classification and safety precautions
  Hazard Dangerous for the environment, Very toxic to aquatic organisms
  Transport code 9
  Packaging group III
  UN Number 3082
  Operator protection A, H [1-4]; U02a [1-3, 5-7]; U05a, U20c [1-7]
  Environmental protection E15a [2]; E15b [1, 3-7]; E16a, E34, E38, H410 [1-7]
  Storage and disposal D01, D02, D05, D09a, D10c, D12a
  Medical advice M03

313   pendimethalin + pyroxsulam

A dinitroaniline + triazolopyrimidine herbicide mixture for use in winter wheat, rye and triticale
HRAC mode of action code: K1 + B

Products

Broadway Sunrise	Dow (Corteva)	314:5.4 g/l	OD	14960

Uses
- Annual dicotyledons in *triticale*, *winter rye*, *winter wheat*
- Blackgrass in *triticale* (MS only), *winter rye* (MS only), *winter wheat* (MS only)
- Ryegrass in *triticale* (from seed), *winter rye* (from seed), *winter wheat* (from seed)
- Sterile brome in *triticale*, *winter rye*, *winter wheat*
- Wild oats in *triticale*, *winter rye*, *winter wheat*

Approval information
- Pendimethalin and pyroxsulam included in Annex I under EC Regulation 1107/2009.

Efficacy guidance
- To avoid the build-up of resistance, do not apply this or any other product containing an ALS inhibitor with claims for control of grass weeds more than once to any crop.

Following crops guidance
- In the event of crop failure, plough and sow only spring barley, spring wheat or maize at least 5 months after application and cultivate soil to at least 15 cm.

Environmental safety
- LERAP Category B

Hazard classification and safety precautions
  Hazard Irritant, Dangerous for the environment
  Transport code 9
  Packaging group III
  UN Number 3082
  Risk phrases H315, H317
  Operator protection A, H; U05a, U14, U15, U20b
  Environmental protection E15b, E16a, E34, E38, H410
  Storage and disposal D01, D02, D05, D10c, D12a

FOR FULL CONDITIONS OF USE ALWAYS READ THE PRODUCT LABEL

314   penflufen

An SDHI seed treatment for use in potatoes
FRAC mode of action code: 7

Products

| Emesto Prime DS | Bayer CropScience | 2% w/w | DP | 17280 |

Uses

- Black scurf in **potatoes**
- Stem canker in **potatoes** *(reduction)*
- Stolon canker in **potatoes** *(reduction)*

Approval information

- Penflufen is included in Annex 1 under EC Regulation 1107/2009

Efficacy guidance

- Allow treated seed coming out of cold store to reach ambient temperature before planting.
- Damage to tubers or sprouts may affect subsequent yields.

Restrictions

- Treated tubers may only be planted in the same field once every 3 years.
- Applications to potato seed must not exceed 100 g penflufen per hectare.
- Sowing of treated seed must not exceed a maximum planting density of 5 tonne per hectare.

Hazard classification and safety precautions

Transport code 9
Packaging group III
UN Number 3077
Operator protection U20a
Environmental protection E15b, H411
Storage and disposal D01, D02, D09a, D11a
Treated seed S01, S02, S04a, S05, S06b, S08a, S09

315   penthiopyrad

An SDHI fungicide for disease control in cereals, apples and pears
FRAC mode of action code: 7

See also cyproconazole + penthiopyrad

Products

| Fontelis | DuPont (Corteva) | 200 g/l | SC | 17465 |

Uses

- Powdery mildew in **apples** *(suppression)*, **pears** *(suppression)*
- Scab in **apples**, **pears**

Approval information

- Penthiopyrad is included in Annex 1 under EC Regulation 1107/2009
- Accepted by BBPA for use on malting barley

Efficacy guidance

- Always use in mixture with another product from a different fungicide resistance group that is used for control of the same target disease.
- Best results achieved from applications made before the disease is established in the crop

Restrictions

- Do not use more than two foliar applications of SDHI fungicides on any cereal crop.
- Must be applied in at least 100 l/ha since it is a skin sensitiser.
- Avoid application in either frosty or hot, sunny weather conditions [1]
- Do not apply to any crop suffering from stress as a result of drought, waterlogging, low temperatures, nutrient or lime deficiency or other factors reducing crop growth.

SEE SECTION 3 FOR PRODUCTS ALSO REGISTERED

SECTION 2

- A maximum total dose of not more than 500g penthiopyrad/hectare may be applied in a two year period on the same field

Environmental safety
- Broadcast air-assisted LERAP 40m [1]
- LERAP Category B

Hazard classification and safety precautions
Hazard Very toxic to aquatic organisms
Transport code 9
Packaging group III
UN Number 3082
Operator protection A
Environmental protection E15b, E16a, E34, H410; E17a (40 m)
Storage and disposal D05, D09a, D12a

316   pepino mosaic virus isolate VX1

For use in tomatoes grown under permanent protection but available only in mixtures
FRAC mode of action code: Unclassified

See also Mild Pepino Mosaic Virus isolate VC1 + VX1

317   phenmedipham

A contact phenyl carbamate herbicide for beet crops and strawberries
HRAC mode of action code: C1

See also desmedipham + ethofumesate + lenacil + phenmedipham
desmedipham + ethofumesate + phenmedipham
desmedipham + phenmedipham
ethofumesate + phenmedipham

Products

1	Beetup Flo	UPL Europe	160 g/l	SC	14328
2	Betasana SC	UPL Europe	160 g/l	SC	14209
3	Corzal SC	UPL Europe	160 g/l	SC	17751

Uses
- Annual dicotyledons in *chard* (off-label), *herbs (see appendix 6)* (off-label), *rocket* (off-label), *seakale* (off-label), *spinach* (off-label), *spinach beet* (off-label) [3]; *fodder beet*, *mangels*, *red beet*, *sugar beet* [1-3]
- Black bindweed in *ornamental plant production* (off-label) [2, 3]
- Chickweed in *fodder beet*, *mangels*, *red beet*, *sugar beet* [1, 2]; *ornamental plant production* (off-label) [2, 3]
- Fat hen in *ornamental plant production* (off-label) [2, 3]
- Field pansy in *ornamental plant production* (off-label) [2, 3]
- Knotgrass in *ornamental plant production* (off-label) [2, 3]
- Penny cress in *ornamental plant production* (off-label) [2, 3]
- Scentless mayweed in *ornamental plant production* (off-label) [2, 3]

Extension of Authorisation for Minor Use (EAMUs)
- *chard* 20172731 [3]
- *herbs (see appendix 6)* 20172731 [3]
- *ornamental plant production* 20152050 [2], 20180376 [3]
- *rocket* 20172731 [3]
- *seakale* 20172731 [3]
- *spinach* 20172731 [3]
- *spinach beet* 20172731 [3]

FOR FULL CONDITIONS OF USE ALWAYS READ THE PRODUCT LABEL

Approval information
- Phenmedipham included in Annex I under EC Regulation 1107/2009

Efficacy guidance
- Best results achieved by application to young seedling weeds, preferably cotyledon stage, under good growing conditions when low doses are effective
- If using a low-dose programme 2-3 repeat applications at 7-10 d intervals are recommended on mineral soils, 3-5 applications may be needed on organic soils
- Addition of adjuvant oil may improve effectiveness on some weeds
- Various tank-mixtures with other beet herbicides recommended. See label for details
- Use of certain pre-emergence herbicides is recommended in combination with post-emergence treatment. See label for details

Restrictions
- Maximum number of treatments and maximum total dose varies with crop and product used. See label for details
- At high temperatures (above 21°C) reduce rate and spray after 5 pm
- Do not apply immediately after frost or if frost expected
- Do not spray wet foliage or if rain imminent
- Do not spray crops stressed by wind damage, nutrient deficiency, pest or disease attack etc. Do not roll or harrow for 7 d before or after treatment
- Do not use on strawberries under cloches or polythene tunnels
- Consult processor before use on crops for processing
- Do not apply to crops with wet leaves

Crop-specific information
- Latest use: before crop leaves meet between rows for beet crops; before flowering for strawberries
- Apply to beet crops at any stage as low dose/low volume spray or from fully developed cotyledon stage with full rate. Apply to red beet after fully developed cotyledon stage
- Apply to strawberries at any time when weeds in susceptible stage, except in period from start of flowering to picking

Following crops guidance
- Beet crops may follow at any time after a treated crop. 3 mth must elapse from treatment before any other crop is sown and must be preceded by mould-board ploughing to 15 cm

Environmental safety
- Dangerous for the environment
- Very toxic to aquatic organisms
- Harmful to fish or other aquatic life. Do not contaminate surface waters or ditches with chemical or used container
- LERAP Category B

Hazard classification and safety precautions
Hazard Irritant [1, 2]; Dangerous for the environment [1-3]
Transport code 9
Packaging group III
UN Number 3082
Risk phrases H317, H319
Operator protection A [1-3]; C, H [1, 2]; U05a [3]; U08 [1-3]; U11, U14, U19a, U20c [1, 2]
Environmental protection E15b, E16a, H411 [1-3]; E16b, E34, E38 [3]
Storage and disposal D01, D02, D09a, D12a [1-3]; D05, D10c [1, 2]; D10b [3]

318   phenothrin

A non-systemic pyrethroid insecticide available only in mixtures
IRAC mode of action code: 3A

SEE SECTION 3 FOR PRODUCTS ALSO REGISTERED

319   phenothrin + tetramethrin

A pyrethroid insecticide mixture for control of flying insects
IRAC mode of action code: 3 + 3

See also tetramethrin

Products

1	Killgerm ULV 1500	Killgerm	4.8:2.4% w/w	UL	H6961
2	Killgerm ULV 500	Killgerm	4.8:2.4% w/w	UL	H4647

Uses

- Flies in *agricultural premises*
- Grain storage mite in *grain stores*
- Mosquitoes in *agricultural premises*
- Wasps in *agricultural premises*

Approval information

- Product approved for ULV application. See label for details
- Phenothrin and tetramethrin are not included in Annex 1 under EC Regulation 1107/2009

Efficacy guidance

- Close doors and windows and spray in all directions for 3-5 sec. Keep room closed for at least 10 min

Restrictions

- For use only by professional operators
- Do not use space sprays containing pyrethrins or pyrethroid more than once per week in intensive or controlled environment animal houses in order to avoid development of resistance. If necessary, use a different control method or product

Crop-specific information

- May be used in the presence of poultry and livestock

Environmental safety

- Dangerous for the environment
- Toxic to aquatic organisms
- Do not apply directly to livestock/poultry
- Remove exposed milk and collect eggs before application. Protect milk machinery and containers from contamination

Hazard classification and safety precautions

Hazard Harmful, Dangerous for the environment, Harmful if swallowed
Transport code 9
Packaging group III
UN Number 3082
Operator protection A, C, D, E, H; U02b, U09b, U14, U19a, U20a, U20b
Environmental protection E05a, E15a, E38, H411
Consumer protection C06, C07, C08, C09, C11, C12
Storage and disposal D01, D05, D06a, D09a, D10a, D12a
Medical advice M05b

320   Phlebiopsis gigantea

A fungal protectant against root and butt rot in conifers

Products

PG Suspension	Forest Research	0.5% w/w	SC	17009

Uses

- Fomes root and butt rot in *farm forestry, forest, forest nurseries*

FOR FULL CONDITIONS OF USE ALWAYS READ THE PRODUCT LABEL

Approval information

- Phlebiopsis gigantea included in Annex 1 under EC Regulation 1107/2009
- Approval expiry 19 Dec 2020 [1]

Efficacy guidance

- Store sachets in the original box in a cool dry place between 2°C and 15°C out of the sun.
- Use the diluted solution within 24 hours of mixing. Ensure complete coverage of stumps for optimum efficacy.

Hazard classification and safety precautions

UN Number N/C
Operator protection A, D, H; U20c
Environmental protection E15a
Storage and disposal D11a

321   physical pest control

Products that work by physical action only

Products

1	SB Plant Invigorator	Fargro	-	SL	-
2	Silico-Sec	Sumitomo	>90% w/w	DS	-
3	Vazor DE	Killgerm	n/a	PO	-
4	Vazor Liquid Mosquito Film	Killgerm	n/a	RH	-

Uses

- Aphids in *all edible crops (outdoor and protected), apples, asparagus, asteraceae, aubergines, banana, beans without pods (fresh), beetroot, broccoli, cabbages, chillies, chives, clovers, cotton, cucumbers, cyclamen, freesias, garlic, geraniums, gladioli, grapevines, hibiscus trionum, honeysuckle, lettuce, lilies, lupins, maize, marrows, melons, morning glory, onions, orchids, ornamental plant production, peas, pelargoniums, peppers, poinsettias, potatoes, primulas, protected ornamentals, prunus, pumpkins, roses, spinach, squashes, strawberries, sugar beet, sunflowers, swedes, sweet potato, tomatoes, turnips* [1]
- Insect control in *agricultural premises* [3]
- Mealybugs in *all edible crops (outdoor and protected), apples, asparagus, asteraceae, aubergines, banana, beans without pods (fresh), beetroot, broccoli, cabbages, chillies, chives, clovers, cotton, cucumbers, cyclamen, freesias, garlic, geraniums, gladioli, grapevines, hibiscus trionum, honeysuckle, lettuce, lilies, lupins, maize, marrows, melons, morning glory, onions, orchids, ornamental plant production, peas, pelargoniums, peppers, poinsettias, potatoes, primulas, protected ornamentals, prunus, pumpkins, roses, spinach, squashes, strawberries, sugar beet, sunflowers, swedes, sweet potato, tomatoes, turnips* [1]
- Mites in *stored grain* [2]
- Mosquitoes in *enclosed waters* [4]
- Powdery mildew in *all edible crops (outdoor and protected), apples, asparagus, asteraceae, aubergines, banana, beans without pods (fresh), beetroot, broccoli, cabbages, chillies, chives, clovers, cotton, cucumbers, cyclamen, freesias, garlic, geraniums, gladioli, grapevines, hibiscus trionum, honeysuckle, lettuce, lilies, lupins, maize, marrows, melons, morning glory, onions, orchids, ornamental plant production, peas, pelargoniums, peppers, poinsettias, potatoes, primulas, protected ornamentals, prunus, pumpkins, roses, spinach, squashes, strawberries, sugar beet, sunflowers, swedes, sweet potato, tomatoes, turnips* [1]
- Scale insects in *all edible crops (outdoor and protected), apples, asparagus, asteraceae, aubergines, banana, beans without pods (fresh), beetroot, broccoli, cabbages, chillies, chives, clovers, cotton, cucumbers, cyclamen, freesias, garlic, geraniums, gladioli, grapevines, hibiscus trionum, honeysuckle, lettuce, lilies, lupins, maize, marrows, melons, morning glory, onions, orchids, ornamental plant production, peas, pelargoniums, peppers, poinsettias, potatoes, primulas, protected ornamentals, prunus,*

SEE SECTION 3 FOR PRODUCTS ALSO REGISTERED

pumpkins, roses, spinach, squashes, strawberries, sugar beet, sunflowers, swedes, sweet potato, tomatoes, turnips [1]

- Spider mites in *all edible crops (outdoor and protected), apples, asparagus, asteraceae, aubergines, banana, beans without pods (fresh), beetroot, broccoli, cabbages, chillies, chives, clovers, cotton, cucumbers, cyclamen, freesias, garlic, geraniums, gladioli, grapevines, hibiscus trionum, honeysuckle, lettuce, lilies, lupins, maize, marrows, melons, morning glory, onions, orchids, ornamental plant production, peas, pelargoniums, peppers, poinsettias, potatoes, primulas, protected ornamentals, prunus, pumpkins, roses, spinach, squashes, strawberries, sugar beet, sunflowers, swedes, sweet potato, tomatoes, turnips* [1]
- Suckers in *all edible crops (outdoor and protected), apples, asparagus, asteraceae, aubergines, banana, beans without pods (fresh), beetroot, broccoli, cabbages, chillies, chives, clovers, cotton, cucumbers, cyclamen, freesias, garlic, geraniums, gladioli, grapevines, hibiscus trionum, honeysuckle, lettuce, lilies, lupins, maize, marrows, melons, morning glory, onions, orchids, ornamental plant production, peas, pelargoniums, peppers, poinsettias, potatoes, primulas, protected ornamentals, prunus, pumpkins, roses, spinach, squashes, strawberries, sugar beet, sunflowers, swedes, sweet potato, tomatoes, turnips* [1]
- Whitefly in *all edible crops (outdoor and protected), apples, asparagus, asteraceae, aubergines, banana, beans without pods (fresh), beetroot, broccoli, cabbages, chillies, chives, clovers, cotton, cucumbers, cyclamen, freesias, garlic, geraniums, gladioli, grapevines, hibiscus trionum, honeysuckle, lettuce, lilies, lupins, maize, marrows, melons, morning glory, onions, orchids, ornamental plant production, peas, pelargoniums, peppers, poinsettias, potatoes, primulas, protected ornamentals, prunus, pumpkins, roses, spinach, squashes, strawberries, sugar beet, sunflowers, swedes, sweet potato, tomatoes, turnips* [1]

Approval information
- Products included in this profile are not subject to the Control of Pesticides Regulations/Plant Protection Products Regulations because they act by physical means only

Efficacy guidance
- Products act by physical means following direct contact with spray. They may therefore be used at any time of year on all pest growth stages
- Ensure thorough spray coverage of plant, paying special attention to growing points and the underside of leaves
- Treat as soon as target pests are seen and repeat as often as necessary

Restrictions
- No limit on the number of treatments and no minimum interval between applications
- Before large scale use on a new crop, treat a few plants to check for crop safety
- Do not treat ornamental crops when in flower

Crop-specific information
- HI zero

Environmental safety
- May be used in conjunction with biological control agents. Spray 24 h before they are introduced

Hazard classification and safety precautions
    Hazard Irritant [1]; Highly flammable [3]
    UN Number N/C
    Risk phrases H319, H336 [3]
    Operator protection A, H [3]; F [2]; U10, U14, U15 [1]
    Storage and disposal D01, D02, D05, D06h [1]

FOR FULL CONDITIONS OF USE ALWAYS READ THE PRODUCT LABEL

322　pinoxaden

A phenylpyrazoline grass weed herbicide for cereals
HRAC mode of action code: A

See also clodinafop-propargyl + pinoxaden
*　　　　florasulam + pinoxaden*

Products

Axial Pro	Syngenta	60 g/l	EC	19010

Uses

- Blackgrass in **spring barley, winter barley**
- Italian ryegrass in **spring barley, spring wheat, winter barley, winter wheat**
- Perennial ryegrass in **spring barley, spring wheat, winter barley, winter wheat**
- Wild oats in **spring barley, spring wheat, winter barley, winter wheat**

Approval information

- Pinoxaden included in Annex 1 under EC Regulation 1107/2009

Efficacy guidance

- Best results obtained from treatment when all grass weeds have emerged. There is no residual activity
- Broad-leaved weeds are not controlled
- Treat before emerged weed competition reduces yield
- Blackgrass in winter and spring barley is also controlled when used as part of an integrated control strategy. Product is not recommended for blackgrass control in winter wheat
- Grass weed control may be reduced if rain falls within 1 hr of application
- Pinoxaden is an ACCase inhibitor herbicide. To avoid the build up of resistance do not apply products containing an ACCase inhibitor herbicide more than twice to any crop. In addition do not use any product containing pinoxaden in mixture or sequence with any other product containing the same ingredient
- Use these products as part of a resistance management strategy that includes cultural methods of control and does not use ACCase inhibitors as the sole chemical method of grass weed control
- Applying a second product containing an ACCase inhibitor to a crop will increase the risk of resistance development; only use a second ACCase inhibitor to control different weeds at a different timing
- Always follow WRAG guidelines for preventing and managing herbicide resistant weeds. See Section 5 for more information

Restrictions

- Maximum number of treatments 1 per crop
- Do not spray crops under stress or suffering from waterlogging, pest attack, disease or frost damage
- Do not spray crops undersown with grass mixtures
- Avoid the use of hormone-containing herbicides in mixture or in sequence. Allow 21 d after, or 7 d before, hormone application
- To avoid the build-up of resistance do not apply products containing an ACCase inhibitor herbicide more than twice to any crop
- This product may only be applied after 1st February in the year of harvest [1]

Crop-specific information

- Latest use: before flag leaf extending stage (GS 41)
- Spray cereals in autumn, winter or spring from the 2-leaf stage (GS 12)

Following crops guidance

- There are no restrictions on succeeding crops in a normal rotation
- In the event of failure of a treated crop ryegrass, maize, oats or any broad-leaved crop may be planted after a minimum interval of 4 wk from application

Environmental safety

- Dangerous for the environment
- Toxic to aquatic organisms
- Do not allow spray to drift onto neighbouring crops of oats, ryegrass or maize

Hazard classification and safety precautions

Hazard Irritant, Dangerous for the environment
Transport code 9
Packaging group III
UN Number 3082
Risk phrases H315, H317
Operator protection A, C, H; U05a, U09a, U20b
Environmental protection E15b, E38, H411
Storage and disposal D01, D02, D05, D09a, D10c, D11a, D12a

323   pirimicarb

A carbamate insecticide for aphid control
IRAC mode of action code: 1A

See also lambda-cyhalothrin + pirimicarb

Products

1 Aphox	Syngenta	50% w/w	WG	17401
2 Aphox	Adama	50% w/w	WG	18562
3 Clayton Pirimicarb	Clayton	50% w/w	WG	17704
4 Pirimate 500	Becesane	50% w/w	WG	17657

Uses

- Aphids in *broad beans, combining peas, spring field beans, vining peas, winter field beans* [1-4]; *celeriac* (off-label), *celery (outdoor)* (off-label), *dwarf beans* (off-label), *edible podded peas* (off-label), *florence fennel* (off-label), *french beans* (off-label), *horseradish* (off-label), *parsley root* (off-label), *protected aubergines* (off-label), *protected baby leaf crops* (off-label), *protected chilli peppers* (off-label), *protected courgettes* (off-label), *protected cress* (off-label), *protected cucumbers* (off-label), *protected edible flowers* (off-label), *protected endives* (off-label), *protected gherkins* (off-label), *protected herbs (see appendix 6)* (off-label), *protected land cress* (off-label), *protected lettuce* (off-label), *protected peppers* (off-label), *protected pumpkins* (off-label), *protected red mustard* (off-label), *protected rocket* (off-label), *protected summer squash* (off-label), *protected tomatoes* (off-label), *radishes* (off-label), *red beet* (off-label), *rhubarb* (off-label), *runner beans* (off-label) [1, 2]; *chillies* (off-label), *peppers* (off-label), *protected lamb's lettuce* (off-label) [2]
- Bean aphid in *dwarf beans* (off-label), *edible podded peas* (off-label), *french beans* (off-label), *runner beans* (off-label) [2]
- Cotton aphids in *celeriac* (off-label), *protected chilli peppers* (off-label), *protected peppers* (off-label) [2]; *chillies* (off-label), *peppers* (off-label) [1]
- Foxglove aphid in *celeriac* (off-label), *protected chilli peppers* (off-label), *protected peppers* (off-label) [2]; *chillies* (off-label), *peppers* (off-label) [1]
- Lettuce root aphid in *protected baby leaf crops* (off-label), *protected cress* (off-label), *protected edible flowers* (off-label), *protected endives* (off-label), *protected herbs (see appendix 6)* (off-label), *protected land cress* (off-label), *protected lettuce* (off-label), *protected red mustard* (off-label), *protected rocket* (off-label) [1]
- Melon aphid in *protected baby leaf crops* (off-label), *protected cress* (off-label), *protected edible flowers* (off-label), *protected endives* (off-label), *protected herbs (see appendix 6)* (off-label), *protected land cress* (off-label), *protected lettuce* (off-label), *protected red mustard* (off-label), *protected rocket* (off-label) [1]
- Pea aphid in *dwarf beans* (off-label), *edible podded peas* (off-label), *french beans* (off-label), *runner beans* (off-label) [2]
- Peach-potato aphid in *celeriac* (off-label), *protected chilli peppers* (off-label), *protected peppers* (off-label) [2]; *chillies* (off-label), *peppers* (off-label), *protected baby leaf crops* (off-label), *protected cress* (off-label), *protected edible flowers* (off-label), *protected endives* (off-

label), **protected herbs (see appendix 6)** *(off-label)*, **protected land cress** *(off-label)*, **protected lettuce** *(off-label)*, **protected red mustard** *(off-label)*, **protected rocket** *(off-label)* [1]

Extension of Authorisation for Minor Use (EAMUs)
- **celeriac** *20161609* [1], *20182151* [2]
- **celery (outdoor)** *20161610* [1], *20182135* [2]
- **chillies** *20181759* [1], *20182149* [2]
- **dwarf beans** *20171445* [1], *20182152* [2]
- **edible podded peas** *20171445* [1], *20182152* [2]
- **florence fennel** *20161606* [1], *20182153* [2]
- **french beans** *20171445* [1], *20182152* [2]
- **horseradish** *20161605* [1], *20182136* [2]
- **parsley root** *20161605* [1], *20182136* [2]
- **peppers** *20181759* [1], *20182149* [2]
- **protected aubergines** *20191392* [1], *20191393* [2]
- **protected baby leaf crops** *20191177* [1], *20191178* [2]
- **protected chilli peppers** *20171444* [1], *20182150* [2]
- **protected courgettes** *20161611* [1], *20182154* [2]
- **protected cress** *20191177* [1], *20191178* [2]
- **protected cucumbers** *20191394* [1], *20191395* [2]
- **protected edible flowers** *20191177* [1], *20191178* [2]
- **protected endives** *20191177* [1], *20191178* [2]
- **protected gherkins** *20161611* [1], *20182154* [2]
- **protected herbs (see appendix 6)** *20191177* [1], *20191178* [2]
- **protected lamb's lettuce** *20191178* [2]
- **protected land cress** *20191177* [1], *20191178* [2]
- **protected lettuce** *20191177* [1], *20191178* [2]
- **protected peppers** *20171444* [1], *20182150* [2]
- **protected pumpkins** *20191390* [1], *20191391* [2]
- **protected red mustard** *20191177* [1], *20191178* [2]
- **protected rocket** *20191177* [1], *20191178* [2]
- **protected summer squash** *20161611* [1], *20182154* [2]
- **protected tomatoes** *20191392* [1], *20191393* [2]
- **radishes** *20161605* [1], *20182136* [2]
- **red beet** *20161605* [1], *20182136* [2]
- **rhubarb** *20161610* [1], *20182135* [2]
- **runner beans** *20171445* [1], *20182152* [2]

Approval information
- Pirimicarb included in Annex I under EC Regulation 1107/2009
- In 2006 CRD required that all products containing this active ingredient should carry the following warning in the main area of the container label: "Pirimicarb is an anticholinesterase carbamate. Handle with care"

Efficacy guidance
- Chemical has contact, fumigant and translaminar activity
- Best results achieved under warm, calm conditions when plants not wilting and spray does not dry too rapidly. Little vapour activity at temperatures below 15°C
- Apply as soon as aphids seen or warning issued and repeat as necessary
- Addition of non-ionic wetter recommended for use on brassicas
- On cucumbers and tomatoes a root drench is preferable to spraying when using predators in an integrated control programme
- Where aphids resistant to pirimicarb occur control is unlikely to be satisfactory

Restrictions
- Contains an anticholinesterase carbamate compound. Do not use if under medical advice not to work with such compounds
- Maximum number of treatments not specified in some cases but normally 2-6 depending on crop

SEE SECTION 3 FOR PRODUCTS ALSO REGISTERED

Crop-specific information
- Latest use in accordance with harvest intervals below
- HI protected courgettes and gherkins 24 h; other edible crops 3 d;

Environmental safety
- Dangerous for the environment
- Very toxic to aquatic organisms
- Dangerous to fish or other aquatic life. Do not contaminate surface waters or ditches with chemical or used container
- Chemical has little effect on bees, ladybirds and other insects and is suitable for use in integrated control programmes on apples and pears
- Keep all livestock out of treated areas for at least 7 d. Bury or remove spillages
- LERAP Category B

Hazard classification and safety precautions
Hazard Toxic, Dangerous for the environment, Toxic if swallowed, Harmful if inhaled
Transport code 6.1
Packaging group III
UN Number 2757
Risk phrases H319
Operator protection A, C, D, H, J, M; U05a, U08, U19a, U20b
Environmental protection E06c (7 d); E15b, E16a, E34, E38, H410
Storage and disposal D01, D02, D09a, D11a, D12a
Medical advice M02, M03, M04a

324   pirimiphos-methyl

A contact, fumigant and translaminar organophosphorus insecticide
IRAC mode of action code: 1B

Products

1 Actellic 50 EC	Syngenta	500 g/l	EC	12726
2 Actellic Smoke Generator No. 20	Syngenta	22.5% w/w	FU	15739
3 Flycatcher	AgChem Access	500 g/l	EC	14135
4 Pan PMT	Pan Agriculture	500 g/l	EC	13996

Uses
- Flour beetle in *crop handling & storage structures* [1]; *stored grain* [1, 3, 4]
- Flour moth in *crop handling & storage structures* [1]; *stored grain* [1, 3, 4]
- Grain beetle in *crop handling & storage structures* [1]; *stored grain* [1, 3, 4]
- Grain storage mite in *crop handling & storage structures* [1]; *stored grain* [1, 3, 4]
- Grain storage pests in *grain stores* [2]
- Grain weevil in *crop handling & storage structures* [1]; *stored grain* [1, 3, 4]
- Insect pests in *grain stores* [2]
- Mites in *grain stores* [2]
- Warehouse moth in *crop handling & storage structures* [1]; *stored grain* [1, 3, 4]

Approval information
- Pirimiphos-methyl included in Annex I under EC Regulation 1107/2009
- In 2006 PSD required that all products containing this active ingredient should carry the following warning in the main area of the container label: "Pirimiphos-methyl is an anticholinesterase organophosphate. Handle with care"
- Accepted by BBPA for use in stores for malting barley

Efficacy guidance
- Chemical acts rapidly and has short persistence in plants, but persists for long periods on inert surfaces
- Best results for protection of stored grain achieved by cleaning store thoroughly before use and employing a combination of pre-harvest and grain/seed treatments [1, 2]

FOR FULL CONDITIONS OF USE ALWAYS READ THE PRODUCT LABEL

- Best results for admixture treatment obtained when grain stored at 15% moisture or less. Dry and cool moist grain coming into store but then treat as soon as possible, ideally as it is loaded [1]
- Surface admixture can be highly effective on localised surface infestations but should not be relied on for long-term control unless application can be made to the full depth of the infestation [1]
- Where insect pests resistant to pirimiphos-methyl occur control is unlikely to be satisfactory

Restrictions

- Contains an organophosphorus anticholinesterase compound. Do not use if under medical advice not to work with such compounds
- Maximum number of treatments 2 per grain store [1, 2]; 1 per batch of stored grain [1]
- Do not apply surface admixture or complete admixture using hand held equipment [1]
- Surface admixture recommended only where the conditions for store preparation, treatment and storage detailed in the label can be met [1]

Crop-specific information

- Latest use: before storing grain [1, 2]
- Disinfect empty grain stores by spraying surfaces and/or fumigation and treat grain by full or surface admixture. Treat well before harvest in late spring or early summer and repeat 6 wk later or just before harvest if heavily infested. See label for details of treatment and suitable application machinery
- Treatment volumes on structural surfaces should be adjusted according to surface porosity. See label for guidelines
- Treat inaccessible areas with smoke generating product used in conjunction with spray treatment of remainder of store
- Predatory mites (*Cheyletus*) found in stored grain feeding on infestations of grain mites may survive treatment and can lead to rejection by buyers. They do not remain for long but grain should be inspected carefully before selling [1, 2]

Environmental safety

- Dangerous for the environment
- Very toxic to aquatic organisms [1]
- Toxic to aquatic organisms [2]
- Highly flammable [2]
- Flammable [1]
- Ventilate fumigated or fogged spaces thoroughly before re-entry
- Unprotected persons must be kept out of fumigated areas within 3 h of ignition and for 4 h after treatment
- Wildlife must be excluded from buildings during treatment [1]
- Keep away from combustible materials [2]

Hazard classification and safety precautions

Hazard Harmful, Dangerous for the environment, Harmful if swallowed [1-4]; Highly flammable [2]; Flammable [1, 3, 4]; Flammable liquid and vapour [1, 3]; Harmful if inhaled [4]; Very toxic to aquatic organisms [1-3]

Transport code 3 [1, 3, 4]; 9 [2]

Packaging group III

UN Number 1993 [1, 3, 4]; 3077 [2]

Risk phrases H304, H317, H318, H336, H370 [1, 3]; H320, H371 [4]; H335 [1, 3, 4]

Operator protection A, C, J, M [1, 3, 4]; D, H [1-4]; U04a, U09a, U16a, U24 [1, 3, 4]; U05a, U19a, U20b [1-4]; U14 [2]; U23a [1, 3]

Environmental protection E02a [2] (4 h); E15a, E34, E38, H410 [1-4]

Consumer protection C09 [1, 3, 4]; C12 [1-4]

Storage and disposal D01, D02, D09a, D12a [1-4]; D05, D10c [1, 3, 4]; D07, D11a [2]

Treated seed S06a [1, 3, 4]

Medical advice M01 [1-4]; M03 [2]; M05b [1, 3, 4]

SEE SECTION 3 FOR PRODUCTS ALSO REGISTERED

325   potassium bicarbonate (commodity substance)

An inorganic fungicide for use in horticulture.

Products

1	Atilla	Belchim	85% w/w	SP	16026
2	Karma	Certis	85.42% w/w	WP	16363
3	potassium bicarbonate	various	100% w/w	AP	00000

Uses

- Botrytis in **courgettes** *(off-label)*, **gherkins** *(off-label)*, **okra** *(off-label)*, **protected aubergines** *(off-label)*, **protected chilli peppers** *(off-label)*, **protected peppers** *(off-label)*, **protected tomatoes** *(off-label)*, **pumpkins** *(off-label)*, **summer squash** *(off-label)*, **sweetcorn** *(off-label)*, **table grapes** *(off-label)*, **wine grapes** *(off-label)*, **winter squash** *(off-label)* [2]
- Disease control in **all edible crops (outdoor and protected)**, **all non-edible crops (outdoor)**, **all protected non-edible crops** [3]
- Downy mildew in **baby leaf crops** *(off-label)*, **herbs (see appendix 6)** *(off-label)*, **lettuce** *(off-label)*, **protected lettuce** *(off-label)*, **table grapes** *(off-label)*, **wine grapes** *(off-label)* [2]
- Fungus diseases in **apples**, **nursery fruit trees and bushes** [2]
- Insect pests in **pears** [1]
- Pear sucker in **pears** *(reduction)* [1]
- Powdery mildew in **baby leaf crops** *(off-label)*, **broccoli** *(off-label)*, **brussels sprouts** *(off-label)*, **cabbages** *(off-label)*, **cauliflowers** *(off-label)*, **choi sum** *(off-label)*, **collards** *(off-label)*, **courgettes** *(off-label)*, **cucumbers** *(off-label)*, **gherkins** *(off-label)*, **herbs (see appendix 6)** *(off-label)*, **kale** *(off-label)*, **kohlrabi** *(off-label)*, **melons** *(off-label)*, **okra** *(off-label)*, **oriental cabbage** *(off-label)*, **ornamental plant production** *(off-label)*, **protected aubergines** *(off-label)*, **protected chilli peppers** *(off-label)*, **protected courgettes** *(off-label)*, **protected cucumbers** *(off-label)*, **protected melons** *(off-label)*, **protected ornamentals** *(off-label)*, **protected peppers** *(off-label)*, **protected tomatoes** *(off-label)*, **pumpkins** *(off-label)*, **strawberries** *(off-label)*, **summer squash** *(off-label)*, **sweetcorn** *(off-label)*, **table grapes** *(off-label)*, **wine grapes** *(off-label)*, **winter squash** *(off-label)* [2]
- Sooty moulds in **apples**, **nursery fruit trees and bushes** [2]; **pears** *(reduction)* [1]

Extension of Authorisation for Minor Use (EAMUs)

- **baby leaf crops** *20182444* [2]
- **broccoli** *20193074* [2]
- **brussels sprouts** *20193074* [2]
- **cabbages** *20193074* [2]
- **cauliflowers** *20193074* [2]
- **choi sum** *20193074* [2]
- **collards** *20193074* [2]
- **courgettes** *20151941* [2], *20192503* [2]
- **cucumbers** *20151941* [2]
- **gherkins** *20192503* [2]
- **herbs (see appendix 6)** *20182444* [2]
- **kale** *20193074* [2]
- **kohlrabi** *20193074* [2]
- **lettuce** *20151940* [2]
- **melons** *20151941* [2]
- **okra** *20192503* [2]
- **oriental cabbage** *20193074* [2]
- **ornamental plant production** *20193338* [2]
- **protected aubergines** *20171299* [2]
- **protected chilli peppers** *20171299* [2]
- **protected courgettes** *20151941* [2]
- **protected cucumbers** *20151941* [2]
- **protected lettuce** *20151940* [2]
- **protected melons** *20151941* [2]
- **protected ornamentals** *20193338* [2]
- **protected peppers** *20171299* [2]

- *protected tomatoes* 20171299 [2]
- *pumpkins* 20192503 [2]
- *strawberries* 20150901 [2]
- *summer squash* 20192503 [2]
- *sweetcorn* 20192503 [2]
- *table grapes* 20161327 [2]
- *wine grapes* 20161327 [2]
- *winter squash* 20192503 [2]

SECTION 2

Approval information
- Approval for the use of potassium bicarbonate as a commodity substance was granted on 26 July 2005 by Ministers under regulation 5 of the Control of Pesticides Regulations 1986
- Potassium bicarbonate is being supported for review in the fourth stage of the EC Review Programme under EC Regulation 1107/2009. It was included in Annex 1 under the alternative name of potassium hydrogen carbonate and now products containing it will need to gain approval in the normal way if they carry label claims for pesticidal activity
- Accepted by BBPA for use on hops
- Approval expiry 31 Aug 2020 [3]

Efficacy guidance
- Adjuvants authorised by CRD may be used in conjunction with potassium bicarbonate
- Crop phytotoxicity can occur

Restrictions
- Food grade potassium bicarbonate must only be used

Hazard classification and safety precautions
UN Number N/C
Operator protection C [2]; U05a [1, 2]; U19a [2]
Environmental protection E15a [1, 2]
Storage and disposal D09b [1, 2]

326   potassium phosphonates

Inorganic fungicide (formerly potassium phosphite) for disease control in grapes
FRAC mode of action code: P07

See also dithianon + potassium phosphonates

Products

1 Frutogard	Certis	342 g/l	SL	19105
2 Soriale	BASF	755 g/l	SC	19166

Uses
- Disease control in *table grapes*, *wine grapes* [1]
- Phytophthora in *table grapes*, *wine grapes* [1]
- Pythium in *table grapes*, *wine grapes* [1]
- Scab in *apples*, *pears* [2]

Approval information
- Potassium phosphonate included in Annex I under EC Regulation 1107/2009

Restrictions
- A minimum interval of 7 days must be observed between applications.
- Do not handle treated foliage until 24 days after treatment.

Environmental safety
- Buffer zone requirement 15 m
- LERAP Category B

Hazard classification and safety precautions
Hazard Very toxic to aquatic organisms [2]

SEE SECTION 3 FOR PRODUCTS ALSO REGISTERED

UN Number N/C
Risk phrases H317, H319, H351 [2]
Operator protection A, H [1, 2]; C [2]; U19a, U20a [2]
Environmental protection E16a [1, 2]; E17a [2] (30 m); H410 [2]
Storage and disposal D01, D02, D05

327   prochloraz

A broad-spectrum protectant and eradicant imidazole fungicide
FRAC mode of action code: 3

See also epoxiconazole + prochloraz
        prochloraz + proquinazid + tebuconazole
        prochloraz + tebuconazole

Products

1	Mirage	Adama	450 g/l	EC	18361
2	Sporgon 50 WP	Sylvan	46% w/w	WP	17700

Uses

- Cobweb in **mushrooms** [2]
- Dry bubble in **mushrooms** [2]
- Eyespot in **rye** *(moderate control)*, **spring wheat** *(moderate control)*, **triticale** *(moderate control)*, **winter wheat** *(moderate control)* [1]
- Glume blotch in **spring wheat**, **winter wheat** [1]
- Leaf spot in **spring wheat**, **winter wheat** [1]
- Powdery mildew in **rye** [1]
- Rhynchosporium in **rye** *(moderate control)*, **triticale** *(moderate control)* [1]
- Wet bubble in **mushrooms** [2]

Approval information

- Prochloraz included in Annex 1 under EC Regulation 1107/2009

Efficacy guidance

- Spray cereals at first signs of disease. Protection of winter crops through season usually requires at least 2 treatments. See label for details of rates and timing. Treatment active against strains of eyespot resistant to benzimidazole fungicides
- Tank mixes with other fungicides recommended to improve control of rusts in wheat. See label for details
- A period of at least 3 h without rain should follow spraying
- Prochloraz is a DMI fungicide. Resistance to some DMI fungicides has been identified in Septoria leaf blotch which may seriously affect performance of some products. For further advice contact a specialist advisor and visit the Fungicide Resistance Action Group (FRAG)-UK website

Restrictions

- No application is permitted to any crop before BBCH 30, i.e. only apply after stem elongation has begun [1]

Crop-specific information

- Application with other fungicides may cause cereal crop scorch

Environmental safety

- Dangerous for the environment
- Very toxic to aquatic organisms
- LERAP Category B [1]

Hazard classification and safety precautions

Hazard Irritant [2]; Very toxic to aquatic organisms [1]
Transport code 9
Packaging group III

FOR FULL CONDITIONS OF USE ALWAYS READ THE PRODUCT LABEL

UN Number 3077 [2]; 3082 [1]
Operator protection A, H [1, 2]; C, D, J [2]; U05a, U09a, U20b [2]
Environmental protection E13b, E34 [2]; E16a [1]; H410 [1, 2]
Storage and disposal D01, D02, D09a, D10b [2]

328   prochloraz + proquinazid + tebuconazole

A broad spectrum systemic fungicide mixture for cereals
FRAC mode of action code: 3 + U7 + 3

See also proquinazid
    tebuconazole

Products

Vareon	DuPont (Corteva)	320:40:160 g/l	EC	14376

Uses

*   Brown rust in **spring barley**, **spring wheat**, **winter barley**, **winter wheat**
*   Crown rust in **spring oats** *(qualified minor use recommendation)*, **winter oats** *(qualified minor use recommendation)*
*   Ear diseases in **spring wheat** *(reduction only)*, **winter wheat** *(reduction only)*
*   Eyespot in **spring barley**, **spring wheat**, **winter barley**, **winter wheat**
*   Glume blotch in **spring wheat**, **winter wheat**
*   Net blotch in **spring barley** *(moderate control)*, **winter barley** *(moderate control)*
*   Powdery mildew in **spring barley**, **spring oats**, **spring wheat**, **winter barley**, **winter oats**, **winter wheat**
*   Rhynchosporium in **spring barley** *(moderate control)*, **winter barley** *(moderate control)*
*   Septoria leaf blotch in **spring wheat** *(moderate control)*, **winter wheat** *(moderate control)*
*   Yellow rust in **spring barley**, **spring wheat**, **winter barley**, **winter wheat**

Approval information

*   Prochloraz, proquinazid and tebuconazole included in Annex 1 under EC Regulation 1107/2009
*   Accepted by BBPA for use on malting barley

Efficacy guidance

*   Resistance to DMI fungicides has been identified in Septoria leaf blotch and may seriously affect the performance of some products. Advice on resistance management is available from the Fungicide Action Group (FRAG) web site

Restrictions

*   Maximum number of applications is 2 per crop
*   Newer authorisations for tebuconazole products require application to cereals only after GS 30 and applications to oilseed rape and linseed after GS20 - check label

Crop-specific information

*   Latest application in wheat is before full flowering (GS65)
*   Latest application in barley is before beginning of heading (GS49)

Environmental safety

*   Avoid spraying within 5 m of the field boundary to reduce the effect on non-target insects or other arthropods
*   LERAP Category B

Hazard classification and safety precautions

Hazard Harmful, Dangerous for the environment
Transport code 9
Packaging group III
UN Number 3082
Risk phrases H317, H351, H361
Operator protection A, C, H
Environmental protection E15a, E16a, E22c, E34, E38, H410
Storage and disposal D01, D02, D09a, D12a

SEE SECTION 3 FOR PRODUCTS ALSO REGISTERED

329   prochloraz + tebuconazole

A broad spectrum systemic fungicide mixture for cereals, oilseed rape and amenity use
FRAC mode of action code: 3 + 3

See also tebuconazole

Products

1 Agate EW	Nufarm UK	267:133 g/l	EW	19000
2 Clayton Prius	Clayton	267:133 g/l	EW	18946
3 Monkey	Nufarm UK	267:133 g/l	EW	18997
4 Orius P	Nufarm UK	267:133 g/l	EW	18994

Uses

- Brown rust in *spring rye, spring wheat, triticale, winter rye, winter wheat*
- Eyespot in *spring rye, spring wheat, triticale, winter rye, winter wheat*
- Glume blotch in *spring wheat, triticale, winter wheat*
- Late ear diseases in *spring wheat, triticale, winter wheat*
- Powdery mildew in *spring rye, spring wheat, triticale, winter rye, winter wheat*
- Rhynchosporium in *spring rye, winter rye*
- Septoria leaf blotch in *spring wheat, triticale, winter wheat*
- Yellow rust in *spring rye, spring wheat, triticale, winter rye, winter wheat*

Approval information

- Prochloraz and tebuconazole included in Annex I under EC Regulation 1107/2009

Efficacy guidance

- Optimum application timing is normally when disease first seen but varies with main target disease - see labels
- Prochloraz and tebuconazole are DMI fungicides. Resistance to some DMI fungicides has been identified in Septoria leaf blotch which may seriously affect performance of some products. For further advice contact a specialist advisor and visit the Fungicide Resistance Action Group (FRAG)-UK website

Restrictions

- Maximum total dose on cereals equivalent to two full dose treatments
- Maximum number of treatments on managed amenity turf 2 per yr
- Newer authorisations for tebuconazole products require application to cereals only after GS 30 and applications to oilseed rape and linseed after GS 20 - check label
- Do not exceed 450 g prochloraz/hectare per application for outdoor uses
- To protect non target insects/arthropods respect an unsprayed buffer zone of 5 m to non-crop land [4]

Crop-specific information

- Latest use: Up to and including flowering (anthesis) just complete (GS 69)
- HI 6 wk for cereals
- Occasionally transient leaf speckling may occur after treating wheat. Yield responses should not be affected

Environmental safety

- Dangerous for the environment
- Very toxic to aquatic organisms
- To protect non target insects/arthropods respect an unsprayed buffer zone of 5 m to non-crop land
- LERAP Category B

Hazard classification and safety precautions

Hazard Harmful, Dangerous for the environment, Harmful if swallowed, Very toxic to aquatic organisms
Transport code 9
Packaging group III
UN Number 3082

FOR FULL CONDITIONS OF USE ALWAYS READ THE PRODUCT LABEL

Risk phrases H319, H361
Operator protection A, C, H; U05a, U09b, U14, U20a [1-4]; U11 [1]
Environmental protection E13b, E16a, E34, E38, H410
Storage and disposal D01, D02, D05, D09a, D10c, D12a
Medical advice M03

330   prohexadione-calcium

A cyclohexanecarboxylate growth regulator for use in apples

See also mepiquat chloride + prohexadione-calcium

Products

1	Bolster	De Sangosse	50 g/l	OD	18773
2	Flagstaff	De Sangosse	50 g/l	OD	18586
3	Kudos	Fine	10% w/w	WG	17578
4	Regalis Plus	BASF	10% w/w	WG	16485

Uses

- Botrytis in **protected ornamentals** *(off-label)*, **table grapes** *(off-label)*, **wine grapes** *(off-label)* [4]
- Control of shoot growth in **apples**, **pears** *(off-label)* [3, 4]; **cherries** *(off-label)* [3]; **hops in propagation** *(off-label)*, **nursery fruit trees** *(off-label)*, **ornamental plant production** *(off-label)*, **plums** *(off-label)* [4]
- Disease control in **protected ornamentals** *(off-label)* [4]
- Growth regulation in **rye**, **spring barley**, **triticale**, **winter barley**, **winter wheat** [1, 2]

Extension of Authorisation for Minor Use (EAMUs)

- **cherries** *20192120* [3]
- **hops in propagation** *20150175* [4]
- **nursery fruit trees** *20150175* [4]
- **ornamental plant production** *20150175* [4]
- **pears** *20171068* [3], *20150182* [4]
- **plums** *20171174* [4]
- **protected ornamentals** *20192153* [4]
- **table grapes** *20150180* [4]
- **wine grapes** *20150180* [4]

Approval information

- Prohexadione-calcium included in Annex I under EC Regulation 1107/2009

Efficacy guidance

- For a standard orchard treat at the start of active growth and again after 3-5 wk depending on growth conditions
- Alternatively follow the first application with four reduced rate applications at two wk intervals

Restrictions

- Maximum total dose equivalent to two full dose treatments
- Do not apply in conjunction with calcium based foliar fertilisers

Crop-specific information

- HI 55 d for apples

Environmental safety

- Harmful to aquatic organisms
- Avoid spray drift onto neighbouring crops and other non-target plants
- Product does not harm natural insect predators when used as directed

Hazard classification and safety precautions

UN Number N/C
Risk phrases H317 [4]
Operator protection A, H [1, 2]; U02a, U08, U19a, U20b [3]; U05a [3, 4]

Environmental protection E15b, H412 [3]; E38 [3, 4]
Storage and disposal D01, D02, D09a, D12a [3, 4]; D10b [3]; D10c [4]
Medical advice M05a [3]

331   prohexadione-calcium + trinexapac-ethyl

A plant growth regulator for use in cereal crops

See also trinexapac-ethyl

Products

Medax Max	BASF	0.5:0.75% w/w	SG	17263

Uses

- Growth regulation in **durum wheat**, **spring barley**, **spring oats**, **spring wheat**, **triticale**, **winter barley**, **winter oats**, **winter rye**, **winter wheat**

Approval information

- Prohexadione-calcium and trinexapac-ethyl included in Annex I under EC Regulation 1107/2009

Efficacy guidance

- Do not use on stressed crops

Following crops guidance

- Any crop can follow a normally harvested or failed cereal crop. Ploughing is not necessary before sowing a following crop.

Hazard classification and safety precautions

Transport code 9
Packaging group III
UN Number 3077
Operator protection A; U08, U19a, U20a
Environmental protection E15b, E34, H412
Storage and disposal D01, D02, D05, D09a, D10c
Medical advice M03

332   propamocarb hydrochloride

A translocated protectant carbamate fungicide
FRAC mode of action code: 28

See also fenamidone + propamocarb hydrochloride
fluopicolide + propamocarb hydrochloride
fosetyl-aluminium + propamocarb hydrochloride
mancozeb + propamocarb hydrochloride

Products

1	Promess	Arysta	722 g/l	SL	16008
2	Proplant	Arysta	722 g/l	SL	15422

Uses

- Damping off in **protected broccoli**, **protected brussels sprouts**, **protected calabrese**, **protected cauliflowers** [1, 2]; **protected tomatoes** *(off-label)* [2]
- Downy mildew in **forest nurseries** *(off-label)*, **ornamental plant production** *(off-label)*, **protected bulbs** *(off-label)*, **protected ornamentals** *(off-label)* [1]; **protected broccoli**, **protected brussels sprouts**, **protected calabrese**, **protected cauliflowers**, **protected watercress** *(off-label)*, **radishes** *(off-label)* [1, 2]
- Peronospora spp in **forest nurseries** *(off-label)*, **ornamental plant production** *(off-label)*, **protected broccoli**, **protected brussels sprouts**, **protected calabrese**, **protected cauliflowers**, **protected ornamentals** *(off-label)* [1, 2]; **protected bulbs** *(off-label)* [1]

FOR FULL CONDITIONS OF USE ALWAYS READ THE PRODUCT LABEL

- Phytophthora in **forest nurseries** *(off-label)*, **ornamental plant production** *(off-label)*, **protected broccoli, protected brussels sprouts, protected calabrese, protected cauliflowers, protected ornamentals** *(off-label)*, **protected tomatoes** *(off-label)*, **protected watercress** *(off-label)* [1, 2]; **protected bulbs** *(off-label)* [1]
- Pythium in **forest nurseries** *(off-label)*, **ornamental plant production** *(off-label)*, **protected broccoli, protected brussels sprouts, protected calabrese, protected cauliflowers, protected ornamentals** *(off-label)*, **protected watercress** *(off-label)* [1, 2]; **protected bulbs** *(off-label)* [1]
- Root rot in **protected broccoli, protected brussels sprouts, protected calabrese, protected cauliflowers, protected tomatoes** *(off-label)* [1, 2]; **protected watercress** *(off-label)* [1]

Extension of Authorisation for Minor Use (EAMUs)
- **forest nurseries** *20160796* [1], *20160785* [2]
- **ornamental plant production** *20160796* [1], *20160785* [2]
- **protected bulbs** *20160796* [1]
- **protected ornamentals** *20160796* [1], *20160785* [2]
- **protected tomatoes** *20160797* [1], *20160784* [2]
- **protected watercress** *20160792* [1], *20160783* [2]
- **radishes** *20160795* [1], *20160782* [2]

Approval information
- Propamocarb hydrochloride included in Annex I under EC Regulation 1107/2009

Efficacy guidance
- Chemical is absorbed through roots and translocated throughout plant
- Incorporate in compost before use or drench moist compost or soil before sowing, pricking out, striking cuttings or potting up
- Concentrated solution is corrosive to all metals other than stainless steel

Restrictions
- Maximum number of treatments 1 per crop for listed brassicas;
- When applied over established seedlings rinse off foliage with water and do not apply under hot, dry conditions
- Do not apply in a recirculating irrigation/drip system
- Store away from seeds and fertilizers
- Consult processor before use on crops grown for processing

Crop-specific information
- Latest use: before transplanting for brassicas

Hazard classification and safety precautions
Hazard Irritant
UN Number N/C
Risk phrases H317
Operator protection A, H, M; U02a, U05a, U08, U19a, U20b
Environmental protection E15a, E34
Storage and disposal D01, D02, D05, D09a, D10c

333   propaquizafop

A phenoxy alkanoic acid foliar acting herbicide for grass weeds in a range of crops
HRAC mode of action code: A

Products

1	Clayton Satchmo	Clayton	100 g/l	EC	17061
2	Cleancrop Peregrine	Clayton	100 g/l	EC	18999
3	Falcon	Adama	100 g/l	EC	16459
4	Longhorn	AgChem Access	100 g/l	EC	16951
5	Shogun	Adama	100 g/l	EC	16527

SEE SECTION 3 FOR PRODUCTS ALSO REGISTERED

SECTION 2

Uses

- Annual grasses in *all edible seed crops grown outdoors* (off-label), *all non-edible seed crops grown outdoors* (off-label), *game cover* (off-label), *mustard* (off-label), *poppies for morphine production* (off-label) [3, 5]; *broad beans, bulb onions, carrots, combining peas, dwarf beans, early potatoes, fodder beet, french beans, linseed, maincrop potatoes, spring field beans, spring oilseed rape, sugar beet, swedes, turnips, winter field beans, winter oilseed rape* [1-5]; *forest* (off-label), *forest nurseries* (off-label), *poppies* (off-label), *poppies grown for seed production* (off-label) [3]; *garlic* (off-label), *lupins* (off-label), *red beet* (off-label), *shallots* (off-label) [5]
- Annual meadow grass in *all edible seed crops grown outdoors* (off-label), *all non-edible seed crops grown outdoors* (off-label), *game cover* (off-label), *poppies for morphine production* (off-label), *red beet* (off-label) [5]; *garlic* (off-label), *lupins* (off-label), *shallots* (off-label) [3, 5]; *poppies* (off-label), *poppies grown for seed production* (off-label) [3]
- Couch in *mustard* (off-label) [3]
- Grass weeds in *garlic* (off-label), *lupins* (off-label), *shallots* (off-label) [3]
- Perennial grasses in *all edible seed crops grown outdoors* (off-label), *all non-edible seed crops grown outdoors* (off-label), *game cover* (off-label), *poppies for morphine production* (off-label) [3, 5]; *broad beans, bulb onions, carrots, combining peas, dwarf beans, early potatoes, fodder beet, french beans, linseed, maincrop potatoes, spring field beans, spring oilseed rape, sugar beet, swedes, turnips, winter field beans, winter oilseed rape* [1-5]; *forest* (off-label), *forest nurseries* (off-label), *poppies* (off-label), *poppies grown for seed production* (off-label) [3]; *garlic* (off-label), *lupins* (off-label), *mustard* (off-label), *red beet* (off-label), *shallots* (off-label) [5]
- Volunteer cereals in *mustard* (off-label), *poppies for morphine production* (off-label), *red beet* (off-label) [3, 5]; *poppies* (off-label), *poppies grown for seed production* (off-label) [3]
- Volunteer rye in *red beet* (off-label) [3]

Extension of Authorisation for Minor Use (EAMUs)

- *all edible seed crops grown outdoors* 20141283 [3], 20141927 [5], 20142509 [5]
- *all non-edible seed crops grown outdoors* 20141283 [3], 20141927 [5], 20142509 [5]
- *forest* 20163458 [3]
- *forest nurseries* 20163458 [3]
- *game cover* 20141283 [3], 20141927 [5], 20142509 [5]
- *garlic* 20141281 [3], 20141926 [5]
- *lupins* 20141282 [3], 20141928 [5]
- *mustard* 20172803 [3], 20180165 [5]
- *poppies* 20180090 [3]
- *poppies for morphine production* 20152056 [3], 20160285 [5]
- *poppies grown for seed production* 20180090 [3]
- *red beet* 20151827 [3], 20160286 [5]
- *shallots* 20141281 [3], 20141926 [5]

Approval information

- Propaquizafop included in Annex 1 under EC Regulation 1107/2009

Efficacy guidance

- Apply to emerged weeds when they are growing actively with adequate soil moisture
- Activity is slower under cool conditions
- Broad-leaved weeds and any weeds germinating after treatment are not controlled
- Annual meadow grass up to 3 leaves checked at low doses and severely checked at highest dose
- Spray barley cover crops when risk of wind blow has passed and before there is serious competition with the crop
- Various tank mixtures and sequences recommended for broader spectrum weed control in oilseed rape, peas and sugar beet. See label for details
- Severe couch infestations may require a second application at reduced dose when regrowth has 3-4 leaves unfolded
- Products contain surfactants. Tank mixing with adjuvants not required or recommended
- Propaquizafop is an ACCase inhibitor herbicide. To avoid the build up of resistance do not apply products containing an ACCase inhibitor herbicide more than twice to any crop. In addition do not

use any product containing propaquizafop in mixture or sequence with any other product containing the same ingredient
- Use these products as part of a resistance management strategy that includes cultural methods of control and does not use ACCase inhibitors as the sole chemical method of grass weed control
- Applying a second product containing an ACCase inhibitor to a crop will increase the risk of resistance development; only use a second ACCase inhibitor to control different weeds at a different timing
- Always follow WRAG guidelines for preventing and managing herbicide resistant weeds. See Section 5 for more information

Restrictions
- Maximum number of treatments 1 or 2 per crop or yr. See label for details
- See label for list of tolerant tree species
- Do not treat seed potatoes

Crop-specific information
- Latest use: before crop flower buds visible for winter oilseed rape, linseed, field beans; before 8 fully expanded leaf stage for spring oilseed rape; before weeds are covered by the crop for potatoes, sugar beet, fodder beet, swedes, turnips; when flower buds visible for peas
- HI onions, carrots, early potatoes, parsnips 4 wk; combining peas, early potatoes 7 wk; sugar beet, fodder beet, maincrop potatoes, swedes, turnips 8 wk; field beans 14 wk
- Application in high temperatures and/or low soil moisture content may cause chlorotic spotting especially on combining peas and field beans
- Overlaps at the highest dose can cause damage from early applications to carrots and parsnips

Following crops guidance
- In the event of a failed treated crop an interval of 2 wk must elapse between the last application and redrilling with winter wheat or winter barley. 4 wk must elapse before sowing oilseed rape, peas or field beans, and 16 wk before sowing ryegrass or oats

Environmental safety
- Dangerous for the environment
- Toxic to aquatic organisms
- Risk to certain non-target insects or other arthropods. See directions for use
- LERAP Category B

Hazard classification and safety precautions
Hazard Harmful, Dangerous for the environment
Transport code 9
Packaging group III
UN Number 3082
Risk phrases H304, H319
Operator protection A, C [1-5]; H [4, 5]; U02a, U05a, U08, U11, U13, U14, U15, U19a, U20b
Environmental protection E15a, E16a, E22b, E34, E38, H411
Storage and disposal D01, D02, D05, D09a, D12a, D14 [1-5]; D10b [1-3]; D12b [4, 5]
Medical advice M05b

334   propoxycarbazone-sodium

A sulfonylaminocarbonyl triazolinone residual grass weed herbicide for winter wheat
HRAC mode of action code: B

See also iodosulfuron-methyl-sodium + propoxycarbazone-sodium
*   mesosulfuron-methyl + propoxycarbazone-sodium*

Products

Attribut	Sumitomo	70% w/w	SG	18894

Uses
- Blackgrass in *winter wheat*
- Couch in *winter wheat*

SEE SECTION 3 FOR PRODUCTS ALSO REGISTERED

Approval information
- Propoxycarbazone-sodium included in Annex I under EC Regulation 1107/2009

Efficacy guidance
- Activity by root and foliar absorption but depends on presence of sufficient soil moisture to ensure root uptake by weeds. Control is enhanced by use of an adjuvant to encourage foliar uptake
- Best results obtained from treatments applied when weed grasses are growing actively. Symptoms may not become apparent for 3-4 wk after application
- Ensure good even spray coverage
- To achieve best control of couch and reduction of infestation in subsequent crop treat between 2 true leaves and first node stage of the weed
- Blackgrass should be treated after using specialist blackgrass herbicides with a different mode of action
- Propoxycarbazone-sodium is a member of the ALS-inhibitor group of herbicides. To avoid the build up of resistance do not use any product containing an ALS-inhibitor herbicide with claims for control of grass weeds more than once on any crop
- Use these products as part of a resistance management strategy that includes cultural methods of control and does not use ALS inhibitors as the sole chemical method of grass weed control

Restrictions
- Maximum total dose equivalent to one full dose treatment
- Do not apply when temperature near or below freezing
- Avoid treatment under dry soil conditions
- To avoid the build up of resistance do not apply this or any other product containing an ALS inhibitor herbicide with claims for control of grass-weeds more than once to any crop

Crop-specific information
- Latest use: third node detectable in wheat (GS 33)

Following crops guidance
- Only winter wheat, field beans or winter barley may be sown in the autumn following spring treatment
- Any crop may be grown in the spring on land treated during the previous calendar yr

Environmental safety
- Dangerous for the environment
- Very toxic to aquatic organisms
- Take extreme care to avoid drift onto adjacent plants or land as this could result in severe damage
- LERAP Category B

Hazard classification and safety precautions
Hazard Dangerous for the environment
Transport code 9
Packaging group III
UN Number 3077
Operator protection A; U05a, U20b
Environmental protection E15b, E16a, E16b, E38, H410
Storage and disposal D01, D02, D09a, D11a, D12a

335   propyzamide

A residual benzamide herbicide for use in a wide range of crops
HRAC mode of action code: K1

See also aminopyralid + propyzamide
clopyralid + propyzamide

Products

1 Barclay Propyz	Barclay	400 g/l	SC	15083
2 Cleancrop Forward	Dow (Corteva)	400 g/l	SC	18145

Products – continued

3	Judo	FMC Agro	400 g/l	SC	18743
4	Kerb Flo	Dow (Corteva)	400 g/l	SC	13716
5	Kerb Flo 500	Dow (Corteva)	500 g/l	SC	15586
6	Kerb Granules	Barclay	4% w/w	GR	14213
7	Levada	Certis	400 g/l	SC	15743
8	Menace 80 EDF	Dow (Corteva)	80% w/w	WG	13714
9	Proper Flo	Belchim	400 g/l	SC	15102
10	PureFlo	Pure Amenity	400 g/l	SC	15138
11	Relva	Belcrop	400 g/l	SC	14873
12	Relva Granules	Belcrop	4% w/w	GR	16744
13	Setanta Flo	Certis	400 g/l	SC	15791
14	Solitaire	Certis	400 g/l	SC	15792
15	Stroller	AgChem Access	400 g/l	SC	14603
16	Stymie	Becesane	400 g/l	SC	16328
17	Zamide 80 WG	Albaugh UK	80% w/w	WG	14723
18	Zamide Flo	Albaugh UK	400 g/l	SC	14679

SECTION 2

Uses

- Annual and perennial weeds in *amenity vegetation* [7, 13, 14]
- Annual dicotyledons in *all edible seed crops grown outdoors* (off-label), *all non-edible seed crops grown outdoors* (off-label), *almonds* (off-label), *bilberries* (off-label), *blueberries* (off-label), *bog myrtle* (off-label), *broccoli* (off-label), *calabrese* (off-label), *cauliflowers* (off-label), *cherries* (off-label), *chestnuts* (off-label), *chicory root* (off-label), *cob nuts* (off-label), *corn gromwell* (off-label), *courgettes* (off-label), *cranberries* (off-label), *cress* (off-label), *endives* (off-label), *frise* (off-label), *game cover* (off-label), *hazel nuts* (off-label), *herbs (see appendix 6)* (off-label), *hops* (off-label), *lamb's lettuce* (off-label), *leaf brassicas* (off-label), *marrows* (off-label), *mirabelles* (off-label), *protected endives* (off-label), *protected forest nurseries* (off-label), *protected herbs (see appendix 6)* (off-label), *protected lettuce* (off-label), *pumpkins* (off-label), *quinces* (off-label), *radicchio* (off-label), *salad brassicas* (off-label), *scarole* (off-label), *squashes* (off-label), *table grapes* (off-label), *walnuts* (off-label), *wine grapes* (off-label) [4]; *amenity vegetation* [1, 2, 4, 6, 8, 10, 12, 15, 16]; *apple orchards, blackberries, blackcurrants, clover seed crops, gooseberries, hedges, loganberries, pear orchards, plums, raspberries* (England only), *redcurrants, strawberries* [1, 2, 4, 7, 8, 10, 13-16]; *brassica seed crops, lettuce* (outdoor crops), *rhubarb* (outdoor), *woody ornamentals* [8]; *farm forestry* [1, 2, 4, 6-8, 10, 12-18]; *fodder rape seed crops, kale seed crops, turnip seed crops* [1, 2, 4, 7, 9, 10, 13-16]; *forest* [1, 2, 4, 6, 7, 9, 10, 12-18]; *forest nurseries* [1, 2, 4, 7, 8, 10, 13-18]; *lettuce* [2, 4, 7, 9, 13-18]; *lucerne* [2, 4, 7, 8, 10, 13-16]; *ornamental plant production* [6, 12]; *red clover, white clover* [9]; *rhubarb* [1, 2, 4, 7, 10, 13-16]; *sugar beet seed crops* [1, 2, 4, 7-10, 13-16]; *winter field beans* [2, 4, 5, 7-10, 13-18]; *winter oilseed rape* [1-5, 7-11, 13-18]
- Annual grasses in *all edible seed crops grown outdoors* (off-label), *all non-edible seed crops grown outdoors* (off-label), *almonds* (off-label), *bilberries* (off-label), *blueberries* (off-label), *bog myrtle* (off-label), *broccoli* (off-label), *calabrese* (off-label), *cauliflowers* (off-label), *cherries* (off-label), *chestnuts* (off-label), *chicory root* (off-label), *cob nuts* (off-label), *corn gromwell* (off-label), *courgettes* (off-label), *cranberries* (off-label), *cress* (off-label), *endives* (off-label), *frise* (off-label), *game cover* (off-label), *hazel nuts* (off-label), *herbs (see appendix 6)* (off-label), *hops* (off-label), *lamb's lettuce* (off-label), *leaf brassicas* (off-label), *marrows* (off-label), *mirabelles* (off-label), *ornamental plant production* (off-label), *protected endives* (off-label), *protected forest nurseries* (off-label), *protected herbs (see appendix 6)* (off-label), *protected lettuce* (off-label), *pumpkins* (off-label), *quinces* (off-label), *radicchio* (off-label), *salad brassicas* (off-label), *scarole* (off-label), *squashes* (off-label), *table grapes* (off-label), *walnuts* (off-label), *wine grapes* (off-label) [4]; *amenity vegetation* [1, 2, 4, 6, 8, 10, 12, 15, 16]; *apple orchards, blackberries, blackcurrants, clover seed crops, gooseberries, hedges, loganberries, pear orchards, plums, raspberries* (England only), *redcurrants, strawberries* [1, 2, 4, 7, 8, 10, 13-16]; *brassica seed crops, lettuce* (outdoor crops), *rhubarb* (outdoor), *woody ornamentals* [8]; *farm forestry* [1, 2, 4, 6-8, 10, 12-18]; *fodder rape seed crops, kale seed crops, turnip seed crops* [1, 2, 4, 7, 9, 10, 13-16]; *forest* [1, 2, 4, 6, 7, 9, 10, 12-18]; *forest nurseries* [1, 2, 4, 7, 8, 10, 13-18]; *lettuce* [2, 4, 7, 9, 13-18]; *lucerne* [2, 4, 7, 8, 10, 13-16]; *ornamental plant production* [6, 12]; *red clover, white clover* [9]; *rhubarb* [1, 2, 4, 7, 10, 13-

16]; *sugar beet seed crops* [1, 2, 4, 7-10, 13-16]; *winter field beans* [2, 4, 7-10, 13-18]; *winter oilseed rape* [1-4, 7-11, 13-18]

- Annual meadow grass in *winter field beans, winter oilseed rape* [5]
- Blackgrass in *winter field beans, winter oilseed rape* [5]
- Fat hen in *ornamental plant production (off-label)* [4]
- Horsetails in *forest* [8]
- Perennial grasses in *all edible seed crops grown outdoors (off-label), all non-edible seed crops grown outdoors (off-label), almonds (off-label), bilberries (off-label), blueberries (off-label), bog myrtle (off-label), broccoli (off-label), calabrese (off-label), cauliflowers (off-label), cherries (off-label), chestnuts (off-label), chicory root (off-label), cob nuts (off-label), corn gromwell (off-label), courgettes (off-label), cranberries (off-label), cress (off-label), endives (off-label), frise (off-label), game cover (off-label), hazel nuts (off-label), herbs (see appendix 6) (off-label), hops (off-label), lamb's lettuce (off-label), leaf brassicas (off-label), marrows (off-label), mirabelles (off-label), protected endives (off-label), protected forest nurseries (off-label), protected herbs (see appendix 6) (off-label), protected lettuce (off-label), pumpkins (off-label), quinces (off-label), radicchio (off-label), salad brassicas (off-label), scarole (off-label), squashes (off-label), table grapes (off-label), walnuts (off-label), wine grapes (off-label)* [4]; *amenity vegetation* [1, 2, 4, 6, 8, 10, 12, 15, 16]; *apple orchards, blackberries, blackcurrants, forest nurseries, gooseberries, hedges, loganberries, pear orchards, plums, raspberries (England only), redcurrants* [1, 2, 4, 7, 8, 10, 13-16]; *clover seed crops, rhubarb, strawberries* [1, 2, 4, 7, 10, 13-16]; *farm forestry* [1, 2, 4, 6-8, 10, 12-16]; *fodder rape seed crops, kale seed crops, sugar beet seed crops, turnip seed crops* [1, 2, 4, 7, 9, 10, 13-16]; *forest* [1, 2, 4, 6-10, 12-16]; *lettuce* [2, 4, 7, 9, 13-16]; *lucerne* [2, 4, 7, 10, 13-16]; *ornamental plant production* [6, 12]; *red clover, white clover* [9]; *rhubarb (outdoor), woody ornamentals* [8]; *winter field beans* [2, 4, 7, 9, 10, 13-16]; *winter oilseed rape* [1-4, 7, 9-11, 13-16]
- Sedges in *forest* [8]
- Volunteer cereals in *sugar beet seed crops* [8]; *winter field beans, winter oilseed rape* [5, 8]
- Wild oats in *sugar beet seed crops* [8]; *winter field beans, winter oilseed rape* [5, 8]

Extension of Authorisation for Minor Use (EAMUs)

- *all edible seed crops grown outdoors* 20082942 [4]
- *all non-edible seed crops grown outdoors* 20082942 [4]
- *almonds* 20082420 [4]
- *bilberries* 20082419 [4]
- *blueberries* 20082419 [4]
- *bog myrtle* 20120509 [4]
- *broccoli* 20091902 [4]
- *calabrese* 20091902 [4]
- *cauliflowers* 20091902 [4]
- *cherries* 20082418 [4]
- *chestnuts* 20082420 [4]
- *chicory root* 20091530 [4]
- *cob nuts* 20082420 [4]
- *corn gromwell* 20122395 [4]
- *courgettes* 20082416 [4]
- *cranberries* 20082419 [4]
- *cress* 20082410 [4]
- *endives* 20082410 [4]
- *frise* 20082411 [4]
- *game cover* 20082942 [4]
- *hazel nuts* 20082420 [4]
- *herbs (see appendix 6)* 20082412 [4]
- *hops* 20082414 [4]
- *lamb's lettuce* 20082410 [4]
- *leaf brassicas* 20082410 [4]
- *marrows* 20082416 [4]
- *mirabelles* 20082418 [4]
- *ornamental plant production* 20130207 [4]
- *protected endives* 20082415 [4]

- ***protected forest nurseries*** *20082942* [4]
- ***protected herbs (see appendix 6)*** *20082412* [4]
- ***protected lettuce*** *20082415* [4]
- ***pumpkins*** *20082416* [4]
- ***quinces*** *20082413* [4]
- ***radicchio*** *20082411* [4]
- ***salad brassicas*** *20082410* [4]
- ***scarole*** *20082411* [4]
- ***squashes*** *20082416* [4]
- ***table grapes*** *20082417* [4]
- ***walnuts*** *20082420* [4]
- ***wine grapes*** *20082417* [4]

Approval information
- Propyzamide included in Annex I under EC Regulation 1107/2009
- Some products may be applied through CDA equipment. See labels for details
- Accepted by BBPA for use on hops

Efficacy guidance
- Active via root uptake. Weeds controlled from germination to young seedling stage, some species (including many grasses) also when established
- Best results achieved by winter application to fine, firm, moist soil. Rain is required after application if soil dry
- Uptake is slow and and may take up to 12 wk
- Excessive organic debris or ploughed-up turf may reduce efficacy
- For heavy couch infestations a repeat application may be needed in following winter
- Always follow WRAG guidelines for preventing and managing herbicide resistant weeds. See Section 5 for more information

Restrictions
- Maximum number of treatments 1 per crop or yr
- Maximum total dose equivalent to one full dose treatment for all crops
- Do not treat protected crops
- Apply to listed edible crops only between 1 Oct and the date specified as the latest time of application (except lettuce)
- Do not apply in windy weather and avoid drift onto non-target crops
- Do not use on soils with more than 10% organic matter except in forestry

Crop-specific information
- Latest use: labels vary but normally before 31 Dec in year before harvest for rhubarb, lucerne, strawberries and winter field beans; before 31 Jan for other crops
- HI: 6 wk for edible crops
- Apply as soon as possible after 3-true leaf stage of oilseed rape (GS 1,3) and seed brassicas, after 4-leaf stage of sugar beet for seed, within 7 d after sowing but before emergence for field beans, after perennial crops established for at least 1 season, strawberries after 1 yr
- Only apply to strawberries on heavy soils. Do not use on matted row crops
- Only apply to field beans on medium and heavy soils
- Only apply to established lucerne not less than 7 d after last cut
- In lettuce lightly incorporate in top 25 mm pre-drilling or irrigate on dry soil
- See label for lists of ornamental and forest species which may be treated

Following crops guidance
- Following an application between 1 Apr and 31 Jul at any dose the following minimum intervals must be observed before sowing the next crop: lettuce 0 wk; broad beans, chicory, clover, field beans, lucerne, radishes, peas 5 wk; brassicas, celery, leeks, oilseed rape, onions, parsley, parsnips 10 wk
- Following an application between 1 Aug and 31 Mar at any dose the following minimum intervals must be observed before sowing the next crop: lettuce 0 wk; broad beans, chicory, clover, field beans, lucerne, radishes, peas 10 wk; brassicas, celery, leeks, oilseed rape, onions, parsley, parsnips 25 wk or after 15 Jun, whichever occurs sooner

SEE SECTION 3 FOR PRODUCTS ALSO REGISTERED

- Cereals or grasses or other crops not listed may be sown 30 wk after treatment up to 840 g.a.i/ha between 1 Aug and 31 Mar or 40 wk after treatment at higher doses at any time and after mouldboard ploughing to at least 15 cm
- A period of at least 9 mth must elapse between applications of propyzamide to the same land

Environmental safety
- Dangerous for the environment
- Very toxic to aquatic organisms
- Some pesticides pose a greater threat of contamination of water than others and propyzamide is one of these pesticides. Take special care when applying propyzamide near water and do not apply if heavy rain is forecast

Hazard classification and safety precautions
> **Hazard** Harmful, Dangerous for the environment [1-18]; Very toxic to aquatic organisms [7]
> **Transport code** 9
> **Packaging group** III
> **UN Number** 3077 [6, 8, 12, 17]; 3082 [1-5, 7, 9-11, 13-16, 18]
> **Risk phrases** H351 [1-5, 7, 9-12, 14, 16]; R40, R53a [6, 8, 13, 15, 17, 18]; R50 [8, 13, 15, 18]; R51 [6, 17]
> **Operator protection** A, H [1-18]; D [1-4, 8, 10, 15-18]; M [1-4, 7, 8, 10, 11, 13-18]; U05a [6, 12]; U20c [1-5, 7-11, 13-18]
> **Environmental protection** E15a [1-6, 8-10, 12, 15-18]; E15b, E39 [7, 11, 13, 14]; E34 [9]; E38 [1-18]; H410 [7, 9, 11, 14]; H411 [1-5, 10, 12, 14, 16]
> **Consumer protection** C02a [1-4, 8-10, 15-18] (6 wk)
> **Storage and disposal** D01, D02 [6, 12]; D05, D11a [1-5, 8-10, 15-18]; D07 [9]; D09a, D12a [1-6, 8-10, 12, 15-18]
> **Medical advice** M03 [6, 12]

336   proquinazid

A quinazolinone fungicide for powdery mildew control in cereals
FRAC mode of action code: U7

See also prochloraz + proquinazid + tebuconazole

Products

1	Justice	DuPont (Corteva)	200 g/l	EC	12835
2	Talius	DuPont (Corteva)	200 g/l	EC	12752

Uses
- Powdery mildew in **apples** *(off-label)*, **pears** *(off-label)*, **protected strawberries** *(off-label)*, **spring barley**, **spring oats**, **spring rye**, **spring wheat**, **triticale**, **winter barley**, **winter oats**, **winter rye**, **winter wheat** [1, 2]; **courgettes** *(off-label)*, **cucumbers** *(off-label)*, **forest nurseries** *(off-label)*, **marrows** *(off-label)*, **protected aubergines** *(off-label)*, **protected courgettes** *(off-label)*, **protected cucumbers** *(off-label)*, **protected marrows** *(off-label)*, **protected pumpkins** *(off-label)*, **protected squashes** *(off-label)*, **protected tomatoes** *(off-label)*, **pumpkins** *(off-label)*, **soft fruit** *(off-label)*, **squashes** *(off-label)*, **top fruit** *(off-label)* [2]; **grapevines** *(off-label)* [1]

Extension of Authorisation for Minor Use (EAMUs)
- **apples** *20170726* [1], *20170725* [2]
- **courgettes** *20152627* [2]
- **cucumbers** *20152627* [2]
- **forest nurseries** *20090420* [2]
- **grapevines** *20111763* [1]
- **marrows** *20152627* [2]
- **pears** *20170726* [1], *20170725* [2]
- **protected aubergines** *20152627* [2]
- **protected courgettes** *20152627* [2]
- **protected cucumbers** *20152627* [2]
- **protected marrows** *20152627* [2]

- *protected pumpkins* *20152627* [2]
- *protected squashes* *20152627* [2]
- *protected strawberries* *20172436* [1], *20170210* [2]
- *protected tomatoes* *20152627* [2]
- *pumpkins* *20152627* [2]
- *soft fruit* *20090420* [2]
- *squashes* *20152627* [2]
- *top fruit* *20090420* [2]

Approval information
- Proquinazid included in Annex 1 under EC Regulation 1107/2009
- Accepted by BBPA on malting barley

Efficacy guidance
- Best results obtained from preventive treatment before disease is established in the crop
- Where mildew has already spread to new growth a tank mix with a curative fungicide with an alternative mode of action should be used
- Use as part of an integrated crop management (ICM) strategy incorporating other methods of control or fungicides with different modes of action

Restrictions
- Maximum number of treatments 2 per crop
- Do not apply to any crop suffering from stress from any cause
- Avoid application in either frosty or hot, sunny conditions
- Must not be applied by hand-held equipment

Crop-specific information
- Latest use: before beginning of heading (GS 49) for barley, oats, rye, triticale; before full flowering (GS 65) for wheat

Environmental safety
- Dangerous for the environment
- Toxic to aquatic organisms
- Dangerous to fish or other aquatic life. Do not contaminate surface waters or ditches with chemical or used container
- LERAP Category B

Hazard classification and safety precautions
Hazard Harmful, Dangerous for the environment
Transport code 9
Packaging group III
UN Number 3082
Risk phrases H315, H318, H351
Operator protection A, C, H; U05a, U11
Environmental protection E13b, E16a, E16b, E34, E38, H410
Storage and disposal D01, D02, D05, D09a, D10b, D11a, D12a
Medical advice M03, M05a

337 prosulfocarb

A thiocarbamate herbicide for grass and broad-leaved weed control in cereals and potatoes
HRAC mode of action code: N

See also clodinafop-propargyl + prosulfocarb

Products

1	Bokken	Syngenta	800 g/l	EC	17557
2	Clayton Comply	Clayton	800 g/l	EC	17702
3	Clayton Obey	Clayton	800 g/l	EC	16626
4	Defy	Syngenta	800 g/l	EC	16202
5	Fidox 800 EC	Belchim	800 g/l	EC	17904

SEE SECTION 3 FOR PRODUCTS ALSO REGISTERED

Products – continued

6	Lees	Rotam	800 g/l	EC	18769
7	Moose 800 EC	Belchim	800 g/l	EC	17968
8	Quidam	UPL Europe	800 g/l	EC	17595
9	Roxy 800 EC	Belchim	800 g/l	EC	17859
10	Spinnaker	Adama	800 g/l	EC	18855
11	Topsail	Adama	800 g/l	EC	19024

Uses

- Annual dicotyledons in *almonds* (off-label), *apples* (off-label), *apricots* (off-label), *cherries* (off-label), *chestnuts* (off-label), *hazel nuts* (off-label), *medlar* (off-label), *nectarines* (off-label), *peaches* (off-label), *pears* (off-label), *plums* (off-label), *quinces* (off-label), *walnuts* (off-label) [2, 10]; *bulb onion sets* (off-label) [10]; *bulb onions* (off-label), *herbs (see appendix 6)* (off-label) [2]; *carrots* (off-label), *celery (outdoor)* (off-label), *forest nurseries* (off-label), *game cover* (off-label), *garlic* (off-label), *horseradish* (off-label), *leeks* (off-label), *miscanthus* (off-label), *ornamental plant production* (off-label), *parsley root* (off-label), *parsnips* (off-label), *poppies for morphine production* (off-label), *rye* (off-label), *salsify* (off-label), *shallots* (off-label), *spring field beans* (off-label), *spring onions* (off-label), *winter field beans* (off-label), *winter linseed* (off-label) [2, 4, 10]; *celeriac* (off-label), *triticale* (off-label) [2, 4]; *onion sets* (off-label), *top fruit* (off-label) [4]; *potatoes, winter barley, winter wheat* [1-11]; *spring barley* (off-label) [4, 10]
- Annual grasses in *almonds* (off-label), *apples* (off-label), *apricots* (off-label), *cherries* (off-label), *chestnuts* (off-label), *hazel nuts* (off-label), *medlar* (off-label), *nectarines* (off-label), *peaches* (off-label), *pears* (off-label), *plums* (off-label), *poppies for morphine production* (off-label), *quinces* (off-label), *spring field beans* (off-label), *walnuts* (off-label), *winter field beans* (off-label) [2, 10]; *bulb onion sets* (off-label), *spring barley* (off-label) [10]; *bulb onions* (off-label), *herbs (see appendix 6)* (off-label) [2]; *carrots* (off-label), *celery (outdoor)* (off-label), *forest nurseries* (off-label), *game cover* (off-label), *garlic* (off-label), *horseradish* (off-label), *leeks* (off-label), *miscanthus* (off-label), *ornamental plant production* (off-label), *parsley root* (off-label), *parsnips* (off-label), *rye* (off-label), *salsify* (off-label), *shallots* (off-label), *spring onions* (off-label), *winter linseed* (off-label) [2, 4, 10]; *celeriac* (off-label), *triticale* (off-label) [2, 4]; *onion sets* (off-label), *top fruit* (off-label) [4]
- Annual meadow grass in *bulb onions* (off-label), *garlic* (off-label), *herbs (see appendix 6)* (off-label), *leeks* (off-label), *shallots* (off-label), *spring onions* (off-label) [8]; *carrots* (off-label), *celeriac* (off-label), *celery (outdoor)* (off-label), *forest nurseries* (off-label), *game cover* (off-label), *horseradish* (off-label), *miscanthus* (off-label), *ornamental plant production* (off-label), *parsley root* (off-label), *parsnips* (off-label), *poppies for morphine production* (off-label), *rye* (off-label), *salsify* (off-label), *top fruit* (off-label), *triticale* (off-label) [4]; *poppies* (off-label), *poppies grown for seed production* (off-label), *spring barley* (off-label) [4, 10]; *potatoes, winter barley, winter wheat* [1-11]; *winter linseed* (off-label) [4, 8]
- Black nightshade in *broad beans* (off-label), *edible podded peas* (off-label), *poppies* (off-label), *poppies grown for seed production* (off-label), *vining peas* (off-label) [4, 10]; *bulb onions* (off-label), *garlic* (off-label), *leeks* (off-label), *shallots* (off-label), *spring onions* (off-label), *winter linseed* (off-label) [8]; *edible flowers* (off-label) [10]; *herbs (see appendix 6)* (off-label) [2, 4, 8, 10]
- Blackgrass in *rye* (off-label) [10]
- Chickweed in *broad beans* (off-label), *edible podded peas* (off-label), *poppies* (off-label), *poppies grown for seed production* (off-label), *vining peas* (off-label) [4, 10]; *bulb onions* (off-label), *garlic* (off-label), *herbs (see appendix 6)* (off-label), *leeks* (off-label), *shallots* (off-label), *spring onions* (off-label), *winter linseed* (off-label) [8]; *celeriac* (off-label) [2, 4, 10]; *potatoes, winter barley, winter wheat* [1-11]
- Cleavers in *bulb onions* (off-label), *garlic* (off-label), *leeks* (off-label), *shallots* (off-label), *spring onions* (off-label), *winter linseed* (off-label) [8]; *carrots* (off-label), *horseradish* (off-label), *parsley root* (off-label), *parsnips* (off-label), *salsify* (off-label) [2, 4, 10]; *edible flowers* (off-label) [10]; *herbs (see appendix 6)* (off-label) [2, 4, 8, 10]; *poppies* (off-label), *poppies grown for seed production* (off-label) [4, 10]; *potatoes, winter barley, winter wheat* [1-11]; *spring field beans* (off-label), *winter field beans* (off-label) [2, 4]
- Crane's-bill in *broad beans* (off-label), *edible podded peas* (off-label), *poppies* (off-label), *poppies grown for seed production* (off-label), *vining peas* (off-label) [4, 10]; *bulb onions* (off-

label), **garlic** *(off-label)*, **herbs (see appendix 6)** *(off-label)*, **leeks** *(off-label)*, **shallots** *(off-label)*, **spring onions** *(off-label)*, **winter linseed** *(off-label)* [8]

- Fat hen in **carrots** *(off-label)*, **celeriac** *(off-label)*, **herbs (see appendix 6)** *(off-label)*, **horseradish** *(off-label)*, **parsley root** *(off-label)*, **parsnips** *(off-label)*, **salsify** *(off-label)* [2, 4, 10]; **edible flowers** *(off-label)* [10]
- Field speedwell in **potatoes, winter barley, winter wheat** [1-11]
- Forget-me-not in **broad beans** *(off-label)*, **edible podded peas** *(off-label)*, **poppies** *(off-label)*, **poppies grown for seed production** *(off-label)*, **vining peas** *(off-label)* [4, 10]; **bulb onions** *(off-label)*, **garlic** *(off-label)*, **herbs (see appendix 6)** *(off-label)*, **leeks** *(off-label)*, **shallots** *(off-label)*, **spring onions** *(off-label)*, **winter linseed** *(off-label)* [8]
- Fumitory in **bulb onion sets** *(off-label)*, **edible flowers** *(off-label)*, **garlic** *(off-label)*, **leeks** *(off-label)*, **shallots** *(off-label)*, **spring onions** *(off-label)* [10]; **herbs (see appendix 6)** *(off-label)* [2, 4, 10]
- Grass weeds in **celeriac** *(off-label)* [10]
- Ivy-leaved speedwell in **poppies** *(off-label)*, **poppies grown for seed production** *(off-label)* [4, 10]; **potatoes, winter barley, winter wheat** [1-11]
- Knotgrass in **carrots** *(off-label)*, **herbs (see appendix 6)** *(off-label)*, **horseradish** *(off-label)*, **parsley root** *(off-label)*, **parsnips** *(off-label)*, **salsify** *(off-label)* [2, 4, 10]; **edible flowers** *(off-label)* [10]
- Loose silky bent in **bulb onions** *(off-label)*, **garlic** *(off-label)*, **herbs (see appendix 6)** *(off-label)*, **leeks** *(off-label)*, **shallots** *(off-label)*, **spring onions** *(off-label)*, **winter linseed** *(off-label)* [8]; **poppies** *(off-label)*, **poppies grown for seed production** *(off-label)* [4, 10]; **potatoes, winter barley, winter wheat** [1-11]
- Mayweeds in **carrots** *(off-label)*, **celeriac** *(off-label)*, **herbs (see appendix 6)** *(off-label)*, **horseradish** *(off-label)*, **parsley root** *(off-label)*, **parsnips** *(off-label)*, **salsify** *(off-label)* [2, 4, 10]; **edible flowers** *(off-label)* [10]
- Nettles in **celeriac** *(off-label)* [10]
- Polygonums in **bulb onion sets** *(off-label)*, **garlic** *(off-label)*, **leeks** *(off-label)*, **shallots** *(off-label)*, **spring onions** *(off-label)* [10]; **celeriac** *(off-label)* [2, 4, 10]
- Red dead-nettle in **broad beans** *(off-label)*, **edible podded peas** *(off-label)*, **poppies** *(off-label)*, **poppies grown for seed production** *(off-label)*, **vining peas** *(off-label)* [4, 10]
- Rough-stalked meadow grass in **bulb onions** *(off-label)*, **garlic** *(off-label)*, **herbs (see appendix 6)** *(off-label)*, **leeks** *(off-label)*, **shallots** *(off-label)*, **spring onions** *(off-label)*, **winter linseed** *(off-label)* [8]; **poppies** *(off-label)*, **poppies grown for seed production** *(off-label)* [4, 10]; **potatoes, winter barley, winter wheat** [1-11]
- Ryegrass in **rye** *(off-label)* [10]
- Speedwells in **broad beans** *(off-label)*, **edible podded peas** *(off-label)*, **poppies** *(off-label)*, **poppies grown for seed production** *(off-label)*, **vining peas** *(off-label)* [4, 10]; **bulb onions** *(off-label)*, **garlic** *(off-label)*, **herbs (see appendix 6)** *(off-label)*, **leeks** *(off-label)*, **shallots** *(off-label)*, **spring onions** *(off-label)*, **winter linseed** *(off-label)* [8]
- Thistles in **celeriac** *(off-label)* [2, 4, 10]

Extension of Authorisation for Minor Use (EAMUs)
- **almonds** *20162930* [2], *20191742* [10]
- **apples** *20162930* [2], *20191742* [10]
- **apricots** *20162930* [2], *20191742* [10]
- **broad beans** *20191041* [4], *20191749* [10]
- **bulb onion sets** *20191740* [10]
- **bulb onions** *20162925* [2], *20181690* [8]
- **carrots** *20162922* [2], *20131354* [4], *20191747* [10]
- **celeriac** *20162933* [2], *20131355* [4], *20191737* [10]
- **celery (outdoor)** *20162924* [2], *20131353* [4], *20191751* [10]
- **cherries** *20162930* [2], *20191742* [10]
- **chestnuts** *20162930* [2], *20191742* [10]
- **edible flowers** *20191750* [10]
- **edible podded peas** *20191041* [4], *20191749* [10]
- **forest nurseries** *20162926* [2], *20131432* [4], *20191741* [10]
- **game cover** *20162931* [2], *20131430* [4], *20191739* [10]
- **garlic** *20162925* [2], *20131357* [4], *20181690* [8], *20191740* [10]

SEE SECTION 3 FOR PRODUCTS ALSO REGISTERED

- **hazel nuts** *20162930* [2], *20191742* [10]
- **herbs (see appendix 6)** *20162923* [2], *20152809* [4], *20181691* [8], *20191750* [10]
- **horseradish** *20162922* [2], *20131354* [4], *20191747* [10]
- **leeks** *20162925* [2], *20131357* [4], *20181690* [8], *20191740* [10]
- **medlar** *20162930* [2], *20191742* [10]
- **miscanthus** *20162931* [2], *20131430* [4], *20191739* [10]
- **nectarines** *20162930* [2], *20191742* [10]
- **onion sets** *20131357* [4]
- **ornamental plant production** *20162927* [2], *20131431* [4], *20191746* [10]
- **parsley root** *20162922* [2], *20131354* [4], *20191747* [10]
- **parsnips** *20162922* [2], *20131354* [4], *20191747* [10]
- **peaches** *20162930* [2], *20191742* [10]
- **pears** *20162930* [2], *20191742* [10]
- **plums** *20162930* [2], *20191742* [10]
- **poppies** *20180615* [4], *20191743* [10]
- **poppies for morphine production** *20162934* [2], *20131385* [4], *20191744* [10]
- **poppies grown for seed production** *20180615* [4], *20191743* [10]
- **quinces** *20162930* [2], *20191742* [10]
- **rye** *20162929* [2], *20131429* [4], *20191754* [10]
- **salsify** *20162922* [2], *20131354* [4], *20191747* [10]
- **shallots** *20162925* [2], *20131357* [4], *20181690* [8], *20191740* [10]
- **spring barley** *20131373* [4], *20191745* [10]
- **spring field beans** *20162932* [2], *20131356* [4], *20191753* [10]
- **spring onions** *20162925* [2], *20131357* [4], *20181690* [8], *20191740* [10]
- **top fruit** *20131433* [4]
- **triticale** *20162929* [2], *20131429* [4]
- **vining peas** *20191041* [4], *20191749* [10]
- **walnuts** *20162930* [2], *20191742* [10]
- **winter field beans** *20162932* [2], *20131356* [4], *20191753* [10]
- **winter linseed** *20162928* [2], *20131428* [4], *20181692* [8], *20191752* [10]

Approval information
- Prosulfocarb included in Annex I under EC Regulation 1107/2009
- Accepted by BBPA for use on malting barley

Efficacy guidance
- Best results obtained in cereals from treatment of crops in a firm moist seedbed, free from clods
- Pre-emergence use will reduce blackgrass populations but the product should only be used against this weed as part of a management strategy involving sequences with products of alternative modes of action
- Always follow WRAG guidelines for preventing and managing herbicide resistant weeds. See Section 5 for more information
- Must be applied in 200L of water [9]

Restrictions
- Maximum number of treatments 1 per crop for winter barley, winter wheat and potatoes
- Do not apply to crops under stress from any cause. Transient yellowing can occur from which recovery is complete
- Winter cereals must be covered by 3 cm of settled soil
- Do not apply by hand held equipment

Crop-specific information
- Latest use: at emergence (soil rising over emerging potato shoots) for potatoes, up to and including early tillering (GS 21) for winter barley, winter wheat
- When applied pre-emergence to cereals crop emergence may occasionally be slowed down but yield is not affected
- For potatoes, complete ridge formation before application and do not disturb treated soil afterwards
- Peas are severely damaged or killed.

FOR FULL CONDITIONS OF USE ALWAYS READ THE PRODUCT LABEL

Following crops guidance
- Do not sow field or broad beans within 12 mth of treatment
- In the event of failure of a treated cereal crop, winter wheat, winter barley or transplanted brassicas may be re-sown immediately. In the following spring sunflowers, maize, flax, spring cereals, peas, oilseed rape or soya beans may be sown without ploughing, and carrots, lettuce, onions, sugar beet or potatoes may be sown or planted after ploughing

Environmental safety
- Dangerous for the environment
- Very toxic to aquatic organisms
- LERAP Category B

Hazard classification and safety precautions
Hazard Irritant, Dangerous for the environment
Transport code 9
Packaging group III
UN Number 3082
Risk phrases H304, H315, H317 [1-4, 8, 10, 11]; H319 [1-5, 7-11]; R38, R43, R50, R53a [6]
Operator protection A, C, H; U02a, U05a, U08, U14, U15, U20b
Environmental protection E15b, E16a, E16b, E38 [1-11]; H410 [1-5, 7-11]
Storage and disposal D01, D02, D05, D09a, D10c, D12a
Medical advice M05a

338   prosulfuron

A contact and residual sulfonyl urea herbicide for use in maize crops
HRAC mode of action code: B

See also bromoxynil + prosulfuron
dicamba + prosulfuron

Products
1	Clayton Kibo	Clayton	75% w/w	WG	15822
2	Peak	Syngenta	75% w/w	WG	15521

Uses
- Annual dicotyledons in *forage maize*, *grain maize* [1, 2]; *game cover* (off-label), *millet* (off-label) [2]
- Black bindweed in *forage maize*, *grain maize* [1, 2]
- Chickweed in *forage maize*, *grain maize* [1, 2]
- Fumitory in *forage maize*, *grain maize* [1, 2]
- Groundsel in *forage maize*, *grain maize* [1, 2]
- Knotgrass in *forage maize*, *grain maize* [1, 2]
- Mayweeds in *forage maize*, *grain maize* [1, 2]
- Redshank in *forage maize*, *grain maize* [1, 2]
- Scarlet pimpernel in *forage maize*, *grain maize* [1, 2]
- Shepherd's purse in *forage maize*, *grain maize* [1, 2]
- Sowthistle in *forage maize*, *grain maize* [1, 2]

Extension of Authorisation for Minor Use (EAMUs)
- *game cover* 20120906 [2]
- *millet* 20172107 [2]

Approval information
- Prosulfuron included in Annex I under EC Regulation 1107/2009

Efficacy guidance
- For optimum efficacy, apply when weeds are at the 2 - 4 leaf stage

Restrictions
- Do not apply to forage maize or grain maize grown for seed production
- Do not apply with organo-phosphate insecticides

SEE SECTION 3 FOR PRODUCTS ALSO REGISTERED

- Do not apply by hand held equipment or in water volumes less than recommended
- A maximum total dose of 15g of prosulfuron per hectare may only be applied every third year on the same field

Following crops guidance
- In the event of crop failure, wait 4 weeks after treatment and then re-sow
- After normal harvest wheat, barley and winter beans may be sown as a following crop in the autumn once the soil has been ploughed to 15 cms. In spring, wheat, barley, peas or beans may be sown but do not sow any other crop at this time.

Environmental safety
- LERAP Category B

Hazard classification and safety precautions
    Hazard Harmful, Dangerous for the environment, Harmful if swallowed, Very toxic to aquatic organisms
    Transport code 9
    Packaging group III
    UN Number 3077
    Operator protection A, H; U05a, U14, U20c
    Environmental protection E15b, E16a, E38, H410
    Storage and disposal D01, D02, D05, D09b, D10c, D12a

339   prothioconazole

A systemic, protectant and curative triazole fungicide
FRAC mode of action code: 3

See also benzovindiflupyr + prothioconazole
        bixafen + fluopyram + prothioconazole
        bixafen + fluoxastrobin + prothioconazole
        bixafen + prothioconazole
        bixafen + prothioconazole + spiroxamine
        bixafen + prothioconazole + tebuconazole
        clothianidin + prothioconazole
        clothianidin + prothioconazole + tebuconazole + triazoxide
        fluopyram + prothioconazole
        fluopyram + prothioconazole + tebuconazole
        fluoxastrobin + prothioconazole
        fluoxastrobin + prothioconazole + tebuconazole
        fluoxastrobin + prothioconazole + trifloxystrobin
        isopyrazam + prothioconazole

Products

1	Decoy 250EC	BASF	250 g/l	EC	18639
2	Proline 275	Bayer CropScience	275 g/l	EC	14790
3	Rudis	Bayer CropScience	480 g/l	SC	14122
4	Traciafin Plus	Clayton	250 g/l	EC	19212

Uses
- Alternaria in *broccoli, brussels sprouts, cabbages, calabrese, cauliflowers, red beet* (off-label) [3]
- Alternaria blight in *carrots, parsnips, swedes, turnips* [3]
- Brown rust in *durum wheat* [1, 4]; *spring barley, spring wheat, winter barley, winter rye, winter wheat* [1, 2, 4]; *triticale* [4]
- Cercospora leaf spot in *red beet* (off-label) [3]
- Crown rust in *spring oats, winter oats* [1, 2, 4]
- Disease control in *linseed* (off-label), *mustard* (off-label) [2]
- Eyespot in *durum wheat* [1, 4]; *spring barley, spring wheat, winter barley, winter rye, winter wheat* [1, 2, 4]; *spring oats, winter oats* [1, 2]; *triticale* [4]

FOR FULL CONDITIONS OF USE ALWAYS READ THE PRODUCT LABEL

- Glume blotch in **durum wheat** [1, 4]; **spring wheat**, **winter wheat** [1, 2, 4]; **triticale** [4]
- Kabatiella lini in **linseed** *(off-label)* [2]
- Late ear diseases in **durum wheat** [1, 4]; **spring barley**, **spring wheat**, **winter barley**, **winter wheat** [1, 2, 4]; **triticale** [4]
- Leaf blotch in **leeks** *(useful reduction)* [3]
- Light leaf spot in **broccoli**, **brussels sprouts**, **cabbages**, **calabrese**, **cauliflowers** [3]; **winter oilseed rape** [2]
- Net blotch in **spring barley**, **winter barley** [1, 2, 4]
- Phoma in **borage** *(off-label)*, **spring corn gromwell** *(off-label)*, **winter corn gromwell** *(off-label)*, **winter oilseed rape** [2]; **broccoli**, **brussels sprouts**, **cabbages**, **calabrese**, **cauliflowers**, **red beet** *(off-label)* [3]
- Powdery mildew in **borage** *(off-label)*, **spring corn gromwell** *(off-label)*, **winter corn gromwell** *(off-label)* [2]; **broccoli**, **brussels sprouts**, **cabbages**, **calabrese**, **carrots**, **cauliflowers**, **parsnips**, **red beet** *(off-label)*, **swedes**, **turnips** [3]; **durum wheat** [1, 4]; **spring barley**, **spring oats**, **spring wheat**, **winter barley**, **winter oats**, **winter rye**, **winter wheat** [1, 2, 4]; **triticale** [4]
- Purple blotch in **leeks** [3]
- Rhynchosporium in **spring barley**, **winter barley**, **winter rye** [1, 2, 4]; **triticale** [4]
- Ring spot in **broccoli**, **brussels sprouts**, **cabbages**, **calabrese**, **cauliflowers** [3]
- Rust in **leeks** [3]
- Sclerotinia in **borage** *(off-label)*, **spring corn gromwell** *(off-label)*, **winter corn gromwell** *(off-label)* [2]
- Sclerotinia rot in **carrots**, **parsnips** [3]
- Sclerotinia stem rot in **spring oilseed rape** [4]; **winter oilseed rape** [1, 2, 4]
- Septoria leaf blotch in **durum wheat** [1, 4]; **spring wheat**, **winter wheat** [1, 2, 4]; **triticale** [4]
- Septoria linicola in **linseed** *(off-label)* [2]
- Stem canker in **winter oilseed rape** [2]
- Stemphylium in **leeks** *(useful reduction)* [3]
- Tan spot in **durum wheat** [1, 4]; **spring wheat**, **winter wheat** [1, 2, 4]
- Yellow rust in **durum wheat** [1, 4]; **spring wheat**, **winter barley**, **winter wheat** [1, 2, 4]; **triticale** [4]

Extension of Authorisation for Minor Use (EAMUs)
- **borage** *20193125* [2]
- **linseed** *20162863* [2]
- **mustard** *20111003* [2]
- **red beet** *20193126* [3]
- **spring corn gromwell** *20193125* [2]
- **winter corn gromwell** *20193125* [2]

Approval information
- Accepted by BBPA for use on malting barley
- Prothioconazole included in Annex I under EC Regulation 1107/2009

Efficacy guidance
- Seed treatments must be applied by manufacturer's recommended treatment application equipment
- Treated cereal seed should preferably be drilled in the same season
- Follow-up treatments will be needed later in the season to give protection against air-borne and splash-borne diseases
- Best results on cereal foliar diseases obtained from treatment at early stages of disease development. Further treatment may be needed if disease attack is prolonged [2]
- Foliar applications to established infections of any disease are likely to be less effective [2]
- Best control of cereal ear diseases obtained by treatment during ear emergence [2]
- Treat oilseed rape at early to full flower [2]
- Prothioconazole is a DMI fungicide. Resistance to some DMI fungicides has been identified in Septoria leaf blotch which may seriously affect performance of some products. For further advice contact a specialist advisor and visit the Fungicide Resistance Action Group (FRAG)-UK website

SEE SECTION 3 FOR PRODUCTS ALSO REGISTERED

Restrictions

- Maximum number of seed treatments one per batch
- Maximum total dose equivalent to two full dose treatments on barley, oats and oilseed rape, three full dose treatments on wheat and rye [2]
- Seed treatment must be fully re-dispersed and homogeneous before use
- Do not use on seed with more than 16% moisture content, or on sprouted, cracked or skinned seed
- All seed batches should be tested to ensure they are suitable for treatment
- Treated winter barley seed must be used within the season of treatment; treated winter wheat seed should preferably be drilled in the same season
- Do not make repeated treatments of the product alone to the same crop against pathogens such as powdery mildew. Use tank mixtures or alternate with fungicides having a different mode of action

Crop-specific information

- Latest use: pre-drilling for seed treatments; before grain milky ripe for foliar sprays on winter rye, winter wheat; beginning of flowering for barley, oats [2]
- HI 56 d for winter oilseed rape [2]

Environmental safety

- Dangerous for the environment [2]
- Toxic to aquatic organisms [2]
- Harmful to aquatic organisms
- LERAP Category B

Hazard classification and safety precautions

Hazard Irritant [1, 2, 4]; Dangerous for the environment, Very toxic to aquatic organisms [1-4]
Transport code 9
Packaging group III
UN Number 3082
Risk phrases H319 [1, 2, 4]; H335 [2, 4]
Operator protection A, H [1-4]; C [1, 2, 4]; U05a, U20b [1-4]; U09b [1, 2, 4]
Environmental protection E15a [1, 2, 4]; E15b [3]; E16a, E34, E38, H410 [1-4]
Storage and disposal D01, D02, D09a, D10b, D12a [1-4]; D05 [1, 2, 4]
Medical advice M03

340 prothioconazole + spiroxamine

A broad spectrum fungicide mixture for cereals
FRAC mode of action code: 3 + 5

See also spiroxamine

Products

Helix	Bayer CropScience	160:300 g/l	EC	18289

Uses

- Brown rust in *durum wheat, spring barley, spring rye, spring wheat, triticale, winter barley, winter rye, winter wheat*
- Crown rust in *spring oats, winter oats*
- Eyespot in *durum wheat, spring barley, spring oats, spring rye, spring wheat, triticale, winter barley, winter oats, winter rye, winter wheat*
- Glume blotch in *durum wheat, spring wheat, triticale, winter wheat*
- Late ear diseases in *durum wheat, spring rye, spring wheat, triticale, winter rye, winter wheat*
- Net blotch in *spring barley, winter barley*
- Powdery mildew in *durum wheat, spring barley, spring oats, spring rye, spring wheat, triticale, winter barley, winter oats, winter rye, winter wheat*
- Rhynchosporium in *spring barley, spring rye, winter barley, winter rye*
- Septoria leaf blotch in *durum wheat, spring wheat, triticale, winter wheat*
- Tan spot in *durum wheat, spring wheat, triticale, winter wheat*

FOR FULL CONDITIONS OF USE ALWAYS READ THE PRODUCT LABEL

- Yellow rust in *durum wheat*, *spring barley*, *spring wheat*, *triticale*, *winter barley*, *winter wheat*

Approval information
- Prothioconazole and spiroxamine included in Annex I under EC Regulation 1107/2009
- Accepted by BBPA for use on malting barley

Efficacy guidance
- Best results obtained from treatment at early stages of disease development. Further treatment may be needed if disease attack is prolonged
- Applications to established infections of any disease are likely to be less effective
- Best control of cereal ear diseases obtained by treatment during ear emergence
- Prothioconazole is a DMI fungicide. Resistance to some DMI fungicides has been identified in Septoria leaf blotch which may seriously affect performance of some products. For further advice contact a specialist advisor and visit the Fungicide Resistance Action Group (FRAG)-UK website

Restrictions
- Maximum total dose equivalent to two full dose treatments on barley and oats and three full dose treatments on wheat and rye
- Maintain an aquatic buffer zone of 6 m [1]

Crop-specific information
- Latest use: before grain watery ripe for rye, oats, winter wheat; up to beginning of anthesis for barley

Environmental safety
- Dangerous for the environment
- Very toxic to aquatic organisms
- Buffer zone requirement 6m [1]
- LERAP Category B

Hazard classification and safety precautions
Hazard Harmful, Dangerous for the environment, Harmful if swallowed, Harmful if inhaled, Very toxic to aquatic organisms
Transport code 9
Packaging group III
UN Number 3082
Risk phrases H315, H319, H335
Operator protection A, C, H; U05a, U09b, U11, U19a, U20b
Environmental protection E15a, E16a, E34, E38
Storage and disposal D01, D02, D05, D09a, D10b, D12a
Medical advice M03

341   prothioconazole + spiroxamine + tebuconazole

A broad spectrum fungicide mixture for cereals
FRAC mode of action code: 3 + 5 + 3

See also spiroxamine
tebuconazole

Products

Cello	Bayer CropScience	100:250:100 g/l	EC	18290

Uses
- Brown rust in *durum wheat*, *spring barley*, *spring wheat*, *triticale*, *winter barley*, *winter rye*, *winter wheat*
- Crown rust in *spring oats*, *winter oats*

SEE SECTION 3 FOR PRODUCTS ALSO REGISTERED

- Eyespot in **durum wheat** *(reduction)*, **spring barley** *(reduction)*, **spring oats**, **spring wheat** *(reduction)*, **triticale** *(reduction)*, **winter barley** *(reduction)*, **winter oats**, **winter rye** *(reduction)*, **winter wheat** *(reduction)*
- Glume blotch in **durum wheat**, **spring wheat**, **triticale**, **winter wheat**
- Late ear diseases in **durum wheat**, **spring barley**, **spring wheat**, **triticale**, **winter barley**, **winter wheat**
- Net blotch in **spring barley**, **winter barley**
- Powdery mildew in **durum wheat**, **spring barley**, **spring oats**, **spring wheat**, **triticale**, **winter barley**, **winter oats**, **winter rye**, **winter wheat**
- Rhynchosporium in **spring barley**, **winter barley**, **winter rye**
- Septoria leaf blotch in **durum wheat**, **spring wheat**, **triticale**, **winter wheat**
- Yellow rust in **durum wheat**, **spring barley**, **spring wheat**, **triticale**, **winter barley**, **winter wheat**

Approval information
- Prothioconazole, spiroxamine and tebuconazole included in Annex I under EC Regulation 1107/2009

Efficacy guidance
- Best results obtained from treatment at early stages of disease development. Further treatment may be needed if disease attack is prolonged
- Applications to established infections of any disease are likely to be less effective
- Best control of cereal ear diseases obtained by treatment during ear emergence
- Prothioconazole and tebuconazole are DMI fungicides. Resistance to some DMI fungicides has been identified in Septoria leaf blotch which may seriously affect performance of some products. For further advice contact a specialist advisor and visit the Fungicide Resistance Action Group (FRAG)-UK website

Restrictions
- Maximum total dose equivalent to two full dose treatments
- Newer authorisations for tebuconazole products require application to cereals only after GS 30 and applications to oilseed rape and linseed after GS 20 - check label
- Maintain an interval of at least 21 days between applications

Crop-specific information
- Latest use: before grain milky ripe stage for rye, wheat; up to beginning of anthesis for barley and oats

Environmental safety
- Dangerous for the environment
- Very toxic to aquatic organisms
- LERAP Category B

Hazard classification and safety precautions
Hazard Harmful, Dangerous for the environment, Harmful if inhaled, Very toxic to aquatic organisms
Transport code 9
Packaging group III
UN Number 3082
Risk phrases H315, H317, H319, H361
Operator protection A, C, H; U05a, U09b, U11, U14, U19a, U20b
Environmental protection E15a, E16a, E34, E38, H410
Storage and disposal D01, D02, D05, D09a, D10b, D12a
Medical advice M03

342   prothioconazole + tebuconazole

A triazole fungicide mixture for cereals
FRAC mode of action code: 3 + 3

See also tebuconazole

Products

1	Corinth	Bayer CropScience	80:160 g/l	EC	16742
2	Harvest Proteb	Harvest	125:125 g/l	EC	17692
3	Kestrel	Bayer CropScience	160:80 g/l	EC	16751
4	Prosaro	Bayer CropScience	125:125 g/l	EC	16732
5	Protefin	Clayton	125:125 g/l	EC	19126
6	Redigo Pro	Bayer CropScience	150:20 g/l	FS	15145

Uses

- Black stem rust in **grass seed crops** *(off-label)* [4]
- Blue mould in **durum wheat**, **rye**, **spring oats**, **spring wheat**, **triticale**, **winter oats**, **winter wheat** [6]
- Brown rust in **spring barley**, **winter barley**, **winter rye**, **winter wheat** [2-5]; **spring wheat** [2-4]
- Bunt in **durum wheat**, **rye**, **spring oats**, **spring wheat**, **triticale**, **winter oats**, **winter wheat** [6]
- Covered smut in **spring barley**, **spring oats**, **winter barley**, **winter oats** [6]
- Crown rust in **spring oats**, **winter oats** [2-5]
- Ergot in **rye** *(qualified minor use)* [6]
- Eyespot in **spring barley**, **winter barley**, **winter rye**, **winter wheat** [3]; **spring barley** *(reduction)*, **winter barley** *(reduction)*, **winter rye** *(reduction)*, **winter wheat** *(reduction)* [2, 4, 5]; **spring oats**, **winter oats** [2-5]; **spring wheat** [2-4]
- Fusarium ear blight in **spring barley**, **spring wheat**, **winter barley**, **winter wheat** [3]
- Fusarium seedling blight in **durum wheat**, **rye**, **spring barley**, **spring oats**, **spring wheat**, **triticale**, **winter barley**, **winter oats**, **winter wheat** [6]
- Glume blotch in **spring wheat** [2-4]; **winter wheat** [2-5]
- Late ear diseases in **spring barley**, **winter barley**, **winter wheat** [2, 4, 5]; **spring wheat** [2, 4]
- Leaf spot in **grass seed crops** *(off-label)* [4]
- Leaf stripe in **spring barley**, **winter barley** [6]
- Light leaf spot in **oilseed rape** [3]; **oilseed rape** *(moderate control only)*, **winter oilseed rape** *(moderate control only)* [1]; **spring oilseed rape**, **winter oilseed rape** [2, 4, 5]
- Loose smut in **durum wheat**, **rye**, **spring barley**, **spring oats**, **spring wheat**, **triticale**, **winter barley**, **winter oats**, **winter wheat** [6]
- Net blotch in **spring barley**, **winter barley** [2-5]
- Phoma in **spring oilseed rape**, **winter oilseed rape** [2, 4, 5]
- Phoma leaf spot in **oilseed rape** [1, 3]; **winter oilseed rape** [1]
- Powdery mildew in **grass seed crops** *(off-label)* [4]; **spring barley**, **spring oats**, **winter barley**, **winter oats**, **winter rye**, **winter wheat** [2-5]; **spring wheat** [2-4]
- Rhynchosporium in **grass seed crops** *(off-label)* [4]; **spring barley**, **winter barley**, **winter rye** [2-5]
- Sclerotinia stem rot in **oilseed rape** [1, 3]; **spring oilseed rape** [2, 4, 5]; **winter oilseed rape** [1, 2, 4, 5]
- Septoria leaf blotch in **spring wheat** [2-4]; **winter wheat** [2-5]
- Sooty moulds in **spring barley**, **spring wheat**, **winter barley**, **winter wheat** [3]
- Stem canker in **oilseed rape**, **winter oilseed rape** [1]
- Tan spot in **spring wheat** [2-4]; **winter wheat** [2-5]
- Yellow rust in **spring barley**, **winter barley**, **winter wheat** [2-5]; **spring wheat** [2-4]

Extension of Authorisation for Minor Use (EAMUs)

- **grass seed crops** *20160083* [4]

Approval information

- Prothioconazole and tebuconazole included in Annex I under EC Regulation 1107/2009
- Accepted by BBPA for use on malting barley

SEE SECTION 3 FOR PRODUCTS ALSO REGISTERED

Efficacy guidance
- Best results on cereal foliar diseases obtained from treatment at early stages of disease development. Further treatment may be needed if disease attack is prolonged
- On oilseed rape apply a protective treatment in autumn/winter for Phoma followed by a further spray in early spring from the onset of stem elongation, if necessary. For control of Sclerotinia apply at early to full flower
- Applications to established infections of any disease are likely to be less effective
- Best control of cereal ear diseases obtained by treatment during ear emergence
- Prothioconazole and tebuconazole are DMI fungicides. Resistance to some DMI fungicides has been identified in Septoria leaf blotch which may seriously affect performance of some products. For further advice contact a specialist advisor and visit the Fungicide Resistance Action Group (FRAG)-UK website
- Treated seed should be drilled to a depth of 40mm. Ensure no seed is left on the soil surface . After drilling if conditions allow, field should be harrowed and then rolled to ensure good incorporation [6]

Restrictions
- Maximum total dose equivalent to two full dose treatments on barley, oats, oilseed rape; three full dose treatments on wheat and rye
- Newer authorisations for tebuconazole products require application to cereals only after GS 30 and applications to oilseed rape and linseed after GS20 - check label
- Treated seed should not be left on the soil surface. Bury or remove spillages [6]

Crop-specific information
- Latest use: before grain milky ripe for rye, wheat; beginning of flowering for barley, oats
- HI 56 d for oilseed rape

Environmental safety
- Dangerous for the environment
- Toxic to aquatic organisms
- LERAP Category B [1-5]

Hazard classification and safety precautions
Hazard Harmful [1, 3]; Irritant, Very toxic to aquatic organisms [2, 4, 5]; Dangerous for the environment [1-5]; Harmful if inhaled [1]

Transport code 9

Packaging group III

UN Number 3082

Risk phrases H315, H361 [1-5]; H317 [1, 3]; H319 [1, 2, 4, 5]; H335 [2, 4, 5]

Operator protection A, H [1-6]; C [1, 3]; D [6]; U05a [1-6]; U09b, U20b [2-5]; U11, U20c [1]; U19a [1, 2, 4, 5]; U20a, U20e [6]

Environmental protection E15a, E16a [1-5]; E15b [6]; E22b [1]; E22c [2-5]; E34, E38 [1-6]; H410 [2, 4-6]; H411 [1, 3]

Storage and disposal D01, D02, D10b [1-5]; D05, D09a, D12a [1-6]; D10d, D21 [6]

Treated seed S01, S02, S03, S04a, S04b, S05, S07, S08 [6]

Medical advice M03 [1-5]

343   prothioconazole + trifloxystrobin

A triazole and strobilurin fungicide mixture for cereals
FRAC mode of action code: 3 + 11

See also trifloxystrobin

Products

1 Mobius	Bayer CropScience	175:150 g/l	SC	13395
2 Zephyr	Bayer CropScience	175:88 g/l	SC	13174

Uses
- Brown rust in *durum wheat*, *rye*, *spring barley*, *triticale*, *winter barley*, *winter wheat* [1, 2]; *spring wheat* [2]

- Ear diseases in *durum wheat, rye, triticale, winter wheat* [1]
- Eyespot in *durum wheat, rye, triticale, winter wheat* [1, 2]; *spring barley (reduction)*, *spring wheat, winter barley (reduction)* [2]; *spring barley (reduction in severity)*, *winter barley (reduction in severity)* [1]
- Glume blotch in *durum wheat, rye, triticale, winter wheat* [1, 2]; *spring wheat* [2]
- Net blotch in *spring barley, winter barley* [1, 2]
- Powdery mildew in *durum wheat, rye, spring barley, triticale, winter barley, winter wheat* [1, 2]; *spring wheat* [2]
- Rhynchosporium in *spring barley, winter barley* [1, 2]
- Septoria leaf blotch in *durum wheat, rye, triticale, winter wheat* [1, 2]; *spring wheat* [2]
- Yellow rust in *durum wheat, rye, spring barley, triticale, winter barley, winter wheat* [1, 2]; *spring wheat* [2]

Approval information
- Prothioconazole and trifloxystrobin included in Annex I under EC Regulation 1107/2009
- Accepted by BBPA for use on malting barley

Efficacy guidance
- Best results obtained from treatment at early stages of disease development. Further treatment may be needed if disease attack is prolonged
- Applications to established infections of any disease are likely to be less effective
- Best control of cereal ear diseases obtained by treatment during ear emergence
- Prothioconazole is a DMI fungicide. Resistance to some DMI fungicides has been identified in Septoria leaf blotch which may seriously affect performance of some products. For further advice contact a specialist advisor and visit the Fungicide Resistance Action Group (FRAG)-UK website
- Trifloxystrobin is a member of the QoI cross resistance group. Product should be used preventatively and not relied on for its curative potential
- Use product as part of an Integrated Crop Management strategy incorporating other methods of control, including where appropriate other fungicides with a different mode of action. Do not apply more than two foliar applications of QoI containing products to any cereal crop
- There is a significant risk of widespread resistance occurring in *Septoria tritici* populations in UK. Failure to follow resistance management action may result in reduced levels of disease control
- Strains of wheat and barley powdery mildew resistant to QoIs are common in the UK. Control of wheat mildew can only be relied on from the triazole component
- Where specific control of wheat mildew is required this should be achieved through a programme of measures including products recommended for the control of mildew that contain a fungicide from a different cross-resistance group and applied at a dose that will give robust control

Restrictions
- Maximum total dose equivalent to two full dose treatments
- To avoid the build-up of resistance do not apply products containing QoI fungicides more than twice per cereal crop

Crop-specific information
- Latest use: before grain milky ripe for wheat; beginning of flowering for barley

Environmental safety
- Dangerous for the environment
- Very toxic to aquatic organisms
- LERAP Category B

Hazard classification and safety precautions
Hazard Irritant, Dangerous for the environment, Very toxic to aquatic organisms
Transport code 9
Packaging group III
UN Number 3082
Risk phrases H317
Operator protection A, C, H; U05a, U09b, U19a [1, 2]; U20b [2]; U20c [1]

SEE SECTION 3 FOR PRODUCTS ALSO REGISTERED

Environmental protection E15a, E16a, E38, H410 [1, 2]; E34 [2]
Storage and disposal D01, D02, D09a, D10c, D12a
Medical advice M03

344   Pseudomonas SP (DSMZ 13134)

For use on potato seed

Products

Proradix	Fargro	0.535% w/w	WP	18115

Uses

* Rhizoctonia in *potatoes* (reduction)

Efficacy guidance

* For the reduction of root rot caused by Rhizoctonia solani in potato crops. It may be applied as spray during sowing of potato or as seed treatment prior to sowing
* Can be stored for 12 months if stored deep frozen or 3 months if stored at room temperature

Restrictions

* Consult processors before using on crops destined for processing

Hazard classification and safety precautions
UN Number N/C
Operator protection A, H; U05a, U19a, U20b
Storage and disposal D01, D12c
Medical advice M04a

345   pyraclostrobin

A protectant and curative strobilurin fungicide for cereals
FRAC mode of action code: 11

See also boscalid + epoxiconazole + pyraclostrobin
boscalid + pyraclostrobin
dimethomorph + pyraclostrobin
dithianon + pyraclostrobin
epoxiconazole + fluxapyroxad + pyraclostrobin
epoxiconazole + kresoxim-methyl + pyraclostrobin
epoxiconazole + pyraclostrobin
fluxapyroxad + pyraclostrobin

Products

1	Comet 200	BASF	200 g/l	EC	12639
2	Eland	Rigby Taylor	20% w/w	WG	14549
3	Flyer 200	BASF	200 g/l	EC	17293
4	Halley	AgChem Access	200 g/l	EC	15916
5	Tucana	BASF	250 g/l	EC	10899
6	Vanguard	Sherriff Amenity	20% w/w	WG	13838
7	Vivid	BASF	250 g/l	EC	10898

Uses

* Brown rust in *spring barley, spring wheat, winter barley, winter wheat* [1, 3-5, 7]
* Crown rust in *spring oats, winter oats* [1, 3-5, 7]
* Disease control in *forage maize, grain maize* [3]
* Dollar spot in *managed amenity turf* (useful reduction) [2, 6]
* Eyespot in *forage maize, grain maize* [1, 3]; *forage maize* (off-label) [7]
* Foliar disease control in *ornamental plant production* (off-label) [7]
* Fusarium patch in *managed amenity turf* (moderate control) [2]; *managed amenity turf* (moderate control only) [6]

FOR FULL CONDITIONS OF USE ALWAYS READ THE PRODUCT LABEL

- Glume blotch in **spring wheat**, **winter wheat** [1, 3-5, 7]
- Net blotch in **spring barley**, **winter barley** [1, 3-5, 7]
- Northern leaf blight in **forage maize**, **grain maize** [1]; **forage maize** *(moderate control)*, **grain maize** *(moderate control)* [3]
- Red thread in **managed amenity turf** [2, 6]
- Rhynchosporium in **spring barley** *(moderate control)*, **winter barley** *(moderate control)* [1, 3-5, 7]
- Rust in **grass seed crops** *(off-label)*, **rye** *(off-label)*, **triticale** *(off-label)* [5, 7]
- Septoria leaf blotch in **spring wheat**, **winter wheat** [1, 3-5, 7]
- Yellow rust in **spring barley**, **spring wheat**, **winter barley**, **winter wheat** [1, 3-5, 7]

Extension of Authorisation for Minor Use (EAMUs)
- **forage maize** *20131684* [7]
- **grass seed crops** *20193266* [5], *20193251* [7]
- **ornamental plant production** *20082884* [7]
- **rye** *20193266* [5], *20193251* [7]
- **triticale** *20193266* [5], *20193251* [7]

Approval information
- Pyraclostrobin included in Annex I under EC Regulation 1107/2009
- Accepted by BBPA for use on malting barley (before ear emergence only) and on hops

Efficacy guidance
- For best results apply at the start of disease attack on cereals [1, 5, 7]
- For Fusarium Patch treat early as severe damage to turf can occur once the disease is established [2, 6]
- Regular turf aeration, appropriate scarification and judicious use of nitrogenous fertiliser will assist the control of Fusarium Patch [2, 6]
- Best results on Septoria glume blotch achieved when used as a protective treatment and against Septoria leaf blotch when treated in the latent phase [1, 5, 7]
- Yield response may be obtained in the absence of visual disease symptoms [1, 5, 7]
- Pyraclostrobin is a member of the QoI cross resistance group. Product should be used preventatively and not relied on for its curative potential
- Use product as part of an Integrated Crop Management strategy incorporating other methods of control, including where appropriate other fungicides with a different mode of action. Do not apply more than two foliar applications of QoI containing products to any cereal crop or to grass
- There is a significant risk of widespread resistance occurring in *Septoria tritici* populations in UK. Failure to follow resistance management action may result in reduced levels of disease control [1, 5, 7]
- On cereal crops product must always be used in mixture with another product, recommended for control of the same target disease, that contains a fungicide from a different cross resistance group and is applied at a dose that will give robust control [1, 5, 7]
- Late application to maize crops can give a worthwhile increase in biomass

Restrictions
- Maximum number of treatments 2 per crop on cereals [1, 5, 7]
- Maximum total dose on turf equivalent to two full dose treatments [2, 6]
- Do not apply during drought conditions or to frozen turf [2, 6]
- The maximum concentration must not exceed 1.25 litre of product per 200 litres water

Crop-specific information
- Latest use: before grain watery ripe (GS 71) for wheat; up to and including emergence of ear just complete (GS 59) for barley and oats [1, 5, 7]
- Avoid applying to turf immediately after cutting or 48 h before mowing [2, 6]

Environmental safety
- Dangerous for the environment
- Very toxic to aquatic organisms
- LERAP Category B

SECTION 2

SEE SECTION 3 FOR PRODUCTS ALSO REGISTERED

Hazard classification and safety precautions

Hazard Harmful, Dangerous for the environment, Harmful if inhaled, Very toxic to aquatic organisms [1-7]; Harmful if swallowed [1, 3-5, 7]
Transport code 6.1 [1, 3-5, 7]; 9 [2, 6]
Packaging group III
UN Number 2902 [1, 3-5, 7]; 3077 [2, 6]
Risk phrases H304, H319 [1, 3, 4]; H315, H317 [1, 3-5, 7]; H335 [5, 7]
Operator protection A [1-7]; D [2, 6]; U05a, U14, U20b [1, 3-5, 7]
Environmental protection E15a [6]; E15b, E16a, E38 [1-7]; E16b [1-5, 7]; E34, H410 [1, 3-5, 7]
Storage and disposal D01, D02 [1, 3-5, 7]; D08, D09a, D10c, D12a [1-7]
Medical advice M03, M05a [1, 3-5, 7]

346   pyraflufen-ethyl

A phenylpyrazole herbicide for potatoes
HRAC mode of action code: E

See also glyphosate + pyraflufen-ethyl

Products

1 Gozai	Belchim	26.5 g/l	EC	17381
2 Kabuki	Belchim	26.5 g/l	EC	18187

Uses

- Annual dicotyledons in *hops (off-label)* [1]; *potatoes* [1, 2]
- Desiccation in *hops (off-label)* [1]; *potatoes* [1, 2]

Extension of Authorisation for Minor Use (EAMUs)

- *hops 20171023* [1]

Approval information

- Pyraflufen-ethyl included in Annex I under EC Regulation 1107/2009

Efficacy guidance

- Use with a methylated vegetable oil adjuvant

Following crops guidance

- After cultivation to at least 20 cms winter cereals can be sown as a following crop but the safety to following broad-leaved crops has not yet been established.

Environmental safety

- Buffer zone requirement 20 m in hops [1]
- LERAP Category B

Hazard classification and safety precautions

Hazard Harmful, Dangerous for the environment, Harmful if inhaled, Very toxic to aquatic organisms
Transport code 9
Packaging group III
UN Number 3082
Risk phrases H304, H315, H317, H318
Operator protection A, C, H; U05a, U09a, U11, U14, U15
Environmental protection E15b, E16a, E34, H410
Storage and disposal D01, D02, D05, D09a, D10a, D12a, D12b
Medical advice M03, M05b

347 pyrethrins

A non-persistent, contact acting insecticide extracted from Pyrethrum
IRAC mode of action code: 3

Products

1 Dairy Fly Spray	B H & B	0.75 g/l	AL	H5579
2 Spruzit	Certis	4.59 g/l	EC	18434

Uses

- Aphids in **asparagus, beans without pods (fresh), broad beans, bulb onions, cabbages, carrots, celeriac, cucumbers, dwarf beans, edible podded peas, french beans, garlic, herbs (see appendix 6), horseradish, jerusalem artichokes, kohlrabi, leeks, lettuce, parsley, parsnips, protected cabbages, protected lettuce, protected ornamentals, protected tomatoes, radishes, red beet, runner beans, salsify, shallots, soya beans, spinach, swedes, turnips, vining peas** [2]
- Caterpillars in **asparagus, beans without pods (fresh), broad beans, bulb onions, cabbages, carrots, celeriac, cucumbers, dwarf beans, edible podded peas, french beans, garlic, herbs (see appendix 6), horseradish, jerusalem artichokes, kohlrabi, leeks, lettuce, parsley, parsnips, protected cabbages, protected lettuce, protected ornamentals, protected tomatoes, radishes, red beet, runner beans, salsify, shallots, soya beans, spinach, swedes, turnips, vining peas** [2]
- Insect pests in **dairies, farm buildings, livestock houses, poultry houses** [1]
- Spider mites in **asparagus, beans without pods (fresh), broad beans, bulb onions, cabbages, carrots, celeriac, cucumbers, dwarf beans, edible podded peas, french beans, garlic, herbs (see appendix 6), horseradish, jerusalem artichokes, kohlrabi, leeks, lettuce, parsley, parsnips, protected cabbages, protected lettuce, protected ornamentals, protected tomatoes, radishes, red beet, runner beans, salsify, shallots, soya beans, spinach, swedes, turnips, vining peas** [2]
- Thrips in **asparagus, beans without pods (fresh), broad beans, bulb onions, cabbages, carrots, celeriac, cucumbers, dwarf beans, edible podded peas, french beans, garlic, herbs (see appendix 6), horseradish, jerusalem artichokes, kohlrabi, leeks, lettuce, parsley, parsnips, protected cabbages, protected lettuce, protected ornamentals, protected tomatoes, radishes, red beet, runner beans, salsify, shallots, soya beans, spinach, swedes, turnips, vining peas** [2]

Approval information

- Pyrethrins have been included in Annex 1 under EC Regulation 1107/2009
- Some products formulated for ULV application [1]. May be applied through fogging machine or sprayer [1]. See label for details
- Accepted by BBPA for use in empty grain stores, malting barley and hops

Efficacy guidance

- For indoor fly control close doors and windows and spray or apply fog as appropriate [1]
- For best fly control outdoors spray during early morning or late afternoon and evening when conditions are still [1]
- For all uses ensure good spray coverage of the target area or plants by increasing spray volume where necessary
- Best results on outdoor or protected crops achieved from treatment at first signs of pest attack in early morning or evening
- To avoid possibility of development of resistance do not spray more frequently than once per week

Restrictions

- Maximum number of treatments on edible crops 3 per crop or season
- For use only by professional operators [1]
- Do not allow spray to contact open food products or food preparing equipment or utensils [1]
- Remove exposed milk and collect eggs before application [1]
- Do not treat plants [1]
- Avoid direct application to open flowers

SEE SECTION 3 FOR PRODUCTS ALSO REGISTERED

- Do not use space sprays containing pyrethrins or pyrethroid more than once per week in intensive or controlled environment animal houses in order to avoid development of resistance. If necessary, use a different control method or product [1]
- Store away from strong sunlight

Crop-specific information
- HI: 24 h for edible crops
- Some plant species including *Ageratum*, ferns, *Ficus, Lantana, Poinsettia, Petroselium crispum*, and some strawberries may be sensitive especially where more than one application is made. Test before large scale treatment

Environmental safety
- Dangerous for the environment
- Very toxic to aquatic organisms [1]
- High risk to bees. Do not apply to crops in flower or to those in which bees are actively foraging. Do not apply when flowering weeds are present
- Risk to non-target insects or other arthropods
- Do not apply directly to livestock and exclude all persons and animals during treatment [1]
- Wash spray equipment thoroughly after use to avoid traces of pyrethrum causing damage to susceptible crops sprayed later
- LERAP Category B [2]

Hazard classification and safety precautions
Hazard Harmful, Irritant [1]; Dangerous for the environment [2]
Transport code 9
Packaging group III
UN Number 3082
Risk phrases R22a, R36, R38 [1]
Operator protection A [1, 2]; B, E [1]; H [2]; U05a, U09a, U19a, U20b [1]; U20d [2]
Environmental protection E05a, E13c [1]; E12c, E15b, E16a, E19b, E38, H410 [2]
Consumer protection C04, C06, C07, C08, C09, C10, C11 [1]
Storage and disposal D01, D02, D09a, D11a [1, 2]; D12a [2]

348   pyridate

A contact phenylpyridazine herbicide for bulb onions and brassicas
HRAC mode of action code: C3

Products

1	Diva	Belchim	600 g/l	EC	17774
2	Gyo	Belchim	600 g/l	EC	17875
3	Lentagran WP	Belchim	45% w/w	WP	14162

Uses
- Annual dicotyledons in *asparagus* (off-label), *broccoli* (off-label), *calabrese* (off-label), *cauliflowers* (off-label), *chives* (off-label), *collards* (off-label), *fodder rape* (off-label), *game cover* (off-label), *garlic* (off-label), *kale* (off-label), *leeks* (off-label), *lupins* (off-label), *oilseed rape* (off-label), *salad onions* (off-label), *shallots* (off-label), *spring greens* (off-label) [3]; *forage maize, grain maize* [1, 2]
- Black nightshade in *asparagus* (off-label), *broccoli* (off-label), *brussels sprouts, bulb onions, cabbages, calabrese* (off-label), *cauliflowers* (off-label), *chives* (off-label), *collards* (off-label), *fodder rape* (off-label), *game cover* (off-label), *garlic* (off-label), *kale* (off-label), *leeks* (off-label), *lupins* (off-label), *oilseed rape* (off-label), *salad onions* (off-label), *shallots* (off-label), *spring greens* (off-label) [3]; *forage maize, grain maize* [1, 2]
- Cleavers in *asparagus* (off-label), *broccoli* (off-label), *brussels sprouts, bulb onions, cabbages, calabrese* (off-label), *cauliflowers* (off-label), *chives* (off-label), *collards* (off-label), *fodder rape* (off-label), *game cover* (off-label), *garlic* (off-label), *kale* (off-label), *leeks* (off-label), *lupins* (off-label), *oilseed rape* (off-label), *salad onions* (off-label), *shallots* (off-label), *spring greens* (off-label) [3]; *forage maize, grain maize* [1, 2]

FOR FULL CONDITIONS OF USE ALWAYS READ THE PRODUCT LABEL

- Fat hen in **asparagus** *(off-label)*, **broccoli** *(off-label)*, **brussels sprouts**, **bulb onions**, **cabbages**, **calabrese** *(off-label)*, **cauliflowers** *(off-label)*, **chives** *(off-label)*, **collards** *(off-label)*, **fodder rape** *(off-label)*, **game cover** *(off-label)*, **garlic** *(off-label)*, **kale** *(off-label)*, **leeks** *(off-label)*, **lupins** *(off-label)*, **oilseed rape** *(off-label)*, **salad onions** *(off-label)*, **shallots** *(off-label)*, **spring greens** *(off-label)* [3]; **forage maize**, **grain maize** [1, 2]
- Fumitory in **asparagus** *(off-label)*, **broccoli** *(off-label)*, **calabrese** *(off-label)*, **cauliflowers** *(off-label)*, **chives** *(off-label)*, **collards** *(off-label)*, **fodder rape** *(off-label)*, **game cover** *(off-label)*, **garlic** *(off-label)*, **kale** *(off-label)*, **leeks** *(off-label)*, **lupins** *(off-label)*, **oilseed rape** *(off-label)*, **salad onions** *(off-label)*, **shallots** *(off-label)*, **spring greens** *(off-label)* [3]; **forage maize**, **grain maize** [1, 2]
- Groundsel in **asparagus** *(off-label)*, **broccoli** *(off-label)*, **calabrese** *(off-label)*, **cauliflowers** *(off-label)*, **chives** *(off-label)*, **collards** *(off-label)*, **fodder rape** *(off-label)*, **game cover** *(off-label)*, **garlic** *(off-label)*, **kale** *(off-label)*, **leeks** *(off-label)*, **lupins** *(off-label)*, **oilseed rape** *(off-label)*, **salad onions** *(off-label)*, **shallots** *(off-label)*, **spring greens** *(off-label)* [3]; **forage maize**, **grain maize** [1, 2]

Extension of Authorisation for Minor Use (EAMUs)
- **asparagus** *20091039* [3]
- **broccoli** *20090786* [3]
- **calabrese** *20090786* [3]
- **cauliflowers** *20090786* [3]
- **chives** *20120267* [3]
- **collards** *20090785* [3]
- **fodder rape** *20093230* [3]
- **game cover** *20090788* [3]
- **garlic** *20092862* [3]
- **kale** *20090785* [3]
- **leeks** *20090784* [3]
- **lupins** *20090787* [3]
- **oilseed rape** *20093230* [3]
- **salad onions** *20120267* [3]
- **shallots** *20092862* [3]
- **spring greens** *20090785* [3]

Approval information
- Pyridate included in Annex I under EC Regulation 1107/2009

Efficacy guidance
- Best results achieved by application to actively growing weeds at 6-8 leaf stage when temperatures are above 8°C before crop foliage forms canopy [3]
- Apply when maize is between 2 and 8 leaves for optimum activity [1, 2]

Restrictions
- Do not apply in mixture with or within 14 d of any other product which may result in dewaxing of crop foliage
- Do not use on crops suffering stress from frost, drought, disease or pest attack
- Horizontal boom sprayers must be fitted with three star drift reduction technology for all uses [1, 2]
- Must not be applied via hand-held equipment
- Low drift spraying equipment must be used for 30 m from the top of the bank of any surface water bodies [1, 2]
- Applications must not be made before 15th May [1, 2]

Crop-specific information
- Latest use: before 6 true leaf stage of Brussels sprouts and cabbage; before 7 true leaves for onions; before flower buds visible for oilseed rape
- Apply to cabbages and Brussels sprouts after 4 fully expanded leaf stage. Allow 2 wk after transplanting before treating

Environmental safety
- Dangerous for the environment

SEE SECTION 3 FOR PRODUCTS ALSO REGISTERED

- Toxic to aquatic organisms
- Buffer zone requirement 12 m [1, 2]
- LERAP Category B [1, 2]

Hazard classification and safety precautions

Hazard Irritant, Dangerous for the environment [3]; Flammable liquid and vapour, Very toxic to aquatic organisms [1, 2]

Transport code 9

Packaging group III

UN Number 3077 [3]; 3082 [1, 2]

Risk phrases H315, H319, H370 [1, 2]; H317 [1-3]

Operator protection A [1-3]; C [3]; H [1, 2]; U05a, U14, U22a

Environmental protection E15a [3]; E16a [1, 2]; E38, H410 [1-3]

Storage and disposal D01, D02, D09a, D10c, D12a

349 pyrimethanil

An anilinopyrimidine fungicide for apples and strawberries
FRAC mode of action code: 9

See also fludioxonil + pyrimethanil

Products

1	Clayton Patriot	Clayton	400 g/l	SL	18418
2	Pyrus 400 SC	Arysta	400 g/l	SL	16254
3	Scala	BASF	400 g/l	SC	15222
4	Triple SC	Pan Agriculture	400 g/l	SL	18669

Uses

- Botrytis in *bilberries (off-label)*, *blackberries (off-label)*, *blackcurrants (off-label)*, *blueberries (off-label)*, *cranberries (off-label)*, *forest nurseries (off-label)*, *gooseberries (off-label)*, *grapevines (off-label)*, *ornamental plant production (off-label)*, *pears (off-label)*, *protected aubergines (off-label)*, *protected baby leaf crops (off-label)*, *protected blackberries (off-label)*, *protected herbs (see appendix 6) (off-label)*, *protected lettuce (off-label)*, *protected raspberries (off-label)*, *protected strawberries*, *protected tomatoes (off-label)*, *quinces (off-label)*, *raspberries (off-label)*, *redcurrants (off-label)*, *top fruit (off-label)*, *vaccinium spp. (off-label)*, *whitecurrants (off-label)* [3]; *blackberries*, *raspberries* [1, 2, 4]; *strawberries* [1-4]
- Disease control in *forest nurseries (off-label)*, *top fruit (off-label)* [3]
- Grey mould in *protected strawberries (moderate control)*, *strawberries* [3]
- Scab in *apples*, *pears (off-label)*, *quinces (off-label)* [3]

Extension of Authorisation for Minor Use (EAMUs)

- *bilberries* 20110291 [3]
- *blackberries* 20110293 [3]
- *blackcurrants* 20110291 [3]
- *blueberries* 20110291 [3]
- *cranberries* 20110291 [3]
- *forest nurseries* 20122054 [3]
- *gooseberries* 20110291 [3]
- *grapevines* 20110283 [3]
- *ornamental plant production* 20111315 [3]
- *pears* 20110295 [3]
- *protected aubergines* 20110282 [3]
- *protected baby leaf crops* 20110287 [3]
- *protected blackberries* 20110292 [3]
- *protected herbs (see appendix 6)* 20110287 [3]
- *protected lettuce* 20110287 [3]
- *protected raspberries* 20110292 [3]
- *protected tomatoes* 20110282 [3]
- *quinces* 20110295 [3]

FOR FULL CONDITIONS OF USE ALWAYS READ THE PRODUCT LABEL

- *raspberries* 20110293 [3]
- *redcurrants* 20110291 [3]
- *top fruit* 20122054 [3]
- *vaccinium spp.* 20110291 [3]
- *whitecurrants* 20110291 [3]

Approval information
- Pyrimethanil included in Annex I under EC Regulation 1107/2009.

Efficacy guidance
- On apples a programme of sprays will give early season control of scab. Season long control can be achieved by continuing programme with other approved fungicides
- In strawberries product should be used as part of a programme of disease control treatments which should alternate with other materials to prevent or limit development of less sensitive strains of grey mould

Restrictions
- Maximum number of treatments 4 per yr for apples and pears; 2 per yr for cane fruit, bush fruit, strawberries, tomatoes
- Product does not taint apples. Processors should be consulted before use on strawberries

Crop-specific information
- Latest use: before end of flowering for apples
- HI protected cane fruit, strawberries 1 d; aubergines, tomatoes 3 d; protected lettuce 14 d; bush fruit, grapevines 21 d
- All varieties of apples and strawberries may be treated
- Treat apples from bud burst at 10-14 d intervals
- In strawberries start treatments at white bud to give maximum protection of flowers against grey mould and treat every 7-10 d. Product should not be used more than once in a 3 or 4 spray programme

Environmental safety
- Harmful to aquatic organisms
- Product has negligible effect on hoverflies and lacewings. Limited evidence indicates some margin of safety to *Typhlodromus pyri*
- Broadcast air-assisted LERAP [3] (20 m); LERAP Category B [3]

Hazard classification and safety precautions
Hazard Dangerous for the environment [1, 2, 4]
Transport code 9 [2, 4]
Packaging group III [2, 4]
UN Number 3082 [2, 4]; N/C [1, 3]
Risk phrases R51 [1, 2, 4]; R52 [3]; R53a [1-4]
Operator protection A, H [1-4]; M [3]; U05a [1, 2, 4]; U08, U20b [3]
Environmental protection E15a, E16a [3]; E17b [3] (20 m); E38 [1, 2, 4]
Storage and disposal D01, D02 [1-4]; D05, D08, D09a, D10b, D12a [3]; D12b, D19 [1, 2, 4]

350   pyriofenone

A protectant fungicide with some curative activity against mildew in the latent phase
FRAC mode of action code: U8

Products

Property 180SC	Belchim	180 g/l	SC	17666

Uses
- Powdery mildew in *spring wheat*, *winter wheat*

Approval information
- Pyrifenone is included in Annex 1 under EC Regulation 1107/2009
- Approval expiry 12 Sep 2020 [1]

SEE SECTION 3 FOR PRODUCTS ALSO REGISTERED

Efficacy guidance

- Best results obtained from application at the start of disease attack. Acts preventatively against powdery mildew with some curative activity during the latent phase. Apply before disease spreads on to new growth with a follow-up after 2 - 4 weeks according to climatic conditions.
- Should be used as part of a mildew control programme that incorporates other methods of control including fungicides active against mildew via a different mode of action.

Restrictions

- Only two applications of benzoylpyridine fungicides are permitted per crop per season. Where a second application is required it should be applied in mixture with fungicides using a different mode of action such as fenpropimorph.

Crop-specific information

- Latest use: mid-flowering (GS65) in spring and winter wheat

Following crops guidance

- Oilseed rape, maize, potato, sugar beet, kidney bean, soybean, pea, onion, turnip, lettuce, flax, oat, barley and wheat may be sown as a following crop.

Hazard classification and safety precautions

Hazard Harmful, Dangerous for the environment
Transport code 9
Packaging group III
UN Number 3082
Risk phrases H351, R53a
Operator protection A, H; U05a, U20b
Environmental protection E15b, E34, E38, H411
Storage and disposal D01, D02, D05, D09a, D12a, D20
Medical advice M05a

351   pyroxsulam

A triazolopyrimidine herbicide for use in cereals
HRAC mode of action code: B

See also florasulam + pyroxsulam
flupyrsulfuron-methyl + pyroxsulam
pendimethalin + pyroxsulam

Products

Avocet	Dow (Corteva)	7.5% w/w	WG	14829

Uses

- Annual dicotyledons in **triticale, winter rye, winter wheat**
- Blackgrass in **triticale, winter rye, winter wheat**
- Cleavers in **triticale, winter rye, winter wheat**
- Ryegrass in **triticale, winter rye, winter wheat**
- Volunteer oilseed rape in **triticale, winter rye, winter wheat**
- Wild oats in **triticale, winter rye, winter wheat**

Approval information

- Pyroxsulam included in Annex 1 under EC Regulation 1107/2009

Efficacy guidance

- Avoid overlapping of spray swaths.
- Rainfast in one hour.
- Do not spray when crops are under stress from cold, drought, water logging, pest damage, nutrient deficiency etc.
- Do not roll or harrow 7 days before or after application.
- Take extreme care to avoid drift onto susceptible crops, non-target plants or waterways. Do not apply directly to, or allow spray drift to come into contact with agricultural or horticultural crops, amenity plantings, gardens, ponds, lakes or watercourses.

FOR FULL CONDITIONS OF USE ALWAYS READ THE PRODUCT LABEL

- To avoid crop injury do not apply in tank mixture with plant growth regulators. A minimum interval of 7 days between applications should be left.
- To avoid crop injury do not apply in tank mix with an organophosphate insecticide. A minimum interval of 14 days between applications should be left.

Restrictions
- To avoid the build-up of resistance, do not apply this or any other product containing an ALS inhibitor herbicide with claims for control of grass-weeds more than once to any crop
- To avoid subsequent injury to crops other than cereals, all spraying equipment must be thoroughly cleaned both inside and out, using All Clear Extra spray cleaner

Crop-specific information
- Winter wheat, winter barley, winter oats, triticale, winter rye, oilseed rape, grass, winter beans or brassicas as transplants may be sown as a following crop in the autumn.
- Following a sequence of pyroxsulam and metsulfuron-methyl, treated ground should be ploughed to a depth of 15 cm before establishing winter oilseed rape
- In spring, spring wheat, spring barley, spring oats, triticale, rye, spring oilseed rape, sugar beet, potatoes, grass, clover (as part of a grass/clover ley), beans, peas, maize, linseed may follow a normally harvested treated crop.

Following crops guidance
- In the event of a crop failure after a crop has been treated before February 1st only the following crops may be planted after ploughing to a depth of at least 15 cm; Spring wheat, spring barley, grass or maize. At least 6 weeks must elapse between treatment and re-drilling.
- Where crop failure occurs following an application after February 1st plough and then allow 6 weeks before planting grass or maize.

Environmental safety
- LERAP Category B

Hazard classification and safety precautions
Hazard Very toxic to aquatic organisms
Transport code 9
Packaging group III
UN Number 3077
Operator protection A, H; U05a, U09b, U14, U15
Environmental protection E16a, E34, H410
Storage and disposal D01, D02, D09a, D12a

352   quinmerac

A residual herbicide available only in mixtures
HRAC mode of action code: O

See also dimethenamid-p + metazachlor + quinmerac
dimethenamid-p + quinmerac
imazamox + quinmerac
metamitron + quinmerac
metazachlor + quinmerac

353   quizalofop-P-ethyl

An aryl phenoxypropionic acid post-emergence herbicide for grass weed control
HRAC mode of action code: A

Products

1 Leopard 5EC	Adama	50 g/l	EC	16915
2 Pilot Ultra	Gowan	50 g/l	SC	17136
3 Quizallo	Harvest	100 g/l	EC	18308
4 Targa Max	Gowan	100 g/l	EC	17135

SEE SECTION 3 FOR PRODUCTS ALSO REGISTERED

Products – continued

5 Targa Super   Gowan   50 g/l   EC   17134

Uses

- Annual grasses in *combining peas, linseed, spring field beans, vining peas, winter field beans* [1, 2]; *fodder beet, mangels, potatoes, red beet, spring oilseed rape, sugar beet, winter oilseed rape* [1-5]; *forest nurseries (off-label)* [1]
- Couch in *combining peas, linseed, spring field beans, vining peas, winter field beans* [1, 2]; *fodder beet, mangels, potatoes, red beet, spring oilseed rape, sugar beet, winter oilseed rape* [1-5]
- Perennial grasses in *combining peas, linseed, spring field beans, vining peas, winter field beans* [1, 2]; *fodder beet, mangels, potatoes, red beet, spring oilseed rape, sugar beet, winter oilseed rape* [1-5]; *forest nurseries (off-label)* [1]
- Volunteer cereals in *combining peas, linseed, spring field beans, vining peas, winter field beans* [1, 2]; *fodder beet, mangels, potatoes, red beet, spring oilseed rape, sugar beet, winter oilseed rape* [1-5]

Extension of Authorisation for Minor Use (EAMUs)

- *forest nurseries 20151181* [1]

Approval information

- Quizalofop-P-ethyl has been included in Annex 1 under EC Regulation 1107/2009

Efficacy guidance

- Best results achieved by application to emerged weeds growing actively in warm conditions with adequate soil moisture
- Weed control may be reduced under conditions such as drought that limit uptake and translocation
- Annual meadow-grass is not controlled
- For effective couch control do not hoe beet crops within 21 d after spraying
- At least 2 h without rain should follow application otherwise results may be reduced
- Quizalofop-P-ethyl is an ACCase inhibitor herbicide. To avoid the build up of resistance do not apply products containing an ACCase inhibitor herbicide more than twice to any crop. In addition do not use any product containing quizalofop-P-ethyl in mixture or sequence with any other product containing the same ingredient
- Use these products as part of a resistance management strategy that includes cultural methods of control and does not use ACCase inhibitors as the sole chemical method of grass weed control
- Applying a second product containing an ACCase inhibitor to a crop will increase the risk of resistance development; only use a second ACCase inhibitor to control different weeds at a different timing
- Always follow WRAG guidelines for preventing and managing herbicide resistant weeds. See Section 5 for more information

Restrictions

- Maximum number of treatments 1 on all recommended crops
- Do not spray crops under stress from any cause or in frosty weather
- Consult processor before use on crops recommended for processing
- An interval of at least 3 d must elapse between treatment and use of another herbicide on beet crops, 14 d on oilseed rape, 21 d on linseed and other recommended crops
- Avoid drift onto neighbouring crops

Crop-specific information

- HI 16 wk for beet crops; 11 wk for oilseed rape, linseed; 8 wk for field beans; 5 wk for peas
- In some situations treatment can cause yellow patches on foliage of peas, especially vining varieties. Symptoms usually rapidly and completely outgrown
- May cause taint in peas

Following crops guidance

- In the event of failure of a treated crop broad-leaved crops may be resown after a minimum interval of 2 wks, and cereals after 2 - 6 wks depending on dose applied
- Onions, leeks and maize are not recommended to follow a failed treated crop

FOR FULL CONDITIONS OF USE ALWAYS READ THE PRODUCT LABEL

Environmental safety
- Dangerous for the environment
- Toxic to aquatic organisms
- Flammable

Hazard classification and safety precautions
Hazard Harmful, Flammable [1]; Dangerous for the environment [1-5]; Harmful if inhaled, Very toxic to aquatic organisms [5]
Transport code 3 [1]; 9 [3-5]
Packaging group III [1, 3-5]
UN Number 1993 [1]; 3082 [3-5]; N/C [2]
Risk phrases H304 [1, 5]; H315, H319, H336 [1]; H317 [5]; H318 [3-5]
Operator protection A, C, H; U05a, U11, U14, U15, U19a [1]; U09a [2-5]; U20b [1-5]
Environmental protection E13b [2-5]; E15a, H411 [1]; E34, E38 [1-5]; H410 [5]
Storage and disposal D01, D02, D05, D10b [1]; D09a, D12a [1-5]; D10c [2-5]
Medical advice M03, M05b [1]

354 quizalofop-P-tefuryl

An aryloxyphenoxypropionate herbicide for grass weed control
HRAC mode of action code: A

Products
1	Panarex	Arysta	40 g/l	EC	17960
2	Rango	Arysta	40 g/l	EC	17959

Uses
- Blackgrass in *combining peas, fodder beet, potatoes, spring field beans, spring linseed, spring oilseed rape, sugar beet, winter field beans, winter linseed, winter oilseed rape*
- Couch in *combining peas, fodder beet, potatoes, spring field beans, spring linseed, spring oilseed rape, sugar beet, winter field beans, winter linseed, winter oilseed rape*
- Italian ryegrass in *combining peas, fodder beet, potatoes, spring field beans, spring linseed, spring oilseed rape, sugar beet, winter field beans, winter linseed, winter oilseed rape*
- Perennial ryegrass in *combining peas, fodder beet, potatoes, spring field beans, spring linseed, spring oilseed rape, sugar beet, winter field beans, winter linseed, winter oilseed rape*
- Volunteer cereals in *combining peas, fodder beet, potatoes, spring field beans, spring linseed, spring oilseed rape, sugar beet, winter field beans, winter linseed, winter oilseed rape*
- Wild oats in *combining peas, fodder beet, potatoes, spring field beans, spring linseed, spring oilseed rape, sugar beet, winter field beans, winter linseed, winter oilseed rape*

Approval information
- Quizalofop-P-tefuryl has been included in Annex 1 under EC Regulation 1107/2009

Efficacy guidance
- Best results on annual grass weeds when growing actively and treated from 2 leaves to the start of tillering
- Treat cover crops when they have served their purpose and the threat of wind blow has passed
- Best results on couch achieved when the weed is growing actively and commencing new rhizome growth
- Grass weeds germinating after treatment will not be controlled
- Treatment quickly stops growth and visible colour changes to the leaf tips appear after about 7 d. Complete kill takes 3-4 wk under good growing conditions
- Quizalofop-P-tefuryl is an ACCase inhibitor herbicide. To avoid the build up of resistance do not apply products containing an ACCase inhibitor herbicide more than twice to any crop. In addition do not use any product containing quizalofop-P-tefuryl in mixture or sequence with any other product containing the same ingredient
- Use these products as part of a resistance management strategy that includes cultural methods of control and does not use ACCase inhibitors as the sole chemical method of grass weed control

SEE SECTION 3 FOR PRODUCTS ALSO REGISTERED

- Applying a second product containing an ACCase inhibitor to a crop will increase the risk of resistance development; only use a second ACCase inhibitor to control different weeds at a different timing
- Always follow WRAG guidelines for preventing and managing herbicide resistant weeds. See Section 5 for more information

Restrictions
- Maximum number of treatments 1 per crop
- Consult processors before use on peas or potatoes for processing
- Do not treat crops and weeds growing under stress from any cause

Crop-specific information
- HI 60 d for all crops
- Treat oilseed rape from the fully expanded cotyledon stage
- Treat linseed, peas and field beans from 2-3 unfolded leaves
- Treat sugar beet from 2 unfolded leaves

Following crops guidance
- In the event of failure of a treated crop any broad-leaved crop may be planted at any time. Cereals may be drilled from 4 wk after treatment

Environmental safety
- Dangerous for the environment
- Very toxic to aquatic organisms
- Risk to certain non-target insects or other arthropods. For advice on risk management and use in Integrated Pest Management (IPM) see directions for use
- Avoid spraying within 6 m of the field boundary to reduce effects on non-target insects and other arthropods

Hazard classification and safety precautions
Hazard Toxic [2]; Irritant [1]; Dangerous for the environment, Very toxic to aquatic organisms [1, 2]
Transport code 9
Packaging group III
UN Number 3082
Risk phrases H304, H315, H317, H319, H336, H341, H360
Operator protection A, C, H; U02a, U04a, U05a, U10, U11, U14, U15, U19a, U20b
Environmental protection E15a, E22b, E38, H410 [1, 2]; H411 [1]
Storage and disposal D01, D02, D09a, D10b, D12a
Medical advice M04a [2]; M05b [1, 2]

355   rimsulfuron

A selective systemic sulfonylurea herbicide
HRAC mode of action code: B

Products
1 Caesar	AgChem Access	25% w/w	SG	15272
2 Clayton Bramble	Clayton	25% w/w	SG	15281
3 Clayton Nero	Clayton	25% w/w	SG	15511
4 Titus	Adama	25% w/w	SG	15050

Uses
- Annual dicotyledons in *forage maize, potatoes* [1-4]; *forest nurseries* (off-label), *ornamental plant production* (off-label) [4]
- Charlock in *forage maize, potatoes* [4]
- Chickweed in *forage maize, potatoes* [4]
- Cleavers in *forage maize, potatoes* [4]
- Hemp-nettle in *forage maize, potatoes* [4]
- Red dead-nettle in *forage maize, potatoes* [4]
- Redshank in *forage maize, potatoes* [4]
- Scentless mayweed in *forage maize, potatoes* [4]

FOR FULL CONDITIONS OF USE ALWAYS READ THE PRODUCT LABEL

- Small nettle in *forage maize*, *potatoes* [4]
- Volunteer oilseed rape in *forage maize*, *potatoes* [1-4]

Extension of Authorisation for Minor Use (EAMUs)
- *forest nurseries 20141912* [4]
- *ornamental plant production 20141912* [4]

Approval information
- Rimsulfuron included in Annex I under EC Regulation 1107/2009

Efficacy guidance
- Product should be used with a suitable adjuvant or a suitable herbicide tank-mix partner. See label for details
- Product acts by foliar action. Best results obtained from good spray cover of small actively growing weeds. Effectiveness is reduced in very dry conditions
- Weed spectrum can be broadened by tank mixture with other herbicides. See label for details
- Susceptible weeds cease growth immediately and symptoms can be seen 10 d later
- Rimsulfuron is a member of the ALS-inhibitor group of herbicides

Restrictions
- Maximum number of treatments 1 per crop
- Do not treat maize previously treated with organophosphorus insecticides
- Do not apply to potatoes grown for certified seed
- Consult processor before use on crops grown for processing
- Avoid high light intensity (full sunlight) and high temperatures on the day of spraying
- Do not treat during periods of substantial diurnal temperature fluctuation or when frost anticipated
- Do not apply to any crop stressed by drought, water-logging, low temperatures, pest or disease attack, nutrient or lime deficiency
- Do not apply to forage maize treated with organophosphate insecticides
- Do not apply to forage maize undersown with grass or clover

Crop-specific information
- Latest use: before most advanced potato plants are 25 cm high; before 4-collar stage of fodder maize
- All varieties of ware potatoes may be treated, but variety restrictions of any tank-mix partner must be observed
- Only certain named varieties of forage maize may be treated. See label

Following crops guidance
- Only winter wheat should follow a treated crop in the same calendar yr
- Only barley, wheat or maize should be sown in the spring of the yr following treatment
- In the second autumn after treatment any crop except brassicas or oilseed rape may be drilled

Environmental safety
- Dangerous for the environment
- Toxic to aquatic organisms
- Extremely dangerous to fish or other aquatic life. Do not contaminate surface waters or ditches with chemical or used container
- Herbicide is very active. Take particular care to avoid drift onto plants outside the target area
- Spraying equipment should not be drained or flushed onto land planted, or to be planted, with trees or crops other than potatoes or forage maize and should be thoroughly cleansed after use - see label for instructions

Hazard classification and safety precautions
Hazard Dangerous for the environment, Very toxic to aquatic organisms
Transport code 9
Packaging group III
UN Number 3077
Operator protection A [3, 4]; U08, U19a, U20b
Environmental protection E13a, E34, E38, H410
Storage and disposal D01, D09a, D11a, D12a

SEE SECTION 3 FOR PRODUCTS ALSO REGISTERED

SECTION 2

356   silthiofam

A thiophene carboxamide fungicide seed dressing for cereals
FRAC mode of action code: 38

Products

Latitude	Certis	125 g/l	FS	18036

Uses

- Seed-borne diseases in **rye** *(off-label)*, **triticale** *(off-label)*
- Take-all in **rye** *(off-label)*, **spring wheat** *(seed treatment)*, **triticale** *(off-label)*, **winter barley** *(seed treatment)*, **winter wheat** *(seed treatment)*

Extension of Authorisation for Minor Use (EAMUs)

- *rye 20193262*
- *triticale 20193262*

Approval information

- Silthiofam included in Annex I under EC Regulation 1107/2009
- Accepted by BBPA for use on malting barley

Efficacy guidance

- Apply using approved seed treatment equipment which has been accurately calibrated
- Apply simultaneously (i.e. not in mixture) with a standard seed treatment using a recommended seed treatment machine
- Drill treated seed in the season of purchase. The viability of treated seed and fungicide activity may be reduced by physical storage
- Drill treated seed at 2.5-4 cm into a well prepared firm seedbed
- As precaution against possible development of disease resistance do not treat more than three consecutive susceptible cereal crops in any one rotation

Restrictions

- Maximum number of treatments one per batch of seed
- Do not use on seed with more than 16% moisture content, or on sprouted, cracked or skinned seed
- Test germination of all seed batches before treatment

Crop-specific information

- Latest use: immediately prior to drilling

Hazard classification and safety precautions
   UN Number N/C
   Operator protection A, H; U20b
   Environmental protection E03, E15a, E34
   Storage and disposal D05, D09a, D10a
   Treated seed S01, S02, S04b, S05, S06a, S07

357   S-metolachlor

A chloroacetamide residual herbicide for use in forage maize and grain maize
HRAC mode of action code: K3

See also mesotrione + S-metolachlor

Products

1	Clayton Smelter	Clayton	960 g/l	EC	14937
2	Dual Gold	Syngenta	960 g/l	EC	14649

Uses

- Annual dicotyledons in **baby leaf crops** *(off-label)*, **forest nurseries** *(off-label)*, **garlic** *(off-label)*, **herbs (see appendix 6)** *(off-label)*, **lettuce** *(off-label)*, **onions** *(off-label)*, **ornamental plant**

SECTION 2

production *(off-label)*, **red beet** *(off-label)*, **shallots** *(off-label)*, **swedes** *(off-label)*, **turnips** *(off-label)* [2]

- Annual grasses in **begonias** *(off-label)*, **broccoli** *(off-label)*, **brussels sprouts** *(off-label)*, **cabbages** *(off-label)*, **calabrese** *(off-label)*, **cauliflowers** *(off-label)*, **chinese cabbage** *(off-label)*, **collards** *(off-label)*, **kale** *(off-label)* [2]; **chicory** *(off-label)*, **dwarf beans** *(off-label)*, **endives** *(off-label)*, **french beans** *(off-label)*, **runner beans** *(off-label)*, **strawberries** *(off-label)* [1, 2]
- Annual meadow grass in **baby leaf crops** *(off-label)*, **edible podded peas** *(off-label)*, **forest nurseries** *(off-label)*, **garlic** *(off-label)*, **herbs (see appendix 6)** *(off-label)*, **lettuce** *(off-label)*, **onions** *(off-label)*, **ornamental plant production** *(off-label)*, **red beet** *(off-label)*, **shallots** *(off-label)*, **spring field beans** *(off-label)*, **swedes** *(off-label)*, **sweetcorn** *(off-label)*, **turnips** *(off-label)*, **vining peas** *(off-label)* [2]; **forage maize, grain maize** [1, 2]
- Barnyard grass in **sweetcorn** *(off-label)* [2]
- Black nightshade in **chicory** *(off-label)*, **endives** *(off-label)* [1, 2]
- Chickweed in **forage maize** *(moderately susceptible)*, **grain maize** *(moderately susceptible)* [1, 2]; **spring field beans** *(off-label)* [2]
- Crane's-bill in **sweetcorn** *(off-label)* [2]
- Fat hen in **forage maize** *(moderately susceptible)*, **grain maize** *(moderately susceptible)* [1, 2]; **spring field beans** *(off-label)* [2]
- Field speedwell in **forage maize, grain maize** [1, 2]; **spring field beans** *(off-label)* [2]
- Groundsel in **sweetcorn** *(off-label)* [2]
- Mayweeds in **forage maize, grain maize** [1, 2]; **spring field beans** *(off-label)*, **sweetcorn** *(off-label)* [2]
- Red dead-nettle in **forage maize, grain maize** [1, 2]
- Sowthistle in **forage maize, grain maize** [1, 2]; **sweetcorn** *(off-label)* [2]

Extension of Authorisation for Minor Use (EAMUs)
- *baby leaf crops 20120594* [2]
- *begonias 20132574* [2]
- *broccoli 20111258* [2]
- *brussels sprouts 20111259* [2]
- *cabbages 20111259* [2]
- *calabrese 20111258* [2]
- *cauliflowers 20111258* [2]
- *chicory 20111262* [1], *20111257* [2]
- *chinese cabbage 20111260* [2]
- *collards 20111260* [2]
- *dwarf beans 20103103* [1], *20132572* [2]
- *edible podded peas 20140164* [2]
- *endives 20111262* [1], *20111257* [2]
- *forest nurseries 20120501* [2]
- *french beans 20103103* [1], *20132572* [2]
- *garlic 20110840* [2]
- *herbs (see appendix 6) 20120594* [2]
- *kale 20111260* [2]
- *lettuce 20120594* [2]
- *onions 20110840* [2]
- *ornamental plant production 20120501* [2]
- *red beet 20111006* [2]
- *runner beans 20103103* [1], *20132572* [2]
- *shallots 20110840* [2]
- *spring field beans 20140163* [2]
- *strawberries 20103104* [1], *20132573* [2]
- *swedes 20111006* [2]
- *sweetcorn 20172834* [2]
- *turnips 20111006* [2]
- *vining peas 20140164* [2]

Approval information
- S-metolachlor included in Annex I under EC Regulation 1107/2009

SEE SECTION 3 FOR PRODUCTS ALSO REGISTERED

Environmental safety
- LERAP Category B

Hazard classification and safety precautions
    Hazard Irritant, Dangerous for the environment, Very toxic to aquatic organisms
    Transport code 9
    Packaging group III
    UN Number 3082
    Risk phrases H317, H319
    Operator protection A [1, 2]; C [2]; H [1]; U05a, U09a, U20b
    Environmental protection E15b, E16a, E16b, E38, H410
    Storage and disposal D01, D02, D09a, D10c, D11a, D12a

358   spearmint oil

A hot fogging treatment for post-harvest sprout suppression in potatoes

Products

Biox-M	Juno	100% w/w	HN	16021

Uses
- Sprout suppression in **potatoes**

Approval information
- Not listed in Annex 1 under EC Regulation 1107/2009 since it is a natural product.

Hazard classification and safety precautions
    Hazard Harmful, Dangerous for the environment
    UN Number N/C
    Risk phrases R22b, R43, R50, R53a
    Operator protection A, C, D, H, J, M
    Storage and disposal D12a

359   spinosad

A selective insecticide derived from naturally occurring soil fungi (naturalyte)
IRAC mode of action code: 5

Products

1	Conserve	Fargro	120 g/l	SC	12058
2	Spindle	Pan Agriculture	120 g/l	SC	18949
3	Tracer	Landseer	480 g/l	SC	12438

Uses
- Asparagus beetle in **asparagus** (off-label) [3]
- Blastobasis lacticoella in **protected strawberries** (off-label) [3]
- Cabbage moth in **broccoli, brussels sprouts, cabbages, calabrese, cauliflowers, chinese cabbage, collards** (off-label), **kale** (off-label), **kohlrabi** (off-label), **oriental cabbage** (off-label) [3]
- Cabbage root fly in **collards** (off-label), **kale** (off-label), **kohlrabi** (off-label), **oriental cabbage** (off-label) [3]
- Cabbage white butterfly in **broccoli, brussels sprouts, cabbages, calabrese, cauliflowers, chinese cabbage, collards** (off-label), **kale** (off-label), **kohlrabi** (off-label), **oriental cabbage** (off-label) [3]
- Caterpillars in **celeriac** (off-label), **celery (outdoor)** (off-label), **chicory** (off-label), **collards** (off-label), **courgettes** (off-label), **fennel** (off-label), **kale** (off-label), **kohlrabi** (off-label), **oriental cabbage** (off-label), **pattisons** (off-label), **protected melons** (off-label), **protected pumpkins** (off-label), **protected strawberries** (off-label), **swedes** (off-label), **turnips** (off-label), **witloof** (off-label) [3]

FOR FULL CONDITIONS OF USE ALWAYS READ THE PRODUCT LABEL

- Celery fly in **celery (outdoor)** *(off-label)*, **fennel** *(off-label)* [3]
- Codling moth in **apples** [3]
- Diamond-back moth in **broccoli**, **brussels sprouts**, **cabbages**, **calabrese**, **cauliflowers**, **chinese cabbage**, **collards** *(off-label)*, **kale** *(off-label)*, **kohlrabi** *(off-label)*, **oriental cabbage** *(off-label)* [3]
- Insect pests in **blackcurrants** *(off-label)*, **bush fruit** *(off-label)*, **cane fruit** *(off-label)*, **grapevines** *(off-label)*, **hops** *(off-label)*, **nursery fruit trees** *(off-label)*, **ornamental plant production** *(off-label)*, **protected strawberries**, **redcurrants** *(off-label)*, **strawberries** *(off-label)*, **whitecurrants** *(off-label)* [3]
- Large white butterfly in **collards** *(off-label)*, **kale** *(off-label)*, **kohlrabi** *(off-label)*, **oriental cabbage** *(off-label)* [3]
- Leaf miner in **celeriac** *(off-label)*, **celery (outdoor)** *(off-label)*, **chicory** *(off-label)*, **courgettes** *(off-label)*, **fennel** *(off-label)*, **pattisons** *(off-label)*, **protected melons** *(off-label)*, **protected pumpkins** *(off-label)*, **swedes** *(off-label)*, **turnips** *(off-label)*, **witloof** *(off-label)* [3]
- Light brown apple moth in **blueberries** *(off-label)*, **cranberries** *(off-label)*, **gooseberries** *(off-label)*, **haskap** *(off-label)*, **protected blackcurrants** *(off-label)*, **protected blueberry** *(off-label)*, **protected gooseberries** *(off-label)*, **protected haskap** *(off-label)*, **protected redcurrants** *(off-label)* [3]
- Moths in **bush fruit** *(off-label)*, **cane fruit** *(off-label)*, **hops** *(off-label)*, **nursery fruit trees** *(off-label)*, **ornamental plant production** *(off-label)*, **strawberries** *(off-label)* [3]
- Onion thrips in **celery leaves** *(off-label)*, **cress** *(off-label)*, **frise** *(off-label)*, **lamb's lettuce** *(off-label)*, **lettuce** *(off-label)*, **radicchio** *(off-label)*, **scarole** *(off-label)* [3]
- Sawflies in **blackcurrants** *(off-label)*, **redcurrants** *(off-label)*, **whitecurrants** *(off-label)* [3]
- Silver Y moth in **celery (outdoor)** *(off-label)*, **celery leaves** *(off-label)*, **cress** *(off-label)*, **dwarf beans** *(off-label)*, **fennel** *(off-label)*, **french beans** *(off-label)*, **frise** *(off-label)*, **lamb's lettuce** *(off-label)*, **lettuce** *(off-label)*, **radicchio** *(off-label)*, **scarole** *(off-label)* [3]
- Small white butterfly in **broccoli**, **brussels sprouts**, **cabbages**, **calabrese**, **cauliflowers**, **chinese cabbage**, **collards** *(off-label)*, **kale** *(off-label)*, **kohlrabi** *(off-label)*, **oriental cabbage** *(off-label)* [3]
- Spotted wing drosophila in **blueberries** *(off-label)*, **cranberries** *(off-label)*, **gooseberries** *(off-label)*, **haskap** *(off-label)*, **protected blackcurrants** *(off-label)*, **protected blueberry** *(off-label)*, **protected gooseberries** *(off-label)*, **protected haskap** *(off-label)*, **protected redcurrants** *(off-label)*, **rubus hybrids** *(off-label)*, **strawberries** *(off-label)* [3]
- Thrips in **asparagus** *(off-label)*, **blackberries** *(off-label)*, **bulb onions**, **celeriac** *(off-label)*, **celery (outdoor)** *(off-label)*, **chicory** *(off-label)*, **courgettes** *(off-label)*, **fennel** *(off-label)*, **garlic**, **leeks**, **pattisons** *(off-label)*, **protected blackberries** *(off-label)*, **protected melons** *(off-label)*, **protected pumpkins** *(off-label)*, **protected raspberries** *(off-label)*, **protected strawberries**, **raspberries** *(off-label)*, **salad onions**, **shallots**, **swedes** *(off-label)*, **turnips** *(off-label)*, **witloof** *(off-label)* [3]
- Tortrix moths in **apples** [3]
- Western flower thrips in **protected aubergines**, **protected cucumbers**, **protected ornamentals**, **protected peppers**, **protected tomatoes** [1, 2]; **protected tomatoes** *(off-label)* [1]
- Winter moth in **blackcurrants** *(off-label)*, **redcurrants** *(off-label)*, **whitecurrants** *(off-label)* [3]

Extension of Authorisation for Minor Use (EAMUs)
- *asparagus* 20181201 [3]
- *blackberries* 20181203 [3]
- *blackcurrants* 20113223 [3]
- *blueberries* 20150775 [3]
- *bush fruit* 20181202 [3]
- *cane fruit* 20181202 [3]
- *celeriac* 20181204 [3]
- *celery (outdoor)* 20181206 [3]
- *celery leaves* 20181205 [3]
- *chicory* 20181204 [3]
- *collards* 20150102 [3]
- *courgettes* 20181204 [3]
- *cranberries* 20150775 [3]

SEE SECTION 3 FOR PRODUCTS ALSO REGISTERED

- ***cress*** *20181205* [3]
- ***dwarf beans*** *20150103* [3]
- ***fennel*** *20181206* [3]
- ***french beans*** *20150103* [3]
- ***frise*** *20181205* [3]
- ***gooseberries*** *20150775* [3]
- ***grapevines*** *20122222* [3]
- ***haskap*** *20150775* [3]
- ***hops*** *20181202* [3]
- ***kale*** *20150102* [3]
- ***kohlrabi*** *20150102* [3]
- ***lamb's lettuce*** *20181205* [3]
- ***lettuce*** *20181205* [3]
- ***nursery fruit trees*** *20181202* [3]
- ***oriental cabbage*** *20150102* [3]
- ***ornamental plant production*** *20181202* [3]
- ***pattisons*** *20181204* [3]
- ***protected blackberries*** *20181207* [3]
- ***protected blackcurrants*** *20150775* [3]
- ***protected blueberry*** *20150775* [3]
- ***protected gooseberries*** *20150775* [3]
- ***protected haskap*** *20150775* [3]
- ***protected melons*** *20181204* [3]
- ***protected pumpkins*** *20181204* [3]
- ***protected raspberries*** *20181207* [3]
- ***protected redcurrants*** *20150775* [3]
- ***protected strawberries*** *20181941* [3]
- ***protected tomatoes*** *20181208* [1]
- ***radicchio*** *20181205* [3]
- ***raspberries*** *20181203* [3]
- ***redcurrants*** *20113223* [3]
- ***rubus hybrids*** *20142018* [3]
- ***scarole*** *20181205* [3]
- ***strawberries*** *20181202* [3], *20192181* [3]
- ***swedes*** *20181204* [3]
- ***turnips*** *20181204* [3]
- ***whitecurrants*** *20113223* [3]
- ***witloof*** *20181204* [3]

Approval information
- Spinosad included in Annex I under EC Regulation 1107/2009

Efficacy guidance
- Product enters insects by contact from a treated surface or ingestion of treated plant material therefore good spray coverage is essential
- Some plants, for example Fuchsia flowers, can provide effective refuges from spray deposits and control of western flower thrips may be reduced [1]
- Apply to protected crops when western flower thrip nymphs or adults are first seen [1]
- Monitor western flower thrip development carefully to see whether further applications are necessary. A 2 spray programme at 5-7 d intervals may be needed when conditions favour rapid pest development [1]
- Treat top fruit and field crops when pests are first seen or at very first signs of crop damage [3]
- Ensure a rain-free period of 12 h after treatment before applying irrigation [3]
- To reduce possibility of development of resistance, adopt resistance management measures. See label and Section 5 for more information

Restrictions
- Maximum number of treatments 4 per crop for brassicas, onions, leeks [3]; 1 pre-blossom and 3 post blossom for apples, pears [3]

FOR FULL CONDITIONS OF USE ALWAYS READ THE PRODUCT LABEL

- Apply no more than 2 consecutive sprays. Rotate with another insecticide with a different mode of action or use no further treatment after applying the maximum number of treatments
- Establish whether any incoming plants have been treated and apply no more than 2 consecutive sprays. Maximum number of treatments 6 per structure per yr
- Avoid application in bright sunlight or into open flowers [1]

Crop-specific information
- HI 3 d for protected cucumbers, brassicas; 7 d for apples, pears, onions, leeks
- Test for tolerance on a small number of ornamentals or cucumbers before large scale treatment
- Some spotting of african violet flowers may occur

Environmental safety
- Dangerous for the environment
- Very toxic to aquatic organisms
- Whenever possible use an Integrated Pest Management system. Spinosad presents low risk to beneficial arthropods
- Product has low impact on many insect and mite predators but is harmful to adults of most parasitic wasps. Most beneficials may be introduced to treated plants when spray deposits are dry but an interval of 2 wk should elapse before introduction of parasitic wasps. See label for details
- Treatment may cause temporary reduction in abundance of insect and mite predators if present at application
- Exposure to direct spray is harmful to bees but dry spray deposits are harmless. Treatment of field crops should not be made in the heat of the day when bees are actively foraging [3]
- Buffer zone requirement 10 m for melon and pumpkin, 5 m for celeriac, swede and turnip
- Broadcast air-assisted LERAP [3] (40 m); LERAP Category B [3]

Hazard classification and safety precautions
Hazard Dangerous for the environment [1-3]; Very toxic to aquatic organisms [3]
Transport code 9
Packaging group III
UN Number 3082
Operator protection A, H [3]; U05a [1, 2]; U08, U20b [3]
Environmental protection E15a, E16a [3]; E15b [1, 2]; E17b [3] (40 m); E34, E38, H410 [1-3]
Storage and disposal D01, D05 [1, 2]; D02, D10b, D12a [1-3]

360   spirodiclofen

A tetronic acid which inhibits lipid biosynthesis
IRAC mode of action code: 23

Products

Envidor	Bayer CropScience	240 g/l	SC	17518

Uses
- Mussel scale in *apples*, *pears*
- Pear sucker in *pears*
- Red spider mites in *apples*, *pears*
- Rust mite in *apples*, *pears*, *plums* (off-label)
- Spider mites in *apricots* (off-label), *blackcurrants* (off-label), *cherries* (off-label), *gooseberries* (off-label), *nectarines* (off-label), *ornamental plant production* (off-label), *peaches* (off-label), *plums* (off-label), *protected chilli peppers* (off-label), *protected cucumbers* (off-label), *protected peppers* (off-label), *protected tomatoes* (off-label), *redcurrants* (off-label), *whitecurrants* (off-label)
- Two-spotted spider mite in *apples*, *hops* (off-label), *nursery fruit trees* (off-label), *pears*, *protected strawberries* (off-label), *strawberries* (off-label)

Extension of Authorisation for Minor Use (EAMUs)
- *apricots* 20160999
- *blackcurrants* 20161000

SEE SECTION 3 FOR PRODUCTS ALSO REGISTERED

SECTION 2

- *cherries* 20160998
- *gooseberries* 20161000
- *hops* 20172203
- *nectarines* 20160999
- *nursery fruit trees* 20172204
- *ornamental plant production* 20160997
- *peaches* 20160999
- *plums* 20160995
- *protected chilli peppers* 20160996
- *protected cucumbers* 20160996
- *protected peppers* 20160996
- *protected strawberries* 20171600
- *protected tomatoes* 20160996
- *redcurrants* 20161000
- *strawberries* 20171600
- *whitecurrants* 20161000

Approval information
- Spirodiclofen included in Annex 1 under EC Regulation 1107/2009

Restrictions
- Consult processor on crops grown for processing or for cider production

Environmental safety
- Broadcast air-assisted LERAP (18 m)

Hazard classification and safety precautions
Hazard Harmful
UN Number N/C
Risk phrases H317, H351
Operator protection A, H; U05a, U20c
Environmental protection E12c, E12f, E15b, E22c, E34, E38, H410; E17b (18 m)
Storage and disposal D01, D02, D09a, D10c, D12a
Medical advice M05a

361   spirotetramat

Tetramic acid derivative that acts on lipid synthesis
IRAC mode of action code: 23

Products

1	Batavia	Bayer CropScience	100 g/l	SC	18449
2	Clayton Occupy	Clayton	150 g/l	OD	18924
3	Movento	Bayer CropScience	150 g/l	OD	18435

Uses
- Aphids in **apples**, **apricots** *(off-label)*, **cherries**, **ornamental plant production** *(off-label)*, **peaches** *(off-label)*, **pears**, **plums**, **protected strawberries**, **strawberries**, **sweetcorn** *(off-label)*, **whitecurrants** *(off-label)* [1]; **baby leaf crops** *(off-label)*, **choi sum** *(off-label)*, **cress** *(off-label)*, **endives** *(off-label)*, **herbs (see appendix 6)** *(off-label)*, **lamb's lettuce** *(off-label)*, **land cress** *(off-label)*, **oriental cabbage** *(off-label)*, **protected baby leaf crops** *(off-label)*, **protected cress** *(off-label)*, **protected endives** *(off-label)*, **protected lamb's lettuce** *(off-label)*, **protected land cress** *(off-label)*, **protected oriental cabbage** *(off-label)*, **protected rocket** *(off-label)*, **protected spinach** *(off-label)*, **rocket** *(off-label)*, **spinach** *(off-label)* [3]; **blackcurrants** *(off-label)*, **blueberries** *(off-label)*, **gooseberries** *(off-label)*, **redcurrants** *(off-label)* [1, 3]; **broccoli**, **brussels sprouts**, **bulb onions**, **cabbages**, **calabrese**, **carrots**, **cauliflowers**, **collards**, **kale**, **lettuce**, **parsnips**, **potatoes**, **shallots**, **swedes**, **turnips** [2, 3]
- Bird-cherry aphid in **sweetcorn** *(off-label)* [1]
- Blackcurrant gall mite in **blackcurrants** *(off-label)*, **blueberries** *(off-label)*, **gooseberries** *(off-label)*, **redcurrants** *(off-label)*, **whitecurrants** *(off-label)* [1]

FOR FULL CONDITIONS OF USE ALWAYS READ THE PRODUCT LABEL

- Blackcurrant leaf midge in **blackcurrants** *(off-label)*, **blueberries** *(off-label)*, **gooseberries** *(off-label)*, **redcurrants** *(off-label)* [3]
- Damson-hop aphid in **hops** *(off-label)* [1]
- Glasshouse whitefly in **protected ornamentals** *(off-label)* [1]
- Lettuce root aphid in **baby leaf crops** *(off-label)*, **choi sum** *(off-label)*, **cress** *(off-label)*, **endives** *(off-label)*, **herbs (see appendix 6)** *(off-label)*, **lamb's lettuce** *(off-label)*, **land cress** *(off-label)*, **oriental cabbage** *(off-label)*, **protected baby leaf crops** *(off-label)*, **protected cress** *(off-label)*, **protected endives** *(off-label)*, **protected lamb's lettuce** *(off-label)*, **protected land cress** *(off-label)*, **protected oriental cabbage** *(off-label)*, **protected rocket** *(off-label)*, **protected spinach** *(off-label)*, **rocket** *(off-label)*, **spinach** *(off-label)* [3]
- Mealy aphid in **baby leaf crops** *(off-label)*, **choi sum** *(off-label)*, **cress** *(off-label)*, **endives** *(off-label)*, **herbs (see appendix 6)** *(off-label)*, **lamb's lettuce** *(off-label)*, **land cress** *(off-label)*, **oriental cabbage** *(off-label)*, **protected baby leaf crops** *(off-label)*, **protected cress** *(off-label)*, **protected endives** *(off-label)*, **protected lamb's lettuce** *(off-label)*, **protected land cress** *(off-label)*, **protected oriental cabbage** *(off-label)*, **protected rocket** *(off-label)*, **protected spinach** *(off-label)*, **rocket** *(off-label)*, **spinach** *(off-label)* [3]
- Mussel scale in **apples**, **pears** [1]
- Peach-potato aphid in **baby leaf crops** *(off-label)*, **choi sum** *(off-label)*, **cress** *(off-label)*, **endives** *(off-label)*, **herbs (see appendix 6)** *(off-label)*, **lamb's lettuce** *(off-label)*, **land cress** *(off-label)*, **oriental cabbage** *(off-label)*, **protected baby leaf crops** *(off-label)*, **protected cress** *(off-label)*, **protected endives** *(off-label)*, **protected lamb's lettuce** *(off-label)*, **protected land cress** *(off-label)*, **protected rocket** *(off-label)*, **protected spinach** *(off-label)*, **rocket** *(off-label)*, **spinach** *(off-label)* [3]; **protected ornamentals** *(off-label)* [1]
- Rose-grain aphid in **sweetcorn** *(off-label)* [1]
- Scale insects in **wine grapes** *(off-label)* [1]
- Soft scales Coccidae in **wine grapes** *(off-label)* [1]
- Tarsonemid mites in **protected strawberries**, **strawberries** [1]
- Vine phylloxera in **wine grapes** *(off-label)* [1]
- Western flower thrips in **protected ornamentals** *(off-label)* [1]
- Whitefly in **baby leaf crops** *(off-label)*, **choi sum** *(off-label)*, **cress** *(off-label)*, **endives** *(off-label)*, **herbs (see appendix 6)** *(off-label)*, **lamb's lettuce** *(off-label)*, **land cress** *(off-label)*, **oriental cabbage** *(off-label)*, **protected baby leaf crops** *(off-label)*, **protected cress** *(off-label)*, **protected endives** *(off-label)*, **protected lamb's lettuce** *(off-label)*, **protected land cress** *(off-label)*, **protected oriental cabbage** *(off-label)*, **protected rocket** *(off-label)*, **protected spinach** *(off-label)*, **rocket** *(off-label)*, **spinach** *(off-label)* [3]; **broccoli**, **brussels sprouts**, **bulb onions**, **cabbages**, **calabrese**, **cauliflowers**, **collards**, **kale**, **shallots**, **swedes**, **turnips** [2, 3]
- Woolly aphid in **apples** *(off-label)* [1]

Extension of Authorisation for Minor Use (EAMUs)
- **apples** *20192594* [1]
- **apricots** *20191055* [1]
- **baby leaf crops** *20183138* [3]
- **blackcurrants** *20191054* [1], *20190245* [3]
- **blueberries** *20191054* [1], *20190245* [3]
- **choi sum** *20180711* [3]
- **cress** *20183138* [3]
- **endives** *20183138* [3]
- **gooseberries** *20191054* [1], *20190245* [3]
- **herbs (see appendix 6)** *20180918* [3]
- **hops** *20191056* [1]
- **lamb's lettuce** *20183138* [3]
- **land cress** *20183138* [3]
- **oriental cabbage** *20180711* [3]
- **ornamental plant production** *20191058* [1]
- **peaches** *20191055* [1]
- **protected baby leaf crops** *20183138* [3]
- **protected cress** *20183138* [3]
- **protected endives** *20183138* [3]
- **protected lamb's lettuce** *20183138* [3]

SEE SECTION 3 FOR PRODUCTS ALSO REGISTERED

- ***protected land cress*** *20183138* [3]
- ***protected oriental cabbage*** *20183226* [3]
- ***protected ornamentals*** *20192597* [1]
- ***protected rocket*** *20183138* [3]
- ***protected spinach*** *20183138* [3]
- ***redcurrants*** *20191054* [1], *20190245* [3]
- ***rocket*** *20183138* [3]
- ***spinach*** *20183138* [3]
- ***sweetcorn*** *20193152* [1]
- ***whitecurrants*** *20191054* [1]
- ***wine grapes*** *20191057* [1]

Approval information
- Spirotetramat is included in Annex 1 under EC Regulation 1107/2009
- Accepted by BBPA for use on hops

Restrictions
- Application to baby leaf crops, cress, endive, lamb's lettuce, land cress, rocket and spinach must only be made between growth stages BBCH 12 and BBCH 49
- Applications to blackcurrants, redcurrants, blueberries and gooseberries must made between March and July and only in a year where there is no harvest of the crop
- DO NOT apply to the following ornamental plant species due to a high risk of phytotoxicity (plant damage): Alstroemeria spp., Begonia spp., Cyclamen spp., Euphorbia spp., Ficus spp., Fuchsia spp., Hedera spp., Hydrangea spp., Impatiens spp., Pelargonium spp., Populus spp., Salix spp., Saintpaulia spp., Tilia spp., Quercus frainetto [1]

Environmental safety
- Aquatic buffer zone 15 m in apples [1]

Hazard classification and safety precautions
Hazard Irritant, Dangerous for the environment [2, 3]
Transport code 9
Packaging group III
UN Number 3082
Risk phrases H317, H361 [1-3]; H319 [2, 3]
Operator protection A [1-3]; C [2, 3]; H [1]; U05a, U20b [1-3]; U14 [2, 3]
Environmental protection E12c, E15b, E34, H411 [1-3]; E17a [1] (10 m); E22c, E38 [2, 3]
Storage and disposal D02, D09a, D10c, D12a
Medical advice M03

362   starch, protein, oil and water

Natural plant extracts for control of moss, liverworts and algae

Products
MossKade	Fargro	n/a g/l	SL	00000

Uses
- Algae in **hard surfaces**
- Liverworts in **hard surfaces**
- Moss in **hard surfaces**

Approval information
- No listing required on Annex 1

Efficacy guidance
- Treat surfaces at first sign of moss, lichen, liverworts or algae since young growth is more susceptible.
- Must dry on the treated surface before it begins to work
- Summer growth is less susceptible than growth developing in autumn, winter or spring
- Repeat applications may be necessary for full efficacy

FOR FULL CONDITIONS OF USE ALWAYS READ THE PRODUCT LABEL

Hazard classification and safety precautions
UN Number N/C
Operator protection U08, U09a, U19a
Storage and disposal D06c, D07

363 sulfosulfuron

A sulfonylurea herbicide for grass and broad-leaved weed control in winter wheat
HRAC mode of action code: B

See also glyphosate + sulfosulfuron

Products

Monitor	Sumitomo	80% w/w	WG	17695

Uses

- Brome grasses in **winter wheat** *(moderate control of barren brome)*
- Chickweed in **winter wheat**
- Cleavers in **winter wheat**
- Couch in **winter wheat** *(moderate control only)*
- Loose silky bent in **winter wheat**
- Mayweeds in **winter wheat**
- Onion couch in **winter wheat** *(moderate control only)*

Approval information

- Sulfosulfuron included in Annex I under EC Regulation 1107/2009

Efficacy guidance

- For best results treat in early spring when annual weeds are small and growing actively. Avoid treatment when weeds are dormant for any reason
- Best control of onion couch is achieved when the weed has more than two leaves. Effects on bulbils or on growth in the following yr have not been examined
- An extended period of dry weather before or after treatment may result in reduced control
- The addition of a recommended surfactant is essential for full activity
- Specific follow-up treatments may be needed for complete control of some weeds
- Use only where competitively damaging weed populations have emerged otherwise yield may be reduced
- Sulfosulfuron is a member of the ALS-inhibitor group of herbicides. To avoid the build up of resistance do not use any product containing an ALS-inhibitor herbicide with claims for control of grass weeds more than once on any crop
- Use these products as part of a resistance management strategy that includes cultural methods of control and does not use ALS inhibitors as the sole chemical method of grass weed control

Restrictions

- Maximum total dose equivalent to one treatment at full dose
- Do not treat crops under stress
- Apply only after 1 Feb and from 3 expanded leaf stage
- Do not treat durum wheat or any undersown wheat crop
- Do not use in mixture or in sequence with any other sulfonyl urea herbicide on the same crop

Crop-specific information

- Latest use: flag leaf ligule just visible for winter wheat (GS 39)

Following crops guidance

- In the autumn following treatment winter wheat, winter rye, winter oats, triticale, winter oilseed rape, winter peas or winter field beans may be sown on any soil, and winter barley on soils with less than 60% sand
- In the next spring following application crops of wheat, barley, oats, maize, peas, beans, linseed, oilseed rape, potatoes or grass may be sown
- In the second autumn following application winter linseed may be sown, and winter barley on soils with more than 60% sand

SEE SECTION 3 FOR PRODUCTS ALSO REGISTERED

- Sugar beet or any other crop not mentioned above must not be drilled until the second spring following application
- Where winter oilseed rape is to be sown in the autumn following treatment soil cultivation to a minimum of 10 cm is recommended

Environmental safety
- Dangerous for the environment
- Very toxic to aquatic organisms
- Extremely dangerous to fish or other aquatic life. Do not contaminate surface waters or ditches with chemical or used container
- Take extreme care to avoid drift onto broad leaved plants or other crops, or onto ponds, waterways or ditches.
- Follow detailed label instructions for cleaning the sprayer to avoid damage to sensitive crops during subsequent use
- LERAP Category B

Hazard classification and safety precautions
Hazard Dangerous for the environment, Very toxic to aquatic organisms
UN Number N/C
Operator protection U20b
Environmental protection E15a, E16a, E16b, E34, E38, H410
Storage and disposal D09a, D11a, D12a

364   sulfoxaflor

An insecticide for use on protected crops
IRAC mode of action code: 4C

Products

Sequoia	Fargro	120 g/l	SC	18677

Uses
- Aphids in **protected aubergines, protected chilli peppers, protected courgettes, protected cucumbers, protected gherkins, protected melons, protected ornamentals, protected peppers, protected pumpkins, protected tomatoes, protected watermelon**
- Whitefly in **protected aubergines, protected chilli peppers, protected courgettes, protected cucumbers, protected gherkins, protected melons, protected ornamentals, protected peppers, protected pumpkins, protected tomatoes, protected watermelon**

Efficacy guidance
- Acts on the insect nicotinic-acetylcholine site at a unique target receptor. Symptoms appear almost immediately and complete mortality occurs within a few hours
- Test the product on a small number of plants before treating the whole crop, especially when treating sensitive species
- May adversely impact on beneficial insects used in IPM as well as bees used for pollination services. For further information and the latest advice on beneficial insects and mites and their integrated use, consult Corteva or Fargro Ltd

Restrictions
- See label for details of recommended ornamentals

Hazard classification and safety precautions
Transport code 9
Packaging group III
UN Number 3082
Operator protection A; U09a, U20b
Environmental protection E15b, E34, H411
Storage and disposal D01, D02, D10b, D11b, D22

FOR FULL CONDITIONS OF USE ALWAYS READ THE PRODUCT LABEL

365   sulphur

A broad-spectrum inorganic protectant fungicide, foliar feed and acaricide
FRAC mode of action code: M2

Products

1	Kumulus DF	BASF	80% w/w	WG	04707
2	Microthiol Special	UPL Europe	80% w/w	SC	16989
3	Thiopron	UPL Europe	82.5% g/l	SC	19147

Uses

- Disease control/foliar feed in **bilberries** *(off-label)*, **blueberries** *(off-label)*, **cranberries** *(off-label)*, **redcurrants** *(off-label)*, **whitecurrants** *(off-label)* [1]
- Foliar feed in **rye**, **triticale** [3]; **spring barley**, **spring wheat**, **winter barley**, **winter wheat** [1-3]; **spring oats**, **sugar beet**, **swedes**, **turnips**, **winter oats** [2, 3]; **winter oilseed rape** [1]
- Gall mite in **blackcurrants** [1]
- Powdery mildew in **apples**, **hops** [1, 2]; **blackcurrants**, **redcurrants**, **spring barley**, **spring oats**, **spring wheat**, **swedes**, **turnips**, **winter barley**, **winter oats**, **winter wheat** [2, 3]; **borage for oilseed production** *(off-label)*, **combining peas** *(off-label)*, **fodder beet** *(off-label)*, **grapevines**, **grass seed crops** *(off-label)*, **parsnips** *(off-label)*, **protected aubergines** *(off-label)*, **protected cayenne peppers** *(off-label)*, **protected chilli peppers** *(off-label)*, **protected cucumbers** *(off-label)*, **protected herbs (see appendix 6)** *(off-label)*, **protected peppers** *(off-label)*, **protected tomatoes** *(off-label)*, **vining peas** *(off-label)* [2]; **gooseberries**, **strawberries**, **sugar beet** [1-3]; **rye**, **table grapes**, **triticale** [3]; **wine grapes** [1, 3]
- Rust mite in **protected aubergines** *(off-label)*, **protected cayenne peppers** *(off-label)*, **protected chilli peppers** *(off-label)*, **protected cucumbers** *(off-label)*, **protected peppers** *(off-label)*, **protected tomatoes** *(off-label)* [2]
- Scab in **apples** [1]

Extension of Authorisation for Minor Use (EAMUs)

- **bilberries** *20061042* [1]
- **blueberries** *20061042* [1]
- **borage for oilseed production** *20151521* [2]
- **combining peas** *20151524* [2]
- **cranberries** *20061042* [1]
- **fodder beet** *20151526* [2]
- **grass seed crops** *20151523* [2]
- **parsnips** *20151527* [2]
- **protected aubergines** *20191907* [2]
- **protected cayenne peppers** *20191907* [2]
- **protected chilli peppers** *20191907* [2]
- **protected cucumbers** *20191907* [2]
- **protected herbs (see appendix 6)** *20151522* [2]
- **protected peppers** *20191907* [2]
- **protected tomatoes** *20191907* [2]
- **redcurrants** *20061042* [1]
- **vining peas** *20151524* [2]
- **whitecurrants** *20061042* [1]

Approval information

- Sulphur is included in Annex 1 under EC Regulation 1107/2009
- Accepted by BBPA for use on malting barley and hops (before burr)

Efficacy guidance

- Apply when disease first appears and repeat 2-3 wk later. Details of application rates and timing vary with crop, disease and product. See label for information
- Sulphur acts as foliar feed as well as fungicide and with some crops product labels vary in whether treatment recommended for disease control or growth promotion
- In grassland best results obtained at least 2 wk before cutting for hay or silage, 3 wk before grazing
- Treatment unlikely to be effective if disease already established in crop

SEE SECTION 3 FOR PRODUCTS ALSO REGISTERED

Restrictions

- Maximum number of treatments normally 2 per crop for grassland, sugar beet, parsnips, swedes, hops, protected herbs; 3 per yr for blackcurrants, gooseberries; 4 per crop on apples, pears but labels vary
- Do not use on sulphur-shy apples (Beauty of Bath, Belle de Boskoop, Cox's Orange Pippin, Lanes Prince Albert, Lord Derby, Newton Wonder, Rival, Stirling Castle) or pears (Doyenne du Comice)
- Do not use on gooseberry cultivars Careless, Early Sulphur, Golden Drop, Leveller, Lord Derby, Roaring Lion, or Yellow Rough
- Do not use on apples or gooseberries when young, under stress or if frost imminent
- Do not use on fruit for processing, on grapevines during flowering or near harvest on grapes for wine-making
- Do not use on hops at or after burr stage
- Do not spray top or soft fruit with oil or within 30 d of an oil-containing spray

Crop-specific information

- Latest use: before burr stage for hops; before end Sep for parsnips, swedes, sugar beet; fruit swell for gooseberries; milky ripe stage for cereals
- HI cutting grass for hay or silage 2 wk; grazing grassland 3 wk

Environmental safety

- Sulphur products are attractive to livestock and must be kept out of their reach
- Do not empty into drains
- To protect non target insects/arthropods respect an unsprayed buffer zone of 5 m to non-crop land [3]

Hazard classification and safety precautions

Transport code 9 [3]
Packaging group III [3]
UN Number 3082 [3]; N/C [1, 2]
Risk phrases H317 [3]
Operator protection A, H [3]; U20b [2, 3]; U20c [1]
Environmental protection E15a
Storage and disposal D01 [2, 3]; D09a, D11a [1-3]

366 tau-fluvalinate

A contact pyrethroid insecticide for cereals and oilseed rape
IRAC mode of action code: 3

Products

Mavrik	Adama	240 g/l	EW	10612

Uses

- Aphids in **borage for oilseed production** *(off-label)*, **evening primrose** *(off-label)*, **grass seed crops** *(off-label)*, **honesty** *(off-label)*, **linseed** *(off-label)*, **mustard** *(off-label)*, **rye** *(off-label)*, **spring barley**, **spring oilseed rape**, **spring wheat**, **triticale** *(off-label)*, **winter barley**, **winter oilseed rape**, **winter wheat**
- Barley yellow dwarf virus vectors in **winter barley**, **winter wheat**
- Cabbage stem flea beetle in **winter oilseed rape**
- Pollen beetle in **spring oilseed rape**, **winter oilseed rape**

Extension of Authorisation for Minor Use (EAMUs)

- **borage for oilseed production** *20141329*
- **evening primrose** *20141329*
- **grass seed crops** *20193244*
- **honesty** *20141329*
- **linseed** *20141329*
- **mustard** *20141329*
- **rye** *20193244*
- **triticale** *20193244*

FOR FULL CONDITIONS OF USE ALWAYS READ THE PRODUCT LABEL

Approval information
- Tau-fluvalinate included in Annex 1 under EC Regulation 1107/2009
- Accepted by BBPA for use on malting barley

Efficacy guidance
- For BYDV control on winter cereals follow local warnings or spray high risk crops in mid-Oct and make repeat application in late autumn/early winter if aphid activity persists
- For summer aphid control on cereals spray once when aphids present on two thirds of ears and increasing
- On oilseed rape treat peach potato aphids in autumn in response to local warning and repeat if necessary
- Best control of pollen beetle in oilseed rape obtained from treatment at green to yellow bud stage and repeat if necessary
- Good spray cover of target essential for best results
- Do not mix with boron or products containing boron [1]

Restrictions
- Maximum total dose equivalent to two full dose treatments on oilseed rape. See label for dose rates on cereals
- Must not be applied to a cereal crop if any other product containing either a pyrethroid or dimethoate has been applied to that crop after the start of ear emergence
- A minimum of 14 d must elapse between applications to cereals

Crop-specific information
- Latest use: before caryopsis watery ripe (GS 71) for barley; before flowering for oilseed rape; before kernel medium milk (GS 75) for wheat

Environmental safety
- Dangerous for the environment
- Very toxic to aquatic organisms
- High risk to non-target insects or other arthropods. Do not spray within 6 m of the field boundary [1]
- Avoid spraying oilseed rape within 6 m of field boundary to reduce effects on certain non-target species or other arthropods
- Must not be applied to cereals if any product containing a pyrethroid insecticide or dimethoate has been sprayed after the start of ear emergence (GS 51)
- LERAP Category A

Hazard classification and safety precautions
Hazard Dangerous for the environment, Very toxic to aquatic organisms
Transport code 9
Packaging group III
UN Number 3082
Operator protection A, C, H; U05a, U10, U11, U19a, U20a
Environmental protection E15a, E16c, E16d, E22a, H410
Storage and disposal D01, D02, D05, D09a, D10c

367   tebuconazole

A systemic triazole fungicide for cereals and other field crops
FRAC mode of action code: 3

See also azoxystrobin + tebuconazole
    bixafen + prothioconazole + tebuconazole
    bromuconazole + tebuconazole
    clothianidin + prothioconazole + tebuconazole + triazoxide
    difenoconazole + fludioxonil + tebuconazole
    fludioxonil + tebuconazole
    fluopyram + prothioconazole + tebuconazole
    fluoxastrobin + prothioconazole + tebuconazole
    fluoxastrobin + tebuconazole
    imidacloprid + tebuconazole + triazoxide
    prochloraz + proquinazid + tebuconazole
    prochloraz + tebuconazole
    prothioconazole + spiroxamine + tebuconazole
    prothioconazole + tebuconazole
    prothioconazole + tebuconazole + triazoxide
    spiroxamine + tebuconazole
    sulphur + tebuconazole

Products

1	Clayton Tebucon 250 EW	Clayton	250 g/l	EW	17823
2	Deacon	Nufarm UK	200 g/l	EW	19003
3	Erase	FMC Agro	200 g/l	EC	18741
4	Fathom	FMC Agro	200 g/l	EC	18737
5	Fezan	Sipcam	250 g/l	EW	17082
6	Harvest Teb 250	Harvest	250 g/l	EW	17706
7	Hecates	Ventura	430 g/l	SC	17403
8	Mitre	Nufarm UK	200 g/l	EW	19002
9	Odin	Rotam	250 g/l	EW	18016
10	Orius	Nufarm UK	200 g/l	EW	18992
11	Rosh	Harvest	250 g/l	EW	17673
12	Savannah	Rotam	250 g/l	EW	18058
13	Spekfree	Rotam	430 g/l	SC	16004
14	Tebucur 250	Belchim	255.1 g/l	EW	13975
15	Tebuzol 250	UPL Europe	250 g/l	EW	17417
16	Teson	Belchim	250 g/l	EW	17128
17	Tesoro	Belchim	250 g/l	EW	17222
18	Toledo	Rotam	430 g/l	SC	18298
19	Tubosan	Belchim	250 g/l	EW	17127
20	Ulysses	Rotam	430 g/l	SC	16052

Uses

- Alternaria in *cabbages*, *carrots*, *horseradish* [3, 4, 7, 9, 12, 13, 17, 18, 20]; *parsnips* [9, 12]; *spring oilseed rape*, *winter oilseed rape* [1-4, 6-20]; *spring wheat*, *winter wheat* [7, 13, 18, 20]
- Botrytis in *linseed* [7]; *linseed* (reduction) [3, 4, 17]
- Brown rust in *rye* [7, 13, 18, 20]; *spring barley*, *spring wheat*, *winter barley*, *winter wheat* [1-20]; *spring oats*, *winter oats* [9, 12]; *spring rye* [2-6, 8-10, 12, 14, 16, 17, 19]; *triticale* [3-5, 7, 9, 12, 13, 18, 20]; *winter rye* [1-6, 8-12, 14-17, 19]
- Canker in *apples* (off-label), *pears* (off-label), *quinces* (off-label) [3, 4]
- Chocolate spot in *spring field beans*, *winter field beans* [3, 4, 7, 9-15, 17, 18, 20]; *spring field beans* (moderate control), *winter field beans* (moderate control) [2]
- Cladosporium in *spring wheat*, *winter wheat* [7, 13, 18, 20]
- Crown rust in *spring oats* [2-5, 7, 8, 13, 14, 18, 20]; *spring oats* (reduction), *winter oats* (reduction) [1, 6, 11, 15-17, 19]; *winter oats* [2-5, 7, 8, 10, 13, 14, 18, 20]
- Dark leaf spot in *spring oilseed rape*, *winter oilseed rape* [7, 13, 18, 20]

FOR FULL CONDITIONS OF USE ALWAYS READ THE PRODUCT LABEL

- Disease control in **borage for oilseed production** *(off-label)*, **canary flower (echium spp.)** *(off-label)*, **evening primrose** *(off-label)*, **grass seed crops** *(off-label)*, **honesty** *(off-label)*, **mallow (althaea spp.)** *(off-label)*, **mustard** *(off-label)*, **parsley root** *(off-label)*, **triticale** *(off-label)* [14]; **linseed** [1, 11, 19]; **spring field beans, winter field beans** [1, 19]
- Ear diseases in **spring wheat, winter wheat** [7, 13, 18, 20]
- Fusarium in **spring wheat, winter wheat** [7, 13, 18, 20]
- Fusarium ear blight in **spring wheat, winter wheat** [1-4, 6, 8, 10, 11, 14-17, 19]; **triticale** [3, 4]
- Glume blotch in **spring wheat, winter wheat** [1-8, 10, 11, 13-20]; **triticale** [3-5, 7, 13, 18, 20]
- Light leaf spot in **cabbages** [3, 4, 7, 9, 12, 13, 17, 18, 20]; **spring oilseed rape, winter oilseed rape** [1-20]
- Lodging control in **spring oilseed rape, winter oilseed rape** [1, 3, 4, 6, 8, 10, 11, 15-17, 19]
- Net blotch in **spring barley, winter barley** [1, 3-20]; **spring oats, spring rye, spring wheat, triticale, winter oats, winter rye, winter wheat** [9, 12]
- Phoma in **spring oilseed rape, winter oilseed rape** [1, 3-6, 8-12, 14-17, 19]
- Phoma leaf spot in **spring oilseed rape, winter oilseed rape** [7, 13, 18, 20]
- Powdery mildew in **apples** *(off-label)*, **pears** *(off-label)*, **quinces** *(off-label)* [3, 4]; **cabbages, carrots, parsnips, swedes, turnips** [3, 4, 7, 9, 12, 13, 17, 18, 20]; **linseed** [3, 4, 7, 14, 17]; **rye** [7, 13, 18, 20]; **spring barley, spring wheat, winter barley, winter wheat** [1, 3-20]; **spring barley** *(moderate control)*, **spring wheat** *(moderate control)*, **winter barley** *(moderate control)*, **winter wheat** *(moderate control)* [2]; **spring oats, winter oats** [1-20]; **spring rye** [2-6, 8-10, 12, 14, 16, 17, 19]; **triticale** [3-5, 9, 12]; **winter rye** [1-6, 8-12, 14-17, 19]
- Rhynchosporium in **rye** [7, 13, 18, 20]; **spring barley, winter barley** [1, 3-20]; **spring barley** *(moderate control)*, **spring rye** *(moderate control)*, **winter barley** *(moderate control)*, **winter rye** *(moderate control)* [2]; **spring oats, spring wheat, triticale, winter oats, winter wheat** [9, 12]; **spring rye** [3-6, 8-10, 12, 14, 16, 17, 19]; **winter rye** [1, 3-6, 8-12, 14-17, 19]
- Ring spot in **cabbages** [3, 4, 7, 9, 12, 13, 17, 18, 20]; **spring oilseed rape, winter oilseed rape** [7, 13, 18, 20]; **spring oilseed rape** *(reduction)*, **winter oilseed rape** *(reduction)* [1, 3, 4, 6, 8, 10, 11, 15-17, 19]
- Rust in **borage for oilseed production** *(off-label)*, **canary flower (echium spp.)** *(off-label)*, **evening primrose** *(off-label)*, **grass seed crops** *(off-label)*, **honesty** *(off-label)*, **mallow (althaea spp.)** *(off-label)*, **parsley root** *(off-label)*, **triticale** *(off-label)* [14]; **dwarf beans** *(off-label)*, **french beans** *(off-label)*, **runner beans** *(off-label)* [3, 4]; **leeks** [7, 17]; **linseed** [1, 11]; **mustard** *(off-label)* [3, 4, 14]; **spring field beans, winter field beans** [1, 3, 4, 7, 9, 11-15, 17-20]
- Sclerotinia in **carrots** [3, 4, 7, 9, 12, 13, 17, 18, 20]; **parsnips** [9, 12]; **spring oilseed rape, winter oilseed rape** [2, 5, 7, 13, 18, 20]
- Sclerotinia stem rot in **spring oilseed rape, winter oilseed rape** [1, 3, 4, 6, 8-12, 14-17, 19]
- Septoria leaf blotch in **spring wheat, winter wheat** [1, 3-8, 10, 11, 13-20]; **triticale** [3-5, 7, 13, 18, 20]
- Sooty moulds in **spring wheat, winter wheat** [1-4, 6, 8, 10, 11, 14-17, 19]; **triticale** [3, 4]
- Stem canker in **spring oilseed rape, winter oilseed rape** [1, 3, 4, 6-20]
- White rot in **bulb onion sets** *(off-label)* [3, 4]
- Yellow rust in **rye** [7, 13, 18, 20]; **spring barley, spring wheat, winter barley, winter wheat** [1-20]; **spring oats, winter oats** [9, 12]; **spring rye** [2-6, 8-10, 12, 14, 16, 17, 19]; **triticale** [3-5, 9, 12]; **winter rye** [1-6, 8-12, 14-17, 19]

Extension of Authorisation for Minor Use (EAMUs)
- **apples** *20190753* [3], *20190834* [4]
- **borage for oilseed production** *20122270* [14]
- **bulb onion sets** *20190754* [3], *20190833* [4]
- **canary flower (echium spp.)** *20122270* [14]
- **dwarf beans** *20190756* [3], *20190831* [4]
- **evening primrose** *20122270* [14]
- **french beans** *20190756* [3], *20190831* [4]
- **grass seed crops** *20122273* [14]
- **honesty** *20122270* [14]
- **mallow (althaea spp.)** *20122272* [14]
- **mustard** *20190755* [3], *20190832* [4], *20122270* [14]
- **parsley root** *20122272* [14]
- **pears** *20190753* [3], *20190834* [4]

SEE SECTION 3 FOR PRODUCTS ALSO REGISTERED

- **quinces** *20190753* [3], *20190834* [4]
- **runner beans** *20190756* [3], *20190831* [4]
- **triticale** *20122271* [14]

Approval information
- Tebuconazole is included in Annex 1 under EC Regulation 1107/2009
- Accepted by BBPA for use on malting barley
- Approval expiry 01 Nov 2020 [1]

Efficacy guidance
- For best results apply at an early stage of disease development before infection spreads to new crop growth
- To protect flag leaf and ear from Septoria diseases apply from flag leaf emergence to ear fully emerged (GS 37-59). Earlier application may be necessary where there is a high risk of infection
- For light leaf spot control in oilseed rape apply in autumn/winter with a follow-up spray in spring/summer if required
- For control of most other diseases spray at first signs of infection with a follow-up spray 2-4 wk later if necessary. See label for details
- For disease control in cabbages a 3-spray programme at 21-28 d intervals will give good control
- Tebuconazole is a DMI fungicide. Resistance to some DMI fungicides has been identified in Septoria leaf blotch which may seriously affect performance of some products. For further advice contact a specialist advisor and visit the Fungicide Resistance Action Group (FRAG)-UK website

Restrictions
- Maximum total dose equivalent to 1 full dose treatment on linseed; 2 full dose treatments on wheat, barley, rye, oats, field beans, swedes, turnips, onions; 2.25 full dose treatments on cabbages; 2.5 full dose treatments on oilseed rape; 3 full dose treatments on leeks, parsnips, carrots
- Do not treat durum wheat
- Apply only to listed oat varieties (see label) and do not apply to oats in tank mixture
- Do not apply before swedes and turnips have a root diameter of 2.5 cm, or before heart formation in cabbages, or before button formation in Brussels sprouts
- Consult processor before use on crops for processing
- Newer authorisations for tebuconazole products require application to cereals only after GS 30 and applications to oilseed rape and linseed after GS 20 - check label
- Must not be applied via hand-held equipment [18]
- Applications to linseed must be made after BBCH 20, applications to field bean must be made after BBCH 40
- For use on cereals a maximum individual dose of 1.25 l/ha applies after BBCH 30 and before early boot stage (BBCH39). A further maximum individual dose of 1.25 l/ha cannot be applied until after BBCH 40 stage [10]
- Applications must only be made in the Spring [10]
- For use on oilseed rape a single application must only be made after BBCH 20 [10]

Crop-specific information
- Latest use: before grain milky-ripe for cereals (GS 71); when most seed green-brown mottled for oilseed rape (GS 6,3); before brown capsule for linseed
- HI field beans, swedes, turnips, linseed, winter oats 35 d; market brassicas, carrots, horseradish, parsnips 21 d; leeks 14 d
- Some transient leaf speckling on wheat or leaf reddening/scorch on oats may occur but this has not been shown to reduce yield response to disease control
- Do not handle treated apples, pears or quince for at least 3 days after treatment [3]

Environmental safety
- Dangerous for the environment
- Toxic to aquatic organisms
- LERAP Category B [1, 3-6, 8-12, 14-19]

FOR FULL CONDITIONS OF USE ALWAYS READ THE PRODUCT LABEL

Hazard classification and safety precautions

Hazard Harmful, Dangerous for the environment [1, 6, 7, 9, 11-20]; Irritant [2-5, 8, 10]; Harmful if swallowed [1, 11, 15]; Harmful if inhaled [11, 15]; Very toxic to aquatic organisms [1, 11]

Transport code 9

Packaging group III

UN Number 3082

Risk phrases H317 [3, 4]; H318 [3-6, 11, 14-17, 19]; H319 [1, 2, 8, 10]; H335 [1, 14, 16, 17]; H361 [1-20]

Operator protection A, C, H; U05a, U20a [1-20]; U09a, U12 [7, 13, 18, 20]; U11 [1-6, 8-12, 14-17, 19]; U14, U15 [2-5, 7, 8, 10, 13, 18, 20]; U19a [9, 12, 14]

Environmental protection E15a [2-16, 18, 20]; E16a [1, 3-6, 8-12, 14-19]; E34 [1-20]; E38 [1, 6, 7, 11, 13, 15-20]; H410 [1-4, 8, 10, 11, 13, 15, 18, 20]; H411 [5-7, 9, 12, 14, 16, 17, 19]

Storage and disposal D01, D02, D09a [1-20]; D05, D10b [1, 6, 9, 11, 12, 14-17, 19]; D10c [7, 13, 18, 20]; D12a [1, 6, 7, 11, 13, 15-20]

Medical advice M03 [1, 6, 7, 9, 11-20]; M05a [7, 11, 13, 15, 18, 20]

368   tebuconazole + trifloxystrobin

A conazole and stobilurin fungicide mixture for wheat and vegetable crops
FRAC mode of action code: 3 + 11

See also trifloxystrobin

Products

1	Clayton Bestow	Clayton	200:100 g/l	SC	17185
2	Dedicate	Bayer CropScience	200:100 g/l	SC	17003
3	Fortify	Rigby Taylor	200:100 g/l	SC	17718
4	Fusion	Rigby Taylor	200:100 g/l	SC	17225
5	Nativo 75 WG	Bayer CropScience	50:25% w/w	WG	16867
6	Oath	Harvest	200:100 g/l	SC	18520

Uses

* Alternaria in **broccoli, brussels sprouts, cabbages, calabrese, carrots, cauliflowers, celeriac** *(off-label)*, **swedes** *(off-label)*, **turnips** *(off-label)* [5]
* Anthracnose in **amenity grassland, managed amenity turf** [4]
* Black blight in **celeriac** *(off-label)*, **celery (outdoor)** *(off-label)*, **swedes** *(off-label)*, **turnips** *(off-label)* [5]
* Black root rot in **ornamental plant production** *(off-label)*, **protected ornamentals** *(off-label)* [5]
* Dollar spot in **amenity grassland, managed amenity turf** [1-4, 6]
* Fusarium in **amenity grassland, managed amenity turf** [1-3, 6]
* Fusarium patch in **amenity grassland, managed amenity turf** [4]
* Leaf spot in **horseradish** *(off-label)*, **parsley root** *(off-label)*, **parsnips** *(off-label)*, **salsify** *(off-label)* [5]
* Light leaf spot in **broccoli, brussels sprouts, cabbages, calabrese, cauliflowers** [5]
* Melting out in **amenity grassland, managed amenity turf** [1-3, 6]
* Mycosphaerella in **celeriac** *(off-label)*, **swedes** *(off-label)*, **turnips** *(off-label)* [5]
* Phoma leaf spot in **broccoli, brussels sprouts, cabbages, calabrese, cauliflowers** [5]
* Powdery mildew in **broccoli, brussels sprouts, cabbages, calabrese, carrots, cauliflowers, celeriac** *(off-label)*, **horseradish** *(off-label)*, **parsley root** *(off-label)*, **parsnips** *(off-label)*, **salsify** *(off-label)*, **swedes** *(off-label)*, **turnips** *(off-label)*, **wine grapes** *(off-label)* [5]
* Red thread in **amenity grassland, managed amenity turf** [1-4, 6]
* Ring spot in **broccoli, brussels sprouts, cabbages, calabrese, cauliflowers** [5]
* Rust in **amenity grassland, managed amenity turf** [1-3, 6]; **leeks** *(off-label)*, **salad onions** *(off-label)* [5]
* Sclerotinia in **horseradish** *(off-label)*, **parsley root** *(off-label)*, **parsnips** *(off-label)*, **salsify** *(off-label)* [5]
* Sclerotinia rot in **carrots** [5]
* White blister in **broccoli, brussels sprouts, cabbages, calabrese, cauliflowers** [5]
* White rust in **ornamental plant production** *(off-label)*, **protected ornamentals** *(off-label)* [5]

Extension of Authorisation for Minor Use (EAMUs)
- *celeriac* *20182969* [5]
- *celery (outdoor)* *20182970* [5]
- *horseradish* *20151959* [5]
- *leeks* *20182971* [5]
- *ornamental plant production* *20192513* [5]
- *parsley root* *20151959* [5]
- *parsnips* *20151959* [5]
- *protected ornamentals* *20192513* [5]
- *salad onions* *20182971* [5]
- *salsify* *20151959* [5]
- *swedes* *20182969* [5]
- *turnips* *20182969* [5]
- *wine grapes* *20182972* [5]

Approval information
- Tebuconazole and trifloxystrobin included in Annex I under EC Regulation 1107/2009

Efficacy guidance
- Best results obtained from treatment at early stages of disease development. Further treatment may be needed if disease attack is prolonged
- Applications to established infections of any disease are likely to be less effective
- Treatment may give some control of *Stemphylium botryosum* and leaf blotch in leeks
- Tebuconazole is a DMI fungicide. Resistance to some DMI fungicides has been identified in Septoria leaf blotch which may seriously affect performance of some products. For further advice contact a specialist advisor and visit the Fungicide Resistance Action Group (FRAG)-UK website
- Trifloxystrobin is a member of the QoI cross resistance group. Product should be used preventatively and not relied on for its curative potential
- Use product as part of an Integrated Crop Management strategy incorporating other methods of control, including where appropriate other fungicides with a different mode of action. Do not apply more than two foliar applications of QoI containing products to any cereal crop, broccoli, calabrese or cauliflower. Do not apply more than three applications to Brussels sprouts, cabbage, carrots or leeks
- Do not apply to grass during drought conditions nor to frozen turf

Restrictions
- Maximum number of treatments 2 per crop for broccoli, calabrese, cauliflowers; 3 per crop for Brussels sprouts, cabbages, carrots and 1 for leeks
- In addition to the maximum number of treatments per crop a maximum of 3 applications may be applied to the same ground in one calendar year
- Consult processor before use on vegetable crops for processing
- Horizontal boom sprayers must be fitted with three star drift reduction technology for all uses [5]
- Low drift spraying equipment must be operated according to the specific conditions stated in the official three star rating for that equipment as published on HSE Chemicals Regulation Division's website. These operating conditions must be maintained until 30m from the top of the bank of any surface water bodies [5]

Crop-specific information
- Latest use: before milky ripe stage on winter wheat
- HI 35 d for wheat; 21 d for all other crops
- Performance against leaf spot diseases of brassicas, *Alternaria* leaf blight of carrots and rust in leeks may be improved by mixing with an approved sticker/wetter

Environmental safety
- Dangerous for the environment
- Very toxic to aquatic organisms
- Buffer zone requirement 12 m, 18 m on protected and outdoor ornamentals [5]
- LERAP Category B [1-4, 6]

Hazard classification and safety precautions
    Hazard Harmful, Dangerous for the environment [1-6]; Very toxic to aquatic organisms [1-4, 6]

FOR FULL CONDITIONS OF USE ALWAYS READ THE PRODUCT LABEL

Transport code 9
Packaging group III
UN Number 3077 [5]; 3082 [1-4, 6]
Risk phrases H317, H319 [5]; H335 [4]; H361 [1-6]
Operator protection A [1-6]; H [4, 5]; P [1-3, 6]; U05a [1-3, 5, 6]; U09a [1-3, 6]; U09b [4, 5]; U19a [1-4, 6]; U20b [1-6]
Environmental protection E15a, E34, E38, H410 [1-6]; E16a [1-4, 6]
Storage and disposal D01, D02, D09a, D10b, D12a [1-6]; D05 [4]
Medical advice M03

369   tebufenpyrad

A pyrazole mitochondrial electron transport inhibitor (METI) aphicide and acaricide
IRAC mode of action code: 21

Products

| Clayton Bonsai | Clayton | 20% w/w | WB | 17983 |

Uses

- Damson-hop aphid in *hops*
- Red spider mites in *apples*, *pears*
- Two-spotted spider mite in *hops*, *protected roses*, *strawberries*

Approval information

- Tebufenpyrad included in Annex 1 under EC Regulation 1107/2009
- Accepted by BBPA for use on hops

Efficacy guidance

- Acts on eggs (except winter eggs) and all motile stages of spider mites up to adults
- Treat spider mites from 80% egg hatch but before mites become established
- For effective control total spray cover of the crop is required
- Product can be used in a programme to give season-long control of damson-hop aphids coupled with mite control
- Where aphids resistant to tebufenpyrad occur in hops control is unlikely to be satisfactory and repeat treatments may result in lower levels of control. Where possible use different active ingredients in a programme

Restrictions

- Maximum total dose equivalent to one full dose treatment on apples, pears, strawberries; 3 full dose treatments on hops
- Other mitochondrial electron transport inhibitor (METI) acaricides should not be applied to the same crop in the same calendar yr either separately or in mixture
- Do not treat apples before 90% petal fall
- Small-scale testing of rose varieties to establish tolerance recommended before use
- Inner liner of container must not be removed
- Must not be applied via hand-held equipment

Crop-specific information

- Latest use: end of burr stage for hops
- HI strawberries 3 d; apples, bilberries, blackcurrants, blueberries, cranberries, gooseberries, pears, redcurrants, whitecurrants 7d; blackberries, plums, raspberries 21 d
- Product has no effect on fruit quality or finish

Environmental safety

- Dangerous for the environment
- Very toxic to aquatic organisms
- High risk to bees. Do not apply to crops in flower or to those in which bees are actively foraging. Do not apply when flowering weeds are present
- Broadcast air-assisted LERAP (18 m); LERAP Category B

SEE SECTION 3 FOR PRODUCTS ALSO REGISTERED

Hazard classification and safety precautions

    Hazard Harmful, Dangerous for the environment

    Transport code 9

    Packaging group III

    UN Number 3077

    Risk phrases R20, R22a, R50, R53a

    Operator protection A; U02a, U05a, U09a, U13, U14, U20b

    Environmental protection E12a, E12e, E15a, E16a, E16b, E38; E17b (18 m)

    Storage and disposal D01, D02, D09a, D11a, D12a

    Medical advice M05a

370   tefluthrin

A soil acting pyrethroid insecticide seed treatment

IRAC mode of action code: 3

See also fludioxonil + tefluthrin

Products

Force ST	Syngenta	200 g/l	CF	19042

Uses

- Millipedes in **fodder beet** *(seed treatment)*, **sugar beet** *(seed treatment)*
- Pygmy beetle in **fodder beet** *(seed treatment)*, **sugar beet** *(seed treatment)*
- Springtails in **fodder beet** *(seed treatment)*, **sugar beet** *(seed treatment)*
- Symphylids in **fodder beet** *(seed treatment)*, **sugar beet** *(seed treatment)*

Approval information

- Tefluthrin included in Annex 1 under EC Regulation 1107/2009

Efficacy guidance

- Apply during process of pelleting beet seed. Consult manufacturer for details of specialist equipment required
- Micro-capsule formulation allows slow release to provide a protection zone around treated seed during establishment

Restrictions

- Maximum number of treatments 1 per batch of seed
- Sow treated seed as soon as possible. Do not store treated seed from one drilling season to next
- If used in areas where soil erosion by wind or water likely, measures must be taken to prevent this happening
- Can cause a transient tingling or numbing sensation to exposed skin. Avoid skin contact with product, treated seed and dust throughout all operations in the seed treatment plant and at drilling
- To protect birds the product must be entirely incorporated in the soil; ensure that the product is fully incorporated at the end of rows. Remove spillages

Crop-specific information

- Latest use: before drilling seed
- Treated seed must be drilled within the season of treatment

Environmental safety

- Dangerous for the environment
- Very toxic to aquatic organisms
- Keep treated seed secure from people, domestic stock/pets and wildlife at all times during storage and use
- Treated seed harmful to game and wild life. Bury spillages
- In the event of seed spillage clean up as much as possible into the related seed sack and bury the remainder completely
- Do not apply treated seed from the air
- Keep livestock out of areas drilled with treated seed for at least 80 d

FOR FULL CONDITIONS OF USE ALWAYS READ THE PRODUCT LABEL

SECTION 2

Hazard classification and safety precautions
Hazard Harmful, Dangerous for the environment, Harmful if inhaled, Very toxic to aquatic organisms
Transport code 9
Packaging group III
UN Number 3082
Risk phrases H317
Operator protection A, D, E, H; U02a, U04a, U05a, U08, U20b
Environmental protection E06a (80 d); E15a, E34, E38, H410
Storage and disposal D01, D02, D05, D09a, D11a, D12a
Treated seed S02, S03, S04b, S05, S07
Medical advice M05b

371   tetradecadienyl + trienyl acetate

A pheromone insecticide for use in glasshouses

Products

Isonet T	Fargro	8.3% w/w	VP	17929

Uses

- Tomato leaf miner (Tuta absoluta) in **protected aubergines**, **protected chilli peppers**, **protected peppers**, **protected tomatoes**

Approval information

- Tetradecadienyl acetate included in Annex I under EC Regulation 1107/2009

Restrictions

- For use in greenhouses only
- Do not contaminate water with the product or its container

Hazard classification and safety precautions
Hazard Irritant, Dangerous for the environment
Transport code 9
Packaging group III
UN Number 3082
Risk phrases H315
Operator protection A; U05a, U09c
Environmental protection H410
Storage and disposal D01, D12a, D12b, D22

372   thiabendazole

A systemic, curative and protectant benzimidazole (MBC) fungicide
FRAC mode of action code: 1

See also imazalil + thiabendazole
metalaxyl + thiabendazole

Products

Storite Excel	Syngenta	500 g/l	SC	12705

Uses

- Dry rot in **seed potatoes** (tuber treatment - post-harvest), **ware potatoes** (tuber treatment - post-harvest)
- Gangrene in **seed potatoes** (tuber treatment - post-harvest), **ware potatoes** (tuber treatment - post-harvest)
- Silver scurf in **seed potatoes** (tuber treatment - post-harvest), **ware potatoes** (tuber treatment - post-harvest)
- Skin spot in **seed potatoes** (tuber treatment - post-harvest), **ware potatoes** (tuber treatment - post-harvest)

SEE SECTION 3 FOR PRODUCTS ALSO REGISTERED

Approval information

- Thiabendazole included in Annex I under EC Regulation 1107/2009
- Use of thiabendazole products on ware potatoes requires that the discharge of thiabendazole to receiving water from washing plants is kept within emission limits set by the UK monitoring authority

Efficacy guidance

- For best results tuber treatments should be applied as soon as possible after lifting and always within 24 hr
- Dust treatments should be applied evenly over the whole tuber surface
- Thiabendazole should only be used on ware potatoes where there is a likely risk of disease during the storage period and in combination with good storage hygiene and maintenance [1]
- Benzimidazole tolerant strains of silver scurf and skin spot are common in UK and tolerant strains of dry rot have been reported. To reduce the chance of such strains increasing benzimidazole based products should not be used more than once in the cropping cycle [1]

Restrictions

- Maximum number of treatments 1 per batch for ware or seed potato tuber treatments [1]
- Treated seed potatoes must not be used for food or feed [1]
- Do not remove treated potatoes from store for sale, processing or consumption for at least 21 d after application [1]
- Do not mix with any other product

Crop-specific information

- Latest use: 21 d before removal from store for sale, processing or consumption for ware potatoes [1]
- Apply to potatoes as soon as possible after harvest using suitable equipment and always within 2 wk of lifting provided the skins are set. See label for details [1]
- Potatoes should only be treated by systems that provide an accurate dose to tubers not carrying excessive quantities of soil [1]

Environmental safety

- Dangerous for the environment
- Toxic to aquatic organisms

Hazard classification and safety precautions

Hazard Irritant, Dangerous for the environment
Transport code 9
Packaging group III
UN Number 3082
Risk phrases H317
Operator protection A, C, D, H; U05a, U14, U19a, U20c
Environmental protection E15b, E38, H410
Storage and disposal D01, D02, D05, D09a, D10c, D12a

373   thiacloprid

A chloronicotinyl insecticide for use in agriculture and horticulture
IRAC mode of action code: 4A

Products

1 Agrovista Reggae	Agrovista	480 g/l	SC	13706
2 Biscaya	Bayer CropScience	240 g/l	OD	15014
3 Calypso	Bayer CropScience	480 g/l	SC	11257
4 Clayton Cayman	Clayton	240 g/l	OD	18386
5 Exemptor	ICL (Everris) Ltd	10% w/w	GR	15615
6 Zubarone	AgChem Access	240 g/l	OD	15798

FOR FULL CONDITIONS OF USE ALWAYS READ THE PRODUCT LABEL

Uses

- Aphids in *all edible seed crops grown outdoors* (off-label), *all non-edible seed crops grown outdoors* (off-label), *celeriac* (off-label), *chinese cabbage* (off-label), *choi sum* (off-label), *pak choi* (off-label), *spring corn gromwell* (off-label), *swedes* (off-label), *tatsoi* (off-label), *turnips* (off-label), *winter corn gromwell* (off-label) [2]; *bedding plants, hardy ornamental nursery stock, ornamental plant production, pot plants, protected ornamentals* [5]; *carrots, parsnips, potatoes, seed potatoes* [2, 4, 6]; *cherries* (off-label), *courgettes* (off-label), *gherkins* (off-label), *hops* (off-label), *leaf brassicas* (off-label), *marrows* (off-label), *mirabelles* (off-label), *protected forest nurseries* (off-label), *soft fruit* (off-label), *top fruit* (off-label) [3]; *combining peas, vining peas, winter wheat* [2, 4]; *leaf brassicas* (off-label - baby leaf production) [1]; *ornamental plant production* (off-label) [2, 3, 5]; *peas, ware potatoes* [6]; *protected herbs (see appendix 6)* (off-label), *protected lamb's lettuce* (off-label), *protected lettuce* (off-label) [1, 3]
- Asparagus beetle in *asparagus* (off-label) [3]
- Beetles in *bedding plants, hardy ornamental nursery stock, ornamental plant production, ornamental plant production* (off-label), *pot plants, protected ornamentals* [5]
- Blueberry gall midge in *protected blueberry* (off-label) [3]
- Bruchid beetle in *spring field beans, winter field beans* [2, 4]
- Bud mite in *almonds* (off-label), *chestnuts* (off-label), *cob nuts* (off-label), *hazel nuts* (off-label), *walnuts* (off-label) [1, 3]; *filberts* (off-label) [3]
- Cabbage aphid in *broccoli, brussels sprouts, cabbages, calabrese, cauliflowers* [2, 4, 6]; *chinese cabbage* (off-label), *choi sum* (off-label), *collards* (off-label), *kale* (off-label), *pak choi* (off-label), *spring greens* (off-label), *tatsoi* (off-label) [2]
- Capsids in *strawberries* (off-label) [1, 3]
- Common green capsid in *blackberries* (off-label), *protected blackberries* (off-label), *protected raspberries* (off-label), *raspberries* (off-label), *rubus hybrids* (off-label) [3]
- Damson-hop aphid in *plums* (off-label) [1, 3]
- Gall midge in *protected blueberry* (off-label) [1]
- Insect pests in *bilberries* (off-label), *blackberries* (off-label), *blackcurrants* (off-label), *blueberries* (off-label), *cherries* (off-label - under temporary protective rain covers), *courgettes* (off-label), *cranberries* (off-label), *gherkins* (off-label), *gooseberries* (off-label), *marrows* (off-label), *mirabelles* (off-label - under temporary protective rain covers), *pears* (off-label), *protected blackberries* (off-label), *protected raspberries* (off-label), *protected strawberries* (off-label), *raspberries* (off-label), *redcurrants* (off-label), *rubus hybrids* (off-label), *whitecurrants* (off-label) [1]; *hops* (off-label), *ornamental plant production* (off-label), *protected forest nurseries* (off-label), *soft fruit* (off-label), *top fruit* (off-label) [1, 3]
- Leaf miner in *hemp* (off-label) [3]; *protected aubergines* (off-label), *protected courgettes* (off-label), *protected cucumbers* (off-label), *protected ornamentals* (off-label), *protected peppers* (off-label), *protected tomatoes* (off-label) [1]
- Light brown apple moth in *protected blueberry* (off-label) [3]
- Mealy aphid in *broccoli, brussels sprouts, cabbages, calabrese, cauliflowers* [2, 4, 6]; *chinese cabbage* (off-label), *choi sum* (off-label), *collards* (off-label), *kale* (off-label), *pak choi* (off-label), *spring greens* (off-label), *swedes* (off-label), *tatsoi* (off-label), *turnips* (off-label) [2]
- Moths in *protected blueberry* (off-label) [1]
- Palm thrips in *protected aubergines* (off-label), *protected courgettes* (off-label), *protected cucumbers* (off-label), *protected ornamentals* (off-label), *protected peppers* (off-label), *protected tomatoes* (off-label) [3]
- Pea aphid in *spring field beans, winter field beans* [2, 4]
- Pea midge in *combining peas, vining peas* [2, 4]; *peas* [6]
- Peach-potato aphid in *leaf brassicas* (off-label) [3]; *leaf brassicas* (off-label - baby leaf production) [1]; *protected herbs (see appendix 6)* (off-label), *protected lamb's lettuce* (off-label), *protected lettuce* (off-label) [1, 3]
- Pear leaf midge in *pears* (off-label) [3]
- Pear midge in *pears* (off-label) [1, 3]
- Pollen beetle in *mustard, spring oilseed rape, winter oilseed rape* [2, 4, 6]
- Rosy apple aphid in *apples* [1, 3]
- Sciarid flies in *ornamental plant production* (off-label) [5]; *protected ornamentals* (off-label) [3]

SEE SECTION 3 FOR PRODUCTS ALSO REGISTERED

- South American leaf miner in **protected aubergines** *(off-label)*, **protected courgettes** *(off-label)*, **protected cucumbers** *(off-label)*, **protected ornamentals** *(off-label)*, **protected peppers** *(off-label)*, **protected tomatoes** *(off-label)* [3]
- Tarnished plant bug in **blackberries** *(off-label)*, **protected blackberries** *(off-label)*, **protected raspberries** *(off-label)*, **protected strawberries** *(off-label)*, **raspberries** *(off-label)*, **rubus hybrids** *(off-label)* [3]
- Thrips in **chives** *(off-label)*, **leeks** *(off-label)*, **salad onions** *(off-label)* [3]; **protected aubergines** *(off-label)*, **protected courgettes** *(off-label)*, **protected cucumbers** *(off-label)*, **protected ornamentals** *(off-label)*, **protected peppers** *(off-label)*, **protected tomatoes** *(off-label)* [1, 3]
- Tobacco whitefly in **protected aubergines** *(off-label)*, **protected courgettes** *(off-label)*, **protected cucumbers** *(off-label)*, **protected ornamentals** *(off-label)*, **protected peppers** *(off-label)*, **protected tomatoes** *(off-label)* [3]
- Vine weevil in **bedding plants**, **hardy ornamental nursery stock**, **ornamental plant production**, **ornamental plant production** *(off-label)*, **pot plants**, **protected ornamentals** [5]; **protected ornamentals** *(off-label)* [3]
- Western flower thrips in **protected aubergines** *(off-label)*, **protected courgettes** *(off-label)*, **protected cucumbers** *(off-label)*, **protected ornamentals** *(off-label)*, **protected peppers** *(off-label)*, **protected tomatoes** *(off-label)* [1, 3]
- Wheat-blossom midge in **spring wheat**, **winter wheat** [2, 4, 6]
- Whitefly in **bedding plants**, **hardy ornamental nursery stock**, **ornamental plant production**, **ornamental plant production** *(off-label)*, **pot plants**, **protected ornamentals** [5]; **protected aubergines** *(off-label)*, **protected courgettes** *(off-label)*, **protected cucumbers** *(off-label)*, **protected ornamentals** *(off-label)*, **protected peppers** *(off-label)*, **protected tomatoes** *(off-label)* [1]
- Willow aphid in **horseradish** *(off-label)*, **parsley root** *(off-label)*, **red beet** *(off-label)*, **salsify** *(off-label)* [2]
- Woolly currant scale in **bilberries** *(off-label)*, **blackcurrants** *(off-label)*, **blueberries** *(off-label)*, **cranberries** *(off-label)*, **gooseberries** *(off-label)*, **redcurrants** *(off-label)*, **whitecurrants** *(off-label)* [3]

Extension of Authorisation for Minor Use (EAMUs)

- **all edible seed crops grown outdoors** *20142041* [2]
- **all non-edible seed crops grown outdoors** *20142041* [2]
- **almonds** *20080471* [1], *20142137* [3]
- **asparagus** *20142142* [3]
- **bilberries** *20080466* [1], *20142133* [3]
- **blackberries** *20080475* [1], *20142138* [3]
- **blackcurrants** *20080466* [1], *20142133* [3]
- **blueberries** *20080466* [1], *20142133* [3]
- **celeriac** *20152075* [2]
- **cherries** *(under temporary protective rain covers)* *20080469* [1], *20142135* [3]
- **chestnuts** *20080471* [1], *20142137* [3]
- **chinese cabbage** *20142063* [2]
- **chives** *20142150* [3]
- **choi sum** *20142063* [2]
- **cob nuts** *20080471* [1], *20142137* [3]
- **collards** *20142249* [2]
- **courgettes** *20102032* [1], *20142140* [3]
- **cranberries** *20080466* [1], *20142133* [3]
- **filberts** *20142137* [3]
- **gherkins** *20102032* [1], *20142140* [3]
- **gooseberries** *20080466* [1], *20142133* [3]
- **hazel nuts** *20080471* [1], *20142137* [3]
- **hemp** *20171455* [3]
- **hops** *20102034* [1], *20142149* [3]
- **horseradish** *20142062* [2]
- **kale** *20142249* [2]
- **leaf brassicas** *(baby leaf production)* *20080470* [1], *20142136* [3]
- **leeks** *20142150* [3]

FOR FULL CONDITIONS OF USE ALWAYS READ THE PRODUCT LABEL

- **marrows** *20102032* [1], *20142140* [3]
- **mirabelles** *(under temporary protective rain covers) 20080469* [1], *20142135* [3]
- **ornamental plant production** *20102034* [1], *20142041* [2], *20142149* [3], *20170555* [5]
- **pak choi** *20142063* [2]
- **parsley root** *20142062* [2]
- **pears** *20080464* [1], *20142130* [3]
- **plums** *20080468* [1], *20142134* [3]
- **protected aubergines** *20080474* [1], *20142151* [3]
- **protected blackberries** *20080467* [1], *20142139* [3]
- **protected blueberry** *20102035* [1], *20142141* [3]
- **protected courgettes** *20080474* [1], *20142151* [3]
- **protected cucumbers** *20080474* [1], *20142151* [3]
- **protected forest nurseries** *20102034* [1], *20142149* [3]
- **protected herbs (see appendix 6)** *20080472* [1], *20142152* [3]
- **protected lamb's lettuce** *20080472* [1], *20142152* [3]
- **protected lettuce** *20080472* [1], *20142152* [3]
- **protected ornamentals** *20080474* [1], *20142151* [3], *20142153* [3]
- **protected peppers** *20080474* [1], *20142151* [3]
- **protected raspberries** *20080467* [1], *20142139* [3]
- **protected strawberries** *20102033* [1], *20142132* [3]
- **protected tomatoes** *20080474* [1], *20142151* [3]
- **raspberries** *20080475* [1], *20142138* [3]
- **red beet** *20142062* [2]
- **redcurrants** *20080466* [1], *20142133* [3]
- **rubus hybrids** *20080475* [1], *20142138* [3]
- **salad onions** *20142150* [3]
- **salsify** *20142062* [2]
- **soft fruit** *20102034* [1], *20142149* [3]
- **spring corn gromwell** *20192821* [2]
- **spring greens** *20142249* [2]
- **strawberries** *20080465* [1], *20142131* [3]
- **swedes** *20142061* [2]
- **tatsoi** *20142063* [2]
- **top fruit** *20102034* [1], *20142149* [3]
- **turnips** *20142061* [2]
- **walnuts** *20080471* [1], *20142137* [3]
- **whitecurrants** *20080466* [1], *20142133* [3]
- **winter corn gromwell** *20192821* [2]

Approval information
- Thiacloprid included in Annex I under EC Regulation 1107/2009

Efficacy guidance
- Best results in apples obtained by a programme of sprays commencing pre-blossom at the first sign of aphids [3]
- For best pear midge control treat under warm conditions (mid-day) when the adults are flying (temperatures higher than 12˚C). Applications on cool damp days are unlikely to be successful [3]
- Best results in field crops obtained from treatments when target pests reach threshold levels. For wheat blossom midge use pheromone or sticky yellow traps to monitor adult activity in crops at risk [2]
- Treatment of ornamentals and protected ornamentals is by incorporation into peat-based growing media prior to sowing or planting using suitable automated equipment [5]
- Ensure thorough mixing in the compost to achieve maximum control. Top dressing is ineffective [5]
- Unless transplants have been treated with a suitable pesticide prior to planting in treated compost, full protection against target pests is not guaranteed [5]
- Minimise the possibility of the development of resistance by alternating insecticides with different modes of action in the programme
- In dense canopies and on larger trees increase water volume to ensure full coverage

SEE SECTION 3 FOR PRODUCTS ALSO REGISTERED

- It can give effective contact activity against adult CSFB "present and active" in oilseed rape at the time of an application to control Myzus persicae [2]

Restrictions
- Maximum number of treatments 2 per yr on apples, seed potatoes; 1 per yr on other field crops and for compost incorporation
- Consult processor before use on crops for processing [2]
- Do not mix treated compost with any other bulky materials such as perlite or bark [5]
- Use treated compost as soon as possible, and within 4 wk of mixing [5]

Crop-specific information
- Latest use: up to and including flowering just complete (GS 69) for wheat; before sowing or planting for treated compost
- HI apples and potatoes 14 d; oilseed rape and mustard 30 d
- Test tolerance of ornamental species before large scale use [5]

Environmental safety
- Dangerous for the environment
- Very toxic or toxic to aquatic organisms
- Risk to certain non-target insects or other arthropods. See directions for use [3]
- Broadcast air-assisted LERAP [1] (18 m); Broadcast air-assisted LERAP [3] (30 m)

Hazard classification and safety precautions
Hazard Harmful, Dangerous for the environment [1-6]; Harmful if swallowed [1-4, 6]; Harmful if inhaled [1]; Very toxic to aquatic organisms [3, 5]
Transport code 6.1 [1, 3]; 9 [5]
Packaging group III [1, 3, 5]
UN Number 2902 [1, 3]; 3077 [5]; N/C [2, 4, 6]
Risk phrases H315, H319 [2-4, 6]; H317 [1]; H336 [1-4, 6]; H351, H360 [1-6]
Operator protection A [1-6]; C [2, 4, 6]; D [5]; H [1, 3]; U05a [1-6]; U11, U16a, U19a [2, 4, 6]; U14 [1, 3]; U20b [1-4, 6]; U20c [5]
Environmental protection E13c, E22b [3]; E15a, E38 [5]; E15b [1, 2, 4, 6]; E17b [1] (18 m); E17b [3] (30 m); E22c [1]; E34 [1-6]; H410 [1, 2, 4-6]
Storage and disposal D01, D02, D09a [1-6]; D10b [1, 3]; D10c [2, 4, 6]; D11a [5]; D12a [1, 2, 4, 6]
Treated seed S04a [5]
Medical advice M03 [1-6]; M05a [1]; M05b [3]

---

374   thifensulfuron-methyl

A translocated sulfonylurea herbicide
HRAC mode of action code: B

See also carfentrazone-ethyl + thifensulfuron-methyl
flupyrsulfuron-methyl + thifensulfuron-methyl
fluroxypyr + metsulfuron-methyl + thifensulfuron-methyl
fluroxypyr + metsulfuron-methyl + thifensulfuron-methyl
fluroxypyr + thifensulfuron-methyl + tribenuron-methyl
metsulfuron-methyl + thifensulfuron-methyl
nicosulfuron + thifensulfuron-methyl
thifensulfuron-methyl + tribenuron-methyl

Products
Pinnacle	FMC Agro	50% w/w	SG	18781

Uses
- Annual dicotyledons in *soya beans* (off-label), *spring barley*, *spring wheat*, *winter barley*, *winter wheat*
- Docks in *grassland*
- Green cover in *land temporarily removed from production*

FOR FULL CONDITIONS OF USE ALWAYS READ THE PRODUCT LABEL

Extension of Authorisation for Minor Use (EAMUs)
- *soya beans* 20190886

Approval information
- Thifensulfuron-methyl included in Annex I under EC Regulation 1107/2009
- Accepted by BBPA for use on malting barley

Efficacy guidance
- Best results achieved from application to small emerged weeds when growing actively. Broad-leaved docks are susceptible during the rosette stage up to onset of stem extension
- Ensure good spray coverage and apply to dry foliage
- Susceptible weeds stop growing almost immediately but symptoms may not be visible for about 2 wk
- Only broad-leaved docks (*Rumex obtusifolius*) are controlled; curled docks (*Rumex crispus*) are resistant
- Docks with developing or mature seed heads should be topped and the regrowth treated later
- Established docks with large tap roots may require follow-up treatment
- High populations of docks in grassland will require further treatment in following yr
- Thifensulfuron-methyl is a member of the ALS-inhibitor group of herbicides and products should be used in a planned Resistance Management strategy. See Section 5 for more information

Restrictions
- Maximum number of treatments 1 per year for grassland. Must only be applied from 1 Feb in year of harvest
- Do not treat new leys in year of sowing
- Do not treat where nutrient imbalances, drought, waterlogging, low temperatures, lime deficiency, pest or disease attack have reduced crop or sward vigour
- Do not roll or harrow within 7 d of spraying
- Do not graze grass crops within 7 d of spraying
- Specific restrictions apply to use in sequence or tank mixture with other sulfonylurea or ALS-inhibiting herbicides. See label for details
- Only one application of a sulfonylurea product may be applied per calendar yr to grassland and green cover on land temporarily removed from production

Crop-specific information
- Latest use: before 1 Aug on grass
- On grass apply 7-10 d before grazing and do not graze for 7 d afterwards
- Product may cause a check to both sward and clover which is usually outgrown

Following crops guidance
- Only grass or cereals may be sown within 4 wk of application to grassland or setaside, or in the event of failure of any treated crop
- No restrictions apply after normal harvest of a treated cereal crop

Environmental safety
- Dangerous for the environment
- Very toxic to aquatic organisms
- Keep livestock out of treated areas for at least 7 d following treatment
- Take extreme care to avoid drift onto broad-leaved plants outside the target area or onto surface waters or ditches, or land intended for cropping
- Spraying equipment should not be drained or flushed onto land planted, or to be planted, with trees or crops other than cereals and should be thoroughly cleansed after use - see label for instructions

Hazard classification and safety precautions
    Hazard Dangerous for the environment, Very toxic to aquatic organisms
    Transport code 9
    Packaging group III
    UN Number 3077
    Operator protection U19a, U20b
    Environmental protection E06a (7 d); E15b, E38, H410
    Storage and disposal D09a, D11a, D12a

SECTION 2

SEE SECTION 3 FOR PRODUCTS ALSO REGISTERED

375   thifensulfuron-methyl + tribenuron-methyl

A mixture of two sulfonylurea herbicides for cereals
HRAC mode of action code: B + B

See also tribenuron-methyl

Products

1 Clayton Rift	Clayton	40:15% w/w	WG	18945
2 Hiatus	Rotam	40:15% w/w	WG	16059
3 Inka SX	FMC Agro	25:25% w/w	SG	18760
4 Ratio SX	FMC Agro	40:10% w/w	SG	18786
5 Seduce	Certis	33.3:16.7% w/w	SG	16261

Uses

- Annual dicotyledons in *game cover* *(off-label)*, *spring oats*, *spring rye* [3]; *spring barley*, *spring wheat*, *winter barley*, *winter wheat* [1-5]; *triticale*, *winter rye* [1-3]; *winter oats* [3, 4]
- Black bindweed in *spring barley*, *spring wheat*, *triticale*, *winter barley*, *winter rye*, *winter wheat* [1, 2]
- Charlock in *spring barley*, *spring wheat*, *winter barley*, *winter wheat* [3-5]; *spring oats*, *spring rye*, *triticale*, *winter rye* [3]; *winter oats* [3, 4]
- Chickweed in *spring barley*, *spring wheat*, *winter barley*, *winter wheat* [1-5]; *spring oats*, *spring rye* [3]; *triticale*, *winter rye* [1-3]; *winter oats* [3, 4]
- Docks in *spring barley*, *spring oats*, *spring rye*, *spring wheat*, *triticale*, *winter barley*, *winter oats*, *winter rye*, *winter wheat* [3]
- Fat hen in *spring barley*, *spring wheat*, *triticale*, *winter barley*, *winter rye*, *winter wheat* [1, 2]
- Mayweeds in *spring barley*, *spring wheat*, *winter barley*, *winter wheat* [1-5]; *spring oats*, *spring rye* [3]; *triticale*, *winter rye* [1-3]; *winter oats* [3, 4]
- Poppies in *spring barley*, *spring wheat*, *triticale*, *winter barley*, *winter rye*, *winter wheat* [1, 2]

Extension of Authorisation for Minor Use (EAMUs)

- *game cover* 20190897 [3]

Approval information

- Thifensulfuron-methyl and tribenuron-methyl included in Annex I under EC Regulation 1107/2009
- Accepted by BBPA for use on malting barley

Efficacy guidance

- Apply when weeds are small and actively growing
- Apply after end of Feb in year of harvest
- Ensure good spray cover of the weeds
- Apply in a volume of 200 - 400 l/ha [2]
- Susceptible weeds cease growth almost immediately after application and symptoms become evident 2 wk later
- Effectiveness reduced by rain within 4 h of treatment and in very dry conditions
- Various tank mixtures recommended to broaden weed control spectrum
- Thifensulfuron-methyl and tribenuron-methyl are members of the ALS-inhibitor group of herbicides and products should be used in a planned Resistance Management strategy. See Section 5 for more information

Restrictions

- Maximum number of treatments 1 per crop
- Do not apply to cereals undersown with grass, clover or other legumes, or any other broad-leaved crop
- Do not apply within 7 d of rolling
- Specific restrictions apply to use in sequence or tank mixture with other sulfonylurea or ALS-inhibiting herbicides. See label for details
- Do not apply to any crop suffering from stress
- Consult contract agents before use on crops grown for seed

FOR FULL CONDITIONS OF USE ALWAYS READ THE PRODUCT LABEL

- Horizontal boom sprayers must be fitted with three star drift reduction technology for all uses [2]
- Low drift spraying equipment must be operated according to the specific conditions stated in the official three star rating for that equipment as published on HSE Chemicals Regulation Division's website. These operating conditions must be maintained until the operator is 30m from the top of the bank of any surface water bodies [2]

Crop-specific information
- Latest use: before flag leaf ligule first visible (GS 39) for all crops

Following crops guidance
- Only cereals, field beans, grass or oilseed rape may be sown in the same calendar year as harvest of a treated crop.
- In the event of failure of a treated crop sow only a cereal crop within 3 mth of product application and after ploughing and cultivating to at least 15 cm. After 3 mth field beans or oilseed rape may also be sown
- Only cereals, oilseed rape or field beans may be sown in the same calendar year after harvest. In the following spring only cereals, oilseed rape or sugar beet may be sown. In case of crop failure for any reason, sow only spring small grain cereal. Before sowing, soil should be ploughed and cultivated to a depth of at least 15 cm[2]

Environmental safety
- Dangerous for the environment
- Very toxic to aquatic organisms
- Buffer zone requirement 6m [2]
- Spraying equipment should not be drained or flushed onto land planted, or to be planted, with trees or crops other than cereals and should be thoroughly cleansed after use - see label for instructions
- Take particular care to avoid damage by drift onto broad-leaved plants outside the target area or onto surface waters or ditches
- LERAP Category B

Hazard classification and safety precautions
Hazard Irritant [4, 5]; Dangerous for the environment [1-5]; Very toxic to aquatic organisms [5]
Transport code 9
Packaging group III
UN Number 3077
Operator protection A, H [4, 5]; U05a, U08, U20a [4, 5]; U14 [4]; U19a [1, 2]; U20c [3]
Environmental protection E15a [4, 5]; E15b [1-3]; E16a, E38, H410 [1-5]; E16b [5]
Storage and disposal D01, D02 [1, 2, 4, 5]; D09a, D11a, D12a [1-5]

376   thiophanate-methyl

A thiophanate fungicide with protectant and curative activity
FRAC mode of action code: 1

Products

1	Cercobin WG	Certis	70% w/w	WG	13854
2	Taurus	Certis	70% w/w	WG	15508
3	Topsin WG	Certis	70% w/w	WG	13988

Uses
- Disease control in *hops* (off-label), *soft fruit* (off-label) [1]
- Fusarium in *durum wheat*, *spring oilseed rape*, *winter oilseed rape* [3]; *durum wheat* (reduction), *spring wheat* (reduction), *triticale* (reduction), *winter wheat* (reduction) [2]; *spring wheat*, *triticale*, *winter wheat* [1, 3]
- Fusarium diseases in *protected ornamentals* (off-label) [1]
- Mycotoxins in *durum wheat* (reduction) [2, 3]; *spring oilseed rape* (reduction), *winter oilseed rape* (reduction) [3]; *spring wheat* (reduction), *triticale* (reduction), *winter wheat* (reduction) [1-3]

SEE SECTION 3 FOR PRODUCTS ALSO REGISTERED

- Root diseases in **container-grown ornamentals** *(off-label)*, **protected ornamentals** *(off-label)* [1]
- Sclerotinia stem rot in **spring oilseed rape**, **winter oilseed rape** [2]
- Verticillium wilt in **protected tomatoes** *(off-label)* [1]

Extension of Authorisation for Minor Use (EAMUs)
- **container-grown ornamentals** *20111655* [1]
- **hops** *20082833* [1]
- **protected ornamentals** *20111887* [1]
- **protected tomatoes** *20091969* [1]
- **soft fruit** *20082833* [1]

Approval information
- Thiophanate-methyl included in Annex I under EC Regulation 1107/2009

Efficacy guidance
- To avoid the development of resistance, a maximum of 2 applications of any MBC product (thiophanate-methyl or carbendazim) are allowed in any one crop. Avoid using MBC fungicides alone.

Environmental safety
- Dangerous for the environment
- Very toxic to aquatic organisms
- LERAP Category B

Hazard classification and safety precautions
Hazard Harmful, Dangerous for the environment, Harmful if swallowed, Harmful if inhaled [2]
Transport code 9
Packaging group III
UN Number 3077
Risk phrases H317, H371 [2]
Operator protection A; U19a [2, 3]
Environmental protection E15b, E16a, E16b [1-3]; E38 [2, 3]; H411 [2]
Storage and disposal D01, D02, D12a [2, 3]; D10b [1-3]
Medical advice M05a [2, 3]

377   tri-allate

A soil-acting thiocarbamate herbicide for grass weed control
HRAC mode of action code: N

Products

1	Avadex Excel 15G	Gowan	15% w/w	GR	17872
2	Avadex Factor	Gowan	450 g/l	CS	17877

Uses
- Annual grasses in **canary seed** *(off-label)*, **linseed** *(off-label)*, **miscanthus** *(off-label)*, **triticale** *(off-label)*, **winter linseed** *(off-label)*, **winter rye** *(off-label)* [1]; **spring barley**, **winter barley**, **winter wheat** [1, 2]
- Annual meadow grass in **canary seed** *(off-label)*, **linseed** *(off-label)*, **miscanthus** *(off-label)*, **winter linseed** *(off-label)* [1]
- Blackgrass in **canary seed** *(off-label)*, **linseed** *(off-label)*, **miscanthus** *(off-label)*, **triticale** *(off-label)*, **winter rye** *(off-label)* [1]; **spring barley**, **winter barley**, **winter linseed** *(off-label)*, **winter wheat** [1, 2]; **spring linseed** *(off-label)* [2]
- Italian ryegrass in **spring linseed** *(off-label)*, **winter linseed** *(off-label)* [2]
- Loose silky bent in **spring linseed** *(off-label)*, **winter linseed** *(off-label)* [2]
- Meadow grasses in **spring barley**, **triticale** *(off-label)*, **winter barley**, **winter rye** *(off-label)*, **winter wheat** [1]
- Wild oats in **canary seed** *(off-label)*, **linseed** *(off-label)*, **miscanthus** *(off-label)*, **spring barley**, **triticale** *(off-label)*, **winter barley**, **winter linseed** *(off-label)*, **winter rye** *(off-label)*, **winter wheat** [1]

FOR FULL CONDITIONS OF USE ALWAYS READ THE PRODUCT LABEL

Extension of Authorisation for Minor Use (EAMUs)
- *canary seed* 20170467 [1]
- *linseed* 20170468 [1]
- *miscanthus* 20170466 [1]
- *spring linseed* 20191053 [2]
- *triticale* 20171361 [1]
- *winter linseed* 20170468 [1], 20191053 [2]
- *winter rye* 20171361 [1]

Approval information
- Tri-allate included in Annex 1 under EC Regulation 1107/2009
- Accepted by BBPA for use on malting barley

Efficacy guidance
- Apply to soil surface pre-emergence
- For maximum activity apply to well-prepared moist seedbeds
- Do not use on soils with more than 10% organic matter
- Wild oats controlled up to 2-leaf stage
- Do not apply with spinning disc granule applicator; see label for suitable types [1]
- Do not apply to cloddy seedbeds
- Use sequential treatments to improve control of barren brome and annual dicotyledons (see label for details)

Restrictions
- Maximum number of treatments 1 per crop
- Drill wheat well below treated layer of soil (see label for safe drilling depths)
- Do not apply to shallow-drilled wheat crops
- Do not undersow grass species. It is safe to undersow with clover and other legumes
- Do not sow oats or grasses within 1 yr of treatment
- Low drift spraying equipment must be operated according to the specific conditions stated in the official three star rating for that equipment as published on HSE Chemicals Regulation Division's website. These operating conditions must be maintained until the operator is 30m from the top of the bank of any surface water bodies [2]

Crop-specific information
- Latest use: pre-drilling for beet crops; before crop emergence for field beans, spring barley, peas, forage legumes; before first node detectable stage (GS 31) for winter wheat, winter barley, durum wheat, triticale, winter rye

Following crops guidance
- Do not sow oat or grass crops within a year of application

Environmental safety
- Irritating to eyes and skin
- May cause sensitization by skin contact
- Harmful to fish or other aquatic life. Do not contaminate surface waters or ditches with chemical or used container
- Buffer zone requirement 10 m [1] and 6 m [2]
- Avoid drift on to non-target plants outside the treated area
- LERAP Category B

Hazard classification and safety precautions
Hazard Harmful, Dangerous for the environment, Very toxic to aquatic organisms [1]
Transport code 9
Packaging group III
UN Number 3077 [1]; 3082 [2]
Risk phrases H317 [2]; H373 [1, 2]
Operator protection A, C, H; U02a, U05a, U19a, U20a [1]
Environmental protection E16a, H410 [1, 2]; H411 [2]
Storage and disposal D01, D02, D09a, D11a, D19 [1]
Medical advice M05a [1]

SEE SECTION 3 FOR PRODUCTS ALSO REGISTERED

378   tribenuron-methyl

A foliar acting sulfonylurea herbicide with some root activity for use in cereals
HRAC mode of action code: B

See also florasulam + tribenuron-methyl
fluroxypyr + thifensulfuron-methyl + tribenuron-methyl
metsulfuron-methyl + tribenuron-methyl
thifensulfuron-methyl + tribenuron-methyl

Products

1	Flame	Albaugh UK	50% w/w	WG	17842
2	Nuance	Nufarm UK	75% w/w	WG	18616
3	Triad	FMC Agro	50% w/w	TB	18699

Uses

- Annual dicotyledons in **grassland**, **undersown spring barley** [3]; **spring barley** [1-3]; **spring oats**, **spring wheat**, **triticale**, **winter oats** [2]; **winter barley**, **winter wheat** [1, 2]
- Chickweed in **grassland**, **spring barley**, **undersown spring barley** [3]

Approval information

- Tribenuron-methyl included in Annex I under EC Regulation 1107/2009
- Accepted by BBPA for use on malting barley

Efficacy guidance

- Best control achieved when weeds small and actively growing
- Good spray cover must be achieved since larger weeds often become less susceptible
- Susceptible weeds cease growth almost immediately after treatment and symptoms can be seen in about 2 wk
- Weed control may be reduced when conditions very dry
- Tribenuron-methyl is a member of the ALS-inhibitor group of herbicides and products should be used in a planned Resistance Management strategy. See Section 5 for more information

Restrictions

- Maximum number of treatments 1 per crop
- Specific restrictions apply to use in sequence or tank mixture with other sulfonylurea or ALS-inhibiting herbicides. See label for details
- Do not apply to crops undersown with grass, clover or other broad-leaved crops
- Do not apply to any crop suffering stress from any cause or not actively growing
- Do not apply within 7 d of rolling
- Must not be applied before end of February in the year of harvest [1-3]

Crop-specific information

- Latest use: up to and including flag leaf ligule/collar just visible (GS 39)
- Apply in autumn or in spring from 3 leaf stage of crop

Following crops guidance

- Only cereals, field beans or oilseed rape may be sown in the same calendar yr as harvest of a treated crop
- In the event of crop failure sow only a cereal within 3 mth of application. After 3 mth field beans or oilseed rape may also be sown

Environmental safety

- Dangerous for the environment
- Very toxic to aquatic organisms
- Take extreme care to avoid drift onto broad-leaved plants outside the target area or onto surface waters or ditches, or land intended for cropping
- Spraying equipment should not be drained or flushed onto land planted, or to be planted, with trees or crops other than cereals and should be thoroughly cleansed after use - see label for instructions

FOR FULL CONDITIONS OF USE ALWAYS READ THE PRODUCT LABEL

Hazard classification and safety precautions

Hazard Irritant, Dangerous for the environment [1-3]; Very toxic to aquatic organisms [1, 2]
Transport code 9
Packaging group III
UN Number 3077
Risk phrases H317 [1, 3]; H319 [1]
Operator protection A [1-3]; C [1]; H [1, 2]; U05a, U08, U20b
Environmental protection E07b [3] (3 weeks); E07e [3]; E15a, H410 [1-3]; E38 [1, 2]
Storage and disposal D01, D02, D09a, D11a [1-3]; D12a [1, 2]

379   Trichoderma asperellum (Strain T34)

A biological control agent

Products

T34 Biocontrol	Fargro	12% w/w	WP	17290

Uses

* Fusarium in **protected baby leaf crops** *(off-label)*, **protected bilberries** *(off-label)*, **protected blackberries** *(off-label)*, **protected blackcurrants** *(off-label)*, **protected blueberry** *(off-label)*, **protected broccoli** *(off-label)*, **protected brussels sprouts** *(off-label)*, **protected cabbages** *(off-label)*, **protected calabrese** *(off-label)*, **protected cauliflowers** *(off-label)*, **protected choi sum** *(off-label)*, **protected collards** *(off-label)*, **protected courgettes** *(off-label)*, **protected cranberries** *(off-label)*, **protected cucumbers** *(off-label)*, **protected elderberries** *(off-label)*, **protected forest nurseries** *(off-label)*, **protected gherkins** *(off-label)*, **protected gooseberries** *(off-label)*, **protected herbs (see appendix 6)** *(off-label)*, **protected kale** *(off-label)*, **protected kohlrabi** *(off-label)*, **protected lettuce** *(off-label)*, **protected loganberries** *(off-label)*, **protected melons** *(off-label)*, **protected mulberry** *(off-label)*, **protected oriental cabbage** *(off-label)*, **protected ornamentals** *(off-label)*, **protected pumpkins** *(off-label)*, **protected raspberries** *(off-label)*, **protected redcurrants** *(off-label)*, **protected rose hips** *(off-label)*, **protected rubus hybrids** *(off-label)*, **protected strawberries** *(off-label)*, **protected summer squash** *(off-label)*, **protected watercress** *(off-label)*, **protected winter squash** *(off-label)*
* Fusarium foot rot and seedling blight in **aubergines, chillies, ornamental plant production, peppers, tomatoes**
* Pythium in **protected baby leaf crops** *(off-label)*, **protected bilberries** *(off-label)*, **protected blackberries** *(off-label)*, **protected blackcurrants** *(off-label)*, **protected blueberry** *(off-label)*, **protected broccoli** *(off-label)*, **protected brussels sprouts** *(off-label)*, **protected cabbages** *(off-label)*, **protected calabrese** *(off-label)*, **protected cauliflowers** *(off-label)*, **protected choi sum** *(off-label)*, **protected collards** *(off-label)*, **protected courgettes** *(off-label)*, **protected cranberries** *(off-label)*, **protected cucumbers** *(off-label)*, **protected elderberries** *(off-label)*, **protected forest nurseries** *(off-label)*, **protected gherkins** *(off-label)*, **protected gooseberries** *(off-label)*, **protected herbs (see appendix 6)** *(off-label)*, **protected kale** *(off-label)*, **protected kohlrabi** *(off-label)*, **protected lettuce** *(off-label)*, **protected loganberries** *(off-label)*, **protected melons** *(off-label)*, **protected mulberry** *(off-label)*, **protected oriental cabbage** *(off-label)*, **protected ornamentals** *(off-label)*, **protected pumpkins** *(off-label)*, **protected raspberries** *(off-label)*, **protected redcurrants** *(off-label)*, **protected rose hips** *(off-label)*, **protected rubus hybrids** *(off-label)*, **protected strawberries** *(off-label)*, **protected summer squash** *(off-label)*, **protected watercress** *(off-label)*, **protected winter squash** *(off-label)*

Extension of Authorisation for Minor Use (EAMUs)

* *protected baby leaf crops 20161809*
* *protected bilberries 20161808*
* *protected blackberries 20161808*
* *protected blackcurrants 20161808*
* *protected blueberry 20161808*
* *protected broccoli 20161809*
* *protected brussels sprouts 20161809*
* *protected cabbages 20161809*
* *protected calabrese 20161809*

SEE SECTION 3 FOR PRODUCTS ALSO REGISTERED

SECTION 2

- *protected cauliflowers* 20161809
- *protected choi sum* 20161809
- *protected collards* 20161809
- *protected courgettes* 20161805
- *protected cranberries* 20161808
- *protected cucumbers* 20161805
- *protected elderberries* 20161808
- *protected forest nurseries* 20161810
- *protected gherkins* 20161805
- *protected gooseberries* 20161808
- *protected herbs (see appendix 6)* 20161809
- *protected kale* 20161809
- *protected kohlrabi* 20161809
- *protected lettuce* 20161809
- *protected loganberries* 20161808
- *protected melons* 20161805
- *protected mulberry* 20161808
- *protected oriental cabbage* 20161809
- *protected ornamentals* 20161810
- *protected pumpkins* 20161805
- *protected raspberries* 20161808
- *protected redcurrants* 20161808
- *protected rose hips* 20161808
- *protected rubus hybrids* 20161808
- *protected strawberries* 20161808
- *protected summer squash* 20161805
- *protected watercress* 20161809
- *protected winter squash* 20161805

Approval information
- Trichoderma asperellum (Strain T34) included in Annex 1 under EC Regulation 1107/2009

Efficacy guidance
- May be applied by spraying, through an irrigation system or by root dipping.

Restrictions
- Efficacy has been demonstrated on peat and coir composts but should be checked on the more unusual composts before large scale use.

Crop-specific information
- Crop safety has been demonstrated on a range of carnation species. It is advisable to check the safety to other species on a sample of the population before large-scale use.

Hazard classification and safety precautions
Hazard Harmful
UN Number N/C
Risk phrases H317
Operator protection A, D, H; U05a, U09a, U19a, U20b
Environmental protection E15a
Storage and disposal D01, D02, D05, D09a, D09b, D10c
Medical advice M03, M04a

380   trifloxystrobin

A protectant strobilurin fungicide for cereals, apples and pears and managed amenity turf
FRAC mode of action code: 11

See also cyproconazole + trifloxystrobin
fluopyram + trifloxystrobin
fluoxastrobin + prothioconazole + trifloxystrobin
prothioconazole + trifloxystrobin
tebuconazole + trifloxystrobin

Products

1	Flint	Bayer CropScience	50% w/w	WG	11259
2	Martinet	AgChem Access	500 g/l	SC	14614
3	PureFloxy	Pure Amenity	50% w/w	WG	15473
4	Scorpio	Bayer CropScience	50% w/w	WG	12293
5	Swift SC	Bayer CropScience	500 g/l	SC	11227

Uses

- Brown rust in **spring barley**, **winter barley**, **winter wheat** [2, 5]
- Disease control in **apples**, **pears** [1]
- Foliar disease control in **ornamental specimens** *(off-label)* [5]
- Fusarium patch in **amenity grassland**, **managed amenity turf** [3, 4]
- Glume blotch in **winter wheat** [2, 5]
- Net blotch in **spring barley**, **winter barley** [2, 5]
- Red thread in **amenity grassland**, **managed amenity turf** [3, 4]
- Rhynchosporium in **spring barley**, **winter barley** [2, 5]
- Rust in **grass seed crops** *(off-label)*, **rye** *(off-label)*, **triticale** *(off-label)* [5]
- Septoria leaf blotch in **winter wheat** [2, 5]

Extension of Authorisation for Minor Use (EAMUs)

- **grass seed crops** *20193265* [5]
- **ornamental specimens** *20082882* [5]
- **rye** *20193265* [5]
- **triticale** *20193265* [5]

Approval information

- Trifloxystrobin included in Annex I under EC Regulation 1107/2009
- Accepted by BBPA for use on malting barley

Efficacy guidance

- Should be used protectively before disease is established in crop. Further treatment may be necessary if disease attack prolonged
- Treat grass after cutting and do not mow for at least 48 h afterwards to allow adequate systemic movement [4]
- Trifloxystrobin is a member of the QoI cross resistance group. Product should be used preventatively and not relied on for its curative potential
- Use product as part of an Integrated Crop Management strategy incorporating other methods of control, including where appropriate other fungicides with a different mode of action. Do not apply more than two foliar applications of QoI containing products to any cereal crop
- There is a significant risk of widespread resistance occurring in *Septoria tritici* populations in UK. Failure to follow resistance management action may result in reduced levels of disease control
- On cereal crops product must always be used in mixture with another product, recommended for control of the same target disease, that contains a fungicide from a different cross resistance group and is applied at a dose that will give robust control
- Strains of barley powdery mildew resistant to QoIs are common in the UK

Restrictions

- Maximum number of treatments 2 per crop per yr
- Do not apply to turf during dought conditions or to frozen turf [4]

Crop-specific information
- HI barley, wheat 35 d [5]

Environmental safety
- Dangerous for the environment
- Very toxic to aquatic organisms
- LERAP Category B [3, 4]

Hazard classification and safety precautions

Hazard Irritant [3, 4]; Dangerous for the environment [2-5]; Very toxic to aquatic organisms [1-5]
Transport code 9
Packaging group III
UN Number 3077 [1, 3, 4]; 3082 [2, 5]
Risk phrases H317 [1, 3, 4]
Operator protection A, H [2-5]; U02a, U09a, U19a [2, 5]; U05a, U20b [1-5]; U11, U13, U14, U15 [1, 3, 4]
Environmental protection E13b [2, 5]; E15b, E16b, E34 [1, 3, 4]; E16a [3, 4]; E17a [1] (15 m); E38, H410 [1-5]
Consumer protection C02a [2, 5] (35 d)
Storage and disposal D01, D02, D09a, D10c, D12a
Medical advice M03 [1, 3, 4]

381   triflusulfuron-methyl

A sulfonyl urea herbicide for beet crops
HRAC mode of action code: B

See also lenacil + triflusulfuron-methyl

Products

1	Debut	FMC Agro	50% w/w	WG	18766
2	Shiro	UPL Europe	50% w/w	WG	17439
3	Upbeet	FMC Agro	50% w/w	WG	18794

Uses
- Annual dicotyledons in **chicory root** *(off-label)*, **endives** *(off-label)* [1]; **fodder beet**, **sugar beet** [1-3]
- Black bindweed in **red beet** *(off-label)* [2]
- Charlock in **red beet** *(off-label)* [1, 2]
- Chickweed in **red beet** *(off-label)* [2]
- Cleavers in **red beet** *(off-label)* [1, 2]
- Fat hen in **red beet** *(off-label)* [2]
- Field pansy in **red beet** *(off-label)* [2]
- Flixweed in **red beet** *(off-label)* [1]
- Fool's parsley in **red beet** *(off-label)* [1, 2]
- Fumitory in **red beet** *(off-label)* [2]
- Knotgrass in **red beet** *(off-label)* [2]
- Nipplewort in **red beet** *(off-label)* [1]
- Red dead-nettle in **red beet** *(off-label)* [2]
- Redshank in **red beet** *(off-label)* [2]
- Scentless mayweed in **red beet** *(off-label)* [2]
- Small nettle in **red beet** *(off-label)* [2]
- Volunteer oilseed rape in **red beet** *(off-label)* [2]
- Wild chrysanthemum in **red beet** *(off-label)* [1]

Extension of Authorisation for Minor Use (EAMUs)
- **chicory root** *20183483* [1]
- **endives** *20183483* [1]
- **red beet** *20183484* [1], *20191915* [2]

Approval information
- Triflusulfuron-methyl included in Annex I under EC Regulation 1107/2009

Efficacy guidance
- Product should be used with a recommended adjuvant or a suitable herbicide tank-mix partner - see label for details
- Product acts by foliar action. Best results obtained from good spray cover of small actively growing weeds
- Susceptible weeds cease growth immediately and symptoms can be seen 5-10 d later
- Best results achieved from a programme of up to 4 treatments starting when first weeds have emerged with subsequent applications every 5-14 d when new weed flushes at cotyledon stage
- Weed spectrum can be broadened by tank mixture with other herbicides. See label for details
- Product may be applied overall or via band sprayer
- Triflusulfuron-methyl is a member of the ALS-inhibitor group of herbicides

Restrictions
- Maximum number of treatments 4 per crop
- Do not apply to any crop stressed by drought, water-logging, low temperatures, pest or disease attack, nutrient or lime deficiency

Crop-specific information
- Latest use: before crop leaves meet between rows
- HI 4 wk for red beet
- All varieties of sugar beet and fodder beet may be treated from early cotyledon stage until the leaves begin to meet between the rows

Following crops guidance
- Only winter cereals should follow a treated crop in the same calendar yr. Any crop may be sown in the next calendar yr
- After failure of a treated crop, sow only spring barley, linseed or sugar beet within 4 mth of spraying unless prohibited by tank-mix partner

Environmental safety
- Dangerous for the environment
- Very toxic to aquatic organisms
- Extremely dangerous to fish or other aquatic life. Do not contaminate surface waters or ditches with chemical or used container
- Take extreme care to avoid drift onto broad-leaved plants outside the target area or onto surface waters or ditches, or land intended for cropping
- Spraying equipment should not be drained or flushed onto land planted, or to be planted, with trees or crops other than sugar beet and should be thoroughly cleansed after use - see label for instructions
- LERAP Category B

Hazard classification and safety precautions
Hazard Irritant, Dangerous for the environment, Very toxic to aquatic organisms
Transport code 9
Packaging group III
UN Number 3077
Risk phrases H351 [2]
Operator protection A; U05a, U08, U19a, U20a
Environmental protection E13a, E15a, E16a, E16b, E38, H410
Storage and disposal D01, D02, D05, D09a, D11a, D12a

SEE SECTION 3 FOR PRODUCTS ALSO REGISTERED

382   trinexapac-ethyl

A novel cyclohexanecarboxylate plant growth regulator for cereals, turf and amenity grassland

See also chlormequat + trinexapac-ethyl
   prohexadione-calcium + trinexapac-ethyl

Products

1	Circle	Syngenta	250 g/l	ME	18381
2	Cleancrop Cutlass	Nufarm UK	175 g/l	EC	18820
3	Clipless NT	Headland Amenity	120 g/l	ME	17558
4	Clipless NT	FMC Agro	120 g/l	ME	18729
5	Confine	Zantra	250 g/l	EC	16109
6	Cutaway	Syngenta	121 g/l	SL	14445
7	Freeze NT	FMC Agro	250 g/l	EC	18732
8	Gyro	UPL Europe	250 g/l	EC	17426
9	Limitar	Belcrop	250 g/l	EC	16301
10	Modan	Belchim	250 g/l	EC	17060
11	Moddus	Syngenta	250 g/l	EC	15151
12	Moxa	Globachem	250 g/l	EC	16105
13	Optimus	Nufarm UK	175 g/l	EC	18817
14	Pan Tepee	Pan Agriculture	250 g/l	EC	15867
15	Primo Maxx II	Syngenta	121 g/l	SL	17509
16	Pure Max	Pure Amenity	121 g/l	SL	18524
17	Shorten	Becesane	250 g/l	EC	15377
18	Shrink	AgChem Access	250 g/l	EC	15660
19	Sudo Mor	Life Scientific	250 g/l	EC	18704
20	Terplex	Belchim	200 g/l	EC	18405
21	Tridus	Globachem	250 g/l	EC	16938
22	Trinexis	Arysta	250 g/l	EC	16429

Uses

* Growth regulation in **canary seed** *(off-label)* [2, 13]; **durum wheat, spring barley, spring oats, triticale, winter barley, winter oats, winter wheat** [8, 9, 17-19]; **forest nurseries** *(off-label)*, **ornamental plant production** *(off-label)* [6, 11, 15]; **grass seed crops** [8, 9, 19]; **grass seed crops** *(off-label)*, **red clover** *(off-label)* [11]; **lettuce** *(off-label)* [6]; **rye** [9]; **ryegrass seed crops** [8, 17, 18]; **spring oilseed rape, winter oilseed rape** [1]; **spring red wheat** *(off-label)* [5]; **spring rye, winter rye** [8, 17-19]; **spring wheat** [9, 19]
* Growth retardation in **amenity grassland** [3, 4, 6, 15, 16]; **managed amenity turf** [3, 4, 15, 16]
* Lodging control in **durum wheat** [2, 5, 7, 11-14, 21, 22]; **grass seed crops** [2, 5, 7, 12, 20, 21]; **grass seed crops** *(off-label)*, **red clover** *(off-label)* [11]; **rye** [7, 12, 21]; **ryegrass seed crops** [11, 13, 14, 22]; **spring barley, spring oats, winter oats** [2, 5, 7, 10-14, 21, 22]; **spring red wheat** *(off-label)* [5, 11]; **spring rye, winter rye** [2, 5, 10, 11, 13, 14, 20, 22]; **spring wheat** [7, 11-13, 20, 21]; **triticale, winter barley, winter wheat** [2, 5, 7, 10-14, 20-22]

Extension of Authorisation for Minor Use (EAMUs)

* **canary seed** *20191183* [2], *20192154* [13]
* **forest nurseries** *20162140* [6], *20122055* [11], *20180621* [15]
* **grass seed crops** *20171035* [11]
* **lettuce** *20162140* [6]
* **ornamental plant production** *20162140* [6], *20103062* [11], *20180621* [15]
* **red clover** *20171035* [11]
* **spring red wheat** *20130322* [5], *20121385* [11]

Approval information

* Trinexapac-ethyl included in Annex I under EC Regulation 1107/2009
* Accepted by BBPA for use on malting barley

Efficacy guidance

* Best results on cereals and ryegrass seed crops obtained from treatment from the leaf sheath erect stage [11, 14]

- Best results on turf achieved from application to actively growing weed free turf grass that is adequately fertilized and watered and is not under stress. Adequate soil moisture is essential
- Turf should be dry and weed free before application
- Environmental conditions, management and cultural practices that affect turf growth and vigour will influence effectiveness of treatment
- Repeat treatments on turf up to the maximum approved dose may be made as soon as growth resumes

Restrictions
- Maximum total dose equivalent to one full dose on cereals and ryegrass seed crops [11, 14]
- Do not apply if rain or frost expected or if crop wet. Products are rainfast after 12 h
- Only use on crops at risk of lodging [11, 14]
- Do not apply within 12 h of mowing turf
- Do not treat newly sown turf
- Not to be used on food crops
- Do not compost or mulch grass clippings
- Some products stipulate that application should not be made before GS30
- Must not be used on grass seed crops that will be grazed by livestock or cut for fodder [2]
- No more than 28 applications must be carried out on managed amenity turf and amenity grassland per year [15]

Crop-specific information
- Latest use: Up to and including flag leaf just visible (GS37) for oats and spring barley, up to and including flag leaf ligule just visible (GS39) for wheat, rye, triticale and winter barley.
- On wheat apply as single treatment between leaf sheath erect stage (GS 30) and flag leaf fully emerged (GS 39) [11, 14]
- On barley, rye, triticale and durum wheat apply as single treatment between leaf sheath erect stage (GS 30) and second node detectable (GS 32), or on winter barley at higher dose between flag leaf just visible (GS 37) and flag leaf fully emerged (GS 39) [11, 14]
- On oats and ryegrass seed crops apply between leaf sheath erect stage (GS 30) and first node detectable stage (GS 32) [11, 14]
- Treatment may cause ears of cereals to remain erect through to harvest [11, 14]
- Turf under stress when treated may show signs of damage
- Any weed control in turf must be carried out before application of the growth regulator

Environmental safety
- Dangerous for the environment [11, 14]
- Toxic to aquatic organisms [11, 14]

Hazard classification and safety precautions
Hazard Irritant, Dangerous for the environment [1, 2, 5, 7, 8, 10-14, 17-22]; Flammable liquid and vapour [1, 12, 21]; Harmful if swallowed [20]; Harmful if inhaled [1, 12, 15, 16, 21]; Very toxic to aquatic organisms [8]
Transport code 9 [1, 3-5, 7, 8, 10-12, 14, 15, 17-22]
Packaging group III [1, 3-5, 7, 8, 10-12, 14, 15, 17-22]
UN Number 3082 [1, 3-5, 7, 8, 10-12, 14, 15, 17-22]; N/C [2, 6, 9, 13, 16]
Risk phrases H315 [2, 8, 13]; H317 [2, 7, 8, 10, 11, 13-19, 22]; H318 [8, 10, 20]; H319 [1, 2, 6, 7, 9, 12, 13, 21]; H320 [1]; H335 [1, 9, 10, 12, 21]; H360 [6]
Operator protection A [1-22]; C [1-7, 9-22]; H [3, 4, 6-8, 10, 15, 16, 20]; K [3, 4, 6, 15, 16]; U05a [1-8, 10-22]; U08, U09c, U20a [9]; U10, U14 [10, 20]; U11 [9, 10, 20]; U15, U20c [1, 2, 5, 7, 8, 10-14, 17-22]; U19a [2, 9, 10, 20]; U20b [3, 4, 6, 15, 16]
Environmental protection E15a [1, 2, 5, 7, 8, 10-14, 17-22]; E15b, E34 [9]; E38 [1, 2, 5, 7-14, 17-22]; H410 [11]; H411 [1, 5, 6, 9, 10, 12, 14, 17-22]; H412 [2-4, 7, 13, 15, 16]
Consumer protection C01 [3, 4, 6, 15, 16]
Storage and disposal D01 [1-22]; D02, D05, D10c [1-8, 10-22]; D09a, D12a [1, 2, 5, 7-14, 17-22]; D11a, D12b [9]
Medical advice M03 [9]

SEE SECTION 3 FOR PRODUCTS ALSO REGISTERED

383 urea

Commodity substance for fungicide treatment of cut tree stumps

384 warfarin

A coumarin anti-coagulant rodenticide

Products

1	Grey Squirrel Bait	Killgerm	0.02% w/w	RB	UK19-1169
2	Sakarat Warfarin Whole Wheat	Killgerm	0.05% w/w	RB	UK17-1059

Uses
- Grey squirrels in **agricultural premises**, **forest**, **trees** (nut crops) [1]
- Mice in **farm buildings/yards** [2]
- Rats in **farm buildings/yards** [2]

Approval information
- Warfarin included in Annex I under EC Regulation 1107/2009

Efficacy guidance
- For rodent control place ready-to-use or prepared baits at many points wherever rats active. Out of doors shelter bait from weather
- Inspect baits frequently and replace or top up as long as evidence of feeding. Do not underbait

Restrictions
- For use only by local authorities, professional operators providing a pest control service and persons occupying industrial, agricultural or horticultural premises

Environmental safety
- Harmful to wildlife
- Prevent access to baits by children and animals, especially cats, dogs and pigs
- Rodent bodies must be searched for and burned or buried, not placed in refuse bins or rubbish tips. Remains of bait and containers must be removed after treatment and burned or buried
- Bait must not be used where food, feed or water could become contaminated
- Warfarin baits must not be used outdoors where pine martens are known to occur naturally

Hazard classification and safety precautions
    UN Number N/C [1]
    Risk phrases H360, H373 [2]
    Operator protection A; U13, U20b
    Environmental protection E10b, E15a [1]
    Storage and disposal D05, D07 [2]; D09a, D11a [1, 2]
    Vertebrate/rodent control products V01b, V02 [1, 2]; V03b, V04b [2]; V04a [1]
    Medical advice M03 [2]

385 zeta-cypermethrin

A contact and stomach acting pyrethroid insecticide
IRAC mode of action code: 3

Products

1	Fury 10 EW	FMC Agro	100 g/l	EW	18711
2	Minuet EW	FMC Agro	100 g/l	EW	18702

Uses
- Aphids in **broad beans** (off-label), **fodder beet** (off-label), **grass seed crops** (off-label), **lupins** (off-label), **rye** (off-label), **triticale** (off-label) [1]; **spring barley**, **spring oats**, **spring wheat**, **winter barley**, **winter oats**, **winter wheat** [1, 2]

FOR FULL CONDITIONS OF USE ALWAYS READ THE PRODUCT LABEL

- Barley yellow dwarf virus vectors in **spring barley, spring wheat, winter barley, winter wheat** [1, 2]
- Cabbage seed weevil in **spring oilseed rape, winter oilseed rape** [1, 2]
- Cabbage stem flea beetle in **spring oilseed rape, winter oilseed rape** [1, 2]
- Cutworms in **potatoes, sugar beet** [1, 2]
- Flax flea beetle in **linseed** [1, 2]
- Flea beetle in **spring oilseed rape, winter oilseed rape** [1, 2]
- Insect pests in **broad beans** *(off-label)*, **fodder beet** *(off-label)*, **grass seed crops** *(off-label)*, **lupins** *(off-label)*, **rye** *(off-label)*, **triticale** *(off-label)* [1]
- Large flax flea beetle in **linseed** [1, 2]
- Pea and bean weevil in **combining peas, spring field beans, vining peas, winter field beans** [1, 2]
- Pea aphid in **combining peas, vining peas** [1, 2]
- Pea moth in **combining peas, vining peas** [1, 2]
- Pod midge in **spring oilseed rape, winter oilseed rape** [1, 2]
- Pollen beetle in **spring oilseed rape, winter oilseed rape** [1, 2]
- Rape winter stem weevil in **spring oilseed rape, winter oilseed rape** [1, 2]

Extension of Authorisation for Minor Use (EAMUs)
- **broad beans** *20190807 expires 31 Jul 2020* [1]
- **fodder beet** *20190805 expires 31 Jul 2020* [1]
- **grass seed crops** *20190804 expires 31 Jul 2020* [1]
- **lupins** *20190806 expires 31 Jul 2020* [1]
- **rye** *20190804 expires 31 Jul 2020* [1]
- **triticale** *20190804 expires 31 Jul 2020* [1]

Approval information
- Zeta-cypermethrin included in Annex I under EC Regulation 1107/2009
- Accepted by BBPA for use on malting barley
- Approval expiry 31 Jul 2020 [1, 2]

Efficacy guidance
- On winter cereals spray when aphids first found in the autumn for BYDV control. A second spray may be required on late drilled crops or in mild conditions
- For summer aphids on cereals spray when treatment threshold reached
- For listed pests in other crops spray when feeding damage first seen or when treatment threshold reached. Under high infestation pressure a second treatment may be necessary
- Best results for pod midge and seed weevil control in oilseed rape obtained from treatment after pod set but before 80% petal fall
- Pea moth treatments should be applied according to ADAS/PGRO warnings or when economic thresholds reached as indicated by pheromone traps
- Treatments for cutworms should be made at egg hatch and repeated no sooner than 10 d later

Restrictions
- Maximum total dose equivalent to two full dose treatments on all crops
- Consult processors before use on crops for processing

Crop-specific information
- Latest use: before 4 true leaves for linseed, before end of flowering for oilseed rape, cereals
- HI potatoes, field beans, peas 14 d; sugar beet 60 d

Environmental safety
- Dangerous for the environment
- Very toxic to aquatic organisms
- High risk to non-target insects or other arthropods. Do not spray within 6 m of the field boundary
- LERAP Category A

Hazard classification and safety precautions
Hazard Harmful, Dangerous for the environment, Harmful if swallowed, Harmful if inhaled [1, 2]; Very toxic to aquatic organisms [1]
Transport code 9
Packaging group III

SEE SECTION 3 FOR PRODUCTS ALSO REGISTERED

UN Number 3082
Risk phrases H317
Operator protection A, C, H; U05a, U08, U14, U15, U19a, U20b
Environmental protection E15b, E16c, E16d, E22a, E34, E38, H410
Storage and disposal D01, D02, D09a, D10b, D12a
Medical advice M05a

386   zoxamide

A substituted benzamide fungicide available only in mixtures
FRAC mode of action code: 22

See also cymoxanil + zoxamide
dimethomorph + zoxamide
mancozeb + zoxamide

SECTION 3
PRODUCTS ALSO REGISTERED

Products also Registered

Products listed in the table below have not been notified for inclusion in Section 2 of this edition of the *Guide*. These products may legally be stored and used in accordance with their label until their approval expires, but they may not still be available for purchase.

Product	Approval holder	MAPP No.	Expiry Date
3 acetamiprid			
Aceta 20 SG	Euro	16919	09 Sep 2099
Acetamex 20 SP	MAC	15888	09 Sep 2099
Antelope	Gemini	18041	09 Sep 2099
Gazelle	Certis	12909	09 Sep 2099
Persist	RAAT	17633	09 Sep 2099
4 acetic acid			
OWK	UK Organic	15363	28 Feb 2025
7 adoxophyes orana gv			
Capex	Andermatt	18258	31 Jul 2025
8 alpha-cypermethrin			
Alert	BASF	16785	31 Jan 2023
Alpha C 6 ED	Techneat	13611	31 Jan 2023
Ceranock	Russell	18604	31 Jan 2023
Contest	BASF	16764	31 Jan 2023
Eribea	Belchim	17270	31 Jan 2023
Fastac	BASF	16761	31 Jan 2023
Fastac ME	BASF	17686	31 Jan 2023
Fasthrin 10EC	Sharda	18320	31 Jan 2023
Hi-Cyrus	Hockley	19018	31 Jan 2023
9 aluminium ammonium sulphate			
Curb Crop Spray Powder	Sphere	02480	30 Sep 2020
Guardsman	Chiltern	05494	30 Nov 2020
Guardsman	Chiltern	19150	28 Feb 2023
Liquid Curb Crop Spray	Sphere	03164	30 Apr 2020
Rezist	Sphere	08576	30 Nov 2020
Rezist Liquid	Sphere	14643	30 Nov 2020
Sphere ASBO	Sphere	15064	30 Nov 2020
10 aluminium phosphide			
Detia Gas Ex-P	Rentokil	17036	28 Feb 2025
Detia Gas-Ex-T	Rentokil	17034	28 Feb 2025
Quickphos Pellets 56% GE	UPL Europe	16987	28 Feb 2025
13 ametoctradin + dimethomorph			
Resplend	BASF	14975	31 Jan 2022
14 ametoctradin + mancozeb			
Diablo	BASF	16084	31 Jul 2022
15 amidosulfuron			
Eagle	Sumitomo	18902	30 Jun 2022
Squire Ultra	Sumitomo	18906	30 Jun 2022
16 amidosulfuron + iodosulfuron-methyl-sodium			
Sekator OD	Sumitomo	18905	09 Sep 2099

Product	Approval holder	MAPP No.	Expiry Date
18 aminopyralid			
Pro-Banish	Dow (Corteva)	14730	31 Dec 2020
19 aminopyralid + fluroxypyr			
Forefront Pro	Dow (Corteva)	19177	04 Jun 2022
Halcyon	Dow (Corteva)	14709	31 Dec 2020
Halcyon	Dow (Corteva)	18561	04 Jun 2022
22 aminopyralid + propyzamide			
Amino-Pro	HMpG GMBH	18948	09 Sep 2099
Clayton Propel Plus	Clayton	17227	09 Sep 2099
Dymid	Dow (Corteva)	17585	31 Jan 2021
Galactic Pro	Chem-Wise	17180	09 Sep 2099
Milepost	Terrechem	17537	09 Sep 2099
Milestone	PSI	17494	09 Sep 2099
Pyzamid Universe	RealChemie	18169	09 Sep 2099
23 aminopyralid + triclopyr			
Garlon Ultra	Nomix Enviro	19172	31 Oct 2022
Speedline Pro	Dow (Corteva)	19173	31 Oct 2022
Speedline Pro	Bayer CropScience	16208	31 Dec 2020
Speedline Pro	Dow (Corteva)	19004	31 Jul 2021
24 amisulbrom			
Gachinko	Nissan	18112	30 Sep 2021
Leimay	Syngenta	18243	28 Nov 2020
Leimay	Nissan	17412	28 Nov 2020
Sanblight	Nissan	17411	28 Nov 2020
29 Aureobasidium pullulans strain DSM 14940 and DSM 14941			
Boni Protect	Nufarm UK	19233	31 Jul 2026
31 azoxystrobin			
5504	Syngenta	12351	30 Jun 2027
Azaka	FMC Agro	16422	30 Sep 2022
Azofin	Finchimica	18193	30 Jun 2027
Azoshy	Sharda	18072	30 Jun 2027
Cleancrop Celeb	Agrii	18040	30 Jun 2027
Cleancrop Cortina	Agrii	18506	30 Jun 2027
Conclude	Belchim	16905	30 Jun 2027
Globaztar AZT 250SC	Belchim	18441	30 Jun 2027
Globaztar SC	Zantra	15575	30 Jun 2027
Legado	Industrias Afrasa	18087	30 Jun 2027
Legate	Novastar	18531	30 Jun 2027
Mirador 250SC	Adama	18292	30 Jun 2027
Ortiva	Syngenta	10542	30 Jun 2027
Oxe	Agform	18941	13 Jun 2022
Pantha	Terrechem	18619	13 Jun 2022
Phloem	Capital CP	17754	30 Jun 2027
Priori	Syngenta	10543	30 Jun 2027
Promesa	Galenika	18840	30 Jun 2027
Puma	Ascot Pro-G	17678	30 Jun 2027
Puma Super	Ascot Pro-G	18977	13 Jun 2022
Quadris	Syngenta	19211	30 Jun 2027
Sinstar Pro	Agrii	18523	30 Jun 2027
Valiant	FMC Agro	16776	30 Sep 2022
Xylem	EuroChem	17677	30 Jun 2027
Zakeo 250 SC	Adama	18379	30 Jun 2027

Product	Approval holder	MAPP No.	Expiry Date
32 azoxystrobin + chlorothalonil			
Amistar Opti	Syngenta	18156	20 May 2020
Curator	Syngenta	18157	20 May 2020
Olympus	Syngenta	18158	20 May 2020
Ortiva Opti	Syngenta	18208	20 May 2020
Perseo	Sipcam	17230	20 May 2020
Quadris Opti	Syngenta	18207	20 May 2020
34 azoxystrobin + difenoconazole			
Priori Gold	Syngenta	19226	30 Jun 2022
36 azoxystrobin + fluazinam			
Vendetta	FMC Agro	17500	31 Aug 2021
37 azoxystrobin + isopyrazam			
Symetra Flex	Adama	19220	30 Sep 2025
38 azoxystrobin + propiconazole			
Headway	Syngenta	14396	19 Mar 2020
PureProgress	Pure Amenity	15471	19 Mar 2020
Quilt Xcel	Syngenta	16450	19 Mar 2020
39 azoxystrobin + tebuconazole			
Seraphin	Adama	16248	28 Feb 2023
41 Bacillus amyloliquefaciens strain FZB24			
Taegro	Syngenta	19204	31 Dec 2034
43 Bacillus firmus I - 1582			
Flocter	Bayer CropScience	16480	31 Mar 2026
Votivo	BASF	19087	31 Mar 2026
44 Bacillus pumilus QST2808			
Sonata	Bayer CropScience	19161	28 Feb 2027
45 Bacillus subtilis			
Solani	Russell	16585	31 Oct 2022
46 Bacillus thuringiensis			
Biocure	Russell	18413	23 Sep 2021
Bruco	Progreen	17919	23 Sep 2021
Clayton Expel	Clayton	18879	23 Sep 2021
Delfin WG	Andermatt	18452	31 Oct 2022
Dipel DF	Sumitomo	18874	23 Sep 2021
47 Bacillus thuringiensis aizawai GC-91			
Agree 50 WG	Certis	17502	31 Oct 2021
49 Bacillus thuringiensis israelensis, strain AM65-52			
Gnatrol SC	Sumitomo	17802	24 Oct 2020
Gnatrol SC	Resource Chemicals	17998	24 Oct 2020
Gnatrol SC	Sumitomo	18875	24 Oct 2020
50 Beauveria bassiana			
Naturalis-L	Belchim	14655	31 Mar 2020
52 benalaxyl + mancozeb			
Galben M	FMC Agro	17247	26 Jul 2020
Intro Plus	FMC Agro	17253	26 Jul 2020
Tairel	FMC Agro	17254	26 Jul 2020

SECTION 3

Product	Approval holder	MAPP No.	Expiry Date
54 bentazone			
Basagran	BASF	00188	31 Dec 2020
Bently	Chem-Wise	14210	31 Dec 2020
Clayton Dent 480	Clayton	15546	31 Dec 2020
Euro Benta 480	Euro	14707	31 Dec 2020
Forban	Novastar	18269	31 Dec 2020
Globagran SL	Globachem	18509	31 Dec 2020
Hockley Bentazone 48	Hockley	15543	31 Dec 2020
IT Bentazone	I T Agro	13132	31 Dec 2020
Mac-Bentazone 480 SL	MAC	13598	31 Dec 2020
RouteOne Benta 48	Albaugh UK	15266	31 Dec 2020
RouteOne Bentazone 48	Albaugh UK	15269	31 Dec 2020
UPL B Zone	UPL Europe	14808	31 Dec 2020
58 bentazone + pendimethalin			
Impuls	BASF	13372	31 Dec 2020
59 benthiavalicarb-isopropyl			
Versilus	Certis	18618	09 Jul 2022
60 benthiavalicarb-isopropyl + mancozeb			
En-Garde	Certis	14901	31 Jan 2022
61 benzoic acid			
MENNO Florades	Brinkman	15091	09 Sep 2099
62 benzovindiflupyr			
Aprovia Plus	Syngenta	19174	02 Sep 2025
63 benzovindiflupyr + prothioconazole			
Elate	Generica	18091	13 Dec 2020
Pro-Benzo	HMpG GMBH	19058	13 Dec 2020
Tacanza Era	Syngenta	19217	13 Dec 2020
Velogy Era	Syngenta	18981	13 Dec 2020
64 benzyladenine			
MaxCel	Sumitomo	18878	30 Nov 2026
65 benzyladenine + gibberellin			
Floralife Bulb 100	Oasis	17995	28 Feb 2023
Promalin	Sumitomo	18325	28 Feb 2023
Promalin	Sumitomo	18892	28 Feb 2023
66 beta-cyfluthrin			
Gandalf	Nufarm UK	12865	30 Apr 2021
Gandalf	Nufarm UK	18993	30 Apr 2022
69 bifenazate			
Inter Bifenazate 240 SC	Iticon	14543	31 Jan 2023
70 bifenox			
Cleancrop Diode	Agrii	14620	31 Dec 2021
Wolf	RAAT	17378	31 Dec 2021
75 bixafen			
Bixafen EC125	Bayer CropScience	15951	31 Mar 2026
Inception Xpro	Bayer CropScience	18312	31 Mar 2026
78 bixafen + prothioconazole			
Saltri	Euro	16999	31 Jan 2023

Product	Approval holder	MAPP No.	Expiry Date
79 bixafen + prothioconazole + spiroxamine			
Boogie Xpro Plus	Bayer CropScience	18835	31 Jan 2023
82 boscalid			
Bonafide	Sharda	18775	31 Jan 2023
Boscler	Life Scientific	19113	31 Jan 2023
Coli	Gemini	17616	31 Jan 2023
Rasput	Globachem	19252	31 Jan 2023
84 boscalid + epoxiconazole			
Whistle	BASF	16093	31 Oct 2022
87 boscalid + pyraclostrobin			
Daisy	RAAT	17492	31 Jul 2022
Insignis	Euro	17116	31 Jul 2022
Pyrabos	HMpG GMBH	19080	31 Jul 2022
89 bromadiolone			
Romax Bromablock	Zapi	UK13-0777	31 Aug 2020
90 bromoxynil			
Akocynil 225 EC	Aako	15688	31 Jan 2023
Alpha Bromolin 225 EC	MAUK	14864	31 Jan 2023
Buctril	Bayer CropScience	16597	31 Jan 2023
Flagon 400 EC	MAUK	14921	31 Jan 2023
Xinca	Nufarm UK	16833	31 Jan 2023
92 bromoxynil + diflufenican			
Nessie	Nufarm UK	17815	30 Jun 2022
99 bromoxynil + prosulfuron			
Jester	Syngenta	13500	31 Jul 2020
102 bromuconazole + tebuconazole			
Djembe	Sumitomo	18323	28 Feb 2023
Djembe	Sumitomo	19223	28 Feb 2023
Sakura	Sumitomo	18327	28 Feb 2023
Sakura	Sumitomo	19224	28 Feb 2023
Soleil	Sumitomo	19222	28 Feb 2023
104 buprofezin			
Applaud 25 WP	Certis	17197	31 Jul 2025
105 Candida oleophila Strain 0			
Nexy 1	BioNext	16961	31 Mar 2026
106 captan			
Akotan 80 WG	Aako	16307	31 Jan 2023
Malvin WG	Arysta	16308	31 Jan 2023
Orthocide WG	Arysta	16309	31 Jan 2023
Ratan 80 WG	RAAT	17252	31 Jan 2023
116 carbetamide			
Kartouch 60 WG	Aako	16897	30 Nov 2022
120 carfentrazone-ethyl			
Aurora	Belchim	11613	31 Jan 2021
Aurora 40 WG	FMC Agro	17278	31 Jan 2021
Carfen 50	HMpG GMBH	18427	31 Jan 2021
Clayton Xenon	Clayton	18278	31 Jan 2021
Harrier	FMC Agro	18803	09 Sep 2099

SECTION 3

Product	Approval holder	MAPP No.	Expiry Date
Harrier	Belchim	14164	31 Jan 2021
Platform	FMC Agro	17809	31 Jan 2021
Shark	FMC Agro	17256	31 Jan 2021
Spotlight Plus	FMC Agro	17241	31 Jan 2021

122 carfentrazone-ethyl + mecoprop-P

Jewel	FMC Agro	18780	31 Jul 2022
Platform S	FMC Agro	17259	31 Jan 2021

125 cerevisane

Romeo	Fargro	18534	08 May 2022

126 chlorantraniliprole

Coragen	FMC Agro	14930	26 Oct 2020

127 chloridazon

Better DF	Sipcam	16101	30 Jun 2020
Better Flowable	Sipcam	15740	30 Jun 2020
Pyramin DF	BASF	16768	30 Jun 2020

130 chloridazon + metamitron

Volcan Combi FL	Sipcam	16681	30 Jun 2020

131 chloridazon + quinmerac

Fiesta T	BASF	17153	30 Jun 2020

132 chlormequat

Adjust	Taminco	17141	31 May 2024
Barleyquat B	Taminco	17204	31 May 2024
BEC CCC 720	Becesane	19009	31 May 2024
Belcocel	Taminco	17773	31 May 2024
Belcocel	Taminco	16652	31 Oct 2020
Belcocel 750	Taminco	19125	31 May 2024
Bettequat B	Taminco	17208	31 May 2024
Cleancrop Conicen	Agrii	18986	31 May 2024
Eddystone	Terrechem	18345	31 May 2024
K2	Taminco	17206	31 May 2024
Manipulator	Taminco	17207	31 May 2024
Plectrum	Gemini	18317	31 May 2024
Selon	Taminco	17209	31 May 2024
Silk 750 SL	Terrechem	17496	31 May 2024

133 chlormequat + 2-chloroethylphosphonic acid

Spatial Plus	Sumitomo	18888	31 Jan 2022
Vivax	Sumitomo	18901	31 Jan 2023

136 chlormequat + imazaquin

Meteor	BASF	16800	30 Jun 2020
Upright	BASF	16884	30 Jun 2020

138 chlormequat + trinexapac-ethyl

Completto	Adama	18448	14 May 2022

142 chlorothalonil

Abringo	Belcrop	15607	20 May 2020
Alternil 500	Arysta	16813	20 May 2020
Alternil Excel 720	Arysta	17434	20 May 2020
Asterix	Barclay	17383	30 Apr 2020
Balear 720 SC	Arysta	15545	20 May 2020
Banko 720	Arysta	17728	20 May 2020
Barclay Avoca	Barclay	15549	20 May 2020

Product	Approval holder	MAPP No.	Expiry Date
Barclay Chloroflash	Barclay	16651	20 May 2020
Bravo 500	Syngenta	14548	20 May 2020
Chlorta 500	HMpG GMBH	18474	30 Apr 2020
Chlorthalis	Arysta	16513	20 May 2020
Claw 500	Gemini	17332	20 May 2020
Clayton Turret	Clayton	17220	20 May 2020
Cleancrop Fortis	Agrii	19098	20 May 2020
Cleancrop Wanderer	Agrii	18355	20 May 2020
Cleancrop Wanderer 2	Agrii	16302	30 Apr 2020
Daconil	Syngenta	17778	20 May 2020
Damocles	AgChem Access	15689	20 May 2020
Doolin	Barclay	17676	20 May 2020
Duster	JT Agro	17517	20 May 2020
FSA Chlorothalonil	Synergy	17182	20 May 2020
GF Chlorothalonil	Genfarm	17011	20 May 2020
Hi-Charlo	Hockley	18621	20 May 2020
Invictus	Ascot Pro-G	18877	20 May 2020
IT Chlorothalonil	I T Agro	15845	20 May 2020
Joules	Nufarm UK	15730	20 May 2020
Life Scientific Chlorothalonil	Life Scientific	15369	30 Apr 2020
Life Scientific Chlorothalonil 500	Life Scientific	15813	20 May 2020
LS Chlorothalonil XL	Agrovista	15443	20 May 2020
MS Catelyn	Micromix	17148	20 May 2020
Multi-Star 500	Albaugh UK	16784	20 May 2020
Mycoguard 720	Arysta	18330	20 May 2020
Optio 500	Novastar	17216	20 May 2020
Ovarb	Life Scientific	17972	20 May 2020
Piper	Nufarm UK	15786	20 May 2020
Premium 500	Agroquimicos	16137	20 May 2020
Pronto	Stefes	17688	20 May 2020
Raindance	Chem-Wise	19100	20 May 2020
Renew Chlorothalonil 500SC	Renew	16706	20 May 2020
Resort	Belchim	17861	20 May 2020
Rover 500	Sipcam	15496	20 May 2020
Sinconil	Agrii	16268	20 May 2020
Soar	Agform	18147	20 May 2020
Spirodor	Belchim	17710	20 May 2020
Stefonil	Stefes	16862	20 May 2020
Sundance	Chem-Wise	19099	20 May 2020
Supreme	Certis	16505	20 May 2020
Synergy Chlorothalonil	Synergy	17215	20 May 2020
Thalo	Synergy	18606	20 May 2020
Thalonil 500	Euro	14926	20 May 2020
Tymoon 500	Novastar	17336	20 May 2020
UPL Chlorothalonil	UPL Europe	15472	20 May 2020
Zambesi	Terrechem	17840	20 May 2020

143 chlorothalonil + cymoxanil

Mixanil	Sipcam	15675	20 May 2020

144 chlorothalonil + cyproconazole

Alto Elite	Syngenta	08467	20 May 2020
Baritone	AgChem Access	15910	20 May 2020
Belimar	AgChem Access	18090	20 May 2020
Bravo Xtra	Syngenta	11824	20 May 2020
Citadelle	Syngenta	17886	20 May 2020
Cleancrop Cyprothal	Agrii	09580	31 Dec 2021
Octolan	Syngenta	11675	20 May 2020
SAN 703	Syngenta	11676	20 May 2020

SECTION 3

Product	Approval holder	MAPP No.	Expiry Date
147 chlorothalonil + fludioxonil + propiconazole			
Instrata	Syngenta	16458	19 Mar 2020
150 chlorothalonil + fluxapyroxad			
Divexo	BASF	17937	20 May 2020
Divexo	BASF	18632	20 May 2020
151 chlorothalonil + mancozeb			
Guru	Sumitomo	15268	20 May 2020
153 chlorothalonil + metalaxyl-M			
Broadsheet	AgChem Access	15906	20 May 2020
Folio Gold	Syngenta	14368	20 May 2020
154 chlorothalonil + penthiopyrad			
Aylora	DuPont (Corteva)	17905	20 May 2020
Intellis Plus	DuPont (Corteva)	17908	20 May 2020
Treoris	DuPont (Corteva)	17849	20 May 2020
Trust	Generica	16561	20 May 2020
157 chlorothalonil + propiconazole			
Barclay Avoca Premium	Barclay	15619	19 Mar 2020
Oxana	Arysta	15177	19 Mar 2020
Prairie	Syngenta	13994	19 Mar 2020
SIP 313	Sipcam	14755	19 Mar 2020
158 chlorothalonil + proquinazid			
Fielder SE	DuPont (Corteva)	15120	20 May 2020
160 chlorothalonil + tebuconazole			
Cigal Plus	Rotam	17229	20 May 2020
Confucius	Rotam	17257	20 May 2020
Crafter	Nufarm UK	15895	20 May 2020
Fezan Plus	Sipcam	15490	20 May 2020
Nectar	Nufarm UK	16872	20 May 2020
Pentangle	Nufarm UK	16792	20 May 2020
Timpani	Nufarm UK	14651	20 May 2020
Tonga	Arysta	18416	20 May 2020
166 chlorpropham			
Aceto Sprout Nip	Aceto	14156	31 Jan 2021
BL 500	UPL Europe	14387	31 Jan 2021
CIPC Gold	Dormfresh	15674	31 Jan 2021
Cleancrop Amigo	Agrii	15292	31 Jan 2022
Gro-Stop Electro	Certis	17445	31 Jan 2021
Gro-Stop HN	Certis	14146	31 Jan 2021
MSS CIPC 50 M	UPL Europe	14388	31 Jan 2021
Neo-Stop 120 RTU	Agriphar	17420	31 Jan 2022
Neo-Stop Starter	Arysta	15651	31 Jan 2022
Prolonger	UPL Europe	17453	31 Jan 2021
Tuberprop Easy	UPL Europe	15701	31 Jan 2022
169 chlorpyrifos			
Akofos 480 EC	Aako	17986	22 Jan 2021
Ballad	FMC Agro	17999	31 Jul 2022
Cyren	FMC Agro	17934	31 Jul 2022
Dursban WG	Dow (Corteva)	17932	22 Jan 2021
Equity	Dow (Corteva)	17933	31 Jul 2022
Pyrinex 48 EC	Adama	17935	22 Jan 2021
Pyrinex 48 EC	Nufarm UK	18893	22 Jan 2021

Product	Approval holder	MAPP No.	Expiry Date
170 chlorpyrifos-methyl			
Garrison	Pan Agriculture	18174	31 Jul 2022
Reldan 22	Dow (Corteva)	16710	31 Jul 2022
174 clethodim			
Centurion Max	Sumitomo	16310	09 Nov 2020
Chellist	Chem-Wise	16263	09 Nov 2020
Clayton Gatso	Clayton	17698	09 Nov 2020
Clayton Gatso	Clayton	17732	09 Nov 2020
Clayton Gatso	Clayton	17746	09 Nov 2020
Mollivar	Generica	17795	09 Nov 2020
Mollivar	Generica	17798	09 Nov 2020
Mollivar	Nouvelle Tec	18590	09 Nov 2020
V-Dim 240 EC	VextaChem	18530	26 Feb 2023
Vextadim 240 EC	Clayton	18465	26 Feb 2023
175 clodinafop-propargyl			
Clodinastar	Life Scientific	17404	31 Oct 2020
Ravena	FMC Agro	17313	31 Oct 2021
Sword	Adama	15099	31 Oct 2021
Viscount	Syngenta	17495	31 Oct 2021
Wildcat	UKChem	17783	30 Jun 2020
Wildcat	Matrix	17317	31 Oct 2021
176 clodinafop-propargyl + cloquintocet-mexyl			
Ciclope	Industrias Afrasa	17226	31 Oct 2022
178 clodinafop-propargyl + pinoxaden			
Traxos	Syngenta	12742	30 Sep 2020
Traxos Pro	Syngenta	16713	30 Sep 2020
179 clodinafop-propargyl + prosulfocarb			
Auxiliary	BASF	14576	30 Apr 2022
Grapple	Syngenta	16523	30 Apr 2022
181 clofentezine			
Acaristop 500 SC	Aako	17232	30 Jun 2022
Ariane	RAAT	17296	30 Jun 2022
182 clomazone			
Centium 360 CS	FMC Agro	17327	30 Apr 2021
Cirrus CS	FMC Agro	17328	30 Apr 2021
Clayton Chrome CS	Clayton	16383	30 Apr 2022
Cleancrop Chicane	FMC Agro	17330	30 Apr 2021
Czar	Terrechem	17423	30 Apr 2022
Evea	Rotam	18511	30 Apr 2022
Gambit	Harvest	17747	31 Mar 2021
Gamit 36 CS	FMC Agro	17314	30 Apr 2021
Hobby	Ascot Pro-G	17667	30 Apr 2022
Noble	Rovogate	17162	30 Apr 2022
Pegasus	Terrechem	18658	30 Apr 2022
Pegasus PC	Terrechem	18927	30 Apr 2021
Regal	UKChem	17782	30 Jun 2020
Regal	Matrix	16617	30 Apr 2021
Retribute	Agform	16739	31 Mar 2020
Standon Soulmate	Standon	16879	30 Apr 2021
Standon Soulmate	Enviroscience	18499	30 Apr 2022
Token	Becesane	18574	30 Apr 2022
Upstage	UPL Europe	18984	30 Apr 2022
Zone 360 CS	HMpG GMBH	18424	30 Apr 2022

SECTION 3

Product	Approval holder	MAPP No.	Expiry Date
Zone 360 CS-II	HMpG GMBH	18437	30 Apr 2022
Zulkon	Unisem	17610	30 Apr 2022

183 clomazone + dimethenamid-p + metazachlor

Nimbus Gold	BASF	18192	30 Apr 2022

185 clomazone + metazachlor

Circuit Synctec	FMC Agro	17118	30 Apr 2022
Nimbus CS	BASF	16573	30 Apr 2022

186 clomazone + metazachlor + napropamide

Colzor SyncTec	FMC Agro	18092	30 Apr 2021
Colzor SyncTec	FMC Agro	18716	05 Jul 2021

188 clomazone + napropamide

Altiplano DAMtec	FMC Agro	17189	30 Apr 2022

189 clomazone + pendimethalin

Stallion Sync TEC	FMC Agro	17243	31 Dec 2020

190 clopyralid

Bariloche	Proplan-Plant	17577	30 Nov 2025
Cliophar 600 SL	Arysta	16735	31 Oct 2022
GF-2895	Dow (Corteva)	16821	31 Oct 2022
Glopyr 400	Globachem	15009	31 Oct 2022
Lontrel 72SG	Dow (Corteva)	15235	31 Oct 2022

194 clopyralid + florasulam + fluroxypyr

Dakota	Dow (Corteva)	16121	31 Jan 2021
Galaxy	Dow (Corteva)	14085	30 Sep 2020
Mogul	Aremie	17289	30 Sep 2020
Polax	Pan Amenity	15915	30 Sep 2020
Praxys	Dow (Corteva)	13912	30 Sep 2020

195 clopyralid + fluroxypyr + MCPA

Brittas	Barclay	18587	30 Apr 2022
Interfix	Iticon	15762	30 Apr 2022

197 clopyralid + halauxifen-methyl

Korvetto	Dow (Corteva)	19111	31 Oct 2022

198 clopyralid + picloram

Barca 334 SI	Hockley	18539	30 Jun 2022
Barca 334 SL	Pestila	19121	30 Jun 2022
Chaco	Globachem	18783	30 Jun 2022
Pyralid Extra	Euro	16606	30 Jun 2022

200 clopyralid + triclopyr

Blaster Pro	Headland Amenity	15752	31 Oct 2020
Headland Flail 2	FMC Agro	16007	31 Oct 2022
Tor	Dow (Corteva)	17777	31 Oct 2022

214 cyantraniliprole

Mainspring	Syngenta	19198	14 Mar 2029
Minecto One	Syngenta	18649	14 Mar 2029
Verimark 20 SC	FMC Agro	17519	31 Oct 2022

215 cyazofamid

Rithfir	DuPont (Corteva)	14816	31 Jan 2023
RouteOne Roazafod	Albaugh UK	15271	31 Jan 2023
Sugoi	ISK Biosciences	19122	31 Jan 2023

Product	Approval holder	MAPP No.	Expiry Date
219 Cydia pomonella GV			
Cyd-X	Certis	17019	31 Oct 2021
Cyd-X Duo	Certis	16779	31 Oct 2021
Cyd-X Xtra	Certis	17020	31 Oct 2021
Madex Top	Andermatt	18227	31 Oct 2021
Madex Top	Andermatt	19213	31 Oct 2022
220 cyflufenamid			
Cyflufen EW	HMpG GMBH	18425	30 Sep 2025
Diego	Star	16980	30 Sep 2025
NF-149 SC	Certis	15835	30 Sep 2025
Sydly	Terrechem	18307	30 Sep 2025
221 cymoxanil			
Cymbal Flow	Belchim	17843	29 Feb 2024
Cymostraight 45	Belchim	17437	29 Feb 2024
Drum Flow	Belchim	17906	29 Feb 2024
Krug Flow	Belchim	17964	29 Feb 2024
222 cymoxanil + famoxadone			
Tanos	DuPont (Corteva)	17160	31 Dec 2022
223 cymoxanil + fluazinam			
Shirlan Forte	Syngenta	15542	31 Aug 2022
225 cymoxanil + mancozeb			
Nautile WP	UPL Europe	16468	31 Jul 2022
Palmas WP	SFP Europe	16632	31 Jul 2022
Solace Max	Nufarm UK	16697	31 Jul 2022
Solution	DuPont (Corteva)	19078	31 Jul 2022
Zetanil WG	Sipcam	15488	31 Jul 2022
226 cymoxanil + mandipropamid			
Bedrock	Syngenta	18426	29 Feb 2024
227 cymoxanil + propamocarb			
Omix Duo	Agria SA	18978	31 Jan 2023
Rival Duo	Agria SA	18442	31 Jan 2023
228 cymoxanil + zoxamide			
Reboot	Gowan	16909	31 Jul 2020
Reboot	Gowan	18202	09 Sep 2099
229 cypermethrin			
Afrisect 500 EC	Arysta	17137	30 Apr 2022
Langis 300 ES	Arysta	18178	30 Apr 2022
Permasect 500 EC	Arysta	17132	30 Apr 2022
Signal 300 ES	Arysta	16481	30 Apr 2022
Supasect 500 EC	Arysta	17131	30 Apr 2022
233 cyproconazole + penthiopyrad			
Cielex	DuPont (Corteva)	16540	30 Sep 2020
Cielex	DuPont (Corteva)	17758	30 Nov 2023
238 cyprodinil			
Unix	Syngenta	14846	31 Oct 2022
239 cyprodinil + fludioxonil			
Botrefin	Clayton	18537	30 Apr 2022
Button	Euro	16842	30 Apr 2022
Lever	HMpG GMBH	18467	30 Apr 2022

SECTION 3

Product	Approval holder	MAPP No.	Expiry Date
Reversal	RAAT	17359	30 Apr 2022
242 2,4-D			
Depitox 500	Nufarm UK	17597	09 Sep 2099
Herboxone 60	FMC Agro	14080	09 Sep 2099
Maton	FMC Agro	13234	09 Sep 2099
244 2,4-D + dicamba			
Landscaper Pro Weed Control + Fertilizer	ICL (Everris) Ltd	18291	30 Jun 2022
247 2,4-D + dicamba + iron sulphate + mecoprop-P			
Elliott's Professional Feed Weed & Moss Killer	Elliott	19205	30 Jun 2022
Feed 'N' Weed Extra	Vitax	19103	30 Jun 2022
GTHMF Pro	Global Crop Care	18443	30 Jun 2022
Professional Turf Feed, Weed And Moss Killer	Proctor	19106	30 Jun 2022
Restore Plus	Coburn	19102	30 Jun 2022
Turfmaster Complete Feed, Weed & Mosskiller	Hygeia	18985	30 Jun 2022
248 2,4-D + dicamba + MCPA + mecoprop-P			
Dicophar	Arysta	16460	30 Apr 2022
252 2,4-D + glyphosate			
Kurtail Evo	Progreen	19020	09 Sep 2099
Rasto	Nufarm UK	16795	09 Sep 2099
253 2,4-D + MCPA			
Agroxone Combi	Nufarm UK	14907	09 Sep 2099
256 2,4-D + triclopyr			
Genoxone ZX EC	Arysta	17016	09 Sep 2099
257 daminozide			
B-Nine SG	Arysta	14435	30 Apr 2022
258 dazomet			
Basamid	Certis	11324	31 Jul 2020
Basamid	Kanesho	12895	31 Jul 2020
259 2,4-DB			
Butoxone DB	Nufarm UK	14840	30 Apr 2020
Butoxone DB	Nufarm UK	18172	09 Sep 2099
CloverMaster	Nufarm UK	17927	30 Apr 2020
DB Straight	UPL Europe	13736	30 Apr 2020
Embutone 24DB	Nufarm UK	17596	30 Apr 2020
Embutone 24DB	Nufarm UK	18255	09 Sep 2099
Headland Spruce	FMC Agro	15089	30 Apr 2020
261 2,4-DB + MCPA			
Butoxone DB Extra	Nufarm UK	15741	09 Sep 2099
262 deltamethrin			
CMI Delta 2.5 EC	CMI	16695	30 Apr 2022
Decis	Bayer CropScience	16124	31 Jul 2020
Deltason-D	DAPT	17735	30 Apr 2022
GAT Decline 2.5 EC	FMC Agro	17703	30 Apr 2021
GAT Decline 2.5 EC	FMC Agro	18715	30 Apr 2022
Grain-Tect ULV	Barrettine	18076	30 Apr 2022

Product	Approval holder	MAPP No.	Expiry Date
Grain-Tect ULV	Limagrain	15253	30 Apr 2020
Tecsis	HMpG GMBH	18439	30 Apr 2022

267 desmedipham + phenmedipham

Betanal Maxxim	Bayer CropScience	14186	01 Jul 2020
Rifle	Adama	15705	01 Jul 2020

271 dicamba + MCPA + mecoprop-P

Nomix Tribute Turf	Nomix Enviro	16313	30 Apr 2022
Relay Turf Elite	FMC Agro	17192	30 Apr 2022

272 dicamba + mecoprop-P

Dimeco XL	UPL Europe	17190	30 Jun 2022
Headland Swift	FMC Agro	11945	30 Jun 2022
Mircam	Nufarm UK	11707	30 Jun 2022
Optica Forte	Nufarm UK	14845	30 Jun 2022

274 dicamba + prosulfuron

Clayton Spook	Clayton	15846	09 Sep 2099
Rosan	FMC Agro	16591	30 Apr 2020

277 dichlorprop-P + ferrous sulphate + MCPA

SHL Granular Feed, Weed & Mosskiller	Sinclair	10972	30 Apr 2020

280 dichlorprop-P + MCPA + mecoprop-P

Hymec Triple	Agrichem	15753	30 Apr 2022

283 difenacoum

Romax D Rat & Mouse Killer	Rentokil	UK14-0808	19 Mar 2023
Romax DP	Rentokil	UK13-0804	31 Aug 2020
Sorexa D Mouse Killer	BASF	UK12-0319	14 Feb 2023

285 difenoconazole

Bogard	Syngenta	17310	30 Jun 2022
Brace	Terrechem	19197	30 Jun 2022
Change	RAAT	17400	30 Jun 2022
Diconazol 250	Euro	16387	30 Jun 2022
Difend	Q-Chem	16316	30 Jun 2022
Difenostar	Life Scientific	17598	28 Jun 2020
Mavita 250EC	Adama	18293	30 Jun 2022
Revel	Life Scientific	17973	28 Jun 2021
Slick	Syngenta	17331	30 Jun 2022
Stratus	B&W Farms	18624	30 Jun 2022

287 difenoconazole + fludioxonil

Celest Extra	Syngenta	16630	30 Apr 2022

288 difenoconazole + fludioxonil + tebuconazole

Celest Trio	Syngenta	15510	30 Apr 2022

292 difenoconazole + propiconazole

Armure	Syngenta	16587	19 Mar 2020
Difure Pro	Belchim	17303	19 Mar 2020

293 diflubenzuron

Diflox Flow	Hockley	15640	31 Dec 2021
Dimilin Flo	Arysta	11056	31 Dec 2021

294 diflufenican

Beeper 500 SC	Aako	15358	30 Jun 2022

SECTION 3

Product	Approval holder	MAPP No.	Expiry Date
Beluga	Ascot Pro-G	17806	30 Jun 2022
Deflect	Sharda	18236	30 Jun 2022
Diecot	Gemini	17191	30 Jun 2022
Diecot	Cropthetics	19243	30 Jun 2022
Diflufen 500	Euro	16687	30 Jun 2022
Diflufenican GL 500	Globachem	17084	30 Jun 2022
Dina 50	Globachem	17083	30 Jun 2022
Flash	Sharda	17943	30 Jun 2022
Flyflo	Gemini	17694	30 Jun 2022
Flyflo	Cropthetics	19242	30 Jun 2022
Inter Diflufenican	UPL Europe	17342	29 Feb 2020
Overlord	MAUK	13521	30 Jun 2022
Prefect	Novastar	17713	30 Jun 2022
Sempra	UPL Europe	16967	30 Jun 2022
Solo D500	Sipcam	16098	30 Jun 2022
Terrier	Terrechem	17761	30 Jun 2022
Tornado	UKChem	17781	30 Jun 2020
Tornado	Matrix	17425	31 Oct 2020
Turnpike	Rotam	18631	30 Jun 2022
Twister	Unique Marketing	16374	30 Jun 2022

295 diflufenican + florasulam

Product	Approval holder	MAPP No.	Expiry Date
Bow	Nufarm UK	16845	30 Jun 2022
Lector Delta	Nufarm UK	16808	30 Jun 2022

296 diflufenican + flufenacet

Product	Approval holder	MAPP No.	Expiry Date
Adept	Agform	17628	30 Apr 2023
Amaranth	PSI	16949	30 Apr 2023
Cleancrop Marauder	Agrii	15933	30 Apr 2023
Deliverer	Euro	16692	30 Apr 2023
Dephend	FMC Agro	18144	30 Jun 2021
Diflufenastar	Life Scientific	17611	30 Apr 2023
Diflufenastar 100	Life Scientific	17475	30 Apr 2021
Duke	Chem-Wise	16738	30 Apr 2023
Fenican	Euro	16790	30 Apr 2023
Feud	Gemini	18384	30 Apr 2023
Firebird	Bayer CropScience	14826	30 Apr 2023
Fludual	Hockley	18554	30 Apr 2023
Flufenican	Euro	15386	30 Apr 2023
Navigate	FMC Agro	17506	30 Apr 2021
Navigate	FMC Agro	18950	30 Apr 2022
Nucleus	FMC Agro	18610	30 Apr 2022
Saviour	Chem-Wise	16702	30 Apr 2023
Sharp XD	Cropthetics	18431	30 Apr 2023
Sharp XD	JT Agro	18268	30 Apr 2021
Terrane	FMC Agro	17619	30 Jun 2021
UPL DFF + FFT	UPL Europe	18503	30 Apr 2022

297 diflufenican + flufenacet + flurtamone

Product	Approval holder	MAPP No.	Expiry Date
Flufenamone	Euro	17260	27 Mar 2020
Movon	Bayer CropScience	14784	27 Mar 2020
Vigon	Bayer CropScience	14785	27 Mar 2020
Voyage	Chem-Wise	17238	27 Mar 2020

300 diflufenican + flurtamone

Product	Approval holder	MAPP No.	Expiry Date
Bacara	Bayer CropScience	16448	27 Mar 2020

301 diflufenican + glyphosate

Product	Approval holder	MAPP No.	Expiry Date
Pistol Rail	Bayer CropScience	17450	14 Aug 2020

Product	Approval holder	MAPP No.	Expiry Date
302 diflufenican + iodosulfuron-methyl-sodium			
LockStar	ICL (Everris) Ltd	19101	30 Jun 2022
Valdor Flex	Bayer CropScience	19033	30 Jun 2022
303 diflufenican + iodosulfuron-methyl-sodium + mesosulfuron-methyl			
Kalenkoa	Bayer CropScience	17733	09 Sep 2099
306 diflufenican + metribuzin			
Tavas	Adama	18213	30 Jun 2022
307 diflufenican + metsulfuron-methyl			
Pelican Delta	Nufarm UK	14312	30 Jun 2021
309 dimethachlor			
Clayton Dimethachlor	Clayton	16188	30 Jun 2024
311 dimethenamid-p + metazachlor			
Caribou	RealChemie	18195	30 Apr 2022
Caribou	RealChemie	18263	30 Apr 2022
Muntjac	BASF	16893	30 Apr 2022
Sika	HMpG GMBH	18428	30 Apr 2022
312 dimethenamid-p + metazachlor + quinmerac			
Luna	Goldengrass	17053	30 Apr 2022
Medirac	Euro	17214	30 Apr 2022
313 dimethenamid-p + pendimethalin			
Dime	Pan Agriculture	17398	09 Sep 2099
Dime	Pan Agriculture	18388	09 Sep 2099
Pendi P	HMpG GMBH	19040	09 Sep 2099
Pendi P	HMpG GMBH	19092	09 Sep 2099
314 dimethenamid-p + quinmerac			
Butisan Pro	BASF	18726	30 Apr 2022
315 dimethoate			
Clayton Tote	Clayton	18423	31 Jan 2023
Danadim Progress	FMC Agro	15890	31 Jan 2021
316 dimethomorph			
Morph	Adama	15121	31 Jan 2022
Murphy 500 SC	Aako	15203	31 Jan 2022
Navio	FMC Agro	17827	31 Jan 2022
Rigel WP	Servem	17347	31 Jan 2022
318 dimethomorph + mancozeb			
Saracen	BASF	15250	31 Jul 2022
319 dimethomorph + propamocarb			
Diprospero	Arysta	19188	31 Jan 2023
320 dimethomorph + pyraclostrobin			
Cassiopeia	BASF	16522	31 Jul 2022
Optimo Tech	BASF	16455	31 Jul 2022
325 diquat			
A1412A2	Syngenta	13440	04 Feb 2020
Balista	Novastar	15186	04 Feb 2020
Barclay DQ-200	Barclay	15260	04 Feb 2020
Barclay D-Quat	Barclay	14833	04 Feb 2020
BD-200	Barclay	14918	04 Feb 2020

SECTION 3

Product	Approval holder	MAPP No.	Expiry Date
Belquat	Agroquimicos	18238	04 Feb 2020
Brogue	Certis	16421	04 Feb 2020
Clayton Diquat	Clayton	12739	04 Feb 2020
Clayton Diquat 200	Clayton	13942	04 Feb 2020
Clayton IQ	Clayton	14519	04 Feb 2020
CleanCrop Flail	Agrii	14870	04 Feb 2020
CropStar Diquat	CropStar	17679	04 Feb 2020
Dessicash 200	Sharda	14848	04 Feb 2020
Di Novia	Agroquimicos	15840	04 Feb 2020
Diqua	Sharda	14032	04 Feb 2020
Diquanet	Belcrop	16868	04 Feb 2020
Diquash	Agrichem	14849	04 Feb 2020
Di-Quattro	Arysta	15712	04 Feb 2020
Dragoon	Hermoo	13927	04 Feb 2020
Dragoon Gold	Belcrop	14973	04 Feb 2020
Hockley Diquat 20	Hockley	15585	04 Feb 2020
I.T. Diquat	UPL Europe	16970	04 Feb 2020
Inter Diquat	UPL Europe	16986	04 Feb 2020
Kalahari	Barclay	14917	04 Feb 2020
Knoxdoon	AgChem Access	13987	04 Feb 2020
Life Scientific Diquat	Life Scientific	15398	04 Feb 2020
Life Scientific Diquat 200	Life Scientific	15811	04 Feb 2020
Mission 200 SL	UPL Europe	16977	04 Feb 2020
Quad S	Q-Chem	13578	04 Feb 2020
Quad-Glob 200 SL	Belchim	14758	04 Feb 2020
Quat 200	Euro	15042	04 Feb 2020
Quazar	Belchim	15357	04 Feb 2020
Quit	Belchim	14874	04 Feb 2020
Reglone	Syngenta	10534	04 Feb 2020
Retro	Syngenta	13841	04 Feb 2020
Roquat 20	Belchim	15703	04 Feb 2020
Standon Googly	Standon	12995	04 Feb 2020
Standon Googly	Enviroscience	18492	04 Feb 2020
Synchem Diquat 200	Synchem	16817	04 Feb 2020
Tuber Mission	UPL Europe	16965	04 Feb 2020
Woomera	Belcrop	14972	04 Feb 2020

326 dithianon

Rathianon 70 WG	RAAT	17471	30 Nov 2026

328 dithianon + pyraclostrobin

Maccani	BASF	18546	31 Jul 2022

334 dodine

Radspor 400	Agriphar	16001	31 Dec 2021

335 epoxiconazole

Amber	Adama	16285	31 Oct 2022
Aska	Terrechem	17461	31 Oct 2021
Barret 125 SL	Agrifarm	16709	31 Oct 2022
Bassoon	BASF	14402	31 Oct 2022
Cibury	HMpG GMBH	18528	31 Oct 2022
Cibury	HMpG GMBH	18553	31 Oct 2022
Corral	FMC Agro	18819	31 Oct 2022
Corral	FMC Agro	16942	31 Oct 2021
Ignite	BASF	15205	31 Oct 2022
Inopia	Chem-Wise	19021	31 Oct 2021
Intercord	Iticon	16155	31 Oct 2021
Opus	BASF	12057	31 Oct 2022
Propov	Syngenta	17926	31 Oct 2022

Product	Approval holder	MAPP No.	Expiry Date
Rapier	Ascot Pro-G	17857	31 Oct 2021
Regent	UKChem	17784	30 Jun 2020
Regent	Matrix	17087	31 Oct 2020
Rubric	FMC Agro	14118	31 Oct 2021
Spike	Q-Chem	17910	31 Oct 2022
Strand	FMC Agro	14261	31 Oct 2021
Voodoo	Sipcam	16072	31 Oct 2021
Warlock	Matrix	17026	31 Oct 2020
Warlock	UKChem	17775	30 Jun 2020

336 epoxiconazole + fenpropimorph

Product	Approval holder	MAPP No.	Expiry Date
Eclipse	BASF	11731	31 Oct 2020
Opus Team	BASF	11759	31 Oct 2020
Oracle Duet	AgChem Access	16011	31 Oct 2020

337 epoxiconazole + fenpropimorph + kresoxim-methyl

Product	Approval holder	MAPP No.	Expiry Date
Asana	BASF	11934	31 Oct 2020
Mantra	BASF	11728	31 Oct 2020

338 epoxiconazole + fenpropimorph + metrafenone

Product	Approval holder	MAPP No.	Expiry Date
Capalo	BASF	13170	31 Oct 2020
Polaca	Euro	16436	31 Oct 2020
Stiletto	BASF	14729	31 Oct 2020

339 epoxiconazole + fenpropimorph + pyraclostrobin

Product	Approval holder	MAPP No.	Expiry Date
Diamant	BASF	14149	31 Oct 2020
Sapphire	Euro	16648	31 Oct 2020

340 epoxiconazole + fluxapyroxad

Product	Approval holder	MAPP No.	Expiry Date
Adexar	PSI	17441	31 Oct 2022
Apex Pro	Chem-Wise	17497	31 Oct 2022
Apex Pro	Chem-Wise	17632	31 Oct 2022
Cougar	UKChem	17785	30 Jun 2020
Cougar	Matrix	17442	31 Oct 2022
Morex	BASF	17307	31 Oct 2022
Standon Coolie	Standon	17382	31 Oct 2022
Standon Coolie	Enviroscience	18494	31 Oct 2022
ZORO	Terrechem	17462	31 Oct 2022

341 epoxiconazole + fluxapyroxad + pyraclostrobin

Product	Approval holder	MAPP No.	Expiry Date
Ceriax	PSI	17527	31 Jul 2022
Chusan	Terrechem	17528	31 Jul 2022
Smaragdin	PSI	17532	31 Jul 2022

343 epoxiconazole + isopyrazam

Product	Approval holder	MAPP No.	Expiry Date
Clayton Kudos	Clayton	16580	31 Oct 2021
Micaraz	Adama	18563	31 Oct 2022
Seguris	Adama	18564	31 Oct 2022

346 epoxiconazole + metconazole

Product	Approval holder	MAPP No.	Expiry Date
Ettu	Euro	16435	31 Oct 2022
Icarus	BASF	14471	31 Oct 2022

347 epoxiconazole + metrafenone

Product	Approval holder	MAPP No.	Expiry Date
Ceando	Dow (Corteva)	16012	31 Oct 2022
Ceando	BASF	19253	31 Oct 2022
Cloister	BASF	19254	31 Oct 2022
Cloister	Dow (Corteva)	15994	31 Oct 2022

348 epoxiconazole + prochloraz

Product	Approval holder	MAPP No.	Expiry Date
Corrib	Barclay	18634	31 Oct 2022

SECTION 3

Product	Approval holder	MAPP No.	Expiry Date
349 epoxiconazole + pyraclostrobin			
Ambassador	Euro	16410	31 Jul 2022
Ibex	BASF	16457	31 Jul 2022
Inter Opal	BASF	19035	31 Jul 2022
Inter Opal	Inter-Phyto	18430	31 Oct 2021
Spike Max	Belchim	18551	31 Jul 2022
350 esfenvalerate			
Barclay Alphasect	Barclay	16053	09 Sep 2099
Clayton Cajole	Clayton	14995	09 Sep 2099
Clayton Slalom	Clayton	15054	09 Sep 2099
Clayton Vindicate	Clayton	15055	09 Sep 2099
Gocha	Belchim	17320	30 Apr 2020
Gocha	Sumitomo	18623	09 Sep 2099
Kingpin	Sumitomo	18626	09 Sep 2099
Kingpin	Belchim	18176	30 Apr 2020
Sumi-Alpha	Sumitomo	18637	09 Sep 2099
Sven	Sumitomo	18627	09 Sep 2099
352 ethanol			
Ethy-Gen II	Ripe Rite	15839	31 Dec 2022
Restrain Fuel	Restrain	14520	31 Dec 2021
353 ethephon			
Cerone	Sumitomo	18903	31 Jan 2023
Coupon	Globachem	19123	31 Jan 2023
Coupon 480	Globachem	18865	31 Jan 2023
Hi-Phone 48	Hockley	15506	31 Jan 2023
Inter Ethephon	Inter-Phyto	18638	31 Jan 2022
354 ethephon + mepiquat chloride			
Gunbar	Terrechem	17460	31 Aug 2022
Mepicame	Unisem	17541	31 Aug 2022
Riggid	Gemini	17522	31 Aug 2022
Terpal	PSI	17440	31 Aug 2022
355 ethofumesate			
Nortron Flo	Bayer CropScience	18526	30 Apr 2034
Princess	Emerald	19191	09 Sep 2099
356 ethofumesate + metamitron			
Goltix Plus	Aako	16538	28 Feb 2020
Goltix Plus	Aako	18665	28 Feb 2025
Metafol Super	UPL Europe	18925	28 Feb 2025
Metafol Super	UPL Europe	16881	30 Jun 2020
Oblix MT	UPL Europe	16487	30 Aug 2020
Volcano	UPL Europe	16532	30 Jun 2020
358 ethofumesate + phenmedipham			
Brigida	Agroquimicos	15658	09 Sep 2099
Gemini	EZCrop	15670	09 Sep 2099
Magic Tandem	Bayer CropScience	17358	09 Sep 2099
Powertwin	Adama	14004	09 Sep 2099
Teamforce	UPL Europe	16978	31 Jul 2020
360 ethylene			
Banarg	BOC	18190	28 Feb 2025
Biofresh Safestore	Biofresh	15729	28 Feb 2025

Product	Approval holder	MAPP No.	Expiry Date
361 etofenprox			
Trebon 30 EC	Certis	16666	30 Jun 2024
362 etoxazole			
Borneo	Sumitomo	18873	31 Jan 2023
365 fatty acids			
Flipper	Fargro	17767	28 Feb 2023
Flipper	Bayer CropScience	19154	28 Feb 2023
Hydro Coco Houseplant Pest & Spider Mite Killer	151 Products Ltd	16898	30 Apr 2020
NEU 1170 H	Sinclair	15754	28 Feb 2023
373 fenazaquin			
Matador 200 SC	Gowan	16875	31 Dec 2020
377 fenhexamid			
Agrovista Fenamid	Agrovista	13733	09 Sep 2099
RouteOne Fenhex 50	Albaugh UK	13665	09 Sep 2099
RouteOne Fenhex 50	Albaugh UK	15282	09 Sep 2099
379 fenoxaprop-P-ethyl			
Foxtrot	FMC Agro	13243	30 Jun 2021
Oskar	FMC Agro	13344	30 Jun 2021
Polecat	Sumitomo	18896	30 Jun 2022
383 fenpropidin + prochloraz + tebuconazole			
Artemis	Adama	14178	30 Jun 2020
Artemis	Adama	18533	30 Jun 2022
387 fenpropimorph			
Clayton Spigot	Clayton	11560	31 May 2020
Clayton Spigot	Clayton	19127	31 Oct 2020
Cleancrop Fenpro	Agrii	09885	31 Oct 2020
Cleancrop Fenpropimorph	Agrii	09445	31 May 2020
Corbel	BASF	00578	31 May 2020
Corbel	BASF	18904	31 Oct 2020
Crebol	AgChem Access	13922	31 May 2020
Marnoch Phorm	Me2	11087	31 May 2020
Standon Fenpropimorph 750	Standon	08965	31 May 2020
391 fenpropimorph + pyraclostrobin			
Jenton	BASF	11898	31 Oct 2020
393 fenpyrazamine			
Clayton Vitis	Clayton	16974	30 Jun 2025
Prolectus	Sumitomo	18891	30 Jun 2025
395 ferric phosphate			
Chicane	Certis	19059	30 Jun 2033
Daxxos	Certis	18809	30 Jun 2033
Derrex	Certis	15351	30 Jun 2033
Ironflexx	Certis	18555	30 Jun 2033
Ironmax Pro	De Sangosse	18839	30 Jun 2033
Limafer	Frunol Delicia	19145	30 Jun 2033
Luminare	Certis	19060	30 Jun 2033
Minixx	Certis	19068	30 Jun 2033
NEU 1181 M	Neudorff	14355	30 Jun 2033
Regiment	Certis	19065	30 Jun 2033
Sluggo	Omex	14788	31 Aug 2020
Spinner	Certis	19067	30 Jun 2033

SECTION 3

Product	Approval holder	MAPP No.	Expiry Date
TurboDisque	Frunol Delicia	19144	30 Jun 2033
TurboPads	Frunol Delicia	19146	31 Dec 2031

396 ferrous sulphate

Product	Approval holder	MAPP No.	Expiry Date
Landscaper Pro Moss Control + Fertiliser	ICL (Everris) Ltd	16723	28 Feb 2023
LSTF Pro	LSTF	16992	28 Feb 2023
Maxicrop No. 2 Moss Killer and Conditioner	Maxicrop	17300	31 May 2022
Maxicrop Professional Moss Killer & Conditioner	Valagro	18514	28 Feb 2023
Moss-Kill Pro	Bioservices	17029	28 Feb 2023
Rigby Taylor Turf Moss Killer	Rigby Taylor	17219	28 Feb 2023
Sinclair Lawn Sand	Westland Horticulture	17562	28 Feb 2023
Sinclair Pro Autumn Feed and Moss Killer	Westland Horticulture	18398	31 Jan 2022

399 flazasulfuron

Product	Approval holder	MAPP No.	Expiry Date
Chikara	Nomix Enviro	13775	09 Sep 2099
Flaza 250	HMpG GMBH	18470	09 Sep 2099
Flazaraat	RAAT	17912	09 Sep 2099
Hinoki	ISK Biosciences	19134	09 Sep 2099
Paradise	Pan Agriculture	14504	09 Sep 2099
Paradise	Pan Agriculture	16829	09 Sep 2099
Railtrax	PSI	15139	09 Sep 2099
Sixte	FMC Agro	18142	31 Jul 2020
Sixte	FMC Agro	18734	31 Jul 2020
Terafit	Syngenta	18280	31 Jul 2020

400 flocoumafen

Product	Approval holder	MAPP No.	Expiry Date
Storm Mini Bits	BASF	UK15-0915	08 Mar 2023

401 flonicamid

Product	Approval holder	MAPP No.	Expiry Date
Hinode	ISK Biosciences	19135	28 Feb 2026
Pekitek	HMpG GMBH	19029	28 Feb 2026
Pekitek	HMpG GMBH	19070	28 Feb 2026
Primeman	RAAT	17656	28 Feb 2026

402 florasulam

Product	Approval holder	MAPP No.	Expiry Date
Barton WG	Dow (Corteva)	13284	30 Jun 2033
Flora 50	Euro	15700	09 Sep 2099
Flora 50 II	Euro	17716	31 Aug 2022
Globus	Globachem	18834	07 Mar 2023
Lector	Nufarm UK	16565	31 Aug 2022
Paramount	Nufarm UK	16452	31 Aug 2022
RouteOne Florasul 50	Albaugh UK	15285	09 Sep 2099
Solstice	Nufarm UK	17164	30 Sep 2022
Suprime	Nufarm UK	18110	09 Sep 2099

403 florasulam + fluroxypyr

Product	Approval holder	MAPP No.	Expiry Date
Flatline	Aremie	17422	09 Sep 2099
Flurostar XL	Belchim	18617	17 Oct 2022
Flurosulam XL	Euro	15236	09 Sep 2099
Gal-Gone XL	Agrovista	18918	17 Oct 2022

404 florasulam + halauxifen-methyl

Product	Approval holder	MAPP No.	Expiry Date
Mattera	Dow (Corteva)	19133	29 Feb 2028
Renitar	Dow (Corteva)	19129	05 Feb 2028

Product	Approval holder	MAPP No.	Expiry Date
405 florasulam + pinoxaden			
Axial One	Syngenta	17262	30 Apr 2020
Axial One	Syngenta	18837	31 Dec 2028
406 florasulam + pyroxsulam			
Capri Duo	Dow (Corteva)	19142	31 Oct 2026
Gyga	Dow (Corteva)	19143	31 Oct 2026
407 florasulam + tribenuron-methyl			
Bolt	Nufarm UK	17363	30 Apr 2021
Flame Duo	Albaugh UK	19097	30 Apr 2022
Paramount Max	Nufarm UK	17271	30 Apr 2021
409 fluazifop-P-butyl			
Clayton Crowe	Clayton	12572	28 Feb 2021
Fluazifop +	RouteOne	13382	28 Feb 2021
Fusilade Max	Nufarm UK	11519	31 Aug 2020
Fusilade Max	Nufarm UK	18815	28 Feb 2021
Greencrop Bantry	Greencrop	12737	28 Feb 2021
Howitzer	ChemSource	14942	28 Feb 2021
RouteOne Fluazifop +	Albaugh UK	15288	28 Feb 2021
RouteOne Fluazifop +	Generica	15679	28 Feb 2021
RouteOne Fluazifop +	Albaugh UK	13641	28 Feb 2021
Salvo	Generica	16228	28 Feb 2021
410 fluazinam			
Boyano	Belchim	16621	31 Aug 2022
Dalimo	ISK Biosciences	19132	31 Aug 2022
Deltic Pro	FMC Agro	18733	31 Aug 2022
Deltic Pro	FMC Agro	17855	31 Aug 2021
Float	FMC Agro	18810	31 Aug 2022
Float	FMC Agro	16718	31 Aug 2021
Fluazifin	Clayton	18549	31 Aug 2022
Fluazinam 500 SC	Finchimica	19261	31 Aug 2022
Flunam 500 III	HMpG GMBH	18480	31 Aug 2022
FOLY 500 SC	Aako	16486	31 Aug 2022
Frowncide	ISK Biosciences	16619	31 Aug 2021
Frowncide	Belchim	18392	31 Aug 2022
Ibiza 500	Belchim	16620	31 Aug 2022
Legacy	Syngenta	16622	31 Aug 2021
Legacy	Belchim	18389	31 Aug 2022
Magic	Cropthetics	18861	31 Aug 2022
Magic	JT Agro	18750	31 Aug 2021
Ohayo	Belchim	18391	31 Aug 2022
Ohayo	Syngenta	16588	31 Aug 2021
Shirlan	Syngenta	16624	31 Aug 2021
Shirlan Programme	Syngenta	16623	31 Aug 2021
Shirlan Programme	Belchim	18407	31 Aug 2022
Smash	RAAT	17488	31 Aug 2022
Tizca	FMC Agro	16289	31 Aug 2021
Winby	ISK Biosciences	16618	31 Aug 2021
Winby	Belchim	18390	31 Aug 2022
Zinam 500	Euro	16831	31 Aug 2021
Zinam II	Euro	16886	31 Aug 2022
412 fludioxonil			
Maxim 480FS	Syngenta	16725	30 Apr 2022
418 fludioxonil + sedaxane			
A20078F	Syngenta	17945	19 Jan 2021

Product	Approval holder	MAPP No.	Expiry Date
421 flufenacet			
Fence	Albaugh UK	17393	30 Apr 2023
Foxton	Terrechem	18595	30 Apr 2023
Glosset SC	Belchim	19265	30 Apr 2023
Macho	Albaugh UK	18122	30 Apr 2023
Osprey	Albaugh UK	18657	30 Apr 2023
Steeple	Albaugh UK	18082	30 Apr 2023
423 flufenacet + isoxaflutole			
Cadou Star	Bayer CropScience	13242	31 Jan 2022
RouteOne Oxanet 481	Albaugh UK	15309	31 Jan 2022
425 flufenacet + pendimethalin			
Cleancrop Hector	Agrii	15932	09 Sep 2099
Ice	BASF	13930	31 Jan 2021
Kruos	Chem-Wise	15880	09 Sep 2099
Latice	Euro	16693	09 Sep 2099
Standon Diwana	Standon	17221	09 Sep 2099
Standon Diwana	Enviroscience	18485	09 Sep 2099
426 flufenacet + picolinafen			
Equs	BASF	17736	09 Sep 2099
Kudu	BASF	17804	09 Sep 2099
427 flumioxazin			
Digital	Sumitomo	13561	31 Dec 2022
Digital	Sumitomo	18885	31 Dec 2022
Guillotine	Sumitomo	18876	31 Dec 2022
Guillotine	Sumitomo	13562	31 Dec 2022
Sumimax	Sumitomo	18884	31 Dec 2022
431 fluopyram + prothioconazole			
Clayton Repel	Clayton	18037	31 Jan 2023
Prolense	HMpG GMBH	18560	31 Jan 2023
434 fluoxastrobin			
Bayer UK 831	Bayer CropScience	12091	31 Jan 2022
435 fluoxastrobin + prothioconazole			
Exploit	Euro	16851	31 Jan 2023
Firefly	Bayer CropScience	13692	31 Jan 2023
Prostrob 20	Chem-Wise	15698	31 Jan 2023
437 fluoxastrobin + prothioconazole + trifloxystrobin			
Haven	Bayer CropScience	18403	09 Sep 2099
446 fluroxypyr			
Arbiter	Barclay	18326	30 Jun 2027
Awac	Globachem	17501	02 Feb 2023
Casino	Certis	17592	30 Jun 2027
Clean Crop Gallifrey 200	Globachem	17503	02 Feb 2023
GF-1784	Dow (Corteva)	16850	30 Jun 2027
Havoc	Gemini	19057	30 Jun 2027
Hyflux GB	Hygeia	19015	31 May 2022
Taipan	Agro Trade	17063	30 Jun 2027
Toska EC	Novastar	17709	30 Jun 2027
Tumu	Terrechem	19032	30 Jun 2027
450 fluroxypyr + metsulfuron-methyl			
Croupier OD	Certis	17630	31 Aug 2020

Product	Approval holder	MAPP No.	Expiry Date
451 fluroxypyr + metsulfuron-methyl + thifensulfuron-methyl			
Omnera LQM	FMC Agro	17769	31 Dec 2020
Omnera LQM	FMC Agro	18758	09 Sep 2099
Provalia LQM	FMC Agro	17846	31 Dec 2020
453 fluroxypyr + triclopyr			
Pas	Dow (Corteva)	17772	31 Oct 2022
456 flutolanil			
NNF-136	Nichino	14302	31 Aug 2022
457 flutriafol			
Consul	FMC Agro	12976	31 Dec 2021
Impact	FMC Agro	12776	31 Dec 2021
Impact	FMC Agro	18789	31 Dec 2021
Topguard	FMC Agro	18723	31 Dec 2021
458 fluxapyroxad			
BAS 700	BASF	17106	30 Nov 2020
Flux	Chem-Wise	17602	30 Jun 2025
459 fluxapyroxad + mefentrifluconazole			
Mivyto XE	BASF	19259	22 Sep 2023
Revystar XE	BASF	19250	22 Sep 2023
Verydor XE	BASF	19251	22 Sep 2023
460 fluxapyroxad + metconazole			
Wolverine	BASF	19181	31 Oct 2022
461 fluxapyroxad + pyraclostrobin			
Syrex	BASF	18908	31 Jul 2022
462 folpet			
Lamast	Syngenta	19140	31 Jan 2023
Mirror	Syngenta	18866	31 Jan 2023
Sesto	Belchim	19148	31 Jan 2023
Sesto	Adama	19219	31 Jan 2023
464 foramsulfuron + iodosulfuron-methyl-sodium			
Logo	Bayer CropScience	17624	09 Sep 2099
466 fosetyl-aluminium			
Plant Trust	ICL (Everris) Ltd	15779	31 Oct 2022
467 fosetyl-aluminium + propamocarb hydrochloride			
Avatar	ICL (Everris) Ltd	16608	31 Oct 2022
472 garlic extract			
Eagle Green Care	ECOspray	14989	31 Mar 2021
EGC Liquid	Rigby Taylor	17852	28 Feb 2023
EGCA Granules	Rigby Taylor	17233	28 Feb 2023
Grenadier	Certis	17617	28 Feb 2023
NEMguard A PCN Granules	Certis	17922	28 Feb 2023
NEMguard granules	ECOspray	15254	28 Feb 2023
Pitcher GR	ECOspray	18126	28 Feb 2023
Pitcher SC	ECOspray	18125	28 Feb 2023
474 gibberellins			
Regulex 10 SG	Sumitomo	18890	28 Feb 2023
Smartgrass	Sumitomo	18883	28 Feb 2023

SECTION 3

Product	Approval holder	MAPP No.	Expiry Date
475 Gliocladium catenulatum			
Prestop Mix	Fargro	15104	31 Jan 2022
477 glyphosate			
Alekto Plus TF	Belchim	17143	09 Sep 2099
Amega Duo	Nufarm UK	14131	30 Jun 2020
Asteroid	FMC Agro	11118	31 Dec 2020
Asteroid Pro	FMC Agro	15373	31 Dec 2020
Asteroid Pro 450	FMC Agro	16384	31 Dec 2020
Barclay Gallup Biograde Amenity	Barclay	12716	30 Jun 2020
Boom Efekt	Albaugh UK	17588	09 Sep 2099
Buggy SL	Sipcam	17962	30 Jun 2020
Buggy XTG	Sipcam	12951	30 Jun 2020
CDA Vanquish Biactive	Bayer CropScience	12586	30 Jun 2020
Cleancrop Bronco	Belchim	17056	09 Sep 2099
Cleancrop Corral	Agrii	17554	30 Jun 2020
Cleancrop Corral 2	Agrii	18023	09 Sep 2099
Cleancrop Tungsten	Agrii	17553	09 Sep 2099
Cleancrop Tungsten Ultra	Agrii	18297	09 Sep 2099
Clinic Grade	Nufarm UK	19028	09 Sep 2099
Clinic TF	Nufarm UK	16716	09 Sep 2099
Clipper	Adama	14820	09 Sep 2099
Cosmic NG	Arysta	15646	30 Jun 2020
Credit DST	Nufarm UK	14066	30 Jun 2020
Crestler	Nufarm UK	16663	30 Jun 2020
Egret	Monsanto	17510	09 Sep 2099
Envision	FMC Agro	10569	31 Dec 2020
Euro Glyfo 450	Euro	14936	30 Jun 2020
Excel DF Gold	Excel	17413	28 Feb 2020
Gallup Hi-Aktiv Amenity	Barclay	12898	30 Jun 2020
Glister Ultra	Sinon EU	15209	09 Sep 2099
Glister Ultramax	Sinon EU	18021	09 Sep 2099
Glyder 450 TF	Belchim	16520	09 Sep 2099
Glyfer	Nouvelle Tec	17895	30 Jun 2020
Glyfer	Nouvelle Tec	19200	09 Sep 2099
Glyfo Star	Q-Chem	17457	30 Jun 2020
Glyfos Dakar Pro	Headland Amenity	13147	31 Dec 2020
Glyfos Dakar Pro	FMC Agro	18812	09 Sep 2099
Glyfos Gold	FMC Agro	18790	09 Sep 2099
Glyfos Gold	Nomix Enviro	10570	31 Dec 2020
Glyfos Gold ECO	FMC Agro	16489	31 Dec 2020
Glyfos Gold ECO	Headland Amenity	18730	09 Sep 2099
Glyfos Monte	FMC Agro	15069	31 Dec 2020
Glyfos Monte	FMC Agro	18807	09 Sep 2099
Glyfos Supreme XL	FMC Agro	17705	31 Dec 2020
Glyfos Supreme XL	FMC Agro	18695	09 Sep 2099
Glypho-Rapid 450	Barclay	17647	09 Sep 2099
Habitat	Barclay	17662	09 Sep 2099
Helosate 450 TF	Helm	16326	09 Sep 2099
Hilite	Nomix Enviro	13871	30 Jun 2020
Jali 450	DAPT	15930	30 Jun 2020
Klean G	Ventura	17645	30 Jun 2020
Kraken	Bayer CropScience	15175	30 Jun 2020
Mascot Hi-Aktiv Amenity	Rigby Taylor	15178	30 Jun 2020
Master Gly	Ventura	18515	30 Jun 2020
MON 76473	Monsanto	15348	09 Sep 2099
MON 79351	Monsanto	12860	09 Sep 2099
MON 79376	Monsanto	12664	09 Sep 2099
MON 79545	Monsanto	12663	09 Sep 2099

Product	Approval holder	MAPP No.	Expiry Date
MON 79991	Monsanto	16300	09 Sep 2099
Monosate G	Monsanto	17449	29 Feb 2020
Monosate G	Monsanto	18651	09 Sep 2099
NASA	Agria SA	18170	11 Jul 2021
Nomix Conqueror Amenity	Nomix Enviro	16039	30 Jun 2020
Nomix Frontclear	Nomix Enviro	15180	30 Jun 2020
Nomix Frontclear	Nomix Enviro	18365	30 Jun 2020
Nomix G	Nomix Enviro	18359	09 Sep 2099
Nomix G	Nomix Enviro	13872	30 Jun 2020
Nomix Nova	Nomix Enviro	13873	30 Jun 2020
Nomix Nova	Nomix Enviro	18360	30 Jun 2020
Nomix Prolite	Nomix Enviro	18346	09 Sep 2099
Nomix Prolite	Frontier	16132	30 Jun 2020
Nomix Revenge	Nomix Enviro	13874	30 Jun 2020
Nomix Revenge	Nomix Enviro	18368	30 Jun 2020
Ormond	Barclay	17712	09 Sep 2099
Ovation	Monsanto	17476	09 Sep 2099
Oxalis NG	Arysta	15612	30 Jun 2020
Palomo Green	ProKlass	17081	09 Sep 2099
Pitch	Nufarm UK	15485	09 Sep 2099
Pitch	Nufarm UK	15693	30 Jun 2020
Preline	Linemark	12756	30 Jun 2020
Pultare	Maxwell	18242	30 Jun 2020
Rattler	Nufarm UK	15692	30 Jun 2020
Roundup Ace	Monsanto	12772	30 Jun 2020
Roundup Advance	Monsanto	15540	30 Jun 2020
Roundup Assure	Monsanto	15537	30 Jun 2020
Roundup Biactive	Monsanto	10320	09 Sep 2099
Roundup Biactive 3G	Monsanto	13409	30 Jun 2020
Roundup Biactive Dry	Monsanto	12646	30 Jun 2020
Roundup Bio	Monsanto	15538	30 Jun 2020
Roundup Express	Monsanto	12526	30 Jun 2020
Roundup Gold	Monsanto	10975	30 Jun 2020
Roundup Klik	Monsanto	12866	09 Sep 2099
Roundup Max	Monsanto	12952	30 Jun 2020
Roundup Pro Biactive	Monsanto	10330	09 Sep 2099
Roundup ProBiactive 450	Monsanto	12778	30 Jun 2020
Roundup Pro-Green	Rigby Taylor	11907	30 Jun 2020
Roundup Provide	Monsanto	12953	30 Jun 2020
Snapper	Nufarm UK	15489	30 Jun 2020
Snapper	Nufarm UK	15695	30 Jun 2020
Surrender T/F	Agrovista	17575	30 Jun 2020
Tangent	FMC Agro	18826	09 Sep 2099
Tangent	FMC Agro	11872	31 Dec 2020
Tanker	Nufarm UK	15694	30 Jun 2020
Touchdown Quattro	Syngenta	10608	09 Sep 2099
Trustee Amenity	Barclay	12897	30 Jun 2020
Typhoon 360	Adama	14817	09 Sep 2099
Ultramax	PSI	15931	30 Jun 2020
Vesuvius Green	Ventura	17650	30 Jun 2020

479 glyphosate + sulfosulfuron

Nomix Blade	Nomix Enviro	13907	30 Jun 2020
Nomix Blade	Nomix Enviro	18358	09 Sep 2099
Nomix Dual	Nomix Enviro	13420	30 Jun 2020
Nomix Duplex	Nomix Enviro	18350	09 Sep 2099
Nomix Duplex	Nomix Enviro	14953	30 Jun 2020

SECTION 3

Product	Approval holder	MAPP No.	Expiry Date
483 halauxifen-methyl			
GF-2573	Dow (Corteva)	17540	05 Feb 2028
486 imazalil			
Fungazil 50LS	Certis	14069	31 Jul 2020
487 imazalil + ipconazole			
Conima	Arysta	18966	28 Feb 2027
Rancona i-MIX	Arysta	16492	28 Feb 2027
493 imazamox + metazachlor + quinmerac			
Clesima	BASF	16275	31 Jan 2024
495 imazamox + quinmerac			
Clentiga	BASF	18121	31 Oct 2026
503 indoxacarb			
Explicit	FMC Agro	15359	30 Apr 2022
Picard 300 WG	Agrifarm	16714	30 Apr 2022
Rumo	FMC Agro	14883	30 Apr 2021
Steward	FMC Agro	13149	30 Apr 2021
504 iodosulfuron-methyl-sodium			
Hussar	Bayer CropScience	12364	30 Apr 2020
505 iodosulfuron-methyl-sodium + mesosulfuron-methyl			
Atlantis Pro	PSI	18402	09 Sep 2099
Atlantis WG	Bayer CropScience	12478	30 Apr 2020
Ficap	Life Scientific	18135	30 Apr 2020
Greencrop Doonbeg 2	Clayton	13025	30 Apr 2020
Idosem 36	Chem-Wise	15913	30 Apr 2020
Iodomeso OD	Euro	16501	09 Sep 2099
Jerico	Terrechem	17477	30 Apr 2020
Lanista	Life Scientific	18118	09 Sep 2099
Lusitania	Life Scientific	17989	30 Apr 2020
Mesiodo 35	Euro	15305	30 Apr 2020
Mesiodos OD	HMpG GMBH	19071	09 Sep 2099
Nemo	Nurture	18929	30 Apr 2020
Nemo	AgChem Access	14140	30 Apr 2020
Ocelot	UKChem	17790	30 Apr 2020
Pacifica	Bayer CropScience	12049	30 Apr 2020
RouteOne Seafarer	Albaugh UK	15317	30 Apr 2020
Standon Mimas WG	Standon	17170	30 Apr 2020
Standon Mimas WG	Enviroscience	18497	30 Apr 2020
506 iodosulfuron-methyl-sodium + propoxycarbazone-sodium			
Caliban Duo	FMC Agro	14283	30 Apr 2020
508 ipconazole			
Rancona 15 ME	Arysta	16500	28 Feb 2027
514 isopyrazam			
A15149W	Syngenta	15046	30 Jun 2022
A15149W	Adama	18571	30 Sep 2025
Zulu	Adama	18575	30 Sep 2025
515 isopyrazam + prothioconazole			
Polarize	Adama	18980	31 Jan 2023
516 isoxaben			
Pan Isoxaben 500	Pan Amenity	17980	31 Dec 2021

Product	Approval holder	MAPP No.	Expiry Date
519 kresoxim-methyl			
Strong WG	RAAT	17470	30 Jun 2027
520 lambda-cyhalothrin			
Clayton Lambada	Clayton	13231	09 Sep 2099
Clayton Lanark	Clayton	12942	09 Sep 2099
Cleancrop Argent	Agrii	18286	09 Sep 2099
CleanCrop Corsair	Agrii	14124	09 Sep 2099
CM Lambaz 50 EC	CMI	16241	09 Sep 2099
Colt 10 CS	FMC Agro	17430	31 Dec 2020
Colt 10 CS	FMC Agro	18708	09 Sep 2099
Eminentos 10 CS	FMC Agro	18712	09 Sep 2099
Eminentos 10 CS	FMC Agro	17690	31 Dec 2020
Euro Lambda 100 CS	Euro	14794	09 Sep 2099
Hockley Lambda 5EC	Hockley	15516	09 Sep 2099
Karis 10 CS	FMC Agro	18707	09 Sep 2099
Karis 10 CS	FMC Agro	16747	31 Dec 2020
Kendo	Syngenta	15562	09 Sep 2099
Kung Fu	Syngenta	18974	09 Sep 2099
Kusti	Syngenta	16656	09 Sep 2099
Laidir 10 CS	FMC Agro	18705	09 Sep 2099
Laidir 10 CS	FMC Agro	17693	31 Dec 2020
Lamdex Extra	Adama	18541	31 May 2020
MAC-Lambda-Cyhalothrin 50 EC	MAC	14941	09 Sep 2099
Ninja 5CS	Syngenta	16417	09 Sep 2099
Reparto	Sipcam	15452	09 Sep 2099
RouteOne Lambda C	Albaugh UK	13663	09 Sep 2099
RVG Lambda-cyhalothrin	Rovogate	16806	31 Mar 2021
Stealth	Syngenta	14551	09 Sep 2099
Triumph	Matrix	16929	09 Sep 2099
Triumph CS	Capital CP	18235	09 Sep 2099
Warrior	Syngenta	13857	09 Sep 2099
524 lenacil			
Venzar 500 SC	FMC Agro	17743	30 Jun 2021
Venzar 80 WP	FMC Agro	17742	30 Jun 2021
Venzar 80 WP	FMC Agro	18757	30 Jun 2022
525 lenacil + triflusulfuron-methyl			
Debut Plus	FMC Agro	17991	30 Jun 2021
Safari Lite WSB	FMC Agro	17954	30 Jun 2021
Upbeet Lite	FMC Agro	17990	30 Jun 2022
Upbeet Lite	FMC Agro	18795	30 Jun 2022
527 magnesium phosphide			
Magtoxin Pellets	Rentokil	17376	28 Feb 2025
528 maleic hydrazide			
Clayton Stun	Clayton	17194	30 Apr 2020
Cleancrop Malahide	Agrii	13629	30 Apr 2020
Fazor	Arysta	13617	31 Oct 2020
Fazor	Arysta	13679	31 Oct 2020
Fazor	Dow (Corteva)	19075	09 Sep 2099
Gro-Slo	Arysta	15851	30 Apr 2020
Himalaya	Arysta	15214	30 Apr 2020
Himalaya XL	Arysta	17511	31 Oct 2020
Himalaya XL	Arysta	19076	09 Sep 2099
Itcan SL 270	Gemini	17957	09 Sep 2099
Magna SL	Certis	18589	09 Sep 2099
Source II	Drexel	17858	30 Apr 2020

SECTION 3

Product	Approval holder	MAPP No.	Expiry Date
Source II	Chiltern	13618	30 Apr 2020
529 maleic hydrazide + pelargonic acid			
Finalsan Plus	Certis	17928	09 Sep 2099
530 maltodextrin			
Eradicoat Max	Certis	18852	15 Nov 2022
Terminus	Certis	17319	03 Aug 2021
531 mancozeb			
Agria Mancozeb 75WDG	Agria SA	16807	31 Jul 2022
Agria Mancozeb 80WP	Agria SA	16636	31 Jul 2022
Avtar 75 NT	Syngenta	18909	31 Jul 2022
Cleancrop Feudal	Agrii	17894	31 Jul 2022
Cleancrop Mandrake	Agrii	14792	31 Jul 2022
Cleancrop Mandrake Plus WG	Agrii	18508	31 Jul 2022
Dithane 945	Indofil	17269	31 Jul 2020
Dithane 945	Sumitomo	18900	31 Jul 2022
Dithane Dry Flowable Neotec	Indofil	17268	31 Jul 2020
Dithane NT Dry Flowable	Sumitomo	18889	31 Jul 2022
Emzeb 80 WP	Sabero	17714	02 Aug 2020
Fortuna	Hockley	18620	31 Jul 2022
Fortuna Globe	Hockley	18612	31 Jul 2022
Karamate Dry Flo Neotec	Landseer	14632	31 Jul 2022
Laminator 75 WG	Sumitomo	19022	31 Jul 2022
Laminator Plus WG	Sumitomo	18907	31 Jul 2022
Manfil 75 WG	Indofil	15958	31 Jul 2022
Manfil 80 WP	Indofil	15956	31 Jul 2022
Manfil WP Plus	Agrii	17939	31 Jul 2022
Quell Flo	Sumitomo	19077	31 Jul 2022
Tridex	UPL Europe	18196	31 Jul 2021
Trimanoc 75 DG	DuPont (Corteva)	18822	31 Jul 2022
Trimanzone	UPL Europe	16759	31 Jul 2022
Zebra WDG	FMC Agro	14875	31 Jul 2022
532 mancozeb + metalaxyl-M			
Clayton Mohawk	Clayton	18086	31 Jul 2022
535 mandipropamid			
Evagio	Syngenta	18279	31 Jan 2026
Mandimid	Euro	17651	31 Jan 2026
Mandoprid	Generica	18182	31 Jan 2026
Manpro	HMpG GMBH	18468	31 Jan 2026
Pergardo Uni	Fargro	17513	31 Jan 2026
Standon Mandor	Standon	17536	31 Jan 2026
Standon Mandor	Enviroscience	18463	31 Jan 2026
Standon Mandor	Enviroscience	18496	31 Jan 2026
Standon Mandor	Standon	17533	31 Mar 2022
537 MCPA			
Agrichem MCPA 500	UPL Europe	16964	30 Apr 2022
Agritox 50	Nufarm UK	14814	30 Apr 2022
Dow MCPA Amine 50	Dow (Corteva)	14911	30 Apr 2022
Go-Low Power	DuPont (Corteva)	13812	30 Apr 2022
Haksar 500 SL	Ciech	18446	30 Apr 2022
Lagoon	Gemini	17856	30 Apr 2022
MCPA 25%	Nufarm UK	14893	30 Apr 2022
MCPA 50	UPL Europe	14908	30 Apr 2022
MCPA 500	Nufarm UK	18415	30 Apr 2022
Nufarm MCPA 750	Nufarm UK	14892	30 Apr 2022

Product	Approval holder	MAPP No.	Expiry Date
POL-MCPA 500 SL	Zaklady	14912	30 Apr 2020
POL-MCPA 500 SL	Clayton	17808	30 Apr 2022
Tasker 75	FMC Agro	14913	30 Apr 2022

539 MCPA + mecoprop-P

Cleanrun Pro	ICL (Everris) Ltd	15073	30 Apr 2022
Greenmaster Extra	ICL (Everris) Ltd	15817	30 Apr 2022

540 MCPB

Bellmac Straight	UPL Europe	14448	30 Apr 2021
Bellmac Straight	UPL Europe	18636	30 Apr 2022
Butoxone	FMC Agro	14406	30 Apr 2021
Tropotox	Nufarm UK	14450	30 Apr 2021

541 mecoprop-P

Clenecorn Super	Nufarm UK	14628	31 Jul 2020
Clenecorn Super	Nufarm UK	18378	31 Jul 2022
Compitox Plus	Nufarm UK	14390	31 Jul 2020
Duplosan KV	Nufarm UK	13971	31 Jul 2020
Headland Charge	FMC Agro	14394	31 Jul 2021
Isomec	Nufarm UK	14385	31 Jul 2020
Isomec	Nufarm UK	18376	31 Jul 2022

544 mefentrifluconazole

Lenvyor Duo	BASF	19255	24 Jun 2023

547 mepiquat chloride + metconazole

Regulate OSR	HMpG GMBH	18495	31 Aug 2022

548 mepiquat chloride + prohexadione-calcium

Medax Top	BASF	16574	31 Aug 2022

549 meptyldinocap

Kindred	Landseer	13891	29 Feb 2020

552 mesotrione

A12739A	Syngenta	16158	09 Sep 2099
Barracuda	Albaugh UK	17433	21 Mar 2020
Callisto	PSI	17512	09 Sep 2099
Clayton Goldcob	Clayton	16299	09 Sep 2099
Clayton Goldcob	Clayton	19247	09 Sep 2099
Cuter	Clayton	17668	09 Sep 2099
Evolya	Syngenta	17490	09 Sep 2099
Greencrop Goldcob	Clayton	12863	31 Mar 2021
Hockley Mesotrione 10	Hockley	15519	09 Sep 2099
Kalypstowe	AgChem Access	13844	09 Sep 2099
Meristo	Syngenta	16865	09 Sep 2099
Meruba	Helm	17961	09 Sep 2099
Minerva	Terrechem	17514	09 Sep 2099
Osorno	Belchim	18250	09 Sep 2099
RouteOne Trione 10	Albaugh UK	14765	09 Sep 2099
Standon Cobmajor	Standon	16721	09 Sep 2099
Standon Cobmajor	Enviroscience	18489	09 Sep 2099

553 mesotrione + nicosulfuron

Choriste	Syngenta	16899	09 Sep 2099
Clayton Aspen	Clayton	16114	09 Sep 2099
Elumis	Syngenta	15800	09 Sep 2099

554 mesotrione + S-metolachlor

Camix	Syngenta	17722	09 Sep 2099

SECTION 3

Product	Approval holder	MAPP No.	Expiry Date
Clarido	Syngenta	17891	09 Sep 2099

555 mesotrione + terbuthylazine

Calaris	Syngenta	12405	29 Feb 2020
Callistar	Syngenta	16864	29 Feb 2020
Clayton Faize	Clayton	13810	29 Feb 2020
Destiny	AgChem Access	14159	29 Feb 2020
Orion	EZCrop	16049	29 Feb 2020
Pan Theta	Pan Agriculture	14501	29 Feb 2020
RouteOne Mesot	Albaugh UK	15302	29 Feb 2020
Standon Cobmaster	Standon	16772	29 Feb 2020

557 metalaxyl-M

Fongarid Gold	Syngenta	12547	31 Dec 2022
Lenomax	ProKlass	18395	31 Dec 2022
Shams 465	DAPT	16107	31 Dec 2022

558 metaldehyde

Allure	Chiltern	12651	31 Dec 2021
Appeal	Chiltern	12022	31 Dec 2021
Appeal	Chiltern	12713	31 Dec 2021
Attract	Chiltern	12023	31 Dec 2021
Attract	Chiltern	12712	31 Dec 2021
Cargo 3 GB	Aako	15784	31 Dec 2021
Desire	Chiltern	14048	31 Dec 2021
Doff Horticultural Slug Killer Blue Mini Pellets	Certis	11463	31 Dec 2021
Slug III	HMpG GMBH	18724	31 Dec 2021
TDS Metarex Amba	De Sangosse	13461	31 Dec 2021
Tempt	Chiltern	14227	31 Dec 2021
Tremolo	Chiltern	16582	31 Dec 2021
Trounce	Chiltern	14222	31 Dec 2021

559 metamitron

Betra SC	Novastar	17244	28 Feb 2025
Bettix WG	UPL Europe	16496	28 Feb 2025
Celmitron 70% WDG	UPL Europe	16990	31 Dec 2021
Clayton Devoid	Agrii	18252	28 Feb 2025
Cleancrop Savant	Agrii	18399	28 Feb 2025
Defiant WG	UPL Europe	16544	28 Feb 2025
Devoid	JT Agro	18067	30 Apr 2022
Devoid	Cropthetics	18512	28 Feb 2025
Goltix 90	Adama	16497	28 Feb 2025
Goltix Compact	Aako	16545	28 Feb 2025
Goltix WG	Adama	16853	28 Feb 2022
Inter Metron 700 SC	UPL Europe	17389	28 Feb 2025
Mitron BX SC	Certiplant NV	19279	28 Feb 2025
MTM 700	Terrechem	19084	28 Feb 2025
Target SC	UPL Europe	16969	31 Aug 2020

560 metamitron + quinmerac

Kezuro	BASF	17988	25 Oct 2021

562 Metarhizium anisopliae

Lycomax	Russell	18860	31 Oct 2022
Met52 OD	Fargro	17367	31 Oct 2022

563 metazachlor

Clayton Buzz	Clayton	16928	31 Jan 2024
Fuego	Adama	16679	31 Jan 2024

Product	Approval holder	MAPP No.	Expiry Date
Makila 500 SC	Novastar	18607	31 Jan 2024
Metaza	Euro	17168	31 Jan 2024
Mikado	Matrix	17004	31 Jan 2021
Mikado	UKChem	17788	30 Jun 2020
Standon New Metazachlor	Enviroscience	18498	31 Jan 2024
Standon New Metazachlor	Standon	17473	30 Apr 2022

564 metazachlor + quinmerac

Katamaran	BASF	16766	31 Jan 2024
Metamerac	Euro	17139	31 Jan 2024
Metamerac II	RealChemie	18146	31 Jan 2022
Metamerac-II	RealChemie	18171	31 Jan 2024
Rocket	Adama	17093	31 Jan 2024

565 metconazole

Artina EC	Globachem	18680	31 Oct 2022
Caramba	BASF	15337	31 Oct 2022
Conatra 60	Globachem	19208	31 Oct 2022
Life Scientific Metconazole	Life Scientific	16022	31 Oct 2022
Life Scientific Metconazole 60	Life Scientific	17037	29 Feb 2020
Metal	Arysta	17265	31 Oct 2022
Metal 60	Arysta	17074	31 Oct 2022
Plexeo 60	Syngenta	18281	31 Oct 2022
Plexeo 90	Globachem	18681	31 Oct 2022
Quickstep	Syngenta	18831	31 Oct 2022
Rambaca 60	Euro	16841	31 Oct 2022
Saloon	Globachem	18923	31 Oct 2022
Turret 90	Globachem	19025	31 Oct 2022

567 methiocarb

Mesurol	Bayer CropScience	15311	31 Jan 2023

571 1-methylcyclopropene

AppleSmart 3.3VP	Innvigo	18968	30 Apr 2022
EthylBloc Tabs	Smither	18288	30 Apr 2022
Ethylene Buster	Chrysal	16222	30 Apr 2022
Fysium	Janssen	17584	03 Mar 2023
Ripelock Tabs	AgroFresh	17480	30 Apr 2022
Ripelock Tabs 2.0	AgroFresh	17486	30 Apr 2022
Ripelock VP	AgroFresh	17484	30 Apr 2022

573 metobromuron

Fresco	Belchim	19264	30 Jun 2027
Lianto	Belchim	18394	30 Jun 2027

575 metrafenone

Attenzo	BASF	11917	31 Oct 2022
Flexity	BASF	11775	31 Oct 2022

576 metribuzin

Acorix 70 WG	Globachem	18755	31 Jan 2023
Clayton Mizen	Clayton	19207	31 Jan 2023
Discus	RAAT	17552	31 Jan 2023
Metricam	Proplan-Plant	18630	31 Jan 2023
Python	Feinchemie	16010	31 Jan 2023
Sencorex Flow	Sumitomo	18895	31 Jan 2023

577 metsulfuron-methyl

Accurate	Nufarm UK	19237	30 Sep 2025
Alias SX	FMC Agro	18602	30 Sep 2022

SECTION 3

Products also Registered

Product	Approval holder	MAPP No.	Expiry Date
Ally SX	FMC Agro	18456	30 Sep 2022
Ally SX	FMC Agro	18675	30 Sep 2025
Answer SX	FMC Agro	18597	30 Sep 2022
Cleancrop Mondial	Agrii	18598	30 Sep 2025
Finy	UPL Europe	16960	31 Jan 2022
Gropper SX	FMC Agro	18599	30 Sep 2022
Jubilee SX	FMC Agro	18600	30 Sep 2022
Laya	Life Scientific	17984	30 Sep 2020
Life Scientific Metsulfuron-Methyl	Life Scientific	16654	29 Sep 2021
Lorate	FMC Agro	18603	30 Sep 2022
Ricorso Premium	Rotam	18527	30 Sep 2025
Simba SX	FMC Agro	18601	30 Sep 2022

578 metsulfuron-methyl + thifensulfuron-methyl

Product	Approval holder	MAPP No.	Expiry Date
Avro SX	FMC Agro	14313	31 Dec 2020
Chimera SX	FMC Agro	13171	31 Dec 2020
Concert SX	FMC Agro	12288	31 Dec 2020
Finish SX	FMC Agro	12259	31 Dec 2020
Harmony M SX	FMC Agro	12258	31 Dec 2020
Mozaic SX	FMC Agro	16108	31 Dec 2020
Pennant	FMC Agro	13350	31 Oct 2020
Presite SX	FMC Agro	12291	31 Dec 2020
Refine Max SX	FMC Agro	15622	31 Dec 2020

579 metsulfuron-methyl + tribenuron-methyl

Product	Approval holder	MAPP No.	Expiry Date
Ali Macx	Terrechem	18408	09 Sep 2099
Ally Max SX	FMC Agro	14835	31 Dec 2020
BiPlay SX	FMC Agro	14836	31 Dec 2020
DP911 SX	FMC Agro	17493	31 Dec 2020
Equator SX	FMC Agro	18455	09 Sep 2099
Traton SX	FMC Agro	14837	31 Dec 2020

580 Mild Pepino Mosaic Virus isolate VC1 + VX1

Product	Approval holder	MAPP No.	Expiry Date
V10	Valto	19124	30 Sep 2033

581 myclobutanil

Product	Approval holder	MAPP No.	Expiry Date
Systhane 20 EW	Landseer	19160	30 Nov 2022

584 napropamide

Product	Approval holder	MAPP No.	Expiry Date
Colzamid	UPL Europe	17967	30 Jun 2026
Naprop 450	Belchim	18682	30 Jun 2026

585 nicosulfuron

Product	Approval holder	MAPP No.	Expiry Date
Alaris	Novastar	18482	30 Jun 2022
Bandera	Rotam	16576	26 Mar 2022
Entail	FMC Agro	15671	30 Jun 2021
Fornet 4 SC	Belchim	16082	01 Feb 2022
Hill-Agro 40	Stefes	17654	30 Jun 2022
Nicoron	Stefes	18025	30 Jun 2022
Nicosh 4% OD	Sharda	19044	30 Jun 2022
Samson	Belchim	16425	01 Feb 2022
Scape	Terrechem	18863	30 Jun 2022
Standon Frontrunner Super 6	Enviroscience	18486	30 Jun 2022
Standon Frontrunner Super 6	Standon	16890	30 Jun 2021
Stretch	Zenith	19162	30 Jun 2022

586 nicosulfuron + thifensulfuron-methyl

Product	Approval holder	MAPP No.	Expiry Date
Collage	FMC Agro	16820	09 Sep 2099

Product	Approval holder	MAPP No.	Expiry Date
589 oxathiapiprolin			
DP 717	DuPont (Corteva)	18667	03 Sep 2029
Hesper	Ceres	18814	03 Sep 2029
590 paclobutrazol			
A10784A	Syngenta	17172	30 Nov 2025
Bonzi	Syngenta Bioline	17095	30 Apr 2020
592 pelargonic acid			
Kalipe	Belchim	18746	31 Jan 2021
Redialo	Belchim	18747	31 Jan 2021
593 penconazole			
Clayton Castile	Clayton	18310	30 Jun 2022
Rapas	RAAT	17740	30 Jun 2022
Topenco	Belchim	17539	30 Jun 2022
Topenco 100 EC	Belchim	17161	17 Jul 2020
594 pencycuron			
Monceren DS	Bayer CropScience	11292	31 Dec 2021
595 pendimethalin			
Akolin 330 EC	Aako	14282	31 Jan 2021
Alpha Pendimethalin 330 EC	MAUK	13815	31 Jan 2021
Aquarius	MAUK	14712	31 Jan 2021
Atlas Pendimethalin 400	Adama	16815	31 Jan 2021
Blazer M	FMC Agro	15084	31 Jan 2021
Bromo	Novastar	17936	31 Jan 2021
Bunker	Feinchemie	13816	31 Jan 2021
Campus 400 CS	Aako	15366	31 Jan 2021
Cinder	MAUK	14526	31 Jan 2021
CleanCrop Stomp	Agrii	13443	31 Jan 2021
Domitrel	Syngenta	19165	09 Sep 2099
Eximus	Gemini	17174	31 Jan 2021
Eximus II	Gemini	18483	31 Jan 2021
Fastnet	Sipcam	14068	31 Jan 2021
Hockley Pendi 330	Hockley	15513	31 Jan 2021
PDM 330 EC	BASF	13406	31 Jan 2021
Pendi 330	Euro	15807	31 Jan 2021
Pendifin	Finchimica	18164	09 Sep 2099
Pendimet 400	Goldengrass	14715	09 Sep 2099
Pendinova	Finchimica	19157	09 Sep 2099
Pendragon	Certis	16262	31 Jan 2021
Penfox	Sharda	18342	09 Sep 2099
Penta	Sharda	17680	09 Sep 2099
Pre-Empt	Euro	16649	09 Sep 2099
Quarry	Gemini	17789	09 Sep 2099
RouteOne Penthal 330	Albaugh UK	15314	31 Jan 2021
Sherman	Adama	13859	31 Jan 2021
Sovereign	BASF	13442	09 Sep 2099
596 pendimethalin + picolinafen			
Flight	BASF	12534	09 Sep 2099
Sienna Pro	BASF	17047	09 Sep 2099
599 penflufen			
Emesto Prime FS	Bayer CropScience	15446	31 Jul 2026
600 penthiopyrad			
Avella	Belchim	19256	31 Oct 2026

SECTION 3

Product	Approval holder	MAPP No.	Expiry Date
Celica	Belchim	19260	31 Oct 2026
DP 747	DuPont (Corteva)	17530	31 Oct 2026
Intellis	DuPont (Corteva)	17520	30 Sep 2023
Sentenza	Terrechem	17529	31 Oct 2026
Vertisan	DuPont (Corteva)	17485	31 Oct 2026

603 pepino mosaic virus strain CH2 isolate 1906

PMV-01	De Ceuster	17587	07 Feb 2033

608 phenmedipham

Agrichem Phenmedipham EC	UPL Europe	16224	31 Jan 2023
Agrichem PMP EC	UPL Europe	16963	31 Jan 2023
Agrichem PMP SE	UPL Europe	16971	31 Jan 2023
Alpha Phenmedipham 320 SC	Feinchemie	14070	31 Jan 2023
Betanal Flow	Bayer CropScience	13893	31 Jan 2023
Corzal	UPL Europe	16952	31 Jan 2023
Dancer Flow	Sipcam	14395	31 Jan 2023
Herbasan Flow	Nufarm UK	13894	31 Jan 2023
Mandolin Flow	Nufarm UK	13895	31 Jan 2023
Parish	Adama	17464	31 Jan 2023
Pump	Feinchemie	14087	31 Jan 2023
Rubie	EZCrop	15655	31 Jan 2023
Shrapnel	Feinchemie	14088	31 Jan 2021
Shrapnel	Feinchemie	18234	31 Jan 2023

614 picolinafen

AC 900001	BASF	10714	09 Sep 2099
Vixen	BASF	13621	09 Sep 2099

616 pinoxaden

A13814D	Syngenta	17386	31 Aug 2020
A13814D	Syngenta	19006	31 Dec 2028
Axial Pro	Syngenta	18275	31 Aug 2020

617 pirimicarb

Clayton Pirimicarb	Clayton	17759	31 Oct 2021

618 pirimiphos-methyl

Clayton Galic	Clayton	15374	31 Dec 2021
Phobi Smoke Pro90	Lodi UK	17117	31 Jan 2023

621 potassium phosphonates

Alginure Bio Schutz	Tilco-Alginure	18659	31 Mar 2026

622 potassium salts of fatty acids

Jaboland	Invest A. B.	17324	28 Feb 2022
Jabolim	Quimicas	17482	28 Feb 2022
Nakar	Seipasa	17483	28 Feb 2022
Tec-bom	Iberfol	17455	28 Feb 2022

624 prochloraz + propiconazole

Bumper P	Adama	08548	19 Mar 2020
Cirkon	Adama	16643	19 Mar 2020
Greencrop Twinstar	Greencrop	09516	19 Mar 2020

626 prochloraz + tebuconazole

Agate EW	Nufarm UK	18383	28 Feb 2022
Amplitude	HMpG GMBH	18488	28 Feb 2023
Monkey	Nufarm UK	18382	28 Feb 2022
Orius P	Nufarm UK	18338	28 Feb 2022

Product	Approval holder	MAPP No.	Expiry Date
629 prohexadione-calcium			
Attraxor	BASF	18939	30 Jun 2025
631 propamocarb hydrochloride			
Edipro	Arysta	15564	31 Jan 2023
Propamex-I 604 SL	MAC	15901	31 Jan 2023
Rival	Agria SA	18177	31 Jan 2023
632 propaquizafop			
Chitral	Terrechem	17655	31 May 2024
Clayton Enigma	Clayton	17391	31 May 2024
Clayton Enigma	Clayton	17435	31 May 2024
Flanoc	RealChemie	18155	31 May 2024
Profop 100	HMpG GMBH	18429	31 May 2024
Zetrola	Syngenta	18272	31 May 2024
633 propiconazole			
Anode	MAUK	14447	19 Mar 2020
Apache 250 EC	Aako	17454	19 Mar 2020
Atlas Propiconazole 250	Whyte Agrochemicals	17416	19 Mar 2020
Atlas Propiconazole 250	Gemini	16377	19 Mar 2020
Banner Maxx	Syngenta	13167	19 Mar 2020
Banner Maxx II	Syngenta	18038	19 Mar 2020
Barclay Propizole	Barclay	15113	19 Mar 2020
Bounty	AgChem Access	14616	19 Mar 2020
Bumper 250 EC	Adama	14399	19 Mar 2020
Exocet II	Pan Amenity	18306	19 Mar 2020
GF Propiconazole 250	AgChem Access	17154	19 Mar 2020
Hockley Propicon 25	Hockley	15517	19 Mar 2020
Mascot Exocet	Rigby Taylor	15598	19 Mar 2020
Matsuri	Sumi Agro	15544	19 Mar 2020
Nimble Pro	Capital CP	18003	19 Mar 2020
Profiol 250	PSI	15393	19 Mar 2020
Propi 25 EC	Sharda	14939	19 Mar 2020
Regalia	Pure Amenity	18576	19 Mar 2020
Span	ProKlass	17165	19 Mar 2020
Spaniel	Pan Amenity	15530	19 Mar 2020
Spaniel II	Pan Amenity	18257	19 Mar 2020
Zolex	ITACA	15713	19 Mar 2020
637 propoxycarbazone-sodium			
Attribut	Sumitomo	14749	30 Jun 2020
638 propyzamide			
Artax Flo	Stefes	16317	31 Jul 2021
Careca	UPL Europe	16976	31 Jul 2021
Cleancrop Rumble	Agrii	17600	09 Sep 2099
Cohort	Adama	15035	31 Jul 2021
Dennis	Sumitomo	15081	31 Jul 2021
Edge 400	Albaugh UK	14683	31 Jul 2021
Engage	Sumitomo	14233	31 Jul 2021
Flomide	Sumitomo	14223	31 Jul 2021
Gem Granules	Agrigem	19182	09 Sep 2099
Gemstone Granules	Agrigem	17898	31 Jul 2022
Hockley Propyzamide 40	Hockley	15449	09 Sep 2099
Judo	FMC Agro	15346	31 Jul 2021
KeMiChem - Propyzamide 400 SC	KeMiChem	14706	09 Sep 2099
Kerb 50 W	Dow (Corteva)	13715	31 Jul 2021
MAC-Propyzamide 400 SC	MAC	14786	09 Sep 2099
MS Eddard	Micromix	17149	31 Mar 2021

Product	Approval holder	MAPP No.	Expiry Date
Pizza 400 SC	Goldengrass	14430	09 Sep 2099
Pizza Flo	Goldengrass	16600	09 Sep 2099
Prova	Frontier	15641	31 Jul 2021
Pyzamid 400 SC	Euro	15257	09 Sep 2099
RouteOne Zamide Flo	Albaugh UK	14559	09 Sep 2099
Setanta 50 WP	Certis	16518	31 Jul 2021
Shamal	Rotam	15686	31 Jul 2021
Solitaire 50 WP	Certis	16521	31 Jul 2021
Standon Santa Fe 50 WP	Standon	14966	31 Jul 2021
Verdah 400	DAPT	14680	09 Sep 2099
Verge 400	Albaugh UK	14682	31 Jul 2021
Zammo	FMC Agro	15313	31 Jul 2021
Zammo	FMC Agro	18744	31 Jul 2021

639 proquinazid

Product	Approval holder	MAPP No.	Expiry Date
Proquin 200	Euro	16019	31 Jan 2025
Quin Pro	HMpG GMBH	18469	31 Jan 2025
Talius	PSI	16912	31 Jan 2025
Zorkem	Unisem	16906	31 Jan 2025

640 prosulfocarb

Product	Approval holder	MAPP No.	Expiry Date
A8545G	Syngenta	16204	30 Apr 2022
Atlas Pro 800	Whyte Agrochemicals	17734	30 Apr 2022
Atlas Pro 800	Cropthetics	19258	30 Apr 2022
Clayton Heed	Clayton	17701	30 Apr 2022
Clayton Obey	Clayton	16405	30 Apr 2022
Cleancrop Carburettor	Agrii	18772	30 Apr 2022
Crozier	Barclay	19052	30 Apr 2022
Dogo	Proplan-Plant	18628	30 Apr 2022
Elude	Chem-Wise	17195	30 Apr 2022
Fade	Life Scientific	17982	30 Apr 2022
Fidox	Syngenta	16209	30 Apr 2022
IC1574	Syngenta	16205	30 Apr 2022
Jade	Syngenta	16203	30 Apr 2022
Krum 800	Proplan-Plant	18420	30 Apr 2022
Nuron	Gemini	17757	30 Apr 2022
Nuron	Cropthetics	19241	30 Apr 2022
NYX	Terrechem	17737	30 Apr 2022
Peloton	Barclay	19049	30 Apr 2022
Penzo	Proplan-Plant	18629	30 Apr 2022
Pro-Star	Cropthetics	19238	30 Apr 2022
Prosulfix	Novastar	17925	30 Apr 2022
Prosulfix EC	Novastar	18774	30 Apr 2022
Prosulfocarbstar	Life Scientific	17431	31 Mar 2020
Prosulfostar	Life Scientific	17542	28 Feb 2021
PSC Pro	CropStar	18957	30 Apr 2022
Wicket	Syngenta	17555	30 Apr 2022
Zen	Terrechem	19115	30 Apr 2022

642 prothioconazole

Product	Approval holder	MAPP No.	Expiry Date
Banguy	PSI	15140	31 Jan 2023
Proline	Bayer CropScience	12084	31 Jan 2023
Prothio	ChemSource	14945	31 Jan 2022
Redigo	Bayer CropScience	12085	31 Jan 2023

643 prothioconazole + spiroxamine

Product	Approval holder	MAPP No.	Expiry Date
Pro-Spirox	HMpG GMBH	19062	31 Jan 2023
Torsion	HMpG GMBH	18476	31 Jan 2023
Torsion II	HMpG GMBH	19061	31 Jan 2023

Product	Approval holder	MAPP No.	Expiry Date
645 prothioconazole + tebuconazole			
Clayton Zorro Pro	Clayton	17073	31 Jan 2023
Pro Fit 25	Chem-Wise	17040	31 Jan 2023
Proteus	Matrix	17210	31 Jan 2021
Proteus	UKChem	17787	30 Jun 2020
Rimtil	HMpG GMBH	18466	31 Jan 2023
Romtil	RealChemie	18117	31 Jan 2023
Sequana	Ascot Pro-G	17818	31 Jan 2023
Standon Cumer	Enviroscience	18484	31 Jan 2023
Standon Cumer	Standon	17848	31 Jan 2022
Standon Mastana	Standon	17039	31 Jan 2022
647 prothioconazole + trifloxystrobin			
Infinity	HMpG GMBH	18475	09 Sep 2099
Jager	Bayer CropScience	18942	31 Jan 2023
648 Pseudomonas chlororaphis MA 342			
Cerall	Chemtura	14546	31 Oct 2021
Cerall	Koppert	19001	31 Oct 2022
651 pyraclostrobin			
BAS 500 06	BASF	12338	31 Jul 2022
BASF Insignia	BASF	11900	31 Jul 2022
Comet	BASF	10875	31 Jul 2022
Flyer	BASF	12654	31 Jul 2022
Insignia	Vitax	11865	31 Jul 2022
Inter Pyrastrobin 200	Inter-Phyto	18188	31 Jul 2022
Inter Pyrastrobin 200	BASF	19039	31 Jul 2022
LEY	BASF	14774	31 Jul 2021
Platoon	BASF	12325	31 Jul 2022
Platoon 250	BASF	12640	31 Jul 2022
Tucana 200	BASF	17294	31 Jul 2022
Vivid 200	BASF	17295	31 Jul 2022
653 pyrethrins			
Pyrethrum 5 EC	Agropharm	12685	31 May 2020
Pyrethrum 5 EC	PelGar	18210	31 May 2020
Pyrethrum 5 EC	PelGar	18532	28 Feb 2025
654 pyridate			
Clayton Viva	Clayton	18264	30 Jun 2033
655 pyrimethanil			
Penbotec 400 SC	Janssen	17384	31 Oct 2022
Precious	Emerald	18134	31 Oct 2022
Pyrimala	Servem	17325	31 Oct 2022
Spectrum	RAAT	17469	31 Oct 2022
656 pyriofenone			
Diopyr	ISK Biosciences	19131	12 Sep 2020
658 Pythium oligandrum M1			
Polyversum	De Sangosse	17456	31 Oct 2022
660 quinoclamine			
Mogeton	Certis	15837	30 Jun 2020
661 quinoxyfen			
Apres	Dow (Corteva)	08881	27 Mar 2020
Fortress	Dow (Corteva)	08279	27 Mar 2020
Me2 After	Me2	10463	27 Mar 2020

SECTION 3

Product	Approval holder	MAPP No.	Expiry Date
662 quizalofop-P-ethyl			
Quizz	Harvest	18254	11 Jun 2020
664 rimsulfuron			
Clayton Rasp	Clayton	15952	31 Oct 2022
Rimsulf 250	Euro	15802	31 Oct 2022
667 Sheep fat			
Trico	Kwizda	18149	28 Feb 2023
668 silthiofam			
Latitude XL	Certis	16467	30 Apr 2020
Latitude XL	Certis	18030	09 Sep 2099
669 S-metolachlor			
Efica 960EC	Adama	18344	31 Jan 2023
671 sodium silver thiosulphate			
Chrysal AVB	Chrysal	19041	31 Oct 2026
Florissant 100	Dejex	19158	31 Oct 2026
673 spinosad			
Exocet	Matrix	17146	31 Jan 2021
Exocet	UKChem	17792	30 Jun 2020
676 spirotetramat			
Cleancrop Sierra	Agrii	19005	31 Oct 2026
Spirotet	HMpG GMBH	18450	31 Oct 2026
Tetramat	Progreen	18947	31 Oct 2026
680 Streptomyces griseoviridis strain K61			
Mycostop	Verdera	16637	31 Oct 2021
Mycostop	Danstar	18613	31 Oct 2021
681 sulfosulfuron			
Monitor	Sumitomo	18887	09 Sep 2099
682 sulfoxaflor			
Sequoia	Fargro	18938	18 Feb 2028
683 sulfuryl fluoride			
ProFume	Douglas	17309	30 Apr 2026
684 sulphur			
POL-Sulphur 80 WG	Ciech	17613	30 Jun 2023
POL-Sulphur 800 SC	Ciech	17614	30 Jun 2023
Solfa WG	Nufarm UK	11602	31 Dec 2021
685 sulphur + tebuconazole			
Unicorn DF	Sulphur Mills	18109	28 Feb 2023
686 tau-fluvalinate			
Clayton Spirit	Clayton	19190	31 Dec 2021
Clayyton Malin	Clayton	19245	31 Dec 2021
Evure	Syngenta	18284	31 Dec 2021
Greencrop Malin	Greencrop	11787	31 Mar 2021
Klartan	Adama	11074	31 Dec 2021
Revolt	Adama	13383	31 Dec 2021
687 tebuconazole			
Buzz Ultra DF	Arysta	17924	20 Oct 2020

Product	Approval holder	MAPP No.	Expiry Date
Buzz Ultra DF	Arysta	18328	20 Oct 2020
Cezix	Rotam	18056	28 Feb 2023
Clayton Ohio	Clayton	18454	31 May 2020
Cleancrop Teboo	Agrii	18111	28 Feb 2023
Deacon	Nufarm UK	17604	28 Feb 2022
Erase	FMC Agro	17507	28 Feb 2022
Erasmus	Rotam	18057	28 Feb 2023
Fathom	FMC Agro	17452	28 Feb 2022
Folicur	Bayer CropScience	16731	28 Feb 2023
Gizmo	Nufarm UK	16502	28 Feb 2021
Hi-Tebura	Hockley	18421	28 Feb 2023
Jackal	Gemini	18912	28 Feb 2023
Legend	JT Agro	18116	28 Feb 2022
Legend	Cropthetics	18432	28 Feb 2023
Life Scientific Tebuconazole	Life Scientific	18548	28 Feb 2022
Mitre	Nufarm UK	17606	28 Feb 2022
Mystic	Nufarm UK	19156	28 Feb 2023
Orius	Nufarm UK	17414	28 Feb 2022
Quarta	Terrechem	17844	28 Feb 2023
Santal	Novastar	17955	28 Feb 2023
Savannah	Rotam	18373	28 Feb 2023
Standon Beamer	Standon	17350	28 Feb 2023
Starpro	Rotam	18367	28 Feb 2023
Starpro	Rotam	18059	31 Dec 2021
Tamok	Matrix	17264	31 Jan 2021
Tamok	UKChem	17786	30 Jun 2020
Tebusha 25 EW	Sharda	18034	28 Feb 2023
Tharsis	Generica	17201	28 Feb 2023
Toledo	Rotam	14036	31 May 2020
Trident	Ascot Pro-G	18858	28 Feb 2023
Zonor	Life Scientific	18650	28 Feb 2023

690 tebuconazole + trifloxystrobin

Defusa	Rigby Taylor	18232	09 Sep 2099
Dualitas	ProKlass	18000	09 Sep 2099
Dualitas	ProKlass	18070	09 Sep 2099
Inter Tebloxy	Iticon	17311	09 Sep 2099

691 tebufenpyrad

Clayton Bonsai	Clayton	18666	30 Apr 2025
Masai	BASF	18287	30 Apr 2025

693 tefluthrin

A13219F	Syngenta	16161	31 Oct 2020
A13219F	Syngenta	19083	30 Jun 2027
Force ST	Syngenta	11752	30 Sep 2020

694 tembotrione

Laudis	Bayer CropScience	17302	31 Oct 2026

697 terpenoid Blend (QRD 460)

Requiem Prime	Bayer CropScience	19112	27 Jun 2023

698 tetraconazole

Eminent 126 ME	Belchim	18864	30 Jun 2024

702 thiacloprid

Pintail	UKChem	17817	30 Jun 2020
Rana SC	Servem	17211	31 Oct 2022
Sonido	Bayer CropScience	16368	31 Oct 2022

Product	Approval holder	MAPP No.	Expiry Date
Standon Zero Tolerance	Standon	17551	31 Oct 2020
Standon Zero Tolerance	Enviroscience	18500	31 Oct 2022
Thia 240	HMpG GMBH	18633	31 Oct 2022
Thiaclomex 480 SC	MAC	15900	31 Oct 2022

705 thifensulfuron-methyl

Product	Approval holder	MAPP No.	Expiry Date
Harmony SX	FMC Agro	12181	31 Dec 2020
Harmony SX	FMC Agro	18761	09 Sep 2099
Pinnacle	FMC Agro	12285	31 Dec 2020
Prospect SX	FMC Agro	12212	31 Dec 2020
Prospect SX	FMC Agro	18777	09 Sep 2099

706 thifensulfuron-methyl + tribenuron-methyl

Product	Approval holder	MAPP No.	Expiry Date
Calibre SX	Certis	15032	09 Sep 2099
Clayton Pause	Clayton	18937	09 Sep 2099
Counter SX	FMC Agro	18976	09 Sep 2099
Inka SX	FMC Agro	13601	31 Dec 2020
Nautius	Rotam	18838	09 Sep 2099
Parana	Certis	16258	09 Sep 2099
Ratio SX	FMC Agro	12601	31 Dec 2020

707 thiophanate-methyl

Product	Approval holder	MAPP No.	Expiry Date
Bull	Gemini	18991	30 Apr 2022
Thiofin WG	Q-Chem	15384	30 Apr 2022

709 tolclofos-methyl

Product	Approval holder	MAPP No.	Expiry Date
Basilex	ICL (Everris) Ltd	16243	30 Apr 2020
Rizolex 10D	Sumitomo	14204	30 Apr 2020
Rizolex 50 WP	Sumitomo	14217	30 Apr 2020
Rizolex Flowable	Sumitomo	14207	30 Apr 2020

713 tri-allate

Product	Approval holder	MAPP No.	Expiry Date
Avadex Excel 15G	Gowan	16998	31 Dec 2020
Avadex Factor	Gowan	17748	17 Oct 2020

716 tribenuron-methyl

Product	Approval holder	MAPP No.	Expiry Date
Cameo SX	FMC Agro	18990	30 Apr 2022
Corida	Zenith	18247	30 Apr 2022
Helmstar A 75 WG	Belchim	17077	30 Apr 2022
Nuance	Nufarm UK	14813	30 Apr 2021
Quantum	FMC Agro	15190	30 Apr 2022
Quantum	FMC Agro	18836	30 Apr 2022
Quantum SX	FMC Agro	18787	30 Apr 2022
Quantum SX	FMC Agro	15189	30 Apr 2021
Taxi	FMC Agro	15733	30 Apr 2022
Taxi	FMC Agro	18995	30 Apr 2022
Thor	Nufarm UK	15239	30 Apr 2022
Toscana	Proplan-Plant	17726	30 Apr 2022
Trailer	Industrias Afrasa	17903	30 Apr 2022
Triad	FMC Agro	12751	30 Apr 2021
Tribenuron-methyl 750 g/kg WDG	Zenith	18160	30 Apr 2020
Tribun 75 WG	Belchim	17090	30 Apr 2022
Trimeo 75 WG	Belchim	17076	30 Apr 2022

718 Trichoderma harzianum (Strain T22)

Product	Approval holder	MAPP No.	Expiry Date
Trianum G	Koppert	16740	31 Oct 2022
Trianum P	Koppert	16741	31 Oct 2021

719 triclopyr

Product	Approval holder	MAPP No.	Expiry Date
Topper	Arysta	15719	31 Oct 2022

Product	Approval holder	MAPP No.	Expiry Date
720 trifloxystrobin			
Action	Pan Agriculture	19072	31 Jan 2021
Mascot Defender	Rigby Taylor	14065	31 Jan 2021
Pan Aquarius	Pan Agriculture	13018	31 Jan 2021
Pan Tees	Pan Agriculture	12894	31 Jan 2021
722 triflusulfuron-methyl			
Debut	FMC Agro	07804	30 Jun 2022
Safari	FMC Agro	17547	30 Jun 2022
Safari	FMC Agro	18796	30 Jun 2022
Trek	Life Scientific	19206	30 Jun 2022
Tricle	UPL Europe	17448	30 Jun 2022
Upbeet	FMC Agro	17544	30 Jun 2022
Zareba	Terrechem	17987	30 Jun 2022
723 trinexapac-ethyl			
A17600C	Syngenta	17548	31 Oct 2022
Cleancrop Alatrin	Agrii	15196	31 Oct 2022
Cleancrop Alatrin Evo	Agrii	17763	31 Oct 2022
Cleancrop Cutlass	Nufarm UK	16046	31 Oct 2021
Confine NT	FMC Agro	17770	09 May 2021
Freeze NT	FMC Agro	17478	30 Sep 2022
Iceni	UPL Europe	16835	31 Oct 2022
Jaguar	Ascot Pro-G	18882	31 Oct 2022
Life Scientific Trinexapac	Life Scientific	17125	31 Oct 2021
Life Scientific Trinexapac 250	Life Scientific	16047	31 Oct 2020
Maintain NT	FMC Agro	18081	27 Jun 2021
Maintain NT	FMC Agro	18736	27 Jun 2021
Moddus ME	Syngenta	17179	31 Oct 2022
Moxa 250 EC	Belchim	16176	31 Oct 2022
Moxa New	Globachem	17830	31 Oct 2020
Next	Sharda	17797	31 Oct 2022
Optimus	Nufarm UK	15249	31 Oct 2021
Paket 250 EC	Arysta	17436	31 Oct 2022
Palisade	Syngenta	17860	31 Oct 2022
Prop	Terrechem	18605	31 Oct 2022
Scitec	Syngenta	15588	31 Oct 2022
Seize	FMC Agro	16780	31 Oct 2022
Seize NT	FMC Agro	17794	09 May 2021
Seize NT	FMC Agro	18735	09 May 2021
Sonis	Syngenta	16891	31 Oct 2022
Staylow	Unique Marketing	16295	31 Oct 2022
Sudo	Life Scientific	17979	31 Oct 2022
Tempest	Ascot Pro-G	18027	31 Oct 2022
Tempo	Syngenta	15170	31 Oct 2022
TP 100	Certis	17605	31 Oct 2022
Trexstar	Belchim	17062	31 Oct 2022
Trexxus	Belchim	17059	31 Oct 2022
Tribune	Novastar	18647	31 Oct 2022
Trinex 222	Euro	15322	31 Oct 2022
Zira	Terrechem	17467	31 Oct 2022
726 Verticillium alobo-atrum			
Dutch Trig	BTL Bomendienst	17481	30 Oct 2021
728 zeta-cypermethrin			
Angri	AgChem Access	13730	31 Dec 2021
Fury 10 EW	FMC Agro	17255	31 Jul 2020
Minuet EW	FMC Agro	17250	31 Jul 2020

SECTION 3

SECTION 4
ADJUVANTS

Adjuvants

Adjuvants are not themselves classed as pesticides and there is considerable misunderstanding over the extent to which they are legally controlled under the Food and Environment Protection Act. An adjuvant is a substance other than water which enhances the effectiveness of a pesticide with which it is mixed. Consent C(i)5 under the Control of Pesticides Regulations allows that an adjuvant can be used with a pesticide only if that adjuvant is authorised and on a list published on the HSE adjuvant database: https//secure.pesticides. gov.uk/adjuvants/search.asp. An authorised adjuvant has an *adjuvant number* and may have specific requirements about the circumstances in which it may be used.

Adjuvant product labels must be consulted for full details of authorised use, but the table below provides a summary of the label information to indicate the area of use of the adjuvant. Label precautions refer to the keys given in Appendix 4, and may include warnings about products harmful or dangerous to fish. The table includes all adjuvants notified by suppliers as available in 2020.

Product	Supplier	Adj. No.	Type
Abacus	De Sangosse	A0543	vegetable oil
Contains	53.43 % w/w oil (rapeseed fatty acid esters), 20.0 % w/w alkoxylated alcohols and 9.0 % w/w oil (tall oil fatty acids)		
Use with	All approved pesticides on all edible crops when used at half their recommended dose or less, and on all non-edible crops up to their full recommended dose. Also at a maximum concentration of 0.1% with approved pesticides on listed crops up to specified growth stages		
Protective clothing	A, C, H		
Precautions	R36, U05a, U11, U14, U19a, U20b, E15a, E19b, E34, D01, D02, D05, D10a, D12a, H04		
Activator 90	De Sangosse	A0547	non-ionic surfactant/wetter
Contains	375 g/kg alkoxylated alcohols, 375 g/kg alkoxylated alcohol, 150 g/kg oil (tall oil fatty acids)		
Use with	All approved pesticides on all edible crops when used at half their recommended dose or less, and on all non-edible crops up to their full recommended dose. Also at a maximum concentration of 0.1% with approved pesticides on listed crops up to specified growth stages		
Protective clothing	A, C, H		
Precautions	R36, R38, R53a, U02a, U04a, U05a, U10, U11, U20b, E15a, E19b, D01, D02, D05, D10a, H04		
Adigor	Syngenta	A0522	wetter
Contains	47 % w/w methylated rapeseed oil		
Use with	Topik, Axial, Trazos and Amazon on cereals in accordance with recommendations on the respective herbicide labels		
Protective clothing	A, C, H		
Precautions	R43, R51, R53a, U02a, U05a, U09a, U20b, E15b, E38, D01, D02, D05, D09a, D10c, D12a, H04, H11		
AdjiFe	Amega	A0797	wetter
Contains	345 g/l ammonium iron (III) citrate, 300 g/l alkyl polyglycosides and 104.5 g/l ethylene oxide-propylene oxide copolymers		
Use with	All authorised plant protection herbicides and fungicides		
Protective clothing	A		
Precautions	U02a, U04a, U05a, U14, U15, U19a, E13b, E34, D01, D02, D09a, D10a, M03		

SECTION 4

Product	Supplier	Adj. No.	Type
AdjiKataSil	Amega	A0844	spreader/wetter
Contains	90.0 % w/w trisiloxane organosilicone copolymers		
Use with	All approved pesticides at half the approved maximum dose on edible crops and at the full approved dose on non-edible crops and non-crop production at 0.25% spray volume.		
Protective clothing	A, C, H		
Precautions	U02a, U04a, U05a, U11, U14, U15, E15a, E34, D01, D02, D09a, D10a, M03, H04		
AdjiMin	Amega	A0812	adjuvant
Contains	920 g/kg oil (petroleum oils)		
Use with	All approved pesticides at half the approved maximum dose on edible crops and at the full approved dose on non-edible crops and non-crop production at 1% spray volume.		
Protective clothing	A		
Precautions	U02a, U04a, U05a, U14, U15, U19a, E13b, E34, D01, D02, D09a, D10a, M03		
AdjiSil	Amega	A0763	spreader/wetter
Contains	83% w/w trisiloxane organosilicone copolymers		
Use with	All approved pesticides at half the approved maximum dose on edible crops and at the full approved dose on non-edible crops and non-crop production at 0.15% spray volume.		
Protective clothing	A, C, H		
Precautions	U02a, U04a, U05a, U14, U15, U19a, E13b, E34, D01, D02, D09a, D10a, M03, H11		
AdjiVeg	Amega	A0756	adjuvant
Contains	90% w/w fatty acid esters		
Use with	All approved pesticides at half the approved maximum dose on edible crops and at the full approved dose on non-edible crops and non-crop production at 1% spray volume.		
Protective clothing	A, C, H		
Precautions	U02a, U04a, U05a, U11, U14, U15, E15a, E34, D01, D02, D09a, D10a, M03, H04		
Adpro Addit	Koppert	A0868	spreader/sticker/wetter
Contains	780.2 g/l oil (rapeseed triglycerides)		
Use with	Mycotal at 0.25% spray solution and all approved pesticides at half or less than half the approved pesticide rate. Name and no changed from Addit to Adpro Addit Aug 2018		
Protective clothing	A, F, H		
Precautions	R20, R21, R36, R51, R53a		
Amber	Interagro	A0367	vegetable oil
Contains	95% w/w methylated rapeseed oil		
Use with	Sugar beet herbicides, oilseed rape herbicides, cereal graminicides and a wide range of other pesticides that have a label recommendation for use with authorised adjuvant oils on specified crops. See label for details		
Protective clothing	A, C		
Precautions	U05a, U20b, E15a, D01, D02, D05, D09a, D10a		

Product	Supplier	Adj. No.	Type
Arma	Interagro	A0306	penetrant
Contains	500 g/l alkoxylated fatty amine + 500 g/l polyoxyethylene monolaurate		
Use with	Cereal growth regulators, cereal herbicides, cereal fungicides, oilseed rape fungicides and a wide range of other pesticides on specified crops		
Protective clothing	A, C		
Precautions	R51, R58, U05a, E15a, E34, E37, D01, D02, D05, D09a, D10a, H11		
Asu-Flex	Greenaway	A0677	spreader/sticker/wetter
Contains	10.0% w/w rapeseed oil		
Use with	Asulox, Greencrop Found, I T Asulam, Inter Asulam and Spitfire		
Protective clothing	A, C		
Precautions	U11, U12, U15, U20b, E15a, E34, D02		
BackRow	Interagro	A0850	mineral oil
Contains	60% w/w refined paraffinic petroleum oil		
Use with	Pre-emergence herbicides. Refer to label or contact supplier for further details		
Protective clothing	A		
Precautions	R22b, R38, U02a, U05a, U08, U20b, E15a, D01, D02, D05, D09a, D10a, M05b, H03		
Banka	Interagro	A0245	spreader/wetter
Contains	14.6 % w/w alkyl pyrrolidone copolymers and 14.6 % w/w alkyl pyrrolidones		
Use with	Potato fungicides and a wide range of other pesticides on specified crops		
Protective clothing	A, C		
Precautions	R38, R41, R52, R58, U02a, U05a, U11, U19a, U20b, E15a, E34, E37, D01, D02, D05, D09a, H04		
Binder	Amega	A0598	spreader/wetter
Contains	30.0 % w/w alkoxylated alcohols		
Use with	All approved pesticides at half or less than half the approved pesticide rate and all approved formulations of glyphosate		
Protective clothing	A		
Precautions	U02a, U04a, U05a, U14, U15, U19a, E13b, E34, D01, D02, D09a, D10a, M03		
Bio Syl	Intracrop	A0773	spreader/sticker/wetter
Contains	32.67% w/w alkoxylated alcohols and 1.0% w/w trisiloxane organosilicone copolymers		
Use with	Recommended rates of approved pesticides on non-crop and non-edible crops; when used on edible crops the dose of the approved pesticide must be half or less than half the approved dose rate		
Protective clothing	A, C		
Precautions	R36, R38, U05a, U08, U20c, E15a, D01, D02, D10a, H04		
Bioduo	Intracrop	A0606	wetter
Contains	700 g/l alkoxylated alcohols and 150 g/l rapeseed fatty acids		
Use with	A wide range of pesticides used in grassland, agriculture and horticulture and with pesticides used in non-crop situations		
Protective clothing	A, C		
Precautions	R22a, R36, R38, U05a, U08, U20c, E13c, E34, D01, D02, D09a, D10a, M03, H03, H04, H08		

SECTION 4

Product	Supplier	Adj. No.	Type
Biofilm	Intracrop	A0634	anti-drift agent/anti-transpirant/sticker/ UV screen/wetter
Contains	96.0% w/w pinene oligomers		
Use with	All approved fungicides and insecticides on edible crops and all approved formulations of glyphosate used pre-harvest on wheat, barley, oilseed rape, stubble, and in non-crop situations and grassland destruction		
Protective clothing	A, C		
Precautions	U19a, U20c, E13c, D09a, D11a		
BioPower	Bayer CropScience	A0617	wetter
Contains	6.7% w/w 3,6-dioxaeicosylsulphate sodium salt and 20.2% w/w 3,6-dioxaoctadecylsulphate sodium salt		
Use with	Atlantis and all other approved cereal herbicides		
Protective clothing	A, C		
Precautions	R36, R38, U02a, U05a, U08, U13, U19a, U20b, E13c, E34, D01, D02, D05, D09a, D10a, H04		
Biothene	Intracrop	A0633	anti-drift agent/anti-transpirant/sticker/ UV screen/wetter
Contains	96.0% w/w pinene oligomers		
Use with	All approved fungicides and insecticides on edible crops up to 30 d before harvest, and all approved formulations of glyphosate used pre-harvest on wheat, barley, oilseed rape, stubble, and in non-crop situations and grassland destruction. Must not be used in mixture with adjuvant oils or surfactants		
Precautions	U19a, U20c, E13c, D09a, D11a		
Bond	De Sangosse	A0556	extender/sticker/wetter
Contains	10% w/w alkoxylated alcohols, 45% w/w styrene-butadiene copolymers		
Use with	All approved potato blight fungicides. Also with all approved pesticides on all edible crops when used at half their recommended dose or less, and on all non-edible crops up to their full recommended dose. Also at a maximum concentration of 0.14% with approved pesticides on listed crops up to specified growth stages		
Protective clothing	A, C, H		
Precautions	R36, R38, U11, U14, U16b, U19a, E15a, E19b, D01, D02, D05, D10a, D12a, H04		
Broad-Flex	Greenaway	A0680	spreader/sticker/wetter
Contains	10.0% w/w rapeseed oil		
Use with	Broad Sword or Green Guard		
Protective clothing	A, C		
Precautions	U11, U12, U15, U20b, E15a, E34, D02		
Buzz	De Sangosse	A0520	wetter
Contains	375g/l alkoxylated alcohols, 250g/l alkoxylated coconut amines 50g/l oil (tall oil fatty acids)		
Use with	All approved pesticides at half or less than half the approved pesticide rate at 0.1% spray solution on edible crops. Restored Mar 2018 since listed by De Sangosse.		
Protective clothing	A, C, H		
Precautions	R36, R38, R53a, U02a, U04a, U05a, U10, U11, U20b, E15a, E19b, D01, D02, D05, D10a, H04		

Product	Supplier	Adj. No.	Type
Byo-Flex	Greenaway	A0545	sticker/wetter
Contains	10.0% w/w rapeseed oil		
Use with	GLY 490 (MAPP 12718)		
Protective clothing	A, C		
Precautions	U11, U12, U15, U20b, E34		
C-Cure	Interagro	A0851	mineral oil
Contains	60% w/w refined mineral oil		
Use with	Pre-emergence herbicides		
Protective clothing	A		
Precautions	R22b, R38, U02a, U05a, U08, U20b, E15a, D01, D02, D05, D09a, D10a, M05b, H03		
Clayton Bower	Clayton	A0861	wetter
Contains	20.1 % w/w 3,6-dioxaoctadecylsulphate, sodium salt and 6.7 % w/w 3,6-dioxaeicosylsulphate, sodium salt		
Use with	all approved products on cereals and all approved herbicides on natural surfaces not intended to bear vegetation, permeable surfaces overlying soil, industrial and amenity areas and ornamental plant production.		
Protective clothing	A, H		
Precautions	H315, H320, U02a, U05a, U08, U13, U19a, U20b, E13c, E34, D01, D02, D05, D09a, D10a, H04		
Codacide Oil	Microcide	A0629	vegetable oil
Contains	95% w/w oil (rapeseed triglycerides)		
Use with	All approved pesticides and tank mixes. See label for details		
Protective clothing	A, C		
Precautions	U20b, D09a, D10b		
Companion Gold	Agrovista	A0723	acidifier/buffering agent/drift retardant/ extender
Contains	16% w/w ammonium sulphate + 0.95% w/w polyacrylamide		
Use with	Diquat or glyphosate on oilseed rape at 0.5% solution; glyphosate on wheat, rye or triticale at 0.5% solution and with all approved pesticides on non-edible crops at 1% solution or all approved pesticides at 50% dose rate or less on edible crops		
Protective clothing	A, C, H		
Precautions	U05a, U14, U15, U19a, U19c, U20b, E15a, D02, D05, D09a, D10a		
Compliment	Agrii	A0705	spreader/vegetable oil/wetter
Contains	75% w/w mixed fatty acid esters of rapeseed oil		
Use with	All approved pesticides in non-edible crops, all approved pesticides on edible crops when used at half or less their recommended rate, morpholine or triazine fungicides on cereals, and with all pesticides on listed crops up to specified growth stages		
Protective clothing	A, C		
Precautions	R43, R52, R53a, U14, E38, D01, D05, D07, D08, H04		

SECTION 4

Product	Supplier	Adj. No.	Type
County Mark	Greenaway	A0689	sticker/wetter
Contains	10% w/w rapeseed oil		
Use with	'Greenaway Gly-490' (MAPP 12718) and all approved 490 g/l glyphosate products		
Protective clothing	A, C		
Precautions	U11, U12, U15, U20b, E34		
Dash HC	BASF	A0729	non-ionic surfactant/wetter
Contains	348.75 g/l oil (fatty acid esters) and 209.25 g/l alkoxylated alcohols-phosphate esters		
Use with	For use with Cleranda at 1.0 l/ha in a water volume of 100 - 400 l/ha on Clearfield oilseed rape		
Protective clothing	A, C, H		
Precautions	U05a, U12, U14, D01, D02, D10c, M05b, H03		
Deploy	De Sangosse	A0801	wetter
Contains	91.04 % w/w rapeseed fatty acid esters, 4.16 % w/w alkoxylated alcohols and 2.07 % w/w oil (tall oil fatty acids)		
Use with	All approved pesticides on all edible crops when used at half their recommended dose or less, and on all non-edible crops up to their full recommended dose. Also at a maximum concentration of 0.5% with approved pesticides on listed crops up to specified growth stages		
Protective clothing	A, H		
Precautions	U14, U15, H411, D01, D09a		
Designer	De Sangosse	A0660	drift retardant/extender/sticker/wetter
Contains	25% w/w styrene-butadiene copolymers, 7.1% w/w trisiloxane organosilicone copolymers		
Use with	A wide range of fungicides, insecticides and trace elements for cereals and specified agricultural and horticultural crops		
Protective clothing	A, C, H		
Precautions	R36, R38, R52, U02a, U11, U14, U15, U19a, E15a, E37, D01, D02, D05, D09a, D10a, H04		
Diagor	AgChem Access	A0671	wetter
Contains	47% w/w methylated rapeseed oil		
Use with	Topik, Axial, Trazos, Amazon and Viscount on cereals in accordance with recommendations on the respective herbicide labels		
Protective clothing	A, C, H		
Precautions	H317, U02a, U05a, U09a, U20b, E15b, E38, H411, D01, D02, D05, D09a, D10c, D12a, H04, H11, H401		
Drill	De Sangosse	A0544	adjuvant
Contains	15% w/w alkoxylated alcohols, 7.5% w/w oil (tall oil fatty acids) and 63.34% w/w oil (rapeseed fatty acid esters)		
Use with	All approved pesticides on all edible crops when used at half their recommended dose or less, and on all non-edible crops up to their full recommended dose. Also at a maximum concentration of 0.1% with approved pesticides on listed crops up to specified growth stages		
Protective clothing	A, C, H		
Precautions	R36, U05a, U11, U14, U19a, U20b, E15a, E19b, E34, D01, D02, D05, D10a, D12a, H04		

Product	Supplier	Adj. No.	Type
Eco-flex	Greenaway	A0696	sticker/wetter
Contains	10% w/w refined rapeseed oil		
Use with	Approved formulations of glyphosate, 2,4-D		
Protective clothing	A, C		
Precautions	U11, U12, U15, U20b, E34		
Elan Xtra	Intracrop	A0735	spreader/wetter
Contains	530 g/l ethylene oxide-propylene oxide copolymers and 415 g/l trisiloxane organosilicone copolymers		
Use with	All approved pesticides on non-edible crops and non-crop production and with half dose of all approved pesticides on edible crops		
Protective clothing	A		
Emerald	Intracrop	A0636	anti-drift agent/anti-transpirant/extender/ UV screen
Contains	96.0% w/w pinene oligomers		
Use with	Recommended rates of approved pesticides up to the growth stage indicated for specified crops, and with half or less than the recommended rate on these crops after the stated growth stages. Also for use alone on transplants, turf, fruit crops, glasshouse crops and Christmas trees. Must not be used in mixture with adjuvant oils or surfactants		
Protective clothing	A, H		
Precautions	U19a, U20c, E13c, E37, D09a, D11a		
Euroagkem Pen-e-trate	EuroAgkem	A0564	spreader/wetter
Contains	350 g/l propionic acid		
Use with	All approved pesticides which have a recommendation for use with a wetting agent; pesticides must be used at half approved dose rate or less on edible crops, up to full dose rate on non-edible crops		
Protective clothing	A, C		
Precautions	U05a, U08, U11, U14, U15, U19a, U20c, E13e, E15a, D01, D02, D09a, D10b, M03, H05		
Felix	Intracrop	A0178	spreader/wetter
Contains	600 g/l alkoxylated alcohols		
Use with	Mecoprop, 2,4-D in cereals and amenity turf, and a range of grass weedkillers in agriculture. See label for details		
Protective clothing	A, C		
Precautions	R36, R38, U05a, U08, U20a, E15a, D01, D02, D09a, D10a, H04, H08		
Firebrand	Barclay	-	fertiliser/water conditioner
Contains	500 g/l ammonium sulphate		
Use with	Use at 0.5% v/v in the spray solution with glyphosate		
Precautions	U08, U20a, E15a, D09a, D10b		
Galion	Amega	A0162	spreader/wetter
Contains	600 g/l alkoxylated alcohols		
Use with	On all specified crops up to the listed growth stage at 0.5% spray volume		
Protective clothing	A, C, H		
Precautions	U02a, U04a, U05a, U11, U14, U15, E15a, E34, D01, D02, D09a, D10a, M03, H04		

SECTION 4

Product	Supplier	Adj. No.	Type
Gateway	Agrii	A0651	extender/sticker/wetter
Contains	73% w/v synthetic latex solution and 8.5% w/v polyether modified trisiloxane		
Use with	All approved pesticides in crops not destined for human or animal consumption, all approved pesticides on edible crops when used at half or less their recommended rate, and with all pesticides on listed crops up to specified growth stages		
Protective clothing	A, C, H		
Precautions	R41, R52, R53a, U11, U15, E37, E38, D01, D05, D07, D08, H04		
Gly-Flex	Greenaway	A0588	spreader/sticker/wetter
Contains	95% w/w refined rapeseed oil		
Use with	GLY-490 (MAPP 12718) or any approved formulations of 490 g/l glyphosate		
Protective clothing	A, C		
Precautions	R36, U11, U12, U15, U20b, E13c, E34, E37, H04		
Gly-Plus A	Greenaway	A0736	spreader/sticker/wetter
Contains	10.0% w/w oil (rapeseed triglycerides)		
Use with	Any approved 360 g/l glyphosate		
Protective clothing	A, C		
Precautions	R36, U05a, U08, U11, U12, U15, U20b, E13c, E34, E37, E38, H04		
Green Gold	Intracrop	A0250	spreader/wetter
Contains	950 g/l oil (rapeseed triglycerides)		
Use with	All pesticides which have a recommendation for the addition of a wetter/spreader		
Precautions	U08, U20b, E13c, E34, D09a, D10b		
Grounded	Helena	A0456	mineral oil
Contains	732 g/l petroleum oils		
Use with	All approved pesticides on edible and non-edible crops when used at half their approved dose or less. Also with approved pesticides on specified crops, up to specified growth stages, at up to their full approved dose		
Protective clothing	A, C		
Precautions	R53a, U02a, U05a, U08, U20b, E15a, E34, E37, D01, D02, D05, D09a, D10c, H11		
Headland Fortune	FMC Agro	A0703	penetrant/spreader/vegetable oil/wetter
Contains	75% w/w mixed methylated fatty acid esters of seed oil and N-butanol		
Use with	Herbicides and fungicides in a wide range of crops. See label for details		
Protective clothing	A, C		
Precautions	R43, U02a, U05a, U14, U20a, E15a, E34, E38, D01, D02, D05, D09a, D10b, H04		
Headland Guard Pro	FMC Agro	A0653	extender/sticker
Contains	10% w/w styrene/butadiene co-polymers		
Use with	All approved pesticides on all edible crops when used at half their recommended dose or less, and on all non-edible crops up to their full recommended dose. Also at a maximum concentration of 0.1% with approved pesticides on listed crops up to specified growth stages		
Protective clothing	A, C		
Precautions	E15a, D01, D02		

Product	Supplier	Adj. No.	Type
Headland Intake	FMC Agro	A0074	penetrant

Contains — 450 g/l propionic acid

Use with — All approved pesticides on any crop not intended for human or animal consumption, and with all approved pesticides on beans, peas, edible podded peas, oilseed rape, linseed, sugar beet, cereals (except triazole fungicides), maize, Brussels sprouts, potatoes, cauliflowers. See label for detailed advice on timing on these crops

Protective clothing — A, C

Precautions — R34, U02a, U05a, U10, U11, U14, U15, U19a, E34, D01, D02, D05, D09b, D10b, M04a, H05

Product	Supplier	Adj. No.	Type
Headland Rheus	FMC Agro	A0841	wetter

Contains — 85% w/w polyalkylene oxide modified heptamethyl siloxane

Use with — Any herbicide, systemic fungicide, systemic insecticide or plant growth regulator (except any product applied in or near water) where the use of a wetting, spreading and penetrating surfactant is recommended to improve foliar coverage

Protective clothing — A, C

Precautions — R21, R22a, R38, R41, R43, R51, R58, U02a, U05a, U08, U11, U14, U19a, U20b, E13c, E34, D01, D02, D05, D09a, D10b, M03, H03, H11

Product	Supplier	Adj. No.	Type
Herbi-AKtiv	Global Adjuvants	A0839	spreader/wetter

Contains — 57.0 % w/w ethylene oxide-propylene oxide copolymers

Use with — All authorised plant protection products at half or less than half the authorised plant protection product rate. Adj No updated due to expiry of A0821 July 2018

Protective clothing — A, C, H, M

Precautions — U05a, U20c, E34, D01, D02, D10a, M04a

Product	Supplier	Adj. No.	Type
Intracrop Agwet GTX	Intracrop	A0646	spreader/wetter

Contains — 500 g/l alkoxylated alcohols

Use with — All approved pesticides on non-edible crops, all approved pesticides on edible crops at half or less than half the pesticide rate and all approved pesticides with a recommendation for use with a non-ionic or wetting agent.

Protective clothing — A, C, H

Precautions — H318, U05a, U11, U14, U15, U19a, U20a, M04c, H226, H302

Product	Supplier	Adj. No.	Type
Intracrop BLA	Intracrop	A0655	anti-drift agent/anti-transpirant/extender/sticker/UV screen

Contains — 22.0 % w/w styrene-butadiene copolymers

Use with — All potato blight fungicides. Also with recommended rates of approved pesticides in certain non-crop situations and up to the growth stage indicated for specified crops, and with half or less than the recommended rate on these crops after the stated growth stages. Also with pesticides in grassland at half or less their recommended rate

Protective clothing — A, C

Precautions — U08, U20b, E13c, E34, E37, D01, D05, D09a, D10a

SECTION 4

Product	Supplier	Adj. No.	Type
Intracrop Bla-Tex	Intracrop	A0656	extender/sticker
Contains	22.0 % w/w styrene-butadiene copolymers		
Use with	All approved pesticides on non-edible crops and non crop production, with half dose of all approved pesticides on edible crops and with blight fungicides in potatoes		
Protective clothing	A, C		
Precautions	U08, U20b, E13c, E34, E37, D01, D05, D09a, D10a		
Intracrop Boost	Intracrop	A0774	spreader/sticker/wetter
Contains	32.67 % w/w alkoxylated alcohols and 1.0% w/w trisiloxane organosilicone copolymers		
Use with	All approved pesticides at half or less than half the approved pesticide rate on edible crops and all approved pesticides on non-edible crops		
Protective clothing	A, C		
Intracrop Cogent	Intracrop	A0775	spreader/sticker/wetter
Contains	32.67 % w/w alkoxylated alcohols and 1.0% w/w trisiloxane organosilicone copolymers		
Use with	All approved pesticides at half or less than half the approved pesticide rate on edible crops and all approved pesticides on non-edible crops		
Protective clothing	A, C		
Intracrop Dictate	Intracrop	A0673	adjuvant
Contains	91.0 % w/w oil (rapeseed fatty acid esters)		
Use with	All approved pesticides on non-edible crops; all approved pesticides applied at half or less than half dose on edible crops		
Protective clothing	A, C		
Precautions	U19a, U20b, D05, D09a, D10b		
Intracrop Evoque	Intracrop	A0754	spreader/wetter
Contains	652 g/l oil (rapeseed fatty acid esters), 112 g/l trisiloxane organosilicone copolymers, 22.5 g/l alkoxylated alcohols and 19.1 g/l alkoxylated alcohols		
Use with	All authorised plant protection products on non-edible crops and all authorised plant protection products at half or less than half the authorised plant protection product rate on edible crops. Max conc is 0.2% of spray volume.		
Protective clothing	A, C		
Intracrop F16	Intracrop	A0752	spreader/wetter
Contains	652 g/l oil (rapeseed fatty acid esters), 112 g/l trisiloxane organosilicone copolymers, 22.5 g/l alkoxylated alcohols and 19.1 g/l alkoxylated alcohols		
Use with	All authorised plant protection products on non-edible crops and all authorised plant protection products at half or less than half the authorised plant protection product rate on edible crops. Max conc is 0.2% of spray volume.		
Protective clothing	A, C		
Intracrop Impetus	Intracrop	A0647	spreader/wetter
Contains	50.0 % w/w alkoxylated alcohols		
Use with	All approved pesticides on non-edible crops, all approved pesticides on edible crops at half or less than half the pesticide rate and all approved pesticides with a recommendation for use with a non-ionic or wetting agent.		
Protective clothing	A, C, H		
Precautions	H318, U05a, U11, U14, U15, U19a, U20a, M04c, H226, H302		

Product	Supplier	Adj. No.	Type
Intracrop Inca	Intracrop	A0784	adjuvant

Contains | 840 g/l oil (rapeseed fatty acid esters)

Use with | All approved pesticides at half or less than half the approved pesticide rate on edible crops and all approved pesticides on non-edible crops

Protective clothing | A, C

Product	Supplier	Adj. No.	Type
Intracrop Incite	Intracrop	A0785	adjuvant

Contains | 840 g/l oil (rapeseed fatty acid esters)

Use with | All approved pesticides at half or less than half the approved pesticide rate on edible crops and all approved pesticides on non-edible crops

Protective clothing | A, C

Product	Supplier	Adj. No.	Type
Intracrop Mica AF	Intracrop	A0808	spreader/sticker/wetter

Contains | 250 g/l ethylene oxide-propylene oxide copolymers, 70.2 g/l styrene-butadiene copolymers and 41.5 g/l trisiloxane organosilicone copolymers

Use with | All approved pesticides at half or less than half the approved rate

Protective clothing | A, C

Precautions | U05a, U08, U20c, E13e, D01, D02, D09a, D10a

Product	Supplier	Adj. No.	Type
Intracrop Neotex	Intracrop	A0657	anti-drift agent/anti-transpirant/extender/sticker/UV screen

Contains | 22.0 % w/w styrene-butadiene copolymers

Use with | Recommended rates of approved pesticides up to specified growth stages of a wide range of agricultural arable and horticultural crops, and for non-crop uses. Use on edible crops beyond specified growth stages, and in grass, should only be with half recommended rates of the pesticide or less. See label for details of growth stage restrictions. In addition may be used with all potato blight fungicides at their recommended rates of use up to the latest recommended timing of the fungicide

Protective clothing | A, C

Precautions | U08, U20b, E13c, E34, E37, D01, D05, D09a, D10a

Product	Supplier	Adj. No.	Type
Intracrop Novatex	Intracrop	A0658	anti-drift agent/anti-transpirant/extender/sticker/UV screen

Contains | 22.0 % w/w styrene-butadiene copolymers

Use with | Recommended rates of approved pesticides up to the growth stage indicated for specified crops, and with half or less than the recommended rate on these crops after the stated growth stages. Also for use with recommended rates of pesticides on grassland and specified non-crop situations

Protective clothing | A, C

Precautions | U08, U20b, E13c, E34, E37, D01, D09a, D10a

Product	Supplier	Adj. No.	Type
Intracrop Perm-E8	Intracrop	A0565	spreader/wetter

Contains | 42.0 % w/w propionic acid

Use with | All approved formulations of chlormequat, all approved formulations of glyphosate, diquat, fenoxaprop-P-ethyl, tralkoxydim, clodinafop-propargyl, fluazifop-P-butyl, cycloxydim and propaquizafop at half or less than half the approved pesticide rate in edible crops, at full rate in non-edible crops.

Protective clothing | A, C

Precautions | U02a, U04a, U05a, U08, U10, U11, U13, U14, U15, U19a, U20b, D01, D02, D09a, D10a, H05

SECTION 4

Product	Supplier	Adj. No.	Type
Intracrop Predict	Intracrop	A0503	adjuvant/vegetable oil
Contains	91.0 % w/w oil (rapeseed fatty acid esters)		
Use with	All approved pesticides on non-edible crops, and all pesticides approved for use on growing edible crops when used at half recommended dose or less. On specified crops product may be used at a maximum spray concentration of 1% with approved pesticides at their full approved rate up to the growth stages shown in the label		
Protective clothing	A, C		
Precautions	U19a, U20b, D05, D09a, D10b		
Intracrop Quad	Intracrop	A0753	spreader/wetter
Contains	652 g/l oil (rapeseed fatty acid esters), 112 g/l trisiloxane organosilicone copolymers, 22.5 g/l alkoxylated alcohols and 19.1 g/l alkoxylated alcohols		
Use with	All approved pesticides on non-edible crops, and all pesticides approved for use on growing edible crops when used at half recommended dose or less.		
Protective clothing	A, C		
Intracrop Quartz	Intracrop	A0776	spreader/sticker/wetter
Contains	32.67 % w/w alkoxylated alcohols and 1.0% w/w trisiloxane organosilicone copolymers		
Use with	All approved pesticides on non-edible crops, and all pesticides approved for use on growing edible crops when used at half recommended dose or less.		
Protective clothing	A, C		
Intracrop Questor	Intracrop	A0495	activator/non-ionic surfactant/spreader
Contains	750 g/l ethylene oxide-propylene oxide copolymers		
Use with	All approved pesticides on non-edible crops and pesticides used in non-crop production, and all pesticides approved for use on growing edible crops when used at half recommended dose or less. On specified crops product may be used at a maximum spray concentration of 0.3% with approved pesticides at their full approved rate up to the growth stages shown in the label		
Protective clothing	A, C		
Precautions	R36, R38, U04a, U05a, U08, U19a, E13b, D01, D02, D09a, D10b, M03, M05a, H04		
Intracrop Rapide Beta	Intracrop	A0672	wetter
Contains	350 g/l propionic acid and 100 g/l alkoxylated alcohols		
Use with	All approved pesticides which have a recommendation for use with a wetting agent; pesticides must be used at half approved dose rate or less on edible crops, up to full dose rate on non-edible crops		
Protective clothing	A, C		
Precautions	U02a, U04a, U05a, U08, U10, U11, U13, U14, U15, U19a, U20b, E15a, D01, D02, D09a, D10a, M05a, H05		
Intracrop Retainer NF	Intracrop	A0711	adjuvant
Contains	91.0 % w/w oil (rapeseed fatty acid esters)		
Use with	All approved pesticides at half or less than half the approved pesticide rate on edible crops and all approved pesticides on non-edible crops		
Protective clothing	A, C		

Product	Supplier	Adj. No.	Type
Intracrop Rigger	Intracrop	A0783	adjuvant/vegetable oil
Contains	840 g/l oil (rapeseed fatty acid esters)		
Use with	All approved pesticides on non-edible crops, and all pesticides approved for use on growing edible crops when used at half recommended dose or less. On specified crops product may be used at a maximum spray concentration of 1.78% with approved pesticides at their full approved rate up to the growth stages shown in the label		
Protective clothing	A, C		
Precautions	U19a, U20b, D05, D09a, D10b		
Intracrop Rustler	Intracrop	A0777	spreader/sticker/wetter
Contains	32.67 % w/w alkoxylated alcohols and 1.0% w/w trisiloxane organosilicone copolymers		
Use with	All approved pesticides on non-edible crops, and all pesticides approved for use on growing edible crops when used at half recommended dose or less.		
Protective clothing	A, C		
Intracrop Salute AF	Intracrop	A0809	spreader/sticker/wetter
Contains	250 g/l ethylene oxide-propylene oxide copolymers, 70.2 g/l styrene-butadiene copolymers and 41.5 g/l trisiloxane organosilicone copolymers		
Use with	All approved pesticides on non-edible crops, and all pesticides approved for use on growing edible crops when used at half recommended dose or less.		
Protective clothing	A, C		
Precautions	U05a, U08, U20c, E13e, E40c, D01, D02, D09a, D10a		
Intracrop Sapper AF	Intracrop	A0810	spreader/sticker/wetter
Contains	250 g/l ethylene oxide-propylene oxide copolymers , 70.2 g/l styrene-butadiene copolymers and 41.5 g/l trisiloxane organosilicone copolymers		
Use with	All approved pesticides on non-edible crops, and all pesticides approved for use on growing edible crops when used at half recommended dose or less.		
Protective clothing	A, C		
Precautions	U05a, U08, U20c, E13e, E40c, D01, D02, D09a, D10a		
Intracrop Saturn	Intracrop	A0494	activator/non-ionic surfactant/spreader
Contains	750 g/l ethylene oxide-propylene oxide copolymers		
Use with	All approved pesticides on non-edible crops and pesticides used in non-crop production, and all pesticides approved for use on growing edible crops when used at half recommended dose or less. On specified crops product may be used at a maximum spray concentration of 0.3% with approved pesticides at their full approved rate up to the growth stages shown in the label		
Protective clothing	A, C		
Precautions	R36, R38, U04a, U05a, U08, U19a, E13b, E40c, D01, D02, D09a, D10b, M03, M05a, H04		
Intracrop Signal XL	Intracrop	A0659	anti-drift agent/anti-transpirant/extender/sticker/UV screen
Contains	22.0 % w/w styrene-butadiene copolymers		
Use with	All approved pesticides on non-edible crops and non crop production uses; All approved pesticides at half or less than half the approved pesticide rate on edible crops		
Protective clothing	A, C		
Precautions	U05a, U19a, U20b, E13c, E34, D01, D02, D09a, D10b, M04a, M05a		

SECTION 4

Product	Supplier	Adj. No.	Type
Intracrop Sprinter	Intracrop	A0513	spreader/wetter
Contains	19.0 % w/v alkoxylated alcohols		
Use with	Recommended rates of approved pesticides up to the growth stage indicated for specified crops, and with half or less than the recommended rate on these crops after the stated growth stages. Also with herbicides on managed amenity turf at recommended rates, and on grassland at half or less than recommended rates		
Protective clothing	A, C		
Precautions	R36, R38, R41, U05a, U08, U11, U14, U15, U20c, E15a, D01, D02, D09a, D10b, M03, H04		
Intracrop Status	Intracrop	A0506	adjuvant/vegetable oil
Contains	91.0 % w/w oil (rapeseed fatty acid esters)		
Use with	All approved pesticides on non-edible crops, and all pesticides approved for use on growing edible crops when used at half recommended dose or less. On specified crops product may be used at a maximum spray concentration of 1.0% with approved pesticides at their full approved rate up to the growth stages shown in the label		
Protective clothing	A, C		
Precautions	U19a, U20b, D05, D09a, D10b		
Intracrop Stay-Put	Intracrop	A0507	vegetable oil
Contains	91.0 % w/w oil (rapeseed fatty acid esters)		
Use with	Recommended rates of approved pesticides for non-crop uses and with recommended rates of approved pesticides up to the growth stage indicated for specified crops, and with half or less than the recommended rate on these crops after the stated growth stages		
Protective clothing	A, C		
Precautions	U19a, U20b, D05, D09a, D10b		
Intracrop Super Rapeze MSO	Intracrop	A0782	adjuvant/vegetable oil
Contains	840 g/l oil (rapeseed fatty acid esters)		
Use with	All approved pesticides on non-edible crops, and all pesticides approved for use on growing edible crops when used at half recommended dose or less. On specified crops product may be used at a maximum spray concentration of 1.78% with approved pesticides at their full approved rate up to the growth stages shown in the label		
Protective clothing	A, C		
Precautions	U19a, U20b, D05, D09a, D10b		
Intracrop Tonto	Intracrop	A0778	spreader/sticker/wetter
Contains	32.67 % w/w alkoxylated alcohols and 1.0% w/w trisiloxane organosilicone copolymers		
Use with	All approved pesticides at half or less than half the approved pesticide rate on edible crops and all approved pesticides on non-edible crops		
Protective clothing	A, C		
Precautions	U05a, U08, U20c, E13e, E40c, D01, D02, D09a, D10a		

Product	Supplier	Adj. No.	Type
Intracrop Warrior	Intracrop	A0514	spreader/wetter
Contains	19.0 % w/v alkoxylated alcohols		
Use with	Recommended rates of approved pesticides up to the growth stage indicated for specified crops, and with half or less than the recommended rate on these crops after the stated growth stages. Also with herbicides on managed amenity turf at recommended rates, and on grassland at half or less than recommended rates		
Protective clothing	A, C		
Precautions	R36, R38, R41, U05a, U08, U11, U14, U15, U20c, E15a, D01, D02, D09a, D10b, M03, H04		
Intracrop Zenith AF	Intracrop	A0811	spreader/sticker/wetter
Contains	250 g/l ethylene oxide-propylene oxide copolymers, 70.2 g/l styrene-butadiene copolymers and 41.5 g/l trisiloxane organosilicone copolymers.		
Use with	All approved pesticides on non-edible crops, all approved pesticides on edible crops at half or less than half the pesticide rate and all approved pesticides with a recommendation for use on non crop production.		
Protective clothing	A, C, H		
Precautions	H318, U05a, U11, U14, U15, U19a, U20a, M04c, H226, H302		
Intracrop Zodiac	Intracrop	A0755	spreader/wetter
Contains	652 g/l oil (rapeseed fatty acid esters), 112 g/l trisiloxane organosilicone copolymers), 22.5 g/l alkoxylated alcohols and 19.1 g/l alkoxylated alcohols		
Use with	All authorised plant protection products on all non edible crops and all authorised plant protection products on edible crops at half or less than half the authorised plant protection product rate		
Protective clothing	A, C		
Precautions	U05a, U08, U20c, E13e, E40c, D01, D02, D09a, D10a		
Kantor	Interagro	A0623	spreader/wetter
Contains	790 g/l alkoxylated triglycerides		
Use with	All approved pesticides		
Protective clothing	A, C		
Precautions	U05a, U20b, E15b, E19b, E34, D01, D02, D09a, D10a, D12a		
Katalyst	Interagro	A0450	penetrant/water conditioner
Contains	90% w/w alkoxylated fatty amine		
Use with	Glyphosate and a wide range of other pesticides on specified crops. Refer to label or contact supplier for further details		
Protective clothing	A, C		
Precautions	R22a, R36, R38, R50, R58, U02a, U05a, U08, U19a, U20b, E15a, E34, E37, D01, D02, D05, D09a, D10a, M03, H03, H11		

SECTION 4

Product	Supplier	Adj. No.	Type
Kinetic	Helena	A0252	spreader/wetter

Contains: 80% w/w ethylene oxide and propylene oxide copolymers, 16% w/w trisiloxane copolymers

Use with: Approved pesticides on non-edible crops and cereals and stubbles of all edible crops when used at full recommended dose, and with pesticides approved for use on growing edible crops when used at half recommended dose or less. On specified crops product may be used at a maximum spray concentration of 0.2% with approved pesticides at their full approved rate up to the growth stages shown in the label

Protective clothing: A, H

Precautions: R38, R41, U02a, U05a, U08, U19a, U20b, E13c, E34, E37, D01, D02, D05, D09a, D10c, M03, H04

Product	Supplier	Adj. No.	Type
Klipper	Amega	A0260	spreader/wetter

Contains: 600 g/l alkoxylated alcohols and isobutanol

Use with: Listed herbicides, fungicides and chlorpyrifos on managed amenity turf or amenity grassland (see label for details).

Protective clothing: A, C, H

Precautions: R22a, R41, R67, U02a, U04a, U05a, U11, U14, U15, E15a, E34, D01, D02, D09a, D10a, M03, H03, H08

Product	Supplier	Adj. No.	Type
Kotek	Agrovista	A0746	spreader/wetter

Contains: 83.0% w/w trisiloxane organosilicone copolymers

Use with: All approved pesticides at half or less than half the approved rate on edible crops; all approved pesticides on non-edible crops or non-production targets. Max conc 0.15% of spray volume

Protective clothing: A, H

Precautions: R20, R41, R51, R53a, U11

Product	Supplier	Adj. No.	Type
Leaf-Koat	Helena	A0511	spreader/wetter

Contains: 80% w/w ethylene oxide and propylene oxide copolymers, 16% w/w trisiloxane copolymers

Use with: Approved pesticides on non-edible crops and cereals and stubbles of all edible crops when used at full recommended dose, and with pesticides approved for use on growing edible crops when used at half recommended dose or less. On specified crops product may be used at a maximum spray concentration of 0.2% with approved pesticides at their full approved rate up to the growth stages shown in the label

Protective clothing: A, H

Precautions: R38, R41, U02a, U05a, U08, U19a, U20b, E13c, E34, E37, D01, D02, D05, D09a, D10c, M03, H04

Product	Supplier	Adj. No.	Type
Level	Agrii	A0654	extender/sticker

Contains: 10% w/w styrene-butadiene copolymers

Use with: All approved pesticides in non-edible crops, all approved pesticides on edible crops when used at half or less of their recommended rate, potato blight fungicides and with all pesticides on listed crops up to specified growth stages

Protective clothing: A, C

Precautions: E37, D01, D02, D05

Product	Supplier	Adj. No.	Type
Li-700	De Sangosse	A0529	acidifier/drift retardant/penetrant
Contains	9.39% w/w alkoxylated alcohols, 35.0% w/w propionic acid and 35.0% w/w soybean phospholipids			
Use with	All approved pesticides on all edible crops when used at half their recommended dose or less, and on all non-edible crops up to their full recommended dose. Also with morpholine fungicides on cereals and with pirimicarb on legumes when used at full dose; with iprodione on brassicas when used at 75% dose, and at a maximum concentration of 0.5% with all pesticides on listed crops at full dose up to specified growth stages			
Protective clothing	A, C, H			
Precautions	R36, R38, U11, U14, U15, U19a, E15a, E19b, D01, D02, D05, D10a, D12a, H04			
Logic	Microcide	A0288	vegetable oil
Contains	95% w/w oil (rapeseed triglycerides)			
Use with	All approved pesticides on edible and non-edible crops for ground or aerial application			
Protective clothing	A, C			
Precautions	U19a, U20b, D09a, D10b			
Logic Oil	Microcide	A0630	vegetable oil
Contains	95% w/w oil (rapeseed triglycerides)			
Use with	All approved pesticides on edible and non-edible crops for ground or aerial application			
Protective clothing	A, C			
Precautions	U19a, U20b, D09a, D10b			
Low Down	Helena	A0459	mineral oil
Contains	732 g/l petroleum oils			
Use with	All approved pesticides on edible and non-edible crops when used at half their approved dose or less. Also with approved pesticides on specified crops, up to specified growth stages, at up to their full approved dose			
Protective clothing	A, C			
Precautions	R53a, U02a, U05a, U08, U20b, E15a, E34, E37, D01, D02, D05, D09a, D10c, H11			
Master Sil	Global Adjuvants	A0822	spreader/wetter
Contains	83% trisiloxane organosilicone copolymers			
Use with	All approved pesticides on all edible crops when used at half their recommended dose or less, and on all non-edible crops up to their full recommended dose.			
Protective clothing	A, C, H			
Precautions	H319, E34, H411, D01, D02, D09a, D10a			
Master Wett	Global Adjuvants	A0826	spreader/sticker/wetter
Contains	84% trisiloxane organosilicone copolymers			
Use with	All approved pesticide products at a dose of 0.25% spray solution.			
Protective clothing	A, C, H			
Precautions	H319, E34, H411, D01, D02, D09a, D10a			

SECTION 4

Product	Supplier	Adj. No.	Type
Meco-Flex	Greenaway	A0678	spreader/sticker/wetter
Contains	10.0% w/w oil (rapeseed triglycerides)		
Use with	Re-Act, Headland Relay Depitox		
Protective clothing	A, C		
Precautions	U11, U12, U15, U20b, E15a, E34, D02		
Mero	Bayer CropScience	A0818	wetter
Contains	81.4% w/w rapeseed fatty acid esters		
Use with	All approved cereal and maize herbicides and all approved brassica insecticides		
Protective clothing	A, C, H		
Precautions	R38, U05a, U09a, U13, U20b, E34, D01, D02, D09a, D10a, M05a		
Mixture B NF	Amega	A0570	non-ionic surfactant/spreader/wetter
Contains	37% w/w alkoxylated alcohols and 41% w/w alkoxylated alcohols and isopropanol		
Use with	All approved pesticides on non-edible crops at their full recommended rate. Also with Timbrel (MAPP 05815) and all approved formulations of glyphosate in non-crop situations. Also with a range of specified herbicides when used at less than half their recommended rate on forest and grassland, and with all approved pesticides at full rate on a wide range of specified edible crops up to specified growth stages		
Protective clothing	A, C, H		
Precautions	U02a, U04a, U05a, U11, U14, U15, E13a, E34, D01, D02, D09a, D10a, M03, H03, H11		
Nelson	Agrovista	A0796	spreader/wetter
Contains	300 g/l alkoxylated alcohols and 300 g/l alkoxylated alcohols		
Use with	All approved pesticides at half the approved maximum dose on edible crops and at the full approved dose on non-edible crops and non-crop production at 0.5% spray volume.		
Protective clothing	A, C, H		
Precautions	H318, U02a, U05a, U08, U09c, U11, U13, U14, U15, U20d, M03a, M05b, H302		
Newman Cropspray 11E	De Sangosse	A0863	adjuvant
Contains	99% w/w oil (petroleum oils)		
Use with	All approved pesticides on all edible crops when used at half their recommended dose or less, and on all non-edible crops up to their full recommended dose. Also with listed herbicides on a range of specified crops and with all pesticides on a range of specified crops up to specified growth stages.		
Protective clothing	A, C		
Precautions	R22a, U10, U16b, U19a, E15a, E19b, E34, E37, D01, D02, D05, D10a, D12a, M05b, H03		
Nion	Amega	A0760	spreader/wetter
Contains	90% w/w alkoxylated alcohols		
Use with	All approved pesticides at half the approved maximum dose on edible crops and at the full approved dose on non-edible crops and non-crop production at 1% spray volume.		
Protective clothing	A, C		
Precautions	U02a, U04a, U05a, U11, U14, U15, E15a, E34, D01, D02, D09a, D10a, M03, H03		

Product	Supplier	Adj. No.	Type
Nu Film P	Intracrop	A0635	anti-drift agent/anti-transpirant/sticker/ UV screen/wetter
Contains	96.0 % w/w pinene oligomers		
Use with	Glyphosate and many other pesticides and growth regulators for which a protectant is recommended. Do not use in mixture with adjuvant oils or surfactants		
Protective clothing	A, C		
Precautions	U19a, U20b, E13c, D09a, D10a		
Pan Oasis	Pan Agriculture	A0411	vegetable oil
Contains	95% w/w methylated rapeseed oil		
Use with	A wide range of pesticides that have a label recommendation for use with authorised adjuvant oils. Contact distributor for further details		
Protective clothing	A		
Precautions	R36, U02a, U05a, U08, U20b, E15a, D01, D02, D05, D09a, D10a, H04		
Pan Panorama	Pan Agriculture	A0412	mineral oil
Contains	95% w/w mineral oil		
Use with	A wide range of pesticides that have a label recommendation for use with adjuvant oils. Contact distributor for details		
Protective clothing	A		
Precautions	R22b, R38, U02a, U05a, U08, U20b, E13c, D01, D02, D05, D09a, D10a, M05b, H03		
Phase II	De Sangosse	A0622	vegetable oil
Contains	95.2% w/w oil (rapeseed fatty acid esters)		
Use with	Pesticides approved for use in sugar beet, oilseed rape, cereals (for grass weed control), and other specified agricultural and horticultural crops		
Protective clothing	A, C, H		
Precautions	U19a, U20b, E15a, E19b, E34, D01, D02, D05, D10a, D12a		
Pin-o-Film	Intracrop	A0637	anti-drift agent/anti-transpirant/sticker/ UV screen/wetter
Contains	96.0 % w/w pinene oligomers		
Use with	Recommended rates of approved pesticides up to the growth stage indicated for specified crops, and with half or less than the recommended rate on these crops after the stated growth stages. Also for use with pesticides on grassland at half or less than their recommended rates. Must not be used in mixture with adjuvant oils or surfactants		
Protective clothing	A, C		
Precautions	U19a, U20c, E13c, C02a, D09a, D11a		
Planet	Intracrop	A0605	non-ionic surfactant/spreader/wetter
Contains	700 g/l alkoxylated alcohols and 150 g/l rapeseed fatty acids		
Use with	Any spray for which additional wetter is recommended		
Protective clothing	A, C		
Precautions	R22a, R36, U05a, U08, U19a, U20b, E13c, D01, D09a, D10a, D11a, H03, H04, H08		

SECTION 4

Product	Supplier	Adj. No.	Type
Pro Wett	Global Adjuvants	A0860	spreader/sticker/wetter
Contains	84% trisiloxane organosilicone copolymers		
Use with	All approved pesticide products at a dose of 0.25% spray solution.		
Protective clothing	A, C, H		
Precautions	H319, E34, H411, D01, D02, D09a, D10a		
Probe	Life Scientific	A0874	wetter
Contains	276.5 g/l alkyl ether sulphates, sodium salts		
Use with	All authorised plant protection products on cereals at 1% solution and all authorised herbicides at 1% solution on Ornamental Plant Production, Amenity Vegetation, Natural Surfaces Not Intended To Bear Vegetation, Permeable Surfaces Overlying Soil, Industrial And Amenity Areas		
Protective clothing	A, H		
Precautions	H318, U05a, U11, U14, U15, U19a, U20a, M04c, H226, H302		
Profit Oil	Microcide	A0631	extender/sticker/wetter
Contains	95% w/w oil (rapeseed triglycerides)		
Use with	All approved pesticides		
Protective clothing	A, C		
Precautions	U19a, U20b, D09a, D10b		
Pryz-Flex	Greenaway	A0679	spreader/sticker/wetter
Contains	10.0% w/w oil (rapeseed triglycerides)		
Use with	Kerb Flo (MAPP 13716 or 15586) and all approved 490 g/l glyphosate products on permeable surfaces overlying soil		
Protective clothing	A, C		
Precautions	U11, U12, U15, U20b, E34, E37		
Remix	Agrovista	A0765	wetter
Contains	732 g/l petroleum (paraffin) oil		
Use with	All approved pesticides at half or less than half the approved rate on edible crops; all approved pesticides on non-edible crops or non-production targets. Can also be used with all approved pesticides pre-emergence on bulb and stem vegetables, legumes, carrots, parsnips, swedes and turnips. Max conc for all uses is 1% of spray solution.		
Protective clothing	A, C, H		
Precautions	U02a, U05a, U19a, E15a, D01, D02, D05, D09a, D10a, D12a		
Respond	Amega	A0836	spreader/wetter
Contains	57.0 % w/w ethylene oxide-propylene oxide copolymers		
Use with	All approved pesticides on all edible crops when used at half their recommended dose or less, and on all non-edible crops up to their full recommended dose. Also at a maximum concentration of 0.5%		
Protective clothing	A		
Precautions	U02a, U04a, U05a, U14, U15, U19a, E13b, E34, D01, D02, D09a, D10a, M03		
Reward Oil	Microcide	A0632	extender/sticker/wetter
Contains	95% w/w oil (rapeseed triglycerides)		
Use with	All approved pesticides		
Protective clothing	A, C		
Precautions	U19a, U20b, D09a, D10b		

Product	Supplier	Adj. No.	Type
Roller	Agrovista	A0748	spreader/wetter

Contains: 832 g/l ethylene oxide-propylene oxide copolymers and 169.3 g/l trisiloxane organosilicone copolymers

Use with: All approved pesticides up to their maximum dose at a max conc of 0.2% v/v.

Protective clothing: A, C, H

Precautions: R36, R38, R41, R52, R53a, U02a, U05a, U08, U11, U20b, E15a, E34, E40c, D01, D02, D05, D09a, D10a, D12b, H04

SAS 90	Intracrop	A0740	spreader/wetter

Contains: 83% w/w trisiloxane organosilicone copolymers

Use with: All approved herbicides at a maximum concentration of 0.025% spray volume and all approved fungicides and insecticides at a maximum concentration of 0.05% spray volume

Protective clothing: A, C

Precautions: U02a, U08, U19a, U20b, E13c, E34, D09a, D10a

Saturn Plus	Intracrop	A0813	spreader/wetter

Contains: 70.0% w/w ethylene oxide-propylene oxide copolymers

Use with: All approved pesticides on non-edible crops, and all pesticides approved for use on growing edible crops when used at half recommended dose or less.

Protective clothing: A, C

Precautions: U19a, U20c, E13c, C02a, D09a, D11a

Siltex AF	Intracrop	A0803	spreader/sticker/wetter

Contains: 250 g/l ethylene oxide-propylene oxide copolymers, 70.2 g/l styrene-butadiene copolymers and 41.5 g/l trisiloxane organosilicone copolymers

Use with: All approved pesticides on non-edible crops, and all pesticides approved for use on growing edible crops when used at half recommended dose or less.

Protective clothing: A, C

Precautions: U05a, U08, U20c, E13c, E37, D01, D02, D09a, D10a

Silwet L-77	De Sangosse	A0640	drift retardant/spreader/wetter

Contains: 80% w/w trisiloxane organosilicone copolymers

Use with: All approved fungicides on winter and spring sown cereals; all approved pesticides applied at 50% or less of their full approved dose. A wide range of other uses. See label or contact supplier for details

Protective clothing: A, C, H

Precautions: R20, R21, R22a, R36, R43, R48, R51, R58, U11, U15, U19a, E15a, E19b, D01, D02, D05, D10a, D12a, H03, H11

Slippa	Interagro	A0206	spreader/wetter

Contains: 64.0 % w/w trisiloxane organosilicone copolymers

Use with: Cereal fungicides and a wide range of other pesticides and trace elements on specified crops

Protective clothing: A, C, H

Precautions: R20, R38, R41, R43, R48, R51, R58, U02a, U05a, U08, U11, U19a, U20a, E15a, E34, E37, D01, D02, D05, D09a, D10a, M03, H03, H11

SECTION 4

Product	Supplier	Adj. No.	Type
Solar Plus	Intracrop	A0802	spreader/wetter
Contains	70.0% w/w ethylene oxide-propylene oxide copolymers		
Use with	All approved pesticides on non-edible crops, and all pesticides approved for use on growing edible crops when used at half recommended dose or less.		
Protective clothing	A, C		
Precautions	R36, R38, U04a, U05a, U08, U20a, E13c, E34, D01, D02, D09a, D10b, D11a, M03, H04		
Speedway Total	ICL (Everris) Ltd	A0820	extender/spreader/wetter
Contains	41.0 % w/w alkoxylated alcohols and 37.0 % w/w alkoxylated alcohols		
Use with	All authorised plant protection products on non-edible crops and non-crop production uses; all authorised plant protection products at half or less than half the authorised plant protection product rate on edible crops.		
Protective clothing	A		
Precautions	U02a, U04a, U05a, U14, U15, U19a, E13b, E34, D01, D02, D09a, D10a, M03		
Spray-fix	De Sangosse	A0559	extender/sticker/wetter
Contains	10% w/w alkoxylated alcohols and 45% w/w styrene-butadiene copolymers		
Use with	All approved potato blight fungicides. Also with all approved pesticides on all edible crops when used at half their recommended dose or less, and on all non-edible crops up their full recommended dose. Also at a maximum concentration of 0.14% with approved pesticides on listed crops up to specified growth stages		
Protective clothing	A, C, H		
Precautions	R36, R38, U11, U14, U16b, U19a, E15a, E19b, D01, D02, D05, D10a, D12a, H04		
Spraymac	De Sangosse	A0549	acidifier/non-ionic surfactant
Contains	100g/l alkoxylated alcohols and 350g/l propionic acid		
Use with	All approved pesticides on all edible crops when used at half their recommended dose or less, and on all non-edible crops up to their full recommended dose. Also at a maximum concentration of 0.5% with approved pesticides on listed crops up to specified growth stages		
Protective clothing	A, C, H		
Precautions	R34, U04a, U10, U11, U14, U19a, U20b, E15a, E19b, E34, D01, D02, D05, D10a, D12a, M04a, H05		
Standon Shiva	Standon	A0816	wetter
Contains	6.7% w/w 3,6-dioxaeicosylsulphate sodium salt and 20.2% w/w 3,6-dioxaoctadecylsulphate sodium salt		
Use with	Standon Mimas WG and all other approved cereal herbicides		
Protective clothing	A, C		
Precautions	R36, R38, U02a, U05a, U08, U13, U19a, U20b, E13c, E34, D01, D02, D05, D09a, D10a, H04		
Standon Wicket	Standon	A0781	wetter
Contains	47% w/w rapeseed fatty acid esters		
Use with	Clodinafop-propargyl, pinoxaden and mixtures of the two actives in cereal crops		
Protective clothing	A, H		
Precautions	R43, R50, R53a, U02a, U05a, U09b, U20b, E15b, E38, D01, D02, D05, D09a, D10c, D12a, H04, H11		

Product	Supplier	Adj. No.	Type
Stika	De Sangosse	A0557	extender/sticker/wetter
Contains	10% w/w alkoxylated alcohols and 22.5% w/w styrene-butadione copolymers		
Use with	All approved potato blight fungicides. Also with all approved pesticides on all edible crops when used at half their recommended dose or less, and on all non-edible crops up to their full recommended dose. Also at a maximum concentration of 0.14% with approved pesticides on listed crops up to specified growth stages		
Protective clothing	A, C, H		
Precautions	R36, R38, U11, U14, U16b, U19a, E15a, E19b, D01, D02, D05, D10a, D12a, H04		
SU Wett	Global Adjuvants	A0840	spreader/wetter
Contains	57.0 % w/w ethylene oxide-propylene oxide copolymers		
Use with	All authorised plant protection products at half or less than half the authorised plant protection product rate. Adj No updated due to expiry of A0814 July 2018		
Protective clothing	A, C, H, M		
Precautions	U05a, U20c, E34, D01, D02, D10a, M04a		
Surfer	Dow (Corteva)	A0800	wetter
Contains	72.0 % w/w alkoxylated alcohols		
Use with	All approved pesticides on all edible crops when used at half their recommended dose or less, and on all non-edible crops up to their full recommended dose.		
Protective clothing	A, C, H		
Precautions	H318, U05a, U11, U14, U15, U20a, E15a, E34, D01, D02, D09a, D10a		
Sward	Amega	A0747	spreader/wetter
Contains	400 g/l ethylene oxide-propylene oxide copolymers and 58.1 g/l trisiloxane organosilicone copolymers		
Use with	For use with all approved pesticides on managed amenity turf and amenity grassland at 0.2% v/v		
Protective clothing	A, C		
Precautions	U02a, U04a, U05a, U11, U14, U15, E15a, E34, D01, D02, D09a, D10a, D12b, M03, H04		
Tempest	Intracrop	A0887	spreader/wetter
Contains	301 g/l alkoxylated alcohols, 299 g/l alkoxylated alcohols and 18.36 g/l trisiloxane organosilicone copolymers		
Use with	All approved pesticides at a maximum concentration of 0.5% v/v		
Protective clothing	A, C, H		
Precautions	H318, U05a, U09b, U20d, D12b, M03, M05b, H302		
TM 1008	Intracrop	A0675	wetter
Contains	750 g/l ethylene oxide-propylene oxide copolymers		
Use with	All approved pesticides on non-edible crops and non crop production uses; All approved pesticides at half or less than half the approved pesticide rate on edible crops		
Protective clothing	A, C		
Precautions	R36, R38, U04a, U05a, U08, U19a, U20b, E13b, E34, D01, D02, D09a, D10a, M03, M05a, H04		

SECTION 4

Product	Supplier	Adj. No.	Type
Toil	Interagro	A0248	vegetable oil
Contains	95% w/w methylated rapeseed oil		
Use with	Sugar beet herbicides, oilseed rape herbicides, cereal graminicides and a wide range of other pesticides that have a label recommendation for use with authorised adjuvant oils on specified crops. Refer to label or contact supplier for further details		
Protective clothing	A		
Precautions	U05a, U20b, E15a, D01, D02, D05, D09a, D10a		
Transact	Agrii	A0584	penetrant
Contains	40% w/w propionic acid		
Use with	Plant growth regulators in cereals, all approved pesticides in non-edible crops, all approved pesticides on edible crops when used at half or less their recommended rate, and with all pesticides on listed crops up to specified growth stages		
Protective clothing	A, C		
Precautions	R34, U10, U11, U14, U15, U19a, D01, D05, D09b, M04a, H05		
Transcend	Helena	A0333	adjuvant/vegetable oil
Contains	80% w/w oil (soybean fatty acid esters) and 12% w/w trisiloxane organosilicone copolymers.		
Use with	Approved pesticides on non-edible crops, cereals and stubbles of all edible crops, grassland (destruction), and pesticides approved for use on growing edible crops when used at half recommended dose or less. On specified crops product may be used at a maximum spray concentration of 0.5% with approved pesticides at their full approved rate up to the growth stages shown in the label		
Protective clothing	A, H		
Precautions	R38, R41, U02a, U05a, U08, U11, U19a, U20b, E13c, E34, E37, D01, D02, D05, D09a, M03, H04		
Validate	De Sangosse	A0500	adjuvant/wetter
Contains	25% w/w alkoxylated alcohols, 25% w/w oil (soybean fatty acid esters) and 50% w/w soybean phospholipids		
Use with	Approved pesticides on non-edible crops where the addition of a wetter/ spreader or adjuvant oil is recommended on the pesticide label, and pesticides approved for use on growing edible crops when used at half recommended dose or less. On specified crops product may be used at a maximum spray concentration of 0.5% with approved pesticides at their full approved rate up to the growth stages shown in the label		
Protective clothing	A, C, H		
Precautions	R51, R53a, U05a, U11, U14, U19a, U20b, E15a, E19b, E34, D01, D02, D10a, D12a, H11		
Velocity	Agrovista	A0697	adjuvant
Contains	745g/l oil (rapeseed fatty acid esters) and 103g/l trisiloxane organosilicone copolymers.		
Use with	All approved pesticides up to their full approved rate at a maximum conc of 0.5% vol/vol.		
Protective clothing	A, C		
Precautions	R41, R52, R53a, U02a, U05a, U08, U11, U14, U15, U20b, E13c, E15a, E34, E40c, D01, D02, D09a, D10a, H04		

Product	Supplier	Adj. No.	Type
Velvet	Global Adjuvants	A0814	spreader/sticker/wetter

Contains	83% trisiloxane organosilicone copolymers
Use with	All approved pesticides on all edible crops when used at half their recommended dose or less, and on all non-edible crops up to their full recommended dose.
Protective clothing	A, C, H
Precautions	H319, E34, H411, D01, D02, D09a, D10a

Product	Supplier	Adj. No.	Type
Verdant	Agrovista	A0714	wetter

Contains	8% w/w alkoxylated alcohol, 63% w/w alkoxylated alkyl polyglucoside and 10% w/w alkyl polyglucoside
Use with	All approved pesticides at 50% or less on edible crops and with all approved pesticides on non-edible crops, non-crop production and stubbles of edible crops at 0.25% spray sloution.
Protective clothing	A, C, H
Precautions	R38, R41, U02a, U05a, U11, U14, U15, U20b, E15a, E34, D02, D05, D09a, D10a, H04

Product	Supplier	Adj. No.	Type
Wetcit	Plant Solutions	A0586	penetrant/wetter

Contains	8.15% w/w alcohol ethoxylate
Use with	All approved pesticides on all edible crops when used at half their recommended dose or less, and on all non-edible crops up to their full recommended dose. Also at a maximum concentration of 0.25% with approved pesticides on listed crops up to specified growth stages at full rate, and at half rate thereafter
Protective clothing	A, C, H
Precautions	R22a, R38, R41, U08, U11, U19a, U20b, E13c, E15a, E34, E37, D05, D08, D09a, D10b, H03

Product	Supplier	Adj. No.	Type
Zarado	De Sangosse	A0516	vegetable oil

Contains	70% w/w oil (rapeseed oil fatty acid esters) as emulsifiable concentrate
Use with	All approved pesticides on edible and non-edible crops when used at half their approved dose or less. Also with approved pesticides on specified crops, up to specified growth stages, at up to their full approved dose
Protective clothing	A, C, H
Precautions	R43, R52, U05a, U13, U19a, U20b, E15a, E19b, E34, D01, D02, D05, D10a, M05a, H04

Product	Supplier	Adj. No.	Type
Zeal	Interagro	A0685	water conditioner/wetter

Contains	60.0 % w/w alkoxylated tallow amines and 37.0 % w/w alkoxylated sorbitan esters
Use with	Trace elements such as calcium, copper, manganese and sulphur and with macronutrients such as phosphates
Protective clothing	A, C, H
Precautions	R22a, R36, R38, R51, R53a, U05a, U09a, U11, U14, U15, U19a, E15b, E34, D01, D02, D09a, D10a, M03, H03, H11

SECTION 4

Product	Supplier	Adj. No.	Type
Zigzag	FMC Agro	A0692	extender/sticker/wetter
Contains	36.9% w/v styrene/butadiene co-polymer and 6.375% w/v polyether-modified trisiloxane		
Use with	Approved glyphosate formulations on cereal crops; All approved pesticides applied at half-rate or less; All approved pesticides for use on crops not destined for human or animal consumption; With all approved pesticides on edible crops up to the latest growth stage specified on the label.		
Protective clothing	C, H		
Precautions	U11, U15, E38, D01		
Zinzan	Certis	A0600	extender/spreader/wetter
Contains	70% w/w 1,2 bis (2-ethylhexyloxycarbonyl) ethanesulphonate		
Use with	All approved pesticides on non-edible crops, and all pesticides approved for use on growing edible crops when used at half recommended dose or less. On specified edible crops product may be used at a maximum spray concentration of 0.075% with approved pesticides at their full approved rate up to the growth stages shown in the label		
Protective clothing	A, C, P		
Precautions	R38, R41, U11, U14, U20c, D01, D09a, D10a, H04		

SECTION 5
USEFUL INFORMATION

Pesticide Legislation

Anyone who advertises, sells, supplies, stores or uses a pesticide is bound by legislation, including those who use pesticides in their own homes, gardens or allotments. There are numerous UK statutory controls, but the major legal instruments are outlined below.

Regulation 1107/2009 — The Replacement for EU 91/414

Regulation 1107/2009 entered into force within the EU on 14 Dec 2009 and was applied to new approval applications from 14th June 2011. Since that date all pesticides that held a current approval under 91/414 have been deemed to be approved under 1107/2009 and the new criteria for approval will only be applied when the active substance comes up for review. The new regulation largely mirrors the previous regulation but requires additional approval criteria to be met to weed out the more hazardous chemicals. These include assessing the effect on vulnerable groups, taking into account known synergistic and cumulative effects, considering the effect on coastal and estuarine waters, effects on the behaviour of non-target organisms, biodiversity and the ecosystem. However, these effects will only be evaluated when scientific methods approved by EFSA (European Food Safety Authority) have been developed.

The additional criteria will require that no CMR's (substances which are Carcinogens, Mutagens or toxic to Reproduction), no endocrine disruptors and no POP (Persistent Organic Pollutants), PBT (chemicals that are Persistent, Bioaccumulative or Toxic) or vPVB (very Persistent and very Bioaccumulative) will gain approval unless the exposure is negligible. However, note that for EFSA the term 'negligible' has yet to receive a precise definition. A list of 66 chemicals that pose an endocrine disruptor risk has been produced. It contained maneb and vinclozolin and both actives have now been withdrawn.

If an active substance passes these hurdles, the first approval for basic substances will be granted for 10 years (15 years for 'low risk' substances, 7 years for candidates for substitution – see below) and then at each subsequent renewal a further 15 years will be granted. Actives will be divided between 5 categories – low risk substances (e.g. pheromones, semiochemicals, micro-organisms and natural plant extracts), basic substances, candidates for substitution and then safeners/synergists and co-formulants. Candidates for substitution will be those chemicals which have just scraped past the approval thresholds but are deemed to still carry some degree of hazard. They will be withdrawn when significantly safer alternatives are available. This includes physical methods of control but any alternative must not have significant economic or practical disadvantages, must minimise the risk of the development of resistance and the consequences for any 'minor uses' of the original product must also be considered.

The Water Framework Directive

This became law in the UK in 2003 but is being introduced over a number of years to allow systems to acclimatise to the new rules and required member states of the EU to achieve 'good status' in ground and surface water bodies by 22 December 2015 with reviews every six years thereafter. To achieve 'good status' as far as pesticides are concerned the concentration of any individual pesticide must not exceed 0.1 micrograms per litre and the total quantity of all pesticides must not exceed 0.5 micrograms per litre. These levels are not based on the toxicity of the pesticides but on the level of detection that can be reliably achieved. Failure to meet these limits could jeopardise the approval of the problem active ingredients and so every effort must be made to avoid contamination of surface and ground water bodies with pesticides. The pesticides that are currently being found in water samples above the 0.1 ppb limit are metaldehyde, carbetamide, propyzamide, MCPA, mecoprop, glyphosate, bentazone, terbutryne, trietazine, clopyralid, metazachlor, chlorotoluron, chloridazon and ethofumesate. These actives are identified as a risk to water in the pesticide profiles in section 2 and great care should be taken when applying them, particularly metaldehyde which is almost impossible to remove from water at treatment plants.

SECTION 5

Classification, Labelling and Packaging (CLP) Regulation

The CLP Regulation (EC No 1272/2008) governs the classification of substances and mixtures and is intended to introduce the United Nations Globally Harmonised System (GHS) of classification to Europe. This new system and the hazard icons that appear on pesticide labels is very similar to the old system with the most obvious difference being that the hazard icons have a white background rather than an orange one. The new system of classification became mandatory for new supplies from 1st June 2015 in the UK but for products already in the supply chain the compliance date was deferred to 1st June 2017. Since that date all products placed in the market MUST comply with the CLP Regulation. The biggest change was the replacement of the old HARMFUL icon **'X'** with a new one **'!'** to notify that caution is required when handling or using a product. This brings Europe and the UK in line with the rest of the world. Details of the new phrases are given in Appendix 4.

The Food and Environment Protection Act 1986 (FEPA)

FEPA introduced statutory powers to control pesticides with the aims of protecting human beings, creatures and plants, safeguarding the environment, ensuring safe, effective and humane methods of controlling pests and making pesticide information available to the public. This was supplemented by Control of Pesticides Regulations 1986 (COPR) which has now been replaced by EC Regulation 1107/2009 – see above. Details are given on the websites of the Chemicals Regulation Directorate (CRD) and the Health and Safety Executive (HSE). Together these regulations mean that:

- Only approved products may be sold, supplied, stored, advertised or used.
- No advertisement may contain any claim for safety beyond that which is permitted in the approved label text.
- Only products specifically approved for the purpose may be applied from the air.
- A recognised Storeman's Certificate of Competence is required by anyone who stores for sale or supply pesticides approved for agricultural use.
- A recognised Certificate of Competence is required by anyone who gives advice when selling or supplying pesticides approved for agricultural use.
- Users of pesticides must comply with the Conditions of Approval relating to use.
- A recognised Certificate of Competence is required for all contractors and persons applying pesticides approved for agricultural use (unless working under direct supervision of a certificate holder).
- Only those adjuvants authorised by CRD may be used.
- For tank-mixes of convenience, when the purpose is to reduce the number of passes in the crop, no efficacy data is required but physical and chemical compatibility data are required in line with EU guidance. For mixtures that claim positive benefits efficacy data remains necessary in addition to the physical and chemical compatibility data. Note however that ALS herbicides (HRAC mode of action code 'B') may only be mixed or applied in sequence with other ALS herbicides when a label statement permits this. Mixtures of anticholinesterase products are not permitted unless CRD has fully assessed the toxicological effects.

Dangerous Preparations Directive (1999/45/EC)

The Dangerous Preparations Directive (DPD) came into force in the UK for pesticide and biocidal products on 30 July 2004. Its aim is to achieve a uniform approach across all Member States to the classification, packaging and labelling of most dangerous preparations, including crop protection products. The Directive is implemented in the UK under the Chemicals (Hazard Information and Packaging for Supply) Regulations 2002, often referred to under the acronym CHIP3. In most cases the Regulations have led to additional hazard symbols, and associated risk and safety phrases relating to environmental and health hazards, appearing on the label. All products affected by the DPD entering the supply chain from the implementation date above must be so labelled. The new environmental hazard classifications and risk phrases are included.

Under the DPD, the labels of all plant protection products now state 'To avoid risks to man and the environment, comply with the instructions for use'. As the instruction applies to all products, it has not been repeated in each fact sheet.

The Review Programme

The review programme of existing active substances will continue under EC Directive 1107/2009. The programme is designed to ensure that all available plant protection products are supported by up-to-date information on safety and efficacy. As this will now be under 1107/2009, the new standards for approval will be applied at the due date and any that are deemed to be hazardous chemicals will be withdrawn. Any that are just within the safety standards yet to be defined will be called 'candidates for substitution' and granted only a seven-year renewal to allow time for significantly safer alternatives to be developed. If these are available at the time of renewal, the approval will be withdrawn. Chemicals deemed to be low risk, such as natural plant extracts, microorganisms, semiochemicals and pheromones, will be renewed for 15 years, while most active substances will be renewed for a further 10 years.

Control of Substances Hazardous to Health Regulations 1988 (COSHH)

The COSHH regulations, which came into force on 1 October 1989, were made under the Health and Safety at Work Act 1974, and are also important as a means of regulating the use of pesticides. The regulations cover virtually all substances hazardous to health, including those pesticides classed as Very Toxic, Toxic, Harmful, Irritant or Corrosive, other chemicals used in farming or industry, and substances with occupational exposure limits. They also cover harmful micro-organisms, dusts and any other material, mixture, or compound used at work which can harm people's health.

The original Regulations, together with all subsequent amendments, have been consolidated into a single set of regulations: The Control of Substances Hazardous to Health Regulations 1994 (COSHH 1994).

The basic principle underlying the COSHH regulations is that the risks associated with the use of any substance hazardous to health must be assessed before it is used, and the appropriate measures taken to control the risk. The emphasis is changed from that pertaining under the Poisonous Substances in Agriculture Regulations 1984 (now repealed) – whereby the principal method of ensuring safety was the use of protective clothing – to the prevention or control of exposure to hazardous substances by a combination of measures. In order of preference the measures should be:

(a) substitution with a less hazardous chemical or product
(b) technical or engineering controls (e.g. the use of closed handling systems, etc.)
(c) operational controls (e.g. operators located in cabs fitted with air-filtration systems, etc.)
(d) use of personal protective equipment (PPE), which includes protective clothing.

Consideration must be given as to whether it is necessary to use a pesticide at all in a given situation and, if so, the product posing the least risk to humans, animals and the environment must be selected. Where other measures do not provide adequate control of exposure and the use of PPE is necessary, the items stipulated on the product label must be used as a minimum. It is essential that equipment is properly maintained and the correct procedures adopted. Where necessary, the exposure of workers must be monitored, health checks carried out, and employees instructed and trained in precautionary techniques. Adequate records of all operations involving pesticide application must be made and retained for at least 3 years.

Biocidal Products Regulations

These regulations concern disinfectants, preservatives and pest control products but only the latter are listed in this book. The regulation aims to harmonise the European market for biocidal products, to provide a high level of safety for humans, animals and the environment and to ensure that the products are effective against the target organisms. Products that have been reviewed and included under these regulations are issued with a new number of the format UK-2012-xxxx and these are summarised in this book as UK??-xxxx.

SECTION 5

Certificates of Competence – the roles of BASIS and NPTC

COPR, COSHH and other legislation places certain obligations on those who handle and use pesticides. Minimum standards are laid down for the transport, storage and use of pesticides, and the law requires those who act as storekeepers, sellers and advisors to hold recognised Certificates of Competence.

BASIS

BASIS is an independent Registration Scheme for the pesticide industry. It is responsible for organising training courses and examinations to enable such staff to obtain a Certificate of Competence.

In addition, BASIS undertakes an annual assessment of pesticide supply stores, enabling distributors, contractors and seedsmen to meet their obligations under the Code of Practice for Suppliers of Pesticides. Further information can be obtained from BASIS. Since 26 November 2015 all businesses must have sufficient number of staff with BASIS certificates to advise customers at **the time of sale** where previously it was just at the point of sale

Certificates of Competence

Storage

* BASIS Certificate of Competence in the Storage and Handling of Crop Protection Products

Sale and supply

* BASIS Certificate in Crop Protection (Agriculture)
* BASIS Certificate in Crop Protection (Commercial Horticulture)
* BASIS Certificate in Crop Protection (Amenity Horticulture)
* BASIS Certificate in Crop Protection (Forestry)
* BASIS Certificate in Crop Protection (Seed Treatment)
* BASIS Certificate in Crop Protection (Seed Sellers)
* BASIS Certificate in Crop Protection (Field Vegetables)
* BASIS Certificate in Crop Protection (Potatoes)
* BASIS Certificate in Crop Protection (Aquatic)
* BASIS Certificate in Crop Protection (Indoor Landscaping)
* BASIS Certificate in Crop Protection (Grassland and Forage Crops)
* BASIS/LEAF ICM Certificate
* BASIS Advanced Certificate
* FACTS (Fertiliser Advisers Certification and Training Scheme) Certificate

NPTC

Certain spray operators also require certificates of competence under the Control of Pesticides Regulations. NPTC's Pesticides Award is a recognised Certificate of Competence under COPR and it is aimed principally at those people who use pesticide products approved for use in agriculture, horticulture (including amenity horticulture) and forestry. All contractors must possess a certificate if they spray such products unless they are working under the direct supervision of a certificate holder.

Because the required spraying skills vary widely among the uses listed above, candidates are assessed under one (or more) modules that are most appropriate for their professional work. Assessments are carried out by an approved NPTC or Scottish Skills Testing Service Assessor. All candidates must first complete a foundation module (PA1), for which a certificate is not issued, before taking one of the specialist modules. Certificate holders who change their work so that a different specialist module becomes more appropriate may need to obtain a new certificate under the new module.

Holders are required to produce on demand their Certificate of Competence for inspection to any authorised person. Further information can be obtained from NPTC.

Certificates of Competence for spray operators

- PA1 Foundation Module
- PA2 Ground Crop Sprayers – mounted or trailed
- PA3 Broadcast Air Blast Sprayer. Variable Geometry Boom Air Assisted Sprayer
- PA4 Granule Applicator – Mounted or Trailed
- PA5 Boat Mounted Applicators
- PA6 Hand Held Applicators
- PA7 Aerial Application
- PA8 Mixer/Loader
- PA9 Fogging, Misting and Smokes
- PA10 Dipping Bulbs, Corms, Plant Material or Containers
- PA11 Seed Treating Equipment
- PA12 Application of Pesticides to Material as a Continuous or Batch Process

Maximum Residue Levels

A small number of pesticides are liable to leave residues in foodstuffs, even when used correctly. Where residues can occur, statutory limits, known as maximum residue levels (MRLs), have been established. MRLs provide a check that products have been used as directed; **they are not safety limits**. However, they do take account of consumer safety because they are set at levels that ensure normal dietary intake of residues presents no risk to health. Wide safety margins are built in, and eating food containing residues above the MRL does not automatically imply a risk to health. Nevertheless, it is an offence to put into circulation any produce where the MRL is exceeded.

The surrounding legislation is complex. MRLs may be specified by several different bodies. The UK has set statutory MRLs since 1988. The European Union intends eventually to introduce MRLs for all pesticide/commodity combinations. These are being introduced initially by a series of priority lists but will subsequently be covered by the review programme under EC Regulation 1107/2009. However, in cases where no information is available, EU Regulation 2000/42/EC requires many MRLs to be set at the limit of determination (LOD). As a result, certain approvals are being withdrawn where such use would leave residues above the MRL set in the Directive.

MRLs apply to imported as well as home-produced foodstuffs. Details of those that have been set have been published in *The Pesticides (Maximum Residue Levels in Crops, Food and Feeding Stuffs) Regulations 1994*, and successive amendments to these Regulations. These Statutory Instruments are available from The Stationery Office (www.tso.co.uk).

MRLs are set for many chemicals not currently marketed in Britain. Because of this and the ever-changing information on MRLs, direct access is provided to the comprehensive MRL databases on the Chemicals Regulation Division website (www.hse.gov.uk/crd/index.htm). This online database sets out in table form the levels specified by UK Regulations, EC Directives, and the Codex Alimentarius for each commodity.

REACH (Registration, Evaluation, Authorisation and restriction of CHemicals)

REACH is a European regulation that came into force on 1 June 2007 to produce a centralized database of all substances made or imported into the EU in quantities greater than 1 tonne per annum. Its aim was to enhance the protection of human health and the environment, to allow free movement of substances within the EU and to promote the use of alternatives to hazardous materials where possible. The data on these substances is held by the European Chemicals Agency (ECHA) and in the UK the competent authority is Health and Safety Executive (HSE). Full details and background information are available of the HSE website at www.hse.gov.uk/reach/index.htm.

SECTION 5

Approval (On-label and Off-label)

Only officially approved pesticides can be marketed and used in the UK. Approvals are granted by UK Government Ministers in response to applications that are supported by satisfactory data on safety, efficacy and, where relevant, humaneness. The Chemicals Regulation Division (CRD), (www.hse.gov.uk/crd/index.htm.uk) comes under the Health and Safety Executive (HSE), (www.hse.gov.uk), and is the UK Government Agency for regulating agricultural pesticides and plant protection products. The HSE currently fulfils the same role for other pesticides, with the two organisations now merged. The main focus of the regulatory process in both bodies is the protection of human health and the environment.

Statutory Conditions of Use

Approvals are normally granted only in relation to individual products and for specified uses. It is an offence to use non-approved products or to use approved products in a manner that does not comply with the statutory conditions of use, except where the crop or situation is the subject of an off-label extension of use (see below).

Statutory conditions have been laid down for the use of individual products and may include:

- field of use (e.g. agriculture, horticulture etc.)
- crop or situations for which treatment is permitted
- maximum individual dose
- maximum number of treatments or maximum total dose
- maximum area or quantity which may be treated
- latest time of application or harvest interval
- operator protection or training requirements
- environmental protection
- any other specific restrictions relating to particular pesticides.

Products must display these statutory conditions in a boxed area on the label entitled 'Important Information', or words to that effect. At the bottom of the boxed area must be shown a bold text statement: **'Read the label before use. Using this product in a manner that is inconsistent with the label may be an offence. Follow the Code of Practice for Using Plant Protection Products'.** This requirement came into effect in October 2006, and replaced the previous 'Statutory Box'.

Types of Approval

Where there were once three levels of approval (full; provisional; experimental permit), there will now be just two levels (authorisation for use; limited approval for research and development). Provisional approval used to be granted while further data were generated to justify label claims, but under the new regulations, decisions on authorisation will be reached much more quickly, and this 'half-way stage' is deemed to be unnecessary. The official list of approved products, excluding those approved for research and development only, are shown on the websites of CRD and the Health and Safety Executive (HSE).

Withdrawal of Approval

Product approvals may be reviewed, amended, suspended or revoked at any time. Revocation may occur for various reasons, such as commercial withdrawal, or failure by the approval holder to meet data requirements.

From September 2007, where an approval is being revoked for purely administrative, 'housekeeping' reasons, the existing approval will be revoked and in its place will be issued:

- an approval for advertisement, sale and supply by any person for 24 months, and
- an approval for storage and use by any person for 48 months.

Where an approval is replaced by a newer approval, such that there are no safety concerns with the previous approval, but the newer updated approval is more appropriate in terms of reflecting the latest regulatory standard, the existing approval will be revoked and in its place will be issued:

- an approval for advertisement, sale and supply by any person for 12 months, and
- an approval for storage and use by any person for 24 months.

Where there is a need for tighter control of the withdrawal of the product from the supply chain, for example failure to meet data submission deadlines, the current timelines will be retained:

- immediate revocation for advertisement, sale or supply for the approval holder
- approval for 6 months for advertisement, sale and supply by 'others', and
- approval for storage and use by any person for 18 months.

Immediate revocation and product withdrawal remain an option where serious concerns are identified, with immediate revocation of all approvals and approval for storage only by anyone for 3 months, to allow for disposal of product.

The expiry date shown in the product fact sheet is the final date of legal use of the product.

Off-label Extension of Use

Products may legally be used in a manner not covered by the printed label in several ways:

- In accordance with an Extension of Authorisation for Minor Use (EAMU). EAMUs are uses for which individuals or organisations other than the manufacturers have sought approval. The Notices of Approval are published by CRD and are widely available from ADAS or NFU offices. Users of EAMUs must first obtain a copy of the relevant Notice of Approval and comply strictly with the conditions laid down therein. Users of this Guide will find details of extant EAMUs and direct links to the CRD website, where EAMU notices can also be accessed.
- In tank mixture with other approved pesticides in accordance with Consent C(i) made under FEPA. Full details of Consent C(i) are given in Annex A of Guide to Pesticides on the CRD website, but there are two essential requirements for tank mixes. First, all the conditions of approval of all the components of a mixture must be complied with. Second, no person may mix or combine pesticides that are cholinesterase compounds unless allowed by the label of at least one of the pesticides in the mixture.
- In conjunction with authorised adjuvants.
- In reduced spray volume under certain conditions.
- In the use of certain herbicides on specified set-aside areas subject to restrictions, which differ between Scotland and the rest of the UK.
- By mutual recognition of a use fully approved in another Member State of the European Union and authorised by CRD.

Although approved, off-label uses are not endorsed by manufacturers and such treatments are made entirely at the risk of the user.

SECTION 5

Using Crop Protection Chemicals

Use of Herbicides In or Near Water

Products in this Guide approved for use in or near water are listed in Table 5.1. Before use of any product in or near water, the appropriate water regulatory body (Environment Agency/Local Rivers Purification Authority; or in Scotland the Scottish Environmental Protection Agency) must be consulted. Guidance and definitions of the situation covered by approved labels are given in the Defra publication Guidelines for the Use of Herbicides on Weeds in or near Watercourses and Lakes. Always read the label before use.

Table 5.1 Products approved for use in or near water

Chemical	Product
glyphosate	Asteroid, Asteroid Pro, Asteroid Pro 450, Barclay Gallup Biograde 360, Barclay Gallup Biograde Amenity, Barclay Gallup Hi-Aktiv, Barclay Glyde 144, Discman Biograde, Envision, Gallup Hi-Aktiv Amenity, Glyfos Dakar, Mascot Hi-Aktiv Amenity, Roundup Biactive GL, Roundup POWERMAX, Surrender T/F, Trustee Amenity, Vesuvius Green

Use of Pesticides in Forestry

Table 5.2 Products in this Guide approved for use in forestry

Chemical	Product	Use
aluminium ammonium sulphate	Asbo, Curb Liquid Crop Spray	Animal deterrent/repellent
cycloxydim	Laser	Herbicide
cypermethrin	Forester	Insecticide
ferric phosphate	Ferrex, Iroxx, Menorexx, Sluxx HP	Molluscicide
fluazifop-P-butyl	Clayton Maximus, Fusilade Forte, Fusilade Max	Herbicide
glyphosate	Amega Duo, Azural, Barclay Gallup Biograde 360, Barclay Gallup Biograde Amenity, Barclay Gallup Hi-Aktiv, Barclay Glyde 144, Clinic UP, Credit, Crestler, Discman Biograde, Ecoplug Max, Envision, Gallup Hi-Aktiv Amenity, Hilite, Liaison, Mascot Hi-Aktiv Amenity, Mentor, Monsanto Amenity Glyphosate, Monsanto Amenity Glyphosate XL, Motif, Nomix Conqueror Amenity, Rattler, Rodeo, Rosate Green, Samurai, Scorpion, Snapper, Surrender T/F, Tanker, Trustee Amenity, Vesuvius Green	Herbicide
metazachlor	Rapsan 500 SC, Stalwart, Sultan 50 SC, Taza 500	Herbicide
Phlebiopsis gigantea	PG Suspension	Biological agent
propaquizafop	Clayton Satchmo, Cleancrop Peregrine, Falcon, Longhorn, Shogun	Herbicide
propyzamide	Barclay Propyz, Kerb Granules, Menace 80 EDF, Proper Flo, PureFlo, Relva Granules, Zamide 80 WG, Zamide Flo	Herbicide
pyrethrins	Spruzit	Insecticide

SECTION 5

Pesticides Used as Seed Treatments

Information on the target for these products can be found in the relevant pesticide profile in Section 2.

Table 5.3 Products used as seed treatments (including treatments on seed potatoes)

Chemical	Product	Formulation	Crop(s)
Bacillus amyloliquefaciens strain MBI600	Integral Pro	FS	Winter oilseed rape
cymoxanil + fludioxonil + metalaxyl-M	Wakil XL	WS	Carrots, Combining peas, Parsnips, Vining peas
cypermethrin	Signal 300 ES	ES	Spring barley, Spring wheat, Winter barley, Winter wheat
difenoconazole + fludioxonil	Difend Extra	FS	Winter wheat
fludioxonil	Beret Gold	FS	Rye, Spring barley, Spring oats, Spring wheat, Triticale, Winter barley, Winter oats, Winter wheat
	Maxim 100FS	FS	Potatoes, Seed potatoes
fludioxonil + metalaxyl-M	Maxim XL	FS	Forage maize
fludioxonil + metalaxyl-M + sedaxane	Vibrance SB	FS	Fodder beet, Sugar beet
fludioxonil + sedaxane	Vibrance Duo	FS	Rye, Spring oats, Triticale, Winter wheat
fludioxonil + tebuconazole	Fountain	FS	Triticale, Winter barley, Winter oats, Winter rye, Winter wheat
fludioxonil + tefluthrin	Austral Plus	FS	Spring barley, Spring oats, Spring wheat, Triticale, Winter barley, Winter oats, Winter wheat
fluopyram + prothioconazole + tebuconazole	Raxil Star	FS	Winter barley
flutolanil	Rhino DS	DS	Potatoes
metalaxyl-M	Apron XL	ES	Beetroot, Brassica leaves and sprouts, Broccoli, Brussels sprouts, Bulb onions, Cabbages, Calabrese, Cauliflowers, Chard, Chinese cabbage, Choi sum, Collards, Herbs (see appendix 6), Kale, Kohlrabi, Ornamental plant production, Radishes, Shallots, Spinach
pencycuron	Solaren	DS	Potatoes

Chemical	Product	Formulation	Crop(s)
physical pest control	Silico-Sec	DS	Stored grain
prothioconazole + tebuconazole	Redigo Pro	FS	Durum wheat, Rye, Spring barley, Spring oats, Spring wheat, Triticale, Winter barley, Winter oats, Winter wheat
silthiofam	Latitude	FS	Rye, Spring wheat, Triticale, Winter barley, Winter wheat

SECTION 5

Aerial Application of Pesticides

New aerial spraying permit arrangements came into force in June 2012 and those undertaking aerial applications must ensure that the spraying is done in line with an Approved Application Plan that is subject to approval by CRD. Plans will only be approved where there is no viable alternative method of application or where aerial application results in reduced impact on human health and/or the environment compared to land-based application. Thus applications to bracken, forestry and possibly blight sprays to potatoes may meet these requirements but few others are likely to do so. Template application plans are available from the CRD web site and once all the necessary details have been received by CRD they undertake to approve or reject the plan within 10 working days.

Table 5.4 Products approved for aerial application

Chemical	Product	Crop(s)
diflubenzuron	Dimilin Flo	Forest
thifensulfuron-methyl + tribenuron-methyl	Ratio SX	Spring barley, Spring wheat, Winter barley, Winter wheat

Resistance Management

Pest species are, by definition, adaptable organisms. The development of resistance to some crop protection chemicals is just one example of this adaptability. Repeated use of products with the same mode of action will clearly favour those individuals in the pest population able to tolerate the treatment. This leads to a situation where the tolerant (or resistant) individuals can dominate the population and the product becomes ineffective. In general, the more rapidly the pest species reproduces and the more mobile it is, the faster is the emergence of resistant populations, although some weeds seem able to evolve resistance more quickly than would be expected. In the UK, key independent research organisations, chemical manufacturers and other organisations have collaborated to share knowledge and expertise on resistance issues through three action groups. Participants include **ADAS**, the **Chemicals Regulation Division (CRD)**, universities, colleges and the **Agriculture and Horticulture Development Board (AHDB)** The groups have a common aim of monitoring resistance in the UK and devising and publishing management strategies designed to combat it where it occurs. The groups are:

- **The Weed Resistance Action Group (WRAG)**, formed in 1989 (Secretary: Richard Hull, Rothamsted Research, Harpenden, Herts AL5 2JQ *Tel: 01954 268219*)
  Web: https://cereals.ahdb.org.uk/WRAG.
- **The Fungicide Resistance Action Group (FRAG)**, formed in 1995 (Secretary: Mr Paul Ashby, HSE, Email: paul.ashby@hse.gsi.gov.uk). Web: https://cereals.ahdb.org.uk/FRAG
- **The Insecticide Resistance Action Group (IRAG)**, formed in 1997 (Secretary: Dr Sacha White, ADAS Boxworth, Battlegate Road, Cambridge, Cambs, CB3 8NN. *Tel: 01954 267666*, Email: sacha.white@adas.co.uk). Web: https://cereals.ahdb.org.uk/IRAG

The above groups publish detailed advice on resistance management relevant for each sector and, in some cases, specific to a pest problem. This information, together with further details about the function of each group, can be obtained from the **Chemicals Regulation Division** website www.hse.gov.uk/crd/index.htm.

The speed of appearance of resistance depends on the mode of action of the crop protection chemicals, as well as the manner in which they are used. Resistance among insects and fungal diseases has been evident for much longer than weed resistance to herbicides, but examples in all three categories are now widespread and increasing. This has created a need for agreement on the advice given for the use of crop protection chemicals in order to reduce the likelihood of the development of resistance and to avoid the loss of potentially valuable products in the chemical armoury. Mixing or alternating modes of action is one of the guiding principles of resistance management. To assist appropriate product choices, mode of action codes, published by the international Resistance Action Committees (see below), are shown in the respective active ingredient fact sheets. The product label and/or a professional advisor should always be consulted before making decisions. The general guidelines for resistance management are similar for all three problem areas:

Preparation in advance

- Be aware of the factors that favour the development of resistance, such as repeated annual use of the same product, and assess the risk.
- Plan ahead and aim to integrate all possible means of control.
- Use cultural measures such as rotations, stubble hygiene, variety selection and, for fungicides, removal of primary inoculum sources, to reduce reliance on chemical control.
- Monitor crops regularly.
- Keep aware of local resistance problems.
- Monitor effectiveness of actions taken and take professional advice, especially in cases of unexplained poor control.

Using crop protection products

- Optimise product efficacy by using it as directed, at the right time, in good conditions.
- Treat pest problems early.
- Mix or alternate chemicals with different modes of action.
- Avoid repeated applications of very low doses.
- Keep accurate field records.

SECTION 5

Label guidance depends on the appropriate strategy for the product. Most frequently it consists of a warning of the possibility of poor performance due to resistance, and a restriction on the number of treatments that should be applied in order to minimise the development of resistance. This information is summarised in the profiles in the active ingredient fact sheets, but detailed guidance must always be obtained by reading the label itself before use.

International Action Committees

Resistance to crop protection products is an international problem. Agrochemical industry collaboration on a global scale is via three action committees whose aims are to support a coordinated industry approach to the management of resistance worldwide. In particular they produce lists of crop protection chemicals classified according to their mode of action. These lists and other information can be obtained from the respective websites:

Herbicide Resistance Action Committee (HRAC) – www.hracglobal.com

Fungicide Resistance Action Committee (FRAC) – www.frac.infosections

Insecticide Resistance Action Committee (IRAC) – www.irac-online.org

Poisons and Poisoning

Chemicals Subject to the Poison Law

Certain products are subject to the provisions of the Poisons Act 1972. The Poisons Rules 1982 have been revoked under the Deregulation Act 2015 (Poisons and Explosive Precursors) and replaced by The Control of Poisons and Explosive Precursors Regulations 2015 with appropriate amendments to The Poisons Act 1972. These Rules include provisions for the storage and sale and supply of listed non-medicine poisons. Details can be accessed on the HSE website. The nature of the formulation and the concentration of the active ingredient allow some products to be exempted from the Rules, while others with the same active ingredient are included (see below). The chemicals approved for use in the UK are specified under Parts I and II of the Poisons List as follows.

Part I Poisons (sale restricted to registered retail pharmacists and to registered non-pharmacy businesses provided sales do not take place on retail premises):

- aluminium phosphide
- chloropicrin
- magnesium phosphide.

Part II Poisons (sale restricted to registered retail pharmacists and listed sellers registered with a local authority):

- formaldehyde
- oxamyl (a).

Notes

(a) Granular formulations that do not contain more than 12% w/w of this or a combination of similarly flagged poisons are exempt.

Occupational Exposure Limits

A fundamental requirement of the COSHH Regulations is that exposure of employees to substances hazardous to health should be prevented or adequately controlled. Exposure by inhalation is usually the main hazard, and in order to measure the adequacy of control of exposure by this route various substances have been assigned occupational exposure limits.

There are two types of occupational exposure limits defined under COSHH: Occupational Exposure Standards (OES) and Maximum Exposure Limits (MEL). The key difference is that an OES is set at a level at which there is no indication of risk to health; for a MEL a residual risk may exist and the level takes socio-economic factors into account. In practice, MELs have been most often allocated to carcinogens and to other substances for which no threshold of effect can be identified and for which there is no doubt about the seriousness of the effects of exposure.

OESs and MELs are set on the recommendation of the Advisory Committee on Toxic Substances (ACTS). Full details are published by HSE in EH 40/2005 available at:

http://www.hse.gov.uk/pubns/books/eh40.htm

As far as pesticides are concerned, OESs and MELs have been set for relatively few active ingredients. This is because pesticide products usually contain other substances in their formulation, including solvents, which may have their own OES/MEL. In practice inhalation of solvent may be at least, or more, important than that of the active ingredient. These factors are taken into account by the regulators when approving a pesticide product under the Control of Pesticides Regulations. This indicates one of the reasons why a change of pesticide formulation usually necessitates a new approval assessment under COPR.

First Aid Measures

If pesticides are handled in accordance with the required safety precautions, as given on the container label, poisoning should not occur. It is difficult, however, to guard completely against the occasional accidental exposure. Thus, if a person handling, or exposed to, pesticides becomes ill, it is a wise precaution to apply first aid measures appropriate to pesticide poisoning even though the cause of illness may eventually prove to have been quite different. An employer has a legal duty to make adequate first aid provision for employees. Regular pesticide users should consider appointing a trained first aider even if numbers of employees are not large, since there is a specific hazard.

The first essential in a case of suspected poisoning is for the person involved to stop work, to be moved away from any area of possible contamination and for a doctor to be called at once. If no doctor is available the patient should be taken to hospital as quickly as possible. In either event it is most important that the name of the chemical being used should be recorded and preferably the whole product label or leaflet should be shown to the doctor or hospital concerned.

Some pesticides, which are unlikely to cause poisoning in normal use, are extremely toxic if swallowed accidentally or deliberately. In such cases get the patient to hospital as quickly as possible, with all the information you have. Some labels now include Material Safety Data Sheets and these contain valuable information for both the first aider and medical staff. If not included on the label, MSDS are available from company websites.

General Measures

Measures appropriate in all cases of suspected poisoning include the following:

- Remove any protective or other contaminated clothing (taking care to avoid personal contamination).
- Wash any contaminated areas carefully with water or with soap and water if available.
- In cases of eye contamination, flush with plenty of clean water for at least 15 minutes.
- Lay the patient down, keep at rest and under shelter. Cover with one clean blanket or coat, etc. Avoid overheating.
- Monitor level of consciousness, breathing and pulse rate.
- If consciousness is lost, place the casualty in the recovery position (on his/her side with head down and tongue forward to prevent inhalation of vomit).

Reporting of Pesticide Poisoning

Any cases of poisoning by pesticides must be reported without delay to an HM Agricultural Inspector of the Health and Safety Executive. In addition any cases of poisoning by substances named in schedule 2 of The Reporting of Injuries, Diseases and Dangerous Occurrences Regulations 1985, must also be reported to HM Agricultural Inspectorate (this includes organophosphorus chemicals, mercury and some fumigants).

Cases of pesticide poisoning should also be reported to the manufacturer concerned.

Additional Information

General advice on the safe use of pesticides is given in a range of Health and Safety Executive leaflets available from HSE Books. The major agrochemical companies are able to provide authoritative medical advice about their own pesticide products. Useful information is now available from the Material Safety Data Sheet available from the manufacturer or the manufacturer's website.

New arrangements for the provision of information to doctors about poisons and the management of poisonings have been introduced as part of the modernisation of the National Poisons Information Service (NPIS). The Service provides a year-round, 24-hour-a-day service for healthcare staff on the diagnosis, treatment and management of patients who may have been poisoned. The new arrangements are aimed at moving away from the telephone as the first point of contact for poisons information, to the use by doctors of an online database, supported by a second-tier, consultant-led information service for more complex clinical advice.

NPIS no longer provides direct information on poisoning to members of the public. **Anyone suspecting poisoning by pesticides should seek professional medical help immediately via their GP or NHS Direct (www.nhsdirect.nhs.uk) or NHS 24 (www.nhs24.com).**

Environmental Protection

Environmental Land Management: 8 simple steps for arable farmers

Choosing the right measures, putting them in the right place, and managing them in the right way will make all the difference to your farm environment. The general principles given here should be considered in conjunction with local priorities for soil and water protection and wildlife conservation. This approach complements best practice in soil, crop, fertiliser and pesticide management.

An improved farm environment can be achieved by good management of around 4% of the arable area where high quality habitats are maintained or created. However, the actual area required on your farm will depend on factors such as the area of vulnerable soils and length of watercourses.

If you need further advice then consult a competent environmental adviser.

What you can do

It is important to have a balance of environmental measures that contribute to each of the relevant points below to achieve improved environmental benefits.

1. **Look after established wildlife habitats**
   Start by assessing what you already have on the farm! Maintaining, or where necessary, restoring any existing wildlife habitats, such as woodland, ponds, flower-rich grassland or field margins, is critical to the survival of much of the wildlife on the farm, and may count towards some of the following measures without the need to create new habitats. Unproductive land can be used to create new habitats to complement what you already have.

2. **Maximise the environmental value of field boundaries**
   Hedgerow management and ditch management on a 2–3 years rotation boosts flowers, fruit and refuges for wildlife. This is most suited to hedges dominated by hawthorn and blackthorn, and ditches where rotational management will not compromise the drainage function. Establish new hedgerow trees to maintain or restore former numbers within the landscape.

3. **Create a network of grass margins**
   The highest priority is to buffer watercourses, ideally with a minimum of 5 m buffer strips. Grass margins can also be used to boost beneficial insects and small mammals, and buffer hedges, ponds and other environmental features. Beetle banks can be used to reduce soil erosion and run-off on slopes greater than 1:20 and boost beneficial insects in fields greater than 20 ha.

4. **Establish flower rich habitats**
   Evidence suggests that a network of flower-rich margins on 1% of arable land will support beneficial insects and a wealth of wildlife that feeds on insects. Assess whether this is best done by allowing arable plants in the seed bank to germinate, establishing perennial margins with a grass and wildflower mix, or using nectar flower mixtures. Improving the linkages between these features on the farm will also help wildlife move across the landscape.

5. **Provide winter food for birds with weedy over-wintered stubbles or wild bird cover**
   Provision of seed for wildlife is best achieved by leaving over-wintered stubbles unsprayed and uncultivated until at least mid-February on at least 5% of arable land, or growing seed-rich crops as wild bird cover on 2% of arable land.

6. **Use of spring cropping or in-field measures to help ground-nesting birds**
   Spring crops provide better habitat for a range of plants and insects, and birds such as lapwings and skylarks. Use rotational fallows, skylark plots in winter cereals or (if breeding lapwings occur) fallow plots to support ground-nesting birds where spring cropping forms less than 25% of the arable area. Fallow plots should not be created on land liable to runoff or erosion. Evidence suggests that at least 20 skylark plots or a 1 ha fallow plot per 100 ha would support ground-nesting birds.

7. Use winter cover crops to protect water
You should consider whether a winter cover crop (e.g. mustard) is necessary to capture residual nitrogen on cultivated land left fallow through the winter. This is not necessary if the stubble is retained until at least mid-February and forms a green cover.

8. Establish in-field grass areas to reduce soil erosion and run-off
Land liable to act as channels for soil erosion or run-off (e.g. steep slopes or field corners) should be converted to in-field grass areas.

Remember

*   Right measures
*   Right place
*   Right management

This guidance has been produced by The Voluntary Initiative and the Campaign for the Farmed Environment in conjunction with RSPB, Natural England, GWCT, Plantlife, Butterfly Conservation, BCPC and Buglife

Protection of Bees

Honey bees

Honey bees are a source of income for their owners and important to farmers and growers as pollinators of their crops. It is irresponsible and unnecessary to use pesticides in such a way that may endanger them. Pesticides vary in their toxicity to bees, but those that present a special hazard carry a specific warning in the precautions section of the label. They are indicated in this Guide in the hazard classification and safety precautions section of the pesticide profile.

Product labels indicate the necessary environmental precautions to take, but where use of an insecticide on a flowering crop is contemplated the British Beekeepers Association have produced the following guidelines for growers:

*   Target insect pests with the most appropriate product.
*   Choose a product that will cause minimal harm to beneficial species.
*   Follow the manufacturer's instructions carefully.
*   Inspect and monitor crops regularly.
*   Avoid spraying crops in flower or where bees are actively foraging.
*   Keep down flowering weeds.
*   Spray late in the day, in still conditions.
*   Avoid excessive spray volume and run-off.
*   Adjust sprayer pressure to reduce production of fine droplets and drift.
*   Give local beekeepers as much warning of your intention as possible.

Wild bees

Wild bees also play an important role. Bumblebees are useful pollinators of spring flowering crops and fruit trees because they forage in cool, dull weather when honey bees are inactive. They play a particularly important part in pollinating field beans, red and white clover, lucerne and borage. Bumblebees nest and overwinter in field margins and woodland edges. Avoidance of direct or indirect spray contamination of these areas, in addition to the creation of hedgerows and field margins, and late cutting or grazing of meadows, all help the survival of these valuable insects.

BeeConnected

It is best practice to notify local beekeepers before using certain crop protection products where there is a risk to bees. The online BeeConnected tool (www.beeconnected.org.uk) can help you do this.

Accidental and Illegal Poisoning of Wildlife

The Campaign Against Accidental or Illegal Poisoning (CAIP) of Wildlife, was launched back in March 1991 and has now been replaced by **The Wildlife Incident Investigation Scheme**.

SECTION 5

This has two objectives:

- To provide information to HSE on hazards to wildlife, pets and beneficial invertebrates:
- To enforce the correct use of pesticides, identifying and penalizing those who deliberately or recklessly misuse and abuse pesticides.

Full details can be found on the HSE website at www.hse.gov.uk/pesticides/topics/reducing-environmental-impact/wildlife.htm

Water Quality

Even when diluted, some pesticides are potentially dangerous to fish and other aquatic life. Not only this, but many watercourses and groundwaters are sources of drinking water, and it requires only a tiny amount of contamination to breach the stringent European Union water quality standards. The EEC Drinking Water Directive sets a maximum admissible level in drinking water for any pesticide, regardless of its toxicity, at 1 part in 10,000 million. As little as 250 grams could be enough to cause the daily supply to a city the size of London to exceed the permitted levels (although it would be very unlikely to present a health hazard to consumers).

The protection of groundwater quality is therefore vital. The Food and Environment Protection Act 1985 (FEPA) places a special obligation on users of pesticides to "safeguard the environment and in particular avoid the pollution of water". Under the Water Resources Act 1991 it is an offence to pollute any controlled waters (watercourses or groundwater), either deliberately or accidentally. Protection of controlled waters from pollution is the responsibility of the Environment Agency (in England and Wales) and the Scottish Environmental Protection Agency. Addresses for both organisations can be found under Useful Contacts.

Users of pesticides therefore have a duty to adopt responsible working practices and, unless they are applying herbicides in or near water, to prevent them getting into water. Guidance on how to achieve this is given in the Defra Code of Good Agricultural Practice for the Protection of Water.

The duty of care covers not only the way in which a pesticide is sprayed, but also its storage, preparation and disposal of surplus, sprayer washings and the container. Products that are a major hazard to fish, other aquatic life or aquatic higher plants carry one of several specific label precautions in their fact sheet, depending on the assessed hazard level.

Where to get information

For general enquiries telephone 03708 506 506 (Environment Agency) or 0131 449 7296 (SEPA). Both Agencies operate a 24-hour emergency hotline for reporting all environmental incidents relating to air, land and water:

0800 80 70 60 (for incidents in England and Wales) 0345 73 72 71 (for incidents in Scotland)

In addition, printed literature concerning the protection of water is available from the Environment Agency and the Crop Protection Association (see Useful Contacts).

Protecting surface waters

Surface waters are particularly vulnerable to contamination. One of the best ways of preventing those pesticides that carry the greatest risk to aquatic wildlife from reaching surface waters is to prohibit their application within a boundary adjacent to the water. Such areas are known as no-spray, or buffer, zones. Certain products are restricted in this way and have a legally binding label precaution to make sure the potential exposure of aquatic organisms to pesticides that might harm them is minimised.

Before 1999 the protected zones were measured from the edge of the water. The distances were 2 metres for hand-held or knapsack sprayers, 6 metres for ground crop sprayers, and a variable distance (but often 18 metres) for broadcast air-assisted applications, such as in orchards. The introduction of LERAPs (see below) has changed the method of measuring buffer zones.

'Surface water' includes lakes, ponds, reservoirs, streams, rivers and watercourses (natural or artificial). It also includes temporarily or seasonally dry ditches, which have the potential to carry water at different times of the year. Buffer zone restrictions do not necessarily apply to all products containing the same active ingredient. Those in formulations that are not likely to contaminate surface water through spray drift do not pose the same risk to aquatic life and are not subject to the restrictions.

Local Environmental Risk Assessment for Pesticides (LERAPs)

Local Environment Risk Assessments for Pesticides (LERAPs) were introduced in March 1999, and revised guidelines were issued in January 2002. They give users of most products currently subject to a buffer zone restriction the option of continuing to comply with the existing buffer zone restriction (using the new method of measurement), or carrying out a LERAP and possibly reducing the size of the buffer zone as a result. In either case, there is a new legal obligation for the user to record his decision, including the results of the LERAP.

The scheme has changed the method of measuring the buffer zone. Previously the zone was measured from the edge of the water, but it is now the distance from the top of the bank of the watercourse to the edge of the spray area. In 2013 new rules were introduced which allowed LERAP buffer zones greater than the normal 5 metres for LERAP B pesticides to avoid the loss of some important actives. The products concerned are identified in the Environmental safety section of the affected pesticide profiles.

The LERAP provides a mechanism for taking into account other factors that may reduce the risk, such as dose reduction, the use of low drift nozzles, and whether the watercourse is dry or flowing. The previous arrangements applied the same restriction regardless of whether there was actually water present. Now there is a standard zone of 1 m from the top of a dry ditch bank.

Other factors to include in a LERAP that may allow a reduction in the buffer zone are:

- The size of the watercourse, because the wider it is, the greater the dilution factor, and the lower the risk of serious pollution.
- The dose applied. The lower the dose, the less is the risk.
- The application equipment. Sprayers and nozzles are star-rated according to their ability to reduce spray drift fallout. Equipment offering the greatest reductions achieves the highest rating of three stars. The scheme was originally restricted to ground crop sprayers; new, more flexible rules introduced in February 2002 included broadcast air-assisted orchard and hop sprayers.
- Other changes introduced in 2002 allow the reduction of a buffer zone if there is an appropriate living windbreak between the sprayed area and a watercourse.

Not all products that had a label buffer zone restriction are included in the LERAP scheme. The option to reduce the buffer zone does not apply to organophosphorus or synthetic pyrethroid insecticides. This group are classified as Category A products. All other products that had a label buffer zone restriction are classified as Category B. In addition some products have a 'Broadcast air-assisted LERAP' classification. The wording of the buffer zone label precautions has been amended for all products to take account of the new method of measurement, and whether or not the particular product qualifies for inclusion in the LERAP scheme.

Products in this Guide that are in Category A or B, or have a broadcast air-assisted LERAP, are identified with an appropriate icon on the product fact sheet. Updates to the list are published regularly by CRD and details can be obtained from its website at www.pesticides.gov.uk.

The introduction of LERAPs was an important step forward because it demonstrated a willingness to reduce the impact of regulation on users of pesticides by allowing flexibility where local conditions make it safe to do so. This places a legal responsibility on the user to ensure the risk assessment is done either by himself or by the spray operator or by a professional consultant or advisor. It is compulsory to record the LERAP and make it available for inspection by enforcement authorities. In 2013 new rules were introduced which allowed LERAP buffer zones greater than 5 m to avoid the loss of some important actives. These products are identified in the section of Environmental safety for the pesticide profiles affected.

More details of the LERAP arrangements and guidance on how to carry out assessments are published in the Ministry booklet PB5621 *Local Environmental Risk Assessment for Pesticides – Horizontal Boom Sprayers*, and booklet PB6533 *Local Environment Risk Assessment for*

SECTION 5

Pesticides – Broadcast Air-Assisted Sprayers. Additional booklets, PB2088 *Keeping Pesticides Out of Water*, and PB3160 *Is Your Sprayer Fit for Work* give general practical guidance.

Groundwater Regulations

Groundwater Regulations were introduced in 1999 to complete the implementation in UK of the EU Groundwater Directive (Protection of Groundwater Against Pollution Caused by Certain Dangerous Substances – 80/68/EEC). These Regulations help prevent the pollution of groundwater by controlling discharges or disposal of certain substances, including all pesticides.

Groundwater is defined under the Regulations as any water contained in the ground below the water table. Pesticides must not enter groundwater unless it is deemed by the appropriate Agency to be permanently unsuitable for other uses. The Agricultural Waste Regulations were introduced in May 2006 and apply to the disposal of all farm wastes, including pesticides. With certain exemptions farmers are required to obtain a waste management licence for most waste disposal activities.

As far as pesticides are concerned the new Regulations made little difference to existing controls, except for the disposal of empty containers and the disposal of sprayer washings and rinsings. It remains an offence to dispose of pesticides onto land without official authorisation. Normal use of a pesticide in accordance with product approval does not require authorisation. This includes spraying the washings and rinsings back on the crop provided that, in so doing, the maximum approved dose for that product on that crop is not exceeded. However, those wishing to use a lined biobed for this purpose need to obtain an exemption from the Agency.

In practice the best advice to farmers and growers is to plan to use all diluted spray within the crop and to dispose of all washings via the same route making sure that they stay within the conditions of approval of the product. The enforcing agencies for these Regulations are the Environment Agency (in England and Wales) and the Scottish Environment Protection Agency. The Agencies can serve notice at any time to modify the conditions of an authorisation where necessary to prevent pollution of groundwater.

Integrated Farm Management (IFM)

Integrated farm management is a method of farming that balances the requirements of running a profitable farming business with the adoption of responsible and sensitive environmental management. It is a whole-farm, long-term strategy that combines the best of modern technology with some basic principles of good farming practice. It is a realistic way forward that addresses the justifiable concerns of the environmental impact of modern farming practices, at the same time as ensuring that the industry remains viable and continues to provide wholesome, affordable food. IFM embraces arable and livestock management. The phrase 'integrated crop management' is sometimes used where a farm is wholly arable.

Pest control is essential in any management system. Much can be done to minimise the incidence and impact of pests, but their presence is almost inevitable. IFM ensures that, where a pest problem needs to be contained, the action taken is the best combination of all available options. Pesticides are one of these options, and form an essential, but by no means exclusive, part of pest control strategy.

Where chemicals are to be used, the choice of product should be made not only with the pest problem in mind, but with an awareness of the environmental and social risks that might accompany its use. The aim should be to use as much as necessary, but as little as possible. A major part of the approval process aims to safeguard the environment, so that any approved product, when used as directed, will not cause long-term harm to wildlife or the environment. This is achieved by specifying on the label detailed rules for the way in which a product may be used.

The skill in implementing an IFM pest control strategy is the decision on how easily these rules may be complied with in the particular situation where use is contemplated. Although many chemical options may be available, some are likely to be more suitable than others. This Guide lists, under the heading Special precautions/Environmental safety, the key label precautions that need to be considered in this context, although the actual product label must always be read before a product is used.

Campaign for the Farmed Environment

With the demise of set-aside, it was realised that the environmental benefits achieved by this scheme could be lost. The Campaign for the Farmed Environment (CFE) scheme seeks to retain or even exceed the environmental benefits achieved by set-aside by voluntary measures.

It has three main themes:

- farming for cleaner water and a healthier soil
- helping farmland birds thrive
- improving the environment for farm wildlife.

The scheme encourages farmers to sign up for Entry Level Stewardship (ELS) schemes and to target their already extensive knowledge of habitat management to greater effect. If successful, it will help to fend off the proposal that farmers of cultivated land in England will have to adopt environmental management options on up to 6% of their land.

SECTION 6
APPENDICES

Appendix 1
Suppliers of Pesticides and Adjuvants

Aceto:
Aceto Agricultural Chemicals Corporation (UK) Ltd
2nd Floor, Refuge House
33-37 Watergate Row
Chester
Cheshire
CH1 2LE
Tel: (516) 627-6000 ext.
Email: ftrudwig@aceto.com
Web: www.aceto.com

Adama:
Adama Agricultural Solutions UK Ltd
Third Floor East
1410 Arlington Business Park
Theale
Reading
Berks.
RG7 4SA
Tel: 01635 860555
Fax: 01635 861555
Email: ukenquiries@adama.com
Web: www.adama.com

AgChem Access: AgChemAccess Limited
Cedar House,
41 Thorpe Road
Norwich
Norfolk
NR1 1ES
Tel: 0845 459 9413
Fax: 0207 149 9815
Email: thomas@agchemaccess.com
Web: www.agchemaccess.com

Agform: Agform Ltd
Hilldale Farm Research Centre
Titchfield Lane
Wickham
Hampshire
PO17 5NZ
Tel: 01329 836933
Email: info@agform.com
Web: www.agform.com

Agrichem: Agrichem (International) Ltd
Arus Hygeia
Oranmore
Co Galway
Ireland
Tel: 01733 204019
Email: info@agrichem.co.uk
Web: www.agrichem.co.uk

Agrii: United Agri Products Ltd
Throws Farm
Stebbing
Great Dunmow
Essex
DM6 3AQ
Web: www.agrii.co.uk

Agroquimicos : Agroquimicos Genericos
Calle San Vicente Martir
no 85, 8, 46007
Valencia,
Spain
Email: thomas@agchemaccess.com

Agrovista: Agrovista UK Ltd
Rutherford House
Nottingham Science and Technology Park
University Bulevard
Nottingham
NG7 2PZ
Tel: (0115) 939 0202
Fax: (0115) 921 8498
Email: enquiries@agrovista.co.uk
Web: www.agrovista.co.uk

Albaugh UK: Albaugh UK Ltd
1 Northumberland Street
Trafalgar Square
London
WC2N 5BW

Amega: Amega Sciences
Lanchester Way
Royal Oak Industrial Estate
Daventry
Northants.
NN11 5PH
Tel: (01327) 704444
Fax: (01327) 71154
Email: admin@amega-sciences.com
Web: www.amega-sciences.com

Arysta: Arysta Life Sciences
Route d'Artix, BP80
64150
Nogueres
France
Email: Don.pendergrast@arysta.com

SECTION 6

B H & B: Battle Hayward & Bower Ltd
Victoria Chemical Works
Crofton Drive
Allenby Road Industrial Estate
Lincoln
LN3 4NP
Tel: (01522) 529206
Fax: (01522) 538960
Email: orders@battles.co.uk
Web: www.battles.co.uk

Barclay:
Barclay Chemicals Manufacturing Ltd
Damastown Way
Damastown Industrial Park
Mulhuddart
Dublin 15
Ireland
Tel: (+353) 1 811 2900
Fax: (+353) 1 822 4678
Email: info@barclay.ie
Web: www.barclay.ie

Barrier: Barrier BioTech Ltd
36/37 Haverscroft Industrial Estate
New Road
Attleborough
Norfolk
NR17 1YE
Tel: (01953) 456363
Fax: (01953) 455594
Email: sales@barrier-biotech.com
Web: www.barrier-biotech.com

BASF: BASF plc
Agricultural Divison
PO Box 4, Earl Road
Cheadle Hulme
Cheshire
SK8 6QG
Tel: (0845) 602 2553
Fax: (0161) 485 2229
Web: www.agricentre.basf.co.uk

Bayer CropScience:
Bayer CropScience Limited
230 Cambridge Science Park
Milton Road
Cambridge
CB4 0WB
Tel: (01223) 226500
Fax: (01223) 426240
Web: www.bayercropscience.co.uk

Becesane : Becesane s.r.o.
Rohacova, 188/377
130 00
PRAHA 3
Czech Republic
Email: thomas@agchemaccess.com

Belchim: Belchim Crop Protection Ltd
Unit 1b, Fenice Court
Phoenix Park
Eaton Socon
St Neots
Cambs.
PE19 8EP
Tel: (01480) 403333
Fax: (01480) 403444
Email: info@belchim.com
Web: www.belchim.co.uk

Belcrop: Belcrop NV
Tiensestraat 300
3400 Landen
Belgium
Tel: +32 (0) 11 59 83 60
Fax: +32 (0) 11 59 83 61
Email: info@belcrop.com
Web: www.BELCROP.com

Capital CP: Capital CP Europe Limited
Gardeners Cottage
Treemans Road
Lewes Road
Haywards Heath
West Sussex
RH17 7EA

Certis: Certis
Suite 5, 3 Riverside
Granta Park
Great Abington
Cambs.
CB21 6AD
Tel: 0845 373 0305
Fax: 01223 891210
Email: infocertisuk@certiseurope.com
Web: www.certiseurope.co.uk

Chem-Wise: Chem-Wise Ltd
Westwood
Rattar Mains
Thurso
KW14 8XW

Chiltern: Chiltern Farm Chemicals Ltd
East Mellwaters
Stainmore
Bowes
Barnard Castle
Co. Durham
DL12 9RH
Tel: (01833) 628282
Email: chilternfarm@aol.com
Web: www.chilternfarm.com

Chrysal : Chrysal UK Limited
Ardsley Mills
Common Lane
East Ardsley
Wakefield
West Yorkshire
WF3 2DW

Clayton: Clayton Plant Protection Ltd
Bracetown Business Park
Clonee
Dublin 15
Ireland
Tel: (+353) 1 821 0127
Fax: (+353) 81 841 1084
Email: info@cpp.ag
Web: www.cpp.ag

Cropco Ltd: Cropco Limited
Northeys Farm, Yeldham Rd
Belchamp Walter
Sudbury
CO10 7BB
Tel: 01787 238200
Fax: 01787 238222
Email: info@cropco.co.uk
Web:
http://www.cropco.co.uk/contact-cropco/

DAPT: DAPT Agrochemicals Ltd
14 Monks Walk
Southfleet
Gravesend
Kent
DA13 9NZ
Tel: (01474) 834448
Fax: (01474) 834449
Email: rkjltd@supanet.com

De Sangosse: De Sangosse Ltd
Hillside Mill
Quarry Lane
Swaffham Bulbeck
Cambridge
CB25 0LU
Tel: (01223) 811215
Fax: (01223) 810020
Email: info@desangosse.co.uk
Web: www.desangosse.co.uk

Dow (Corteva): Dow AgroSciences Ltd
CPC2,
Capital Park
Fulbourn
Cambridge
Cambridgeshire
CB21 5XE
Tel: 0800 689 8899
Email: ukhotline@corteva.com
Web: www.corteva.co.uk

DuPont (Corteva): DuPont (UK) Ltd
CPC2
Capital Park
Fulbourn
Cambridge
Cambs.
CB21 5XE
Tel: 0800 689 8899
Email: hotline@corteva.com
Web: www.corteva.co.uk

Enviroscience : Enviroscience Limited
Godwin House
Mullbry Business Park
Shakespeare Way
Whitchurch
Shropshire
SY13 1LJ

EuroAgkem: EuroAgkem Ltd
Park Chambers
10 Hereford Road
Abergavenney
NP7 5PR
Tel: (01926) 634801
Fax: (01926) 634798

Fargro: Fargro Ltd
Vinery Fields
Arundel Road
Poling
Arundel
West Sussex
BN18 9PY
Tel: (01903) 721591
Fax: (01903) 883303
Email: info@fargro.co.uk
Web: www.fargro.co.uk

Fine: Fine Agrochemicals Ltd
Hill End House
Whittington
Worcester
WR5 2RQ
Tel: (01905) 361800
Fax: (01905) 361810
Email: enquire@fine.eu
Web: www.fine.eu

FMC Agro: FMC Agro Ltd
Rectors Lane
Pentre
Deeside
Flintshire
CH5 2DH
Tel: 01423 205011
Fax: 01244 532097
Web: http://cp.fmc-agro.co.uk

SECTION 6

Forest Research: Forest Research
Alice Holt Lodge
Farnham
Surrey
GU10 4LH
Tel: 01420 22255
Fax: 01420 23653
Email:
Katherine.tubby@forestry.gsi.gov.uk
Web: www.forestry.gov.uk

FytoFend: FytoFend S.A.
Rue Phocas Lejeune
25-6, 5032 Isnes
Belgium
Tel: 0032 81 72 88 40
Email: info@fytofend.com
Web: http://www.fytofend.com/en

Gemini: Gemini Agriculture Ltd
8 Mythop Road
Lytham St. Annes
Lancs
FY8 4JD

Globachem: Globachem NV
Brustem Industriepark
Lichtenberglaan 2019
3800 Sint-Truiden
Belgium
Tel: (+32) 11 69 01 73
Fax: (+32) 11 68 15 65
Email: globachem@globachem.com
Web: www.globachem.com

Global Adjuvants :
Global Adjuvants Company Ltd
20-22 Wenlock Road
London
N1 7GU
Email: office@global-adjuvants.com
Web: www.global-adjuvants.com

Gowan: Gowan Crop Protection Ltd
Innovation Hub
Rothamsted Centre for Research &
Enterprise
West Common
Harpenden
Herts
AL5 2JQ
Tel: 07584 052323
Email: DLamb@GOWANCO.com

Greenaway: Greenaway Amenity Ltd
7 Browntoft Lane
Donington
Spalding
Lincs.
PE11 4TQ
Tel: (01775) 821031
Fax: (01775) 821034
Email: greenawayamenity@aol.com
Web: www.greenawaycda.com

Greencrop: Greencrop Technology Ltd
c/o Clayton Plant Protection Ltd
Bracetown Business Park
Clonee
Dublin 15
Ireland
Tel: (+353) 1 821 0127
Fax: (+353) 81 841 1084
Email: info@cpp.ag
Web: www.cpp.ag

Harvest: Harvest Agrochemicals Ltd
Carpenter Court,
1 Maple Road,
Bramhall,
Stockport,
Cheshire,
SK7 2DH
United Kingdom

Headland Amenity:
Headland Amenity Limited
1 Burr Elm Court
Main Street
Caldecote
Cambs.
CB23 7NU
Tel: (01223) 481090
Fax: (01223) 491091
Email: info@headlandamenity.com
Web: www.headlandamenity.com

Helena: Helena Chemical Company
Cambridge House
Nottingham Road
Stapleford
Nottingham
NG9 8AB
Tel: (0115) 939 0202
Fax: (0115) 939 8031

Hermoo: Hermoo Belgium NV
Brustem Industriepark
Lichtenberglaan 2045
B-3800 Sint-Truiden
Belgium
Tel: +32 (0) 11 68 68 66
Fax: +32 (0) 11 67 12 05
Email: hermoo@hermoo.be
Web: www.hermoo.com

ICL (Everris) Ltd: Everris Limited
Epsilon House
West Road
Ipswich
Suffolk
IP3 9FJ
Tel: 01473 237111
Fax: 01473 237150
Email: prof.sales@icl-group.com

Industrias Afrasa: Industrias Afrasa, S.A
Ciudad de Sevilla, 53
Pol. Industrial Fuenta del Jarro
46988 - Paterna
Valencia
Spain

Interagro: Interagro (UK) Ltd
Thorley Wash Barn
London Road
Thorley
Bishop's Stortford
Hertfordshire
CM23 4AT
Tel: (01279) 714970
Fax: (01279) 758227
Email: info@interagro.co.uk
Web: www.interagro.co.uk

Inter-Phyto: Inter-Phyto
1 Aps, Sct
Knuds Alle 39
9800 Hjoerring
Denmark

Intracrop: Intracrop
Park Chambers,
10 Hereford Road,
Abergavenny,
NP7 5PR
Tel: (01926) 634801
Fax: (01926) 634798
Email: admin@intracrop.co.uk
Web: www.intracrop.co.uk

Juno : Juno (Plant Protection) Ltd
Little Mill Farm
Underlyn Lane
Marden
Kent
TN12 9AT
Tel: 01580 765079
Email: nick@junopp.com

Killgerm: Killgerm Chemicals Ltd
115 Wakefield Road
Flushdyke
Ossett
W. Yorks.
WF5 9AR
Tel: (01924) 268400
Fax: (01924) 264757
Email: info@killgerm.com
Web: www.killgerm.com

Koppert: Koppert (UK) Ltd
Unit 8, Tudor Rose Court
53 Hollands Road
Haverhill
Suffolk
CB9 8PJ
Tel: (01440) 704488
Fax: (01440) 704487
Email: info@koppert.co.uk
Web: www.koppert.com

Landseer: Landseer Ltd
Lodge Farm
Goat Hall Lane
Galleywood
Chelmsford
Essex
CM2 8PH
Tel: (01245) 357109
Web: www.lanfruit.co.uk

Life Scientific: Life Scientific Limited
Block 4
Belfield Office Park
Beech Hill Road
Dublin 4
Ireland
Tel: 00353 1 283 2024
Web: www.lifescientific.com

Microcide: Microcide Ltd
Shepherds Grove
Stanton
Bury St. Edmunds
Suffolk
IP31 2AR
Tel: (01359) 251077
Fax: (01359) 251545
Email: microcide@microcide co.uk
Web: www.microcide.co.uk

Monsanto: Monsanto (UK) Ltd
PO Box 663
Cambourne
Cambridge
CB1 0LD
Tel: (01954) 717575
Fax: (01954) 717579
Email:
technical.helpline.uk@monsanto.com
Web: www.monsanto-ag.co.uk

SECTION 6

Nomix Enviro: Nomix Enviro Ltd.
The Grain Silos
Weyhill Road
Andover
Hampshire
SP10 3NT
Tel: 01264 388050
Fax: 01522 866176
Email: nomixenviro@frontierag.co.uk
Web: www.nomix.co.uk

Novastar: Novastar Link Limited
17 St Ann's Square,
Manchester,
M2 7PW
UK

NucleateB: NucleateB Limited
57/59 High Street
Dunblane
Perthshire
FK15 0EE

Nufarm UK: Nufarm UK Ltd
Wyke Lane
Wyke
West Yorkshire
BD12 9EJ
Tel: 01274 69 1234
Fax: 01274 69 1176
Email: infouk@uk.nufarm.com
Web: www.nufarm.co.uk

Nurture: Nurture Crop Chemicals Ltd
27 Old Gloucester Street
London
WC1N 3AX

Omex: Omex Agriculture Ltd
Bardney Airfield
Tupholme
Lincoln
LN3 5TP
Tel: (01526) 396000
Fax: (01526) 396001
Email: enquire@omex.com
Web: www.omex.co.uk

Pan Agriculture: Pan Agriculture Ltd
8 Cromwell Mews
Station Road
St Ives
Huntingdon
Cambs.
PE27 5HJ
Tel: (01480) 467790
Fax: (01480) 467041
Email: info@panagriculture.co.uk

Pan Amenity: Pan Amenity Ltd
8 Cromwell Mews
Station Road
St Ives
Cambs.
PE27 5HJ
Tel: 01480 467790
Fax: 01480 467041

Plant Solutions: Plant Solutions
Pyports
Downside Bridge Road
Cobham
Surrey
KT11 3EH
Tel: (01932) 576699
Fax: (01932) 868973
Email: sales@plantsolutionsltd.com
Web: www.plantsolutionsltd.com

Progreen:
Progreen Weed Control Solutions Limited
Kellington House
South Fen Business Park
South Fen Road
Bourne
Lincolnshire
PE10 0DN
Email: info@progreen.co.uk
Web: www.progreen.co.uk

ProKlass: ProKlass Products Limited
145-157 St John Street
London
EC1V 4PW
UK
Tel: 01480 810137
Email: office@proklass-products.com

Pure Amenity : Pure Amenity Limited
Pure House
64-66 Westwick Str.
Norwich
Norfolk
NR2 4SZ
Tel: 0845 257 4710
Email: martin@pureamenity.co.uk
Web: www.pureamenity.co.uk

Rentokil: Rentokil Initial 1927 plc
The Power Centre
Units A1 & A2
Link 10 Napier Way
Crawley
W. Sussex
RH10 9RA
Tel: 01293 858306
Email: dawn.kirby@rentokil-initial.com
Web: www.rentokil.com

Resource Chemicals:
Resource Chemicals Ltd
   Resource House
   76 High Street
   Brackley
   Northants
   NN13 7DS
   Tel: 01280 843800

Rigby Taylor: Rigby Taylor Ltd
   Rigby Taylor House
   Crown Lane
   Horwich
   Bolton
   Lancs.
   BL6 5HP
   Tel: (01204) 677777
   Fax: (01204) 677715
   Email: info@rigbytaylor.com
   Web: www.rigbytaylor.com

Rotam : Rotam Europe Ltd
   Hamilton House
   Mabledon Place
   London
   WC1H 9BB
   Tel: 0207 953 0447
   Web: www.rotam.com/uk

SFP Europe: SFP Europe SA
   11 Boulevard de la Grand Thumine
   Parc d'Ariane - Bat B
   13090 Aix en Provence
   France

Sherriff Amenity:
Sherriff Amenity Services
   The Pines
   Fordham Road
   Newmarket
   Cambs
   CB8 7LG
   Tel: (01638) 721888
   Fax: (01638) 721815
   Web: www.sherriffamenity.com

Sipcam: Sipcam UK Ltd
   4C Archway House
   The Lanterns
   Melbourn Street
   Royston
   Herts.
   SG8 7BX
   Tel: (01763) 212100
   Fax: (01763) 212101
   Email: stewart@sipcamuk.co.uk
   Web: sipcamuk.co.uk

Sphere: Sphere Laboratories (London) Ltd
   c/o Mainswood
   Putley Common
   Ledbury
   Herefordshire
   HR8 2RF
   Tel: 01684 899306
   Fax: 01684 893322
   Email:
   Fiona.Homes@Sphere-London.co.uk

Standon: Standon Chemicals Ltd
   48 Grosvenor Square
   London
   W1K 2HT
   Tel: (020) 7493 8648
   Fax: (020) 7493 4219

Stefes: Stefes GMBH
   Wendenstrasse 21b
   20097 Hamburg
   Germany
   Tel: +49 (0)40 533083-0
   Fax: +49 (0)40 5330833-29
   Email: info@stefes.eu
   Web: www.stefes.eu

Sumi Agro: Sumi Agro Europe Limited
   Vintners' Place
   68 Upper Thames Street
   London
   EC4V 3BJ
   Tel: (020) 7246 3697
   Fax: (020) 7246 3799
   Email: summit@summit-agro.com

Sumitomo :
Sumitomo Chemical Company Ltd
   Kinghams's Place
   36 Newgate Street
   Doddington
   Cambs.
   PE15 0SR
   Tel: (01354) 741414
   Fax: (01354) 741004
   Email: technical@interfarm.co.uk
   Web: www.interfarm.co.uk

Synergy : Synergy Generics Limited
   Market House
   Church Street
   Harlston
   Norfolk
   IP20 9BB
   Tel: 07884 435 282
   Email: thomas@agchemaccess.com

SECTION 6

Syngenta: Syngenta UK Limited
CPC4
Capital Park
Fulbourn
Cambridge
CB21 5XE
Tel: 01223 883400
Fax: (01223) 493700
Web: www.syngenta-crop.co.uk

Top Crop : Top Crop Chem Ltd
27 Old Gloucester Street
London
WC1N 3AX

Unique Marketing : Unique Marketing SL
Calle Marques Del Duero 67 Planta 1
Puerta A
San Pedro Alcantara,
29670 Marbella,
Malaga
Spain
Tel: +34 674 514 601

UPL Europe: UPL Europe Ltd
The Centre
Birchwood Park
Warrington
Cheshire
WA3 6YN
Tel: (01925) 819999
Fax: (01925) 856075
Email: pam.chambers@uniphos.com
Web: www.Uk.uplonline.com

Valto: Valto BV
Leehove 81
2678 MB De Lier
The Netherlands
Tel: +31 174 51 45 19
Email: info@valto.nl
Web: www.valto.nl

Ventura: Ventura Agroscience Ltd
20-22 Wenlock Road
London
N1 7GU
Tel: 01480 810137
Email: office@ventura-agroscience.com

Zantra: Zantra Limited
Westwood
Rattar Mains
Scarfskerry
Thurso
Caithness
KW14 8XW
Tel: 01480 861066
Fax: 01480 861099

Appendix 2
Useful Contacts

Agriculture Industries Confederation Ltd
Confederation House
East of England Showground
Peterborough PE2 6XE
Tel: (01733) 385230
Web: www.agindustries.org.uk

AHDB Horticulture
Stoneleigh Park
Kenilworth
Warwickshire CV8 2TL
Tel: (024) 7669 2051
Web: www.horticulture.ahdb.org.uk

BASIS Ltd
St Monica's House Business Centre
39 Windmill Lane
Ashbourne
Derbyshire DE6 1EY
Tel: (01335) 343945
Fax: (01335) 301205
Web: www.basis-reg.co.uk

British Beekeepers' Association
National Beekeeping Centre
Stoneleigh
Kenilworth
Warwickshire CV8 2LG
Tel: (0871) 811 2282
Web: www.bbka.org.uk

British Beer & Pub Association
Ground Floor
Brewers' Hall
Aldermanbury Square
EC2V 7HR
Tel: (0207) 627 9191
Web: www.beerandpub.com

British Crop Production Council (BCPC)
Garden Studio
4 Hillside
Aldershot
Surrey GU11 3NB
Tel: (01252) 285223
Web: www.bcpc.org

British Pest Control Association (BPCA)
4A Mallard Way
Pride Park
Derby DE24 8GX
Tel: (01332) 294288
Fax: (01332) 225101
Web: www.bpca.org.uk

Chemicals Regulation Division
Room 1A, Mallard House
King's Pool
3 Peasholme Green
York YO1 7PX
Tel: (01904) 640500 or (03459) 335577
Fax: (01904) 455733
Web: www.hse.gov.uk/crd

Crop Protection Association Ltd
2 Swan Court
Cygnet Park
Hampton
Peterborough PE7 8GX
Tel: (01733) 355370
Web: www.cropprotection.org.uk

CropLife International
326 Avenue Louise Box 35
B-1050 Brussels
Belgium
Tel: (+32) 2 542 0410
Fax: (+32) 2 542 0419
Web: www.croplife.org

Department of Agriculture and Rural Development (Northern Ireland)
Pesticides Section
Dundonald House
Upper Newtownards Road
Belfast BT4 3SB
Tel: (028) 9052 4704 or (0300) 200 7850
Fax: (028) 9052 4059
Web: www.daera-ni.gov.uk

Department of Environment, Food and Rural Affairs (Defra)
Nobel House
17 Smith Square
London SW1P 3JR
Tel: (020) 7238 6000 / 03459 335577
Web: www.defra.gov.uk

Environment Agency
National Customer Contact Centre
PO Box 544
Rotherham S60 1BY
Tel: (03708) 506 506
Email: enquiries@environment-agency.gov.uk
Web: www.environment-agency.gov.uk

SECTION 6

European Crop Protection Association (ECPA)
Avenue E van Nieuwenhuyse 6
B-1160 Brussels
Belgium
Tel: (+32) 2 663 1550
Fax: (+32) 2 663 1560
Web: www.ecpa.eu

European Chemicals Agency (ECHA)
PO Box 400
00121 Helsinki
Finland
Tel: +358 9 686180
Web: www.echa.europa.eu

Farmers' Union of Wales
Llys Amaeth
Plas Gogerddan
Aberystwyth
Ceredigion SY23 3BT
Tel: (01970) 820820
Fax:(01970) 820821
Web: www.fuw.org.uk

Forestry Commission
620 Bristol Business Pk
Coldharbour Lane
Bristol BS16 1EJ
Tel: (0300) 067 4000
Web: www.forestry.gov.uk

Game and Wildlife Conservation Trust (GWCT)
Head Office
Burgate Manor
Fordingbridge
Hampshire
SP6 1EF
Tel: 01425 652381
Email: info@gwct.org.uk
Web: www.gwct.org.uk

Health and Safety Executive
Product Approvals
Chemical Regulation
Redgrave Court
Merton Road
Bootle
Merseyside L20 7HS
Tel: (0151) 951 4000
Web: www.hse.gov.uk

Health and Safety Executive – Books
TSO Customer Services
PO Box 29
Norwich NR3 1GN

Lantra
National Agricultural Centre
Lantra House
Stoneleigh
Kenilworth
Warwickshire CV8 2LG
Tel: (024) 7669 6996
Fax: (024) 7669 6732
Web: www.lantra.co.uk

National Association of Agricultural Contractors (NAAC)
The Old Cart Shed
Easton Lodge Farm
Old Oundle Road, Wansford
Peterborough PE8 6NP
Tel: (01780) 784631
Fax: (01780) 784933
Web: www.naac.co.uk

National Chemicals Emergency Centre
The Gemini Building
Fermi Avenue
Harwell
Didcot
Oxfordshire OX11 0RG
Tel: (01235) 753654
Fax: (01235) 753656
Email: ncec@ricardo-aea.com
Web: www.the-ncec.com

National Farmers' Union
Agriculture House
Stoneleigh Park Stoneleigh
Warwickshire CV8 2TZ
Tel: (024) 7685 8500
Web: www.nfu.org.uk

National Poisons Information Service
(Birmingham Unit)
City Hospital, Dudley Road
Birmingham B18 7QH
Tel: (0121) 507 4123
Fax: (0121) 507 5580
Web: www.npis.org

Natural England (enquiries)
County Hall
Spetchley Road
Worcester
WR5 2NP
Tel: 0300 060 3900
Suspected wildlife/pesticide poisoning
Tel: 0800 321 600
Email: enquiries@naturalengland.org.uk
Web: www.gov.uk/government/organisations/natural-england

NIAB

Huntingdon Road
Cambridge, Cambs
CB3 0LE
Tel: (01223) 342200
Email: info@niab.com
Web: www.niab.com

Processors and Growers Research Organisation (PGRO)

The Research Station
Great North Road, Thornhaugh
Peterborough, Cambs. PE8 6HJ
Tel: (01780) 782585
Web: www.pgro.org

Scottish Beekeepers' Association

Cowiemuir
Fochabers
Moray IV32 7PS
Email: secretary@scottishbeekeepers.
org.uk
Web: www.scottishbeekeepers.org.uk

Scottish Environmental Protection Agency

Edinburgh Office, Silvan House
SEPA 3rd Floor
231 Corstorphine Road
Edinburgh EH12 7AT
Tel: (0131) 449 7296

TSO (The Stationary Office)

18 Central Avenue
St Andrews Business Park
Norwich NR7 0HR
Tel: (0333) 202 5070
Email: support@tso.co.uk
Web: www.tsoshop.co.uk

SECTION 6

Appendix 3
Keys to Crop and Weed Growth Stages

Decimal Code for the Growth Stages of Cereals
Illustrations of these growth stages can be found in the reference indicated below and in some company product manuals.

0 Germination

00	Dryseed
03	Imbibition complete
05	Radicle emerged from caryopsis
07	Coleoptile emerged from caryopsis
09	Leaf at coleoptile tip

1 Seedling growth

10	First leaf through coleoptile
11	First leaf unfolded
12	2 leaves unfolded
13	3 leaves unfolded
14	4 leaves unfolded
15	5 leaves unfolded
16	6 leaves unfolded
17	7 leaves unfolded
18	8 leaves unfolded
19	9 or more leaves unfolded

2 Tillering

20	Main shoot only
21	Main shoot and 1 tiller
22	Main shoot and 2 tillers
23	Main shoot and 3 tillers
24	Main shoot and 4 tillers
25	Main shoot and 5 tillers
26	Main shoot and 6 tillers
27	Main shoot and 7 tillers
28	Main shoot and 8 tillers
29	Main shoot and 9 or more tillers

3 Stem elongation

30	Ear at 1 cm
31	1st node detectable
32	2nd node detectable
33	3rd node detectable
34	4th node detectable
35	5th node detectable
36	6th node detectable
37	Flag leaf just visible
39	Flag leaf ligule/collar just visible

4 Booting

41	Flag leaf sheath extending
43	Boots just visibly swollen
45	Boots swollen
47	Flag leaf sheath opening
49	First awns visible

5 Inflorescence

51	First spikelet of inflorescence just visible
52	1/4 of inflorescence emerged
55	1/2 of inflorescence emerged
57	3/4 of inflorescence emerged
59	Emergence of inflorescence completed

6 Anthesis

60	
61	Beginning of anthesis
64	
65	Anthesis half way
68	
69	Anthesis complete

7 Milk development

71	Caryopsis watery ripe
73	Early milk
75	Medium milk
77	Late milk

8 Dough development

83	Early dough
85	Soft dough
87	Hard dough

9 Ripening

91	Caryopsis hard (difficult to divide by thumb-nail)
92	Caryopsis hard (can no longer be dented by thumb-nail)
93	Caryopsis loosening in daytime

(From Tottman, 1987. *Annals of Applied Biology*, **110**, 441–454)

Stages in Development of Oilseed Rape

Illustrations of these growth stages can be found in the reference indicated below and in some company product manuals.

0 Germination and emergence

1 Leaf production

1,0   Both cotyledons unfolded and green
1,1   First true leaf
1,2   Second true leaf
1,3   Third true leaf
1,4   Fourth true leaf
1,5   Fifth true leaf
1,10 About tenth true leaf
1,15 About fifteenth true leaf

2 Stem extension

2,0   No internodes ('rosette')
2,5   About five internodes

3 Flower bud development

3,0   Only leaf buds present
3,1   Flower buds present but enclosed by leaves
3,3   Flower buds visible from above ('green bud')
3,5   Flower buds raised above leaves
3,6   First flower stalks extending
3,7   First flower buds yellow ('yellow bud')

4 Flowering

4,0   First flower opened
4,1   10% all buds opened
4,3   30% all buds opened
4,5   50% all buds opened

5 Pod development

5,3   30% potential pods
5,5   50% potential pods
5,7   70% potential pods
5,9   All potential pods

6 Seed development

6,1   Seeds expanding
6,2   Most seeds translucent but full size
6,3   Most seeds green
6,4   Most seeds green-brown mottled
6,5   Most seeds brown
6,6   Most seeds dark brown
6,7   Most seeds black but soft
6,8   Most seeds black and hard
6,9   All seeds black and hard

7 Leaf senescence

8 Stem senescence

8,1   Most stem green
8,5   Half stem green
8,9   Little stem green

9 Pod senescence

9,1   Most pods green
9,5   Half pods green
9,9   Few pods green

(From Sylvester-Bradley, 1985. *Aspects of Applied Biology*, **10**, 395–400)

SECTION 6

Stages in Development of Peas

Illustrations of these growth stages can be found in the reference indicated below and in some company product manuals.

0 Germination and emergence

000 Dry seed
001 Imbibed seed
002 Radicle apparent
003 Plumule and radicle apparent
004 Emergence

1 Vegetative stage

101 First node (leaf with one pair leaflets, no tendril)
102 Second node (leaf with one pair leaflets, simple tendril)
103 Third node (leaf with one pair leaflets, complex tendril)
•
l0x X nodes (leaf with more than one pair leaflets, complex tendril)
•
•
10n Last recorded node

2 Reproductive stage (main stem)

201 Enclosed buds
202 Visible buds
203 First open flower
204 Pod set (small immature pod)
205 Flat pod
206 Pod swell (seeds small, immature)
207 Podfill
208 Pod green, wrinkled
209 Pod yellow, wrinkled (seeds rubbery)
210 Dry seed

3 Senescence stage

301 Desiccant application stage. Lower pods dry and brown, middle yellow, upper green. Overall moisture content of seed less than 45%
302 Pre-harvest stage. Lower and middle pods dry and brown, upper yellow. Overall moisture content of seed less than 30%
303 Dry harvest stage. All pods dry and brown, seed dry

(From Knott, 1987. *Annals of Applied Biology*, **111**, 233–244)

Stages in Development of Faba Beans

Illustrations of these growth stages can be found in the reference indicated below and in some company product manuals.

0   Germination and emergence

000   Dry seed
001   Imbibed seed
002   Radicle apparent
003   Plumule and radicle apparent
004   Emergence
005   First leaf unfolding
006   First leaf unfolded

1   Vegetative stage

101   First node
102   Second node
103   Third node
•
•
l0x   X nodes
•
•
10n   N, last recorded node

2   Reproductive stage (main stem)

201   Flower buds visible
203   First open flowers
204   First pod set
205   Pods fully formed, green
207   Pod fill, pods green
209   Seed rubbery, pods pliable, turning black
210   Seed dry and hard, pods dry and black

3   Pod senescence

301   10% pods dry and black
•
•
305   50% pods dry and black
•
•
308   80% pods dry and black, some upper pods green
309   90% pods dry and black, most seed dry. Desiccation stage.
310   All pods dry and black, seed hard. Pre-harvest (glyphosate application stage)

4   Stem senescence

401   10% stem brown/black
•
•
405   50% stem brown/black
•
•
409   90% stem brown/black
410   All stems brown/black. All pods dry and black, seed hard.

(From Knott, 1990. *Annals of Applied Biology,* **116**, 391–404)

Stages in Development of Potato

Illustrations of these growth stages can be found in the reference indicated below and in some company product manuals.

0   Seed germination and seedling emergence

000   Dry seed
001   Imbibed seed
002   Radicle apparent
003   Elongation of hypocotyl
004   Seedling emergence
005   Cotyledons unfolded

1   Tuber dormancy

100   Innate dormancy (no sprout development under favourable conditions)
150   Enforced dormancy (sprout development inhibited by environmental conditions)

2   Tuber sprouting

200   Dormancy break, sprout development visible
21x   Sprout with 1 node
22x   Sprout with 2 nodes
•
•
29x   Sprout with 9 nodes
21x(2) Second generation sprout with 1 node
22x(2) Second generation sprout with 2 nodes
•
•
29x(2) Second generation sprout with 9 nodes

Where x = 1, sprout >2 mm;
2, 2-5 mm; 3, 5-20 mm;
4, 20-30 mm; 5, 50-100 mm;
6, 100-150 mm long

3   Emergence and shoot expansion

300   Main stem emergence
301   Node 1
302   Node 2
•
•
319   Node 19
Second order branch
321   Node 1
•
•
Nth   order branch
3N1   Node 1
•
•
3N9   Node 9

4   Flowering

Primary flower
400   No flowers
410   Appearance of flower bud
420   Flower unopen
430   Flower open
440   Flower closed
450   Berry swelling
460   Mature berry
Second order flowers
410(2) Appearance of flower bud
420(2) Flower unopen
430(2) Flower open
440(2) Flower closed
450(2) Berry swelling
460(2) Mature berry

5   Tuber development

500   No stolons
510   Stolon initials
520   Stolon elongation
530   Tuber initiation
540   Tuber bulking (>10 mm diam)
550   Skin set
560   Stolon development

6   Senescence

600   Onset of yellowing
650   Half leaves yellow
670   Yellowing of stems
690   Completely dead

(From Jefferies & Lawson, 1991. *Annals of Applied Biology,* **119**, 387–389)

Stages in Development of Linseed

Illustrations of these growth stages can be found in the reference indicated below and in some company product manuals.

0   Germination and emergence

00	Dry seed
01	Imbibed seed
02	Radicle apparent
04	Hypocotyl extending
05	Emergence
07	Cotyledon unfolding from seed case
09	Cotyledons unfolded and fully expanded

1   Vegetative stage (of main stem)

10	True leaves visible
12	First pair of true leaves fully expanded
13	Third pair of true leaves fully expanded
$1n$	n leaf fully expanded

2   Basal branching

21	One branch
22	Two branches
23	Three branches
$2n$	n branches

3   Flower bud development (on main stem)

31	Enclosed bud visible in leaf axils
33	Bud extending from axil
35	Corymb formed
37	Buds enclosed but petals visible
39	First flower open

4   Flowering (whole plant)

41	10% of flowers open
43	30% of flowers open
45	50% of flowers open
49	End of flowering

5   Capsule formation (whole plant)

51	10% of capsules formed
53	30% Of capsules formed
55	50% of capsules formed
59	End of capsule formation

6   Capsule senescence (on most advanced plant)

61	Capsules expanding
63	Capsules green and full size
65	Capsules turning yellow
67	Capsules all yellow brown but soft
69	Capsules brown, dry and senesced

7   Stem senescence (whole plant)

71	Stems mostly green below panicle
73	Most stems 30% brown
75	Most stems 50% brown
77	Stems 75% brown
79	Stems completely brown

8   Stems rotting (retting)

81	Outer tissue rotting
85	Vascular tissue easily removed
89	Stems completely collapsed

9   Seed development (whole plant)

91	Seeds expanding
92	Seeds white but full size
93	Most seeds turning ivory yellow
94	Most seeds turning brown
95	All seeds brown and hard
98	Some seeds shed from capsule
99	Most seeds shed from capsule

(From Jefferies & Lawson, 1991. *(From Freer, 1991. Aspects of Applied Biology,* **28***, 33–40)*

SECTION 6

Stages in Development of Annual Grass Weeds

Illustrations of these growth stages can be found in the reference indicated below and in some company product manuals.

0 Germination and emergence

00 Dry seed
01 Start of imbibition
03 Imbibition complete
05 Radicle emerged from caryopsis
07 Coleoptile emerged from caryopsis
09 Leaf just at coleoptile tip

1 Seedling growth

10 First leaf through coleoptile
11 First leaf unfolded
12 2 leaves unfolded
13 3 leaves unfolded
14 4 leaves unfolded
15 5 leaves unfolded
16 6 leaves unfolded
17 7 leaves unfolded
18 8 leaves unfolded
19 9 or more leaves unfolded

2 Tillering

20 Main shoot only
21 Main shoot and 1 tiller
22 Main shoot and 2 tillers
23 Main shoot and 3 tillers
24 Main shoot and 4 tillers
25 Main shoot and 5 tillers
26 Main shoot and 6 tillers
27 Main shoot and 7 tillers
28 Main shoot and 8 tillers
29 Main shoot and 9 or more tillers

3 Stem elongation

31 First node detectable
32 2nd node detectable
33 3rd node detectable
34 4th node detectable
35 5th node detectable
36 6th node detectable
37 Flag leaf just visible
39 Flag leaf ligule just visible

4 Booting

41 Flag leaf sheath extending
43 Boots just visibly swollen
45 Boots swollen
47 Flag leaf sheath opening
49 First awns visible

5 Inflorescence emergence

51 First spikelet of inflorescence just visible
53 1/4 of inflorescence emerged
55 1/2 of inflorescence emerged
57 3/4 of inflorescence emerged
59 Emergence of inflorescence completed

6 Anthesis

61 Beginning of anthesis
65 Anthesis half-way
69 Anthesis complete

(From Lawson & Read, 1992. *Annals of Applied Biology,* **12**, 211–214)

Growth Stages of Annual Broad-leaved Weeds

Preferred Descriptive Phrases

Illustrations of these growth stages can be found in the reference indicated below and in some company product manuals.

Pre-emergence
Early cotyledons
Expanded cotyledons
One expanded true leaf
Two expanded true leaves
Four expanded true leaves
Six expanded true leaves
Plants up to 25 mm across/high

Plants up to 50 mm across/high
Plants up to 100 mm across/high
Plants up to 150 mm across/high
Plants up to 250 mm across/high
Flower buds visible
Plant flowering
Plant senescent

(From Lutman & Tucker, 1987. *Annals of Applied Biology*, **110**, 683–687)

SECTION 6

Appendix 4
Key to Hazard Classifications and Safety Precautions

Every product label contains information to warn users of the risks from using the product, together with precautions that must be followed in order to minimise the risks. A hazard classification (if any) and the symbol are shown with associated risk phrases, followed by a series of safety numbers under the heading **Hazard classification and safety precautions**.

The codes are defined below, under the same sub-headings as they appear in the pesticide profiles.

Where a product label specifies the use of personal protective equipment (PPE), the requirements are listed under the sub-heading **Operator protection**, using letter codes to denote the protective items, e.g. handling the concentrate, cleaning equipment etc., but it is not possible to list them separately. The lists of PPE are therefore an indication of what the user may need to have available to use the product in different ways. **When making a COSHH assessment it is therefore essential that the product label is consulted for information on the particular use that is being assessed.**

Where the generalised wording includes a phrase such as '... for xx days', the specific requirement for each pesticide is shown in brackets after the code.

Hazard
H01	Very toxic
H02	Toxic
H03	Harmful
H04	Irritant
H05	Corrosive
H06	Extremely flammable
H07	Highly flammable
H08	Flammable
H09	Oxidising agent
H10	Explosive
H11	Dangerous for the environment
H220	Extremely flammable gas
H225	Highly flammable liquid and vapour
H226	Flammable liquid and vapour
H228	Flammable solid
H260	In contact with water releases flammable gases which may ignite spontaneously
H300	Fatal if swallowed
H301	Toxic if swallowed
H302	Harmful if swallowed
H311	Toxic in contact with skin
H330	Fatal if inhaled
H331	Toxic if inhaled
H332	Harmful if inhaled
H400	Very toxic to aquatic organisms
H401	Toxic to aquatic organisms

Risk phrases
H280	Contains gas under pressure; may explode if heated
H290	May be corrosive to metals
H304	May be fatal if swallowed and enters airways
H310	Fatal in contact with skin
H312	Harmful in contact with skin

H314	Causes severe skin burns and eye damage
H315	Causes skin irritation
H317	May cause an allergic skin reaction
H318	Causes serious eye damage
H319	Causes serious eye irritation
H320	Causes eye irritation
H334	May cause allergy or asthma symptoms or breathing difficulties if inhaled
H335	May cause respiratory irritation
H336	May cause drowsiness or dizziness
H340	May cause genetic defects
H341	Suspected of causing genetic defects
H351	Suspected of causing cancer
H360	May damage fertility or the unborn child
H361	Suspected of damaging fertility or the unborn child
H370	Causes damage to organs
H371	May cause damage to organs
H372	Causes damage to organs through prolonged or repeated exposure
H373	May cause damage to organs through prolonged or repeated exposure
R08	Contact with combustible material may cause fire
R09	Explosive when mixed with combustible material
R15	Contact with water liberates extremely flammable gases
R16	Explosive when mixed with oxidising substances
R19	Fatal if swallowed
R20	Harmful by inhalation
R21	Harmful in contact with skin
R22a	Harmful if swallowed
R22b	May cause lung damage if swallowed
R22c	May be fatal if swallowed and enters airways
R23a	Toxic by inhalation
R23b	Fatal if inhaled
R24	Toxic in contact with skin
R25	Toxic if swallowed
R25b	Toxic; Danger of serious damage to health by prolonged exposure if swallowed
R26	Very toxic by inhalation
R27	Very toxic in contact with skin
R28	Very toxic if swallowed
R31	Contact with acids liberates toxic gas
R34	Causes burns
R35	Causes severe burns
R36	Irritating to eyes
R37	Irritating to respiratory system
R38	Irritating to skin
R39	Danger of very serious irreversible effects
R40	Limited evidence of a carcinogenic effect
R40b	Suspected of causing cancer
R41	Risk of serious damage to eyes
R42	May cause sensitization by inhalation
R43	May cause sensitization by skin contact
R45	May cause cancer
R46	May cause heritable genetic damage
R48	Danger of serious damage to health by prolonged exposure
R50	Very toxic to aquatic organisms
R51	Toxic to aquatic organisms
R52	Harmful to aquatic organisms
R53a	May cause long-term adverse effects in the aquatic environment
R53b	Dangerous to aquatic organisms
R54	Toxic to flora
R55	Toxic to fauna
R56	Toxic to soil organisms
R57	Toxic to bees
R58	May cause long-term adverse effects in the environment

SECTION 6

R60	May impair fertility
R60b	May cause heritable genetic damage
R61	May cause harm to the unborn child
R62	Possible risk of impaired fertility
R63	Possible risk of harm to the unborn child
R64	May cause harm to breast-fed babies
R65	Harmful: May cause lung damage if swallowed
R66	Repeated exposure may cause skin dryness or cracking
R66b	May cause damage to organs through prolonged or repeated exposure
R67	Vapours may cause drowsiness and dizziness
R68	Possible risk of irreversible effects
R69	Danger of serious damage to health by prolonged oral exposure.
R70	May produce an allergic reaction

Operator protection

A	Suitable protective gloves (the product label should be consulted for any specific requirements about the material of which the gloves should be made)
B	Rubber gauntlet gloves
C	Face-shield
D	Approved respiratory protective equipment
E	Goggles
F	Dust mask
G	Full face-piece respirator
H	Coverall
J	Hood
K	Apron/Rubber apron
L	Waterproof coat
M	Rubber boots
N	Waterproof jacket and trousers
P	Suitable protective clothing
U01	To be used only by operators instructed or trained in the use of chemical/product/type of produce and familiar with the precautionary measures to be observed
U02a	Wash all protective clothing thoroughly after use, especially the inside of gloves
U02b	Avoid excessive contamination of coveralls and launder regularly
U02c	Remove and wash contaminated gloves immediately
U03	Wash splashes off gloves immediately
U04a	Take off immediately all contaminated clothing
U04b	Take off immediately all contaminated clothing and wash underlying skin. Wash clothes before re-use
U04c	Wash clothes before re-use
U05a	When using do not eat, drink or smoke
U05b	When using do not eat, drink, smoke or use naked lights
U06	Handle with care and mix only in a closed container
U07	Open the container only as directed (returnable containers only)
U08	Wash concentrate/dust from skin or eyes immediately
U09a	Wash any contamination/splashes/dust/powder/concentrate from skin or eyes immediately
U09b	Wash any contamination/splashes/dust/powder/concentrate from eyes immediately
U09c	If on skin, wash with plenty of soap and water
U10	After contact with skin or eyes wash immediately with plenty of water
U11	In case of contact with eyes rinse immediately with plenty of water and seek medical advice
U12	In case of contact with skin rinse immediately with plenty of water and seek medical advice
U12b	After contact with skin, take off immediately all contaminated clothing and wash immediately with plenty of water
U13	Avoid contact by mouth
U14	Avoid contact with skin
U15	Avoid contact with eyes
U16a	Ensure adequate ventilation in confined spaces
U16b	Use in a well ventilated area

U18	Extinguish all naked flames, including pilot lights, when applying the fumigant/dust/ liquid/product
U19a	Do not breathe dust/fog/fumes/gas/smoke/spray mist/vapour. Avoid working in spray mist
U19b	Do not work in confined spaces or enter spaces in which high concentrations of vapour are present. Where this precaution cannot be observed distance breathing or self-contained breathing apparatus must be worn, and the work should be done by trained operators
U19c	In case of insufficient ventilation, wear suitable respiratory equipment
U19d	Wear suitable respiratory equipment during bagging and stacking of treated seed
U19e	During fumigation/spraying wear suitable respiratory equipment
U19f	In case of accident by inhalation: remove casualty to fresh air and keep at rest
U20a	Wash hands and exposed skin before eating, drinking or smoking and after work
U20b	Wash hands and exposed skin before eating and drinking and after work
U20c	Wash hands before eating and drinking and after work
U20d	Wash hands after use
U20e	Wash hands and exposed skin after cleaning and re-calibrating equipment
U21	Before entering treated crops, cover exposed skin areas, particularly arms and legs
U22a	Do not touch sachet with wet hands or gloves/Do not touch water soluble bag directly
U22b	Protect sachets from rain or water
U23a	Do not apply by knapsack sprayer/hand-held equipment
U23b	Do not apply through hand held rotary atomisers
U23c	Do not apply via tractor mounted horizontal boom sprayers
U24	Do not handle grain unnecessarily
U25	Open the container only as directed
U26	Keep unprotected workers out of treated areas for at least 5 days after treatment.

Environmental protection

E02a	Keep unprotected persons/animals out of treated/fumigation areas for at least xx hours/days
E02b	Prevent access by livestock, pets and other non-target mammals and birds to buildings under fumigation and ventilation
E02c	Vacate treatment areas before application
E02d	Exclude all persons and animals during treatment
E03	Label treated seed with the appropriate precautions, using the printed sacks, labels or bag tags supplied
E05a	Do not apply directly to livestock/poultry
E05b	Keep poultry out of treated areas for at least xx days/weeks
E05c	Do not apply directly to animals
E06a	Keep livestock out of treated areas for at least xx days/weeks after treatment
E06b	Dangerous to livestock. Keep all livestock out of treated areas/away from treated water for at least xx days/weeks. Bury or remove spillages
E06c	Harmful to livestock. Keep all livestock out of treated areas/away from treated water for at least xx days/weeks. Bury or remove spillages
E06d	Exclude livestock from treated fields. Livestock may not graze or be fed treated forage nor may it be used for hay, silage or bedding
E07a	Keep livestock out of treated areas for up to two weeks following treatment and until poisonous weeds, such as ragwort, have died down and become unpalatable
E07b	Dangerous to livestock. Keep livestock out of treated areas/away from treated water for at least xx weeks and until foliage of any poisonous weeds, such as ragwort, has died and become unpalatable
E07c	Harmful to livestock. Keep livestock out of treated areas/away from treated water for at least xx days/weeks and until foliage of any poisonous weeds such as ragwort has died and become unpalatable
E07d	Keep livestock out of treated areas for up to 4-6 weeks following treatment and until poisonous weeds, such as ragwort, have died down and become unpalatable
E07e	Do not take grass crops for hay or silage for at least 21 days after application
E07f	Keep livestock out of treated areas for up to 3 days following treatment and until poisonous weeds, such as ragwort, have died
E07g	Keep livestock out of treated areas for at least 50 days following treatment
E08a	Do not feed treated straw or haulm to livestock within xx days/weeks of spraying

E08b   Do not use on grassland if the crop is to be used as animal feed or bedding

E09   Do not use straw or haulm from treated crops as animal feed or bedding for at least xx days after last application

E10a   Dangerous to game, wild birds and animals

E10b   Harmful to game, wild birds and animals

E10c   Dangerous to game, wild birds and animals. All spillages must be buried or removed

E11   Paraquat can be harmful to hares; spray stubbles early in the day

E12a   High risk to bees

E12b   Extremely dangerous to bees

E12c   Dangerous to bees

E12d   Harmful to bees

E12e   Do not apply to crops in flower or to those in which bees are actively foraging. Do not apply when flowering weeds are present

E12f   Do not apply to crops in flower, or to those in which bees are actively foraging, except as directed on [crop]. Do not apply when flowering weeds are present

E12g   Apply away from bees

E13a   Extremely dangerous to fish or other aquatic life. Do not contaminate surface waters or ditches with chemical or used container

E13b   Dangerous to fish or other aquatic life. Do not contaminate surface waters or ditches with chemical or used container

E13c   Harmful to fish or other aquatic life. Do not contaminate surface waters or ditches with chemical or used container

E13d   Apply away from fish

E13e   Harmful to fish or other aquatic life. The maximum concentration of active ingredient in treated water must not exceed XX ppm or such lower concentration as the appropriate water regulatory body may require.

E14a   Extremely dangerous to aquatic higher plants. Do not contaminate surface waters or ditches with chemical or used container

E14b   Dangerous to aquatic higher plants. Do not contaminate surface waters or ditches with chemical or used container

E15a   Do not contaminate surface waters or ditches with chemical or used container

E15b   Do not contaminate water with product or its container. Do not clean application equipment near surface water. Avoid contamination via drains from farmyards or roads

E15c   To protect groundwater, do not apply to grass leys less than 1 year old

E16a   Do not allow direct spray from horizontal boom sprayers to fall within 5 m of the top of the bank of a static or flowing waterbody, unless a Local Environment Risk Assessment for Pesticides (LERAP) permits a narrower buffer zone, or within 1 m of the top of a ditch which is dry at the time of application. Aim spray away from water

E16b   Do not allow direct spray from hand-held sprayers to fall within 1 m of the top of the bank of a static or flowing waterbody. Aim spray away from water

E16c   Do not allow direct spray from horizontal boom sprayers to fall within 5 m of the top of the bank of a static or flowing waterbody, or within 1m of the top of a ditch which is dry at the time of application. Aim spray away from water. This product is not eligible for buffer zone reduction under the LERAP horizontal boom sprayers scheme.

E16d   Do not allow direct spray from hand-held sprayers to fall within 1 m of the top of the bank of a static or flowing waterbody. Aim spray away from water. This product is not eligible for buffer zone reduction under the LERAP horizontal boom sprayers scheme.

E16e   Do not allow direct spray from horizontal boom sprayers to fall within 5 m of the top of the bank of a static or flowing water body or within 1 m from the top of any ditch which is dry at the time of application. Spray from hand held sprayers must not in any case be allowed to fall within 1 m of the top of the bank of a static or flowing water body. Always direct spray away from water. The LERAP scheme does not extend to adjuvants. This product is therefore not eligible for a reduced buffer zone under the LERAP scheme.

E16f   Do not allow direct spray/granule applications from vehicle mounted/drawn hydraulic sprayers/applicators to fall within 6 m of surface waters or ditches/Do not allow direct spray/granule applications from hand-held sprayers/applicators to fall within 2 m of surface waters or ditches. Direct spray/applications away from water

E16g	Do not allow direct spray from train sprayers to fall within 5 m of the top of the bank of any static or flowing waterbody.
E16h	Do not spray cereals after 31st March in the year of harvest within 6 metres of the edge of the growing crop
E16i	Do not allow direct spray from horizontal boom sprayers to fall within 12 metres of the top of the bank of a static or flowing water body.
E16j	Do not allow direct spray from horizontal boom sprayers to fall within the specified distance of the top of the bank of a static or flowing waterbody,
E17a	Do not allow direct spray from broadcast air-assisted sprayers to fall within xx m of surface waters or ditches. Direct spray away from water
E17b	Do not allow direct spray from broadcast air-assisted sprayers to fall within xx m of the top of the bank of a static or flowing waterbody, unless a Local Environmental Risk Assessment for Pesticides (LERAP) permits a narrower buffer zone, or within 5 m of the top of a ditch which is dry at the time of application. Aim spray away from water
E18	Do not spray from the air within 250 m horizontal distance of surface waters or ditches
E19a	Do not dump surplus herbicide in water or ditch bottoms
E19b	Do not empty into drains
E20	Prevent any surface run-off from entering storm drains
E21	Do not use treated water for irrigation purposes within xx days/weeks of treatment
E22a	High risk to non-target insects or other arthropods. Do not spray within 6 m of the field boundary
E22b	Risk to certain non-target insects or other arthropods. For advice on risk management and use in Integrated Pest Management (IPM) see directions for use
E22c	Risk to non-target insects or other arthropods
E23	Avoid damage by drift onto susceptible crops or water courses
E34	Do not re-use container for any purpose/Do not re-use container for any other purpose
E35	Do not burn this container
E36a	Do not rinse out container (returnable containers only)
E36b	Do not open or rinse out container (returnable containers only)
E37	Do not use with any pesticide which is to be applied in or near water
E38	Use appropriate containment to avoid environmental contamination
E39	Extreme care must be taken to avoid spray drift onto non-crop plants outside the target area
E40	To protect groundwater/soil organisms the maximum total dose of this or other products containing ethofumesate MUST NOT exceed 1.0 kg ethofumesate per hectare in any three year period
E40b	To protect aquatic organisms respect an unsprayed buffer zone to surface waters in line with LERAP requirements
E40c	This product must not be used with any pesticide to be applied in or near water
E40d	To protect non-target arthropods respect an untreated buffer zone of 5m to non crop land
E40e	To protect aquatic organisms respect an unsprayed buffer zone of 20 metres to surface water bodies
E41	Hay for silage must not be cut from treated crops for at least 21 days after treatment
H410	Very toxic to aquatic life with long-lasting effects
H411	Toxic to aquatic life with long lasting effects
H412	Harmful to aquatic life with long lasting effects
H413	May cause long lasting harmful effects to aquatic life

Consumer protection

C01	Do not use on food crops
C02a	Do not harvest for human or animal consumption for at least xx days/weeks after last application
C02b	Do not remove from store for sale or processing for at least 21 days after application
C02c	Do not remove from store for sale or processing for at least 2 days after application
C04	Do not apply to surfaces on which food/feed is stored, prepared or eaten
C05	Remove/cover all foodstuffs before application
C06	Remove exposed milk before application

SECTION 6

C07	Collect eggs before application
C08	Protect food preparing equipment and eating utensils from contamination during application
C09	Cover water storage tanks before application
C10	Protect exposed water/feed/milk machinery/milk containers from contamination
C11	Remove all pets/livestock/fish tanks before treatment/spraying
C12	Ventilate treated areas thoroughly when smoke has cleared/Ventilate treated rooms thoroughly before occupying

Storage and disposal

D01	Keep out of reach of children
D02	Keep away from food, drink and animal feeding-stuffs
D03	Store away from seeds, fertilizers, fungicides and insecticides
D04	Store well away from corms, bulbs, tubers and seeds
D05	Protect from frost
D06a	Store away from heat
D06b	Do not store near heat or open flame
D06c	Do not store in direct sunlight
D06d	Do not store above 30/35 °C
D06e	Store in a well-ventilated place. Keep container tightly closed
D06f	Keep away from sources of ignition - No smoking
D06g	Take precautionary measures against static discharges
D06h	Store at 10 - 25 °C
D07	Store under cool, dry conditions
D08	Store in a safe, dry, frost-free place designated as an agrochemical store
D09a	Keep in original container, tightly closed, in a safe place
D09b	Keep in original container, tightly closed, in a safe place, under lock and key
D09c	Store unused sachets in a safe place. Do not store half-used sachets
D10a	Wash out container thoroughly and dispose of safely
D10b	Wash out container thoroughly, empty washings into spray tank and dispose of safely
D10c	Rinse container thoroughly by using an integrated pressure rinsing device or manually rinsing three times. Add washings to sprayer at time of filling and dispose of container safely
D10d	Do not rinse out container
D11a	Empty container completely and dispose of safely/Dispose of used generator safely
D11b	Empty container completely and dispose of it in the specified manner
D11c	Ventilate empty containers until no phosphine is detected and then dispose of as hazardous waste via an authorised waste-disposal contractor
D12a	This material (and its container) must be disposed of in a safe way
D12b	This material and its container must be disposed of as hazardous waste
D12c	Dispose of contents/container to a licensed hazardous waste disposal contractor or collection site except for empty, clean containers which can be disposed of normally
D13	Treat used container as if it contained pesticide
D14	Return empty container as instructed by supplier (returnable containers only)
D15	Store container in purpose built chemical store until returned to supplier for refilling (returnable containers only)
D16	Do not store below 4 degrees Centigrade
D17	Use immediately on removal of foil
D18	Place the tablets whole into the spray tank - do not break or crumble the tablets
D19	Do not empty into drains
D20	Clean all equipment after use
D21	Open the container only as directed
D22	Collect spillage

Treated Seed

S01	Do not handle treated seed unnecessarily
S02	Do not use treated seed as food or feed
S03	Keep treated seed secure from people, domestic stock/pets and wildlife at all times during storage and use
S04a	Bury or remove spillages

S04b	Harmful to birds/game and wildlife. Treated seed should not be left on the soil surface. Bury or remove spillages
S04c	Dangerous to birds/game and wildlife. Treated seed should not be left on the soil surface. Bury or remove spillages
S04d	To protect birds/wild animals, treated seed should not be left on the soil surface. Bury or remove spillages
S05	Do not reuse sacks or containers that have been used for treated seed for food or feed
S06a	Wash hands and exposed skin before meals and after work
S06b	Wash hands and exposed skin after cleaning and re-calibrating equipment
S07	Do not apply treated seed from the air
S08	Treated seed should not be broadcast
S08a	Treated seed should not be stored longer than 1 month before planting
S09	Label treated seed with the appropriate precautions using printed sacks, labels or bag tags supplied
S10	Only use automated equipment for planting treated seed potatoes

Vertebrate/Rodent control products

V01a	Prevent access to baits/powder by children, birds and other animals, particularly cats, dogs, pigs and poultry
V01b	Prevent access to bait/gel/dust by children, birds and non-target animals, particularly dogs, cats, pigs, poultry
V02	Do not prepare/use/lay baits/dust/spray where food/feed/water could become contaminated
V03a	Remove all remains of bait, tracking powder or bait containers after use and burn or bury
V03b	Remove all remains of bait and bait containers/exposed dust/after treatment (except where used in sewers) and dispose of safely (e.g. burn/bury). Do not dispose of in refuse sacks or on open rubbish tips.
V04a	Search for and burn or bury all rodent bodies. Do not place in refuse bins or on rubbish tips
V04b	Search for rodent bodies (except where used in sewers) and dispose of safely (e.g. burn/bury). Do not dispose of in refuse sacks or on open rubbish tips
V04c	Dispose of safely any rodent bodies and remains of bait and bait containers that are recovered after treatment (e.g. burn/bury). Do not dispose of in refuse sacks or on open rubbish tips
V05	Use bait containers clearly marked POISON at all surface baiting points

Medical advice

M01	This product contains an anticholinesterase organophosphorus compound. DO NOT USE if under medical advice NOT to work with such compounds
M02	This product contains an anticholinesterase carbamate compound. DO NOT USE if under medical advice NOT to work with such compounds
M03	If you feel unwell, seek medical advice immediately (show the label where possible)
M03a	If exposed or concerned, get medical advice or attention
M04a	In case of accident or if you feel unwell, seek medical advice immediately (show the label where possible)
M04b	In case of accident by inhalation, remove casualty to fresh air and keep at rest
M04c	If inhaled, remove victim to fresh air and keep at rest in a position comfortable for breathing. IMMEDIATELY call Poison Centre or doctor/physician
M05a	If swallowed, seek medical advice immediately and show this container or label
M05b	If swallowed, do not induce vomiting: seek medical advice immediately and show this container or label
M05c	If swallowed induce vomiting if not already occurring and take patient to hospital immediately
M05d	If swallowed, rinse the mouth with water but only if the person is conscious
M06	This product contains an anticholinesterase carbamoyl triazole compound. DO NOT USE if under medical advice NOT to work with such compounds

SECTION 6

Appendix 5
Key to Abbreviations and Acronyms

The abbreviations of formulation types in the following list are used in Section 2 (Pesticide Profiles) and are derived from the *Catalogue of Pesticide Formulation Types and International Coding System* (CropLife International Technical Monograph 2, 5th edn, March 2002).

1 Formulation Types

AB	Grain bait
AE	Aerosol generator
AL	Other liquids to be applied undiluted
AP	Any other powder
AS	Aqeous suspension
BB	Block bait
BR	Briquette
CB	Bait concentrate
CC	Capsule suspension in a suspension concentrate
CF	Capsule suspension for seed treatment
CG	Encapsulated granule (controlled release)
CL	Contact liquid or gel (for direct application)
CP	Contact powder (for direct application)
CR	Crystals
CS	Capsule suspension
DC	Dispersible concentrate
DP	Dustable powder
DS	Powder for dry seed treatment
EC	Emulsifiable concentrate
EG	Emulsifiable granule
EO	Water in oil emulsion
ES	Emulsion for seed treatment
EW	Oil in water emulsion
FG	Fine granules
FP	Smoke cartridge
FS	Flowable concentrate for seed treatment
FT	Smoke tablet
FU	Smoke generator
FW	Smoke pellets
GA	Gas
GB	Granular bait
GE	Gas-generating product
GG	Macrogranules
GL	Emulsifiable gel
GP	Flo-dust (for pneumatic application)
GR	Granules
GS	Grease
GW	Water soluble gel
HN	Hot fogging concentrate
KK	Combi-pack (solid/liquid)
KL	Combi-pack (liquid/liquid)
KN	Cold-fogging concentrate
KP	Combi-pack (solid/solid)
LA	Lacquer
LI	Liquid, unspecified
LS	Solution for seed treatment
ME	Microemulsion
MG	Microgranules
MS	Microemulsion for seed treatment

OD	Oil dispersion
OL	Oil miscible liquid
PA	Paste
PC	Gel or paste concentrate
PO	Powder
PS	Seed coated with a pesticide
PT	Pellet
RB	Ready-to-use bait
RC	Ready-to-use low volume CDA sprayer
RH	Ready-to-use emulsion
RH	Ready-to-use spray in hand-operated sprayer
SA	Sand
SC	Suspension concentrate (= flowable)
SE	Suspo-emulsion
SG	Water soluble granules
SL	Soluble concentrate
SP	Water soluble powder
SS	Water soluble powder for seed treatment
ST	Water soluble tablet
SU	Ultra low-volume suspension
TB	Tablets
TC	Technical material
TP	Tracking powder
UL	Ultra low-volume liquid
VP	Vapour releasing product
WB	Water soluble bags
WG	Water dispersible granules
WP	Wettable powder
WS	Water dispersible powder for slurry treatment of seed
WT	Water dispersible tablet
XX	Other formulations
ZZ	Not Applicable

2 Other Abbreviations and Acronyms

ACP	Advisory Committee on Pesticides
ACTS	Advisory Committee on Toxic Substances
ADAS	Agricultural Development and Advisory Service
a.i.	active ingredient
AIC	Agriculture Industries Confederation
BBPA	British Beer and Pub Association
CDA	Controlled droplet application
CPA	Crop Protection Association
cm	centimetre(s)
COPR	Control of Pesticides Regulations 1986
COSHH	Control of Substances Hazardous to Health Regulations
CRD	Chemicals Regulation Directorate
d	day(s)
Defra	Department for Environment, Food and Rural Affairs
EA	Environment Agency
EBDC	ethylene-bis-dithiocarbamate fungicide
EAMU	Extension of Authorisation for Minor Use
FEPA	Food and Environment Protection Act 1985
g	gram(s)
GS	growth stage (unless in formulation column)
h	hour(s)
ha	hectare(s)
HBN	hydroxybenzonitrile herbicide
HI	harvest interval
HSE	Health and Safety Executive
ICM	integrated crop management
IPM	integrated pest management

kg	kilogram(s)
l	litre(s)
LERAP	Local Environmental Risk Assessments for Pesticides
m	metre(s)
MBC	methyl benzimidazole carbamate fungicide
MEL	maximum exposure limit
min	minute(s)
mm	millimetre(s)
MRL	maximum residue level
mth	month(s)
NA	Notice of Approval
NFU	National Farmers' Union
OES	Occupational Exposure Standard
OLA	off-label approval
PPE	personal protective equipment
PPPR	Plant Protection Products Regulations
SOLA	specific off-label approval
ULV	ultra-low volume
VI	Voluntary Initiative
w/v	weight/volume
w/w	weight/weight
wk	week(s)
yr	year(s)

3 UN Number details

UN No.	Substance	EAC	APP	Hazards Class	Sub risks	HIN
1692	Strychnine or Strychnine salts	2X		6.1		66
1760	Corrosive Liquid, N.O.S., Packing groups II & III	2X	B	8		88
1830	Sulphuric acid, with more than 51% acid	2P		8		80
1993	Flammable Liquid, N.O.S., Packing group III	•3Y		3		30
2011	Magnesium Phosphide	4W[(1)]		4.3	6.1	
2588	Pesticide, Solid, TOXIC, N.O.S.	2X		6.1		66/60
2757	Carbamate Pesticide, Solid, Toxic	2X		6.1		66/60
2783	Organophosphorus Pesticide, Solid, TOXIC	2X		6.1		66/60
2902	Pesticide, Liquid, TOXIC, N.O.S., Packing Groups I, II & III	2X	B	6.1		66/60
3077	Environmentally hazardous substance, Solid, N.O.S.	2Z		9		90
3082	Environmentally hazardous substance, Solid, N.O.S.	•3Z		9		90
3265	Corrosive Liquid, Acidic, Organic, N.O.S., Packing group I, II or III	2X	B	8		80/88
3351	Pyrethroid Pesticide, Liquid, Toxic, Flammable, Flash Point 23°C or more	•3W	A(fl)	6.1	3	663/63
3352	Pyrethroid Pesticide, Liquid, TOXIC, Packing groups I, II & III	2X	B	6.1		66/60

Ref: Dangerous Goods Emergency Action Code List 2009 (TSO).
[(1)] Not applicable to the carriage of dangerous goods under RID or ADR.

Appendix 6
Definitions

The descriptions used in this *Guide* for the crops or situations in which products are approved for use are those used on the approved product labels. These are now standardised in a Crop Hierarchy published by the Chemicals Regulation Directorate in which definitions are given. To assist users of this *Guide* the definitions of some of the terminology where misunderstandings can occur are reproduced below.

Rotational grass: Short-term grass crops grown on land that is likely to be growing different crops in future years (*e.g. short-term intensively managed leys for one to three years that may include clover*)

Permanent grassland: Grazed areas that are intended to be permanent in nature (*e.g. permanent pasture and moorland that can be grazed*).

Ornamental Plant Production: All ornamental plants that are grown for sale or are produced for replanting into their final growing position (*e.g. flowers, house plants, nursery stock, bulbs grown in containers or in the ground*).

Managed Amenity Turf: Areas of frequently mown, intensively managed, turf that is not intended to flower and set seed. It includes areas that may be for intensive public use (*e.g. all types of sports turf*).

Amenity Grassland: Areas of semi-natural or planted grassland subject to minimal management. It includes areas that may be accessed by the public (*e.g. railway and motorway embankments, airfields, and grassland nature reserves*). These areas may be managed for their botanical interest, and the relevant authority should be contacted before using pesticides in such locations.

Amenity Vegetation: Areas of semi-natural or ornamental vegetation, including trees, or bare soil around ornamental plants, or soil intended for ornamental planting. It includes areas to which the public have access. It does NOT include hedgerows around arable fields.

Natural surfaces not intended to bear vegetation: Areas of soil or natural outcroppings of rock that are not intended to bear vegetation, including areas such as sterile strips around fields. It may include areas to which the public have access. It does not include the land between rows of crops.

Hard surfaces: Man-made impermeable surfaces that are not intended to bear vegetation (*e.g. pavements, tennis courts, industrial areas, railway ballast*).

Permeable surfaces overlying soil: Any man-made permeable surface (excluding railway ballast) such as gravel that overlies soil and is not intended to bear vegetation

Green Cover on Land Temporarily Removed from Production: Includes fields covered by natural regeneration or by a planted green cover crop that will not be harvested (*e.g. green cover on setaside*). It does NOT include industrial crops.

Forest Nursery: Areas where young trees are raised outside for subsequent forest planting.

Forest: Groups of trees being grown in their final positions. Covers all woodland grown for whatever objective, including commercial timber production, amenity and recreation, conservation and landscaping, ancient traditional coppice and farm forestry, and trees from natural regeneration, colonisation or coppicing. Also includes restocking of established woodlands and new planting on both improved and unimproved land.

Farm forestry: Groups of trees established on arable land or improved grassland including those planted for short rotation coppicing. It includes mature hedgerows around arable fields.

Indoors (for rodenticide use): Situations where the bait is placed within a building or other enclosed structure, and where the target is living or feeding predominantly within that building or structure.

SECTION 6

Herbs: Reference to Herbs or Protected Herbs when used in Section 2 may include any or all of the following. The particular label or EAMU Notice will indicate which species are included in the approval.

Agastache spp.	Lemon thyme
Angelica	Lemon verbena
Applemint	Lovage
Balm	Marigold
Basil	Marjoram
Bay	Mint
Borage (except when grown for oilseed)	Mother of thyme
Camomile	Nasturtium
Caraway	Nettle
Catnip	Oregano
Chervil	Origanum heracleoticum
Clary	Parsley root
Clary sage	Peppermint
Coriander	Pineapplemint
Curry plant	Rocket
Dill	Rosemary
Dragonhead	Rue
English chamomile	Sage
Fennel	Salad burnet
Fenugreek	Savory
Feverfew	Sorrel
French lavender	Spearmint
Gingermint	Spike lavender
Hyssop	Tarragon
Korean mint	Thyme
Land cress	Thymus camphoratus
Lavandin	Violet
Lavender	Winter savory
Lemon balm	Woodruff
Lemon peppermint	

Herbs for Medicinal Uses: Reference to Herbs for Medicinal Uses when used in Section 2 may include any or all of the following. The particular label or EAMU Notice will indicate which species are included in the approval

Black cohosh	Goldenseal
Burdock	Liquorice
Dandelion	Nettle
Echinacea	Valerian
Ginseng	

Appendix 7
References

The information given in *The UK Pesticide Guide* provides some of the answers needed to assess health risks, including the hazard classification and the level of operator protection required. However, the Guide cannot provide all the details needed for a complete hazard assessment, which must be based on the product label itself and, where necessary, the Health and Safety Data Sheet and other official literature.

Detailed guidance on how to comply with the Regulations is available from several sources.

Pesticides: Code of Practice

The sale, supply, storage and use of pesticides are strictly controlled by EU law. The key acts, Regulations and Codes of Practice setting out the duties of employers, supervisors and operators are:

- The *Plant Protection Products Regulations 2011*;
- The *Plant Protection Products (Sustainable Use) Regulations 2012*;
- The *Control of Substances Hazardous to Health Regulations 2002* (COSHH) (as amended);
- Pesticides: *Code of Practice for Using Plant Protection Products 2006*, (incorporating the former 'Green and Orange Codes') with specific Codes of Practice for Scotland and Northern Ireland;
- Classification, Labelling and Packaging (CLP) Regulations.

The Code is currently being updated (as of May 2016) and may be replaced by a series of guidance notes as opposed to a Code in future. The current Code should be read in conjunction with guidance published on the HSE website on how new pesticide legislation has affected the responsibilities of the regulated community.

Further information and details of legislation relating to water, environment protection, waste management and transportation can be found in the UK National action Plan for the Sustainable Use of Pesticides (www.gov.uk/government/publications/pesticides-uk-national-action-plan).

Other Codes of Practice

The Food and Environment Protection Act and its regulations aim to protect the health of human beings, creatures and plants, safeguard the environment and secure safe, efficient and humane methods of controlling pests.

Code of Practice for Suppliers of Pesticides to Agriculture, Horticulture and Forestry (the 'Yellow Code') (Defra Booklet PB 0091)

Code of Good Agricultural Practice for the Protection of Soil (Defra Booklet PB 0617)

Code of Good Agricultural Practice for the Protection of Water (Defra Booklet PB 0587)

Code of Good Agricultural Practice for the Protection of Air (Defra Booklet PB 0618)

Approved Code of Practice for the Control of Substances Hazardous to Health in Fumigation Operations. Health and Safety Commission (ISBN 0-717611-95-7)

Code of Best Practice: Safe use of Sulphuric Acid as an Agricultural Desiccant. National Association of Agricultural Contractors, 2002. (Also available at www.naac.co.uk/?Codes/acidcode.asp)

Safe Use of Pesticides for Non-agricultural Purposes. HSE Approved Code of Practice. HSE L21 (ISBN 0-717624-88-9)

SECTION 6

Other Guidance and Practical Advice

HSE (by mail order from HSE Books – See Appendix 2)

COSHH – A brief guide to the Regulations 2003 (INDG136)

A Step by Step Guide to COSHH Assessment, 2004 (HSG97) (ISBN 0-717627-85-3)

Defra (from The Stationery Office – see Appendix 2)

Local Environment Risk Assessments for Pesticides (LERAP): Horizontal Boom Sprayers.

Local Environment Risk Assessments for Pesticides (LERAP): Broadcast Air-assisted Sprayers.

Crop Protection Association and the Voluntary Initiative (see Appendix 2)

Every Drop Counts: Keeping Water Clean

Best Practice Guides. A range of leaflets giving guidance on best practice when dealing with pesticides before, during and after application.

H2OK? Best Practice Advice and Decision Trees - July 2009/10

BCPC (British Crop Production Council – see Appendix 2)

ukpesticideguide.co.uk (*The UK Pesticide Guide* online) – subscription resource

The Pesticide Manual (18th edition) (ISBN 978-1-9998966-1-4)

The Pesticide Manual online (www.bcpcdata.com) – subscription resource

The GM Crop Manual (ISBN 978-1-901396-20-1)

GM Crop Manual online (www.bcpcdata.com/gm) – subscription resource

The Manual of Biocontrol Agents (5th edition) (ISBN 978-1-901396-87-4)

The Manual of Biocontrol Agents online (www.bcpcdata.com/mba) – subscription resource.

IdentiPest www.identipest.co.uk

Small Scale Spraying (ISBN 1-901396-07-X)

Field Scale Spraying – 2016 (ISBN 978-1-901396-89-8)

Using Pesticides (ISBN 978-1-901396-10-2)

Safety Equipment Handbook (ISBN 1-901396-06-1)

Spreading Fertilisers and Applying Slug Pellets (ISBN 978-1-901396-15-7)

The Environment Agency (see Appendix 2)

Best Farming Practices: Profiting from a Good Environment

Use of Herbicides in or Near Water

SECTION 7
INDEX

Index of Proprietary Names of Products

The references are to entry numbers, not to pages. Adjuvant names are referred to as 'Adj' and are listed separately in Section 4. Products also registered, i.e. not actively marketed, are entered as *PAR* and are listed by their active ingredient in Section 3.

REFERENCES ARE TO ENTRY NUMBERS NOT PAGES

REFERENCES ARE TO ENTRY NUMBERS NOT PAGES

REFERENCES ARE TO ENTRY NUMBERS NOT PAGES

REFERENCES ARE TO ENTRY NUMBERS NOT PAGES

REFERENCES ARE TO ENTRY NUMBERS NOT PAGES

NOTES

NOTES

NOTES

NOTES

THE UK PESTICIDE GUIDE 2020

RE-ORDERS

☐ Please send me _____ more copies of *The UK Pesticide Guide 2020* at £58.00 each

Postage: single copy £6.95; orders up to £200 £9.95... orders over £200 £12.50

Outside the UK please add £10 per order for delivery.

Name _____ Position _____

Institution _____ Department _____

Address _____

City _____ Region _____ Postcode_____ Country_____

Tel_____ Fax_____ E-mail_____

EU countries except UK – VAT No:_____

Payment (pre-payment is required):

☐ I enclose a cheque/draft for £_____ payable to BCPC. Please send me a receipt.

☐ I wish to pay by credit card: ☐ Visa ☐ Mastercard ☐ Amex ☐ Switch

Please charge to my card £_____ and send me a receipt. Name of issuing bank_____

Card no. ☐☐☐☐ ☐☐☐☐ ☐☐☐☐ ☐☐☐☐

Expiry date ☐☐/☐☐ Security code ☐☐☐   Switch cards only: Start date ☐☐/☐☐ Issue no. ☐

Signature_____ Date_____

Name and address of cardholder if different from above:_____

_____

Please photocopy and return to:
BCPC Publications Sales, Garden Studio, 4 Hillside, Aldershot, GU11 3NB, UK.
BCPC Tel: 01252 285 223, Email: publications@bcpc.org, Web: www.bcpc.org

FUTURE EDITIONS

☐ I wish to take out an annual order for _____ copies of each new edition of *The UK Pesticide Guide*

☐ Please send me advance price details for the 2021 edition of *The UK Pesticide Guide* when available

Order by phone: 01252 285 223 or online: www.bcpc.org

	Bulk discount:	
	100+ copies	Contact BCPC
	50–99	10%
	10–49	5%
	List price £58.00	

Thank you for your order